T0139798

Graduate Texts in Physics

Graduate Texts in Physics publishes core learning/teaching material for graduate- and advanced-level undergraduate courses on topics of current and emerging fields within physics, both pure and applied. These textbooks serve students at the MS- or PhD-level and their instructors as comprehensive sources of principles, definitions, derivations, experiments and applications (as relevant) for their mastery and teaching, respectively. International in scope and relevance, the textbooks correspond to course syllabi sufficiently to serve as required reading. Their didactic style, comprehensiveness and coverage of fundamental material also make them suitable as introductions or references for scientists entering, or requiring timely knowledge of, a research field.

More information about this series at
http://www.springer.com/series/8431

Raza Tahir-Kheli

General and Statistical Thermodynamics

Second Edition

Raza Tahir-Kheli
Physics Department
Temple University
Philadelphia, PA, USA

ISSN 1868-4513 ISSN 1868-4521 (electronic)
Graduate Texts in Physics
ISBN 978-3-030-20702-1 ISBN 978-3-030-20700-7 (eBook)
https://doi.org/10.1007/978-3-030-20700-7

Printed on acid-free paper

This Springer imprint is published by the registered company Springer Nature Switzerland AG
The registered company address is: Gewerbestrasse 11, 6330 Cham, Switzerland

Dedicated to my children
Shehra J. Boldt and Kazim R. Tahir-Kheli
and (in memoriam) to my teacher
Martin J. Aitken, F.R.S.

Preface to the Second Edition

Except for a few minor corrections, Chaps. 1–11 are identical to those in the first edition. The second edition adds Chaps. 12–16.

Chapter 12 offers a brief introduction to Landau's theory of second order phase transitions. Cooper pairs, which play a vital role in superconductivity, are discussed in Chap. 13. Bogolyubov's important work is described in Chap. 14. London brothers' theory, which is analyzed in Chap. 15, was the first to provide a viable theoretical explanation for the Meissner effect in the superconducting state. Chapter 16 presents a theoretical description of superconductivity. Several topics mentioned in the text are discussed in the Appendix.

I am greatly indebted to miss Adelheid Duhm and Dr. Zachary Evenson of Springer Verlag for their superlative support in putting together this book. Special thanks are due to Vygintas Vilimas for many insightful suggestions. Gratitude is also owed to Temple University Department of Physics for help and encouragement.

Philadelphia, Pennsylvania
March 3, 2020

Raza Tahir-Kheli

Preface to the First Edition

This book is intended for use both as a textbook and as a source for self-study.

As a textbook: My experience teaching courses related to Thermodynamics and Statistical Mechanics at Temple University has guided its writing. The Thermodynamics course is offered to undergraduates in their junior or senior year. These undergraduates have either a single or a double major in Physics, Biology, Chemistry, Engineering, Earth Sciences, or Mathematics. The course on Statistical Mechanics is attended by graduate students in their second year. Feeling that a robust study of statistical thermodynamics appropriately belongs in a senior level graduate course, only relatively simple aspects of Statistical Mechanics are included here.

As a source for self-study: Years of teaching have taught me a thing or two about what works for students and what does not. In particular, I have learned that the more attention a student pays to taking notes, the less he understands of the subject matter of the lecture being delivered. Further, I have noticed that when, a few days in advance of the delivery of the lecture, a student is provided with details of the algebra to be used, solutions to the problems to be discussed, and some brief information about the physics that is central to the lecture to be given, it obviates much of the need for note-taking during the delivery of the lecture. Another important experience that has guided the writing of this textbook is the pedagogical benefit that accrues from an occasional, quick, recapitulation of the relevant results that have already been presented in an earlier lecture. All this usually results in a better comprehension of the subject matter. Both for purposes of elucidation of the concepts introduced in the text and for providing practical problem solving support, a large number of solved problems have been included. Many of the solutions provided include greater detail than would be necessary for presentation in a lecture itself or needed by teachers or more advanced practitioners. They are there in the given form to offer encouragement and support both for self-study and indeed also to allow for fuller understanding of the subject matter. Therefore, it is as important to read through and understand these solutions as it is to learn the rest of the text.

Part of the vocabulary of thermodynamics are such terms as a thermodynamic system, an adiabatic enclosure, an adiabatically isolated and adiabatically enclosed system, conducting and diathermal walls, isobaric, isochoric, isothermal, quasistatic, nonquasistatic, reversible, and irreversible processes. Also, there are the concepts of thermal and thermodynamic equilibria. These terms and concepts are commonly used in thermodynamic analyses. The study of thermodynamics is greatly assisted by standard mathematical techniques. In particular, exact differentials, and some well-known identities of differential calculus, are central to the theoretical description of the subject. Temperature is not only a word in normal daily use, it also sits at the core of thermodynamics. The zeroth law brings to fore one of the many reasons for its relevance to the physics of thermal equilibria. The foregoing are discussed in Chap. 1.

Arguably, the emphasis on constructing simple models with the hope that they may pertain to real—and necessarily more complicated—systems of interest is unique to the discipline of Physics. Usually, these models are uncomplicated and lacking the complexity of the real systems. But because they can often be exactly solved and understood, the hope always is that the predicted results for the model systems will give insight into the physics of real systems.

A "perfect gas" represents an idealized model system. The model was originally motivated by experimental observation, and constructed to understand the behavior of real gases. It has had a long and illustrious history and a great deal of success. We describe and discuss its implications in Chap. 2.

Caloric was thought to be a massless fluid whose increase warmed an object. Just like a fluid, it was thought to flow from a hot object—where there was more of it—to a cold one until the amounts of caloric in the two objects equalized. Also, like fluids, caloric could neither be created nor destroyed. Therefore, the reason a cannon muzzle got hot when it was being bored is that the chips that were created carried less caloric so more was left behind for the muzzle. While supervising the boring of a cannon, the Count Rumford of Bavaria noticed that the duller the boring bits the hotter the muzzles. In other words, less chips, but more caloric! Rumford came to the obvious conclusion: Heat energy is not exchanged by the transfer of caloric but rather by the expenditure of work that has to be done for the drilling of the cannon. Thus work was empirically observed to be related to heat energy.

Conservation of energy is a concept as old as Aristotle.[1] The first law of thermodynamics recognizes this concept, as well as Count Rumford's observations, that heat energy and work are related. If some work is performed upon adding a given amount of heat energy to a system, the portion of heat energy that is left over is called the change in the internal energy of the system. And while both the amounts of heat energy added to a system and the work done by it depend on the details of how they were carried out[2] the first law asserts that their difference, which is the change in the internal energy, is completely path independent. These issues are discussed in detail in Chap. 3 and solutions to many helpful problems are provided.

Clearly, the hallmark of the first law of thermodynamics is its recognition of the path independent function, the "internal energy." Arguably, the second law has even more to boast about: namely, the identification of the state function, the "entropy,"

[1] Approximately 350 BC.

[2] Meaning they both depend on the physical paths that were taken in executing the two processes.

and its familial relationship with Carnot's ideas about maximum possible efficiency of heat engines. These ideas and their various aftereffects are discussed and treated in Chap. 4.

In Chap. 5, we marry the first and second laws of thermodynamics in a manner that achieves "a perfect union" of the two: a union that provides great insights into the workings of thermodynamics.

Many hold to the view that Johannes Diderik Van der Waals' derivation of a possible equation of state for imperfect gases was the first ever significant contribution to predictive statistical thermodynamic. He noted that any interaction between microscopic constituents of a body must have two distinct features. Because these constituents congregate to form macroscopic entities, the overall interparticle interaction must be attractive. Yet, because matter does condense to finite densities, at small enough distances the interaction must become strong and repulsive, meaning it must have a hard core.

He assumed both these interactions to be "short ranged." It turns out, however, that his equation of state is somewhat more meaningful for a gas that has both a hard core—much as he assumed—but, unlike his assumption, has attractive interaction which is long ranged. These ideas, as well as the many physical consequences of the Van der Waals equation of state, are extensively studied and analyzed in Chap. 6.

State variables such as the volume, pressure, and temperature are easy to measure. In contrast, thermodynamic state functions such as the internal energy and the enthalpy cannot be measured by straightforward procedures. Generally, therefore, one needs to follow somewhat circuitous routes for their measurement.

The famous Gay–Lussac–Joule experiment attempted to measure the internal energy. The experiment was based on determining the amount of heat energy that is produced by free-expansion of a gas enclosed in vessels submerged in water. Unfortunately, the heat capacities of the water that needed to be used plus that of the enclosing core were vastly greater than that of the gas being used. As a result no reliable measurement could be made.

Joule and Kelvin devised an experiment that overcame this difficulty. The experiment shifted the focus from the internal energy to the enthalpy. Additionally, rather than dealing with a fixed amount of gas that is stationary, it used a procedure that involved "steady-flow." As a result, reliable measurements could be made. The related issues and results are presented in Chap. 7.

Chapter 8 deals with the Euler equation—namely the complete fundamental equation—and its dependence on the chemical potential, the Gibbs–Duhem relation in both the energy and the entropy representations, and the three possible equations of state for the ideal gas, again in both the entropy and energy representations.

Several important issues are broached in Chap. 9. For instance, we recall that all spontaneous processes in isolated thermodynamic systems increase their total entropy and this fact follows from the second law. Of course, spontaneous processes continue until the system ceases to change, that is, until it achieves thermal equilibrium. Accordingly, subject to the constraints under which the system has been maintained, and for the given value of its various extensive properties, its total entropy in thermodynamic equilibrium is the maximum possible. With this knowledge in hand, coupled with the concept of the fundamental equation, we are able to more fully examine the nature of the zeroth law.

Entropy extremum also helps explain the direction of thermodynamic motive forces. Namely, why heat energy flows from the warm to the cold, why, at constant temperature and pressure, molecules flow from regions of higher chemical potential to those of lower chemical potential, and why the fact that if two macroscopic systems in thermal contact are placed together in an adiabatically isolated chamber, and if their total volume is constant, their chemical potentials are equal, they are at the same temperature, and they can freely affect each other's volume then the requirement that the total entropy increase in any isothermal spontaneous process ensures that the side with originally lower pressure will shrink in volume? And, why if the process is allowed to continue, the shrinking of the side with originally lower pressure, and the expansion in volume of the side with originally greater pressure, continues until the two pressures become equal?

Much like the entropy, the system energy also obeys an extremum principle. For a given value of the entropy and the equilibrium values of various extensive parameters, the energy of a system is a minimum. And it turns out that the energy extremum leads to the same physical predictions as does the entropy extremum.

These extremum principles make important statements about issues that relate to intrinsic stability of thermodynamic systems. In particular, they help determine the requirement that for intrinsic stability the specific heat C_V and the isothermal compressibility χ_T must always be greater than zero.

Also discussed in Chap. 9 is the Le Châtelier principle which asserts that spontaneous processes caused by displacements from equilibrium help restore the system back to equilibrium.

The analyses presented in Chap. 9 were guided by the use of the extremum principles obeyed by the internal energy and entropy. An important consequence of either of these two extremum principles is the occurrence of other extrema that are related to the Helmholtz potential, Gibbs free energy, and enthalpy. These are identified and some of their consequences predicted in Chap. 10. Legendre transformations provide an essential tool for these studies. The issue of meta-stable equilibrium is commented upon. Also Maxwell relations, the Clausius–Clapeyron differential equation and its use in the study of thermodynamic phases, and the Gibbs phase rule are described in Chap. 10.

Much of what is written in the preceding chapters has been gleaned from standard procedures which do not attempt to carry out exact numerical calculation of state functions such as the entropy and internal energy. Indeed, the focus mainly has been on understanding the rates of change of state functions and their interrelationships. This is because, unlike statistical mechanics, thermodynamics itself does not have any convenient method for performing these calculations.

Chapter 11 deals with statistical thermodynamics. Analyzed first are the classical monatomic perfect gases with constant numbers of particles. Results of their mixing under a variety of ambient circumstances are described. Diatomic gases are treated next. Statistical mechanics of harmonic and anharmonic simple oscillators, the Langevin paramagnet, extremely relativistic ideal gas, gases with interaction, and issues relating to Mayer's cluster expansion and the Lennard-Jones potential are analyzed. Studied next are the differences between the thermodynamics of classical and the quasiclassical quantum systems. Quasiclassical analyses of diatoms with rigid bonds, and others that admit vibrational motion, are described. Nernst's heat theorem, unattainability of zero temperature, the third law, and indeed the concept of negative temperature are described next. Finally, the Richardson effect, the Fermi–Dirac and Bose–Einstein quantum gases, "black body radiation," and the thermodynamics of phonons are presented.

Most users of this textbook should read the relevant appendices. The usefulness of an appendix lies in it containing more details than are originally provided in the body of the text. For instance, the notion that thermodynamics refers to systems that contain very large numbers of atoms is explained with great simplicity in Appendix A, as also is the rationale for the validity of the Gaussian approximation. Appendix B is equally helpful. The rederivation of the perfect gas law is done with straightforward argumentation: this time by using elementary statistical mechanics. Appendix C provides details of an important argument that asserts that the Carnot version of the second law leads to the Clausius version. Other appendices provide solutions to various problems that are identified in the text.

Unlike a novel, which is often read continuously—and the reading is completed within a couple of days—this book is likely to be read piecemeal—a chapter or so a week. At such a slow rate of reading, it is often hard to recall the precise form of a relationship that appeared in a previous chapter or sometimes even in the earlier part of the same chapter. To help relieve this difficulty, when needed, the most helpful explanation of the issue at hand is repeated briefly and the most relevant expressions are mentioned by their equation numbers. Throughout the book, for efficient reading, most equations are numbered in seriatim. When needed, they can be accessed quickly.

A fact well known to researchers is that when a physics problem is analyzed in more ways than one, its understanding is often greatly enhanced. Similar improvement in comprehension is achieved when students are also provided access to alternate, yet equivalent, explanations of the subject matter relating to important physical concepts. To this end, effort has been made to provide—wherever possible—additional, alternate solutions to given problems and also to the derivation of noteworthy physical results. Occasionally, brief historical references have also been included in the text.

Most of the current knowledge of thermodynamics is much older than the students who study it. Numerous books have been written on the subject. While the current book owes greatly to three well-known texts on thermodynamics[3] and one on statistical mechanics,[4] its import is different: it is offered as much for use in formal lectures as for self-study.

[3]Namely, Herbert B. Callen, John Wiley Publishers (1960); D. ter Haar and H. Wergeland, Addison-Wesley Publishing Company (1966); F.W. Sears and G.L. Salinger, Addison-Wesley Publishing Company—Third Edition (1986).

[4]Namely, R.K. Pathria, Pergamon Press (1977).

Many thanks are due to Dr. Claus Ascheron for suggestions and advice. Finally, but for the help and support of my colleagues, Robert Intemann and Peter Riseborough, this book could not have been written.

Philadelphia, Pennsylvania Raza Tahir-Kheli
April 2011

Contents

Definitions and the Zeroth Law

1

Much like other scientific disciplines, thermodynamics also has its own vocabulary. For instance, while dealing with objects composed of very large numbers of particles, one might use terms like thermodynamic system; adiabatically isolating, and adiabatically enclosing, walls; adiabatic and nonadiabatic enclosures; conducting and/or diathermal walls; isothermal, isobaric, isochoric, quasistatic, reversible, and irreversible processes; state functions and state variables; thermodynamic equilibrium and, of course, temperature in its various representations.

Central to the understanding of temperature and its relationship to thermodynamic equilibrium is the zeroth law of thermodynamics.

Just as driving an automobile gets one much farther than walking, thermodynamics is greatly helped by the use of mathematics. And there are a few simple mathematical techniques that are particularly useful.

An attempt at treating the foregoing issues is made in the current chapter. For instance, Sect. 1.1 deals with definitions of the terms that are often used. Section 1.2 deals with the need for large numbers in thermodynamic systems and their effect on the most probable state. Section 1.3 deals with the zeroth law; Sect. 1.4 with some mathematical procedures; Sect. 1.5 deals with the cyclic identity, with exact and inexact differentials and their relevance to state variables and state functions; Sect. 1.6 with the use of simple Jacobian techniques; and finally, Sect. 1.7 with additional helpful identities, the rederivation of the cyclic identity and the introduction of other well-known identities.

1.1 Some Definitions

Let us begin with defining thermodynamic systems.

Thermodynamic System A thermodynamic system, sometimes also called an "object," comprises a collection of very large number of atoms and/or electromagnetic field quanta. In general, in addition to being subject to internal effects, a system may also be affected from the outside. Further, it may be of liquid, solid, gaseous, or even some other exotic form.

R. Tahir-Kheli, *General and Statistical Thermodynamics*,
https://doi.org/10.1007/978-3-030-20700-7_1

Adiabatic Enclosure Generally, the walls of an enclosure in which a system is placed allow for the transfer and exchange of translational momentum, electromagnetic and gravitational fields, heat energy, and sometime even molecules, etc. It is, however, possible to build walls that greatly reduce such transfers and exchanges. Exceptionally, one can imagine the construction of walls that reduce all of the aforementioned transfers and exchanges by 100%. These are called "adiabatic walls."

Built entirely from such adiabatic walls, an enclosure, that completely encloses a (thermodynamic) system, is called an "adiabatic" enclosure.

Isolated System An adiabatically isolated system exists within an adiabatic enclosure and it interchanges no energy, and no information, with the environment (meaning, the rest of the universe).

Enclosed System While an adiabatically enclosed system does not interchange any energy with the environment, unlike an adiabatically isolated system it may be subject to interchanging some information with the environment.

Conducting Walls Walls that are not adiabatic and freely allow for the transfer of momentum, energy, etc., are called "conducting." In a word, they are "open." This openness can take several forms. Particularly relevant to the study of thermodynamics is the openness to the transfer and exchange of "thermal energy," "mass," "electromagnetic fields and charges," etc., and momentum.

Diathermal Walls If a wall allows for the transfer of heat energy, it is called a "diathermal," or equivalently, a "heat energy conducting," or a "thermally open," wall.

Isobaric Process A process that occurs while the pressure remains unchanged is called an "isobaric process."

Isochoric Process An "isochoric process" occurs at constant volume.

Thermal Equilibrium We all know when we feel "hot" or "cold." Also that given two objects, one hot and the other cold, bringing them into contact generally cools-down the hot object and warms-up the cold object.

When a thermometer is used to take a reading of the hot object h, it registers a number, T_{h}, that is higher than that, i.e., T_{c}, registered for the cold object c.

Let identical thermometers be placed in two objects as they are brought into "thermal contact." Now let both objects, with their thermometers, be placed inside an adiabatic enclosure. Assume that the thermometer readings can be observed. These readings, T_{h} and T_{c}, begin to move toward an intermediate number. Indeed, as the contact time increases, the readings move ever closer. Eventually, they stop changing and attain—what should be—the same reading. When that happens we say that the thermometers and the systems have reached—both by themselves and with each-other—a state of "thermal equilibrium."

Quasistatic Process A quasistatic process proceeds extremely—in principle, infinitely—slowly, almost as if it were static when in fact it is proceeding. Generally, it passes through a very long—in principle, infinitely long—series of equilibrium states that are infinitesimally close to each other. In contrast, real processes proceed at finite speeds and pass through states that depart from the equilibrium.

For instance, during a quasistatic transfer of heat energy, both the relevant thermodynamic systems—i.e., the one delivering the heat energy and the one receiving it—pass through only those states that are in thermodynamic equilibrium.

In contrast, a real process both proceeds at nonzero speed and the intermediate states that it passes through may often depart from the equilibrium. Thus a quasistatic process is an idealization and, in practice, is at best achieved only approximately. Yet, in thermodynamics, the concept is of great theoretical value.

Reversible and Irreversible Processes A reversible process is one that occurs with such little "enthusiasm" that it can be reversed with merely an infinitesimal amount of effort. A signature example of such a process would be provided by quasistatic transfer of heat energy from a reservoir to an object that is almost exactly at the same temperature. In this way, the process can proceed in either direction with only an infinitesimal change in the temperature. Note, a reversible process is necessarily quasistatic.

1.2 Large Numbers

Thermodynamics deals with macroscopic systems. Any such system comprises a very large number of particles. For instance, one gram of hydrogen has approximately 6×10^{23} atoms. Considering that the age of the Universe is thought to be less than $\approx 10^{18}$ seconds, 6×10^{23} is a very large number.

Owing to interparticle interaction, theoretical analysis of most such systems is very complicated and can at best be carried out only approximately. Indeed, unlike statistical mechanics, thermodynamics itself does not even attempt to perform a priori theoretical calculations. Rather, it deals with interrelationships of observed physical properties of macroscopic systems. In so far as such knowledge can often help relate easily measurable physical properties to those that are hard to measure, thermodynamics plays an important role in scientific disciplines.

Most Probable State As elucidated in Appendix A, large numbers are fundamental to the accuracy of thermodynamic relationships. Indeed, systems with small numbers of atoms do not satisfy thermodynamic identities.

Macroscopic systems contain very large numbers of particles. For large numbers, the most probable occurrence is overwhelmingly so. As such, the result of any thermodynamic—i.e., a "macroscopic"—measurement is extremely well described by the configuration that refers to the most probable state. Therefore, quite appropriately, thermodynamics focuses primarily on the most probable state.

1.3 The Zeroth Law

Insert identical thermometers into three different objects (i.e., systems) called A, B, and C. Place the trio in an adiabatic enclosure.

Bring A into thermal contact separately with B and C. But make sure that the objects B and C are not placed in direct mutual thermal contact.

Let the two contacts—namely $A \rightarrow B$ and $A \rightarrow C$—last for an extended period of time. As noted before, this causes the objects A and B, as well as the duo A and C, to reach mutual thermal equilibrium.

The zeroth law now makes a seemingly unsurprising prediction, namely that the above two equilibrating processes also ensure that B and C—which are not in direct thermal contact with each other—will also have reached mutual thermal equilibrium. Considering that elementary rules of algebra require B to be equal to C whenever $A = B$ and $A = C$, one might laughingly assert that there is nothing special about this prediction!

But this assertion is fallacious because we are not dealing with the rules of algebra here. For instance, consider two persons, "a" and "b," who are good friends. If "a" is also a good friend of another person named "c," then is it always the case that "b" is a good fiend of "c"? Indeed, if that should be the case, then the trio may be thought to have some common "chemistry" together!

Below we follow an argument, given by Pippard, to show that the zeroth law predicts the trio A, B, and C to have a common state variable. The relevant common state variable is normally called the "temperature" [1].

Empirical Temperature A "simple" system is defined such that a given amount—say, a single mol[1]—can be[2] completely specified by two state variables, pressure p and volume v. Further, with appropriate effort, the magnitude of either, or indeed both, of these variables may be changed.[3]

Consider three such systems—one mol each—labeled A, B, and C. What happens if all three are placed within the same adiabatic enclosure in such a way that A is separately in thermal contact with B, on one side, and C, on the other. In this fashion, because B and C themselves remain physically separated from each other, they are not in direct thermal contact.

Over time, the pressures and volumes settle down at three pairs of values—say, (p_A, v_A), (p_B, v_B), and (p_C, v_C)—that are relevant to the achieved state of mutual thermal equilibrium of A separately with B and C.

An interesting result is that now the values (p_A, v_A), (p_B, v_B), and (p_C, v_C) cannot be arbitrarily adjusted. It is important to note that such an inability did not exist when A and B, and A and C, were not in thermal contact. In fact, it is now

[1]By international agreement, the relative atomic mass of N_A carbon-12 atoms is chosen to be exactly equal to 12. Note that a carbon-12, i.e., $^{12}C_6$, atom has 6 protons and 6 neutrons. The Avogadro's number, N_A, is so chosen that the mass of N_A carbon-12 atoms is exactly equal to 12 g. Measured thus, N_A is equal to $6.02214179(30) \times 10^{23}$ mol^{-1}. References to one mol always specify N_A particles whether they be atoms or molecules.

[2]Note that both carbon and helium molecules are monatomic.

[3]Avogadro's number $N_A = 6.02214179(30) \times 10^{23}$ mol^{-1}.

found that only three of the four variables of a pair of systems in thermal equilibrium can be chosen arbitrarily. The fourth variable is then completely specified.

In other words, the two pairs (p_A, v_A) and (p_B, v_B)—or similarly, the pairs (p_A, v_A) and (p_C, v_C)—have only three independent variables. The fourth variable is dependent on the other three.

Mathematically this fact can be expressed as follows:[4]

$$p_A = f_1(v_A, p_B, v_B) . \tag{1.1}$$

Let us treat next the second pair of systems in thermal equilibrium, namely, A and C. Again, noting that of the four variables (p_A, v_A) and (p_C, v_C) only three are independent, we can write[5]

$$p_A = f_2(v_A, p_C, v_C) . \tag{1.2}$$

Equating p_A in (1.1) and (1.2) yields

$$f_1(v_A, p_B, v_B) = f_2(v_A, p_C, v_C) . \tag{1.3}$$

Let us now remind ourselves of the assertion made by the zeroth law. Because the objects A and B, as well as A and C, are in mutual thermal equilibrium, therefore it is asserted that B and C must also be in mutual thermal equilibrium.

Should the above assertion—namely that B and C are also in mutual thermal equilibrium—be correct, then much as noted earlier, of the four variables (p_B, v_B) and (p_C, v_C), only three would be independent. Thus any of these four variables depends on the other three, e.g.,[6]

$$p_B = f_3(v_B, p_C, v_C) . \tag{1.4}$$

Equation (1.4) is more conveniently written as follows:

$$G(p_B, v_B, p_C, v_C) = f_3(v_B, p_C, v_C) - p_B = 0 . \tag{1.5}$$

Taking Stock Let us take stock of what has been achieved so far. Equation (1.3) acknowledges the fact that the pairs, (A, B) and (A, C), have been brought into mutual thermal equilibrium while the systems B and C have been kept physically separated. Consequent to this happenstance, (1.5) records the prediction—which is actually an assertion—of the zeroth law.

In other words, the zeroth law asserts that (1.5)—which signifies mutual thermal equilibrium for systems B and C—follows from (1.3).

---

[4]Although we have chosen to represent p_A as a function of (v_A, p_B, v_B), any other of these four variables could equally well have been chosen as a function of the remaining three.

[5]It bears noting that $f_1(x, y, z)$ and $f_2(x, y, z)$ are, in all likelihood, not the same functions. This is especially true if B and C are physically different systems.

[6]Note $f_3(x, y, z)$, a function of the three variables x, y, and z, is in all likelihood different from $f_1(x, y, z)$ and $f_2(x, y, z)$ encountered in (1.3).

But how can this be true considering (1.3) is a function of five variables, v_A, (p_B, v_B), and (p_C, v_C), while (1.5) depends only on the four variables (p_B, v_B) and (p_C, v_C)? In order for this to happen, in (1.3), there must occur a complete self-cancelation of the fifth variable, v_A. The most general choice for f_1 and f_2 that satisfies this requirement is the following:

$$\begin{aligned} f_1(v_A, p_B, v_B) &= \alpha(v_A) J(p_B, v_B) + \beta(v_A)\,, \\ f_2(v_A, p_C, v_C) &= \alpha(v_A) K(p_C, v_C) + \beta(v_A)\,. \end{aligned} \tag{1.6}$$

It is important to note that the functions $J(x, y)$ and $K(x, y)$ are neither required, nor are they expected, to be the same.

The equality of $f_1(v_A, p_B, v_B)$ and $f_2(v_A, p_C, v_c)$, demanded by (1.3), i.e.,

$$\alpha(v_A) J(p_B, v_B) + \beta(v_A) = \alpha(v_A) K(p_C, v_C) + \beta(v_A)\,, \tag{1.7}$$

yields

$$J(p_B, v_B) = K(p_C, v_C) \equiv T_{B-C}\,(\text{empirical})\,. \tag{1.8}$$

Equation (1.8) is of central importance. The assertion of the zeroth law that systems B and C are also in thermal equilibrium demands that both B and C lead to a common value of a parameter, T_{B-C}(empirical), that we shall call their empirical temperature.[7]

Analogously, we can conclude that because we started with the knowledge that A and B are in thermal equilibrium, they too must lead to the existence of a common empirical temperature. Let us call that T_{A-B}(empirical).

Equivalently, the same must be true for the duo A and C which also were stated to be in thermal equilibrium. As a result we must have an empirical temperature, T_{A-C}(empirical), that is common to the two systems A and C.

Therefore, using basic rules of algebra, all the three systems, A, B, and C, have in common the same empirical temperature T(empirical). That is,

$$\begin{aligned} T(\text{empirical}) &= T_{A-B}(\text{empirical}) \\ &= T_{A-C}(\text{empirical}) \\ &= T_{B-C}(\text{empirical})\,. \end{aligned} \tag{1.9}$$

Isothermal Process A process that occurs at constant temperature is called an "isothermal process."

Equations of State Relationships of the type given in (1.8), e.g.,

$$J(p, v) = T(\text{empirical})\,, \tag{1.10}$$

where $J(p, v)$ represents a function of the pressure p and the volume v, are often referred to as equations of state of a simple thermodynamic system.

---

[7] Such an empirical temperature could be defined by any appropriate thermometric property that both systems share.

Remark The zeroth law leads to a result that equates a function of the pressure and the volume of a "simple"[8] thermodynamic system to a single parameter that is the same for any two systems in thermal equilibrium. This parameter can be labeled the "empirical temperature." Note that there are also other important consequences of the zeroth law. These consequences are discussed in detail in Sect. 9.1.

1.4 Some Mathematical Procedures

Readers of this text are likely to be familiar with elementary differential and integral calculus. Many will also have been introduced to partial differentiation and possibly also to the use of Jacobian. Experience suggests that at least for some the knowledge has become rusty. Therefore, a quick review of the mathematical procedures, that are most needed for an adequate study of thermodynamics, is often helpful. Salient features of such a review are recorded below. In order to keep the review simple, proofs are not provided and issues of mathematical rigor are not tackled.

1.4.1 Exact Differential

Thermodynamics deals with macroscopic systems that can be described in terms of state variables. For a given amount—such as one mol, for example—of a simple system, there are three such variables: pressure p, volume v, and temperature T. As mentioned above, in general these three variables are related through an equation of state of the form

$$f(p, v, T) = 0\,, \tag{1.11}$$

making only two of them independent. For notational convenience, let us denote the two independent variables X and Y.

Given Z is a function of X and Y,

$$Z = Z(X, Y)\,,$$

and $\mathrm{d}Z$ is an exact differential,[9] then the line integral in the (X, Y) plane,

$$\int_{(X_\mathrm{i}, Y_\mathrm{i})}^{(X_\mathrm{f}, Y_\mathrm{f})} \mathrm{d}Z(X, Y)\,,$$

depends only on the initial, $i \equiv (X_\mathrm{i}, Y_\mathrm{i})$, and the final, $f \equiv (X_\mathrm{f}, Y_\mathrm{f})$, positions and thereby is totally independent of the path traversed between i and f.

[8] It should be noted that the implications of the zeroth law are not limited to just simple thermodynamic systems. Systems with an arbitrary number of thermodynamic state variables also obey the zeroth law equally well.

[9] Note: some authors—e.g., D. ter Haar and H. Wergeland [2]—prefer to use the term total differential.

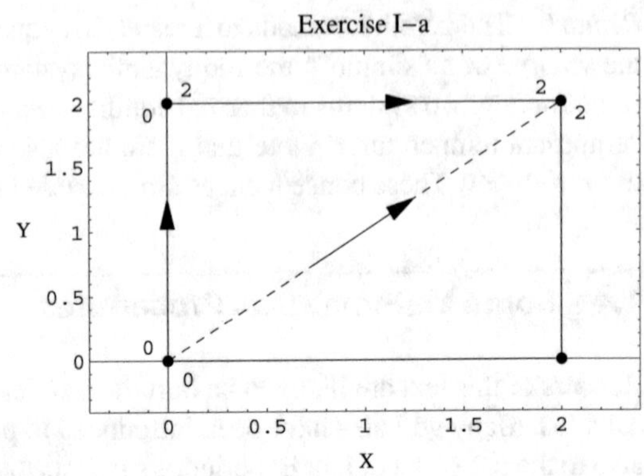

Fig. 1.1 The paths traveled are: (first) from the point (0, 0) up to (0, 2) and then to (2, 2); (second) another possibility is to travel directly from (0, 0) across to the point (2, 2)

1.4.2 Problems

Problem 1.1 Show that

$$\mathrm{d}Z(X, Y) = \big(2AXY^2 + BY^2 + 2CXY + D\big)\mathrm{d}X + \big(2AX^2Y + 2BXY + CX^2 + E\big)\mathrm{d}Y \tag{1.12}$$

is an exact differential.

Solution 1.1 Let us try to determine whether the line-integral,

$$\int_{(X_1,Y_1)}^{(X_2,Y_2)} \mathrm{d}Z(X, Y),$$

between any two points, (X_1, Y_1) and (X_2, Y_2), is path independent. To this end, set $(X_1, Y_1) = (0, 0)$ and $(X_2, Y_2) = (2, 2)$. A simple check is provided by choosing two different paths—as in Fig. 1.1—that use straight-lines. Because, in principle, there are infinitely many different paths in the (X, Y) plane that can connect the given two points, the use of only two paths, while being indicative of the result, can by no means be considered a proof. A convincing proof is provided in Problem 1.1b below.

Route 1 for Problem 1.1 Let us first travel from (0, 0) to (0, 2) and then from (0, 2) to (2, 2). According to (1.12), we get

$$\begin{aligned}\int_{(0,0)}^{(0,2)} \mathrm{d}Z(X, Y) &= \int_{(0,0)}^{(0,2)} (2AXY^2 + BY^2 + 2\,C\;XY + D)\mathrm{d}X \\ &+ \int_{(0,0)}^{(0,2)} (2AX^2Y + 2BXY + C\;X^2 + E)\mathrm{d}Y \\ &= 0 + \int_0^2 (E)\,\mathrm{d}Y = 2E\,.\end{aligned} \tag{1.13}$$

Note, in the above, $X = 0$, and also $dX = 0$.

Next, we need the integral from (0, 2) to (2, 2), i.e.,

$$\int_{(0,2)}^{(2,2)} \mathrm{d}Z(X,Y) = \int_{(0,2)}^{(2,2)} (2AXY^2 + BY^2 + 2\,C\;XY + D)\mathrm{d}X$$
$$+ \int_{(0,2)}^{(2,2)} (2AX^2Y + 2BXY + C\;X^2 + E)\mathrm{d}Y$$
$$= \int_0^2 (8AX + 4B + 4C\;X + D)\,\mathrm{d}X + 0$$
$$= (16A + 8B + 8C + 2D)\,. \tag{1.14}$$

Note, in the above, $Y = 2$. Additionally, $dY = 0$. Thus the total value of the line integral along route I is

$$\int_{(0,0)}^{(2,2)} \mathrm{d}Z(X,Y) = \int_{(0,0)}^{(0,2)} \mathrm{d}Z(X,Y) + \int_{(0,2)}^{(2,2)} \mathrm{d}Z(X,Y)$$
$$= (2E + 16A + 8B + 8C + 2D)\,. \tag{1.15}$$

Route 2 for Problem 1.1 Travel directly along the straight line from (0, 0) to (2, 2). In order to work out the line integral—as always—we need to arrange things so that any of the integrals used involves only one variable. Because the equation for the relevant path is $X = Y$, and also because both X and Y extend from 0 to 2, one gets

$$\int_{(0,0)}^{(2,2)} dZ(X,Y) = \int_{(0,0)}^{(2,2)} (2AXY^2 + BY^2 + 2\,C\;XY + D)\mathrm{d}X$$
$$+ \int_{(0,0)}^{(2,2)} (2AX^2Y + 2BXY + C\;X^2 + E)\mathrm{d}Y$$
$$= \int_0^2 (2AX^3 + BX^2 + 2\,C\;X^2 + D)\,\mathrm{d}X$$
$$+ \int_0^2 (2AY^3 + 2BY^2 + C\;Y^2 + E)\,\mathrm{d}Y$$
$$= (16A + 8B + 8C + 2D + 2E)\,. \tag{1.16}$$

As expected, this result is identical to that noted in (1.15).

Notation Due to an exact differential

$$\int_{(X_\mathrm{i},Y_\mathrm{i})}^{(X_\mathrm{f},Y_\mathrm{f})} \mathrm{d}Z(X,Y) = \alpha(X_\mathrm{f},Y_\mathrm{f}) - \alpha(X_\mathrm{i},Y_\mathrm{i})\,,$$

and similarly

$$\int_{(X_\mathrm{f},Y_\mathrm{f})}^{(X_\mathrm{i},Y_\mathrm{i})} \mathrm{d}Z(X,Y) = \alpha(X_\mathrm{i},Y_\mathrm{i}) - \alpha(X_\mathrm{f},Y_\mathrm{f})\,,$$

adding the two gives

$$\int_{i\to f\to i} dZ(X,Y) = \int_{(X_i,\,Y_i)}^{(X_f,Y_f)} dZ(X,Y) + \int_{(X_f,Y_f)}^{(X_i,Y_i)} dZ(X,Y)$$
$$= \alpha(X_f,Y_f) - \alpha(X_i,Y_i) + \alpha(X_i,Y_i) - \alpha(X_f,Y_f)$$
$$= 0\,. \tag{1.17}$$

This is valid for all paths of the integration which lie within the (X, Y) plane and form a closed loop. Such a loop starts off at some arbitrary initial location i within the (X, Y) plane, travels within the plane to some arbitrary final point f, and at the end returns to the initial location i.

Accordingly, the usual notation for displaying an exact differential dZ is the following:

$$\oint dZ(X,Y) = \oint dZ = 0\,. \tag{1.18}$$

In general, such an exact differential can be expressed as

$$dZ = dZ(X,Y) = N(X,Y)dX + M(X,Y)dY\,, \tag{1.19}$$

where $N(X, Y)$ and $M(X, Y)$ are functions of X and Y and obey the so-called "integrability" requirement

$$N(X,Y) = \left(\frac{\partial Z}{\partial X}\right)_Y, \quad M(X,Y) = \left(\frac{\partial Z}{\partial X}\right)_Z, \tag{1.20}$$

which holds if the second mixed derivatives of Z are equal. That is,[10]

$$\frac{\partial^2 Z}{\partial Y \partial X} = \left(\frac{\partial N}{\partial Y}\right)_X = \left(\frac{\partial M}{\partial X}\right)_Y = \frac{\partial^2 Z}{\partial X \partial Y}\,. \tag{1.21}$$

Thus the standard representation for an exact differential—in two dimensions—is

$$dZ = \left(\frac{\partial Z}{\partial X}\right)_Y dX + \left(\frac{\partial Z}{\partial Y}\right)_X dY\,. \tag{1.22}$$

Problem 1.2 A rigorous method to show

$$dZ(X,Y) = \left(2AXY^2 + BY^2 + 2CXY + D\right)dX$$
$$+ \left(2AX^2Y + 2BXY + CX^2 + E\right)dY \tag{1.23}$$

is an exact differential.

[10]Beginners usually benefit by being reminded that the operation $(\frac{\partial Z}{\partial X})_Y$ consists in finding a derivative of Z with respect to X while holding the variable Y constant.

Solution 1.2 Calculus tells us that a necessary and sufficient condition for

$$dZ(x, y) = M(x, y)dx + N(x, y)dy \tag{1.24}$$

to be an exact differential is the requirement that the following equality holds:

$$\left(\frac{\partial M(x, y)}{\partial y}\right)_x = \left(\frac{\partial N(x, y)}{\partial x}\right)_y .$$

Thus, for $dZ(X, Y)$ to be an exact differential, we must have

$$\left(\frac{\partial(2AXY^2 + BY^2 + 2CXY + D)}{\partial Y}\right)_X = 4AXY + 2BY + 2CX$$

equal to

$$\left(\frac{\partial(2AX^2Y + 2BXY + CX^2 + E)}{\partial X}\right)_Y = 4AXY + 2BY + 2CX ,$$

which is the case.

Inexact Differential

Problem 1.3 Let us look at the differential

$$\begin{aligned} dZ(X, Y) &= \left(BY^2 + 2C\,XY + D\right)dX \\ &\quad + \left(2AX^2Y + 2BXY + C\,X^2 + E\right)dY \\ &\equiv N(X, Y)\,dX + M(X, Y)\,dY . \end{aligned} \tag{1.25}$$

Solution 1.3 Clearly, this differential is inexact because

$$\begin{aligned} &\left(\frac{\partial N}{\partial Y}\right)_X = 2BY + 2CX \\ \neq &\left(\frac{\partial M}{\partial X}\right)_Y = 4AXY + 2BY + 2CX . \end{aligned} \tag{1.26}$$

Of course, information about the inexactness of this differential can also be obtained by integrating along route I—similar to the way it was done in (1.15)—and route II—much as was done in (1.16). As behooves an inexact differential, these two results are not the same. Being equal to $(8B + 8C + 2D + 2E)$ and $(8A + 8B + 8C + 2D + 2E)$, respectively, they differ by an amount equal to $8A$.

1.4.3 State Functions and Variables

Any thermodynamics function, Z, which admits of an exact differential, dZ, is vested with the special title of state function. As mentioned earlier, for a "simple" thermodynamic system, variables X and Y may represent any of the three pairs,

(p, v), (p, t), or (v, t). These pairs are called state variables. An obvious property of these variables is that their differentials, $\mathrm{d}p = \mathrm{d}p(v, t)$, $\mathrm{d}v = \mathrm{d}v(p, t)$, or $\mathrm{d}t = \mathrm{d}t(p, v)$, are also exact.

An essential attribute of a state function $Z(X, Y)$ (for a simple thermodynamic system) therefore is that for any thermodynamic equilibrium state, the state function depends only on the value of the given pair of state variables, X and Y. As a result, when a thermodynamic system travels from an initial position specified by a given initial value of the relevant state variables, namely $(X_\mathrm{i}, Y_\mathrm{i})$, to another position specified by a given final value, say $(X_\mathrm{f}, Y_\mathrm{f})$, of the same two state variables, the resultant change in any state function, namely $Z(X_\mathrm{f}, Y_\mathrm{f}) - Z(X_\mathrm{i}, Y_\mathrm{i})$, is independent of the path taken. Furthermore, as a necessary consequence, any travel that takes the system completely around a closed loop leaves the value of the state function, Z, unchanged.

1.5 Cyclic Identity

Much of the work involved in the study of elementary thermodynamics consists in deriving relationships between quantities that are easy to access experimentally and others that are less so. This is not an idle exercise. Specific heat, compressibility, expansivity, etc., are quantities that are readily measurable under most ambient experimental environments. Their relationships to various derivatives of important thermodynamic state functions, that may not be so easy to measure, help in the evaluation of these functions. Examples of these are the system internal energy, enthalpy, entropy, etc.

The establishment of such relationships can often be accomplished by the use of exact differentials of the state functions themselves as well as those of the relevant state variables. Such procedures involve the use of (1.22).

Noting that $\mathrm{d}p(v, t)$ and $\mathrm{d}v(p, t)$ are exact differentials, we can use (1.22) and write

$$\mathrm{d}p = \left(\frac{\partial p}{\partial v}\right)_\mathrm{t} \mathrm{d}v + \left(\frac{\partial p}{\partial t}\right)_\mathrm{v} \mathrm{d}t\,,$$
$$\mathrm{d}v = \left(\frac{\partial v}{\partial p}\right)_\mathrm{t} \mathrm{d}p + \left(\frac{\partial v}{\partial t}\right)_\mathrm{p} \mathrm{d}t\,. \tag{1.27}$$

Eliminating $\mathrm{d}v$ between these two equations yields

$$\left[1 - \left(\frac{\partial p}{\partial v}\right)_\mathrm{t}\left(\frac{\partial v}{\partial p}\right)_\mathrm{t}\right]\mathrm{d}p = \left[\left(\frac{\partial p}{\partial v}\right)_\mathrm{t}\left(\frac{\partial v}{\partial t}\right)_\mathrm{p} + \left(\frac{\partial p}{\partial t}\right)_\mathrm{v}\right]\mathrm{d}t\,. \tag{1.28}$$

Consider two neighboring equilibrium states of a simple thermodynamic system at temperature t. Recall that the thermodynamics of a simple system depends only on three state variable p, v, and t. The fixing of one of these specifies the interdependence of the other two. For example, given any two neighboring equilibrium states that are both at the same temperature, i.e., $\mathrm{d}t = 0$, then according to (1.28),

$$\left[1 - \left(\frac{\partial p}{\partial v}\right)_{\mathrm{t}} \left(\frac{\partial v}{\partial p}\right)_{\mathrm{t}}\right] \mathrm{d}p = 0 .$$

And unless the two "neighboring" states are identical—in which case $\mathrm{d}p$, $\mathrm{d}v$ are both vanishing—generally, $\mathrm{d}p$ and $\mathrm{d}v$ are $\neq 0$. Therefore, in general we must have

$$1 - \left(\frac{\partial p}{\partial v}\right)_{\mathrm{t}} \left(\frac{\partial v}{\partial p}\right)_{\mathrm{t}} = 0 .$$

This, of course, is the unsurprising statement that

$$\left(\frac{\partial p}{\partial v}\right)_{\mathrm{t}} = \frac{1}{(\frac{\partial v}{\partial p})_{\mathrm{t}}} . \tag{1.29}$$

Next consider two neighboring states with the same pressure. Here $\mathrm{d}p = 0$ but $\mathrm{d}v$ and $\mathrm{d}t$ are in general not equal to zero. Therefore, (1.28), namely

$$\left[1 - \left(\frac{\partial p}{\partial v}\right)_{\mathrm{t}} \left(\frac{\partial v}{\partial p}\right)_{\mathrm{t}}\right] \times 0 = \left[\left(\frac{\partial p}{\partial v}\right)_{\mathrm{t}} \left(\frac{\partial v}{\partial t}\right)_{\mathrm{p}} + \left(\frac{\partial p}{\partial t}\right)_{\mathrm{v}}\right] \mathrm{d}t ,$$

leads to the result

$$0 = \left(\frac{\partial p}{\partial v}\right)_{\mathrm{t}} \left(\frac{\partial v}{\partial t}\right)_{\mathrm{p}} + \left(\frac{\partial p}{\partial t}\right)_{\mathrm{v}} . \tag{1.30}$$

With help from (1.29), (1.30) can also be written as

$$\left(\frac{\partial p}{\partial v}\right)_{\mathrm{t}} \left(\frac{\partial v}{\partial t}\right)_{\mathrm{p}} \left(\frac{\partial t}{\partial p}\right)_{\mathrm{v}} = -1 . \tag{1.31}$$

Equation (1.31)—or, equivalently (1.30)—will be referred to as the "cyclic identity."

Exercise 1.1 Instead of $\mathrm{d}v$, eliminate $\mathrm{d}p$ from the two equations in (1.27) and rederive the cyclic identity. Also, do the same by eliminating $\mathrm{d}t$.

Remark In the following chapters, we shall make much use of the cyclic identity and the formula introduced in (1.31). Also, occasionally we shall work with an easy Jacobian determinant (JD) procedure.

1.6 Jacobian

1.6.1 A Simple Technique

The (JD) offers a simple, and often an efficient, procedure for establishing thermodynamic interrelationships.

Below we describe its use in treating state functions $A(X, Y)$ and $B(X, Y)$ of a simple thermodynamic system where X and Y are any two of the three state variables. The relevant (JD) makes use of a 2×2 Jacobian of the general form $\frac{\partial(A,B)}{\partial(X,Y)}$. That is,

$$\frac{\partial(A, B)}{\partial(X, Y)} = \begin{vmatrix} (\frac{\partial A}{\partial X})_Y & (\frac{\partial A}{\partial Y})_X \\ (\frac{\partial B}{\partial X})_Y & (\frac{\partial B}{\partial Y})_X \end{vmatrix} = \begin{vmatrix} (\frac{\partial A}{\partial X})_Y & (\frac{\partial B}{\partial X})_Y \\ (\frac{\partial A}{\partial Y})_X & (\frac{\partial B}{\partial Y})_X \end{vmatrix}$$
$$= \left(\frac{\partial A}{\partial X}\right)_Y \left(\frac{\partial B}{\partial Y}\right)_X - \left(\frac{\partial A}{\partial Y}\right)_X \left(\frac{\partial B}{\partial X}\right)_Y . \tag{1.32}$$

Clearly, by exchanging A and B, and/or X and Y, we find the relationships

$$\frac{\partial(A, B)}{\partial(X, Y)} = -\frac{\partial(B, A)}{\partial(X, Y)} = \frac{\partial(B, A)}{\partial(Y, X)} = -\frac{\partial(A, B)}{\partial(Y, X)} . \tag{1.33}$$

In other words, a single reversal in the order of the entries at the top or the bottom of $\frac{\partial(A,B)}{\partial(X,Y)}$ causes a (single) change of sign of the Jacobian. Similarly, if the number of reversals is two, it causes (two changes in sign amounting to) no change in sign. An important property of the Jacobian determinant, that will be found very useful, is the following:

$$\frac{\partial(A, B)}{\partial(X, Y)} = \frac{\partial(A, B)}{\partial(C, D)} \cdot \frac{\partial(C, D)}{\partial(X, Y)} . \tag{1.34}$$

An easy way to remember this is the following: what you add at the bottom here, you add "next-door" at the top.

It is clear that in a fashion similar to (1.34), we can extend the process further. That is,

$$\frac{\partial(A, B)}{\partial(C, D)} \cdot \frac{\partial(C, D)}{\partial(X, Y)} = \frac{\partial(A, B)}{\partial(C, D)} \cdot \frac{\partial(C, D)}{\partial(E, F)} \cdot \frac{\partial(E, F)}{\partial(X, Y)}$$
$$= \frac{\partial(A, B)}{\partial(C, D)} \cdot \frac{\partial(D, C)}{\partial(E, F)} \cdot \frac{\partial(E, F)}{\partial(G, H)} \cdot \frac{\partial(G, H)}{\partial(X, Y)}$$
$$= \text{etc.} \tag{1.35}$$

Problem 1.4 Prove the following equality (which is similar to (1.34)):

$$\frac{\partial(x_1, x_2)}{\partial(z_1, z_2)} = \frac{\partial(x_1, x_2)}{\partial(y_1, y_2)} \cdot \frac{\partial(y_1, y_2)}{\partial(z_1, z_2)} . \tag{1.36}$$

Solution 1.4 Here x_1 and x_2 depend on two variables z_1 and z_2, each of which also happens to depend on two other variables y_1 and y_2. For such a case, the "chain rule" of differential calculus tells us that[11]

$$\left(\frac{x_i}{z_1}\right)_{z_2} = \left(\frac{x_i}{y_1}\right)_{y_2} \left(\frac{y_1}{z_1}\right)_{z_2} + \left(\frac{x_i}{y_2}\right)_{y_1} \left(\frac{y_2}{z_1}\right)_{z_2} \tag{1.37}$$

[11] See any text on Differential Calculus, or [3].

and

$$\left(\frac{x_i}{z_2}\right)_{z_1} = \left(\frac{x_i}{y_1}\right)_{y_2}\left(\frac{y_1}{z_2}\right)_{z_1} + \left(\frac{x_i}{y_2}\right)_{y_1}\left(\frac{y_2}{z_2}\right)_{z_1}, \tag{1.38}$$

where $i = 1$ or 2. Equations (1.37) and (1.38) prove the desired equality (1.36).[12]

1.6.2 Jacobian Employed

As noted earlier, partial derivatives of the form $(\frac{\partial Z}{\partial X})_Y$ play an important role in the study of thermodynamics. The (JD) provides a convenient representation for such derivatives. For example, in (1.32), simple notational change A to Z and B to Y yields

$$\frac{\partial(Z,Y)}{\partial(X,Y)} = \begin{vmatrix} (\frac{\partial Z}{\partial X})_Y & (\frac{\partial Z}{\partial Y})_X \\ (\frac{\partial Y}{\partial X})_Y & (\frac{\partial Y}{\partial Y})_X \end{vmatrix}. \tag{1.39}$$

Because

$$\left(\frac{\partial Y}{\partial X}\right)_Y = 0, \quad \left(\frac{\partial Y}{\partial Y}\right)_X = 1,$$

(1.39) gives

$$\frac{\partial(Z,Y)}{\partial(X,Y)} = \begin{vmatrix} (\frac{\partial Z}{\partial X})_Y & (\frac{\partial Z}{\partial Y})_X \\ 0 & 1 \end{vmatrix} = \left(\frac{\partial Z}{\partial X}\right)_Y. \tag{1.40}$$

The significance of (1.40) should be emphasized: it connects the typical partial derivative, $(\frac{\partial Z}{\partial X})_Y$, that is often used in thermodynamics, with a simple (JD), $\frac{\partial(Z,Y)}{\partial(X,Y)}$, which is easy to manipulate. An aid to memory is the following:

The essential feature of the derivative $(\frac{\partial Z}{\partial X})_Y$ is that Z is being varied as a function of X, while Y is being kept constant. Accordingly, in $\frac{\partial(Z,Y)}{\partial(X,Y)}$, Z occurs at the top left, X at the bottom left, and Y occurs to the right, both at the top and the bottom.

1.7 Additional Helpful Identities

Consider variables X, Y, and Z where only two are independent. According to (1.40),

$$\frac{\partial(X,Z)}{\partial(X,Z)} = \left(\frac{\partial X}{\partial X}\right)_Z. \tag{1.41}$$

[12]Note, in order to convince oneself of the above statement, one needs first to write down the two 2 × 2 determinants on the right-hand side of (1.36); next to multiply them and write the result naturally as a 2 × 2 determinant. This resultant determinant consists of the four terms given as (1.37) and (1.38). It should be identical to the 2 × 2 determinant on the left-hand side of (1.36).

Following the procedure described in (1.34), the left-hand side of (1.41) can be extended to

$$\frac{\partial(X,Z)}{\partial(Y,Z)} \cdot \frac{\partial(Y,Z)}{\partial(X,Z)} = \frac{\partial(X,Z)}{\partial(X,Z)} = \left(\frac{\partial X}{\partial X}\right)_Z = 1\,. \tag{1.42}$$

Indeed, we can continue to extend the left-hand side and write

$$\frac{\partial(X,Z)}{\partial(Y,Z)} \cdot \frac{\partial(Y,Z)}{\partial(X,Y)} \cdot \frac{\partial(X,Y)}{\partial(X,Z)} = 1\,.$$

Now, as explained earlier, if we reverse the order in three of the six factors, the overall sign will get reversed three times—which is equivalent to a single reversal of sign, i.e.,

$$\frac{\partial(X,Z)}{\partial(Y,Z)} \cdot \frac{\partial(Z,Y)}{\partial(X,Y)} \cdot \frac{\partial(Y,X)}{\partial(Z,X)} = -1\,. \tag{1.43}$$

Because of its importance in the following, the cyclic identity is derived by a different procedure.

1.7.1 Cyclic Identity Rederived

Using the transliteration embodied in (1.40), the above can be represented as follows:

$$\left(\frac{\partial X}{\partial Y}\right)_Z \cdot \left(\frac{\partial Z}{\partial X}\right)_Y \cdot \left(\frac{\partial Y}{\partial Z}\right)_X = -1\,.$$

The mutual interchange of the last two terms makes this equation easier to memorize. That is,

$$\left(\frac{\partial X}{\partial Y}\right)_Z \cdot \left(\frac{\partial Y}{\partial Z}\right)_X \cdot \left(\frac{\partial Z}{\partial X}\right)_Y = -1\,. \tag{1.44}$$

Aid to memory: think of the cyclical order

$$(X \to Y \to Z \to X \to Y).$$

Often, it is helpful to recast the cyclic identity in the following form:

$$\left(\frac{\partial X}{\partial Y}\right)_Z = -\left(\frac{\partial X}{\partial Z}\right)_Y \cdot \left(\frac{\partial Z}{\partial Y}\right)_X\,. \tag{1.45}$$

The cyclic identity given in (1.44) and (1.45) is identical to that given earlier in (1.31) and (1.30). Recall that this identity represents an important relationship and will be put to good use in this text.

1.7.2 Simple Identity

Another identity that is worth noting is obtained straightforwardly when Jacobian representation is employed:

$$\frac{\partial(A,G)}{\partial(B,G)} = \frac{\partial(A,G)}{\partial(\Sigma,G)} \cdot \frac{\partial(\Sigma,G)}{\partial(B,G)} \,. \tag{1.46}$$

Equivalently, it reads

$$\left(\frac{\partial A}{\partial B}\right)_{\mathrm{G}} = \left(\frac{\partial A}{\partial \Sigma}\right)_{\mathrm{G}} \cdot \left(\frac{\partial \Sigma}{\partial B}\right)_{\mathrm{G}} \,. \tag{1.47}$$

We shall call this the simple identity.[13]

1.7.3 Mixed Identity

As stated earlier, all thermodynamic functions that refer to (a given quantity of) a simple system depend on the variables P, V, and T. While only two of these three variables are linearly independent, occasionally it is useful to work with all three. For instance, consider a function A,

$$A = A(X, Y, Z)\,.$$

Treating the pair X and Y as the independent variables, we have

$$\mathrm{d}A = \left(\frac{\partial A}{\partial X}\right)_{\mathrm{Y}} \mathrm{d}X + \left(\frac{\partial A}{\partial Y}\right)_{\mathrm{X}} \mathrm{d}Y\,. \tag{1.48}$$

Because Y and Z can also, just as well, be treated as the independent pair, X can be represented in terms of them as

$$X = X(Y, Z)\,.$$

Thus

$$\mathrm{d}X = \left(\frac{\partial X}{\partial Y}\right)_{\mathrm{Z}} \mathrm{d}Y + \left(\frac{\partial X}{\partial Z}\right)_{\mathrm{Y}} \mathrm{d}Z\,.$$

When introduced on the right-hand side of (1.48), this gives

$$\mathrm{d}A = \left[\left(\frac{\partial A}{\partial X}\right)_{\mathrm{Y}}\left(\frac{\partial X}{\partial Y}\right)_{\mathrm{Z}} + \left(\frac{\partial A}{\partial Y}\right)_{\mathrm{X}}\right]\mathrm{d}Y + \left(\frac{\partial A}{\partial Z}\right)_{\mathrm{Y}} \mathrm{d}Z\,.$$

[13] Some students may prefer to use (1.47) as a calculus identity and work backwards to (1.46) as a useful Jacobian identity.

Comparing this relationship with one where Y, Z are treated as the independent pair, i.e.,

$$\mathrm{d}A = \left(\frac{\partial A}{\partial Y}\right)_Z \mathrm{d}Y + \left(\frac{\partial A}{\partial Z}\right)_Y \mathrm{d}Z\,,$$

we get an identity

$$\left(\frac{\partial A}{\partial Y}\right)_Z = \left(\frac{\partial A}{\partial Y}\right)_X + \left(\frac{\partial A}{\partial X}\right)_Y \left(\frac{\partial X}{\partial Y}\right)_Z . \tag{1.49}$$

By making a cyclic change of the variables—such that, Y goes to Z, Z goes to X, and X goes to Y—we get two other equivalent relationships:

$$\left(\frac{\partial A}{\partial Z}\right)_X = \left(\frac{\partial A}{\partial Z}\right)_Y + \left(\frac{\partial A}{\partial Y}\right)_Z \left(\frac{\partial Y}{\partial Z}\right)_X \tag{1.50}$$

and

$$\left(\frac{\partial A}{\partial X}\right)_Y = \left(\frac{\partial A}{\partial X}\right)_Z + \left(\frac{\partial A}{\partial Z}\right)_X \left(\frac{\partial Z}{\partial X}\right)_Y . \tag{1.51}$$

For future reference, these will be called mixed identities.

References

1. A.B. Pippard, *Elements of Classical Thermodynamics* (Cambridge University Press, Cambridge, 1957), pp. 7–11
2. D. ter Haar, H. Wergeland, *Elements of Thermodynamics* (Addison-Wesley, Reading, 1966)
3. M.L. Boas, *Mathematical Methods in the Physical Sciences*, 3rd edn. (Wiley, New York, 2005)

Perfect Gas

2

Physics is arguably unique in its emphasis on constructing models that may pertain to systems of interest. Often these models are simple and lacking the complexity of the real systems. But the payoff of such simplicity is the ability to fully predict and understand the model system. The hope always is that such understanding will give insight into the physics of the real system of interest. Section 2.1 refers to perfect gases.

"Perfect gas" is a model of an idealized gas. As is customary, it was constructed to understand the behavior of real gases. Originally motivated by experimental observation, the model has had a long and illustrious history, and a great deal of success. A perfect gas of volume V consists of a large number of atoms. There is no interatomic coupling, the size of the atoms is vanishingly small, and the containing walls of the vessel are smooth and featureless. All collisions between the atoms and walls are perfectly elastic, effects of gravity are absent, also no other external forces are present and the gas is in thermal equilibrium. Further, all N atoms are in random motion. Using standard thermodynamics techniques, the role of pressure, temperature, and volume in perfect gases is analyzed. Section 2.2 deals with issues relating to statistical techniques. These include internal energy, attainment of equilibrium, equation of state, as well as intensive and extensive parameters. In Sect. 2.3 monatomic and diatomic perfect gases are analyzed. A mixture of perfect gases is treated in Sect. 2.4. Dalton's law of partial pressure is studied in Sect. 2.5. Perfect gas atmospheres of various varieties are discussed in Sect. 2.6. The energy of isothermal atmosphere is calculated in Sect. 2.7. A perfect gas of extremely relativistic particle is treated in Sect. 2.8. Seven problems are solved and their import is analyzed in Sect. 2.8.1.

2.1 Model

A perfect gas of volume V consists of a large number of atoms. There is no interatomic coupling, the size of the atoms is vanishingly small, and the containing walls of the vessel are smooth and featureless. All collisions between the atoms and walls

R. Tahir-Kheli, *General and Statistical Thermodynamics*,
https://doi.org/10.1007/978-3-030-20700-7_2

are perfectly elastic, the effects of gravity are absent, also no other external forces are present. The gas is in thermal equilibrium. Further, all N atoms are in random motion.

2.1.1 Pressure

Imagine that the gas consists only of atoms that have the same mass, m. It is contained in a vessel that is shaped as a perfect cube.[1] For convenience, arrange the sides of the cube, each of length L, to lie along the x, y, z axes of a Cartesian coordinate system.

Set the origin of the Cartesian coordinates at the bottom left-hand corner of the cube—i.e., at the position $(0, 0, 0)$—and the positive direction of the axes along the three edges. As such, the top corner, diagonally opposite to the origin, is at the point (L, L, L).

Examine the course of events involved in atomic collisions against the two walls that are perpendicular to the x-axis. Denote the x-component of the velocity of the ith atom, for $i = 1, 2, \ldots, N$, as $v_{\mathrm{i,x}}$.

Perfect elasticity of collisions requires that upon striking the right-hand wall with the x-component of momentum, $m \cdot v_{\mathrm{i,x}}$, the atom gets perfectly reflected. As a result, the x component of its momentum becomes $-m \cdot v_{\mathrm{i,x}}$. Accordingly, the change in the x component of momentum of the atom after one collision of the right hand wall is

$$\begin{aligned} &\text{final momentum of colliding molecule} - \text{its initial momentum} \\ &\quad = [-m \cdot v_{\mathrm{i,x}}] - [m \cdot v_{\mathrm{i,x}}] \\ &\quad = -2m \cdot v_{\mathrm{i,x}} \,. \end{aligned} \tag{2.1}$$

Because there are no external forces, the total momentum in any direction is conserved. Invoking this fact for the x-direction leads to the following requirement:

$$\begin{aligned} &\text{change in total momentum} \\ &\quad = \text{change in particle momentum} + \text{change in momentum of wall} \\ &\quad = -2m \cdot v_{\mathrm{i,x}} + (\text{change in momentum of wall})_{\mathrm{i,x}} \\ &\quad = 0 \,. \end{aligned} \tag{2.2}$$

That is, the ith atom, by a single collision of the wall perpendicular to the x-axis, causes a change in the x-component of the momentum of the wall equal to

$$(\text{change in momentum of wall})_{\mathrm{i,x}} = 2m \cdot v_{\mathrm{i,x}} \,. \tag{2.3}$$

[1] Gas in a more generally shaped vessel is analyzed in Appendix B. A beneficial payback of that analysis is its agreement with Pascal's law. From Pascal's law it can be concluded that, in the absence of external, space- and direction-dependent forces, the pressure of a fluid is constant throughout the vessel.

The absence of slowing down mechanisms insures that after traversing across the cube to the left-hand side wall placed at $x = 0$ this atom returns for another collision against the original wall at $x = L$. Such a round-trip—from the right-hand side wall to the wall on the left and then back to the wall on the right—is of length $2L$. Further, it is traversed at constant speed $|v_{i,x}|$. Therefore, the time t, taken by the atom for the round trip travel, is

$$t = \frac{\text{distance traveled}}{\text{speed of travel}} = 2L/|v_{i,x}| \,. \tag{2.4}$$

As a result, the rate of transfer of momentum perpendicular to the wall by a collision with one atom is as follows:

$$\frac{(\text{change in momentum of wall})_{i,x}}{t} = \frac{2m \cdot |v_{i,x}|}{2L/|v_{i,x}|} = \left(\frac{m}{L}\right) v_{i,x}^2 \,. \tag{2.5}$$

Summing this over all the atoms—that is, for $i = 1 \rightarrow N$—within the cube gives the total transfer rate of momentum to the right-hand side wall perpendicular to the x-axis.

According to Newton's second law of motion, such a transfer rate of momentum is equal to the force, F, exerted by the gas on the relevant wall of the cube:

$$F = \sum_{i=1}^{N} \frac{(\text{change in momentum of wall})_{i,x}}{t} = \left(\frac{m}{L}\right) \sum_{i=1}^{N} v_{i,x}^2 \,. \tag{2.6}$$

The force F acts normal to the wall under consideration. Accordingly, it exerts pressure P, defined as the perpendicular force per unit area,

$$P = F/(\text{area of the wall}) = \frac{F}{L^2} \,. \tag{2.7}$$

Combining this with (2.6) yields

$$P = \frac{1}{L^2}\left(\frac{m}{L}\right) \sum_{i=1}^{N} v_{i,x}^2 = \left(\frac{m}{V}\right) \sum_{i=1}^{N} v_{i,x}^2 = \left(\frac{m}{V}\right) N\langle v_x^2 \rangle \,. \tag{2.8}$$

Here V is the volume of the cubic container that encloses the gas—that is, $V = L^3$—and $N\langle v_x^2 \rangle$ is the average—meaning, the observed—value of the sum, $\sum_{i=1}^{N} v_{i,x}^2$. That is,

$$N\langle v_x^2 \rangle = \sum_{i=1}^{N} v_{i,x}^2 \,. \tag{2.9}$$

The gas is isotropic. Therefore

$$\langle v_x^2 \rangle = \langle v_y^2 \rangle = \langle v_z^2 \rangle = \frac{1}{3}\left[\langle v_x^2 \rangle + \langle v_y^2 \rangle + \langle v_z^2 \rangle\right] = \frac{1}{3}\langle \boldsymbol{v}^2 \rangle \,. \tag{2.10}$$

In the above, $\boldsymbol{v}$ represents the three-dimensional vector

$$\boldsymbol{v} = \boldsymbol{i}v_{\mathrm{x}} + \boldsymbol{j}v_{\mathrm{y}} + \boldsymbol{k}v_{\mathrm{z}}\,,$$

and the pair of pointed brackets, i.e., $\langle\cdot\rangle$, represents an average over all the N atoms. Equation (2.8) can be recast as

$$PV = mN\langle v_{\mathrm{x}}^2\rangle = \frac{m}{3}N\langle \boldsymbol{v}^2\rangle\,. \tag{2.11}$$

Note that, with the system being isotropic, all four quantities—V, m, N, and $\langle \boldsymbol{v}^2\rangle$—are independent of the direction x, y, or z. It is custom to specify the quantity of gas in mol numbers, n, that is, by measuring the total number of molecules[2] N in units of the Avogadro's number,[3] N_{A}, i.e.,

$$N = nN_{\mathrm{A}}\,, \quad N_{\mathrm{A}} = 6.022\,141\,79(30) \times 10^{23}\ \mathrm{mol}^{-1}\,. \tag{2.12}$$

Having derived (2.11), we are confronted with a hurdle that disappoints the users of thermodynamics. Namely, even if the interparticle interaction is exactly known,[4] standard thermodynamics does not provide tools for calculating the state functions. And because it does not have, as it were, an appropriate "arrow in its quiver," one seeks the assistance of "statistical thermodynamics" to possibly hit the target.

2.1.2 Temperature

Atoms in a perfect gas are all of infinitesimal size and are totally noninteracting. Physically, this is equivalent to saying that no atom is aware of the presence of the others.

In three dimensions, the location and momentum of an infinitesimally sized atom is specified by 3 position coordinates, e.g., $(q_{\mathrm{x}}, q_{\mathrm{y}}, q_{\mathrm{z}}) = \boldsymbol{q}$, and 3 components of its momentum vector, $\boldsymbol{p} = (p_{\mathrm{x}}, p_{\mathrm{y}}, p_{\mathrm{z}})$. The relevant Boltzmann–Maxwell–Gibbs (BMG) distribution factor, $f(\boldsymbol{q}, \boldsymbol{p})$, for the given atom is the following:[5]

$$f(\boldsymbol{q}, \boldsymbol{p}) = \frac{\exp[-\beta\mathcal{H}_{\mathcal{O}}(\boldsymbol{q}, \boldsymbol{p})]}{\int_{\boldsymbol{q}'}\int_{\boldsymbol{p}'}\exp[-\beta\mathcal{H}_{\mathcal{O}}(\boldsymbol{q}', \boldsymbol{p}')]\cdot \mathrm{d}\boldsymbol{q}'\cdot \mathrm{d}\boldsymbol{p}'}\,. \tag{2.13}$$

Here $\mathcal{H}_{\mathcal{O}}(\boldsymbol{q}, \boldsymbol{p})$ is the Hamiltonian, i.e., the functional form of the energy of the given atom in terms of its 6 variables $\boldsymbol{q}$ and $\boldsymbol{p}$, and

$$\beta = \frac{1}{k_{\mathrm{B}}T} = \frac{N_{\mathrm{A}}}{RT}\,. \tag{2.14}$$

[2]The molecule being considered here is monatomic, that is, a single atom constitutes a molecule.

[3]For values of the physical constants, see http://physics.nist.gov/cuu/Constants. Note that the numbers in the parentheses represent the standard uncertainty corresponding to the last digits shown.

[4]As is the case here: there is no interaction!

[5]A more complete analysis is given in (11.80)–(11.81).

The parameter T represents the statistical-thermodynamical temperature—usually called the Kelvin temperature and labeled as K; N_A is a constant, called the Avogadro number which has already been defined in (2.12). Additionally, k_B, and therefore R, are also constants. That is,

$$R = 8.3144\,72(15)\ \mathrm{J\,mol^{-1}\,K^{-1}}\,,$$
$$k_B = 1.38065\,04(24) \times 10^{-23}\ \mathrm{J\,K^{-1}}\,. \tag{2.15}$$

Note that R is called the "molar gas constant" and k_B is known as the Boltzmann constant.

In accordance with the BMG postulates, in thermodynamic equilibrium the normalized average (i.e., the observed value $\langle\Omega\rangle$) of any thermodynamic function, $\Omega(\boldsymbol{q}, \boldsymbol{p})$, for the specified single atom is given by the following integral:[6]

$$\langle\Omega\rangle = \int_{\boldsymbol{q}}\int_{\boldsymbol{p}} \Omega(\boldsymbol{q},\boldsymbol{p}) \cdot f(\boldsymbol{q},\boldsymbol{p}) \cdot \mathrm{d}\boldsymbol{q} \cdot \mathrm{d}\boldsymbol{p}\,. \tag{2.16}$$

Note that $f(\boldsymbol{q},\boldsymbol{p})$ is as defined in (2.13). The integrations over the three position variables, $\boldsymbol{q}$, occur over the maximum (three-dimensional) volume V available to the given atom. Each of the three momentum variables in $\boldsymbol{p}$ is integrated over the infinite range from $-\infty$ to $+\infty$.

2.2 Statistical Techniques

Statistical techniques require the use of a Hamiltonian. The Hamiltonian $\mathcal{H}_O(\boldsymbol{q}, \boldsymbol{p})$ does not depend on the position vector $\boldsymbol{q}$. Rather, it contains just the kinetic energy of the given atom. Therefore, it depends only on its mass m and the square of its momentum. That is,

$$\mathcal{H}_O(\boldsymbol{q}, \boldsymbol{p}) = \frac{1}{2m}\left(p_x^2 + p_x^2 + p_x^2\right) = \frac{\boldsymbol{p}^2}{2m}\,. \tag{2.17}$$

Because there are no direction-dependent forces[7] present, the atom behaves isotropically. As a result, we have

$$\langle(p_x^2)\rangle = \langle(p_y^2)\rangle = \langle(p_z^2)\rangle = \frac{\langle(p_x^2)\rangle + \langle(p_y^2)\rangle + \langle(p_z^2)\rangle}{3}\,. \tag{2.18}$$

2.2.1 Internal Energy

According to the argument given earlier, and (2.17) and (2.18), the average value of the energy—to be called the internal energy and denoted as U—of a perfect gas of N noninteracting infinitesimally sized atoms, each of mass m, is given by the

[6]Note that a normalized average of any constant, say α, is equal to itself, that is, $\langle\alpha\rangle = \alpha$.
[7]For example, gravity.

relationship

$$U = N\langle \mathcal{H}_O \rangle = N\frac{\langle \boldsymbol{p}^2 \rangle}{2\,m} = 3\,N\frac{\langle (p_x^2) \rangle}{2m}\,. \tag{2.19}$$

According to (2.13) and (2.16), the observed value $\langle (p_x^2) \rangle$ is the following:

$$\langle (p_x^2) \rangle = \frac{\int_{\boldsymbol{q}} \mathrm{d}\boldsymbol{q} \cdot \int_{\boldsymbol{p}} p_x^2 \cdot \exp[-\beta \mathcal{H}_O(\boldsymbol{q}, \boldsymbol{p})] \cdot \mathrm{d}p_x \cdot \mathrm{d}p_y \cdot \mathrm{d}p_z}{\int_{\boldsymbol{q}'} \mathrm{d}\boldsymbol{q}' \cdot \int_{\boldsymbol{p}} \exp[-\beta \mathcal{H}_O(\boldsymbol{q}', \boldsymbol{p}')] \cdot \mathrm{d}p_x' \cdot \mathrm{d}p_y' \cdot \mathrm{d}p_z'}\,. \tag{2.20}$$

Because $\mathcal{H}_O$ and p_x^2 do not depend on $\boldsymbol{q}$, the integral $\int_{\boldsymbol{q}} \mathrm{d}\boldsymbol{q}$ is equal to just the maximum volume V available for the motion of the molecule. Therefore the integral $\int_{\boldsymbol{q}} \mathrm{d}\boldsymbol{q}$ in the numerator and $\int_{\boldsymbol{q}'} \mathrm{d}\boldsymbol{q}'$ in the denominator are, of course, the same. As a result, they cancel out, and we are left with the relationship

$$\langle (p_x^2) \rangle = \frac{\int_{\boldsymbol{p}} p_x^2 \cdot \exp[-\beta \mathcal{H}_O(\boldsymbol{q}, \boldsymbol{p})] \cdot \mathrm{d}p_x \cdot \mathrm{d}p_y \cdot \mathrm{d}p_z}{\int_{\boldsymbol{p}} \exp[-\beta \mathcal{H}_O(\boldsymbol{q}', \boldsymbol{p}')] \cdot \mathrm{d}p_x' \cdot \mathrm{d}p_y' \cdot \mathrm{d}p_z'}\,. \tag{2.21}$$

The remaining integrals in (2.21) are of a standard form and are worked out in detail in (B.1)–(B.5). In particular, it is shown that the following is true:

$$\begin{aligned} \int_{-\infty}^{+\infty} \exp(-\alpha \eta^2)\mathrm{d}\eta &= \sqrt{\frac{\pi}{\alpha}}\,, \\ \int_{-\infty}^{+\infty} \eta^2 \exp(-\alpha \eta^2)\mathrm{d}\eta &= \frac{1}{2}\sqrt{\frac{\pi}{\alpha^3}}\,. \end{aligned} \tag{2.22}$$

Now let us first look at the denominator in (2.21), namely

$$\begin{aligned} &\int_{\boldsymbol{p}} \exp[-\beta \mathcal{H}_O(\boldsymbol{q}, \boldsymbol{p})] \cdot \mathrm{d}p_x \cdot \mathrm{d}p_y \cdot \mathrm{d}p_z \\ &\quad = \int_{-\infty}^{\infty} \mathrm{d}p_x \int_{-\infty}^{\infty} \mathrm{d}p_y \int_{-\infty}^{\infty} \mathrm{d}p_z \cdot \exp\left[-\frac{\beta}{2m}(p_x^2 + p_y^2 + p_z^2)\right]. \end{aligned} \tag{2.23}$$

The three integrals in (2.23) that are being multiplied together are all equal. Therefore, we have

$$\begin{aligned} &\int_{\boldsymbol{p}} \exp[-\beta \mathcal{H}_O(\boldsymbol{q}, \boldsymbol{p})] \cdot \mathrm{d}p_x \cdot \mathrm{d}p_y \cdot \mathrm{d}p_z \\ &\quad = \left[\int_{-\infty}^{\infty} \mathrm{d}p \cdot \exp\left(\frac{-\beta p^2}{2m}\right)\right]^3 = \left(\frac{2m\pi}{\beta}\right)^{\frac{3}{2}}\,. \end{aligned} \tag{2.24}$$

Equation (2.24) is the denominator of the right-hand side of (2.21). To deal with the numerator of (2.21), let us separate the integral over the variable p_x, i.e.,

$$\int_{-\infty}^{\infty} p_x^2 \exp\left(\frac{-\beta p_x^2}{2m}\right)\mathrm{d}p_x\,,$$

from the rest of the two integrals:

$$\int_p p_x^2 \cdot \exp[-\beta \mathcal{H}_O(\boldsymbol{q}, \boldsymbol{p})] \mathrm{d}p_x \cdot \mathrm{d}p_y \cdot \mathrm{d}p_z$$

$$= \left(\int_{-\infty}^{\infty} p_x^2 \exp\left[\frac{-\beta p_x^2}{2m}\right] \mathrm{d}p_x\right) \cdot \left[\int_{-\infty}^{\infty} \exp\left(\frac{-\beta p_y^2}{2m}\right) \mathrm{d}p_y \int_{-\infty}^{\infty} \exp\left(\frac{-\beta p_z^2}{2m}\right) dp_z\right]$$

$$= \left(\frac{1}{2}\sqrt{\frac{\pi}{(\frac{\beta}{2m})^3}}\right) \cdot \left[\frac{\pi}{\frac{\beta}{2m}}\right]^{\frac{2}{2}} = \left(\frac{1}{2}\sqrt{\pi}\left(\frac{2m}{\beta}\right)^{\frac{3}{2}}\right)\left[\frac{2m\pi}{\beta}\right]. \tag{2.25}$$

Equation (2.25) gives the numerator of the right-hand side of (2.21). In order to determine the thermodynamic average $\langle p_x^2 \rangle$, we need to divide the result obtained in (2.25) by that found in (2.24), where

$$\langle p_x^2 \rangle = \frac{(\frac{1}{2}\sqrt{\pi}(\frac{2m}{\beta})^{\frac{3}{2}}) \cdot [\frac{2m\pi}{\beta}]}{(\frac{2m\pi}{\beta})^{\frac{3}{2}}} = \frac{m}{\beta} = k_B T m . \tag{2.26}$$

The internal energy U, of the perfect gas consisting of N atoms that are all identical to the given atom being studied here, is specified by (2.19) and (2.26). Accordingly, we have

$$U = mN\frac{\langle \boldsymbol{v}^2 \rangle}{2} = N\left[3\frac{\langle p_x^2 \rangle}{2m}\right] = \frac{3N}{2} k_B T . \tag{2.27}$$

2.2.2 Equation of State

According to (2.11) and (2.27), for a perfect gas the product of the gas pressure P and volume V is directly related to the parameter T as

$$PV = \frac{m}{3} N \langle \boldsymbol{v}^2 \rangle = N k_B T = nRT . \tag{2.28}$$

Remember that the number of moles of the perfect gas being treated here is n.

Let us recall a finding of the zeroth law; see (1.10). For a simple thermodynamic system, an equation of state can be defined which relates a function of its pressure and volume to its empirical temperature. In the absence of magnetic and electric effects, and gravity, etc., for simple isotropic systems composed of a fixed number of moles, only two of the three parameters P, V, and T that occur in (2.28) are independent. In that spirit, this defines the equation of state of the perfect gas. There are other equations that also qualify for the title "Equations of State." A more general discussion of this subject is given in Chap. 8.

Legendre transformations and Helmhotz free energy are discussed in Sects. 10.7 and 10.8.

The parameters P, V, and T that appear in the equation of state (2.28) are examples of state variables. Similarly, the function U is an example of a state function.

Intensive and Extensive Properties

For a macroscopic system, any quantity that is proportional to the system size is called *extensive*.

For a macroscopic system, those thermodynamic properties that are independent of the system size are called *intensive*.

To determine whether we are dealing with an extensive or intensive property, we need to consider whether it doubles or remains largely unchanged when the system size—i.e., the number of particles in the system—is doubled. Clearly, all things being equal, a system with twice as many particles will have twice the volume. Also it will have twice the mass and therefore twice the kinetic energy.

Thus the volume V and internal energy U are extensive.

On the other hand, a consequence of the zeroth-law is that once an isolated macroscopic system has reached thermodynamic equilibrium, its temperature becomes uniform.[8] Similarly, while in thermodynamic equilibrium, in the absence of gravity or other external fields, we are assured by Pascal's law that a fluid at rest has uniform pressure. The same applies to isotropic solids under "hydrostatic" pressure.[9] Thus, a thought experiment that divides such systems into two parts would leave both T and the pressure P unaffected.

Statistical-mechanical considerations helped define a precise quantity T. Additionally, a purely thermodynamic description of the temperature is available from the work of N.L. Sadi Carnot[10] and Lord Kelvin.[11]

Therefore, the temperature and pressure are examples of intensive properties.

Identification of Temperature

Also, there is substantial historical information that thermodynamics, in different ways, has played an important role in the identification of the temperature, and indeed that the temperature so identified is identical to the parameter T suggested by statistical mechanics.

Robert Boyle, working with a given amount of gas—that is, keeping the number of atoms N, or equivalently the mole number n, constant—at room temperature, observed that the product of the pressure P and volume V remained unchanged as one or the other, or both, were varied.

Let us look at the equation of state (2.28) above, and note the fact that when measurements are made for a given number of moles at some given room temperature, it is found that the product of the pressure and volume, i.e., PV, remains unchanged. As a result the parameter T that occurs in (2.28), is constant. Clearly, the constancy of T must then be identified with the constancy of the room temperature. And the temperature T—estimated initially in the classic work attributed

---

[8] Different macroscopic parts of a thermodynamic system are in equilibrium. Therefore they have the same temperature. See, for example, Sect. 9.1.

[9] Of course, gravity and other external fields are assumed to be absent.

[10] See, for example, (4.2) and the description provided in the following parts of Chap. 4.

[11] See the related (7.56) and the associated discussion in Sect. 7.2.

to J.A.C. Charles—was found to depend on what—originally proposed by Anders Celsius—is now known as the Celsius temperature θ_c. That is,

$$T = \theta_c + T_0 ,$$

where T_0 is a constant.

For air—assumed here to be a perfect gas—kept in a vessel (that is, at constant volume), a plot of pressure P versus the Celsius temperature θ_c is a straight line that, when lowering the pressure, heads toward lower temperature. In this spirit, zero pressure for a gas implied that it had reached "zero temperature," that is, $T = 0$. When experimental results on different low-density gases were extrapolated far down to zero pressure, they tended to a result for the Celsius temperature which was $\approx -270°$. Interpreting Charles' observations, this would indicate that $\approx -270°$ is the value, of the Celsius temperature, θ_c, when the true thermodynamic temperature, T, is equal to zero, that is, $0 \approx -270° + T_0$, or equivalently, $T_0 \approx 270°$. In recent years, by formal international agreement, the value of T_0 has been fixed at exactly 273.15°. A schematic description of the process which led to this determination is as follows.

Because only two parameters are needed to specify a straight line, both the dictates of the Celsius scale and experiment can be fixed by the satisfaction of two constraints. These constraints arise because the 100° difference on the Celsius scale is equal to the temperature difference between the so-called "ice point"[12] and "steam point."[13] Accordingly, we proceed as follows:

(i) At fixed volume V, let the pressure in a perfect gas, for the ice and steam points be P_i and P_s, respectively. Then, according to the ideal gas equation of state

$$\left(\frac{P_s}{P_i}\right)_V = \frac{T_0 + 100}{T_0} , \quad (2.29)$$

or equivalently,

$$T_0 = \frac{100}{(\frac{P_s}{P_i})_V - 1} , \quad (2.30)$$

where T_0 is the perfect gas temperature at the ice point.

(ii) Experimentally measure, as accurately as possible, the ratio $(\frac{P_s}{P_i})_V$. The choice $T_0 = 273.15$ corresponds to $(\frac{P_s}{P_i})_V$ being equal to 1.3661.

[12]On the Celsius scale, the temperature of the ice point T is 0°. Note that the ice point is represented by the equilibrium state of a mixture of pure water, fully saturated with air at pressure of exactly one atmosphere, and pure ice.

[13]The steam point refers to the equilibrium state of pure water boiling under one atmosphere of air. On the Celsius scale, the temperature of the steam point is set at 100°.

It is worth mentioning that rather than the ice point—the exact reproducibility of which is hard to achieve—the triple point[14] of pure water is a more reliable reference point to use.

Attainment of Thermodynamic Equilibrium

Implicit in the specification of a "temperature," or the quantity T if we prefer, is the assumption of thermodynamic equilibrium. Given the vanishingly small particle size and absence of mutual interaction in a perfect gas, one may legitimately ask how such an equilibrium can be brought about?

Clearly, interparticle communication has to be mediated by the walls of the container with which the particles collide. Because such equilibrating tendency is proportional to the area of the containing walls, its rate is necessarily very slow. For instance, if L is a typical dimension of the containing vessel, then its volume V and area A have the following dependence on L:

$$V \propto L^3 , \qquad A \propto L^2 .$$

For any given particle density,

$$V \propto N .$$

Hence,

$$A \propto N^{\frac{2}{3}} .$$

Therefore for large N, such equilibrating is much slower than that involving direct interparticle communication which is $\propto N$.

2.3 Monatomic and Diatomic Perfect Gases

2.3.1 Monatomic Perfect Gas

Quite clearly, perfect gas is an idealization. Indeed, it is often called an "ideal gas."[15] In practice, gases are "real" and there is enough interaction to obtain thermal equilibrium fairly quickly. For instance, in the limit of low densities, many gases are found to behave much like a perfect gas that has achieved thermal equilibrium.

Because there are no external and internal forces, the physical energy—to be called the internal energy U—consists only of the kinetic energy of the N monatomic molecules,

$$U = \frac{1}{2} m N \langle \boldsymbol{v}^2 \rangle = \left(\frac{3}{2} n R \right) T = \left(\frac{3}{2} N k_{\mathrm{B}} \right) T \equiv C_{\mathrm{V}} T . \tag{2.31}$$

[14]The triple point is where water vapor, pure liquid water, and pure ice all coexist in thermodynamic equilibrium. At this temperature, defined to be exactly equal to 273.16 K, the sublimation pressure of pure ice equals the vapor pressure of pure water.

[15]For this reason, when referring to a gas we shall use the terms "perfect" and "ideal" synonymously.

In the above we have used (2.28). Also we have introduced the notation $C_{\rm V}$ which stands for the specific heat at constant volume of n moles of a monatomic ideal gas. While a more complete description of specific heat is best deferred till later, it is convenient to refer to a single mole and instead represent the molar specific heat $C_{\rm v}$ as[16]

$$C_{\rm v} = \frac{C_{\rm V}}{n} \equiv \frac{f}{2} R\,. \tag{2.32}$$

The number of degrees of freedom of a molecule is denoted as f. As is clear from (2.31), for a monatomic molecule $f = 3$. If an atom that constitutes a monatomic molecule is itself of infinitesimal size—as it is supposed to be in a perfect gas—by fiat any notion of it rotating around its own center can be dropped. Legitimately, therefore, it can possess only three translational degrees of freedom. Accordingly, the motion of such a monatomic molecule is fully described by its momentum vector which has three components in the three-dimensional physical space.

2.3.2 Diatomic Perfect Gas

A monatomic molecule in a perfect gas—i.e., a single zero-sized atom—by definition, cannot have any intraatomic (i.e., intramolecular) vibrational or rotational motion.

On the other hand, if the gas consisted of diatomic molecules, this would not be the case. Here, depending on the temperature, both molecular rotation and intramolecular vibration would occur.

Indeed, owing to the presence of two atoms, each of which has—under the above assumption—only three degrees of freedom, a diatomic molecule in a perfect, totally noninteracting gas—with no associated electronic or nuclear dynamics whatever—can, in principle, have up to six degrees of freedom. Alternatively, one can also represent six degrees of freedom in terms of three translational degrees of freedom of the center of mass of the diatom, two mutually perpendicular rotational degrees of freedom for rotation around the center of mass, and one intramolecular—that is, intradiatomic—vibrational degree of freedom (see Sect. 11.3, "Perfect Gas of Classical Diatoms", (11.34)–(11.71), for a more complete discussion of this subject).

At laboratory temperatures, light diatomic ideal gases generally show only five degrees of freedom, that is, the five degrees without the intradiatomic vibration. This is so because, in practice, contribution of the sixth degree of freedom—namely the intradiatomic vibrational degree of freedom—in typical, nearly perfect, light diatomic gases appears only when the temperature rises well above the room temperature.

---

[16]Usually, thermodynamic quantities for one mole will be denoted by lower case subscripts while upper case subscripts will refer to systems of general size. Thus the specific heat $C_{\rm V}$ is n times the molar specific heat, which in turn is denoted as $C_{\rm v}$. The same applies to $C_{\rm P}$ and $C_{\rm p}$. While we shall make an effort to follow this rule about the subscripts, often, for convenience, $C_{\rm v}$ will equivalently be denoted as $c_{\rm v}$, and $C_{\rm p}$ as $c_{\rm p}$. And occasionally—hopefully not often—we may mistakenly even forget to follow this rule!

Accordingly, when the intramolecular vibrational energies for light diatoms are measured in units of the Boltzmann constant $k_B = (\frac{R}{N_A})$, they correspond to values of the temperature T that are much higher than those normally used in the laboratory. In contrast, heavy diatoms get their vibrational modes excited at much lower temperatures.

2.4 Mixture of Perfect Gases

Let us pose the question: When a quantity of a perfect gas—say, n_2 mol—is introduced into a thermally isolated—i.e., adiabatic—chamber that already contains some amount—say, n_1 mol—of a different perfect gas, how are the temperature and pressure affected? For notational convenience, let us denote the parameters of the original gas with index 1, e.g., m_1, T_1, $C_v(1)$, etc., and those referring to the additional gas with index 2.

For a given perfect gas, the presence of additional molecules of another perfect gas gets noticed only through the intermediation of the containing walls. If $T_2 > T_1$, the newly added molecules will heat up the walls. And for the same reason, if $T_2 < T_1$, their net effect will be to cool the walls down. In due course, this process will affect the average kinetic energies of both the original and the newly introduced molecules, and when thermal equilibrium is reached there will be a common final temperature, say T. Furthermore, because the chamber containing the gases is thermally isolated and does not permit exchange of energy with the environment, there will be no net change in the total energy of the two gases contained therein. Therefore, the final value, U_f, of the internal energy of the mixture will be equal to the sum of the energies of the components (1) and (2). Thus, according to (2.31) and (2.32), we have

$$U_f = n_1 C_v(1) T_1 + n_2 C_v(2) T_2 = \left[n_1 C_v(1) + n_2 C_v(2)\right] T , \tag{2.33}$$

where the common final temperature T is

$$T = \frac{n_1 C_v(1) T_1 + n_2 C_v(2) T_2}{n_1 C_v(1) + n_2 C_v(2)} . \tag{2.34}$$

2.5 Dalton's Law of Partial Pressure

After the temperature has equilibrated to its final value T, let the volume occupied by the mixture of the two gases be V. Because the gases are perfect, molecules roam around, in volume V, independently of each other. As such the pressure exerted on the enclosing walls is caused by impacts of all individual molecules, that is, pressure P_1 is caused by N_1 molecules of variety (1) and P_2 by N_2 molecules of variety (2). These pressures act independently and additively. Thus the total pressure, P, experienced by the enclosing walls is the sum, i.e.,

$$P = P_1 + P_2 . \tag{2.35}$$

The pressure P_1 is specified according to the relationship

$$P_1 V = \left(\frac{m_1}{3}\right) N_1 \langle \boldsymbol{v}_1^2 \rangle = \left(\frac{m_1}{3}\right) n_1 N_A \langle \boldsymbol{v}_1^2 \rangle$$
$$= N_1 k_B T = n_1 N_A k_B T = n_1 RT . \tag{2.36}$$

Similarly, the corresponding relationship for the molecules of variety (2) is

$$P_2 V = \left(\frac{m_2}{3}\right) N_2 \langle \boldsymbol{v}_2^2 \rangle = \left(\frac{m_2}{3}\right) n_2 N_A \langle \boldsymbol{v}_2^2 \rangle$$
$$= N_2 k_B T = n_2 N_A k_B T = n_2 RT . \tag{2.37}$$

Adding the two together gives

$$(P_1 + P_2)V = PV = (n_1 + n_2)RT . \tag{2.38}$$

From (2.35)–(2.38), we have

$$\frac{P_1}{n_1} = \frac{P_2}{n_2} = \frac{RT}{V} = \frac{P}{n_1 + n_2} ,$$

which leads to the useful result:

$$P_1 = \left(\frac{n_1}{n_1 + n_2}\right) P , \quad P_2 = \left(\frac{n_2}{n_1 + n_2}\right) P . \tag{2.39}$$

To recapitulate: The total pressure, exerted by two different perfect gases when placed together in a given volume and kept at a given temperature, is the sum of the partial pressures, P_1 and P_2. These pressures would have been exerted individually by the two perfect gases if they had separately been placed in the same volume V and kept at the same temperature T. Another important feature of the above set of results is that on average the kinetic energy per molecule is the same for the two gases, that is,

$$\left(\frac{m_1}{2}\right) \langle \boldsymbol{v}_1^2 \rangle = \left(\frac{m_2}{2}\right) \langle \boldsymbol{v}_2^2 \rangle = \left(\frac{3}{2}\right) \left(\frac{R}{N_A}\right) T = \left(\frac{3}{2}\right) k_B T . \tag{2.40}$$

Finally, we note that the above analysis can readily be extended to the case where more than two different perfect gases are mixed together. Thus

$$\left(\frac{m_1}{2}\right) \langle \boldsymbol{v}_1^2 \rangle = \left(\frac{m_2}{2}\right) \langle \boldsymbol{v}_2^2 \rangle = \left(\frac{m_3}{2}\right) \langle \boldsymbol{v}_3^2 \rangle = \cdots = \left(\frac{3}{2}\right) \left(\frac{R}{N_A}\right) T = \left(\frac{3}{2}\right) k_B T , \tag{2.41}$$

and

$$P = P_1 + P_2 + P_3 + \cdots , \tag{2.42}$$

where

$$P_i = \left(\frac{n_i}{n_1 + n_2 + n_3 + \cdots}\right) P . \tag{2.43}$$

2.6 Perfect Gas Atmosphere

As long as the gases being mixed are perfect, (2.41), (2.42), and (2.43) hold. This is true irrespective of the complexity of the molecules, or the differences in their masses.

It is important to note that on average the translational kinetic energy of all molecules is equal. To a layman, this result might appear counterintuitive, or at a minimum somewhat surprising. This would especially be so if the molecular masses of the gases being mixed were very different.

2.6.1 Barometric Equation

Treat air in the atmosphere as perfect diatomic gas. The mass of a single molecule is equal to m. Assume that the air temperature, T K, and the acceleration due to gravity, g m/s^2, do not depend on the altitude. With these assumptions, the relationship between the air pressure at height h and the temperature is

$$P_{\mathrm{h}} = P_0 \exp\left(-\frac{N_{\mathrm{A}} mgh}{RT}\right). \tag{2.44}$$

Here P_0 and P_{h} are the atmospheric pressures at sea level and at height h meters, respectively, and (2.44) is the so-called "barometric equation" for the isothermal atmosphere.

Imagine a massless, elementary tablet of base area A placed horizontally at height y in the atmosphere. The thickness of the tablet is Δy. Although Δy is considered to be very small, the volume of the tablet, $A \cdot \Delta y$, is assumed to be sufficiently large so that it encloses enough air molecules that their number is large compared to unity.

Assume the density of the gas at this height is $\rho(y)$. The part of the downward force provided by the air contained within volume $A \cdot \Delta y$ is its weight, which is equal to (density) × (volume) × (acceleration due to gravity), that is, $\rho(y) \cdot A\Delta y \cdot g$.

The total downward force, F_{down}, on the tablet is, of course, caused by the gas pressure P acting on the top horizontal surface of area A plus the weight of the air contained inside the elementary tablet. Consequently,

$$F_{\mathrm{down}} = PA + \rho(y)gA \cdot \Delta y\,. \tag{2.45}$$

The upward force, F_{up}, is caused only by the gas pressure $(P - \Delta P)$ acting at the bottom of the tablet, i.e.,

$$F_{\mathrm{up}} = (P - \Delta P)A\,.$$

In dynamical equilibrium, the upward force acting at the base of the tablet must exactly counterbalance the downward force acting at the top of the tablet. That is,

$$F_{\mathrm{up}} = F_{\mathrm{down}}\,,$$

or equivalently,

$$(P - \Delta P)A = PA + \rho(y)gA \cdot \Delta y\,.$$

Therefore, upon canceling PA and dividing the remainder by A on both sides, we get

$$-\,\Delta P = \rho(y)g \cdot \Delta y\,. \tag{2.46}$$

Remember that $\Delta y \ll 1$. Therefore, without too much loss of generality, we can replace Δy, and as a result ΔP, by $\mathrm{d}y$ and $\mathrm{d}P$, respectively. That is, we can write

$$-\,\mathrm{d}P = \rho(y)g \cdot \mathrm{d}y\,. \tag{2.47}$$

Because of the occurrence of $\mathrm{d}y$ and $\mathrm{d}P$ in the above equation, it is clear that in order to make any headway towards a solution, one needs to know $\rho(y)$ as a function either of y or P. Fortunately, the ideal gas equation of state allows $\rho(y)$ to be expressed in terms of the relevant pressure P. To this end, consider some arbitrary volume $V(y)$ at height y. Then according to (2.28),

$$PV(y) = N(y)k_{\mathrm{B}}T\,, \tag{2.48}$$

where $N(y)$ is the total number of molecules contained in the chosen volume $V(y)$. The mass density for this volume is $\rho(y)$, where

$$\rho(y) = \frac{mN(y)}{V(y)}\,. \tag{2.49}$$

Using (2.48), we can write this as

$$\rho(y) = \frac{mP}{k_{\mathrm{B}}T}\,. \tag{2.50}$$

Thus from (2.47) we have

$$-\,\mathrm{d}P = P \cdot \left(\frac{mg}{k_{\mathrm{B}}T}\right)\mathrm{d}y\,. \tag{2.51}$$

Now divide both sides by P and set the integration from $0 \to h$, which gives

$$-\int_{P_0}^{P_{\mathrm{h}}} \frac{\mathrm{d}P}{P} = \int_0^h \left(\frac{mg}{k_{\mathrm{B}}T}\right)\mathrm{d}y\,. \tag{2.52}$$

Note that the pressures at heights 0 and h are denoted as P_0 and P_{h}, respectively.

Carrying out the above integration, multiplying both sides by -1, i.e.,

$$\ln\left[\frac{P_{\mathrm{h}}}{P_0}\right] = -\left(\frac{mgh}{k_{\mathrm{B}}T}\right), \tag{2.53}$$

and exponentiating both sides leads to the desired barometric equation:

$$P_h = P_0 \exp\left(-\frac{mgh}{k_B T}\right) = P_0 \exp\left(-\frac{N_A mgh}{RT}\right). \tag{2.54}$$

2.6.2 A Related Calculation

Air is composed mostly of diatomic nitrogen and oxygen gases with molar masses of 28 and 32 g, respectively. Therefore, with appropriate ratio of nitrogen and oxygen present, its molar mass M is $\sim 29 \times 10^{-3}$ kg. Assume that the average temperature in the atmosphere is T. Empirically, we know that T decreases with increasing altitude. An average of the ground temperature $\sim$ 300 °K and the temperature $\sim$ 240 K at height $\sim$ 10 km suggests a value for $T \sim 270$ K.

In terms of the molar mass, $M = N_A m$, (2.54) can be written as

$$\frac{P_h}{P_0} = \exp\left(-\frac{N_A mgh}{RT}\right) = \exp\left(-\frac{Mgh}{RT}\right), \tag{2.55}$$

and we get[17] for $h = 10$ km, which is equal to 10^4 m,

$$\frac{P_h}{P_0} = \exp\left[-\frac{(29 \times 10^{-3}) \times (9.8) \times 10^4}{8.3 \times 270}\right] = 0.28\,. \tag{2.56}$$

For an ideal gas at temperature T, we have

$$\left(\frac{P}{RT}\right) = \left(\frac{n}{V}\right).$$

The volume V contains n mol of air each of mass M. Hence its density is

$$\rho = \left(\frac{nM}{V}\right) = \left(\frac{PM}{RT}\right). \tag{2.57}$$

Note that, in the present problem, $(\frac{M}{RT})$ is a constant.

As a result, the pressure P_y at any height y is directly proportional to the density ρ_y at that height. Therefore, we have for the ratio of the densities at heights 10^4 and 0 the following result:

$$\rho_{10^4}/\rho_0 = P_{10^4}/P_0 = 0.28\,. \tag{2.58}$$

Note that the relevant factor 0.28 is given in (2.56).

[17] When mass M is measured in kg, g in m/s^2, and h in m, then the dimensions of the numerator of the exponent are $Mgh = \text{kg} \times \text{m}^2/\text{s}^2 = \text{J}$.

The denominator, i.e., RT, should be considered to be nRT where the number of moles n is equal to 1. Accordingly, the denominator translates into the following units: mol for 1, JK^{-1} mol^{-1} for R, and K for the temperature T. Thus the dimensions of the denominator are mol $\times$ JK^{-1} mol^{-1} $\times$ K = J.

Specified Percentage of Molecules Found Below Given Height

Let us calculate the altitude h_c, below which a specified fraction—say, 90%—of the air molecules in the earth's atmosphere are found. To this end, we use the notation described above. In accord with (2.55) and (2.58), we have the relationship

$$\frac{P_y}{P_0} = \frac{\rho_y}{\rho_0} = \exp\left[-\frac{Mgy}{RT}\right]. \tag{2.59}$$

Thus the total mass of air contained in a cylinder of base area A rising all the way to infinity is

$$\begin{aligned} \text{mass}(\infty) &= \int_0^\infty \mathrm{d}m = \int_0^\infty \rho_y \cdot A\mathrm{d}y = A\rho_0 \int_0^\infty \exp\left[-\frac{Mgy}{RT}\right]\mathrm{d}y \\ &= A\rho_0 (RT/Mg)\,. \end{aligned} \tag{2.60}$$

Similarly, the mass of the column rising to height h is

$$\begin{aligned} \text{mass}(h) &= A\int_0^h \rho_y \mathrm{d}y = A\rho_0 \int_0^h \exp\left[-\frac{Mgy}{RT}\right]\mathrm{d}y \\ &= A\rho_0 (RT/Mg)\left[1 - \exp(-Mgh/RT)\right] \\ &= \text{mass}(\infty)\left[1 - \exp(-Mgh/RT)\right]. \end{aligned} \tag{2.61}$$

When

$$\text{mass}(h_c)/\text{mass}(\infty) = 0.90\,,$$

(2.61) leads to the result

$$\exp\left(-\frac{Mgh_c}{RT}\right) = 0.10\,.$$

Therefore

$$h_c = (RT/Mg)\ln(1/0.10) = \frac{8.3 \times 270}{29 \times 10^{-3} \times 9.8}\ln(10) = 18\ \text{km}\,. \tag{2.62}$$

2.7 Energy of Isothermal Atmosphere

Treat, as before, the atmosphere as an ideal gas and assume that the acceleration due to gravity g is independent of the height h. Then calculate the total energy of a column of area A and height h of the isothermal atmosphere at temperature T_0 K.

To this end, we consider an elementary tablet of gas with base area A, a tiny thickness, Δy, and mass density $\rho(y)$ at altitude y. Its mass, Δm, being proportional to Δy, is also tiny, i.e.,

$$\Delta m = A\rho(y)\Delta y\,, \tag{2.63}$$

as is its gravitational potential energy,

$$\Delta U_{\text{grav}} = gy \cdot \Delta m = gy \cdot A\rho(y)\Delta y\,.$$

In the limit $\Delta y \ll 1$, the sum where y goes from 0 to h is well approximated by appropriate integrals on both sides of the equation, i.e.,

$$\int_0^{U_{\text{grav}}(h)} \mathrm{d}U_{\text{grav}} = Ag \int_0^h y\rho(y)\mathrm{d}y\,,$$

which leads to

$$U_{\text{grav}}(h) = Ag \int_0^h y\rho(y)\mathrm{d}y\,. \tag{2.64}$$

According to (2.59),

$$\rho(y) = \rho(0)\exp\left[-\left(\frac{Mg}{RT}\right)y\right].$$

Therefore, setting $T = T_0$, (2.64) can be written as

$$U_{\text{grav}}(h) = Ag\rho(0)\int_0^h y\exp(-\alpha y)\mathrm{d}y\,, \tag{2.65}$$

where

$$\alpha = \frac{Mg}{RT_0}\,. \tag{2.66}$$

Also, because $P_0V_0 = nRT_0$, we can write

$$P_0M = RT_0\left(\frac{nM}{V_0}\right) = RT_0\,\rho(0)\,.$$

Thus the atmospheric density at the ground level is

$$\rho(0) = (P_0M/RT_0)\,, \tag{2.67}$$

where P_0 is the corresponding pressure (compare with (2.57)).

The integral in (2.65) can be evaluated by parts as follows:

$$\begin{aligned} U_{\text{grav}}(h)/[Ag\rho(0)] &= \left(-\frac{\exp(-\alpha y)y}{\alpha}\bigg|_0^h\right) - \left(\frac{\exp(-\alpha y)}{\alpha^2}\bigg|_0^h\right) \\ &= -\exp(-\alpha h)\cdot\frac{h}{\alpha} + \frac{1-\exp(-\alpha h)}{\alpha^2}\,. \end{aligned} \tag{2.68}$$

Let us next look at the internal energy U_{int} of the same column. For the elementary tablet of air, of mass Δm, the internal energy is ΔU_{int}, which is

$$\Delta U_{\text{int}} = C_{\text{v}}T_0\Delta n\,, \tag{2.69}$$

where Δn counts the number of moles of air contained in the tablet,

$$\Delta n = \frac{\Delta m}{M} = \frac{A}{M}\rho(y)\Delta y\,. \tag{2.70}$$

Thus using (2.70), (2.69), and (2.59), we can write

$$\begin{aligned} U_{\rm int}(h) &= \int_0^h {\rm d}U_{\rm int} = C_{\rm V}T_0\int_0^h {\rm d}n = C_{\rm V}T_0\frac{A}{M}\rho(0)\int_0^h {\rm d}y\exp(-\alpha y) \\ &= \left(\frac{AC_{\rm V}T_0\rho(0)}{M\alpha}\right)\big[1-\exp(-\alpha h)\big]\,. \end{aligned} \tag{2.71}$$

Before we enter any numerical values, it is helpful to organize the expressions for $U_{\rm grav}(h)$ and $U_{\rm int}(h)$ into simpler form. To this end, introduce the notation $\alpha h = \omega$ and write

$$U_{\rm grav}(h) = \left(\frac{AP_0T_0}{Mg}\right)R\big[-\omega\exp(-\omega)+1-\exp(-\omega)\big], \tag{2.72}$$

$$U_{\rm int}(h) = \left(\frac{AP_0T_0}{Mg}\right)C_{\rm V}\big[1-\exp(-\omega)\big] = \frac{f}{2}\cdot\left(\frac{AP_0T_0}{Mg}\right)R\big[1-\exp(-\omega)\big]\,. \tag{2.73}$$

Here f is the number of degrees of freedom of a single molecule. Note that, for diatomic atmosphere at temperature well below 5000 K, f is $= 5$.

Let us calculate the result for $A = 10^6\ {\rm m}^2$, $h = 10^4$ m, $T_0 = 290$ K, and P_0 being $10^5\ {\rm N\cdot m^{-2}}$. We get

$$\omega = \alpha h = \frac{Mgh}{RT_0} = \frac{29\times10^{-3}\times9.8\times10^4}{8.3\times290} = \frac{9.8}{8.3} = 1.2\,, \tag{2.74}$$

and

$$\frac{AP_0T_0R}{Mg} = \frac{10^6\times10^5\times290\times8.3}{29\times10^{-3}\times9.8} = 8.5\times10^{14}\ {\rm J}\,. \tag{2.75}$$

Using the results of the above equations, namely (2.74) and (2.75), the following are readily calculated:

$$U_{\rm grav}(h) = 2.9\times10^{14}\ {\rm J}\,, \qquad U_{\rm int}(h) = 1.5\times10^{15}\ {\rm J}\,.$$

It is interesting to note that if h were very large—still assuming that g is constant—then $\exp(-\alpha h) \ll 1$ and we would have

$$U_{\rm int}(h\gg1) \approx \left(\frac{AP_0T_0}{Mg}\right)C_{\rm V}\,, \tag{2.76}$$

and

$$U_{\rm grav}(h\gg1) \approx \left(\frac{AP_0T_0}{Mg}\right)R\,. \tag{2.77}$$

2.7.1 Atmosphere with Height-Dependent Temperature

Again treat the acceleration due to gravity as being independent of the altitude h. Ask the question: How does the barometric equation change if the atmospheric temperature is not constant but varies with the height h? Assume the atmospheric temperature T is 300 K at the ground level and 240 K at a height of 10^4 m. Also assume that the decrease in T is linear with the rise in the altitude. Begin with (2.50) but in addition to P also treat T as being dependent on the altitude y. That is,

$$\rho(y) = \frac{mP(y)}{k_B T(y)} = \frac{N_A m P(y)}{N_A k_B T(y)} = \frac{MP(y)}{RT(y)} . \tag{2.78}$$

Thus

$$P(y) = \frac{R\rho(y)T(y)}{M} . \tag{2.79}$$

Differentiating with respect to y and using (2.47) gives

$$\frac{dP}{dy} = \frac{R}{M}\left[\rho(y)\frac{dT(y)}{dy} + T(y)\frac{d\rho(y)}{dy}\right] = -\rho(y)g ,$$

which is readily reorganized as

$$-\frac{dT(y)}{dy} = \frac{gM}{R} + \frac{T(y)}{\rho(y)} \cdot \frac{d\rho(y)}{dy} . \tag{2.80}$$

According to the description provided, α, the rate of decrease of temperature per unit increase in the altitude, is constant. Thus

$$-\frac{dT(y)}{dy} = \alpha . \tag{2.81}$$

Combining (2.80) and (2.81) yields

$$\frac{1}{\rho(y)} \cdot d\rho(y) = \left[\alpha - (gM/R)\right]\frac{dy}{T(y)} .$$

Upon integration, we get

$$\ln\frac{\rho(y)}{\rho(0)} = \left[1 - (Mg/R\alpha)\right] \cdot \ln\left(\frac{T(0)}{T(0) - \alpha y}\right) . \tag{2.82}$$

If $T(0) = 300$ K and $T(10^4) = 240$ K,

$$\alpha = 6 \times 10^{-3}\ \text{K}\,\text{m}^{-1} .$$

Therefore, we get

$$\ln\frac{\rho(10^4)}{\rho(0)} = \left[1 - \left(\frac{29 \times 10^{-3} \times 9.8}{8.3 \times 6 \times 10^{-3}}\right)\right] \cdot \ln\left(\frac{300}{240}\right) ,$$

which yields

$$\frac{\rho(10^4)}{\rho(0)} = 0.35\,.$$

Indeed, at any height above sea level, the current atmosphere is cooler, and denser, than the isothermal atmosphere.

2.8 Perfect Gas of Extremely Relativistic Particles

A crude, approximate treatment of such a gas is presented below but for a more complete analysis, which also includes calculation of thermodynamic potentials, see (11.100)–(11.108).

According to the special theory of relativity, a particle with rest mass m_0 and velocity $\boldsymbol{v}_\mathrm{i}$ has energy

$$E_\mathrm{i} = m_\mathrm{i}c^2 = \left(\frac{m_0}{\sqrt{1-\frac{\boldsymbol{v}_\mathrm{i}^2}{c^2}}}\right)c^2 \tag{2.83}$$

and momentum

$$\boldsymbol{p}_\mathrm{i} = m_\mathrm{i}\boldsymbol{v}_\mathrm{i} = \left(\frac{m_0}{\sqrt{1-\frac{\boldsymbol{v}_\mathrm{i}^2}{c^2}}}\right)\boldsymbol{v}_\mathrm{i}\,, \tag{2.84}$$

where c is the velocity of light. Combining these equations leads to

$$E_\mathrm{i} = \sqrt{c^2\boldsymbol{p}_\mathrm{i}^2 + m_0^2c^4}\,. \tag{2.85}$$

In the extreme relativistic limit

$$c^2\boldsymbol{p}_\mathrm{i}^2 \gg m_0^2c^4\,. \tag{2.86}$$

Accordingly, in (2.85) we can ignore $m_0^2c^4$ and retain only the much larger term $c^2\boldsymbol{p}_\mathrm{i}^2$. This leads to the result

$$E_\mathrm{i} \approx cp_\mathrm{i}\,. \tag{2.87}$$

Consider N such particles, enclosed in a box of volume V. For simplicity, assume the box to be a cube of side length L. Perpendicular collision of the ith particle with a wall transfers momentum $\approx 2p_\mathrm{i}$. Because the particle is moving almost as fast as c, it returns for another collision in a time

$$\tau \approx 2\left(\frac{L}{c}\right).$$

Thus the rate of transfer of momentum to the wall per particle moving in the x-direction, i.e.,

$$\approx \frac{2p_i}{\tau},$$

is equal to the force exerted on the wall that is perpendicular to the x-direction. The corresponding contribution to the pressure is therefore

$$P_i = \frac{\text{force}}{\text{area}} \approx \left(\frac{\frac{2p_i}{\tau}}{L^2}\right) = \frac{cp_i}{L^3} = \frac{cp_i}{V}. \tag{2.88}$$

Assuming the particles are evenly distributed along the three orthogonal directions, on average one-third of the particles would be traveling in any one direction. Summing P_i over $(N/3)$ particles gives the total pressure P, on the wall perpendicular to the x-axis,

$$P = \sum_i^{\frac{N}{3}} P_i = V^{-1}\left(\sum_i^{\frac{N}{3}} cp_i\right) = V^{-1}\left(\sum_i^{\frac{N}{3}} E_i\right) = V^{-1}\left(\frac{1}{3}\sum_i^{N} E_i\right). \tag{2.89}$$

Of course, in equilibrium this pressure is uniform throughout the gas.

Thus for an extreme relativistic ideal gas our simple minded analysis leads to the result

$$(PV)_{\text{extreme-relativistic}} = \frac{1}{3}\sum_i^{N} E_i = \frac{U}{3}. \tag{2.90}$$

This result should be contrasted with that for an ordinary monatomic ideal gas for which the multiplying factor on the right-hand side of (2.90) is two-thirds rather than one-third,[18] i.e.,

$$(PV)_{\text{nonrelativistic}} = \frac{2}{3}U. \tag{2.91}$$

2.8.1 Problems 2.1–2.7

Seven different problems are worked out in detail. They include the study of Gibbs potential and partial pressure of mixtures, dissociating triatomic ozone, energy change in leaky container, mixture of carbon and oxygen, and study of pressure, volume, and temperature.

Problem 2.1 (Gibbs potential and partial pressure of mixtures)
At room temperature, $\sim$ 20 °C, and under atmospheric pressure, $P_0 \sim 10^5$ pascal, N_2 and O_2 may be treated as diatomic ideal gases.

[18] Compare with (2.31). Indeed, all nonrelativistic monatomic ideal gases, whether they be classical or quantum, obey this two-thirds U relationship.

(a) Calculate their partial pressures in a mixture containing 56 g/mol of nitrogen and 80 g/mol of oxygen.
(b) What is the volume V that the mixture occupies?

Solution 2.1 The mixture contains n_1 mol of nitrogen and n_2 of oxygen. These are calculated as follows:

$$n_1 = \frac{56}{28} = 2\,, \quad n_2 = \frac{80}{32} = 2.5\,. \tag{2.92}$$

Their partial pressures, P_1 and P_2, are:

$$P_1 = \left(\frac{n_1}{n_1+n_2}\right)P_0 = 0.444P_0\,, \qquad P_2 = \left(\frac{n_2}{n_1+n_2}\right)P_0 = 0.556P_0\,. \tag{2.93}$$

The mixture satisfies the ideal gas equation of state. Therefore, it occupies volume V given below

$$V = (n_1+n_2)\cdot R\cdot\left(\frac{T}{P_0}\right) = (4.5)\cdot 8.31\cdot\left(\frac{273+20}{10^5}\right) = 0.110\ \mathrm{m}^3\,. \tag{2.94}$$

Problem 2.2 (Dissociating triatomic ozone)
A mass M_{total} of ozone gas, which is triatomic oxygen, is contained in a chamber of volume V. The temperature in the chamber is raised to some high temperature T_{H}. As a result, some of the triatomic ozone dissociates into diatomic and monatomic oxygen. The mass of one mole of monatomic oxygen is M_0. At temperature T_{H} it is found that within the chamber the masses of monatomic, diatomic, and triatomic components are M_1, M_2, and M_3, respectively.

Defining the ratios $M_{\mathrm{i}}/M_{\mathrm{total}} = \epsilon_{\mathrm{i}}$, where $i = 1, 2, 3$, and treating the mixture as an ideal gas, show that

$$PV = RT_{\mathrm{H}}\left(\frac{M_{\mathrm{total}}}{3M_0}\right)\left(1 + 2\epsilon_1 + \frac{1}{2}\epsilon_2\right).$$

Solution 2.2 We are given

$$\sum_{\mathrm{i}=1}^{3}\left(\frac{M_{\mathrm{i}}}{M_{\mathrm{total}}}\right) = \sum_{\mathrm{i}=1}^{3}\epsilon_{\mathrm{i}} = 1\,. \tag{2.95}$$

If the number of moles of the component i is n_{i}, then

$$n_1 = \frac{M_1}{M_0}\,, \qquad n_2 = \frac{M_2}{2M_0}\,, \qquad n_3 = \frac{M_3}{3M_0}\,. \tag{2.96}$$

Also, if the partial pressure of the ith component is P_{i}, then the equation of state of the ith component is

$$P_{\mathrm{i}}V = n_{\mathrm{i}}RT_{\mathrm{H}}\,, \quad i = 1, 2, 3\,. \tag{2.97}$$

Adding these and using (2.42), (2.95), and (2.96) gives

$$
\begin{aligned}
(P_1 + P_2 + P_3)V &= PV \\
&= RT_{\mathrm{H}}(n_1 + n_2 + n_3) = \frac{RT_{\mathrm{H}}}{3M_0}\left(3M_1 + \frac{3M_2}{2} + M_3\right) \\
&= RT_{\mathrm{H}}\left(\frac{M_{\mathrm{total}}}{3M_0}\right)\left(3\epsilon_1 + \frac{3}{2}\epsilon_2 + \epsilon_3\right) \\
&= RT_{\mathrm{H}}\left(\frac{M_{\mathrm{total}}}{3M_0}\right)\left(1 + 2\epsilon_1 + \frac{1}{2}\epsilon_2\right). \qquad (2.98)
\end{aligned}
$$

Problem 2.3 (Energy change in leaky container)

(a) A leaky container of volume V_0 is placed in contact with a thermal reservoir that contains a large amount of air at temperature T_0 and pressure P_0. After the pressure and temperature inside the vessel have equilibrated to those of the reservoir, what is the internal energy, U_0, of the air inside the vessel? Treat air as a diatomic ideal gas and assume that each molecule has $f = 5$ degrees of freedom.

Note that part (b) is a slightly different problem.

(b) Air at pressure P_0 and temperature T_0 is enclosed in a container of volume V_0. All the walls of the container, except for one, are made of heat energy nonconducting material. The remaining wall, which is a conductor of heat energy and also leaks air—both in and out of the vessel—is brought into contact with a thermal air reservoir at pressure P_0 but a lower temperature T_c. Calculate the resulting change in the internal energy.

Solution 2.3 (a) The ideal gas equation of state tells us that

$$P_0 V_0 = n_0 R T_0 , \qquad (2.99)$$

where n_0 is the number of moles initially present inside the vessel. Accordingly, if the number of degrees of freedom f is $= 5$, the initial value of the internal energy is

$$U_0 = \frac{f}{2} n_0 R T_0 = \frac{f}{2} P_0 V_0 = \frac{5}{2} P_0 V_0 . \qquad (2.100)$$

(b) When the heat energy conducting wall is brought into contact with the reservoir at the lower temperature T_c, upon equilibration the temperature of the air inside the vessel falls to that of the reservoir. That is, it becomes equal to T_c. If, as a result, the number of moles of air inside the vessel changes to n_c, the equation of state becomes

$$P_0 V_0 = n_c R T_c . \qquad (2.101)$$

Note that the volume of the vessel, V_0, is fixed and because of the leakiness the pressure inside the vessel equilibrates back to the pressure P_0 of the air outside in the reservoir. Thus, despite the cooling, the internal energy

$$U_c = \frac{f}{2} n_c R T_c = \frac{f}{2} P_0 V_0 = \frac{5}{2} P_0 V_0 \quad (2.102)$$

is unchanged and U_c is the same as U_0 given in (2.100).

Remark 2.1 To a laymen, the above result would appear counterintuitive. But, on closer examination, one should notice that while the temperature inside the vessel has fallen, because of its leakiness additional molecules—carrying their kinetic energy with them—have entered the vessel. As a result the total kinetic energy, $\frac{5}{2} \cdot n_c \cdot RT_c$, of all the molecules that are now inside has remained constant at its initial value $\frac{5}{2} \cdot n_0 \cdot RT_0$. Hence, the result $n_c T_c = n_0 T_0$, or equivalently, $U_c = U_0$.

Problem 2.4 (Mixture of carbon and oxygen)
Burning of 1.20 kg of carbon[19] in an atmosphere of 1.92 kg of pure oxygen[20] creates a mixture of n_1 mol of carbon monoxide and n_2 mol of carbon dioxide. In the process, all 1.92 kg of oxygen is used up.

(a) Calculate n_1 and n_2.
(b) If the total pressure of the mixture is 1 bar, and its volume 3.00 m^3, what is its temperature?

Solution 2.4 (a) The mixture, i.e., n_1 mol of carbon monoxide and n_2 mol of carbon dioxide contains only 1200 g/mol—which is equal to n mol—of carbon where n is given as follows:

$$n = \frac{1200 \text{ g/mol}}{12 \text{ g/mol}} = 100\,. \quad (2.103)$$

Because carbon is monatomic and all atoms of carbon monoxide and carbon dioxide each contain only one atom of carbon, the total number of moles of the mixture of the two gases is equal to the number of moles of their constituent element carbon. That is,

$$n_1 + n_2 = n = 100\,. \quad (2.104)$$

Next we note that the total mass of the oxygen, i.e., 1920 g, used in the mixture is made up as follows: $n_1 \times 16$ g in carbon monoxide and $n_2 \times 32$ g in carbon dioxide. Thus

$$16n_1 + 32n_2 = 1920\,. \quad (2.105)$$

Solving (2.104) and (2.105) together readily leads to the result

$$n_1 = 80\,, \qquad n_2 = 20\,.$$

[19]One mole of the abundant isotope contains exactly 12 g/mol of carbon.
[20]The mass of 1 mol of diatomic oxygen is 16 g/mol.

(b) Treating the mixture as an ideal gas, its pressure, volume, and temperature are related through the equation of state

$$PV = (n_1 + n_2)RT = 100RT\,.$$

Therefore

$$T = \frac{PV}{100R} = \frac{10^5 \times 3.00}{100 \times 8.31} = 361\ ^\circ\mathrm{K}\,. \tag{2.106}$$

Problem 2.5 (Burning carbon)
Upon burning 1.44 kg of carbon in an atmosphere of pure oxygen, a mixture of n_1 mol of carbon monoxide and n_2 mol of carbon dioxide is created. The total mass of the mixture is 4.08 kg.

(a) Calculate n_1 and n_2.
(b) If the total pressure of the mixture is 1.5 bar, and its volume 2.00 m^3, what is its temperature T?

Solution 2.5 Using the notation of the preceding problem, the total number of moles $n_1 + n_2$ of carbon is

$$n_1 + n_2 = \frac{1440}{12} = 120\,.$$

Also because the mixture is constituted of n_1 mol of carbon monoxide and n_2 mol of carbon dioxide, its total mass can be expressed as follows:

$$4.08\ \mathrm{kg} = 4080\ \mathrm{g/mol} = n_1 \times (12 + 16)\ \mathrm{g/mol} + n_2 \times (12 + 32)\ \mathrm{g/mol}\,.$$

That is,

$$28\, n_1 + 34\, n_2 = 4080\,.$$

Solving the above two linear equations in the variables n_1 and n_2 readily gives:

$$n_1 = 75\,, \qquad n_2 = 45\,,$$

and

$$T = \frac{PV}{120R} = \frac{(1.5 \times 10^5) \times 2}{120 \times 8.31} = 301\ ^\circ\mathrm{K}\,.$$

Problem 2.6 (Pressure, volume, and temperature)
A vertical cylinder with base area $A = 0.01$ m^2 has a freely moveable piston of mass $m_1 = 10$ kg and encloses $n = 20$ mol of air, which may be treated as a perfect gas of molal mass $M = 29$ g, at initial volume $V_1 = 0.44$ m^3. The cylinder and its contents are thermally isolated. The atmospheric pressure is $P_0 = 10^5$ N/m^2.

(a) Calculate the pressure P_1 and the corresponding temperature T_1 of the gas.
(b) What is the root-mean-square velocity, v_{rms}, of the molecules?

Solution 2.6 (a) Pressure P_1 is the sum of the pressure created by the weight of the piston and the atmospheric pressure being exerted on top of the piston,

$$P_1 = P_0 + \frac{m_1 g}{A} = 10^5 + \frac{10 \times 9.8}{0.01} = 1.1 \times 10^5 \text{ N/m}^2 . \tag{2.107}$$

The corresponding temperature of the enclosed air, i.e., T_1, is now readily found

$$T_1 = \frac{P_1 V_1}{nR} = \frac{(1.1 \times 10^5) \times 0.44}{20 \times 8.3} = 29 \times 10 \text{ K} . \tag{2.108}$$

(b) We have the relationships:

$$\frac{1}{2} m N_A \langle \boldsymbol{v}^2 \rangle = \frac{M}{2} \langle \boldsymbol{v}^2 \rangle = \frac{3}{2} N_A k_B T_1 = \frac{3}{2} R T_1 . \tag{2.109}$$

Since $M = 29 \times 10^{-3}$ kg and $R = 8.3 \text{ J k}^{-1} \text{ mol}^{-1}$, (2.109) gives

$$\langle \boldsymbol{v}^2 \rangle = \frac{3RT_1}{M} = \frac{3 \times 8.3 \times 290}{(29 \times 10^{-3})} = 25 \times 10^4 \text{ (m/s)}^2 , \tag{2.110}$$

and

$$v_{\text{rms}} = \sqrt{\langle \boldsymbol{v}^2 \rangle} = 5 \times 10^2 \text{ m/s} . \tag{2.111}$$

Problem 2.7 (Addition to Problem 2.6)
After doing (a) and (b) of Problem 2.6, do the part (c) described below:

(c) A mass $m_2 = 150$ kg, initially at temperature T_1, is placed on top of the piston. The total heat energy capacity of the cylinder, piston, and the mass m_2 (excluding the enclosed air) is $C_{\text{etc}} = 500$ J/K. Assuming the piston moves inside the cylinder without friction, calculate the final volume V_2 and the temperature T_2 of the enclosed air.

Solution 2.7 (c) When the mass m_2 is added it increases the pressure felt by the contained air(gas) from P_1 to P_2, where

$$P_2 = P_1 + \frac{m_2 g}{A} = 1.1 \times 10^5 + \frac{150 \times 9.8}{0.01} = 2.6 \times 10^5 \text{ N/m}^2 . \tag{2.112}$$

Under this increased pressure, the volume of the enclosed gas decreases from V_1 to V_2. Consequently, the piston moves down a distance Δy where

$$\Delta y = \frac{(V_1 - V_2)}{A} . \tag{2.113}$$

This motion occurs under the influence of the downward force

$$F_{\text{down}} = P_2 A ,$$

and as a result work ΔW is done *on* the gas (or equivalently, work $-\Delta W$ is done *by* the gas)

$$\Delta W = F_{\text{down}} \cdot \Delta y = P_2 A \cdot \Delta y = P_2(V_1 - V_2) . \tag{2.114}$$

In turn, ΔW causes the system temperature to rise from T_1 to T_2. Here, one part of ΔW is used up in increasing the energy of the gas by an amount

$$(\mathrm{d}U)_{\text{gas}} = nc_v(T_2 - T_1) ,$$

and the other,

$$(\mathrm{d}U)_{\text{etc}} = c_{\text{etc}}(T_2 - T_1) ,$$

in increasing the temperature of the cylinder, piston, and the weights. Thus, because the system is thermally isolated, the work done, ΔW, is equal to the sum of the two increases in the energy. That is,

$$P_2(V_1 - V_2) = (\mathrm{d}U)_{\text{gas}} + (\mathrm{d}U)_{\text{etc}} = nc_{\text{v}}(T_2 - T_1) + C_{\text{etc}}(T_2 - T_1) . \tag{2.115}$$

In the above, the only unknowns are V_2 and T_2. Another linear relationship involving these two is available from the equation of state,

$$P_2 V_2 = nRT_2 . \tag{2.116}$$

Adding the left- and right-hand sides of (2.115) and (2.116) gives

$$P_2 V_1 = (nR + nc_{\text{v}} + C_{\text{etc}})T_2 - (nc_{\text{v}} + C_{\text{etc}})T_1 .$$

This, in turn, leads to

$$T_2 = \frac{P_2 V_1 + (nc_{\text{v}} + C_{\text{etc}})T_1}{n(R + c_{\text{v}}) + C_{\text{etc}}} . \tag{2.117}$$

Because $c_{\text{v}} = (5R/2)$ per mole per degree Kelvin, we have

$$T_2 = \frac{2.6 \times 10^5 \times 0.44 + [20 \times (5/2) \times 8.3 + 500] \times 290}{[20 \times (7/2) \times 8.3 + 500]} = 35 \times 10\,\text{K} . \tag{2.118}$$

Using the above value of T_2, V_2 is calculated in the usual fashion. That is,

$$V_2 = nRT_2/P_2 = 20 \times 8.3 \times 35 \times 10/\left(2.6 \times 10^5\right) = 0.22\,\text{m}^3 . \tag{2.119}$$

2.8.2 Exercises for the Student

Exercise 2.1 Location of the 90% of Molecules in an Atmosphere with Decreasing Temperature

Calculate the effective altitude in the atmosphere below which 90% of the air molecules are found. Assume the temperature decreases linearly with the altitude.

Exercise 2.2 Total Energy of a Column in an Atmosphere with Decreasing Temperature

Assuming the temperature decreases linearly with the altitude, construct an expression for the total energy of a column of area A and height h.

3 The First Law

That mechanical work generated heat energy was empirically known to the cave-man who rubbed rough dry kindling against each other to produce fire. Yet an understanding of what constituted heat energy did not come for aeons.

The so-called caloric theory of "heat" held sway until the eighteenth century. Caloric was thought to be a massless fluid whose increase warmed an object. Just like a fluid, it was allowed to flow from a hot object—where there was more of it—to a cold object until the amounts in the two objects equalized. Also, like fluids, caloric could neither be created nor destroyed. The invisible particles of the caloric were self-repellent but were positively attracted by the constituents of the system to which the caloric was added. These properties ensured that the caloric spread evenly, etc.

All this offered a handy explanation for the empirical fact that two objects in contact equalize their level of warmth. Yet, for inexplicable reasons, nobody thought of how the cave-man had created fire! Because caloric could not be created, when he rubbed his sticks of dry wood he would clearly have to have been transferring it from the thinner wood to the thicker one, or perhaps vice versa!

The chapter begins with a discussion of the heat, work, and internal energy in Sect. 3.1. Specific heat is examined in Sect. 3.2 while Sect. 3.3 is devoted to introducing the notation used in the text. Some applications of the first law are described in Sect. 3.4. For the independent variables t, v, and p the first law is used to work out solutions to several problems that relate to systems with two of the three independent variables. These solutions are provided in three sections. Section 3.5 relates to the case where t and v are the two independent variables. And Sects. 3.6 and 3.7 are devoted to discussing the cases when the two independent variables are t and p, and p and v, respectively. Enthalpy is discussed in Sect. 3.8. Hess' rules for chemothermal reactions are mentioned in Sect. 3.9, while Sect. 3.10 contains solved problems that relate to oxidation and latent heat of vaporization, as well as a variety of matters relating to adiabatic processes in ideal gases and nonconducting cylinders. Ideal gas polytropics are treated in Sect. 3.11 and problems related to interrelationships between several different thermodynamic equations are analyzed in Sect. 3.12. The derivation of equation of state from the knowledge of bulk and elastic moduli is analyzed in several solved problems in Sect. 3.13. Section 3.14 deals with Newton's

R. Tahir-Kheli, *General and Statistical Thermodynamics*,
https://doi.org/10.1007/978-3-030-20700-7_3

law of cooling. Section 3.15 refers to internal energy in noninteracting monatomic gases. The volume dependence of single particle energy levels is discussed in the concluding Sect. 3.16.

3.1 Heat, Work, and Internal Energy

Long ago there was a notion that in all substances there existed a mythical, massless substance named caloric whose total was conserved. And when hot and cold objects came into contact enough caloric got transferred from the hot to the cold to equalize their temperatures.

3.1.1 Heat

Indeed, it was not until the end of the eighteenth century that an observant military commander, Count Rumford of Bavaria (Germany), noticed something badly amiss with the caloric theory. While supervising the drilling of cannon muzzles, he noticed that they got hot. The erstwhile explanation of that phenomenon would be that the small chips—i.e., the shavings—that were bored off the cannon lost their caloric to the large remaining part. But not all the boring bits were sharp. And, indeed, the duller the bits—i.e., the greater effort it took to bore through the muzzle—the hotter the muzzles got. In other words, less chips, but more caloric!

Rumford's obvious conclusion was that heat energy was not exchanged by the transfer of the caloric but rather by the expenditure of the work that had to be done for the drilling of the cannon.

Later Ideas It was later, somewhere in the middle of the nineteenth century, that J.P. Joule was able to demonstrate that a given amount of work—electrical and/or mechanical—could be used to cause a reproducible change in the state of a thermally isolated system.

3.1.2 Work

Apparently, Joule's earliest experimentation employed the electric heating effects of resistors immersed in various quantities of water in an adiabatically isolated calorimeter. The current was supplied by a generator. Joule estimated the mechanical work needed to operate the electric generator and measured the consequent rise in the temperature of the contents of the calorimeter. For obvious reasons, such experimentation was subject to a variety of errors. Thus it is remarkable that he was able to estimate that approximately 4.6 J generate one calorie[1] equivalent of heat energy.

[1] One "thermochemical" calorie is exactly equal to 4.184 J.

Joule's later experiments utilized purely mechanical means. Known weights were dropped from measured heights. The decrease in their gravitational potential energy was converted into kinetic energy of churning paddles placed in given amounts of water initially at room temperature. The entire system was thermally isolated.

3.1.3 Internal Energy

The kinetic energy imparted to the paddles is transferred to water. There it is dissipated through viscous forces in the body of the water and drag at the surfaces in contact with the container walls. The process results in raising the temperature of the system.

The temperature rise was carefully measured. Similar experiments were also performed in which mercury was substituted for water.

These later experiments were subject to less uncertainty and yielded a result for the one-gram water/degree calorie equivalent closer to its currently accepted value of 4.184000 J.

Observations made in the foregoing experiments need to be formalized. When a thermally isolated system in equilibrium is acted upon by externally supplied work, its internal energy increases by an equivalent amount. And this increase is independent of any intermediate states that the system may have to pass through.

Outwardly, therefore, it would appear that both work and internal energy are thermodynamic state functions. A closer examination does suggest that internal energy is indeed a state function. But the same is not true for work.

Consider a vanishingly small, quasistatic,[2] volume expansion $\mathrm{d}V$ that occurs under pressure P. In the process, the system *does* an infinitesimal amount of work equal to[3]

$$\mathrm{d}W = P\mathrm{d}V\,. \tag{3.1}$$

In order to determine whether W is a state function, we ask the question: Is $\mathrm{d}W$ an exact differential? The reason we ask this question is that, according to (1.18), a function Z is a *state function* if and only if $\mathrm{d}Z$ is an exact differential.

[2]Quasistatic processes have already been described in the introductory chapter. They proceed extremely slowly and result in long series of equilibrium states. In contrast, real processes proceed at finite speeds and result in states that depart from the equilibrium.

[3]For example, consider a friction-free piston, with base area A, placed on top of some liquid of volume V contained in a cylinder. Assume that just under the piston the liquid is under pressure P and the system is in thermodynamic and mechanical equilibrium. If the pressure of the liquid should increase by an infinitesimal amount $\mathrm{d}P$, the piston will ever so slowly move upwards, say by an infinitesimal distance $\mathrm{d}Z$, thereby increasing the volume, say by an infinitesimal amount $\mathrm{d}V$. Because the pressure is defined as the perpendicular force per unit area, therefore the upward force on the piston is $F_{\mathrm{up}} = (P + dP)A$. Accordingly, the work done in the extension of the piston upwards is $\mathrm{d}W = F_{\mathrm{up}} \times \mathrm{d}Z = (P + dP)A \times \mathrm{d}Z = (PA\mathrm{d}Z + A\mathrm{d}P\mathrm{d}Z)$. Because the increase in the volume is $\mathrm{d}V = A\mathrm{d}Z$, the work done $\mathrm{d}W = (P\mathrm{d}V + \mathrm{d}P \cdot \mathrm{d}V \approx P\mathrm{d}V)$. Integrating the two sides leads to the total work done, "by the liquid," during a finite expansion, $W = \int \mathrm{d}W = \int P\mathrm{d}V$. Note that this is equal to the work that is done "on the piston" as it is raised upwards.

In "simple systems" being treated here, the variables V and P can always be considered as the two independent variables. Thus if $\mathrm{d}W$ were an exact differential, and it explicitly involved $\mathrm{d}V$ and $\mathrm{d}P$, it would have had the following form:

$$\mathrm{d}W \to \left(\frac{\partial W}{\partial V}\right)_{\mathrm{P}} \mathrm{d}V + \left(\frac{\partial W}{\partial P}\right)_{\mathrm{V}} \mathrm{d}P .$$

Accordingly, (3.1) would imply

$$\left(\frac{\partial W}{\partial V}\right)_{\mathrm{P}} = P , \quad \left(\frac{\partial W}{\partial P}\right)_{\mathrm{V}} = 0 . \tag{3.2}$$

Because

$$\frac{\partial^2 W}{\partial P \partial V} = \left(\frac{\partial (\frac{\partial W}{\partial V})_{\mathrm{P}}}{\partial P}\right)_{\mathrm{V}} = 1$$

and

$$\frac{\partial^2 W}{\partial V \partial P} = \left(\frac{\partial (\frac{\partial W}{\partial P})_{\mathrm{V}}}{\partial V}\right)_{\mathrm{P}} = 0 ,$$

the integrability requirement

$$\frac{\partial^2 W}{\partial P \partial V} = \frac{\partial^2 W}{\partial V \partial P} ,$$

which is necessary for an exact differential, would not be satisfied.

Thus the irrefutable conclusion that $\mathrm{d}W$ is not an exact differential. Yet, the experiments of Joule have demonstrated that for thermally isolated systems, in proceeding from the initial state, *initial*, to the final state, *final*, the work done "on the system" is equal to the corresponding increase in the internal energy, i.e.,

$$-\int_{\mathrm{initial}}^{\mathrm{final}} \mathrm{d}W = \int_{\mathrm{initial}}^{\mathrm{final}} \mathrm{d}U = U_{\mathrm{final}} - U_{\mathrm{initial}} . \tag{3.3}$$

Consequently,

$$\begin{aligned} -\int_{\mathrm{initial}}^{\mathrm{final}} \mathrm{d}W - \int_{\mathrm{final}}^{\mathrm{initial}} \mathrm{d}W &= U_{\mathrm{final}} - U_{\mathrm{initial}} + U_{\mathrm{initial}} - U_{\mathrm{final}} = 0 \\ &= \int_{\mathrm{initial}}^{\mathrm{final}} \mathrm{d}U + \int_{\mathrm{final}}^{\mathrm{initial}} \mathrm{d}U = \oint \mathrm{d}U . \end{aligned} \tag{3.4}$$

This result appears to be paradoxical. Despite the fact that $\mathrm{d}W$ is not an exact differential, here we have found a relationship of the form

$$-\int_{\mathrm{initial}}^{\mathrm{final}} \mathrm{d}W - \int_{\mathrm{final}}^{\mathrm{initial}} \mathrm{d}W = 0 , \tag{3.5}$$

which is normally satisfied only by an exact differential. But a close consideration shows that the essential requirement for dW to be an exact differential is indeed not satisfied. For an exact differential, (3.5) is required to be valid for all cyclic paths. But (3.5) is not so valid. The paths traveled during the round-trip integration of dW given in (3.4) and (3.5) are required to be "quasistatic." Thus, their sum is vanishing for only a limited group of paths traveled from *initial* to *final* and *final* to *initial* states. Because all the required conditions for dW to be an exact differential, and consequently for W to be a state function, are not satisfied, we can confidently state that in general dW is not exact differential. But the given work dW has been done in the purest of all possible ways, that is, it was done quasistatically. Therefore, any nonquasistatic work[4] done, namely dW', also cannot possibly be an exact differential. That is,

$$\oint dW' \neq 0 . \tag{3.6}$$

As usual the prime on dW indicates that the work dW may have been done wholly or partially nonquasistatically.

We note that immersion of a hot object into a fluid raises the temperature of the fluid and this can happen without any application of work. Therefore, it is clear that in addition to the input of work, the exchange of heat energy must also affect the change in the internal energy.

To investigate this point further let us add, possibly nonquasistatically, some heat energy to the system and also do some work on the system. Armed with a conversion factor that relates work to heat energy, this can be displayed in the form of an equation. To this end, heat energy $\int_{\text{initial}}^{\text{final}} dQ'$ is added to the work $-\int_{\text{initial}}^{\text{final}} dW'$ that has been done on the system.[5] Assuming conservation of energy, we have

$$\int_{\text{initial}}^{\text{final}} dQ' - \int_{\text{initial}}^{\text{final}} dW' = \int_{\text{initial}}^{\text{final}} dU .$$

Thus

$$\int_{\text{initial}}^{\text{final}} \left(dQ' - dW' - dU\right) = 0 .$$

Because the initial and final positions—called *initial* and *final*—are arbitrary, the above equation can hold only if the integrand itself is vanishing. This leads to the result

$$dQ' - dW' = dU . \tag{3.7}$$

It should be noted that because the right-hand side of (3.7) contains dU, which is an exact differential, and the left-hand side has dW' which is not, the additional quantity, namely dQ', on the left-hand side cannot be an exact differential. Rather,

[4]When any given item refers to a fully or partially nonquasistatic process, it will be denoted with a prime.

[5]The negative sign in $-\int_{\text{initial}}^{\text{final}} d'$ is needed for indicating that rather than being done by the system, which would have carried a positive sign, the work is being done ON the system.

it is the difference of the two inexact differentials—which appear on the left-hand side—that equals an exact differential, which appears on the right-hand side.

Equation (3.7) embodies the first law of thermodynamics. It bears repeating: The first law makes two important statements. It is imperative to recognize that these two statements are unrelated and have different import. The statements are:

(i) Energy is conserved.[6]
(ii) There exists a state function U, that is referred to as the system "internal energy." The sum of the heat energy $\mathrm{d}Q'$ *added to*, and the work $-\mathrm{d}W'$ *done on*, the system equal to the increase $\mathrm{d}U$ in the internal energy.

Consequently, while both the heat energy added to and the work done on the system, as it moves from an initial to a final equilibrium state, depend on the details of the paths taken, their sum equals the change in the system internal energy which has no such dependence. Rather, the change in the internal energy depends only on the location[7] of both the initial and final states. The important thing to note is that this remains true regardless of the nature of the intermediate states encountered en route.

Much of the understanding of the first law of thermodynamics has been gleaned from empirical observations of conversion of work into heat energy. The reader needs to be cautioned that the first law—see (3.7)—does not guarantee the reverse—meaning the conversion of "heat energy" into "work"—can also just as easily be accomplished. Indeed, there are restrictions on the occurrence of any such reverse process. After the introduction of the second law, a complete analysis of all such restrictions will be undertaken.

Perpetual Machines Because internal energy is a state function, it does not change when a system[8] undergoes a complete cycle. Therefore, machines that work in complete cycles leave the total internal energy of the working substance unchanged. While a long tail hangs by this story, at the very least it can be said that any work done must be paid for by an input of energy.

Consequently, according to the first law—see (3.7)—a cyclic machine that continually produces useful work must at the very least undergo a continual input of energy from the outside. A "perpetual machine of the first kind" is a hypothetical machine that does not follow this rule. If it ever existed, it would produce useful work without any input of energy.

[6]Occasionally, just the conservation of energy is referred to as the first law of thermodynamics. In fact, the conservation of energy has been known since the ancient Greek times and the real value of the first law lies in part (ii) which identifies an important state function, the internal energy.

[7]"Location" is defined in terms of its coordinates in the space of thermodynamic variables. We shall learn later that for a simple system the most appropriate coordinates for this representation are the system "volume" and what will be called the "entropy." For the present purposes it suffices to use instead any one of the following three pairs of coordinates: (P, V), (V, T), or (P, T).

[8]Perpetual machine of the first kind.

3.2 Specific Heat

When an amount of heat energy is imparted to an object at temperature T, generally its temperature rises. Such rise, ΔT, is observed to be in direct proportion to the heat energy input, $\Delta Q'$, but in inverse proportion to the mass, M, of the object, i.e.,

$$\Delta T \propto \frac{\Delta Q'}{M}.$$

Moreover, it is specific to the physical and chemical nature of the object. In other words,

$$\frac{\Delta Q'}{\Delta T} = C'M. \tag{3.8}$$

The proportionality parameter C', in addition to depending on the chemical and physical nature of the object, is path dependent and is also found to depend on the system temperature T. It is usually referred to as the "specific heat."[9] Owing to the fact that heat energy exchange, $\Delta Q'$, has these dependencies, C' also displays the same features. While many different paths can be chosen, frequently C' is measured for two types of quasistatic paths: one at constant volume, and the other at constant pressure. Moreover, explicit mention of the mass M is often avoided by particularizing the specific heat to the molar mass.

3.3 Notation

Most of the time, an attempt will be made to use notation that follows a simple rule, namely that when the system size is n mol, and n is not necessarily equal to 1, upper-case letters are employed. That is, unless otherwise helpful, the symbols to be used for the pressure, volume, internal energy, enthalpy, entropy, and temperature, etc., will be P, V, U, H, S, and T, etc., respectively.

On the other hand, if the system size is exactly equal to a single mole, lower-case symbols would be deemed preferable. Thus when $n = 1$, the pressure, volume, internal energy, and temperature would be denoted as p, v, u, and t. Also sometimes, when not occurring as a subscript and not causing confusion, for the sake of convenience of display, both upper- and lower-case symbols may be used with equivalent meaning.

Thus for the specific heat energy at constant volume and constant pressure, we write:

$$\left(\frac{\partial Q}{\partial T}\right)_{\mathrm{V}} = C_{\mathrm{V}} = n\left(\frac{\partial q}{\partial t}\right)_{\mathrm{v}} = nC_{\mathrm{v}} \equiv nc_{\mathrm{v}} \tag{3.9}$$

and

$$\left(\frac{\partial Q}{\partial T}\right)_{\mathrm{P}} = C_{\mathrm{P}} = n\left(\frac{\partial q}{\partial t}\right)_{\mathrm{p}} = nC_{\mathrm{p}} \equiv nc_{\mathrm{p}}. \tag{3.10}$$

[9]However, where needed, we shall make an attempt—which may not always be successful—to refer to it as "specific heat energy."

Note that for convenience of display we may use either C_{P} or, equivalently, c_{P}. Similarly, $C_{\mathrm{V}} \equiv c_{\mathrm{V}}$. The SI units for these are joules per mole per (degree) Kelvin, i.e., J/mol K.

3.4 Some Applications of the First Law

When a simple thermodynamic system expands, work is done by the system. In particular, a single mole, upon quasistatic expansion by an infinitesimal volume $\mathrm{d}v$, under pressure p, *does* an infinitesimal amount of work equal to

$$\mathrm{d}w = p\mathrm{d}v\,. \tag{3.11}$$

For such cases, the first law given in (3.7) requires that the heat energy, $\mathrm{d}q$, added to the system and the work, $-\mathrm{d}w = -p\mathrm{d}v$, done on the system must equal the increase, $\mathrm{d}u$, in the internal energy of the system. That is,

$$\mathrm{d}q - \mathrm{d}w = \mathrm{d}q - p\mathrm{d}v = \mathrm{d}u\,. \tag{3.12}$$

In a simple thermodynamic system, which can be described in terms of the three state variables p, v, and t, the internal energy u can be considered to be a function of any of the three independent pairs, i.e., (t, v), (t, p), and (p, v). In the following we shall treat the three cases separately. We hasten to add, however, that the inclusion of the dictates of the second law much improves the analysis and often leads to a more elegant explanation, and a more compact description, of the results recorded in this chapter. Nevertheless, the analysis appended below should be of benefit to a beginner who wishes to become conversant with some of the simple procedures used for establishing interrelationships between quantities of physical interest.

In Sects. 3.2–3.4 problems are worked out where a pair of variables t, v, and p are independent.

3.5 Independent t and v

Consider a single mole of a simple thermodynamic system where t and v act as the independent pair of variables. Next, denote the internal energy u as a function of t and v,

$$u = u(t, v)\,. \tag{3.13}$$

Physically speaking, $\mathrm{d}u$ represents the difference in internal energies of two neighboring equilibrium states. Mathematically speaking, u is differentiable in terms of the chosen two variables. And because of the physical fact that $u(t, v)$ can be treated as a state function, $\mathrm{d}u$ is an exact differential. Accordingly,

$$\mathrm{d}u = \left(\frac{\partial u}{\partial t}\right)_{\mathrm{V}} \mathrm{d}t + \left(\frac{\partial u}{\partial v}\right)_{\mathrm{t}} \mathrm{d}v\,. \tag{3.14}$$

Thus the first law, i.e., (3.12), becomes

$$\begin{aligned} dq &= du + pdv \\ &= \left(\frac{\partial u}{\partial t}\right)_{\mathrm{v}} dt + \left(\frac{\partial u}{\partial v}\right)_{\mathrm{t}} dv + pdv \\ &= \left(\frac{\partial u}{\partial t}\right)_{\mathrm{v}} dt + \left[p + \left(\frac{\partial u}{\partial v}\right)_{\mathrm{t}}\right] dv\,. \end{aligned} \tag{3.15}$$

Let us work with a process at constant volume, that is, where $dv = 0$. For such a process, (3.15) leads to the relationship

$$(dq)_{\mathrm{v}} = \left(\frac{\partial u}{\partial t}\right)_{\mathrm{v}} (dt)_{\mathrm{v}}\,.$$

Note the subscript v under the differentials (dq) and (dt) which denotes the fact that these differentials are calculated while the volume v is kept constant.

Now divide both sides of the above relationship by $(dt)_{\mathrm{v}}$. Using the definition of c_{v}, we can write[10]

$$c_{\mathrm{v}} = \frac{(dq)_{\mathrm{v}}}{(dt)_{\mathrm{v}}} = \left(\frac{\partial q}{\partial t}\right)_{\mathrm{v}} = \left(\frac{\partial u}{\partial t}\right)_{\mathrm{v}}\,. \tag{3.16}$$

For use in (3.15), we have found it helpful[11] to introduce the notation:

$$\Delta_{\mathrm{v}} = p + \left(\frac{\partial u}{\partial v}\right)_{\mathrm{t}}\,.$$

Therefore, (3.15) can be rewritten in the following convenient form:

$$dq = c_{\mathrm{v}} dt + \Delta_{\mathrm{v}} dv\,. \tag{3.17}$$

Physically, Δ_{v} can be described as follows: Consider an isothermal process, i.e., one that occurs at constant temperature. For such a process, $dt = 0$. Therefore, (3.17) can be written as

$$(dq)_{\mathrm{t}} = \Delta_{\mathrm{v}} (dv)_{\mathrm{t}}\,.$$

Next, divide both sides by $(dv)_{\mathrm{t}}$ to get

$$\left(\frac{\partial q}{\partial v}\right)_{\mathrm{t}} = \Delta_{\mathrm{v}} = p + \left(\frac{\partial u}{\partial v}\right)_{\mathrm{t}}\,. \tag{3.18}$$

[10]An interesting exercise is to show that $0 = c_{\mathrm{v}}(\frac{\partial t}{\partial v})_{\mathrm{u}} + (\frac{\partial u}{\partial v})_{\mathrm{t}}$. This is readily done by working at constant internal energy where $du = 0$. Then, according to (3.14), $0 = (\frac{\partial u}{\partial t})_{\mathrm{v}}(dt)_{\mathrm{u}} + (\frac{\partial u}{\partial v})_{\mathrm{t}}(dv)_{\mathrm{u}}$. Dividing both sides by $(dv)_{\mathrm{u}}$ leads to the desired result.

[11]Also, the following is an amusing comment about (3.16): Because the word "heat" is often imprecisely understood in thermodynamics, it is perhaps wiser to rephrase the word "specific heat at constant volume" to read "specific internal energy at constant volume." Thus the symbol $(\frac{\partial u}{\partial t})_{\mathrm{v}}$ looks more appropriate than the usual symbol $(\frac{\partial q}{\partial t})_{\mathrm{v}}$.

It is interesting to note that owing to the existence of the relationship

$$\Delta_{\mathrm{v}} = t\left(\frac{\partial p}{\partial t}\right)_{\mathrm{v}}, \tag{3.19}$$

—whose proof is given in (5.17) in the chapter on the first-second law—we can relate Δ_{v} to the constant pressure volume expansion coefficient[12] α_{p},

$$\alpha_{\mathrm{p}} = \frac{1}{v}\left(\frac{\partial v}{\partial t}\right)_{\mathrm{p}}, \tag{3.20}$$

and the isothermal compressibility[13] χ_{t},

$$\chi_{\mathrm{t}} = -\frac{1}{v}\left(\frac{\partial v}{\partial p}\right)_{\mathrm{t}}. \tag{3.21}$$

To see how this may be done, first use the cyclic identity—that was given in (1.30) and (1.31)—and then the mathematical definition of the parameters α_{p} and χ_{t}. In this fashion, we can write

$$\begin{aligned}\Delta_{\mathrm{v}} &= t\left(\frac{\partial p}{\partial t}\right)_{\mathrm{v}} = -t\left(\frac{\partial v}{\partial t}\right)_{\mathrm{p}}\left(\frac{\partial p}{\partial v}\right)_{\mathrm{t}} = t\left(\frac{1}{v}\left(\frac{\partial v}{\partial t}\right)_{\mathrm{p}}\right)\left[-v\left(\frac{\partial p}{\partial v}\right)_{\mathrm{t}}\right] \\ &= t\left(\alpha_{\mathrm{p}}\right)\left[\frac{1}{\chi_{\mathrm{t}}}\right] = \left(\frac{t\alpha_{\mathrm{p}}}{\chi_{\mathrm{t}}}\right).\end{aligned} \tag{3.22}$$

Both parameters, α_{p} and χ_{t}, have been measured extensively and the results tabulated. Moreover, these parameters can also be calculated theoretically from an appropriate equation of state whenever the latter is available. By replacing Δ_{v} as above, a more useful form for (3.17) can now be presented. This is version I of the first law,

$$\mathrm{d}q = c_{\mathrm{v}}\mathrm{d}t + \left(\frac{t\alpha_{\mathrm{p}}}{\chi_{\mathrm{t}}}\right)\mathrm{d}v\,. \tag{3.23}$$

Another important point to note is that in view of the foregoing we can state that given one mole of a substance, each of the quantities $(\frac{\partial q}{\partial v})_{\mathrm{t}}$, Δ_{v}, $p + (\frac{\partial u}{\partial v})_{\mathrm{t}}$, and $(\frac{t\alpha_{\mathrm{p}}}{\chi_{\mathrm{t}}})$ represents the amount of heat energy absorbed for a unit increase in the volume at constant temperature.

Let us consider next an isobaric path—i.e., one that is traced while the pressure remains constant—for inputting heat energy. To indicate the constancy of the pressure, we label the heat energy input, as well as the corresponding temperature and

[12]The expansion coefficient α_{p} refers to the rate at which a unit volume expands with rise in the temperature while the pressure is held constant.

[13]Similarly, the isothermal compressibility χ_{t}, is the rate at which a unit volume gets compressed with increase in the pressure while the temperature is held constant.

volume differentials,[14] with subscript p and use (3.23) to write

$$(\mathrm{d}q)_\mathrm{p} = c_\mathrm{v}(\mathrm{d}t)_\mathrm{p} + \left(\frac{t\alpha_\mathrm{p}}{\chi_\mathrm{t}}\right)(\mathrm{d}v)_\mathrm{p}\,. \tag{3.24}$$

Dividing both sides by $(\mathrm{d}t)_\mathrm{p}$ and recalling that $(\frac{\partial q}{\partial t})_\mathrm{p} = c_\mathrm{p}$, we get

$$c_\mathrm{p} = c_\mathrm{v} + \left(\frac{t\alpha_\mathrm{p}}{\chi_\mathrm{t}}\right)\left(\frac{\partial v}{\partial t}\right)_\mathrm{p}\,. \tag{3.25}$$

As noted in (3.20), $\alpha_\mathrm{p} = \frac{1}{v}(\frac{\partial v}{\partial t})_\mathrm{p}$. Therefore, (3.25) becomes

$$c_\mathrm{p} - c_\mathrm{v} = \left(\frac{tv\alpha_\mathrm{p}^2}{\chi_\mathrm{t}}\right). \tag{3.26}$$

This is an important relationship because all quantities on the right-hand side can be measured with relative ease. Also because c_p is generally easier to measure than c_v, this provides a convenient route to the measurement of c_v.

We consider next an adiabatic process, that is, where $\mathrm{d}q = 0$. It is convenient to denote such a process, when it occurs quasistatically, by using a subscript s. In other words,

$$(\mathrm{d}q)_\mathrm{s} = 0\,,$$

and therefore, after dividing both sides by $(\mathrm{d}v)_\mathrm{s}$, and recalling (3.12) which is

$$(\mathrm{d}q)_\mathrm{s} = 0 = (\mathrm{d}u)_\mathrm{s} + p(\mathrm{d}v)_\mathrm{s}\,,$$

leads to the result

$$\left(\frac{\partial u}{\partial v}\right)_\mathrm{s} = -p\,. \tag{3.27}$$

In the t, v representation currently under study, the adiabatic process, i.e., $\mathrm{d}q = 0$, also yields an expression that involves c_v. For instance, setting $(\mathrm{d}q)_\mathrm{s} = 0$ in (3.17) and dividing both sides by $(\mathrm{d}t)_\mathrm{s}$, we get

$$c_\mathrm{v} = -\left(\frac{\partial v}{\partial t}\right)_\mathrm{s}\cdot\Delta_\mathrm{v} = -v\alpha_\mathrm{s}\cdot\Delta_\mathrm{v} = -\left(\frac{tv\alpha_\mathrm{s}\alpha_\mathrm{p}}{\chi_\mathrm{t}}\right). \tag{3.28}$$

In the above, first we have introduced the notation

$$\alpha_\mathrm{s} = \frac{1}{v}\left(\frac{\partial v}{\partial t}\right)_\mathrm{s}\,, \tag{3.29}$$

[14]These differentials represent the physical difference in the corresponding state variables of two neighboring states.

and second we have used (3.22) which says $\Delta_v = (\frac{t\alpha_p}{\chi_t})$. Note that $\alpha_s = (\frac{(\frac{\partial v}{\partial t})_s}{v})$ is the isentropic volume expansion coefficient. (Notice the notational similarity with the isobaric volume expansion coefficient $\alpha_p = (\frac{(\frac{\partial v}{\partial t})_p}{v})$ that is given in (3.20).)

3.5.1 Problems 3.1–3.32

In the following, solutions to 32 problems is provided.

Problem 3.1 (Heat energy needed for raising temperature)
The Debye theory predicts that the constant volume specific heat, C_v, of solids at low temperature obeys the following law:

$$C_v = D\left(\frac{T}{T_d}\right)^3 .$$

(a) Calculate the specific heat at 20 K if the Debye coefficient D is equal to $2 * 10^4$ J per kilomole per degree Kelvin and the Debye temperature, T_d, is 350 K.
(b) What is the heat energy input needed for raising the temperature from 10 to 30 K at constant volume?

Solution 3.1 (a)

$$C_v = 2 * 10^4 * \left(\frac{20}{350}\right)^3 = 3.73 \text{ J kmol}^{-1}\text{ K}^{-1} .$$

(b) At constant volume, the input of heat energy, dQ, is related to the increase in temperature $\mathrm{d}T$, that is, $\mathrm{d}Q = C_v\mathrm{d}T$. Therefore, the heat energy used for raising the prescribed temperature is

$$\begin{aligned} \int \mathrm{d}Q = Q &= \int_{10\text{ K}}^{30\text{ K}} C_v \mathrm{d}T \\ &= D\int_{10\text{ K}}^{30\text{ K}} \left(\frac{T}{T_D}\right)^3 \mathrm{d}T = \left(\frac{\mathrm{D}}{4T_D^3}\right)[(30)^4 - (10)^4] \\ &= 93.3 \text{ J kmol}^{-1} . \end{aligned} \tag{3.30}$$

Problem 3.2 (Heat energy needed for raising temperature)
Here we need to also calculate work done and increase in internal energy due to changes in pressure, volume, and temperature.

(a) An ideal gas undergoes isothermal compression to one-third its initial volume. During this process its internal energy increases by ΔU_a and it does work W_a.
Given the quantity of the gas is $n = 2$ kmol, the pressure P is[15] equal to three atmosphere, and its volume V is equal to 4 m^3, calculate ΔU_a and W_a.

[15] Mole: 1 mol has N_A particles.

(b) Then the same gas undergoes an isochoric process which drops its pressure down to the original value P. In this process its internal energy increases by ΔU_b and it does W_b amount of work. Calculate both of these.

(c) Finally, the gas is expanded back to its original volume in an isobaric process. In this process the work done by the gas is W_c and the increase in its internal energy is ΔU_c. Calculate them both.

(d) What is the total work done and the total change in the internal energy in the three processes (a) + (b) + (c)? Could we have anticipated these sums?

Solution 3.2 (a) Work $dW = p dv$ is done by the gas when it expands an infinitesimal amount dv at pressure p. Therefore, because $pv = nRT$, the work done by the gas during the process (a) is

$$
\begin{aligned}
W_a &= \int_V^{\frac{V}{3}} p dv = -nRT \int_{\frac{V}{3}}^{V} \frac{dv}{v} \\
&= -nRT \ln(3) = -PV \ln(3) \\
&= -3 \text{ atm} * 4 \text{ m}^3 * \ln(3) = -3 * 1.01325 * 10^5 \text{ N m}^{-2} * 4 \text{ m}^3 * \ln(3) \\
&= -3 * 1.01325 * 4 * 10^5 * \ln(3) \text{ J} = -13.358 * 10^5 \text{ J}.
\end{aligned}
$$

The internal energy for the monatomic ideal gas is equal to $\frac{3}{2}nRT$. Because the temperature does not change during the process (a), therefore, there is no change in the internal energy. That is,

$$\Delta U_a = 0.$$

(b) Because the volume of the gas stays constant at the value $\frac{V}{3}$ throughout this process, $p dv = 0$ and no work is done. That is,

$$W_b = 0.$$

The internal energy, on the other hand, changes from $\frac{3}{2}nRT$ to $\frac{3}{2}nR\frac{T}{3}$. Therefore,

$$\Delta U_b = \frac{3}{2}nR\frac{T}{3} - \frac{3}{2}nRT = -nRT = -PV.$$

(c) Finally, in proceeding from $(P, \frac{V}{3}, \frac{T}{3})$ to (P, V, T), the pressure is constant. Therefore, the work done by the gas is

$$W_c = P\left(V - \frac{V}{3}\right) = \left(\frac{2}{3}\right)PV.$$

Inserting the numbers, we get

$$W_c = \left(\frac{2}{3}\right) * (3 * 1.01325 * 10^5 \text{ N m}^{-2}) * (4 \text{ m}^3) = 8.106 * 10^5 \text{ J}.$$

The change in the internal energy is also easily found. The temperature varies from $\frac{T}{3}$ to T, therefore we have

$$\Delta U_{\mathrm{c}} = \frac{3}{2} nR\left(T - \frac{T}{3}\right) = nRT = PV\,.$$

(d) The total work done is

$$W_{\mathrm{a}} + W_{\mathrm{b}} + W_{\mathrm{c}} = -5.252 * 10^5\ \mathrm{J}\,,$$

and the total change in the internal energy is

$$\Delta U_{\mathrm{a}} + \Delta U_{\mathrm{b}} + \Delta U_{\mathrm{c}} = 0 - PV + PV = 0\,.$$

We note that $\mathrm{d}U$ is an exact differential. Therefore the total change in internal energy for a complete round trip must be zero. On the other hand, $\mathrm{d}W$ is not an exact differential. Therefore, nothing definite can be predicted for the work that is done in a complete cycle.

Problem 3.3 (Work done by expanding Van der Waals gas)
The equation of state for one kmol of Van der Waals gas is

$$p = \frac{R_0 t}{v - b} - \frac{a}{v^2}\,.$$

Here

$$\begin{aligned} R_0 &= 8.3143 * 10^3\ \mathrm{J}\,(\mathrm{kmol})^{-1}\,\mathrm{K}^{-1} \\ a &= 580 * 10^3\ \mathrm{J}\,\mathrm{m}^3\,(\mathrm{kmol})^{-2} \\ b &= 0.0319 * \mathrm{m}^3\,(\mathrm{kmol})^{-1}\,. \end{aligned}$$

Following an expansion protocol whereby the gas expands quasistatically from an initial volume $v_1 = 15\ \mathrm{m}^3\,(\mathrm{kmol})^{-1}$ to a final volume $v_2 = 2v_1\,(\mathrm{kmol})^{-1}$, how much work would be done by n mol of such a Van der Waals gas?

Solution 3.3 Let us first calculate the work done by 1 kmol of the gas. We have

$$\begin{aligned} W &= \int_{v_1}^{2v_1} p\mathrm{d}v = R_0 t \int_{v_1}^{2v_1} \frac{\mathrm{d}v}{v - b} - \int_{v_1}^{2v_1} \frac{a}{v^2}\mathrm{d}v \\ &= R_0 t \ln\left[\frac{(2v_1 - b)}{(v_1 - b)}\right] + \frac{a}{2v_1} - \frac{a}{v_1} \\ &= 8.3143 * 10^3\ \mathrm{J}\,\mathrm{K}^{-1} * 373.15\ \mathrm{K} * \ln\left(\frac{30 - 0.0319}{15 - 0.0319}\right) \\ &\quad + 580 * 10^3\ \mathrm{J}\,\mathrm{m}^3 \left(\frac{1}{30} - \frac{1}{15}\right)\frac{1}{\mathrm{m}^3} \\ &= 2.1344 * 10^6\ \mathrm{J}\,. \end{aligned}$$

Clearly, therefore, n mol of gas, when it follows a similar expansion protocol, would do work equal to $(\frac{n}{1000})2.1344 * 10^6$ J $= n * 2.1344 * 10^3$ J.

Problem 3.4 (Metal versus gas regarding work done and volume change) Consider an ideal gas of volume $V_i = 0.1$ m^3 and a block of metal which has twice that volume. They are both at room temperature, $T = 300$ K, and under atmospheric pressure, $P = 1$ atm $= 1.01325 * 10^5$ N m^{-2}.

The pressure is increased quasistatically to 3 atm.

(a) Calculate the work done on the gas in the expansion.
(b) Given the compressibility $\chi_T = -\frac{1}{v}(\frac{\partial v}{\partial p})_T$ of the metal at that temperature is $= 0.8 * 10^{-6}$ atm^{-1}, find the work done on the metal.
(c) What is the change in the volume of the metal and the gas?

Solution 3.4 Assuming there are n moles of the ideal gas present, the work done on the gas in the above process is

$$-\int_{V_i}^{V_f} p\mathrm{d}v = -nRT\int_{V_i}^{V_f}\frac{\mathrm{d}v}{v} = nRT\ln\left(\frac{V_i}{V_f}\right).$$

Clearly, in order to complete this calculation, we need to determine both n and the ratio of the initial and final volume, i.e., $(\frac{V_i}{V_f})$.

To this end, let us look at the equation of state. Because the temperature is constant, we have $P_i V_i = P_f V_f = nRT$. As a result $\frac{V_i}{V_f} = \frac{P_f}{P_i} = \frac{3\text{ atm}}{1\text{ atm}} = 3$. Next, let us introduce the rest of the information that has been provided. We have

$$n = \frac{P_i V_i}{RT} = \frac{(1.01325 * 10^5\,\mathrm{N\,m^{-2}}) * (0.1\,\mathrm{m^3})}{(8.3143\,\mathrm{J\,mol^{-1}\,K^{-1}}) * (300\,\mathrm{K})} = 4.0623\,\mathrm{mol}\,. \tag{3.31}$$

The work done on the ideal gas is

$$\begin{aligned}\int_{V_f}^{V_i} p\mathrm{d}v = nRT\int_{\frac{V_i}{3}}^{V_i}\frac{\mathrm{d}v}{v} &= nRT\ln(3)\\ &= 4.0623 * 8.3143 * 300 * \ln(3)\,\mathrm{J} = 1.1132 * 10^4\,\mathrm{J}\,.\end{aligned} \tag{3.32}$$

Work done on the metal when, at the given temperature T, the pressure increases from 1 to 3 atm is

$$\begin{aligned}\int_{P_f}^{P_i}(p\mathrm{d}v)_T &= \int_{P_f}^{P_i} p\left(\frac{\partial v}{\partial p}\right)_T \mathrm{d}p = \int_{P_f}^{P_i} p[-\chi_T v]\mathrm{d}p\\ &\approx (V)_{\mathrm{metal}}\chi_T\int_{1\,\mathrm{atm}}^{3\,\mathrm{atm}} p\mathrm{d}p = (V)_{\mathrm{metal}}\chi_T\left[\frac{9}{2}-\frac{1}{2}\right] * (\mathrm{atm})^2\\ &= \left(0.2\,\mathrm{m^3}\right)\left(0.8 * 10^{-6}\,\mathrm{atm^{-1}}\right) * 4 * \left(\mathrm{atm^2}\right)\\ &= 0.64 * 10^{-6}\,\mathrm{m^3} * 1.013 * 10^5\,\mathrm{N\,m^{-2}} = 0.065\,\mathrm{J}\,.\end{aligned} \tag{3.33}$$

Note that the work done on the metal is a very small fraction of that done on the gas. Also note that the above integral over p involved the volume v. Therefore, for perfect accuracy, v also should have been expressed as a function of the pressure. In practise, however, much like the work done on the metal during the compression, we can expect the volume of the metal also to change little during the three-fold increase in the pressure. Therefore, extracting v out of the integral is an acceptable approximation.

(c) The change in volume of the gas is easily found. We know that $V_i = 0.1$ m^3 and $V_f = \frac{1}{3} * 0.1$ m^3. Therefore the change in the volume is $V_f - V_i = -0.1 * \frac{2}{3}$ m^3.

The change in the volume of the metal at the given temperature T is

$$\int_{initial}^{final} (dv)_{metal} = \int_{initial}^{final} \left[\left(\frac{\partial v}{\partial p}\right)_T\right]_{metal} dp = -\int_{initial}^{final} \chi_T v dp$$
$$\approx -\chi_T V_{metal}[P_f - P_i] = -\left[0.8 * 10^{-6}\ \text{atm}^{-1}\right]\left(0.2\ \text{m}^3\right) * 2\ \text{atm}$$
$$= -0.32 * 10^{-6}\ \text{m}^3 . \tag{3.34}$$

Clearly, the change in the volume of the metal is only a very small fraction of that of the ideal gas. This retroactively supports our removing $v \approx V_{metal}$ out of the integral wherever it occurs.

Problem 3.5 (Equation of state when given temperature change)
Even though dw—which represents an infinitesimal amount of work done—is not an exact differential itself, for a gas it can be expanded in terms of the exact differentials involving the pressure and the temperature.

(a) Show how that may be done.
(b) Given $(\frac{\partial v}{\partial t})_p = (\frac{R}{p})$ and $(\frac{\partial v}{\partial p})_t = -(\frac{v}{p})$, calculate the equation of state of the gas.

Solution 3.5 (a) Under pressure p, the work done, dw, by a gas when it undergoes a microscopic expansion, dv, is d$w = p$dv. Because dv is an exact differential, for a simple gas—whose thermodynamics depends only on p, v, and t—we have

$$dw = p dv = p\left[\left(\frac{\partial v}{\partial p}\right)_t dp + \left(\frac{\partial v}{\partial t}\right)_p dt\right] = -v dp + R dt . \tag{3.35}$$

(b) The above equation can be rewritten as

$$dt = \left(\frac{v}{R}\right) dp + \left(\frac{p}{R}\right) dv . \tag{3.36}$$

Now let us integrate at constant volume. That is,

$$\int (dt)_v = \int \frac{v}{R} (dp)_v , \quad \text{which gives} \quad t = \left(\frac{v}{R}\right) p + f(v) , \tag{3.37}$$

where $f(v)$ is still to be determined. Now differentiate the above with respect to v and hold the pressure constant. We get

$$\left(\frac{\partial t}{\partial v}\right)_{\mathrm{p}} = \frac{p}{R} + f'(v). \tag{3.38}$$

The statement of the problem tells us that $(\frac{\partial v}{\partial t})_{\mathrm{p}}$ is equal to $\frac{R}{p}$. Upon inversion, it gives $(\frac{\partial t}{\partial v})_{\mathrm{p}} = \frac{p}{R}$. As a result, (3.38) can be written as

$$f'(v) = \left(\frac{\partial t}{\partial v}\right)_{\mathrm{p}} - \frac{p}{R} = \frac{p}{R} - \frac{p}{R} = 0. \tag{3.39}$$

Thus $f(v)$ is a constant, say, equal to t_{c}. Inserting this result into (3.37) gives the desired equation of state,

$$pv = R(t - t_{\mathrm{c}}). \tag{3.40}$$

Problem 3.6 (Spreading gas among three compartments)
A thermally isolated cylinder has three evacuated compartments separated by two massless pistons which move freely without friction. The central compartment has volume V_{i}. Both the right- and left-hand compartments have identical relaxed springs extending from the piston to the end walls of the cylinder. Initially, the pistons are locked in place. The central compartment is now filled with $n = 5$ moles of monatomic ideal gas which has molar specific heat c_{v}. When thermal equilibrium is reached, the gas is at temperature $T_{\mathrm{i}} = 300$ K and pressure $P_{\mathrm{i}} = 2$ atmospheres.

The pistons are now released and the final volume V_{f} of the gas—which is still in the central compartment—settles at $4V_{\mathrm{i}}$. Given the heat energy capacity of the apparatus—excluding the gas—is $C_{\mathrm{a}} = 4c_{\mathrm{v}}$ where $c_{\mathrm{v}} = \frac{3}{2}R$ is the heat capacity of the monatomic ideal gas, find the final temperature T_{f} and the pressure P_{f} of the gas.

Given:

$$n = 5\,\text{mol}, \quad P_{\mathrm{i}} = 2\,\text{atm}, \quad T_{\mathrm{i}} = 300\,\text{K}, \quad c_{\mathrm{v}} = \frac{3}{2}R, \quad C_{\mathrm{a}} = 4c_{\mathrm{v}}, \quad V_{\mathrm{f}} = 4V_{\mathrm{i}}.$$

To be determined:

$$P_{\mathrm{f}}, \quad T_{\mathrm{f}}.$$

Solution 3.6 Let us first analyze the problem in terms of the symbols T_{i}, P_{i}, V_{i}, etc. The numerical values can be introduced afterwards.

We note that three different sources contribute to the total energy of the system. There is the internal energy of the gas, the potential energy due to the compression of the springs, and the internal energy associated with the heat energy content of the apparatus.

The change in the internal energy of the gas due to the change in its temperature is

$$(\mathrm{d}U)_{\mathrm{gas}} = nc_{\mathrm{v}}(T_{\mathrm{f}} - T_{\mathrm{i}}), \tag{3.41}$$

where $c_v = (3/2)R$ is the molar specific heat of the monatomic ideal gas. Because of thermal isolation, the total energy of the system is conserved. The increase in the volume of the central compartment from V_i to V_f results in corresponding decrease in the total volume of the side compartments. As a result, both the springs get compressed by an amount x each. Therefore

$$Ax = \frac{(V_f - V_i)}{2} = \frac{(4V_i - V_i)}{2} = \frac{3V_i}{2}, \tag{3.42}$$

where A is the cross-sectional area of the cylinders.

Let us assume that the springs are "simple." As a result, they obey the Hooke's law, that is, when compressed a small amount x, a spring pushes back with a force proportional to x. Further, because the two springs are identical, and are covered by massless pistons of area A, they both get exerted upon by the same force $P_f A$, where P_f is the pressure in the gas (in the central compartment) at the time that both of the springs have been compressed. As a result, each of the springs experiences the same compression and thus pushes back with the same compressional force kx. Here k is a constant specific to the springs.

The total increase in the compressional potential energy of the two springs is twice that of either of the springs. That is,

$$(\mathrm{d}U)_{\text{two springs}} = 2 \cdot \frac{1}{2}kx^2 = kx^2. \tag{3.43}$$

The compressional force on either of the cylinders is

$$P_f \cdot A = \left(\frac{nRT_f}{V_f}\right) A = kx. \tag{3.44}$$

Multiplying both sides by x and using (3.42) gives

$$\begin{aligned}(\mathrm{d}U)_{\text{springs}} &= (kx)x = \left[\left(\frac{nRT_f}{V_f}\right) \cdot A\right] x \\ &= \left(\frac{nRT_f}{V_f}\right) \cdot \frac{3}{2}V_i = \left(\frac{nRT_f}{4V_i}\right) \cdot \frac{3}{2}V_i = \frac{3}{8}(nRT_f).\end{aligned} \tag{3.45}$$

Finally, the increase in the internal energy of the apparatus is

$$(\mathrm{d}U)_a = C_a(T_f - T_i). \tag{3.46}$$

Because of thermal isolation, the total increase in the system energy must be zero. Thus, using (3.41), (3.45), and (3.46) yields

$$\begin{aligned}0 &= (\mathrm{d}U)_{\text{gas}} + (\mathrm{d}U)_{\text{springs}} + (\mathrm{d}U)_a \\ &= nc_v(T_f - T_i) + \frac{3}{8}(nRT_f) + C_a(T_f - T_i).\end{aligned} \tag{3.47}$$

We are thus led to the result

$$T_{\mathrm{f}} = \left[\frac{nc_{\mathrm{v}} + C_{\mathrm{a}}}{nc_{\mathrm{v}} + (\frac{3}{8})nR + C_{\mathrm{a}}}\right] T_{\mathrm{i}} \,. \tag{3.48}$$

Dividing the top and bottom by c_{v} and inserting the relevant numerical values, we get

$$T_{\mathrm{f}} = \left[\frac{5+4}{5+(5/4)+4}\right] 300 = 263\ \mathrm{K}\,.$$

Now P_{f} is readily found from the equation of state. Because the amount of gas is preserved during the experiment, we have the relationship

$$\frac{P_{\mathrm{f}} V_{\mathrm{f}}}{T_{\mathrm{f}}} = \frac{P_{\mathrm{i}} V_{\mathrm{i}}}{T_{\mathrm{i}}}\,.$$

Therefore,

$$P_{\mathrm{f}} = P_{\mathrm{i}} \left(\frac{V_{\mathrm{i}}}{V_{\mathrm{f}}}\right) \frac{T_{\mathrm{f}}}{T_{\mathrm{i}}} = 2\left(\frac{1}{4}\right)\frac{263}{300} = 0.438\ \mathrm{atm}\,. \tag{3.49}$$

3.6 Independent t and p

In this case

$$u = u(t, p)\,. \tag{3.50}$$

We follow a somewhat similar procedure to that used in the preceding section. Inserting the appropriate representation for the perfect differential du,

$$\mathrm{d}u = \left(\frac{\partial u}{\partial t}\right)_{\mathrm{p}} \mathrm{d}t + \left(\frac{\partial u}{\partial p}\right)_{\mathrm{t}} \mathrm{d}p\,, \tag{3.51}$$

into (3.12) leads to

$$\mathrm{d}q = \mathrm{d}u + p\mathrm{d}v = \left(\frac{\partial u}{\partial t}\right)_{\mathrm{p}} \mathrm{d}t + \left(\frac{\partial u}{\partial p}\right)_{\mathrm{t}} \mathrm{d}p + p\mathrm{d}v\,. \tag{3.52}$$

Keeping p constant, that is, setting $\mathrm{d}p = 0$, i.e.,

$$(\mathrm{d}q)_{\mathrm{p}} = \left(\frac{\partial u}{\partial t}\right)_{\mathrm{p}} (\mathrm{d}t)_{\mathrm{p}} + p(\mathrm{d}v)_{\mathrm{p}}\,,$$

and dividing both sides by $(\mathrm{d}t)_\mathrm{p}$ gives

$$\left(\frac{\partial q}{\partial t}\right)_\mathrm{p} = c_\mathrm{p} = \left(\frac{\partial u}{\partial t}\right)_\mathrm{p} + p\left(\frac{\partial v}{\partial t}\right)_\mathrm{p} = \left(\frac{\partial u}{\partial t}\right)_\mathrm{p} + \left(\frac{\partial (pv)}{\partial t}\right)_\mathrm{p}$$
$$= \left(\frac{\partial (u+pv)}{\partial t}\right)_\mathrm{p} = \left(\frac{\partial h}{\partial t}\right)_\mathrm{p} . \tag{3.53}$$

The last term on the right-hand side of (3.53) looks interesting because it identifies a quantity h, called "enthalpy," which is of great importance in thermodynamics,

$$h = (u + pv) . \tag{3.54}$$

We notice[16] that the temperature derivative of h at constant pressure is equal to the specific heat c_p. (Compare with (3.16) that refers to c_v.) We shall show a little later that $\mathrm{d}h$ is a perfect differential. Hence, the enthalpy h, like the internal energy u, is a state function. While in this section our primary variables are t and p, (3.52) involves three differentials, $\mathrm{d}t$, $\mathrm{d}p$, and $\mathrm{d}v$. Therefore, we need to exclude the presence of $\mathrm{d}v$. To this end, let us express v in terms of t and p. That is,

$$v = v(t, p) . \tag{3.55}$$

Hence

$$\mathrm{d}v = \left(\frac{\partial v}{\partial t}\right)_\mathrm{p} \mathrm{d}t + \left(\frac{\partial v}{\partial p}\right)_\mathrm{t} \mathrm{d}p . \tag{3.56}$$

Inserting this expression for $\mathrm{d}v$ into (3.52),

$$\mathrm{d}q = \left[\left(\frac{\partial u}{\partial t}\right)_\mathrm{p} + p\left(\frac{\partial v}{\partial t}\right)_\mathrm{p}\right]\mathrm{d}t + \left[\left(\frac{\partial u}{\partial p}\right)_\mathrm{t} + p\left(\frac{\partial v}{\partial p}\right)_\mathrm{t}\right]\mathrm{d}p ,$$

achieves the desired end. That is,

$$\mathrm{d}q = c_\mathrm{p}\mathrm{d}t + \Delta_\mathrm{p}\mathrm{d}p , \tag{3.57}$$

where we have used the expression for c_p given in (3.53) and defined a new variable Δ_p as noted below:

$$\Delta_\mathrm{p} = \left[\left(\frac{\partial u}{\partial p}\right)_\mathrm{t} + p\left(\frac{\partial v}{\partial p}\right)_\mathrm{t}\right] = \left(\frac{\partial (u+pv)}{\partial p}\right)_\mathrm{t} - v = \left(\frac{\partial h}{\partial p}\right)_\mathrm{t} - v . \tag{3.58}$$

We shall show in (5.74) in the chapter on the first-second law that, much like Δ_v, Δ_p can also be expressed in terms of parameters that are both easily measured experimentally and calculated theoretically from an equation of state, i.e.,

$$\Delta_\mathrm{p} = -tv\alpha_\mathrm{p} . \tag{3.59}$$

[16] As with (3.16), an interesting comment can also be made about (3.53). Because the word "heat" is often imprecisely understood in thermodynamics, it is perhaps wiser to rephrase the word "specific heat at constant pressure" to read "specific enthalpy at constant pressure." Then the symbol $(\frac{\partial h}{\partial t})_\mathrm{p}$ would look more natural than the usual symbol $(\frac{\partial q}{\partial t})_\mathrm{p}$.

Thus (3.57) can be rewritten as

$$dq = c_p dt - (tv\alpha_p)dp\,. \tag{3.60}$$

This is version II of the first law. For constant temperature, i.e., $dt = 0$, this leads to the result

$$\left(\frac{dq}{dp}\right)_t = \Delta_p = -tv\alpha_p\,.$$

Note that given one mole of a substance, each of the quantities $(\frac{dq}{dp})_t$, Δ_p, and $-tv\alpha_p$ represents the amount of heat energy absorbed during a unit increase in the pressure at constant temperature.

Finally, consider an adiabatic process that proceeds quasistatically. Because no exchange of heat energy occurs, the left-hand side of (3.57) vanishes,

$$(dq)_s = 0 = c_p(dt)_s + \Delta_p(dp)_s\,. \tag{3.61}$$

Separating the right-hand side and dividing both sides by $(dt)_s$ yields

$$\begin{aligned} c_p &= -\left(\frac{\partial p}{\partial t}\right)_s \Delta_p = -\left(\frac{\partial p}{\partial v}\right)_s \left(\frac{\partial v}{\partial t}\right)_s \Delta_p \\ &= \left(\frac{\partial v}{\partial t}\right)_s \frac{-1}{(\frac{\partial v}{\partial p})_s} \Delta_p = \left[\frac{1}{v}\left(\frac{\partial v}{\partial t}\right)_s\right]\left[\frac{-v}{(\frac{\partial v}{\partial p})_s}\right]\Delta_p \\ &= \left(\frac{\alpha_s}{\chi_s}\right)\Delta_p = -\left(\frac{tv\alpha_s\alpha_p}{\chi_s}\right), \end{aligned} \tag{3.62}$$

where we have used the value $\Delta_p = -tv\alpha_p$ given in (3.59).

The result for c_p is completely analogous to that for c_v—given in (3.28)—which, for the reader's convenience, is again recorded below:

$$c_v = -v\alpha_s \cdot \Delta_v = -\left(\frac{tv\alpha_s\alpha_p}{\chi_t}\right). \tag{3.63}$$

Note that α_s and χ_s are respectively the "quasistatic, adiabatic"—or equivalently, the so-called isentropic—volume expansion coefficient and compressibility.

An interesting check on the above results, i.e., (3.62) and (3.63), is provided by—also see (3.73) in the succeeding subsection—the confirmation of the well-known relationship

$$\left[\frac{c_p}{c_v}\right] = \left(\frac{\chi_t}{\chi_s}\right). \tag{3.64}$$

3.7 Independent p and v

This time

$$u = u(p, v)\,. \tag{3.65}$$

We follow the usual format and insert the appropriate representation for the exact differential du,

$$\mathrm{d}u = \left(\frac{\partial u}{\partial v}\right)_\mathrm{p} \mathrm{d}v + \left(\frac{\partial u}{\partial p}\right)_\mathrm{v} \mathrm{d}p\,, \tag{3.66}$$

into (3.12), which in turn leads to

$$\mathrm{d}q = \left(\frac{\partial u}{\partial p}\right)_\mathrm{v} \mathrm{d}p + \left[\left(\frac{\partial u}{\partial v}\right)_\mathrm{p} + p\right] \mathrm{d}v\,. \tag{3.67}$$

First, we note that

$$\left(\frac{\partial u}{\partial p}\right)_\mathrm{v} = \left(\frac{\partial u}{\partial t}\right)_\mathrm{v} \left(\frac{\partial t}{\partial p}\right)_\mathrm{v} = c_\mathrm{V} \left(\frac{\partial t}{\partial p}\right)_\mathrm{v}. \tag{3.68}$$

Next, we write

$$\left(\frac{\partial u}{\partial v}\right)_\mathrm{p} + p = \left(\frac{\partial (u + pv)}{\partial v}\right)_\mathrm{p} = \left(\frac{\partial h}{\partial v}\right)_\mathrm{p} = \left(\frac{\partial h}{\partial t}\right)_\mathrm{p} \left(\frac{\partial t}{\partial v}\right)_\mathrm{p} = c_\mathrm{P} \left(\frac{\partial t}{\partial v}\right)_\mathrm{p}. \tag{3.69}$$

Thus (3.67) becomes

$$\mathrm{d}q = c_\mathrm{V} \left(\frac{\partial t}{\partial p}\right)_\mathrm{v} \mathrm{d}p + c_\mathrm{P} \left(\frac{\partial t}{\partial v}\right)_\mathrm{p} \mathrm{d}v\,. \tag{3.70}$$

Remember that these processes are quasistatic. Let us travel along an adiabatic path. That is, we use the subscript s and write

$$(\mathrm{d}q)_\mathrm{s} = 0 = c_\mathrm{V} \left(\frac{\partial t}{\partial p}\right)_\mathrm{v} (\mathrm{d}p)_\mathrm{s} + c_\mathrm{P} \left(\frac{\partial t}{\partial v}\right)_\mathrm{p} (\mathrm{d}v)_\mathrm{s}\,. \tag{3.71}$$

Dividing both sides by

$$c_\mathrm{V} \left(\frac{\partial t}{\partial v}\right)_\mathrm{p} (\mathrm{d}v)_\mathrm{s}$$

gives

$$\left[\frac{(\frac{\partial t}{\partial p})_\mathrm{V}}{(\frac{\partial t}{\partial v})_\mathrm{P}}\right] \left(\frac{\partial p}{\partial v}\right)_\mathrm{s} + \left(\frac{c_\mathrm{P}}{c_\mathrm{V}}\right) = 0\,. \tag{3.72}$$

Using the cyclic identity in the form

$$\frac{(\frac{\partial t}{\partial p})_\mathrm{V}}{(\frac{\partial t}{\partial v})_\mathrm{P}} = -\left(\frac{\partial v}{\partial p}\right)_\mathrm{t},$$

we can write for the ratio $\frac{c_p}{c_v} = \gamma$ the result

$$\gamma = \left(\frac{c_p}{c_v}\right) = \left(\frac{\partial v}{\partial p}\right)_t \left(\frac{\partial p}{\partial v}\right)_s = \left[\frac{-(\frac{\partial v}{\partial p})_t}{v}\right] / \left[\frac{-(\frac{\partial v}{\partial p})_s}{v}\right] = \left(\frac{\chi_t}{\chi_s}\right), \tag{3.73}$$

where χ_t is the quasistatic isothermal, and χ_s the quasistatic adiabatic, compressibility.

3.7.1 The First Law: Another Version

Equation (3.70), which represents dq in terms of dp and dv can be recast in the following form:

$$dq = c_v\left(\frac{\partial t}{\partial p}\right)_v dp + c_p\left(\frac{\partial t}{\partial v}\right)_p dv = c_v\left(\frac{\chi_t}{\alpha_p}\right)dp + c_p\left(\frac{1}{v\alpha_p}\right). \tag{3.74}$$

Here we have used the identity

$$\left(\frac{\partial t}{\partial p}\right)_v = -\left(\frac{\partial v}{\partial p}\right)_t \left(\frac{\partial t}{\partial v}\right)_p = \left(\frac{-(\frac{\partial v}{\partial p})_t}{(\frac{\partial v}{\partial t})_p}\right) = \frac{\chi_t}{\alpha_p}.$$

3.8 Enthalpy

In (3.54), a quantity $(u + pv)$ was introduced, whose derivative with respect to the temperature, taken at constant pressure, is the specific heat c_p. We assert[17] that this quantity, denoted h and named the enthalpy, is a state function. Of its many uses, some relate to the heat energy of transformation that accompanies an isothermal–isobaric—occurring at constant temperature and pressure—change of phase. Similarly, it is often useful in determining the heat energy produced in chemical reactions. (See, for instance, examples relating to the application of Hess' rules for chemothermal reactions.)

3.8.1 Enthalpy is a State Function

For $h = u + pv$ to be a state function, its differential

$$dh = du + pdv + vdp \tag{3.75}$$

[17]For a proof of this assertion, see Sect. 3.8.1 *below*.

must be exact. Such is the case if and only if the integrability requirement for dh, i.e.,

$$\frac{\partial^2 h}{\partial v \partial p} = \frac{\partial^2 h}{\partial p \partial v},$$

is satisfied. To check this, du in (3.75) must first be expressed in terms of the independent differentials dv and dp as

$$\mathrm{d}u = \left(\frac{\partial u}{\partial v}\right)_{\mathrm{p}} \mathrm{d}v + \left(\frac{\partial u}{\partial p}\right)_{\mathrm{v}} \mathrm{d}p\,. \tag{3.76}$$

Thus

$$\mathrm{d}h = \left[\left(\frac{\partial u}{\partial v}\right)_{\mathrm{p}} + p\right] \mathrm{d}v + \left[\left(\frac{\partial u}{\partial p}\right)_{\mathrm{v}} + v\right] \mathrm{d}p\,. \tag{3.77}$$

A comparison with

$$\mathrm{d}h = \left(\frac{\partial h}{\partial v}\right)_{\mathrm{p}} \mathrm{d}v + \left(\frac{\partial h}{\partial p}\right)_{\mathrm{v}} \mathrm{d}p \tag{3.78}$$

leads to the identification

$$\left(\frac{\partial h}{\partial v}\right)_{\mathrm{p}} = \left[\left(\frac{\partial u}{\partial v}\right)_{\mathrm{p}} + p\right] \tag{3.79}$$

and

$$\left(\frac{\partial h}{\partial p}\right)_{\mathrm{v}} = \left[\left(\frac{\partial u}{\partial p}\right)_{\mathrm{v}} + v\right]. \tag{3.80}$$

Differentiating (3.79) with respect to p while holding v constant gives

$$\frac{\partial^2 h}{\partial p \partial v} = \frac{\partial^2 u}{\partial p \partial v} + 1\,. \tag{3.81}$$

Similarly, differentiating (3.80) with respect to v while p is kept constant gives

$$\frac{\partial^2 h}{\partial v \partial p} = \frac{\partial^2 u}{\partial v \partial p} + 1\,. \tag{3.82}$$

Because u is known to be a state function, du is an exact differential. Therefore, the following relationship must hold:

$$\frac{\partial^2 u}{\partial p \partial v} = \frac{\partial^2 u}{\partial v \partial p}\,.$$

As a result the right-hand sides of (3.81) and (3.82) are identical. Therefore their left-hand sides are equal. This confirms the satisfaction of the integrability requirement for dh.

Thus, much like the internal energy u, which plays an important role in determining the t and v dependent properties,[18] the enthalpy h is a state function. Accordingly, during thermodynamic processes, all changes in enthalpy are determined entirely by the initial and final locations and are independent of the routes taken.

The knowledge of enthalpy is central to the understanding of the p and t dependent behavior of systems in equilibrium. Indeed, aspects of chemical reactions are often described in relation to their "reaction enthalpies."

3.8.2 Enthalpy and the First Law

The first law can equally well be expressed in terms of the enthalpy. To this end, let us introduce (3.75) into (3.12) and get

$$\mathrm{d}q = \mathrm{d}u + p\mathrm{d}v = \mathrm{d}h - v\mathrm{d}p\,. \tag{3.83}$$

Consider a quasistatic, adiabatic process, i.e.,

$$(\mathrm{d}q)_\mathrm{s} = 0 = (\mathrm{d}h)_\mathrm{s} - v(\mathrm{d}p)_\mathrm{s}\,. \tag{3.84}$$

Dividing the above by $(\mathrm{d}p)_\mathrm{s}$ leads to a useful relationship

$$\left(\frac{\partial h}{\partial p}\right)_\mathrm{s} = v\,. \tag{3.85}$$

Next, let us represent the enthalpy as a function of pressure and temperature,

$$h = h(p, t)\,. \tag{3.86}$$

Then

$$\mathrm{d}h = \left(\frac{\partial h}{\partial p}\right)_\mathrm{t} \mathrm{d}p + \left(\frac{\partial h}{\partial t}\right)_\mathrm{p} \mathrm{d}t\,. \tag{3.87}$$

As a result, (3.83) becomes

$$\mathrm{d}q = \left(\frac{\partial h}{\partial t}\right)_\mathrm{p} \mathrm{d}t + \left[\left(\frac{\partial h}{\partial p}\right)_\mathrm{t} - v\right]\mathrm{d}p\,. \tag{3.88}$$

Keeping the pressure constant, i.e., setting $\mathrm{d}p = 0$, and dividing both sides by $(\mathrm{d}t)_\mathrm{p}$ leads to the relationship

$$\left(\frac{\partial q}{\partial t}\right)_\mathrm{p} = c_\mathrm{p} = \left(\frac{\partial h}{\partial t}\right)_\mathrm{p}\,. \tag{3.89}$$

Of course, this relationship was already inferred in (3.53).

[18] That is what we have said here. But strictly speaking, u is best expressed in terms of its natural variables s and v. This subject is visited in detail later.

Thus, in terms of the variables p and t, the first law becomes

$$\mathrm{d}q = c_\mathrm{p}\mathrm{d}t + \left[\left(\frac{\partial h}{\partial p}\right)_\mathrm{T} - v\right]\mathrm{d}p = c_\mathrm{p}\mathrm{d}t + \Delta_\mathrm{p}\mathrm{d}p\,. \tag{3.90}$$

This is the same result that was derived in the preceding section and recorded in (3.57) and (3.58). As mentioned already, similar to Δ_v, Δ_p—the latter may be represented as $(-tv\alpha_\mathrm{p})$—can also be readily measured. And, if the equation of state is available, it is easily calculated.

Particularizing the heat energy input to a path traveled at constant volume,

$$(\mathrm{d}q)_\mathrm{v} = c_\mathrm{p}(\mathrm{d}t)_\mathrm{v} + \Delta_\mathrm{p}(\mathrm{d}p)_\mathrm{v}\,, \tag{3.91}$$

and dividing both sides by $(\mathrm{d}t)_\mathrm{v}$, we get

$$\left(\frac{\partial q}{\partial t}\right)_\mathrm{v} = c_\mathrm{p} + \Delta_\mathrm{p}\left(\frac{\partial p}{\partial t}\right)_\mathrm{v}\,,$$

which is best written as

$$\begin{aligned}
c_\mathrm{p} - c_\mathrm{v} &= c_\mathrm{p} - \left(\frac{\partial q}{\partial t}\right)_\mathrm{v} = -\Delta_\mathrm{p}\left(\frac{\partial p}{\partial t}\right)_\mathrm{v} \\
&= tv\alpha_\mathrm{p}\left(\frac{\partial p}{\partial t}\right)_\mathrm{v} = -tv\alpha_\mathrm{p}\left(\frac{\partial p}{\partial v}\right)_\mathrm{t}\left(\frac{\partial v}{\partial t}\right)_\mathrm{p} \\
&= tv\alpha_\mathrm{p}\left[-v\left(\frac{\partial p}{\partial v}\right)_\mathrm{t}\right]\left[\frac{(\frac{\partial v}{\partial t})_\mathrm{p}}{v}\right] = \left(\frac{tv{\alpha_\mathrm{p}}^2}{\chi_\mathrm{t}}\right).
\end{aligned} \tag{3.92}$$

Note that an earlier derivation of this result[19]—see (3.26)—had traveled a different rout. Also see Problem 3.7 below.

Problem 3.7 Prove

$$c_\mathrm{p} - c_\mathrm{v} = -\Delta_\mathrm{p}\left(\frac{\partial p}{\partial t}\right)_\mathrm{v} = \Delta_\mathrm{v}\left(\frac{\partial v}{\partial t}\right)_\mathrm{p}.$$

Solution 3.7 To show that the expression for $c_\mathrm{p} - c_\mathrm{v}$ given in (3.92), i.e.,

$$c_\mathrm{p} - c_\mathrm{v} = -\Delta_\mathrm{p}\left(\frac{\partial p}{\partial t}\right)_\mathrm{v}\,, \tag{1}$$

is identical to that given earlier in (3.25), i.e.,

$$c_\mathrm{p} - c_\mathrm{v} = \Delta_\mathrm{v}\left(\frac{\partial v}{\partial t}\right)_\mathrm{p}\,, \tag{2}$$

[19] $c_\mathrm{p} - c_\mathrm{v}$ as above by a different rout.

proceed as follows. Noting that

$$\left(\frac{\partial p}{\partial t}\right)_{\rm v} = -\left(\frac{\partial p}{\partial v}\right)_{\rm t}\left(\frac{\partial v}{\partial t}\right)_{\rm p} = \frac{\alpha_{\rm p}}{\chi_{\rm t}}$$

and

$$-\Delta_{\rm p} = tv\alpha_{\rm p} ,$$

equation (1) becomes

$$c_{\rm p} - c_{\rm v} = \left(tv\frac{\alpha_{\rm p}^2}{\chi_{\rm t}}\right).$$

Similarly, noting that

$$\Delta_{\rm v} = t\left(\frac{\partial p}{\partial t}\right)_{\rm v} = \frac{t\alpha_{\rm p}}{\chi_{\rm t}}$$

and

$$\left(\frac{\partial v}{\partial t}\right)_{\rm p} = v\alpha_{\rm p} ,$$

equation (2) also yields the same result

$$c_{\rm p} - c_{\rm v} = \frac{t\alpha_{\rm p}}{\chi_{\rm t}} v\alpha_{\rm p} = \left(tv\frac{\alpha_{\rm p}^2}{\chi_{\rm t}}\right).$$

Enthalpy, Heat of Transformation, and the Internal Energy

Below we present an example of how the enthalpy may sometimes be used to actually get an estimate for the internal energy.

Problem 3.8 (Internal energy from latent heat energy of vaporization)

(a) At temperature $T = 373$ K, the latent heat energy of vaporization, $L_{\rm w\to g}(T)$, of water at atmospheric pressure, P_0, is ≈ 540 calories/g/mol. Estimate its internal energy, $U_{\rm w}(T)$, at that temperature.

(b) The latent heat energy of vaporization, $L_{\rm w\to g}(T)$, is known to decrease with rise in temperature, and at 453 K it is ≈ 479 calories/g/mol. What is the internal energy at 453 K? Does it too decrease with temperature?

(c) At $T = T_{\rm c} \approx 647$ K, normally called the critical temperature, the distinction between liquid and vapor phases disappears. How is this case treated within the framework of (a) and (b) above?

Solution 3.8 (a) In liquid form, one mole of water at 373 K has volume $v_{\rm w}$. Upon conversion to gaseous state, the relevant volume becomes $v_{\rm g}$. At temperatures which are moderately high, but are many degrees below the critical temperature,

$T_c \approx 647$ K, for water— note that 373 and 453 K both satisfy this requirement— we generally have $v_g(T) \gg v_w(T)$.

Because the transition occurs at constant pressure, P_0, the energy difference can conveniently be represented as enthalpy difference. To see how this is done, proceed as follows.

According to the first law, see (3.12), we have

$$\begin{aligned} Q_{w\to g}(T) &= \int_w^g dq(T) = \int_w^g du(T) + \int_w^g dw(T) \\ &= \int_w^g du(T) + \int_w^g P dv(T) = \int_w^g du(T) + P_0 \int_w^g dv(T) \\ &= u_g(T) - u_w(T) + P_0\big[v_g(T) - v_w(T)\big] \\ &= h_g(T) - h_w(T)\,, \end{aligned} \tag{3.93}$$

where $h_g(T) - h_w(T)$ represents the enthalpy difference between the gaseous and the water states. In the following let us use the approximation $v_g(T) \gg v_w(T)$. As a result

$$Q_{w\to g}(T) \approx u_g(T) - u_w(T) + P_0 v_g(T)\,. \tag{3.94}$$

Further, as an additional approximation, treat the water vapor, at temperatures like 373 K or a bit higher, as a diatomic ideal gas.[20] Therefore, write

$$P_0 v_g(T) \approx RT\,.$$

Consistent with this rough approximation, the internal energy of the vapor is

$$u_g(T) \approx C_v T\,,$$

where

$$C_v \sim \frac{5}{2} R\,.$$

Thus from (3.94)

$$-u_w(T) \approx Q_{w\to g} - u_g(T) - P_0 v_g(T) = Q_{w\to g} - (C_v + R)T\,. \tag{3.95}$$

Therefore, using $Q_{w\to g} \approx 540$ calories/g/mol or $540 \times 18 \times 4.18$ J/mol, we get the result

$$\begin{aligned} -u_w(373) &\approx 18 \times 540 \times 4.18 - \frac{7}{2} \times 8.31 \times 373 \\ &= 40,630 - 10,849 \approx 29.8\ \text{kJ mol}^{-1}\,. \end{aligned}$$

[20]Water molecule, H_2O, is triatomic. Here we assume that at the boiling temperature 373 K, effectively two such triatomic molecules act as a single diatom.

Correctly, the above estimate for the internal energy of water is negative. This is so because, at the average intermolecular spacing that exists in water, the intermolecular force is attractive. Accordingly, positive energy has to be expended to tear apart a molecular pair in the liquid phase.

Note that this is true both for separating two water molecules apart which are close to each other—that is what we need here—as well as to tear open a single H_2O molecule so that H_2 and O are separated! Of course, here we are not talking about the latter case.

(b) When following the same procedure as was described above, an estimate for the internal energy of water at 453 K is made, i.e.,

$$-u_{\mathrm{w}}(453) \approx 18 \times 479 \times 4.18 - \frac{7}{2} \times 8.31 \times 453 \approx 22.9\,\mathrm{kJ\,mol^{-1}}\,,$$

it is found to have (algebraically) increased with the rise in temperature.

(c) At the critical point, any distinction between water and its vapor disappears. Therefore the specific volumes of water and gas approach each other. That is, $v_{\mathrm{w}}(T_{\mathrm{c}}) = v_{\mathrm{g}}(T_{\mathrm{c}})$. Also, the internal energies of the two phases become the same,

$$u_{\mathrm{w}}(T_{\mathrm{c}}) = u_{\mathrm{g}}(T_{\mathrm{c}})\,.$$

Thus the latent heat energy of vaporization approaches zero, signalling the onset of a so-called second order phase transition.

3.9 Hess' Rules

Hess' rule can be obtained by making simple deductions from the first law. Heat energy produced in a chemical reaction that occurs at constant pressure is given by the change—i.e., decrease—in the state function called enthalpy. And for a reaction that occurs at constant volume, the heat energy produced is equal to the change—i.e., decrease—in the state function called internal energy. In other words, because the enthalpy and internal energy are both state functions, for both cases the amount of heat energy produced is dependent only on the initial and final states.

3.9.1 Chemothermal Reactions

For a chemical reaction that occurs at constant pressure, say p_{const}, the first law, see (3.12), yields

$$\begin{aligned}
\text{(heat energy produced)} = -q_{\mathrm{added}} &= -\int_{\mathrm{i}}^{\mathrm{f}} \mathrm{d}q \\
&= -\int_{\mathrm{i}}^{\mathrm{f}} [\mathrm{d}u + \mathrm{d}w] = -\big[(u_{\mathrm{f}} - u_{\mathrm{i}}) + p_{\mathrm{const}}(v_{\mathrm{f}} - v_{\mathrm{i}})\big] \\
&= h_{\mathrm{i}} - h_{\mathrm{f}}\,.
\end{aligned} \tag{3.96}$$

Here the indices 'i' and 'f' signify the initial and the final states. On the other hand, if both the pressure and volume are constant—that is, when we also have the relationship $v_f = v_i$—this equation would give

$$(\text{heat energy produced}) = u_i - u_f\,. \tag{3.97}$$

Appended below are two well-known examples where Hess' rules are used to infer useful relationships between the heat energy that is produced for different, but related chemical reactions.

Problem 3.9 (Oxidation of CO to CO_2)
Complete oxidation of one mole of carbon, C, to form one mole of carbon dioxide, CO_2, is known to produce 393.5 kJ of heat energy. On the other hand, partial oxidation to form carbon monoxide, CO, produces only 110.5 kJ of heat energy. Estimate the heat energy produced by complete oxidation of one mole of CO to one mole of CO_2. Note that all these processes occur at the temperature $T_{\text{ref}} = 298$ K.

Solution 3.9 The correct answer is instinctively given. Complete burning of one mole of CO to CO_2, at the specified temperature T_{ref}, would produce $393.5 - 110.5 = 283$ kJ of heat energy.

Still, for pedagogic reasons, it is useful to establish a book-keeping procedure which can, when necessary, deal with more involved accounting of energy differences between the intermediate processes involved in going from the initial constituents to the final product.

Extensive tables of enthalpy—generally given as a difference between the enthalpy of a compound and that of its chemical constituents in their pure form at some specified temperature—are available, see, for instance, [1]. In order to make use of these tables, we set the temperature equal to the reference temperature, $T_{\text{ref}} = 298$ K, for which the necessary tables are available. We then have

$$h_{\text{initial}} = h_C + h_{O_2} = 0 + 0\,,$$

where the initial pure state of carbon is that of graphite. Also, the tables inform us that

$$h_{\text{final}} = h_{CO_2} = -393.5 \text{ kJ mol}^{-1}\,.$$

This information is consistent with the descriptive statement of the current example, namely, that the heat energy produced by complete combustion of carbon, i.e.,

$$h_{\text{initial}} - h_{\text{final}} = h_C + h_{O_2} - h_{CO_2}$$

is $393.5 \text{ kJ mol}^{-1}$.

Next we calculate the heat energy produced by incomplete oxidation of pure carbon when it forms CO. Because of the simplicity of the problem posed, the answer was instinctively guessed. A formal description of the intermediate steps is recorded below.

The statement of the example tells us that partial oxidation to form CO results in the production of 110.5 kJ of heat energy. That is,

$$h_C + \frac{1}{2}h_{O_2} - h_{CO} = 110.5\ \text{kJ mol}^{-1}\,.$$

Subtracting the second equation from the first gives the desired result for the heat energy produced by oxidation of CO to form one mole of CO_2:

$$h_C + h_{O_2} - h_{CO_2} - h_C - \frac{1}{2}h_{O_2} + h_{CO} = h_{CO} + \frac{1}{2}h_{O_2} - h_{CO_2}$$
$$= 393.5 - 110.5 = 283\ \text{kJ mol}^{-1}\,.$$

3.10 Oxidation, Heat of Vaporization

Problem 3.10 At 298 K when one mole of hydrogen and a half mole of oxygen gases combine to form one mole of water vapor, 241.8 kJ of heat energy is released. In contrast, when the same reaction produces one mole of liquid water at 298 K the heat energy released is equal to 285.8 kJ. Calculate the latent heat energy of vaporization of water at 298 K.

Solution 3.10 At 298 K we are told that when measured in molal units

$$h_{H_2-\text{gas}} + \frac{1}{2}h_{O_2-\text{gas}} - h_{H_2O-\text{vapor}} = 241.8\ \text{kJ}\,.$$

Similarly,

$$h_{H_2-\text{gas}} + \frac{1}{2}h_{O_2-\text{gas}} - h_{H_2O-\text{liquid}} = 285.8\ \text{kJ}\,.$$

Subtracting the first equation from the second gives

$$h_{H_2O-\text{vapor}} - h_{H_2O-\text{liquid}} = 285.8 - 241.8 = 44\ \text{kJ}\,,$$

which is the latent heat energy of vaporization at 298.15 K.

It is interesting to recast this result in terms of calories per gram—i.e., write it as $44 \times 1000/(4.186 \times 18) = 584$ calories/gram—and compare it with the corresponding value at the higher temperature 373 K. The latter was quoted in an another example as being equal to 540 calories/gram. Thus, the latent heat energy of vaporization of water decreases with increase in temperature.

3.11 Ideal Gas Adiabatics and Polytropics

First we treat the adiabatic process.

3.11.1 Ideal Gas Adiabatics

For a single mole of an ideal gas with f degrees of freedom per molecule, we have:

$$c_{\rm v}t = u = \frac{f}{2}Rt\,, \quad pv = Rt\,,$$
$$h = u + pv\,, \quad c_{\rm p}t = (c_{\rm v} + R)t\,.$$

The first law for a quasistatic process in one mole of an ideal gas can be written in either of the following two convenient forms:

$$dq = du + pdv = c_{\rm v}dt + pdv\,,$$

or

$$dq = dh - vdp = c_{\rm p}dt - vdp\,.$$

Note: In the above we have used the fact that for an ideal gas $(\frac{\partial u}{\partial v})_{\rm t} = 0$ and $(\frac{\partial h}{\partial p})_{\rm t} = 0$.

Denoting, as usual, the changes occasioned during a quasistatic, adiabatic process, i.e., where $(dq)_{\rm s} = 0$, with subscript s and slightly rearranging the above, we can write

$$(dt)_{\rm s} = \frac{v}{c_{\rm p}}(dp)_{\rm s} = -\frac{p}{c_{\rm v}}(dv)_{\rm s}\,,$$

which leads to the equation[21]

$$\frac{(dp)_{\rm s}}{p} = -\frac{c_{\rm p}}{c_{\rm v}} \cdot \frac{(dv)_{\rm s}}{v}\,,$$

that is readily integrated as

$$\ln(p) = -\gamma \ln(v) + C_1\,.$$

Here C_1 is a constant of integration and $\gamma = (\frac{c_{\rm p}}{c_{\rm v}}) = (\frac{f+2}{f})$ is $\frac{5}{3}$ or $\frac{7}{5}$ depending on whether the gas is monatomic or diatomic.

We can recast the above as

$$\ln(p) + \gamma \ln(v) = \ln\big(pv^{\gamma}\big) = C_1\,,$$

and write more conveniently the final result—that is valid only for adiabatic processes—as

$$pv^{\gamma} = \exp(C_1)\,. \tag{3.98}$$

The constant C_1 is determined from the initial condition.

[21]Note: As always, the subscript s is used only as a reminder that the relevant differential change in the state variables has occurred under quasistatic adiabatic conditions.

Thus, a quasistatic, adiabatic transformation from $(p_\mathrm{i}, v_\mathrm{i})$ to $(p_\mathrm{f}, v_\mathrm{f})$ obeys the relationship

$$p_\mathrm{i} v_\mathrm{i}^\gamma = p_\mathrm{f} v_\mathrm{f}^\gamma \,. \tag{3.99}$$

Because of the quasistatic nature of the transformation process, the gas remains in equilibrium all along the route from the initial to final state, therefore its usual equation of state remains valid. In particular, it is valid at both the initial, $(p_\mathrm{i}, v_\mathrm{i})$, and final, $(p_\mathrm{f}, v_\mathrm{f})$, locations. That is,

$$p_\mathrm{i} v_\mathrm{i} = R t_\mathrm{i} \,, \qquad p_\mathrm{f} v_\mathrm{f} = R t_\mathrm{f} \,.$$

Accordingly, (3.99) can be transformed to a (p, t) or a (t, v) representation to yield

$$\left(\frac{p_\mathrm{i}^{\gamma-1}}{t_\mathrm{i}^\gamma} \right) = \left(\frac{p_\mathrm{f}^{\gamma-1}}{t_\mathrm{f}^\gamma} \right), \tag{3.100}$$

or

$$t_\mathrm{i} v_\mathrm{i}^{\gamma-1} = t_\mathrm{f} v_\mathrm{f}^{\gamma-1} \,. \tag{3.101}$$

Equations (3.99), (3.100), and (3.101) are generally known as the Poisson equations. It should be emphasized that these equations hold only for those adiabatic transformations that are strictly quasistatic—meaning, the adiabatic transformations must be carried out extremely slowly and in such a manner that the system stays in thermodynamic equilibrium.

Problem 3.11 (Work done in adiabatic expansion and isothermal compression of ideal gas)
At point (a), a quantity of monatomic ideal gas at temperature $T_\mathrm{a} = 300$ K and under pressure $P_\mathrm{a} = 5 \times 10^5$ Pa occupies a volume $V_\mathrm{a} = 10^{-3}$ m^3.

In an adiabatic process, it is expanded quasistatically until its temperature drops to $T_\mathrm{b} = 200$ K. This is point (b). Next the gas is compressed isothermally to reach a point (c) where $V_\mathrm{c} = V_\mathrm{a}$. Finally, an isochoric compression brings it back to point (a). (See Fig. 3.1 and note $\gamma = \frac{5}{3}$.)

Calculate the work done by the heat energy put into, and the change in the internal energy of the gas for each of the three legs of its journey.

Given: $V_\mathrm{a} = 10^{-3}$ m^3, $T_\mathrm{a} = 300$ K, $P_a = 5 \times 10^5$ Pa;
(a)→(b) adiabatic, quasistatic expansion, $T_\mathrm{b} = 200$ K;
(b)→(c) isothermal compression, $T_\mathrm{c} = T_\mathrm{b}$, $V_\mathrm{c} = V_\mathrm{a}$;
(c)→(a) isochoric compression, $V_\mathrm{c} = V_\mathrm{a}$.

Solution 3.11 At point (a) the equation of state is

$$P_\mathrm{a} V_\mathrm{a} = 5 \times 10^5 \times 10^{-3} = 500 = nRT_\mathrm{a} = nR \times 300 \,.$$

Thus, in joules/degree Kelvin,

$$nR = \frac{5}{3} \,.$$

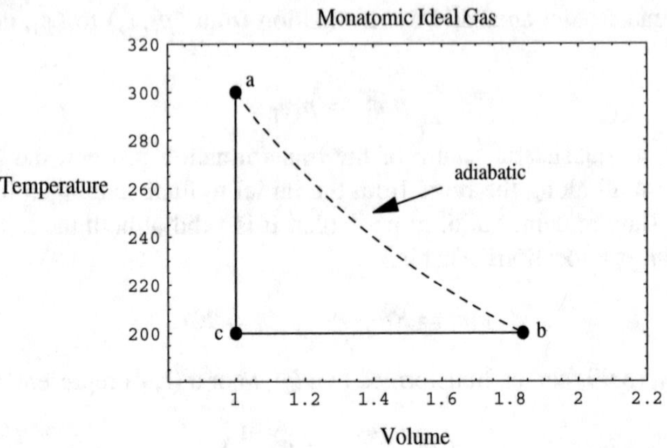

Fig. 3.1 Temperature versus volume

The path from (a) to (b) is quasistatic, adiabatic. Consequently, (3.101) applies and we have

$$T_a V_a^{(\gamma-1)} = 300 \times \left(10^{-3}\right)^{(\gamma-1)} = 300 \times \left(10^{-3}\right)^{(2/3)}$$
$$= 3.00 = T_b V_b^{(\gamma-1)} = 200 \times V_b^{(2/3)} ,$$

which yields

$$V_b = 1.84 \times 10^{-3}\,\mathrm{m}^3 .$$

Because the path from (a) to (b) is adiabatic, there is no exchange of heat energy. Therefore, the heat energy introduced in going from (a) to (b) is zero, i.e.,

$$Q_{a\to b} = 0 .$$

Hence, according to (3.103), the work done by the system is equal to the decrease in its internal energy. That is,

$$W_{a\to b} = U_a - U_b = n c_V (T_a - T_b) = n\frac{3}{2}R(T_a - T_b)$$
$$= \frac{3}{2} \times \frac{5}{3}(300 - 200) = 250\,\mathrm{J} .$$

Note that the system *does* positive work during the expansion.

Next, because temperature is constant from (b) to (c), so is the internal energy. Therefore, any work done along this path is owed to a transfer of heat energy into the system.

At $T_c = T_b = 200$ K, $V_c = V_a$. Thus, from the equation of state

$$P_c V_c = P_c V_a = nRT_c = nRT_b ,$$

we find

$$P_c = nR\left(\frac{T_b}{V_a}\right) = \frac{5}{3} \cdot \frac{200}{10^{-3}} = 3.33 \times 10^5 \text{ N/m}^2 .$$

Because heat energy introduced along this path is equal to the corresponding work done by the system,

$$Q_{b\to c} = W_{b\to c} = \int_b^c P dV = nRT_b \int_{V_b}^{V_c} \frac{dV}{V} = nRT_b \ln\left(\frac{V_c}{V_b}\right).$$

Note that $V_c = V_a$ is less than V_b, therefore the logarithm on the right-hand side is negative. Thus, we get

$$W_{b\to c} = \frac{5}{3} \times 200 \times \ln\left(\frac{1}{1.84}\right) = -\frac{5}{3} \times 200 \times \ln(1.84) = -203 \text{ J},$$

making heat energy input $Q_{b\to c}$ negative.

Clearly, the system discards heat energy as it gets isothermally compressed along $b \to c$.

Finally, the path from (c) to (a) is isochoric. Therefore, no work is done, i.e.,

$$W_{c\to a} = 0 ,$$

and the heat energy input is exactly equal to the increase in the internal energy

$$Q_{c\to a} = U_a - U_c = nc_v(T_a - T_c) = \frac{5}{3} \times \frac{3}{2}(300 - 200) = 250 \text{ J}.$$

Thus, as expected, there is no net change in the internal energy at the completion of the cycle (a) $\to$ (b) $\to$ (c) $\to$ (a). Also, the total heat energy *input* is equal to the total work done *by* the system. That is,

$$Q_{a\to b\to c\to a} = W_{a\to b\to c\to a} = 250 - 203 = 47 \text{ J}.$$

Problem 3.12 (Nonquasistatic free adiabatic expansion of ideal gas)
Two evacuated vessels of volumes V_1 and V_2 are connected across a tube with a stopcock that is closed. The entire system is adiabatically enclosed and its heat energy capacity is assumed to be negligible. Vessel 1 is filled with an ideal gas at temperature T_i. The stopcock is opened and the gas expands to fill both vessels. Determine the final temperature, T_f, of the gas.

Solution 3.12 Here, one is sorely tempted to use (3.101) that would readily lead to a value for T_f because T_i and $V_i = V_1$, $V_f = V_1 + V_2$ are known. Unfortunately, the use of (3.101) would be erroneous because it applies only to processes that are both adiabatic and quasistatic.

During the expansion described in the given example the gas passes through states that are not in thermodynamic equilibrium. As such, the process is not quasistatic. Therefore the prerequisites for (3.101) are not satisfied.

No matter, the first law still applies. Indeed, its general statement—given in (3.7)—comes to the needed rescue. That is,

$$dQ - dW = dU .$$

Because the process occurs in a thermally isolated system, $dQ = 0$. Also, because the gas expands into an evacuated vessel, it does so against zero pressure. Thus it does no work during the expansion, that is, $dW = 0$. Consequently, $dU = 0$. Because for an ideal gas U depends only on the temperature, the temperature remains constant, leading to the result

$$T_f = T_i .$$

Problem 3.13 (Quasistatic adiabatic compression of ideal gas)
A quantity of ideal gas is quasistatically compressed in an adiabatic process from a state (P_i, V_i) to (P_f, V_f). Calculate the work done, $W_{i\to f}$, *on* the gas.

Solution 3.13

$$W_{i\to f} = -\int_i^f P dV .$$

Anywhere en route from $i \to f$ we have

$$PV^\gamma = P_i V_i^\gamma = P_f V_f^\gamma = D ,$$

hence we can write

$$\begin{aligned} W_{i\to f} &= -D \int_i^f V^{-\gamma} dV \\ &= \frac{D}{1-\gamma}\left[V_i^{1-\gamma} - V_f^{1-\gamma}\right] = \frac{1}{1-\gamma}[P_i V_i - P_f V_f] . \end{aligned} \tag{3.102}$$

Recalling that the symbol γ stands for the ratio $(\frac{c_p}{c_v})$, $PV = nRT$, and $R = c_p - c_v$, (3.102) can be recast as

$$W_{i\to f} = nR\left[\frac{T_i - T_f}{1-(\frac{c_p}{c_v})}\right] = nc_v(T_f - T_i) , \tag{3.103}$$

demonstrating that the quasistatic, adiabatic work done on the system is equal to the corresponding increase in the internal energy.

Problem 3.14 (Isobaric, isothermal, or adiabatic expansion of diatomic ideal gas)
One mole of[22] a diatomic ideal gas has volume v_0 under pressure p_0. Upon expansion to volume $2v_0$ it does work w. Calculate w for the following three processes: (a) isobaric, (b) isothermal, and (c) adiabatic. How would these results differ if the gas were monatomic?

[22] Quasistatic adiabatic work equals the change in internal energy.

Solution 3.14 Let t_0 be the initial temperature of the gas, which gives

$$p_0 v_0 = R t_0 .$$

Thus, at constant pressure

$$(w)_{\text{isobaric}} = \int_{v_0}^{2v_0} p_0 \mathrm{d}v = p_0(2v_0 - v_0) = R t_0 . \tag{3.104}$$

(b) For the isothermal process, $Pv = RT_0$. Therefore, at constant temperature

$$(w)_{\text{isothermal}} = \int_{v_0}^{2v_0} p \mathrm{d}v = R t_0 \int_{v_0}^{2v_0} \left(\frac{\mathrm{d}v}{v}\right) = R t_0 \ln 2 = 0.693\, R t_0 . \tag{3.105}$$

(c) The work done in an adiabatic expansion is different for the diatomic and monatomic gases. Here the process follows the relationship

$$p v^\gamma = p_0 v_0^\gamma .$$

In a diatomic *ideal* gas, molecules have five degrees of freedom: three for the translational motion of the center of mass, and two for rotational motion. Thus $c_\mathrm{v} = 5R/2$ and $c_\mathrm{p} = 7R/2$. As a result $c_\mathrm{p}/c_\mathrm{v} = \gamma = (7/5)$. When the gas is monatomic, it only has the three translational degrees of freedom. Thus $c_\mathrm{v} = 3R/2$ and $c_\mathrm{p} = 5R/2$, yielding $\gamma = (5/3)$. Furthermore,

$$\begin{aligned}(w)_\mathrm{s} &= \int_{v_0}^{2v_0} p \mathrm{d}v = p_0 v_0^\gamma \int_{v_0}^{2v_0} \left(\frac{\mathrm{d}v}{v^\gamma}\right) \\ &= \left(\frac{p_0 v_0}{\gamma - 1}\right) \cdot \left[\frac{(2)^{\gamma-1} - 1}{(2)^{\gamma-1}}\right] = \left(\frac{R t_0}{\gamma - 1}\right) \cdot \left[\frac{(2)^{\gamma-1} - 1}{(2)^{\gamma-1}}\right] .\end{aligned} \tag{3.106}$$

Using the appropriate γ's, we get

$$\left[(\mathrm{d}W)_\mathrm{s}\right]_{\text{diatomic}} = 0.605\, R t_0 , \qquad \left[(\mathrm{d}W)_\mathrm{s}\right]_{\text{monatomic}} = 0.555\, R t_0 .$$

It is clear that the larger the number of atoms in the molecule, the closer γ is to 1 and therefore the adiabatic process begins more closely to resemble the isothermal process.

Problem 3.15 (Conducting and nonconducting cylinders in contact)
A cylinder is composed of two sections, each of volume V_0, divided by a freely moving, nonconducting diaphragm. One section is composed of diathermal walls that conduct heat energy from an infinite thermal reservoir at some higher temperature. The walls of the other section are nonconducting. As such, the gas in that section is thermally isolated.

One mole of monatomic ideal gas is introduced into each of the two cylinders. Initially, after introduction, the temperature and pressure of the gas are T_0 and P_0, respectively.

After thermal equilibrium has been established, the pressure in each section is $10P_0$.

Calculate the following: (a) the final volume, V, and temperature, T, of the gas in the nonconducting section, (b) the temperature, T_H, of the thermal reservoir, (c) the amount of heat energy, Q, transferred to the conducting section, and (d) the work, W, done in compressing the gas in the nonconducting section.

Solution 3.15 (a) For monatomic ideal gas, the ratio of the specific heats γ is equal to 5/3. Because the compression of the gas in the nonconducting section happens adiabatically, we have

$$P_0 V_0^\gamma = (10P_0)V^\gamma ,$$

where V is the final volume in this section. Thus

$$V = 0.251\, V_0 .$$

Next, the final temperature in this section is found from the relation

$$PV = (10P_0)\,(0.251V_0) = 2.51(P_0V_0) = 2.51RT_0 = RT ,$$

which yields

$$T = 2.51T_0 .$$

(b) The temperature of the reservoir can be found from the fact that under the final pressure, $10P_0$, the volume of the conducting section of the cylinder is V_H where

$$V_H = 2V_0 - V = V_0(2 - 0.251) = 1.749V_0 ,$$

and therefore

$$10\, P_0(1.749V_0) = RT_H = 17.49RT_0 ,$$

which gives

$$T_H = 17.49T_0 .$$

(c) The heat energy inflow into the conducting section is

$$Q = c_V(T_H - T_0) = \frac{3}{2}R(16.49T_0) = 24.73RT_0 .$$

(d) Finally, because the compression occurs adiabatically, the work done on the gas in the nonconducting section is equal to increase in its internal energy. That is,

$$W = c_V(T - T_0) = \frac{3}{2}R(2.51 - 1)T_0 = 2.27RT_0 .$$

Ideal Gas Polytropics, See Fig. 3.2

In the foregoing we studied a number of examples involving isothermal and/or quasistatic adiabatic processes in ideal gases. Their hallmark was either the use of the equation of state at a given temperature or a transformation that is described by the Poisson equations

$$pv^{\gamma} = \text{const.}\,, \qquad tv^{\gamma-1} = \text{const.}\,, \qquad t^{-\gamma}p^{\gamma-1} = \text{const.} \tag{3.107}$$

However, these two are not the only types of processes that can be obtained. Indeed, there are a variety of applications where the processes utilized lie somewhere between the isothermal and adiabatic. Of course, these two are the extremes represented by perfect contact with a heat energy reservoir at a given temperature or complete thermal isolation.

For the in-between processes, we need to use a more general index ν in place of γ. In this manner, the Poisson equations are replaced by the following:

$$pv^{\nu} = \text{const.}\,, \qquad tv^{\nu-1} = \text{const.}\,, \qquad t^{-\nu}p^{\nu-1} = \text{const.} \tag{3.108}$$

Notice that the process is isothermal when the index $\nu = 1$. Similarly, when $\nu = \gamma$, the process is quasistatic adiabatic. (See, for instance, (3.107).) For other values of ν, we have a so-called polytropic process.

To get some feel for the physics of such processes, we return to the first law for quasistatic processes in an ideal gas whereby du is simply represented by $c_V dt$ and (3.12) gives $dq = c_V dt + p dv$. Using the second of the three forms of (3.108), we can translate $p dv$ into a term that involves dt. Here

$$d\left(tv^{\nu-1}\right) = dt v^{\nu-1} + t(\nu-1)v^{\nu-2}dv = 0\,,$$

or equivalently,

$$dv = -\left(\frac{dt}{t}\right)\cdot\frac{v}{\nu-1}\,.$$

Whence

$$p dv = -\frac{dt}{(\nu-1)}\cdot\left(\frac{pv}{t}\right) = -dt\cdot\frac{R}{(\nu-1)}\,. \tag{3.109}$$

As a result, the equation for the first law attains a simple form

$$(dq)_{\text{polytropic}} = \left(c_V - \frac{R}{\nu-1}\right)(dt)_{\text{polytroic}}\,. \tag{3.110}$$

We can now define a sort of specific heat for this polytropic process

$$c_{\text{polytropic}} = \left(\frac{dq}{dt}\right)_{\text{polytropic}} = c_V - \frac{R}{(\nu-1)}\,. \tag{3.111}$$

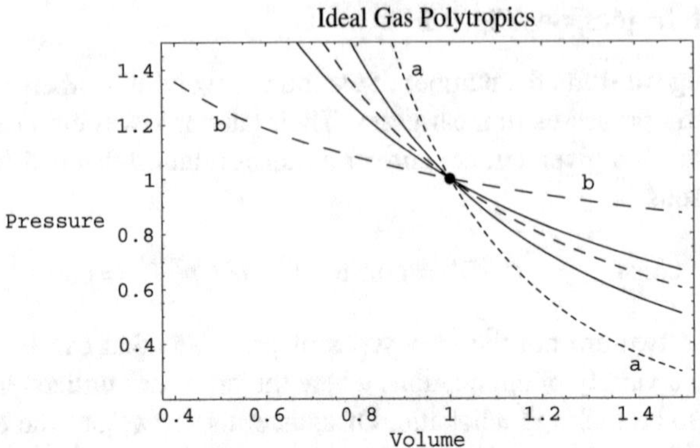

Fig. 3.2 Ideal Gas Polytropics. Plot for a monatomic ideal gas where $\gamma = \frac{5}{3}$. Curve a refers to $\nu = 3$ and curve b to $\nu = 1/3$. In-between these curves, in seriatim lie curves for $\nu = 5/3$, referring to the isentropic case and shown here as the full curve nearest to a; $\nu = 1.25$, dashed curve that lies between the two full curves; and $\nu = 1$, referring to the isothermal case and shown as a full curve nearest to b. Note that specific heat cannot be defined for the cases represented by the full curves. Also note that the polytropic specific heat would be negative for all cases represented by curves that lie in-between the two full curves. On the other hand, for all those polytropic processes that can be represented by curves that lie outside the two full curves, the polytropic specific heat would be positive

Because $R = c_{\rm p} - c_{\rm v} = c_{\rm v}(\gamma - 1)$, we have

$$c_{\rm polytropic} = c_{\rm v}\left[1 - \frac{\gamma - 1}{\nu - 1}\right] = c_{\rm v}\left[\frac{\nu - \gamma}{\nu - 1}\right]. \tag{3.112}$$

When $\nu = \gamma$, the process proceeds as a reversible-adiabatic process for which dq is zero. Hence a specific heat cannot usefully be defined. Similarly, when $\nu = 1$, the process is isothermal and specific heat is not a meaningful concept.

In-between the end-points of the range

$$\gamma > \nu > 1\,,$$

a specific heat can be defined. It turns out that such a polytropic specific heat has the quaint property of being negative!

On the other hand, when[23] either $\nu > \gamma$ or $\nu < 1$, the numerator and denominator have the same sign. Consequently, $c_{\rm polytropic}$ is positive for both such cases.

3.12 Some Interrelationships

Problems 3.16–3.21 deal with interrelated issues.

Problem 3.16 Work out an alternate proof, by a different method, of the relationship $c_{\rm v}(\frac{\partial t}{\partial v})_{\rm u} = -(\frac{\partial u}{\partial v})_{\rm t}$.

[23]Note that γ is greater than 1.

Solution 3.16 Although this relationship has already been derived (see footnote after (3.15)), it is interesting to derive it by a different procedure. Also, it is instructive to see the usefulness of the cyclic identity.

Because $c_{\mathrm{v}} = (\frac{\partial u}{\partial t})_{\mathrm{v}}$, the left-hand side is $(\frac{\partial u}{\partial t})_{\mathrm{v}}(\frac{\partial t}{\partial v})_{\mathrm{u}}$. Multiplying both sides by the inverse of the right-hand side leads to the cyclic identity

$$\left(\frac{\partial u}{\partial t}\right)_{\mathrm{v}}\left(\frac{\partial t}{\partial v}\right)_{\mathrm{u}}\left(\frac{\partial v}{\partial u}\right)_{\mathrm{t}} = -1,$$

which we know is correct. Q.E.D.

Problem 3.17 Show $c_{\mathrm{v}}(\frac{\partial t}{\partial p})_{\mathrm{v}} = (\frac{\partial u}{\partial p})_{\mathrm{v}}$.

Solution 3.17

$$c_{\mathrm{v}}\left(\frac{\partial t}{\partial p}\right)_{\mathrm{v}} = \left(\frac{\partial u}{\partial t}\right)_{\mathrm{v}}\left(\frac{\partial t}{\partial p}\right)_{\mathrm{v}} = \left(\frac{\partial u}{\partial p}\right)_{\mathrm{v}}.$$ Q.E.D.

Problem 3.18 Prove the relationship $c_{\mathrm{p}}(\frac{\partial t}{\partial v})_{\mathrm{p}} = (\frac{\partial h}{\partial v})_{\mathrm{p}}$.

Solution 3.18

$$c_{\mathrm{p}}\left(\frac{\partial t}{\partial v}\right)_{\mathrm{p}} = \left(\frac{\partial h}{\partial t}\right)_{\mathrm{p}}\left(\frac{\partial t}{\partial v}\right)_{\mathrm{p}} = \left(\frac{\partial h}{\partial v}\right)_{\mathrm{p}}.$$ Q.E.D.

Problem 3.19 Show $c_{\mathrm{p}} = (\frac{\partial u}{\partial t})_{\mathrm{p}} + pv\alpha_{\mathrm{p}}$.

Solution 3.19 Consider the relationships $h = u + pv$, $c_{\mathrm{p}} = (\frac{\partial h}{\partial t})_{\mathrm{p}}$, and note that the difference in the enthalpy of two neighboring equilibrium states with the same value of pressure is

$$(\mathrm{d}h)_{\mathrm{p}} = (\mathrm{d}u)_{\mathrm{p}} + \mathrm{d}(pv)_{\mathrm{p}} = (\mathrm{d}u)_{\mathrm{p}} + p(\mathrm{d}v)_{\mathrm{p}}.$$

Dividing both sides by $(\mathrm{d}t)_{\mathrm{p}}$ gives

$$\left(\frac{\partial h}{\partial t}\right)_{\mathrm{p}} = c_{\mathrm{p}} = \left(\frac{\partial u}{\partial t}\right)_{\mathrm{p}} + p\left(\frac{\partial v}{\partial t}\right)_{\mathrm{p}} = \left(\frac{\partial u}{\partial t}\right)_{\mathrm{p}} + pv\alpha_{\mathrm{p}}.$$ Q.E.D.

Problem 3.20 Prove the relationship $(\frac{\partial h}{\partial t})_{\mathrm{v}} = c_{\mathrm{v}} + v(\frac{\alpha_{\mathrm{p}}}{\chi_{\mathrm{t}}})$.

Solution 3.20 Similar to the preceding example, we can write

$$(\mathrm{d}h)_{\mathrm{v}} = (\mathrm{d}u)_{\mathrm{v}} + \mathrm{d}(pv)_{\mathrm{v}} = (\mathrm{d}u)_{\mathrm{v}} + v(\mathrm{d}p)_{\mathrm{v}}.$$

Dividing both sides by $(\mathrm{d}t)_{\mathrm{v}}$ and using the cyclic identity yields

$$\left(\frac{\partial h}{\partial t}\right)_{\mathrm{v}} = \left(\frac{\partial u}{\partial t}\right)_{\mathrm{v}} + v\left(\frac{\partial p}{\partial t}\right)_{\mathrm{v}} = c_{\mathrm{v}} - v\left(\frac{\partial p}{\partial v}\right)_{\mathrm{t}}\left(\frac{\partial v}{\partial t}\right)_{\mathrm{p}} = c_{\mathrm{v}} + v\left(\frac{\alpha_{\mathrm{p}}}{\chi_{\mathrm{t}}}\right).$$ Q.E.D.

Problem 3.21 Prove the relationship $(\frac{\gamma}{v\chi_{\mathrm{t}}}) = (\frac{\partial h}{\partial v})_{\mathrm{p}}(\frac{\partial p}{\partial u})_{\mathrm{v}}$.

Solution 3.21 From Problems 3.17 and 3.18, we have

$$\frac{c_p(\frac{\partial t}{\partial v})_p}{c_v(\frac{\partial t}{\partial p})_v} = \frac{(\frac{\partial h}{\partial v})_p}{(\frac{\partial u}{\partial p})_v} = \left(\frac{\partial h}{\partial v}\right)_p \left(\frac{\partial p}{\partial u}\right)_v .$$

This is the right-hand side of the required result. Next, we need to prove its equality with the left-hand side of the required result. That is, we need to prove the equality

$$\frac{c_p(\frac{\partial t}{\partial v})_p}{c_v(\frac{\partial t}{\partial p})_v} = \left(\frac{\gamma}{v\chi_t}\right). \tag{3.113}$$

Noting that

$$\left(\frac{\partial t}{\partial v}\right)_p = \frac{1}{v\alpha_p}, \qquad \left(\frac{\partial t}{\partial p}\right)_v = \frac{\chi_t}{\alpha_p}, \qquad \text{and} \qquad \frac{c_p}{c_v} = \gamma ,$$

the left-hand side of (3.113) is

$$\gamma\left[\frac{\frac{1}{v\alpha_p}}{\frac{\chi_t}{\alpha_p}}\right] = \left(\frac{\gamma}{v\chi_t}\right). \qquad \text{Q.E.D.}$$

3.13 Equation of State from Bulk and Elastic Moduli

By now[24] we well know that if the equation of state is available, results for certain moduli can usually be predicted. Unfortunately, while extensive experimental data for bulk and elastic moduli are often available—and, indeed, occasionally their dependence on state variables has also been established—detailed knowledge of the equation of state is often hard to come by. Therefore, it is important to study the manner in which information about the moduli can be used to learn about the equation of state itself.

A few examples of how this may be done in practice are given below. While solving these problems, an important thing to remember is that it is always helpful to arrange things so that the left-hand side depends only on the dependent variable and the right-hand side on the independent variables.

Problem 3.22 (χ_t and α_p and the equation of state)
Upon measurement, the isothermal compressibility, χ_t, and the volume expansion coefficient, α_p, of one mole of a certain substance are found to be related to its volume v and temperature t as follows:

$$\alpha_p = \frac{R\chi_t}{(v-d)} \tag{3.114}$$

[24]Construction of equation of state from knowledge of bulk and elastic moduli.

and

$$\chi_t^{-1} = \frac{Rtv}{(v-b)^2} - \left(\frac{2a}{v^2}\right), \tag{3.115}$$

where R, b, and d are constants. Construct an equation of state for the substance and indicate the required relationships between the constants.

Solution 3.22 Because both α_p and χ_t depend only on the two (independent) variables v and t, we make p the dependent variable and express it as a function of v and t, i.e., $p = p(v, t)$. Then

$$\mathrm{d}p = \left(\frac{\partial p}{\partial v}\right)_t \mathrm{d}v + \left(\frac{\partial p}{\partial t}\right)_v \mathrm{d}t\,. \tag{3.116}$$

Next we employ the cyclic identity and express (3.114) in the form

$$\begin{aligned}\left(\frac{\partial p}{\partial t}\right)_v &= -\left(\frac{\partial p}{\partial v}\right)_t\left(\frac{\partial v}{\partial t}\right)_p = \left[-v\left(\frac{\partial p}{\partial v}\right)_t\right]\cdot\left[\frac{(\frac{\partial v}{\partial t})_p}{v}\right] \\ &= \left[(\chi_t)^{-1}\right]\cdot\alpha_p = \left(\frac{R}{v-d}\right).\end{aligned} \tag{3.117}$$

Additionally, we express (3.115) as

$$\left(\frac{\partial p}{\partial v}\right)_t = -\frac{\chi_t^{-1}}{v} = \frac{2a}{v^3} - \frac{Rt}{(v-b)^2}\,. \tag{3.118}$$

Because $\mathrm{d}p$ is an exact differential,

$$\frac{\partial^2 p}{\partial v \partial t} = \frac{\partial^2 p}{\partial t \partial v}\,.$$

Employing (3.117) and (3.118), we get

$$\begin{aligned}\frac{\partial^2 p}{\partial v\partial t} &= \left(\frac{\partial(\frac{\partial p}{\partial t})_v}{\partial v}\right)_t = \left(\frac{\partial(\frac{R}{v-d})}{\partial v}\right)_t = -\frac{R}{(v-d)^2}\,, \\ \frac{\partial^2 p}{\partial t\partial v} &= \left(\frac{\partial(\frac{\partial p}{\partial v})_t}{\partial t}\right)_v = \left(\frac{\partial[\frac{2a}{v^3} - \frac{RT}{(v-b)^2}]}{\partial T}\right)_v = -\frac{R}{(v-b)^2}\,.\end{aligned} \tag{3.119}$$

Because v is arbitrary, the following must hold:

$$d = b\,. \tag{3.120}$$

The integration of the partial derivatives $(\frac{\partial p}{\partial t})_v$ and $(\frac{\partial p}{\partial v})_t$ needs to be done next. According to (3.116), for a given fixed volume v, $\mathrm{d}v = 0$. Therefore we have

$$(\mathrm{d}p)_v = \left(\frac{\partial p}{\partial t}\right)_v (\mathrm{d}t)_v\,.$$

Integrating both sides gives

$$p(v,t) = \int \left(\frac{\partial p}{\partial t}\right)_{\mathrm{v}} (\mathrm{d}t)_{\mathrm{v}} = \int \frac{R}{(v-b)}(\mathrm{d}t)_{\mathrm{v}} = \frac{R}{(v-b)}t + f(v)\,, \tag{3.121}$$

where $f(v)$ is the constant of integration.

A first order linear differential equation for the still to be determined function $f(v)$ can now be found by differentiating the above expression for p with respect to v while t is kept constant. That is,

$$\left(\frac{\partial p}{\partial v}\right)_{\mathrm{t}} = -\frac{Rt}{(v-b)^2} + \frac{\mathrm{d}f(v)}{\mathrm{d}v}\,. \tag{3.122}$$

Equating this with the prior value for $(\frac{\partial p}{\partial v})_{\mathrm{T}}$ given in (3.118), we get

$$\frac{\mathrm{d}f(v)}{\mathrm{d}v} = \frac{2a}{v^3}\,.$$

Thus

$$f(v) = \int \left[\frac{\mathrm{d}f(v)}{\mathrm{d}v}\right]\mathrm{d}v = \int \left(\frac{2a}{v^3}\right)\mathrm{d}v = -\frac{a}{v^2} + c\,, \tag{3.123}$$

where c is a constant to be determined by the initial condition. Combining (3.121) and (3.123) and assuming the initial condition refers to p_0, v_0, t_0, we can write the equation of state as

$$p + \frac{a}{v^2} - \frac{Rt}{(v-b)} = p_0 + \frac{a}{v_0^2} - \frac{Rt_0}{(v_0-b)}\,. \tag{3.124}$$

In Chap. 6 on imperfect gases we shall learn that the above equation is strongly suggestive of what is called the Van der Waals gas but with a subtle difference. Here the right-hand side of the equation is given as a constant which can—but is not necessarily required to—be equal to zero. Unfortunately, this is as far as mathematics can lead us. We need the help of physics to set the constant on the right-hand side equal to zero.

Problem 3.23 Work out alternate solution of Problem 3.22.

Solution 3.23 In solving[25] problems similar to that in Problem 3.22, it may sometime be preferable to more directly exploit the exactness of a differential such as $\mathrm{d}p$ which allows one the freedom to use any arbitrary integration path between the beginning and ending points, i.e., p_0, v_0, t_0 and p, v, t. To this purpose, let us start with (3.116), (3.117), and (3.118), and integrate the perfect differential

$$\mathrm{d}p = \left(\frac{\partial p}{\partial v}\right)_{\mathrm{t}} \mathrm{d}v + \left(\frac{\partial p}{\partial t}\right)_{\mathrm{v}} \mathrm{d}t = \left(\frac{R}{v-d}\right)\mathrm{d}t + \left[\frac{2a}{v^3} - \frac{Rt}{(v-b)^2}\right]\mathrm{d}v\,. \tag{3.125}$$

[25] Equation of state from χ_{t} and α_{p}.

The first issue to deal with is the integrability requirement

$$\frac{\partial^2 p}{\partial v \partial t} = \frac{\partial^2 p}{\partial t \partial v},$$

which leads to the result

$$-\frac{R}{(v-d)^2} = -\frac{R}{(v-b)^2}.$$

Because this is to be true for arbitrary v, the following equality must hold:

$$d = b. \tag{3.126}$$

The boundary conditions to use are the initial values for the pressure, volume and temperature, namely p_0, v_0, t_0, and their final values p, v, and t.

As mentioned earlier, it is always helpful to arrange things so that the left-hand side depends only on the dependent variable—in this case, p—and the right-hand side on the independent ones—in this case, v and t. Again this happens already to be the case here.

For convenience, we choose a path that takes us along the following route: For the left-hand side—which does not depend on v and t—clearly, we can go all the way from $(p_0, v = v_0, t = t_0) \to (p, v, t)$ in one fell swoop. That is, as

$$\int_{p_0}^{p} \mathrm{d}p = p - p_0 = (\mathrm{I}) + (\mathrm{II}). \tag{3.127}$$

For the right-hand side of (3.125), the integration over p is not needed. Therefore, no contribution is recorded for the initial part of the route $(p_0, v_0, t_0) \to (p, v_0, t_0)$.

Accordingly, we start the integration of the right-hand side with journey from $(p, v_0, t = t_0) \to (p, v, t = t_0)$.

The first leg (I) of the integration on the right-hand side is therefore represented as follows:

$$\begin{aligned}
(\mathrm{I}) &= \left(\frac{R}{v-b}\right)\int_{p,v_0,t=t_0}^{p,v,t=t_0} \mathrm{d}t + \int_{p,v_0,t=t_0}^{p,v,t=t_0} \mathrm{d}v\left[\frac{2a}{v^3} - \frac{Rt_0}{(v-b)^2}\right] \\
&= 0 + 2a\left[\frac{v^{-2}}{-2} - \frac{{v_0}^{-2}}{-2}\right] + Rt_0\left[(v-b)^{-1} - (v_0-b)^{-1}\right] \\
&= a\left[{v_0}^{-2} - v^{-2}\right] + Rt_0\left[(v-b)^{-1} - (v_0-b)^{-1}\right].
\end{aligned} \tag{3.128}$$

Note the integral

$$\int_{p,v_0,t=t_0}^{p,v,t=t_0} \mathrm{d}t$$

is zero because it is supposed to extend only from $t = t_0$ to $t = t_0$.

Similarly, the second leg is represented as follows:

$$
\begin{aligned}
(\mathrm{II}) &= \int_{p,v,t_0}^{p,v,t} \left(\frac{R}{v-b}\right) \mathrm{d}t + \int_{p,v,t_0}^{p,v,t} \mathrm{d}v \left[\frac{2a}{v^3} - \frac{Rt}{(v-b)^2}\right] \\
&= \int_{v,t_0}^{v,t} \mathrm{d}t \left[\frac{R}{v-b}\right] + 0 = R\frac{(t-t_0)}{v-b}\,. \qquad (3.129)
\end{aligned}
$$

It is important to note that in the above integral over v, i.e.,

$$
\int_{p,v,t_0}^{p,v,t} \mathrm{d}v \left[\frac{2a}{v^3} - \frac{Rt}{(v-b)^2}\right],
$$

is vanishing because both the initial and final values of the integration variable, v, are the same.

The result of the two integrals (I) and (II) is:

$$
p - p_0 = \mathrm{I} + \mathrm{II} = -a\left(\frac{1}{v^2} - \frac{1}{v_0^2}\right) + R\left(\frac{t}{v-b} - \frac{t_0}{v_0-b}\right). \qquad (3.130)
$$

Hence the equation of state is the following:

$$
p + \frac{a}{v^2} - \frac{Rt}{(v-b)} = p_0 + \frac{a}{v_0^2} - \frac{Rt_0}{(v_0-b)}\,. \qquad (3.131)
$$

Problem 3.24 Find the equation of state of a substance that has isothermal compressibility, $\chi_t = (\frac{b}{v})$ and volume expansion coefficient, $\alpha_p = (\frac{at}{v})$.

Solution 3.24 First, let us organize the above information. We are given

$$
\alpha_p = \frac{1}{v}\left(\frac{\partial v}{\partial t}\right)_p = \left(\frac{at}{v}\right),
$$

which means

$$
\left(\frac{\partial v}{\partial t}\right)_p = at\,. \qquad (3.132)
$$

Also, we are told that

$$
\chi_t = -\frac{1}{v}\left(\frac{\partial v}{\partial p}\right)_t = \left(\frac{b}{v}\right),
$$

which gives

$$
\left(\frac{\partial v}{\partial p}\right)_t = -b\,. \qquad (3.133)
$$

Because

$$
\frac{\partial^2 v}{\partial t \partial p} = \frac{\partial^2 v}{\partial p \partial t} = 0\,,
$$

the integrability requirement is satisfied for arbitrary values of a and b, as long as they are both constants.

We are given information about two partial differential coefficients, $(\frac{\partial v}{\partial t})_{\rm p}$ and $(\frac{\partial v}{\partial p})_{\rm t}$. Therefore, we start with

$$v = v(t, p)\,,$$

so that these two differentials can be utilized. That is, we write

$$\mathrm{d}v = \left(\frac{\partial v}{\partial t}\right)_{\rm p} \mathrm{d}t + \left(\frac{\partial v}{\partial p}\right)_{\rm t} \mathrm{d}p = at\mathrm{d}t - b\mathrm{d}p\,, \tag{3.134}$$

and integrate both sides along the route $(v_0, t_0, p_0) \to (v, t_0, p_0) \to (v, t, p_0) \to (v, t, p)$.

On the left-hand side, only the path from $(v_0, t_0, p_0) \to (v, t_0, p_0)$ makes a contribution, which is equal to $v - v_0$.

On the right-hand side, on the other hand, the path from $(v_0, t_0, p_0) \to (v, t_0, p_0)$ makes no contribution.

Next we look at the contribution, to the right-hand side, from the path $(v, t_0, p_0) \to (v, t, p_0)$. One readily finds this contribution, i.e., $\frac{a}{2}(t^2 - t_0^2)$.

Finally, on the right-hand side, we proceed along the path $(v, t, p_0) \to (v, t, p)$. This gives $-b(p - p_0)$.

Thus the desired equation of state is

$$v - v_0 = \frac{a}{2}(t^2 - t_0^2) - b(p - p_0)\,. \tag{3.135}$$

Problem 3.25 Find[26] the equation of state of one mole of a substance that has isothermal compressibility, $\chi_{\rm t} = a/(p^3t^2)$ and volume expansion coefficient, $\alpha_{\rm p} = b/(p^2t^3)$.

Are constants a and b related?

Solution 3.25 Because information about $\chi_{\rm t}$ and $\alpha_{\rm p}$ has been provided, we write

$$\begin{aligned} \mathrm{d}v &= \left(\frac{\partial v}{\partial t}\right)_{\rm p} \mathrm{d}t + \left(\frac{\partial v}{\partial p}\right)_{\rm t} \mathrm{d}p = v\alpha_{\rm p}\mathrm{d}t - v\chi_{\rm t}\mathrm{d}p \\ &= \left(\frac{bv}{p^2t^3}\right)\mathrm{d}t - \left(\frac{av}{p^3t^2}\right)\mathrm{d}p\,. \end{aligned} \tag{3.136}$$

The integrability requirement,

$$\left(\frac{\partial(\frac{\partial v}{\partial t})_{\rm p}}{\partial p}\right)_{\rm t} = \left(\frac{\partial(\frac{\partial v}{\partial p})_{\rm t}}{\partial t}\right)_{\rm p},$$

leads to the equality

$$\left(\frac{-2bv}{p^3t^3}\right) = \left(\frac{2av}{p^3t^3}\right). \tag{3.137}$$

[26]Equation of state using $\chi_{\rm t}$ and $\alpha_{\rm p}$.

Thus

$$b = -a\,. \tag{3.138}$$

It is interesting to note because thermodynamic stability requires the compressibility, χ_t, to be positive, the constant $a > 0$. Therefore, because of the requirement $b = -a$ for this example, the isobaric volume expansion coefficient, α_p, is negative.

At first sight, the above problem (3.136) appears to be quite complicated. This is because the right-hand side contains all three variables, v, t, and p. It turns out that upon dividing both sides by v, the right-hand side no longer involves v and the left-hand side, $\frac{dv}{v}$, still can be expressed as an exact differential, $\mathrm{d}(\ln v)$. Accordingly, we can write

$$\begin{aligned}\frac{\mathrm{d}v}{v} = \mathrm{d}\ln(v) &= \left(\frac{\partial \ln(v)}{\partial t}\right)_{\mathrm{p}} \mathrm{d}t + \left(\frac{\partial \ln(v)}{\partial p}\right)_{\mathrm{t}} \mathrm{d}p \\ &= \left(\frac{-a}{p^2 t^3}\right)\mathrm{d}t + \left(\frac{-a}{p^3 t^2}\right)\mathrm{d}p\,. \end{aligned}\tag{3.139}$$

The integration of (3.139), between the original location ($v = v_0$, $p = p_0$, $t = t_0$) and the final location ($v = v_1, p = p_1, t = t_1$), can be done along a path of our choice.

To this end, we integrate both sides along the route $(v = v_0,\ p = p_0, t = t_0) \rightarrow (v = v_1, p = p_0, t = t_0) \rightarrow (v = v_1, p = p_1, t = t_0) \rightarrow (v = v_1, p = p_1, t = t_1)$.

On the left-hand side, only the path from $(v = v_0,\ p = p_0, t = t_0) \rightarrow (v = v_1, p = p_0, t = t_0)$ makes a contribution, which is equal to $\ln(\frac{v_1}{v_0})$. For the right-hand side, integrals along the entire length of the path are displayed below:

$$\begin{aligned}
\ln\left(\frac{v_1}{v_0}\right) &= -a\int_{v=v_0,t=t_0,p=p_0}^{v=v_1,t=t_0,p=p_0}\left(\frac{\mathrm{d}t}{p^2t^3}\right) -a\int_{v=v_0,t=t_0,p=p_0}^{v=v_1,t=t_0,p=p_0}\left(\frac{\mathrm{d}p}{p^3t^2}\right)\\
&\quad -a\int_{v=v_1,t=t_0,p=p_0}^{v=v_1,t=t_1,p=p_0}\left(\frac{\mathrm{d}t}{p^2t^3}\right) -a\int_{v=v_1,t=t_0,p=p_0}^{v=v_1,t=t_1,p=p_0}\left(\frac{\mathrm{d}p}{p^3t^2}\right)\\
&\quad -a\int_{v=v_1,t=t_1,p=p_0}^{v=v_1,t=t_1,p=p_1}\left(\frac{\mathrm{d}t}{p^2t^3}\right) -a\int_{v=v_1,t=t_1,p=p_0}^{v=v_1,t=t_1,p=p_1}\left(\frac{\mathrm{d}p}{p^3t^2}\right);\\
\ln\left(\frac{v_1}{v_0}\right) &= 0 \qquad -0\\
&\quad -a\int_{v=v_1,t=t_0,p=p_0}^{v=v_1,t=t_1,p=p_0}\left(\frac{\mathrm{d}t}{p_0^2t^3}\right) \qquad -0\\
&\quad -0 \qquad -a\int_{v=v_1,t=t_1,p=p_0}^{v=v_1,t=t_1,p=p_1}\left(\frac{\mathrm{d}p}{p^3t_1^2}\right);\\
\ln\left(\frac{v_1}{v_0}\right) &= \left(\frac{a}{2{p_0}^2}\right)\left(t_1^{-2} - t_0^{-2}\right) + \left(\frac{a}{2{t_1}^2}\right)\left(p_1^{-2} - p_0^{-2}\right).
\end{aligned}\tag{3.140}$$

Therefore, the desired equation of state is

$$\ln\left(\frac{v_1}{v_0}\right) = \left(\frac{a}{2\, p^2\, t_1{}^2}\right) - \left(\frac{a}{2\, p_0{}^2\, t_0{}^2}\right). \tag{3.141}$$

Problem 3.26 (Alternate solution of Problem 3.14)
It is instructive to retravel the above via the differential equation route. To this end, let us begin with (3.139), but for notational convenience set $v = v_1$ and $t = t_1$.

Solution 3.26 We have

$$\begin{aligned} \mathrm{d}\ln(v_1) &= \left(\frac{\partial \ln(v_1)}{\partial t_1}\right)_{\mathrm{p}} \mathrm{d}t_1 + \left(\frac{\partial \ln(v_1)}{\partial p}\right)_{\mathrm{t_1}} \mathrm{d}p \\ &= \left(\frac{-a}{p^2 t_1^3}\right)\mathrm{d}t_1 + \left(\frac{-a}{p^3 t_1^2}\right)\mathrm{d}p\,. \end{aligned} \tag{3.142}$$

Now let us integrate it for a given fixed value of p:

$$\int \mathrm{d}(\ln v_1) =(\text{for } p \text{ constant}) -a \int \frac{\mathrm{d}t_1}{p^2 t_1^3}\,. \tag{3.143}$$

We get

$$\ln v_1 = \left(\frac{a}{2p^2 t_1^2}\right) + f(p)\,, \tag{3.144}$$

where $f(p)$ is as yet an unknown function.

In order to determine $f(p)$, we differentiate $\ln v_1$ with respect to p while holding t_1 constant:

$$\left(\frac{\partial \ln v_1}{\partial p}\right)_{\mathrm{t_1}} = -\frac{a}{p^3 t_1^2} + \frac{\mathrm{d}f(p)}{\mathrm{d}p}\,. \tag{3.145}$$

Comparing the values of $(\frac{\partial \ln v_1}{\partial p})_{\mathrm{t_1}}$ given in (3.142) and (3.145), we are presented with the equality

$$\left(\frac{\partial \ln v_1}{\partial p}\right)_{\mathrm{t_1}} = -\frac{a}{p^3 t_1^2} = -\frac{a}{p^3 t_1^2} + \frac{\mathrm{d}f(p)}{\mathrm{d}p}\,.$$

This equality can be satisfied only if

$$\frac{\mathrm{d}f(p)}{\mathrm{d}p} = 0\,,$$

which requires that $f(p)$ be a constant. With this information, (3.144) leads to the final result

$$\ln v_1 - \left(\frac{a}{2p^2 t_1^2}\right) = A\,, \tag{3.146}$$

where A is a constant. Note that A can be determined from the initial condition, $v_1 = v_0$, $t_1 = t_0$, and $p = p_0$. Therefore

$$A = \ln v_0 - \left(\frac{a}{2p_0^2 t_0^2}\right),$$

and as a result (3.146) yields the equation of state

$$\ln\left[\frac{v_1}{v_0}\right] = \left(\frac{a}{2p^2 t_1^2}\right) - \left(\frac{a}{2p_0^2 t_0^2}\right). \tag{3.147}$$

The results of the two different procedures—that is, (3.141) and (3.147)—are identical.

Problem 3.27 Find the equation of state of one mole of a certain gas at temperature t for which

$$\left(\frac{\alpha_p}{\chi_t}\right) = \left(\frac{R}{v-b}\right)\left[1 + \frac{a}{Rtv}\right]\exp\left[-\frac{a}{Rtv}\right] \tag{3.148}$$

and

$$\chi_t^{-1} = -v\left(\frac{Rt}{v-b}\right)\left[\frac{a}{Rtv^2} - \frac{1}{v-b}\right]\exp\left[-\frac{a}{Rtv}\right], \tag{3.149}$$

where as usual R is the molar gas constant, χ_t the isothermal compressibility, and α_p the expansion coefficient.

Solution 3.27 Note

$$\left(\frac{\alpha_p}{\chi_t}\right) = \frac{1}{v}\left(\frac{\partial v}{\partial t}\right)_p \cdot \left[-v\left(\frac{\partial p}{\partial v}\right)_t\right] = -\left(\frac{\partial p}{\partial v}\right)_t \left(\frac{\partial v}{\partial t}\right)_p = \left(\frac{\partial p}{\partial t}\right)_v \tag{3.150}$$

and

$$\chi_t^{-1} = -v\left(\frac{\partial p}{\partial v}\right)_t. \tag{3.151}$$

Therefore, we can write

$$\begin{aligned} dp &= \left(\frac{\partial p}{\partial t}\right)_v dt + \left(\frac{\partial p}{\partial v}\right)_t dv \\ &= \left(\frac{R}{v-b}\right)\left[1 + \frac{a}{Rtv}\right]\exp\left[-\frac{a}{Rtv}\right]dt \\ &\quad + \left(\frac{Rt}{v-b}\right)\left[\frac{a}{Rtv^2} - \frac{1}{v-b}\right]\exp\left[-\frac{a}{Rtv}\right]dv \\ &= (\mathrm{I})dt + (\mathrm{II})dv\,. \end{aligned} \tag{3.152}$$

Integrating along the path $(p_0, t_0, v_0) \to (p, t_0, v_0) \to (p, t, v_0) \to (p, t, v)$ yields

$$p - p_0 = \int (\text{I})\mathrm{d}t + \int (\text{II})\mathrm{d}v\,, \tag{3.153}$$

where

$$\begin{aligned}
\int (\text{I})\mathrm{d}t &= \int_{p_0,t_0,v_0}^{p,t_0,v_0} (\text{I})\mathrm{d}t + \int_{p,t_0,v_0}^{p,t,v_0} (\text{I})\mathrm{d}t + \int_{p,t,v_0}^{p,t,v} (\text{I})\mathrm{d}t \\
&= 0 + \int_{p,t_0,v_0}^{p,t,v_0} \left(\frac{R}{v_0 - b}\right)\left[1 + \frac{a}{Rtv_0}\right]\exp\left[-\frac{a}{Rtv_0}\right]\mathrm{d}t + 0 \\
&= \left(\frac{Rt}{v_0 - b}\right)\exp\left(\frac{-a}{Rtv_0}\right)\Bigg|_{t_0}^{t} \\
&= \left(\frac{Rt}{v_0 - b}\right)\exp\left(\frac{-a}{Rtv_0}\right) - \left(\frac{Rt_0}{v_0 - b}\right)\exp\left(\frac{-a}{Rt_0v_0}\right),
\end{aligned} \tag{3.154}$$

and

$$\begin{aligned}
\int (\text{II})\mathrm{d}v &= \int_{p_0,t_0,v_0}^{p,t_0,v_0} (\text{II})\mathrm{d}v + \int_{p,t_0,v_0}^{p,t,v_0} (\text{II})\mathrm{d}v + \int_{p,t,v_0}^{p,t,v} (\text{II})\mathrm{d}v \\
&= 0 + 0 + \int_{p,t,v_0}^{p,t,v} \left(\frac{Rt}{v - b}\right)\left[\frac{a}{Rtv^2} - \frac{1}{v - b}\right]\exp\left[-\frac{a}{Rtv}\right]\mathrm{d}v \\
&= \left(\frac{Rt}{v - b}\right)\exp\left(\frac{-a}{Rtv}\right)\Bigg|_{v_0}^{v} \\
&= \left(\frac{Rt}{v - b}\right)\exp\left(\frac{-a}{Rtv}\right) - \left(\frac{Rt}{v_0 - b}\right)\exp\left(\frac{-a}{Rtv_0}\right).
\end{aligned} \tag{3.155}$$

Therefore, according to (3.153), (3.154), and (3.155), the equation of state is as follows:

$$p - p_0 = \left(\frac{Rt}{v - b}\right)\exp\left(\frac{-a}{Rtv}\right) - \left(\frac{Rt_0}{v_0 - b}\right)\exp\left(\frac{-a}{Rt_0v_0}\right). \tag{3.156}$$

Problem 3.28 Find the equation of state of a single mole of a fluid whose isothermal susceptibility χ_{t} and isobaric volume expansion coefficient α_{p} are well described by the following equations:

$$p\chi_{\mathrm{t}} = 1 + \frac{1}{v}\left(\frac{c}{t^2}\right)$$

and

$$t\alpha_{\mathrm{p}} = 1 + \frac{1}{v}\left(\frac{d}{t^2}\right).$$

Show that for the sake of consistency, d must be equal to $3\,c$.

Solution 3.28 As usual, we write

$$\mathrm{d}v = \left(\frac{\partial v}{\partial t}\right)_{\mathrm{p}} \mathrm{d}t + \left(\frac{\partial v}{\partial p}\right)_{\mathrm{t}} \mathrm{d}p = v\alpha_{\mathrm{p}}\mathrm{d}t - v\chi_{\mathrm{t}}\mathrm{d}p$$
$$= \left(\frac{v}{t} + \frac{d}{t^3}\right)\mathrm{d}t - \left(v + \frac{c}{t^2}\right)\frac{\mathrm{d}p}{p}. \tag{3.157}$$

Next we rearrange (3.157) so that its left-hand side contains a function of only a single variable and the right-hand side only that of the remaining two variables:

$$\frac{\mathrm{d}p}{p} = -\frac{\mathrm{d}v}{v + \frac{c}{t^2}} + \frac{1}{t}\left[\frac{v + \frac{d}{t^2}}{v + \frac{c}{t^2}}\right]\mathrm{d}t\,,$$
$$\mathrm{d}(\ln p) = -\frac{\mathrm{d}v}{v + \frac{c}{t^2}} + \frac{\mathrm{d}t}{t} + \frac{\frac{\mathrm{d}-c}{t^3}}{(v + \frac{c}{t^2})}\mathrm{d}t\,. \tag{3.158}$$

We find that the differential $\mathrm{d}(\ln p)$ satisfies the integrability requirement, i.e.,

$$\frac{\partial^2(\ln p)}{\partial t \partial v} = \frac{\partial^2(\ln p)}{\partial v \partial t} = \frac{\frac{-2c}{t^3}}{(v + \frac{c}{t^2})^2} = \frac{\frac{c-d}{t^3}}{(v + \frac{c}{t^2})^2}\,,$$

only if $(-2c) = (c - d)$, or equivalently, if $d = 3c$. When this is the case, $\mathrm{d}(\ln p)$ is an exact differential, and we may choose an integration path of our choice.

The chosen path proceeds from $(p_0, v_0, t_0) \rightarrow (p, v_0, t_0) \rightarrow (p, v, t_0) \rightarrow (p, v, t)$. For the left-hand side of (3.158), only the portion from $(p_0, v_0, t_0) \rightarrow (p, v_0, t_0)$ contributes. And the result is $\ln[\frac{p}{p_0}]$.

For the right-hand side, the term with $\mathrm{d}v$ contributes only for that portion of the path that extends from $(p, v_0, t_0) \rightarrow (p, v, t_0)$. Similarly, the two terms with $\mathrm{d}t$ contribute only for the portion of the path that extends from $(p, v, t_0) \rightarrow (p, v, t)$.

Consequently, we get

$$\ln\left[\frac{p}{p_0}\right] = -\int_{p,v_0,t_0}^{p,v,t_0} \frac{\mathrm{d}v}{v + \frac{c}{t^2}} + \int_{p,v,t_0}^{p,v,t} \frac{\mathrm{d}t}{t} + \int_{p,v,t_0}^{p,v,t} \frac{\frac{2c}{t^3}}{(v + \frac{c}{t^2})}\mathrm{d}t$$
$$= -\ln\left[\frac{v + \frac{c}{t_0^2}}{v_0 + \frac{c}{t_0^2}}\right] + \ln\left[\frac{t}{t_0}\right] - \ln\left[\frac{(v + \frac{c}{t^2})}{v + \frac{c}{t_0^2}}\right]$$
$$= -\ln\left[\frac{(v + \frac{c}{t^2})t_0}{(v_0 + \frac{c}{t_0^2})t}\right]. \tag{3.159}$$

Transferring the right-hand side term to the left and combining them into a single logarithm, we get

$$\ln\left[\frac{p(v + \frac{c}{t^2})t_0}{t(v_0 + \frac{c}{t_0^2})p_0}\right] = 0\,,$$

which leads to the equation of state

$$\left(v+\frac{c}{t^2}\right)\cdot\frac{p}{t}=\left(v_0+\frac{c}{t_0^2}\right)\cdot\frac{p_0}{t_0}\,. \tag{3.160}$$

Problem 3.29 Work out an alternate solution of Problem 3.28.

Solution 3.29 From (3.158), we have

$$\left(\frac{\partial \ln p}{\partial v}\right)_{\mathrm{t}}=-\frac{1}{v+\frac{c}{t^2}}\,, \tag{3.161}$$

$$\left(\frac{\partial \ln p}{\partial t}\right)_{\mathrm{v}}=\frac{1}{t}\left[\frac{v+\frac{3c}{t^2}}{v+\frac{c}{t^2}}\right]. \tag{3.162}$$

For given fixed value of t, integrating (3.161) gives

$$\int\left(\frac{\partial \ln p}{\partial v}\right)_{\mathrm{t}}\mathrm{d}v=\ln(p)=-\int\frac{1}{v+\frac{c}{t^2}}\mathrm{d}v=-\ln\left(v+\frac{c}{t^2}\right)+\phi(t)\,. \tag{3.163}$$

The unknown function $\phi(t)$ can be determined by performing the operation indicated in (3.162). This process proceeds by differentiating $\ln(p)$ given in (3.163) as follows:

$$\left(\frac{\partial \ln p}{\partial t}\right)_{\mathrm{v}}=\frac{\frac{2c}{t^3}}{v+\frac{c}{t^2}}+\frac{d\phi(t)}{\mathrm{d}t}\,. \tag{3.164}$$

Next we compare the value of $(\frac{\partial \ln p}{\partial t})_{\mathrm{v}}$ given in (3.162) and (3.164). That is,

$$\left(\frac{\partial \ln p}{\partial t}\right)_{\mathrm{v}}=\frac{1}{t}\left[\frac{v+\frac{3c}{t^2}}{v+\frac{c}{t^2}}\right]\quad\text{and}$$
$$\left(\frac{\partial \ln p}{\partial t}\right)_{\mathrm{v}}=\frac{\frac{2c}{t^3}}{v+\frac{c}{t^2}}+\frac{d\phi(t)}{\mathrm{d}t}\,. \tag{3.165}$$

This gives

$$\frac{\mathrm{d}\phi(t)}{\mathrm{d}t}=\frac{1}{t}\,.$$

Upon integration, we get

$$\phi(t)=\ln(t)+C\,, \tag{3.166}$$

where C is a constant. Introducing this into the right-hand side of (3.163) gives

$$\ln(p)=-\ln\left(v+\frac{c}{t^2}\right)+\ln(t)+C\,.$$

Exponentiating both sides leads to the desired equation of state

$$\left(v + \frac{c}{t^2}\right) \cdot \frac{p}{t} = \exp C = \text{const.} \tag{3.167}$$

Again, as required, this result is the same as that obtained by the direct integration method. (Compare (3.160) and (3.167).)

Problem 3.30 Upon measurement, the isothermal compressibility, χ_t, and the volume expansion coefficient, α_p, of one mole of a certain substance are found to be related to its volume v and pressure p as follows: $\alpha_\mathrm{p} = \frac{R}{v(p+b)}$ and $\chi_\mathrm{t}^{-1} = (p + b)$, where R and b are constants.

Construct an equation of state for the substance.

Is any particular relationship between the constants R and b required?

See below for an answer.

Solution 3.30 We have

$$\begin{aligned} \left(\frac{\partial p}{\partial t}\right)_\mathrm{v} &= \frac{\alpha_\mathrm{p}}{\chi_\mathrm{t}} = \frac{\frac{R}{v(p+b)}}{\frac{1}{p+b}} = \left(\frac{R}{v}\right), \\ \left(\frac{\partial p}{\partial v}\right)_\mathrm{t} &= -\frac{1}{v\chi_\mathrm{t}} = -\left(\frac{p+b}{v}\right). \end{aligned} \tag{3.168}$$

If we should try the relationship

$$\mathrm{d}p = \left(\frac{\partial p}{\partial t}\right)_\mathrm{v} \mathrm{d}t + \left(\frac{\partial p}{\partial v}\right)_\mathrm{t} \mathrm{d}v\,,$$

which is

$$\mathrm{d}p = \left(\frac{R}{v}\right)\mathrm{d}t - \left(\frac{p+b}{v}\right)\mathrm{d}v\,, \tag{3.169}$$

we find that the right-hand side contains all the three variables. Things do not improve much if we work instead with equation

$$\mathrm{d}v = \left(\frac{\partial v}{\partial p}\right)_\mathrm{t} \mathrm{d}p + \left(\frac{\partial v}{\partial t}\right)_\mathrm{p} \mathrm{d}t\,.$$

So we look for the third option, namely

$$\mathrm{d}t = \left(\frac{\partial t}{\partial p}\right)_\mathrm{v} \mathrm{d}p + \left(\frac{\partial t}{\partial v}\right)_\mathrm{p} \mathrm{d}v\,.$$

Rather than calculating *ab initio* the right-hand side of the above equation, we can conveniently use the existing information in (3.169) by multiplying both sides with $(\frac{v}{R})$. Accordingly we get

$$\mathrm{d}t = \left(\frac{v}{R}\right)\mathrm{d}p + \left(\frac{p+b}{R}\right)\mathrm{d}v = \left(\frac{\partial t}{\partial p}\right)_\mathrm{v} \mathrm{d}p + \left(\frac{\partial t}{\partial v}\right)_\mathrm{p} \mathrm{d}v\,. \tag{3.170}$$

This satisfies the requirement that the terms involving the dependent variable alone should be on the left-hand side while those involving the independent variables—here they are p and v— are on the right-hand side. We now follow the usual procedure and begin by writing the above equation for the case where v is constant and then integrating it. We get

$$\int (\mathrm{d}t)_\mathrm{v} = \int \left(\frac{v}{R}\right)(\mathrm{d}p)_\mathrm{v}\,, \quad \text{therefore,} \quad t = \left(\frac{v}{R}\right)p + f(v)\,, \tag{3.171}$$

where $f(v)$ is some as yet to be determined function of v. Next we calculate the remaining differential $(\frac{\partial t}{\partial v})_\mathrm{p}$. That is, using (3.171) and (3.170), we get:

$$\begin{aligned} \left(\frac{\partial t}{\partial v}\right)_\mathrm{p} &= \frac{p}{R} + f'(v) = \left(\frac{p+b}{R}\right), \quad \text{which gives} \\ f'(v) &= \frac{df(v)}{\mathrm{d}v} = \left(\frac{b}{R}\right), \quad \text{and, upon integration, we get} \\ f(v) &= \left(\frac{b}{R}\right)v + C\,, \end{aligned} \tag{3.172}$$

where C is some unknown constant. Combining this result with (3.171), we get the desired equation of state

$$\begin{aligned} t &= \left(\frac{v}{R}\right)p + f(v)\,, \quad \text{which gives} \\ t &= \left(\frac{v(p+b)}{R}\right) + C\,. \end{aligned} \tag{3.173}$$

We may also write the above as

$$t - t_0 = \left(\frac{v(p+b)}{R}\right) - \left(\frac{v_0(p_0+b)}{R}\right), \tag{3.174}$$

where the subscript 0 refers to the result at some initial measurement.

Regarding any mandated relationship between b and R, we look at the identity $(\frac{\partial(\frac{\partial t}{\partial v})_\mathrm{p}}{\partial p})_\mathrm{v} = (\frac{\partial(\frac{\partial t}{\partial p})_\mathrm{v}}{\partial v})_\mathrm{t}$. We get

$$\begin{aligned} \left(\frac{\partial(\frac{\partial t}{\partial v})_\mathrm{p}}{\partial p}\right)_\mathrm{v} &= \left(\frac{\partial(\frac{p+b}{R})}{\partial p}\right)_\mathrm{v} = \frac{1}{R}\,, \\ \left(\frac{\partial(\frac{\partial t}{\partial p})_\mathrm{v}}{\partial v}\right)_\mathrm{t} &= \left(\frac{\partial(\frac{v}{R})}{\partial v}\right)_\mathrm{t} = \frac{1}{R}\,. \end{aligned} \tag{3.175}$$

The identity is satisfied for all values of R and b. So they are not required to be interdependent.

Problem 3.31 (Equation of state for constant χ and α)
Find the equation of state of a substance whose isothermal compressibility and volume expansion coefficient are both constant, equal respectively to χ and α.

Solution 3.31 Because $\mathrm{d}v$ is an exact differential, we have

$$\mathrm{d}v = \left(\frac{\partial v}{\partial t}\right)_{\mathrm{p}} dt + \left(\frac{\partial v}{\partial p}\right)_{\mathrm{t}} \mathrm{d}p = v\alpha_{\mathrm{p}}\mathrm{d}t - v\chi_{\mathrm{t}}\mathrm{d}p = v\alpha\mathrm{d}t - v\chi\mathrm{d}p\,. \tag{3.176}$$

Dividing both sides by v gives

$$\frac{\mathrm{d}v}{v} = \alpha\mathrm{d}t - \chi\mathrm{d}p\,. \tag{3.177}$$

Also because $\mathrm{d}(\ln v)$ is an exact differential, we have

$$\frac{dv}{v} = \mathrm{d}(\ln v) = \left(\frac{\partial(\ln v)}{\partial t}\right)_{\mathrm{p}} \mathrm{d}t + \left(\frac{\partial(\ln v)}{\partial p}\right)_{\mathrm{t}} \mathrm{d}p\,. \tag{3.178}$$

This leads to two relationships. Let us begin with the first one,

$$\left(\frac{\partial(\ln v)}{\partial t}\right)_{\mathrm{p}} = \alpha\,,$$

which upon integration gives

$$\ln v = \alpha t + f(p)\,, \tag{3.179}$$

where $f(p)$ is as yet an unknown function of the pressure p.

The second relationship given by (3.178) is

$$\left(\frac{\partial(\ln v)}{\partial p}\right)_{\mathrm{t}} = -\chi\,.$$

Using (3.179), we can write the above as

$$\frac{\mathrm{d}f}{\mathrm{d}p} = -\chi\,,$$

which leads to

$$f(p) = -p\chi + \mathrm{const.}$$

Therefore, (3.179) becomes

$$\ln v - \alpha t + p\chi = \mathrm{const.} \tag{3.180}$$

The constant can be determined by invoking the position of the beginning point p_0, v_0, t_0. This readily leads to the relationship

$$p + \chi^{-1}\ln\left(\frac{v}{v_0}\right) = p_0 + \frac{\alpha}{\chi}(t - t_0)\,. \tag{3.181}$$

3.14 Newton's Law of Cooling

A rule of thumb—sometime referred to as Newton's law of cooling—suggests that the rate of loss of heat energy, $-\Delta Q/\Delta t$, from an object at temperature T placed in calm atmosphere at temperature T_0 is approximately proportional to the temperature difference $(T - T_0)$, i.e.,

$$\Delta Q/\Delta t = -\lambda\left[T(t) - T_0\right]. \tag{3.182}$$

Here λ is a constant that depends on the physical characteristics of the object.

Problem 3.32 (Illustration of Newton's law of cooling)
A small block of mass M kg, with an embedded resistor, has an effective specific heat C_P measured in the units of J/(kg K). It is hung from the ceiling by electrical wiring connected to the resistor. Although the electrical wiring will conduct some heat energy, we shall assume that little of that happens. The air in the room is calm and its temperature $T_0 = 300$ K. The resistor is supplied 350 watts of power for 100 s. As a result the temperature of the block rises from $T_i = 310$ to $T_f = 320$ K. When the current is switched off, the block cools down to the initial temperature 310 K. The cooling takes 30 s. Give estimates for (a) the parameter λ and (b) the parameter MC_P.

Solution 3.32 Newton's heat energy loss law is only moderately accurate. Thus, for temperature differences that are small, i.e., $[(T_i + T_f)/2 - T_0] \ll T_0$, and changes that are small compared to the ambient temperature, i.e., $T_f - T_i \ll T_0$, a linear averaging approximation should be adequate. For instance, given $T_0 = 300$ °K, $T_i = 310$, and $T_f = 320$, the heat energy lost in time Δt is approximately as noted below:

$$\begin{aligned}(\Delta Q)_{\text{heat energy lost}} &= -\lambda \int_0^{\Delta t} dt\left[T(t) - T_0\right] \\ &\approx -\lambda\left[\frac{310+320}{2} - 300\right]\Delta t. \end{aligned} \tag{3.183}$$

Note that here we have replaced the time-dependent temperature by its average in the specified narrow interval $T_f > T(t) > T_i$. Therefore, we can write the difference between the heat energy input, 350×100 J, and that lost from the Newton cooling during the 100 s that the system is being heated, $\lambda[(\frac{310+320}{2}) - 300] \times 100$, to be equal to the heat energy needed, $MC_P(320 - 310)$, to heat the block from 310 to 320 K. That is,

$$350 \times 100 - \lambda\left[\frac{310+320}{2} - 300\right] \times 100 = MC_P(320 - 310). \tag{3.184}$$

The above provides one relationship, but in order to determine both λ and MC_P, two relationships are needed.

The second relationship is obtained from the rate of cooling after the power is turned off. Again, noting that the object cools down according to Newton's law, the heat energy lost in 30 s of cooling from 320 °K to 310 °K is described as follows:

$$\lambda(315 - 300) \times 30 = MC_P(320 - 310)\,. \tag{3.185}$$

These two equations give

$$\lambda \approx 18\ \mathrm{J/(K\,s)}\,, \quad \text{and}$$
$$MC_P \approx 8.1 \times 10^2\ \mathrm{J/K}\,.$$

3.15 Internal Energy in Noninteracting Monatomic Gases Equals $\frac{3}{2}PV$

Irrespective of[27] whether a noninteracting gas is composed of particles that obey classical, or quantum—namely, Bose or Fermi—statistics, the product of its pressure P and volume V is directly related to its internal energy U. The relationship is

$$U = \frac{3}{2}PV\,. \tag{3.186}$$

3.16 Volume Dependence of Single Particle Energy Levels

According to (3.27),

$$\left(\frac{\partial U}{\partial V}\right)_S = -P\,,$$

therefore, for the equality

$$PV = \frac{2}{3}U\,,$$

we should have

$$\left(\frac{\partial U}{\partial V}\right)_S = -\left(\frac{2}{3}\frac{U}{V}\right).$$

Accordingly, we can write

$$\frac{(\mathrm{d}U)_S}{U} = -\frac{2}{3}\frac{(\mathrm{d}V)_S}{V}\,. \tag{3.187}$$

[27] Internal energy in noninteracting monatomic gases is $\frac{3}{2}PV$.

Upon integration, one gets

$$\ln (U)_S = -\frac{2}{3} \ln (V)_S + \text{const.} \tag{3.188}$$

Hence the relationship

$$(U)_S = \text{const.} \times (V)_S^{-\frac{2}{3}} . \tag{3.189}$$

Those who are familiar with elementary Quantum Mechanics will recognize a "connection" here with the result for the single particle energy levels ϵ_i of a particle i, of mass M, confined in a cubic box of volume, $V = L^3$,

$$\epsilon_i = \frac{L^{-2}\pi^2\hbar^2}{2M}\left(l^2 + m^2 + n^2\right) = \frac{V^{-\frac{2}{3}}\pi^2\hbar^2}{2M}\left(l^2 + m^2 + n^2\right), \tag{3.190}$$

where l, m, n are integers and $2\pi\hbar = h$ is Planck's constant.[28]

Consider a generalized ideal gas composed of atoms that have single particle energies comprising both the translational motion and the atomic levels. Generally, these composite energy levels depend on adiabatically invariant parameters—such as quantum numbers, for example—and can be characterized as having a special dependence on the system volume, e.g.,

$$\epsilon_i = a_i V^{\sigma} , \tag{3.191}$$

where i refers to the ith particle which has energy ϵ_i and a_i is the corresponding adiabatic invariant. Note that the objective of the current exercise is to calculate σ.

By summing ϵ_i in (3.191) over all the particles, we get the total internal energy

$$U = \sum_{i=1}^{N} \epsilon_i = V^{\sigma} \sum_{i=1}^{N} a_i . \tag{3.192}$$

Treating the sum

$$\left(\sum_{i=1}^{N} a_i\right)_S = C_S$$

as an invariant for adiabatic processes, we get

$$(U)_S = (V)_S^{\sigma} \times C_S .$$

The following choice for σ reproduces the exact result given in (3.189):

$$\sigma = -\frac{2}{3} . \tag{3.193}$$

[28] Here we have used the subscript S because the quantum numbers are adiabatic invariants. This is the case in the sense that if the length L is varied ever so slowly, the quantum numbers remain unchanged. Note that the subscript S signifies a quasistatic adiabatic process.

Thus the volume dependence of the single particle energy levels has to be the following:

$$\epsilon_{\mathrm{i}} = a_{\mathrm{i}} V^{-\frac{2}{3}}, \tag{3.194}$$

which is clearly the case—see (3.190) above.

References

1. W.M. Haynes, D.R. Lyde, T.J. Bruno, *CRC Handbook of Chemistry and Physics*, 74th edn. (1994)

The Second Law

4

Much of the formulation of the first law of thermodynamics has been based on empirical observations of conversion of work into heat energy. But the reader needs to be cautioned that the insignia of the first law as displayed in (3.7) is not an ordinary mathematical equation because it does not guarantee that the reverse—i.e., the conversion of heat energy into work—can also be fully accomplished. Indeed, such a reverse process is a vexed undertaking whose analysis lies at the very heart of thermodynamics. Not only do Carnot's ideas about perfect heat engines—and how they produce work with maximum possible efficiency that is related to the temperatures of the warm and cold external heat energy—provide a basis for the second law, they also help identify an important thermodynamic state function, the entropy. These ideas, their relevant subject matter, and many associated issues are treated in the form of solved problems. After some introductory remarks about ideal gas heat engines in Sect. 4.1, a perfect Carnot engine that runs on ideal gas as its working substance is described in Sect. 4.2. The Kelvin description of the absolute temperature is discussed in Sect. 4.3. Infinitesimal and finite Carnot cycles are studied in Sect. 4.4. Carnot version of the second law and calculation of the entropy are discussed in Sect. 4.5. Perfect Carnot engine with arbitrary working substance is studied in Sect. 4.6. Different statements of the second law are enunciated in Sect. 4.7. The fact that entropy increases in all spontaneous processes is noted and analyzed in Sects. 4.8 and 4.9. Kelvin–Planck version of the second law and its prediction that entropy always increases in irreversible–adiabatic processes is discussed in Sect. 4.10. Sections 4.10.1 and 4.10.2 respectively deal with the fact that while entropy increases in irreversible adiabats, it remains constant in reversible adiabats. In Sect. 4.11 non-Carnot heat engines and the Clausius inequality are studied. Integral and differential forms of Clausius inequality are reported in Sects. 4.11.1 and 4.11.2, respectively. Section 4.12 contains many solved problems that relate to thermal contact and the entropy change. Carnot refrigerator is described in Sect. 4.13. Section 4.14 deals with idealized version of realistic engine cycles, including the Stirling cycle, perfect Carnot cycle, diesel cycle, Otto and Joule cycles. Some cursory remarks about negative temperature are appended at the end of the chapter in Sect. 4.15.

R. Tahir-Kheli, *General and Statistical Thermodynamics*,
https://doi.org/10.1007/978-3-030-20700-7_4

4.1 Ideal Gas Heat Engines

Let us first yield to some musing.

The first law asserts that the difference between the heat energy, $\Delta Q'$, introduced into, and the corresponding increase in the internal energy ΔU of, the system is equal to the work output—i.e., the work done by the system—which we denote as $\Delta W'$,

$$\Delta W' = \Delta Q' - \Delta U . \tag{4.1}$$

The use of the prime on ΔW and ΔQ indicates that this relationship is true irrespective of whether these processes proceed quasistatically or not.

Let us construct a simple machine that uses heated ideal gas as the working substance contained in an expandable enclosure. The engine withdraws heat energy from a single thermal reservoir, at some temperature T_{H}, and in the process the gas expands and does some useful work.

Clearly, for optimal work output, any increase in the internal energy of the gas should be kept to a minimum. Because for an ideal gas, the increase in the internal energy is directly proportional to its temperature rise, it would be wise to keep the temperature increase as small as possible. Thus, one must begin with the working substance at a temperature only infinitesimally lower than that of the reservoir which provides the heat energy. Also one would need to minimize the friction. In an idealized scenario, we assume all this to have happened.

So far this process appears to have been a great success. It has managed not to cause any change in the internal energy and thus to convert all of the heat energy input into work with 100% efficiency. Unfortunately, this is the end of the story.

In order to reuse the above process, the system needs to be brought back to the condition it was in at the beginning of the cycle. And, even with our best efforts, the return journey will use up all of the work produced!

Engines are cyclic. They use energy—which has to be purchased—and produce work that we need to have done. While a well designed electric engine might produce work that is equal to 99% or higher of the electrical energy it consumes, the situation is quite different for heat engines.

4.1.1 Nonexistence of Perpetual Machines of the Second Kind

Much like the impossibility of construction of a "perpetual machine of the first kind"—which in violation of the first law would continue to produce useful work without any input of energy from an outside source—it is also impossible, as is implicit from the comment made above, to construct a "perpetual machine of the second kind." Such a machine would withdraw heat energy at a single temperature and convert it all into useful work.

A perpetual machine of the second kind is, in principle, far less restrictive. Unlike a machine of the first kind,[1] a perpetual machine of the second kind is allowed to withdraw heat energy from an outside source. Accordingly, it does not fall afoul of the first law.

[1] A perpetual machine of the first kind was described in the preceding chapter.

If a perpetual machine of the second kind existed, it could in principle be built in two alternative ways:

(1) A machine that would operate for only part of a cycle. Such a machine would produce a positive amount of work by extracting energy from a single large heat energy reservoir. Because there is to be no heat energy dump at a lower temperature, heat energy extracted from the reservoir would be available for conversion into useful work. In order not to waste energy, no rise in internal energy would be accepted. If an ideal gas could be used as the working substance, this would demand the temperatures of the reservoir and the working substance to be essentially identical.

Practical considerations rule out the construction of a viable machine of this variety. Unmanageably large changes in the working substance—e.g., increase in its volume, or dangerously high pressure, etc.—would ensue for any sizeable production of work. Also because the temperatures of the source and the working substance would be almost the same, it would take a very long time to introduce any sizable amount of heat energy into the working substance.

(2) On the other hand, if such a machine were to function in a cyclical fashion, a noteworthy thing would happen. Because the differential of the internal energy, i.e., dU, is exact, a complete cycle would leave the internal energy unchanged. Thus, a perpetual cyclic operation, sustained by continual withdrawal of energy from a single thermal reservoir, could result in a complete conversion of the withdrawn energy into useful work. This means that in principle a 100% efficiency would be obtained!

Unfortunately, physics, through the second law of thermodynamics as its intermediary, proscribes such an happenstance. The proscription follows directly from the Kelvin–Planck statement and is to be discussed later in this chapter.

The impossibility of construction of the machines of the first and second kind has important consequences. Work producing engines that would continually withdraw energy from such vast sources of internal energy as described below cannot be constructed. Some of these sources are: (1) The atmosphere, which refers to a quiet atmosphere with no wind; (2) The oceans. Of course, this means oceans without tides and waves, and at temperature no higher than the atmosphere; and (3) The earth itself, assuming, of course, that there is no access to subterranean heat energy sources that are often at higher temperature than the atmosphere.

4.2 Perfect Carnot Engine

Engines that operate on heat energy are generically called Carnot engines. Their hallmark is the existence of one or more hot temperature sources and one or more cold temperature dumps. In cyclic operation, heat energy is withdrawn from the hot sources and transferred to the working substance of the engine. After some of the heat energy has been converted into work, the remainder is discharged into the colder dumps. How well any such engine performs is, of course, dependent on its operational details.[2]

The most elegant, and also the most efficient, of all such engines is the one called a perfect Carnot engine. Such an engine operates in perfect Carnot cycles and, furthermore, is perfectly engineered.

For convenience, a perfect Carnot engine will henceforth be referred to just as a Carnot engine and the perfect Carnot cycle will, more simply, be called the Carnot cycle. All legs of the Carnot cycle are traversed reversibly.[3] Note that reversible processes are globally isentropic. Also a reversible process is quasistatic. And quasistatic processes are often, but not necessarily always, reversible. Indeed, as is to be proven later, the physical properties of a reversible process are independent of the working substance. Therefore, for purposes of illustration, it suffices to employ the simplest of all working substances, namely the ideal gas.

The cycle starts at position "1" and, as indicated by arrows in Fig. 4.1.a, proceeds via positions "2", "3", and "4", back to the starting position.

The leg $1 \rightarrow 2$ describes an isothermal process. It involves a reversible transfer of heat energy equal to, $Q^{\text{rev}}(T_{\text{H}})$, out of a reservoir maintained at temperature T_{H}, and into the working substance of the engine also at the same temperature. Note that to allow for the exchange of heat energy between the reservoir and the working substance—even if the exchange rate should be infinitely slow—their temperatures will have to differ by at least an infinitesimal amount.

During this isothermal process $1 \rightarrow 2$, the gas expands from volume $V_1 \rightarrow V_2$ while the pressure drops from $P_1 \rightarrow P_2$. Also see Figs. 4.1.b–c.

The second leg, $2 \rightarrow 3$, is moved quasistatically and adiabatically—meaning, very slowly and with no exchange of heat energy. Thus, it is a reversible–adiabatic expansion. A reversible–adiabatic expansion occurs quasistatically and without any exchange of heat energy. In principle, like all reversible processes, such an expansion can be reversed with just an infinitesimal amount of effort. Note that all reversible–adiabatic processes are isentropic, meaning they do not cause any change in the entropy of the system that is adiabatically isolated. The reader should note that this statement is distinct from the previous one which referred, rather than to a "reversible–adiabatic process," to just a reversible process. The latter does not imply local isentropy. Rather, it implies only global isentropy.

[2]The operational details include such things as the nature of the cycle of operation and the quality of the engineering, etc.

[3]As we know from the introductory chapter, a reversible process is necessarily quasistatic. There are, however, two additional points to note. First, the entropy of the universe remains unchanged during all reversible processes. This means that any loss in the entropy of the working substance is exactly counterbalanced by the increase in the entropy of the reservoir, or vice versa. Thus, globally speaking, a reversible process is isentropic. Second, despite the fact that all reversible processes are quasistatic, the reverse is not always true. One can imagine a quasistatic process that is not reversible in the sense that globally it is not isentropic. However, for our purposes in this book, essentially all of the quasistatic processes—unless otherwise stated—are likely to be reversible. Therefore, the statement quasistatic may be considered synonymous with the word reversible.

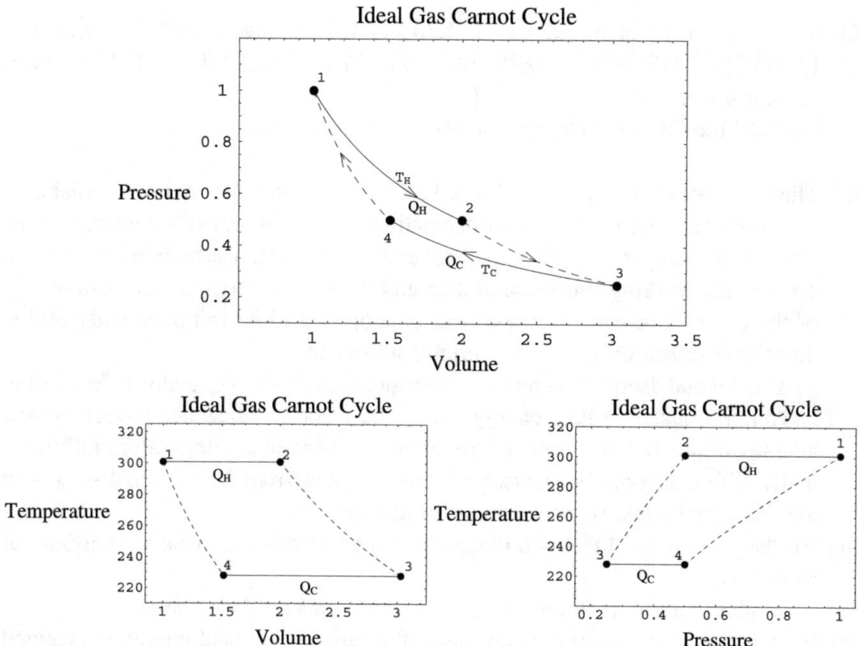

Fig. 4.1 (**a**) Pressure/(5×10^5) Pa versus volume/(10^{-2} m^3). Here $Q_H = Q^{rev}(T_H)$ and $Q_C = Q^{rev}(T_C)$. *Full lines* are isotherms and *broken lines* adiabats. (**b**) Temperature/(K) versus volume/(10^{-2} m^3). Here $Q_H = Q^{rev}(T_H)$ and $Q_C = Q^{rev}(T_C)$. *Full lines* are isotherms and *broken lines* adiabats. (**c**) Temperature/(K) versus pressure/(5×10^5) Pa. Here $Q_H = Q^{rev}(T_H)$ and $Q_C = Q^{rev}(T_C)$. *Full lines* are isotherms and *broken lines* adiabats

Moving from volume $V_2 \rightarrow V_3$, the temperature drops from T_H to T_C and the pressure drops from P_2 to P_3.

The leg $3 \rightarrow 4$ describes a reversible decrease of volume, $V_3 \rightarrow V_4$, and increase of pressure, $P_3 \rightarrow P_4$. Both of these occur at constant temperature T_C. The process involves a transfer of heat energy, $Q^{rev}(T_C)$, between the working substance inside the engine and an external dump maintained at temperature T_C.

Recognize that at temperature T_C, in order to preserve a consistent notation, we have denoted $Q^{rev}(T_C)$ to be the heat energy transferred reversibly "into the working substance" and "out of the dump."[4] In practice, the opposite is the case. Therefore, $Q^{rev}(T_C)$ will in general be a negative quantity.

Note that the point 4 is so chosen that a reversible–adiabatic compression from volume V_4 to V_1 returns the system to its starting position 1—that is, to pressure P_1 which implies temperature T_H.

It is helpful to illustrate the above cycle in all three of its available forms. That is, as plots in the (P, V), (T, V), or (T, P) planes. The essential physics of the cycle described in all the three Figs. 4.1.a–c is the same. These figures were drawn

[4]The similarity between the two notations lies in the fact that both $Q^{rev}(T_H)$ and $Q^{rev}(T_C)$ refer to the amounts of heat energy transferred "into" the working substance. The latter, occurring at T_C, is of course negative.

for two moles of a monatomic gas. In terms of the displayed units, $(P_i, V_i, T_i) = (1, 1, 300.7), (0.5, 2, 300.7), (0.25, 3.031, 227.8)$, and $(0.5, 1.5156, 227.8)$ for $i = 1, 2, 3, 4$, respectively.

As a rule, the Carnot cycle, has the following features:

(a) The cycle consists of a chain of four links: two "isotherms" and two "adiabats."
 The first isotherm—that is, from position $1 \to 2$—refers to a reversible transfer of heat energy, out of the very large external thermal reservoir at temperature T_H, into the working substance at temperature T_H. Note that the temperature T_H of the external heat energy reservoir is supposed to be infinitesimally higher than the temperature T_H of the working substance.
 The second isotherm—that is, from position $3 \to 4$—again refers to the reversible transfer of heat energy "into" the working substance (which is now at temperature T_C) and "out" of the dump maintained at temperature infinitesimally different from T_C. The two adiabats—that is, from $2 \to 3$ and $4 \to 1$—on the other hand, involve no exchange of heat energy.
(b) All the processes—links—in the chain that constitute the cycle are carried out reversibly.
(c) An isothermal link is followed by an adiabatic link, and vice versa.
(d) Irrespective of the working substance, the ratio of the heat energy exchanged with the cold dump and that exchanged with the hot reservoir is a function only of the relevant temperatures.

Indeed, thc Carnot statement is quite explicit. It asserts

$$\frac{Q^{rev}(T_C)}{Q^{rev}(T_H)} = -\left(\frac{T_C}{T_H}\right). \tag{4.2}$$

Simple rearrangement yields[5]

$$\frac{Q^{rev}(T_H)}{T_H} + \frac{Q^{rev}(T_C)}{T_C} = 0. \tag{4.3}$$

Equation (4.3), or its equivalent, will often be referred to as the "Carnot statement" or the "Carnot assertion."

4.3 Kelvin Description of Absolute Temperature

Lord Kelvin appears to have been the first to observe if the Carnot statement embodied in (4.2), or (4.3), is indeed independent of the working substance used in the Carnot engine, then it—i.e., the Carnot statement—can be used to define a universal—or equivalently, an absolute—temperature scale. In honor of Lord

[5] As noted earlier, the heat energy, $Q^{rev}(T_C)$, transferred (supposedly) out of the dump and into the working substance is actually negative!

Kelvin, this absolute scale is often called the "Kelvin scale." Temperatures measured on the Kelvin scale are denoted by the symbol K, or, occasionally, by the symbol k.

Efficiency of an engine represents the amount of (useful) work done for unit input of energy. For the Carnot engine described above, the input of heat energy occurs reversibly. Let us denote this energy as $Q^{\text{rev}}(T_{\text{H}})$. When an engine withdraws heat energy, $Q^{\text{rev}}(T_{\text{H}})$, from a source at temperature T_{H}, the energy withdrawn has to be paid for, much like the gasoline that one buys. On the other hand, the discarded heat energy is energy lost—much like the heat energy contained in the exhaust that emanates from the back of an automobile. So a proper definition of the efficiency would have to be the following: the useful work produced, divided by the energy $Q^{\text{rev}}(T_{\text{H}})$ that we have had to pay for in producing the work. The question then is: What is the useful work produced?

Because the differential of the internal energy is exact, the internal energy of the working substance is unchanged when a complete cycle is performed. Therefore, according to the first law, the work done per cycle is equal to the total input of energy during the cycle.

To this purpose, in addition to the input of heat energy $Q^{\text{rev}}(T_{\text{H}})$ into the working substance from the hot external energy reservoir, we must also include the heat energy input, $Q^{\text{rev}}(T_{\text{C}})$, from the cold dump. The work done during the cycle equals $Q^{\text{rev}}(T_{\text{H}}) + Q^{\text{rev}}(T_{\text{C}})$.[6]

The working efficiency of a Carnot engine is given by the following relationship:

$$\begin{aligned}\epsilon_{\text{carnot}} &= \frac{\text{total work done in the cycle}}{\text{heat energy withdrawn from the hot source}} \\ &= \frac{[Q^{\text{rev}}(T_{\text{H}}) + Q^{\text{rev}}(T_{\text{C}})]}{Q^{\text{rev}}(T_{\text{H}})} = 1 + \left[\frac{Q^{\text{rev}}(T_{\text{C}})}{Q^{\text{rev}}(T_{\text{H}})}\right].\end{aligned} \tag{4.4}$$

Using Carnot's assertion recorded in (4.2), the working efficiency of a Carnot engine operating between a hot reservoir and a cold dump, at temperatures T_{H} and T_{C}, respectively, is

$$\epsilon_{\text{carnot}} = 1 + \left[\frac{Q^{\text{rev}}(T_{\text{C}})}{Q^{\text{rev}}(T_{\text{H}})}\right] = 1 - \left(\frac{T_{\text{C}}}{T_{\text{H}}}\right). \tag{4.5}$$

In order to check the validity of Carnot's assertion—from which (4.5) follows—let us revisit the Carnot engine that uses the ideal gas as its working substance.

Begin by examining the first leg—call it leg "I"—of the cycle. Here the temperature is kept constant. Because the working substance is an ideal gas—whose internal energy is directly proportional to its temperature—there is no net change in the internal energy along this link. Hence, according to the first law, the heat energy withdrawn from the hot reservoir—that is, the heat energy introduced into the

[6]Remember that the two adiabatic legs, $2 \to 3$ and $4 \to 1$, do not entail any heat energy transfer. Also do not forget that positive heat energy is actually discarded out of the working substance into the cold dump. Therefore, $Q^{\text{rev}}(T_{\text{C}})$, as described here, is in fact a negative quantity.

working substance at temperature T_H—is equal to the work done during the prescribed reversible expansion of the ideal gas that constitutes the working substance. That is,

$$Q^{rev}(T_H) = W^{rev}_{1\to 2}\,. \tag{4.6}$$

In view of the quasistatic, isothermal nature of the expansion from $V_1 \to V_2$, the work done by n moles of ideal gas in so expanding is

$$W^{rev}_{1\to 2} = \int_{V_1}^{V_2} P\mathrm{d}V = nRT_H \int_{V_1}^{V_2} \frac{\mathrm{d}V}{V} = nRT_H \ln\left(\frac{V_2}{V_1}\right). \tag{4.7}$$

Similarly, during the reversible isothermal travel from $3 \to 4$—call it leg "III"—the work done by the gas is

$$W^{rev}_{3\to 4} = \int_{V_3}^{V_4} P\mathrm{d}V = nRT_C \int_{V_3}^{V_4} \frac{\mathrm{d}V}{V} = nRT_C \ln\left(\frac{V_4}{V_3}\right). \tag{4.8}$$

Again, because the working substance is an ideal gas and the travel is at constant temperature, there is no change in the internal energy along the leg III—that is, the route $3 \to 4$. Accordingly, the work $W_{3\to 4}$ done (by the working substance) along this path has to be equal to the heat energy $Q^{rev}(T_C)$ that has been introduced by the dump (into the working substance) during the isothermal leg at temperature T_C. That is,

$$W_{3\to 4} = Q^{rev}(T_C) = nRT_C \ln\left(\frac{V_4}{V_3}\right) = -nRT_C \ln\left(\frac{V_3}{V_4}\right). \tag{4.9}$$

Note, because $V_3 > V_4$, the heat energy introduced by the dump into, and therefore the work $W_{3\to 4}$ done by, the working substance are both negative. Equations (4.6), (4.8), and (4.9) lead to the following result:

$$\frac{Q^{rev}(T_C)}{Q^{rev}(T_H)} = -\left(\frac{T_C}{T_H}\right)\left(\frac{\ln(\frac{V_3}{V_4})}{\ln(\frac{V_2}{V_1})}\right). \tag{4.10}$$

Because the links $2 \to 3$ and $4 \to 1$—that is, legs "II" and "IV", respectively—are quasistatic adiabats, according to (3.101), we have the relationships:

$$T_H (V_2)^{\gamma-1} = T_C (V_3)^{\gamma-1}\,, \tag{4.11}$$

$$T_H (V_1)^{\gamma-1} = T_C (V_4)^{\gamma-1}\,. \tag{4.12}$$

On dividing (4.11) by (4.12), we get

$$\left(\frac{V_2}{V_1}\right)^{\gamma-1} = \left(\frac{V_3}{V_4}\right)^{\gamma-1}. \tag{4.13}$$

The specific heat, C_P, measured at constant pressure, is always greater than C_V measured at constant volume. Thus $\gamma > 1$. But more importantly, $\gamma \neq 1$. Thus, the relevant solution of (4.13) is the equality

$$\left(\frac{V_2}{V_1}\right) = \left(\frac{V_3}{V_4}\right). \tag{4.14}$$

Whence, (4.10) becomes

$$\frac{Q^{rev}(T_C)}{Q^{rev}(T_H)} = -\left(\frac{T_C}{T_H}\right), \tag{4.15}$$

which is identical to the Carnot assertion recorded in (4.2). As a result, the efficiency of a perfect Carnot engine that uses the ideal gas as its working substance is the same as asserted by Carnot. (Compare with (4.5).)

$$\begin{aligned}\epsilon_{carnot}(\text{ideal gas}) &= \frac{Q^{rev}(T_H) + Q^{rev}(T_C)}{Q^{rev}(T_H)} \\ &= 1 + \left[\frac{Q^{rev}(T_C)}{Q^{rev}(T_H)}\right] = 1 - \left(\frac{T_C}{T_H}\right).\end{aligned} \tag{4.16}$$

Later we show that (4.5) also holds true when the working substance of the Carnot engine is arbitrary.

The following Carnot assertion can be interpreted as a statement of "The Second Law of Thermodynamics."

"The efficiency of the Carnot engine is either greater than or equal to the efficiency of any cyclic engine operating between the same set of temperatures." That is,

$$\epsilon_{carnot} = 1 + \left[\frac{Q^{rev}(T_C)}{Q^{rev}(T_H)}\right] \geq \epsilon_0 , \tag{4.17}$$

where ϵ_0 is the corresponding efficiency of any cyclic engine operating possibly nonquasistatically between the same set of temperatures. The equality is obtained if and only if the "any given cyclic engine" referred to here is itself a Carnot engine.

While the second law will be described in its various traditional forms a little later, we hasten to add that unless we should be dealing with the exotic phenomenon of "negative temperatures"—see the succeeding subsection—the physical implications of (4.17) will be found to be equivalent to those of the other statements of the second law.[7] Note, however, that much like the usual statements of the second law, (4.17) also is an assertion. It should be added that no known violation of (4.17) exists. Indeed, using the techniques of statistical mechanics, it is possible to estimate that the likelihood of its violation in a thermodynamic system—which generally has an exceedingly large number of particles—is exceedingly small. Therefore, (4.17), for all intents and purposes, is exact.

[7] Indeed, in my experience, what we have dubbed here as the Carnot version of the second law, is as simple—if not simpler—to comprehend by a beginning student as any of the usual statements of the second law.

Problem 4.1 (Two adiabatic legs ignored)
In discussing the ideal gas Carnot engine, we calculated the work output only along the two isothermal legs I and III (see Fig. 4.1.a). Why did we ignore doing the same for the two adiabatic legs II and IV?

Solution 4.1 Because the adiabatic links do not involve any heat energy exchange, according to the first law, the work done by the engine in going from $2 \to 3$ is equal to the corresponding decrease in the internal energy. That is, for the leg II,

$$W^{rev}_{2\to3} = -(U_3 - U_2) = C_V(T_2 - T_3)\,.$$

Similarly, the work done in going from $4 \to 1$—that is, for leg IV—is

$$W^{rev}_{4\to1} = -(U_1 - U_4) = C_V(T_4 - T_1)\,.$$

We know from Fig. 4.1.b that $T_4 = T_3$ and $T_2 = T_1$. This makes

$$W^{rev}_{4\to1} = C_V(T_3 - T_2)\,.$$

Clearly, the sum of the work output for travel through the two legs II and IV is equal to zero. That is,

$$W^{rev}_{2\to3} + W^{rev}_{4\to1} = C_V(T_2 - T_3) + C_V(T_3 - T_2) = 0\,.$$

4.4 Infinitesimal and Finite Carnot Cycles

Infinitesimal Carnot Cycles

In the foregoing, we did not pay any special attention to the size of the heat energy exchanges—with either the hot reservoir or the cold dump. In the present section, we shall work both with finite and infinitesimal Carnot cycles. The latter involve very small—indeed, infinitesimal—exchanges of heat energy.

To this purpose, we shall use the notation, $\Delta Q^{rev}(T)$, for the reversible transfer of a small amount of heat energy *into* the working substance at temperature T. As a result, (4.3) is recast as

$$\frac{\Delta Q^{rev}(T_H)}{T_H} + \frac{\Delta Q^{rev}(T_C)}{T_C} = 0\,. \tag{4.18}$$

This equation makes explicit reference to only the two isothermal legs, I and III, in the cycle, namely those legs that are traversed at temperatures T_H and T_C, respectively. Both these legs involve heat energy transfers between the working substance and the outside. Of course, there are no heat energy transfers involved for the other two legs of the cycle—namely legs II and IV—because they are both adiabats.

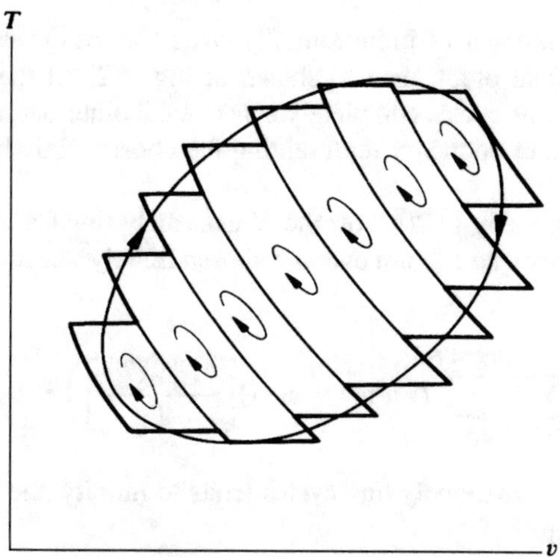

Fig. 4.2 Any arbitrary shaped, finitely sized, reversible cycle can be approximated by a very large number of extremely tiny Carnot cycles. In the figure above, the adiabats in successive contiguous tiny Carnot cycles are seen approximately to cancel each other out—that is, adiabat IV from cycle i cancels the adiabat II in cycle $i+1$. Note that such cancelation becomes perfect—meaning, exact—when the cycles are infinitesimally small. Furthermore, the jagged structure of the uncanceled parts smoothes out to the desired arbitrary shaped, finitely sized, reversible cycle. (This sketch is similar to that given in [1])

Therefore, the above equation can equivalently be represented as a sum over all the four legs, I $\rightarrow$ IV, of the Carnot cycle. That is,

$$\sum_{\text{leg=I}}^{\text{leg=IV}} (\text{Carnot cycle } i)\left[\frac{\Delta Q^{\text{rev}}(T)}{T}\right] = 0\,. \tag{4.19}$$

It is convenient to work with a very large number of very small—indeed, infinitesimal—energy transfers and label one of these as the ith working cycle. With this change and such labeling, (4.19) gives

$$\sum_{\text{leg=I}}^{\text{leg=IV}} (\text{Carnot cycle } i)\left[\frac{\mathrm{d}Q_{\text{i}}^{\text{rev}}(T)}{T}\right] = 0\,. \tag{4.20}$$

A smooth curve, that may represent a finite sized reversible cycle of arbitrary shape, can be approximated to any desired accuracy by putting together a very large number of tiny sized cycles of the type represented by (4.20). Indeed, the approximation is perfect—meaning, it is exact—if the cycles are of infinitesimal size and an infinite number of them are added together in the manner demonstrated in Fig. 4.2. Here an infinitesimal Carnot cycle i is followed by—meaning that in Fig. 4.2 it is placed contiguous to—another infinitesimal Carnot cycle $i + 1$, and that in turn is contiguous to another such cycle numbered $i + 2$, and so on. In this fashion, when an infinitely large number of infinitesimally sized Carnot cycles are all traversed in the same rotational order, then as shown in Fig. 4.2, all the adiabatic legs in successive, contiguous cycles completely cancel each other out, thus leaving intact only the smooth outer boundary representing the chosen, finitely sized, reversible cycle.

Accordingly, summing (4.20) over the N extremely tiny Carnot cycles approximates well the corresponding sum over the chosen finitely[8] sized reversible cycle of arbitrary shape:

$$\sum_{i=1}^{N}\left(\sum_{\text{leg=I}}^{\text{leg=IV}} (\text{Carnot cycle } i)\left[\frac{\mathrm{d}Q_{\mathrm{i}}^{\mathrm{rev}}(T)}{T}\right]\right) = 0\,. \tag{4.21}$$

When the number of extremely tiny cycles tends to infinity, the above sum can be replaced by an integral

$$\oint \frac{\mathrm{d}Q^{\mathrm{rev}}(T)}{T} = 0\,. \tag{4.22}$$

The integral is taken over the whole length of the given, smooth, closed loop—or loops—representing the original, arbitrary shaped, finitely sized, reversible cycle.

Recalling the definition of an exact differential given in (1.18), the integrand in (4.22) must necessarily be a linear combination of exact differentials. It is convenient to use a notation which represents such linear combination as a single exact differential $\mathrm{d}S$. That is,

$$\frac{\mathrm{d}Q^{\mathrm{rev}}(T)}{T} = \mathrm{d}S\,. \tag{4.23}$$

Equations (1.18), (4.22), and (4.23) dictate

$$\int_{\text{initial}}^{\text{final}} \mathrm{d}S = S_{\text{final}} - S_{\text{initial}}\,. \tag{4.24}$$

Note that in the above the initial and final equilibrium states are signified by their relevant suffices. More generally, we can use (4.22), (4.23), and (4.24) together. That is,

$$\oint \mathrm{d}S = \int_{\text{initial}}^{\text{final}} \mathrm{d}S + \int_{\text{final}}^{\text{initial}} \mathrm{d}S = 0\,. \tag{4.25}$$

Clearly, therefore, S is a state function and, as is customary, we shall call it the "entropy."

4.5 Entropy Calculation

How should entropy be calculated?

[8] Meaning, very large compared to the very tiny Carnot cycles mentioned above.

Use Reversible Paths for Entropy Calculation

Whereas each extremely tiny bit of heat energy, i.e., $dQ_i^{rev}(T)$, described in (4.21) and (4.23), was reversibly added (to the working substance), the integrals of dS, given in (4.24) and (4.25), do not demand reversibility. Indeed, they place no requirement on how the path is traversed! Therefore, the most important thing to note about the integrals in (4.24) and (4.25) is that they are path independent.

This fact has important physical implications. Often, an experiment involves an irreversible transfer of heat energy—whose exact details are dependent uniquely on the extremely many nonequilibrium states through which the system passes during the course of the experiment. Therefore, any direct attempt at their evaluation is utterly futile. Fortunately, the fact that entropy S is a function of state, and therefore its change between the initial and final equilibrium states is independent of the path the system may take in going from one to the other, provides the needed key for its calculation.

And the key is to use a reversible path for the calculation of the integral over dS. That is, disregard any irreversible processes encountered in the course of the actual experiment and instead use the good offices of (4.23) and (4.24) for doing the calculation. For a successful calculation, all that is needed is a reversible path that is both convenient for calculational purposes and connects the specified initial and final equilibrium states.

Accordingly, here and in all other cases—involving entropy changes—that arise in the following set of problems and exercises, we shall use appropriate reversible paths for the relevant calculations. Additionally, for reversible paths, whose characteristics can be embedded in the choice of the integrands, the integrals of heat energy transfer, $dQ^{rev}(T)$, as well as that of the work done, dW^{rev}, can generally be treated as ordinary integrals.

Note that normally these integrals would be path dependent. But because the path is specified, that is, it is the chosen reversible path connecting the two specified end points, $dQ^{rev}(T)$ and dW^{rev} behave as though they were exact differentials.

4.6 Perfect Carnot Engine with Arbitrary Working Substance

We now describe a perfect Carnot engine with arbitrary working substance.

As mentioned earlier, consistent with the Carnot's assertion, as long as the working substance allows for a successful traversal of the four legs of the perfect Carnot cycle, the operating efficiency of the resultant cycle is independent of the physical properties of the working substance. This assertion is readily tested by analyzing the cycle in the "entropy–temperature" plane.[9]

[9]Recall that for the given three state variables, P, V, and T, a simple thermodynamic system is completely specified by any of the three pairs. Furthermore, we have seen that the entropy S is also a state variable and thus can be represented by any of the given three pairs of variables. As a result, any thermodynamic function of a simple system can be represented in terms of S and one of the above mentioned three variables, say the temperature T.

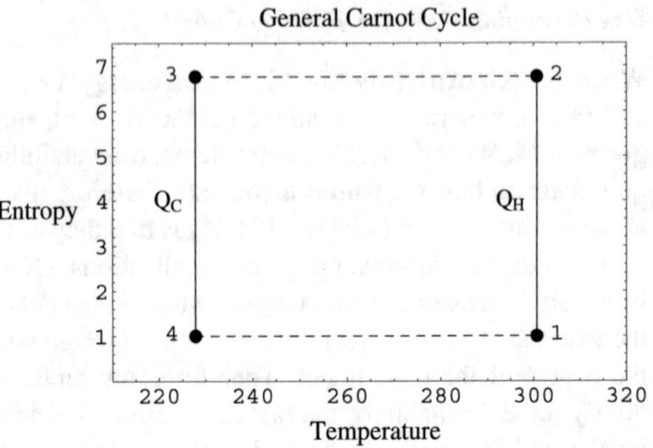

Fig. 4.3 Entropy/(J K^{-1}) versus temperature/K. For convenience, entropy is set at 2 J/K at point 1. Here $Q_\mathrm{H} = Q^\mathrm{rev}(T_\mathrm{H})$ and $Q_\mathrm{C} = Q^\mathrm{rev}(T_\mathrm{C})$. Full lines are isotherms and broken lines adiabats

The Carnot cycle in the entropy–temperature plane is a rectangle (see Fig. 4.3). Recall that the two legs II and IV—that is, the legs $2 \to 3$ and $4 \to 1$, respectively—are reversible adiabats, meaning, they are both reversible and occur when there is no exchange of heat energy. Therefore, they are both isentropes, meaning they both represent states of constant entropy.[10] Accordingly, they appear as horizontal straight lines on the "entropy versus temperature plot."

The remaining two straight lines I and III—meaning the legs $1 \to 2$ and $3 \to 4$ of the rectangle referred to above—are vertical straight lines that represent the two temperatures, T_H and T_C, of the hot reservoir and the cold dump, respectively. Remember that all four legs of such a rectangular cycle are traversed reversibly.

Let us start the cycle at point 1 specified by the coordinates $(T_\mathrm{H}, S_\mathrm{i})$. During isothermal travel, at temperature T_H, from point 1 to point 2, the working substance reversibly absorbs heat energy, $Q^\mathrm{rev}(T_\mathrm{H})$, that is supplied by a thermal reservoir maintained at temperature T_H. As a result, the entropy of the working substance increases from S_i to S_f. That is,

$$S_\mathrm{f} - S_\mathrm{i} = \frac{Q^\mathrm{rev}(T_\mathrm{H})}{T_\mathrm{H}} . \tag{4.26}$$

The coordinates of point 2, therefore, are $(T_\mathrm{H}, S_\mathrm{f})$.

The link $2 \to 3$ is traversed without any exchange of heat energy. Furthermore, the process is quasistatic—which, of course, is always the case if it is reversible. Therefore, the entropy does not change. Accordingly, the entropy of the working substance stays at S_f. Note that the traversal is performed in such a manner that the temperature, T_C, of the working substance at point 3 is lower than T_H.[11] The coordinates of point 3 are $(T_\mathrm{C}, S_\mathrm{f})$.

Next, during the traversal of the leg $3 \to 4$, that occurs at constant temperature T_C, heat energy $Q^\mathrm{rev}(T_\mathrm{C})$ is "supplied by the dump" to the working substance. This is done in such a manner that the entropy of the working substance again becomes equal to S_i. In other words, the entropy changes from S_f *to* S_i as the link III

[10] Constancy of entropy is ensured if the process occurs adiabatically and quasistatically.

[11] This can be arranged by letting the working substance adiabatically perform some work.

is traversed from 3 → 4. Thus

$$S_{\text{i}} - S_{\text{f}} = \frac{Q^{\text{rev}}(T_{\text{C}})}{T_{\text{C}}}. \tag{4.27}$$

Now, an isentropic traversal from the point 4—which is at $(T_{\text{C}}, S_{\text{i}})$—can be so arranged that it brings the working system back to the position 1 that is at $(T_{\text{H}}, S_{\text{i}})$.

Adding (4.26) and (4.27) leads to the perfect Carnot cycle equality

$$\frac{Q^{\text{rev}}(T_{\text{H}})}{T_{\text{H}}} + \frac{Q^{\text{rev}}(T_{\text{C}})}{T_{\text{C}}} = 0\,. \tag{4.28}$$

This equality is identical to that asserted by Carnot, see (4.3).

In other words, the efficiency of the perfect Carnot cycle is independent of the working substance and is as given in (4.5).

4.7 Statements of the Second Law

There have been many equivalent enunciations of the second law.

4.7.1 The Carnot Statement

Owing to the fact that in their original form these statements can appear somewhat abstruse, we develop them via—what we have decided to call—the Carnot version of the second law, namely that

"The perfect Carnot engine operating between any hot and cold temperatures, say T_{H} and T_{C}, is the most efficient." That is,

$$\epsilon_{\text{carnot}} = \frac{Q^{\text{rev}}(T_{\text{H}}) + Q^{\text{rev}}(T_{\text{C}})}{Q^{\text{rev}}(T_{\text{H}})} = \frac{T_{\text{H}} - T_{\text{C}}}{T_{\text{H}}} \geq \epsilon_0\,. \tag{4.29}$$

Recall that ϵ_0 is the efficiency of an ordinary engine. In terms of the heat energy, $Q'(T_{\text{H}})$, added by the high temperature reservoir, and $Q'(T_{\text{C}})$, added by the low temperature dump—to the working substance of such an engine—all or part of which may have been done irreversibly, the efficiency ϵ_0 is specified by the relationship

$$\epsilon_0 = \left[\frac{Q'(T_{\text{H}}) + Q'(T_{\text{C}})}{Q'(T_{\text{H}})}\right]. \tag{4.30}$$

Note that while reversible heat energy transfers and work output/input have been denoted with the superscript "rev," the corresponding irreversible—or, partially irreversible—processes are denoted with a "prime" (′).

4.7.2 Clausius Statement

"Without assistance it is impossible to withdraw positive amount of heat energy from a colder object and transfer the same to a warmer object."

In other words, heat energy does not spontaneously get transferred from a colder object to a warmer one. Therefore, we need to run a refrigerator to affect such a transfer, and running a refrigerator costs money.

Carnot Version of the Second Law Leads to Clausius Version of the Second Law

In an appendix—see (E.7)–(E.11) and the related comments—we examine how a violation of the Carnot version of the second law relates to a violation of the Clausius version given above. Our findings are summarized as follows:

A violation of the Carnot version of the second law leads to a physically unacceptable conclusion, namely that without external assistance a positive amount of heat energy can be extracted from an object that is colder and all of it transferred to an object that is warmer. Accordingly, the following is the case:

"A violation of the Carnot version of the second law results in a violation of the Clausius statement of the second law."

4.8 Entropy Increase in Spontaneous Processes

An important consequence of the Clausius statement of the second law is that a spontaneous process, even if it occurs within adiabatic walls, results in increasing the system entropy. Indeed, entropy always increases in spontaneous processes. To see how and why this is the case, consider two thermodynamic systems placed inside a perfectly enclosed chamber. Assume that at some instant, one system is hot and the other cold, and their temperatures are $T_{\rm H}$ and $T_{\rm C}$, respectively.[12] Note that both systems are assumed to be completely isolated from the rest of the universe, meaning they are placed together within the same adiabatic walls, are not subject to effects of external fields, and do not exchange any matter with the universe outside the walls.

Let the two systems—or equivalently, two macroscopic parts of a single system—be put in complete physical-thermal contact with each other for an infinitesimal length of time. As a result, an infinitesimal amount of heat energy will be exchanged between the two systems/parts. Experience tells us that heat energy spontaneously gets transferred from the warmer system/part to the colder one and the process cannot be reversed without external assistance. Despite the fact that overall such transfer of heat energy will be an irreversible process, we can, in the following fashion, treat the process as being equivalent to the sum of two separate reversible "inputs" of heat energy.

[12] Although, one might imagine that the following applies only to two separate systems that are in physical contact, the argument equally well applies to a single thermodynamic system because it can be "imagined" to have two macroscopic parts.

One such “input” would consist of $\Delta Q^{\text{rev}}(T_{\text{H}})$ amount of heat energy being quasistatically “added” to the warm system/part, occurring exactly at its existing temperature T_{H}. This addition would result in an increase in the entropy of the warm system/part by an amount

$$\Delta(T_{\text{H}}) = \frac{\Delta Q^{\text{rev}}(T_{\text{H}})}{T_{\text{H}}} . \tag{4.31}$$

Similarly, the second reversible “input” would refer to the quasistatic addition of heat energy $\Delta Q^{\text{rev}}(T_{\text{C}})$ to the cold system/part at exactly its existing temperature T_{C}. This would result in increasing the entropy of the cold system/part by an amount

$$\Delta S(T_{\text{C}}) = \frac{\Delta Q^{\text{rev}}(T_{\text{C}})}{T_{\text{C}}} . \tag{4.32}$$

Because the two systems/parts lie within an adiabatic enclosure, the total amount of heat energy thus added to the two must be zero,

$$\Delta Q^{\text{rev}}(T_{\text{H}}) + \Delta Q^{\text{rev}}(T_{\text{C}}) = 0 . \tag{4.33}$$

Using (4.31)–(4.33), we can write for the total increase in the entropy

$$\begin{aligned} \Delta S_{\text{total}}(\text{spontaneous}) &= \Delta S(T_{\text{H}}) + \Delta S(T_{\text{C}}) \\ &= \frac{\Delta Q^{\text{rev}}(T_{\text{H}})}{T_{\text{H}}} + \frac{\Delta Q^{\text{rev}}(T_{\text{C}})}{T_{\text{C}}} \\ &= \Delta Q^{\text{rev}}(T_{\text{H}}) \left[\frac{1}{T_{\text{H}}} - \frac{1}{T_{\text{C}}} \right] && (4.34) \\ &= \Delta Q^{\text{rev}}(T_{\text{C}}) \left[\frac{1}{T_{\text{C}}} - \frac{1}{T_{\text{H}}} \right] . && (4.35) \end{aligned}$$

4.9 Energy Exchange Increases Total Entropy

According to the Clausius enunciation of the second law, positive heat energy cannot spontaneously be transferred from a colder object to a warmer object. Therefore, the energy transferred to the warmer object, $\Delta Q^{\text{rev}}(T_{\text{H}})$, cannot be positive. Furthermore, because in (4.34) the factor multiplying $\Delta Q^{\text{rev}}(T_{\text{H}})$, i.e.,

$$\left[\frac{1}{T_{\text{H}}} - \frac{1}{T_{\text{C}}} \right] ,$$

is negative,[13] the product

$$\Delta Q^{\text{rev}}(T_{\text{H}}) \left[\frac{1}{T_{\text{H}}} - \frac{1}{T_{\text{C}}} \right]$$

[13] This is so because $T_{\text{H}} > T_{\text{C}}$.

cannot be negative. This means that $\Delta S_{\text{total}}(\text{spontaneous})$ cannot be negative, i.e.,

$$\Delta S_{\text{total}}(\text{spontaneous}) \not< 0 . \tag{4.36}$$

Similarly, looking at (4.35), we can say that because the heat energy added to the colder system is positive,

$$\Delta Q^{\text{rev}}(T_{\text{c}}) > 0 ,$$

as is the corresponding multiplying factor in (4.35), i.e., $[\frac{1}{T_{\text{C}}} - \frac{1}{T_{\text{H}}}]$, their product is positive, i.e.,

$$\Delta Q^{\text{rev}}(T_{\text{C}}) \left[\frac{1}{T_{\text{C}}} - \frac{1}{T_{\text{H}}} \right] > 0 . \tag{4.37}$$

As a result of the two inequalities (4.36) and (4.37), we can state with confidence that

$$\Delta S_{\text{total}}(\text{spontaneous}) > 0 . \tag{4.38}$$

For the trivial case $T_{\text{H}} = T_{\text{C}}$, there is no spontaneous heat energy transfer. Accordingly, there is no change in the entropy.

Conclusion In a single isolated system—or a collection of two systems that are in thermodynamic contact with each other and are placed together in a completely isolated adiabatically enclosed chamber—any spontaneous process resulting in internal exchange of heat energy causes the total entropy to increase. Of course, if the exchange of energy should be equal to zero the total entropy stays constant.

4.10 Kelvin–Planck Version of the Second Law

The following statement is attributed to Kelvin and Planck:

"In a cyclic process, it is impossible to extract heat energy from a thermodynamic system and convert it completely into positive work without simultaneously causing other changes in the system or its environment."

Fortunately, to demonstrate the derivation of the above statement from the Carnot version of the second law, we do not need to expend much effort. Because, rather than offering a detailed argument as to how a violation of the Carnot version of the second law would lead to a violation of the Kelvin–Planck version given above, we can instead show that the Clausius statement of the second law in fact leads to that of Kelvin–Planck. As a consequence, a violation of the Carnot statement would also result in a violation of the Kelvin–Planck statement.

Let us assume we have encountered a cyclic process which violates the Kelvin–Planck version of the second law. The process causes a positive amount of heat energy to be extracted from a body at some arbitrary temperature—which for convenience we assume is T_{C}—and completely converts it to positive work, W, without affecting any other changes.

The work W so produced can then be used to run a perfect Carnot refrigerator which is described in Sect. 4.13.

Indeed, such a refrigerator can be made to extract an even more positive amount of heat energy from the body at the given temperature T_C. Such extracted energy can be added to the work W and the sum transferred to a body at a higher temperature. The upshot of such an exercise would be that positive heat energy is transferred from a colder body to a warmer one without any outside assistance. Such an happenstance is forbidden by the Clausius statement.

Thus a violation of the Carnot statement not only causes a violation of the Clausius version of the second law, it also violates the Kelvin–Planck statement of the second law.

It needs to be mentioned that, rather than a Carnot refrigerator, if a perfect Carnot engine were used directly to convert heat energy into work, it would do so while affecting a simultaneous change in its environment. In other words, the engine would necessarily discard part of the heat energy it extracted from the hot reservoir, to the cold reservoir—unless, of course, the temperature of the cold reservoir should be zero. But, alas, that cannot be! Why can it not be, one might ask? Indeed, thereby hangs an important, if not a long, tail and we shall return to it on a later occasion when we discuss the third law of thermodynamics.

4.10.1 Entropy Always Increases in Irreversible Adiabats

We know that an adiabatic process involves no heat energy transfer. And if such a process is completely reversible, it does not cause any change in the entropy.

Such, however, is not the case for an adiabatic process that is even partly irreversible. Indeed, consistency with the Kelvin–Planck's version of the second law ordains a net increase in the entropy of a system that undergoes an adiabatic process which is not fully reversible.

Consider a system that—except for the possibility of contact with an external heat energy reservoir—is fully isolated. Let the system undergo a complete cycle of processes consisting of four traversals that are represented by a schematic plot—see Fig. 4.4—in the T–S plane.

Before any discussion of the round-trip travel—from point 0 all the way around and back to point 0—is given, it is helpful to note that whenever a complete cycle is performed, there is zero net change in the internal energy of the working substance. Accordingly, the total work done by the system while performing the cycle is equal to the net input of heat energy from the external reservoir.

Let us begin with the system in a state represented by the point 0 whose coordinates are (T_0, S_0). Imagine that an irreversible adiabatic process takes the system from the point 0 to point 1, which is represented by the coordinates (T_1, S_1). En route from 0 to 1, the system passes through nonequilibrium states that cannot be displayed in Fig. 4.4 because it refers only to equilibrium states. Note that because the travel from 0 to 1 is adiabatic, no exchange of heat energy has been allowed while traveling from 0 to 1.

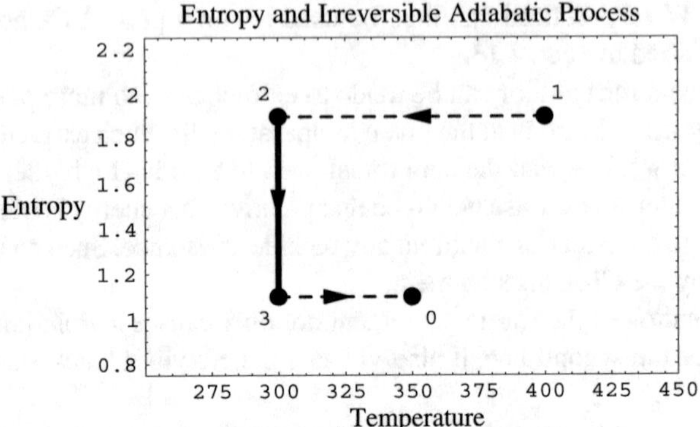

Fig. 4.4 Entropy/(J K^{-1}) versus temperature/K. The scale shown is arbitrary. *Broken lines* are reversible adiabats and the *full line* is a reversible thermostat. The irreversible adiabatic process, that takes the system from the equilibrium state 0 to 1, passes through nonequilibrium states that cannot be displayed in the above plot. Note that $S_3 = S_0$ and $S_2 = S_1$. Also, $T_3 = T_2$

The other three traversals refer to equilibrium processes that are reversible. We shall call these traversals "links."

Let the first of these links represent a process that takes the system from point 1 to point 2. Because this link is both adiabatic and reversible, it occurs along constant entropy. Thus the coordinates of the point 2 are $(T_2, S_2 = S_1)$. Note that again no exchange of heat energy has been allowed during the traversal of this link.

The second link, from point 2 to point 3, is a reversible isotherm at temperature $T_2 = T_3$. In traversing the $2 \to 3$ link, let the system "absorb" heat energy, $Q^{rev}(T_2) = Q^{rev}(T_3)$, from the external reservoir maintained at temperature T_2. Because this heat energy is being added to the system reversibly at one fixed temperature $T_2 = T_3$, we have the equality[14]

$$\frac{Q^{rev}(T_2)}{T_2} = \frac{Q^{rev}(T_3)}{T_3} = S_3 - S_2 = S_0 - S_1 . \tag{4.39}$$

Note that starting at point 2, where the entropy reading is $S_2 = S_1$, and ending at the point 3, where the entropy reads $S_3 = S_0$, the increase in the entropy is equal to $S_3 - S_2 = S_0 - S_1$.

The third link is set to connect point 3 to point 0. Because the entropy is to remain constant all along this link, we must arrange this link to be a reversible adiabat. Thus, there is no interchange of heat energy during this part of the travel.

4.10.2 Entropy Stays Constant in Reversible Adiabats

Let the total amount of work done during the traversal of the full cycle be W. Remember that only the link from point 2 to point 3 involves any heat energy exchange, and also that this heat energy, $Q^{rev}(T_2)$, is reversibly added to the system, see (4.39).

[14]Note that the entropy at point 3, i.e., S_3, can be arranged to be exactly equal to S_0 because we are free to choose the amount of heat energy, $Q^{rev}(T_2)$, to be added.

As mentioned before, because for the full round trip no change in the internal energy occurs, we can equate the total work done during the cycle to the total amount of heat energy that has been added to the working substance. That is,

$$\begin{aligned} W &= Q^{rev}(T_2) \\ &= T_2(S_3 - S_2) = T_2(S_0 - S_1) \\ &= T_3(S_3 - S_2) = T_3(S_0 - S_1)\,. \end{aligned} \tag{4.40}$$

It is important to understand the significance of this observation. What is observed here is that heat energy, $Q^{rev}(T_2)$, is added to the system and all of it has been converted into work W. Note that this energy has been reversibly withdrawn from some very large heat energy reservoir that stays at temperature $T_2 = T_3$.

Consequently, in the above cycle, the Kelvin–Planck assertion, that in a cyclic process heat energy cannot be withdrawn from a source and completely converted into positive work without simultaneously producing other changes, appears to have been violated!

Before we get too alarmed by this seemingly momentous occurrence, we should remember that Kelvin–Planck's stricture against converting heat energy into work in the above fashion only refers to positive work. Thus if the Kelvin–Planck statement is correct, the work W so produced is either equal to zero or is less than zero. That is, according to Kevin–Planck,

$$Q^{rev}(T_2) = W = T_2(S_0 - S_1) \leq 0\,. \tag{4.41}$$

Because T_2 is positive, this means that

$$(S_1 - S_0) \geq 0\,. \tag{4.42}$$

If the transit from 0 to 1 were fully reversible, only the equality would be obtained in the above relationship. Thus the inequality specified in (4.42) must necessarily refer to the transit from 0 to 1 being either wholly or partially irreversible.

This fact confirms the Kelvin–Planck assertion that entropy always increases in irreversible–adiabatic processes.

4.11 Non-Carnot Heat Cycle and Clausius Inequality

Reversible transfer of heat energy is an idealization that is well nigh impossible to achieve in "real" as distinct from "*gedanken*"—that is "thought only"—experiments.

4.11.1 Integral Form of Clausius Inequality

Realistically, energy transfers often occur either wholly or partially nonreversibly. And according to the Carnot version of the second law, the efficiency, ϵ_0, of a realistic, non-Carnot heat engine is always less or equal to the efficiency ϵ_{carnot} of a Carnot engine when both are operating between the same two temperatures.

We shall consider here a series of cycles. During the ith cycle, let an infinitesimal amount of heat energy, $\mathrm{d}Q_i'(T_H)$, be added nonquasistatically to the working substance at temperature T_H and heat energy $dQ_i'(T_C)$ be similarly added at a lower temperature T_C. Then as shown in (4.29) and (4.30), we have

$$\epsilon_{\text{carnot}} \equiv \frac{T_H - T_C}{T_H} \geq \epsilon_0 \equiv \frac{\mathrm{d}Q_i'(T_H) + \mathrm{d}Q_i'(T_C)}{\mathrm{d}Q_i'(T_H)} . \tag{4.43}$$

The relationship (4.43) can readily[15] be reexpressed as follows:

$$0 \geq \frac{\mathrm{d}Q_i'(T_C)}{T_C} + \frac{\mathrm{d}Q_i'(T_H)}{T_H} . \tag{4.44}$$

Of course, the equality in the relationship (4.44) holds only when the flow of heat energy is wholly quasistatic. The relationship (4.44) makes explicit reference to only the two isothermal legs in the cycle, which we number I and III, as before. But here these legs are either wholly or partially traversed nonquasistatically. And as usual, legs I and III involve heat energy transfers—at temperatures T_H and T_C, respectively—between the working substance and the two different heat energy reservoirs. There are no heat energy transfers involved in the other two legs of the cycle—namely the legs that we number II and IV—because they are both adiabats. Therefore, the above relationship can equivalently be represented as the following sum over all the four legs, I → IV, of the ith non-Carnot heat cycle. That is,

$$0 \geq \sum_{\text{leg=I}}^{\text{leg=IV}} (\text{Carnot cycle } i) \left[\frac{\mathrm{d}Q_i'(T)}{T} \right] . \tag{4.45}$$

A smooth curve, that may represent a wholly or partially irreversible finitely sized cycle of arbitrary shape, can be approximated to any desired accuracy by putting together[16] a very large number of tiny cycles of the type represented by (4.45). Indeed, the approximation is perfect—meaning, it is exact—if the cycles are of infinitesimal size and an infinite number of them are added together in the manner described as follows: The infinitesimal non-Carnot cycle i is to be followed by—meaning that it is to be placed contiguous to—another infinitesimal non-Carnot cycle $i + 1$, and that in turn is to be set contiguous to another such cycle numbered $i + 2$, and so on. In this fashion, when an infinitely large number of infinitesimal sized non-Carnot cycles are all traversed in the same rotational order, then all the adiabatic legs, such as II and IV, in successive contiguous cycles completely cancel each other out. As such, this process leaves intact only the smooth outer boundary representing the chosen, finitely sized, partially or wholly irreversible cycle.

[15]To check this, write it as $1 - \frac{T_C}{T_H} \geq 1 + \frac{\mathrm{d}Q_i'(T_C)}{\mathrm{d}Q_i'(T_H)}$. Then subtract 1 from both sides, and multiply both sides by $\frac{\mathrm{d}Q_i'(T_H)}{T_C}$ to get $-\frac{\mathrm{d}Q_i'(T_H)}{T_H} \geq \frac{\mathrm{d}Q_i'(T_C)}{T_C}$. Now transfer $-\frac{\mathrm{d}Q_i'(T_H)}{T_H}$ to the right-hand side.

[16]Compare with Fig. 4.2. However, note that unlike the first time we used Fig. 4.2 where all heat energy transfers were reversible, this time we expect the finite cycle external curve to represent energy transfers which are either wholly or partially irreversible.

Accordingly, for $N \gg 1$, summing (4.45) over $i = 1 \to N$ extremely tiny non-Carnot cycles approximates well the corresponding sum over the chosen finitely sized wholly or partially irreversible cycle of arbitrary shape:

$$0 \geq \sum_{i=1}^{N} \left(\sum_{\text{leg=I}}^{\text{leg=IV}} (\text{Carnot cycle } i) \left[\frac{\mathrm{d}Q_i'(T)}{T} \right] \right). \tag{4.46}$$

When the number of extremely tiny cycles tends to infinity, the above sum can be replaced by an integral

$$0 \geq \oint \frac{\mathrm{d}Q'(T)}{T}. \tag{4.47}$$

The integral is taken over the whole length of the given smooth closed loop—or loops—representing the original, arbitrary shaped, finitely sized, partly or wholly irreversible cycle.

It is important to pay some attention to the detail of the integrand. Unless the transfer of heat energy, $\mathrm{d}Q'$, occurs quasistatically throughout the round trip, say from position a back to the same position a, the inequality will hold. It is, however, convenient to work with a situation in which the heat energy transferred is wholly or partially irreversible for only a part of the loop, say from a to b, while the remaining path in the loop, extending from b to a, is traversed quasistatically. For this scenario, the above inequality becomes

$$0 \geq \int_{\mathrm{a}}^{\mathrm{b}} \frac{\mathrm{d}Q'(T)}{T} + \int_{\mathrm{b}}^{\mathrm{a}} \frac{\mathrm{d}Q^{\mathrm{rev}}(T)}{T}. \tag{4.48}$$

Because $\int_{\mathrm{b}}^{\mathrm{a}} \frac{\mathrm{d}Q^{\mathrm{rev}}(T)}{T}$ is equal to $S_{\mathrm{a}} - S_{\mathrm{b}}$, upon transferring the second term on the right-hand side of (4.48) to the left-hand side, we get

$$S_{\mathrm{b}} - S_{\mathrm{a}} \geq \int_{\mathrm{a}}^{\mathrm{b}} \frac{\mathrm{d}Q'(T)}{T}. \tag{4.49}$$

Equation (4.49) will be called the integral version of the Clausius inequality.[17] Often times, this inequality is treated as an important mathematical result of the second law of thermodynamics.[18]

4.11.2 Differential Form of Clausius Inequality

Imagine a system is brought into contact with a heat energy reservoir that is at temperature T. Assume that the reservoir has infinite heat energy capacity. Let an amount of heat energy equal to $Q'(T)$ flow out of the reservoir. This will change

[17] As mentioned before, while reversible heat energy transfers and work output/input have been denoted with the superscript "rev," the corresponding irreversible—or, partially irreversible—processes are denoted with a "prime" ($'$).

[18] The equality holds only if the path a $\to$ b is traversed quasistatically.

the entropy of the reservoir by an amount $-\frac{Q'(T)}{T}$.[19] Heat flows into the system and causes it to transform from its initial state i, with entropy S_i, into its final state f with entropy S_f. As a result of this spontaneous flow of heat the total change in the entropy of the universe—that is, $\Delta S_{universe}$—must be positive.

The change in the entropy of the universe is equal to that of the system plus that of the reservoir. That is,

$$\Delta S_{universe} = (S_f - S_i) - \frac{Q'(T)}{T} \geq 0 \,. \tag{4.50}$$

Therefore,

$$(S_f - S_i) \geq \frac{Q'(T)}{T} \,. \tag{4.51}$$

A better looking representation of this result is to assume $Q'(T)$ is small and is equal to $dQ'(T)$. As a result the increase, $S_f - S_i$, will also be small and may be represented as dS. Then the above result can be written as

$$dS \geq \frac{dQ'(T)}{T} \,. \tag{4.52}$$

We shall call it the differential form of the Clausius inequality. If we integrate both sides of this inequality around a closed loop, we shall get

$$\oint dS = 0 \geq \oint \frac{dQ'(T)}{T} \,. \tag{4.53}$$

Clearly, this process reproduces the integral inequality that was obtained in (4.47).

Heat Transfer Always Increases Total Entropy

A process in which nonzero, positive heat energy Q' is transferred out from a region "a" at a fixed temperature, T_{high}, the change in its entropy, ΔS_a, is negative. That is, $\Delta S_a = -\frac{Q'}{T_{high}}$. If the same heat energy is transferred into a region "b" at some lower fixed temperature, T_{low}, it increases its entropy by $\Delta S_b = \frac{Q'}{T_{low}}$. As a result the entropy of the universe changes by an amount

$$\Delta S_a + \Delta S_b = Q'\left(-\frac{1}{T_{high}} + \frac{1}{T_{low}}\right) . \tag{4.54}$$

We note that Q' is positive, as is $(-\frac{1}{T_{high}} + \frac{1}{T_{low}}) > 0$. Therefore, we have

$$\Delta S_a + \Delta S_b = Q'\left(-\frac{1}{T_{high}} + \frac{1}{T_{low}}\right) > 0 \,. \tag{4.55}$$

[19]It is not absolutely necessary that the heat energy must flow out of the reservoir quasistatically. Because the reservoir is infinitely large, all heat flows out at exactly the same temperature T, namely the temperature of the reservoir. Thus, even if the entropy change occurs via irreversible processes, one can use reversible paths for its computation, and we can correctly calculate the entropy loss of the reservoir by assuming it occurred quasistatically.

Note that if the regions a and b were of finite size and as a result the temperatures T_{high} and T_{low} were not fixed, then we could work with quasistatic infinitesimal heat energy transfer dQ^{rev}. As a result the increases dS_a and dS_b in the entropy of regions a and b would be $dS_a = -\frac{dQ^{rev}}{T_{high}}$ and $dS_b = \frac{dQ^{rev}}{T_{low}}$. This time the entropy of the universe will increase by

$$dS_a + dS_b = dQ^{rev}\left(-\frac{1}{T_{high}} + \frac{1}{T_{low}}\right) > 0 . \tag{4.56}$$

This relationship will have to be integrated when the heat energy transfer is finitely sized. Then we shall need the temperature dependence of $\frac{dQ^{rev}(T)}{T}$ for each of the two regions a and b and be prepared for the fact that they will depend on the peculiarities of the thermal properties of the two regions.

$$dQ' - dW' = dU . \tag{4.57}$$

Here dQ' and dW' may or may not be quasistatic. Combining (4.52) and (4.57) leads to the Clausius version of the first–second law (compare with (5.6)). That is,

$$TdS \geq dU + dW' . \tag{4.58}$$

The equality in (4.58) is obtained only when the work dW' is performed quasistatically. When that is the case, $dW' \to dW^{rev}$.

An important conclusion is the following: The first law asserts that heat energy, dQ', added to an object plus the work, i.e., $-dW'$, "done on it" are equal to the net increase, dU, in the object's internal energy.

4.12 Solved Problems 4.2–4.24

Entropy Change upon Thermal Contact Discussed in the following are several solved problems that relate to entropy change upon thermal contact.

Problem 4.2 (A finite object in contact with a constant temperature reservoir) A finite sized object is brought in contact with a very large reservoir that maintains its temperature.

Solution 4.2 The reservoir is very large, is maintained at temperature T_{res}, and put in thermal contact with n moles of an object initially at temperature T_{cold}. Let us assume that the reservoir and the object are placed in an adiabatically enclosed large chamber—i.e., a chamber that does not permit any exchange of heat energy with the "universe" outside.

Heat energy is transferred from the hot reservoir to the cold object. The process raises the temperature of the object to that of the reservoir itself. Assuming the specific heat of the object, C_p, per mole is constant, the total amount of heat energy transferred into the object is

$$nC_p(\text{final temp of the system} - \text{its initial temp}) = nC_p(T_{res} - T_{cold}) .$$

Clearly, this heat energy is transferred out of the reservoir while its temperature remains constant at T_{res}. As explained before, because for the reservoir the transfer process begins, continues, and ends at exactly the same temperature, for the purpose of computation of the resulting entropy change of the reservoir, this extraction of heat energy from the reservoir may be treated as though it occurred reversibly.

Therefore, the change in the entropy of the reservoir, ΔS_{res}, is negative and is as given below:

$$\Delta S_{res} = -\frac{\text{heat energy transferred out of the reservoir}}{T_{res}} = -\frac{nC_p(T_{res} - T_{cold})}{T_{res}}. \tag{4.59}$$

Clearly, on the other hand, the entropy of the object increases as its temperature rises from T_{cold} to T_{res}. While this rise in the temperature occurs irreversibly, it can nevertheless be related to the corresponding increase in the entropy by a reversible process in the following manner.

The differential of the entropy is exact. Therefore, it can be integrated via any path that we choose subject only to the requirement that the path must start and end at the specified initial and final locations.

To this end, let us choose a reversible path. Let an infinitesimal amount of heat energy $dQ = nC_p dT$ be added to the object at some temperature T where $T_{res} \geq T \geq T_{cold}$. Note that heat thus added raises the temperature of n mol of the object by an amount dT. Here, C_p is the specific heat energy per mole when the pressure is kept constant. Therefore the total increase in the entropy of the object is

$$\Delta S_{system} = \int_{T_{cold}}^{T_{res}} dS = \int_{T_{cold}}^{T_{res}} \frac{dQ}{T} = \int_{T_{cold}}^{T_{res}} \frac{nC_p dT}{T} = nC_p \ln\left[\frac{T_{res}}{T_{cold}}\right]. \tag{4.60}$$

Being the sum of changes in the entropy of the reservoir and the working substance, total change in the entropy, ΔS_{total}, is

$$\begin{aligned}\Delta S_{total} &= \Delta S_{res} + \Delta S_{system} \\ &= -nC_p\left(\frac{T_{res} - T_{cold}}{T_{res}}\right) + nC_p \ln\left(\frac{T_{res}}{T_{cold}}\right) \\ &= nC_p\left[\ln\left(\frac{T_{res}}{T_{cold}}\right) - \left(\frac{T_{res} - T_{cold}}{T_{res}}\right)\right].\end{aligned} \tag{4.61}$$

Because nC_p is positive, in order to show that the total change in the entropy, i.e., ΔS_{total}, is greater than or equal to zero, all we need to demonstrate is the following:

$$\ln\left(\frac{T_{res}}{T_{cold}}\right) \geq \left(\frac{T_{res} - T_{cold}}{T_{res}}\right). \tag{4.62}$$

Let us introduce the notation

$$y = \left(\frac{T_{\text{cold}}}{T_{\text{res}}}\right).$$

Because $\ln(1/y) = -\ln(y)$, the inequality (4.62) can be written as

$$-\ln(y) \geq 1 - y\,. \tag{4.63}$$

Let us multiply both sides by -1, and note that multiplication by -1 reverses the direction of the inequality. We get

$$\ln(y) \leq y - 1\,. \tag{4.64}$$

Thus in order to prove that ΔS_{total} is indeed greater than or equal to zero, we need to show that the inequality (4.64) holds. To this purpose, let us compare it with a well-known inequality that is known to be true. For all real values of x,

$$(1 + x) \leq \exp(x)\,. \tag{4.65}$$

Therefore, taking natural logarithm of both sides gives

$$\ln(1 + x) \leq x\,. \tag{4.66}$$

Setting

$$x = y - 1\,,$$

gives

$$\ln(y) \leq y - 1\,. \tag{4.67}$$

This proves that the inequality (4.64), and therefore the inequality (4.62), holds. Accordingly, the positivity of the entropy change, ΔS_{total}, given in (4.61) is assured.

Problem 4.3 (Two finite objects in contact with a constant temperature reservoir) Two finite sized objects are brought in contact with a very large reservoir that maintains its temperature.

Solution 4.3 A system consists of two finite masses, m_{h} and m_{c}, originally at temperatures, T_{h} and T_{c}, that have specific heats, C_{h} and C_{c}, respectively. They are both placed in an adiabatic chamber. There the masses are brought into contact until thermal equilibrium is reached. Calculate the change in the entropy of the system and show that when $m_{\text{h}}C_{\text{h}} = m_{\text{c}}C_{\text{c}} = M$, it is given by the expression

$$2M\ln\left(\frac{T_{\text{h}} + T_{\text{c}}}{2\sqrt{T_{\text{h}}T_{\text{c}}}}\right).$$

For convenience, in the following we shall use the notation:

$$m_{\text{h}}C_{\text{h}} = M_{\text{h}}\,, \qquad m_{\text{c}}C_{\text{c}} = M_{\text{c}}\,.$$

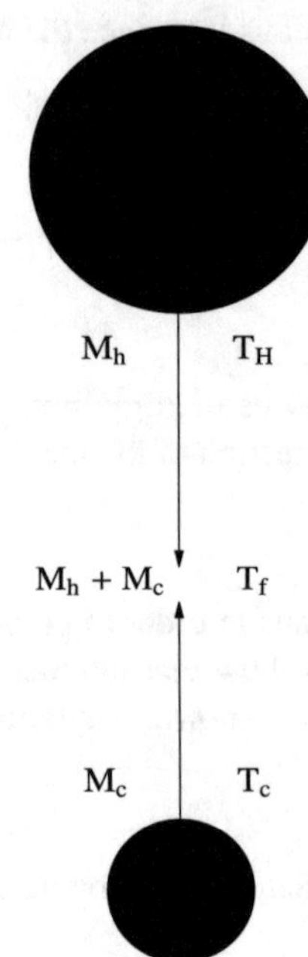

Fig. 4.5 Schematic plot for Problem 4.3. A mass m_h, with specific heat C_h, at temperature T_h is brought into contact with a mass m_c, with specific heat C_c, at a lower temperature T_c. At equilibrium the two masses together reach temperature T_f. For convenience of display, $m_h C_h$ and $m_c C_c$ are represented as M_h and M_c

Upon reaching thermal equilibrium—see Fig. 4.5—the masses will reach a common temperature T_f which must satisfy the requirement that no heat energy has been added to the system during this process:

$$M_h(T_f - T_h) + M_c(T_f - T_c) = 0\,.$$

This yields

$$T_f = \frac{(M_c T_c + M_h T_h)}{(M_c + M_h)}\,. \tag{4.68}$$

It is clear that irreversible processes are involved in the transfer of heat energy from the hot to the cold body. Because entropy is a function of state, for carrying out its calculation, the path followed in going from the initial to the final state is immaterial. Thus we can do this calculation using an idealized path between the initial and final states. Such a path is wholly reversible.

The entropy gains, ΔS_c and ΔS_h, by the cold and hot masses are

$$\Delta Sc = M_c \int_{T_c}^{T_f} \frac{dT}{T} = M_c \ln\left(\frac{T_f}{T_c}\right) \tag{4.69}$$

and[20]

$$\Delta S_h = M_h \int_{T_h}^{T_f} \frac{dT}{T} = M_h \ln\left(\frac{T_f}{T_h}\right). \tag{4.70}$$

[20]Notice that the entropy gain by the hot object is negative because it cools down from T_h to T_f.

Total increase in the entropy of the system is the sum of these two:

$$\Delta S_{\text{total}} = \Delta S_c + \Delta S_h = \ln\left(\frac{T_f}{T_c}\right)^{M_c} + \ln\left(\frac{T_f}{T_h}\right)^{M_h} . \tag{4.71}$$

The proof that $\Delta S_{\text{total}} > 0$ for general values of M_h and M_c is a little involved. As such it is deferred to an appendix. (See Appendix E.) For the special case where $M_h = M_c = M$, the statement

$$\Delta S_{\text{total}} > 0$$

can be shown to be true relatively simply. This is what is do below.

When $M_h = M_c = M$, (4.68) yields the relationship

$$T_f = \frac{T_h + T_c}{2} .$$

Therefore, for this case (4.71) gives

$$\begin{aligned} \Delta S_{\text{total}} &= M\left[\ln\left(\frac{T_f}{T_c}\right) + \ln\left(\frac{T_f}{T_h}\right)\right] = M\ln\left(\frac{T_f^2}{T_c T_h}\right) \\ &= 2M\ln\left(\frac{T_f}{\sqrt{T_c T_h}}\right) = 2M\ln\left[\frac{(T_h + T_c)}{2\sqrt{T_h T_c}}\right] . \end{aligned} \tag{4.72}$$

To prove the positivity of ΔS_{total}, consider the square

$$(\sqrt{T_h} - \sqrt{T_c})^2 = (T_h + T_c) - 2\sqrt{T_h T_c} \geq 0 , \tag{4.73}$$

which leads to the inequality

$$(T_h + T_c) \geq 2\sqrt{T_h T_c} . \tag{4.74}$$

This assures the positivity of the total entropy gain for the special case $M_h = M_c = M$. A demonstration of the positivity of the total increase in the entropy for the general case is given in Appendix E.

Problem 4.4 (Reservoir and mass with temperature-dependent specific heat)
A finite sized mass with temperature-dependent specific heat is brought in contact with a very large reservoir at a lower temperature.

Solution 4.4 A body of mass m has temperature-dependent specific heat given by the relationship $C_V(T) = A + BT + DT^2 + ET^3$ where the coefficients A, B, D, E are all positive.[21] The body is initially at temperature T_C K before it is brought into

[21] Although our choice of constants A, B, D, E is arbitrary, in crystalline solids the occurrence of the ET^3 term at low temperatures is owed to temperature induced lattice vibrations. Similarly, the term linear in temperature—BT—is the electronic contribution to the specific heat in metals. In amorphous solids, the contribution to low temperature specific heat can sometime be quadratic in form, e.g., DT^2.

contact with an infinitely large thermal reservoir at a higher temperature T_H K. The entire process occurs at constant volume. Calculate the change in the entropy of the body, reservoir, and universe composed of the body and reservoir.

At this temperature, T, let the object reversibly contribute an infinitesimal amount of heat energy, $dQ_{obj}^{rev}(T)$, to the working substance of the engine. Similarly, let dQ_{res}^{rev} be the infinitesimal amount of heat energy reversibly contributed by the reservoir—which is at a lower temperature—to the working substance. Note that the reservoir is very large. Therefore, its temperature, T_R, remains essentially unchanged.

As a result of these two infinitesimal additions of heat energy to the working substance, according to the first law and the general properties of the Carnot cycle, and because a complete cycle entails no change in the internal energy—meaning $dU = 0$—the work dW_{done} done by the engine in one complete cycle is equal to the total heat energy input—meaning $dQ_{obj}^{rev}(T) + dQ_{res}^{rev}$—per cycle into the working substance. That is,

$$dW_{done} = dQ_{obj}^{rev}(T) + dQ_{res}^{rev} . \tag{4.75}$$

According to (4.18), the Carnot requirement for this infinitesimal cycle is

$$\frac{dQ_{obj}^{rev}(T)}{T} + \frac{dQ_{res}^{rev}}{T_R} = 0 . \tag{4.76}$$

Using (4.75), the Carnot requirement for the given infinitesimal cycle becomes

$$\frac{dQ_{obj}^{rev}(T)}{T} + \left[\frac{dW_{done} - dQ_{obj}^{rev}(T)}{T_R}\right] = 0 . \tag{4.77}$$

Clearly, withdrawal of heat energy $dQ_{obj}^{rev}(T)$ from the object, causes its temperature to decrease by an infinitesimal amount dT.

For simplicity let us introduce the notation $M = mC$. As a result, we can write

$$dQ_{obj}^{rev}(T) = -M dT . \tag{4.78}$$

Using above two relationships, we get

$$\begin{aligned}\frac{dW_{done}}{T_R} &= -\frac{dQ_{obj}^{rev}(T)}{T} + \frac{dQ_{obj}^{rev}(T)}{T_R}\\ &= -dQ_{obj}^{rev}(T)\left(\frac{1}{T} - \frac{1}{T_R}\right) = M\left(\frac{dT}{T} - \frac{dT}{T_R}\right) .\end{aligned} \tag{4.79}$$

Because the path is reversible we can integrate dW_{done} to calculate the total work done.

$$\int_{initial}^{final} dW_{done} = W_{done} = M T_R \int_{T_{initial}}^{T_{final}} \left(\frac{dT}{T}\right) - M \int_{T_{initial}}^{T_{final}} dT . \tag{4.80}$$

Recalling that the variable T denotes the temperature of the object M, $T_{\text{initial}} = T_{\text{H}}$ and $T_{\text{final}} = T_{\text{R}}$, the total work done is

$$
\begin{aligned}
W_{\text{done}} &= MT_{\text{R}} \ln\left(\frac{T_{\text{R}}}{T_{\text{H}}}\right) - M(T_{\text{R}} - T_{\text{H}}) \\
&= M\left[(T_{\text{H}} - T_{\text{R}}) - T_{\text{R}} \ln\left(\frac{T_{\text{H}}}{T_{\text{R}}}\right)\right]. \qquad (4.81)
\end{aligned}
$$

The total amount of heat energy withdrawn from the object is

$$Q_{\text{object}} = M(T_{\text{H}} - T_{\text{R}}) . \qquad (4.82)$$

The overall efficiency, ε, of the process can now be calculated. It is equal to the ratio of the total work done to the amount of heat energy withdrawn from the hot source. That is,

$$\varepsilon = \frac{W_{\text{done}}}{Q_{\text{object}}} = \frac{W_{\text{done}}}{M(T_{\text{H}} - T_{\text{R}})} . \qquad (4.83)$$

Using (4.81) and (4.82), we get

$$\varepsilon = 1 - \left(\frac{T_{\text{R}}}{T_{\text{H}} - T_{\text{R}}}\right) \ln\left(\frac{T_{\text{H}}}{T_{\text{R}}}\right) . \qquad (4.84)$$

For $T_{\text{H}} = 2T_{\text{R}}$, this gives

$$\varepsilon = 1 - \ln(2) = 0.307 , \qquad (4.85)$$

or 30.7%.

(b) This case is equivalent to that of a Carnot engine operating between two fixed temperatures, T_{H} and T_{R}. (Note that $T_{\text{H}} > T_{\text{R}}$.) Therefore, its efficiency is

$$\varepsilon = 1 - \left(\frac{T_{\text{R}}}{T_{\text{H}}}\right) , \qquad (4.86)$$

which for $T_{\text{H}} = 2T_{\text{R}}$ is 50%.

(c) Because of the use of a Carnot engine, there is no net change in the entropy of the universe for either of the two cases, (a) or (b). Changes in the entropy of the object and the reservoir for case (a) can be calculated as follows:

Note that all of the heat energy added to the reservoir is added reversibly and at a fixed temperature T_{R}. Also that the amount of heat energy, $\Delta Q_{\text{add}}^{\text{res}}$, that actually gets added to the reservoir is the difference between the heat energy it receives from the hot object, i.e., $M(T_{\text{H}} - T_{\text{R}})$, and the work done, W_{done}. That is,

$$\Delta Q_{\text{add}}^{\text{res}} = M(T_{\text{H}} - T_{\text{R}}) - W_{\text{done}} . \qquad (4.87)$$

Using the W_{done} that is recorded in (4.81), the above equation becomes

$$\Delta Q_{\text{add}}^{\text{res}} = M(T_{\text{H}} - T_{\text{R}}) - W_{\text{done}} = M T_{\text{R}} \ln\left(\frac{T_{\text{H}}}{T_{\text{R}}}\right). \tag{4.88}$$

Therefore the increase in the entropy of the reservoir, ΔS_{res}, is

$$\Delta S_{\text{res}} = \frac{\Delta Q_{\text{add}}^{\text{res}}}{T_{\text{R}}} = M \ln\left(\frac{T_{\text{H}}}{T_{\text{R}}}\right). \tag{4.89}$$

Also the increase in the entropy, ΔS_{object}, of the object is readily calculated:

$$\Delta S_{\text{object}} = M \int_{T_{\text{H}}}^{T_{\text{R}}} \frac{\mathrm{d}T}{T} = M \ln\left(\frac{T_{\text{R}}}{T_{\text{H}}}\right). \tag{4.90}$$

Just as one would expect, as a result of the loss in its temperature from T_{H} to T_{R}, the entropy of the object in fact decreases.

It is interesting to check whether the prediction that the Carnot engine conserves the entropy of the universe is satisfied. Its satisfaction would require $\Delta S_{\text{universe}}$ to be equal to zero. We find this is indeed the case. That is,

$$\begin{aligned} \Delta S_{\text{universe}} &= \Delta S_{\text{res}} + \Delta S_{\text{object}} \\ &= M \ln\left(\frac{T_{\text{H}}}{T_{\text{R}}}\right) + M \ln\left(\frac{T_{\text{R}}}{T_{\text{H}}}\right) = 0. \end{aligned} \tag{4.91}$$

The corresponding analysis for the case (b) is easy. Let $Q^{\text{rev}}(T_{\text{H}})$ and $Q^{\text{rev}}(T_{\text{R}})$ be the heat energy introduced reversibly into the working substance at temperatures T_{H} and T_{R}. Clearly, these amounts of heat energy have to be extracted from the corresponding reservoirs. Consequently, the entropies of the reservoirs decrease by the amounts

$$\Delta S_{\text{H}} = -\frac{Q^{\text{rev}}(T_{\text{H}})}{T_{\text{H}}} \tag{4.92}$$

and

$$\Delta S_{\text{R}} = -\frac{Q^{\text{rev}}(T_{\text{R}})}{T_{\text{R}}}. \tag{4.93}$$

But because of the use of the Carnot cycle, we have the relationship

$$\frac{Q^{\text{rev}}(T_{\text{H}})}{T_{\text{H}}} + \frac{Q^{\text{rev}}(T_{\text{R}})}{T_{\text{R}}} = 0, \tag{4.94}$$

therefore

$$\Delta S_{\text{R}} = -\Delta S_{\text{H}}. \tag{4.95}$$

Problem 4.5 (Maximum work and entropy change)
Show that the maximum amount of work, W_{done}, that can be performed, when heat energy is extracted from the object (of mass times specific heat $= M$) and rejected

to the large reservoir as described in the preceding two problems, is given by the relationship

$$W_{\text{done}} = Q_{\text{total}} - T_{\text{R}}(S_{\text{initial}} - S_{\text{final}})\,. \tag{4.96}$$

Here Q_{total} is the total heat energy lost by the object and S_{initial} and S_{final} are the initial and final values of its entropy.

Solution 4.5 The total amount of heat energy lost by the object in cooling from its initial temperature T_{H} to its final temperature T_{R} (equal to that of the reservoir that the Carnot engine uses for discarding heat energy) is

$$Q_{\text{total}} = M(T_{\text{H}} - T_{\text{R}})\,. \tag{4.97}$$

The increase in its entropy is determined in the usual fashion. We have

$$\int_{\text{initial}}^{\text{final}} \mathrm{d}S = \int_{T_{\text{initial}}}^{T_{\text{final}}} \left(\frac{-\mathrm{d}Q(T)}{T}\right) = M \int_{T_{\text{H}}}^{T_{\text{R}}} \frac{\mathrm{d}T}{T}\,, \tag{4.98}$$

therefore

$$S_{\text{final}} - S_{\text{initial}} = -M \ln\left(\frac{T_{\text{H}}}{T_{\text{R}}}\right). \tag{4.99}$$

Note that because the temperature of the mass falls from T_{H} to T_{R}, its entropy decreases in the process. Thus, $S_{\text{final}} - S_{\text{initial}}$ is, as expected, negative.

Combining (4.96) and (4.97), multiplying (4.99) by T_{R}, etc., reproduces the earlier result given in (4.81). That is,

$$\begin{aligned} W_{\text{done}} &= Q_{\text{total}} - T_{\text{R}}(S_{\text{initial}} - S_{\text{final}}) = M T_{\text{R}} \ln\left(\frac{T_{\text{R}}}{T_{\text{H}}}\right) - M(T_{\text{R}} - T_{\text{H}}) \\ &= M\left[(T_{\text{H}} - T_{\text{R}}) - T_{\text{R}} \ln\left(\frac{T_{\text{H}}}{T_{\text{R}}}\right)\right]. \end{aligned} \tag{4.100}$$

Problem 4.6 (Two finite masses in contact)
A system consisting of two finite masses m_{h} and m_{c}, of specific heats C_{h} and C_{c}, that are originally at temperatures T_{h} and T_{c}, is isolated in an adiabatic chamber. These masses are used as the hot and cold finite sources for a perfect Carnot engine which is employed to extract maximum amount of work, W, from the system. Calculate (a) the final equilibrium temperature, T_0, of the masses and (b) the total amount of work extracted.

Solution 4.6 It is convenient to utilize some of the results obtained in the preceding problem. Further, we continue to use the notation $M_{\text{h}} = m_{\text{h}} C_{\text{h}}$ and $M_{\text{c}} = m_{\text{c}} C_{\text{c}}$.

Recognize that the temperature, T_{f}, that was reached by the two masses when they came into thermal equilibrium, i.e.,

$$T_{\text{f}} = \frac{(M_{\text{c}} T_{\text{c}} + M_{\text{h}} T_{\text{h}})}{(M_{\text{h}} + M_{\text{c}})}\,,$$

as recorded in (4.68), is not relevant to the present problem. Therefore, let us simply call the final temperature T_0. Once this is done, we note that the procedure for the evaluation of the total entropy increase, Δ_{total}, that followed (4.68) remains valid and therefore (4.71), with T_f replaced by T_0, still holds. But there is an important caveat. In the present problem, because of the use of the perfect Carnot engine, the total increase in the entropy of the universe—that is, the isolated system consisting of the two masses— is zero. Thus, from (4.71) we get

$$\Delta S_{\text{total}} = \ln\left(\frac{T_0}{T_c}\right)^{M_c} + \ln\left(\frac{T_0}{T_h}\right)^{M_h} = 0\,. \tag{4.101}$$

Adding the logarithms together, i.e.,

$$\ln\left[\frac{T_0^{M_c+M_h}}{T_c^{M_c}T_h^{M_h}}\right] = 0\,,$$

and noting that the solution of $\ln x = 0$ is $x = 1$, we get

$$T_0 = T_c^{\frac{M_c}{\mu}} \cdot T_h^{\frac{M_h}{\mu}}\,, \quad \mu = M_c + M_h\,. \tag{4.102}$$

The total work output, W, of the perfect Carnot engine—which, of course, is the maximum possible work output— is, according to the first law, equal to the heat energy expended. And the heat energy so expended is equal to the difference between the initial and the final heat energy content of the masses. That is,

$$W = (M_c T_c + M_h T_h) - (M_c + M_h)T_0 = (M_c T_c + M_h T_h) - \mu T_c^{\frac{M_c}{\mu}} \cdot T_h^{\frac{M_h}{\mu}}\,. \tag{4.103}$$

Problem 4.7 (Alternate solution for Problem 4.6)
Provide alternate solution for Problem 4.6.

Solution 4.7 Upon thermal contact the body warms up from its initial temperature T_C to T_H, the temperature of the reservoir. As usual, we employ a reversible path to calculate the change in the entropy of the body as it travels from temperature T_C to T_H:

$$\begin{aligned}\Delta S_{\text{body}} &= m\int_{T_C}^{T_H} dT\,\frac{C_V(T)}{T}\\ &= m\left[A\ln\left(\frac{T_H}{T_C}\right) + B(T_H - T_C)\right] + m\left[\frac{D(T_H^2 - T_C^2)}{2} + \frac{E(T_H^3 - T_C^3)}{3}\right].\end{aligned} \tag{4.104}$$

Note that the rise in temperature—meaning that $T_H > T_C$—ensures that the change in the entropy is positive.

Because the heat capacity of the reservoir is very large, its temperature during the entire process hardly changes. Therefore, the (negative) heat energy added to the reservoir can be treated as though it has been added reversibly at temperature T_H.

The resultant entropy increase[22] of the reservoir is

$$\Delta S_{\text{reservoir}} = -\left(\frac{\Delta Q^{\text{rev}}_{\text{body}}}{T_{\text{H}}}\right), \tag{4.105}$$

where $\Delta Q^{\text{rev}}_{\text{body}}$ is the positive amount of heat energy added to the body to raise its temperature from T_{C} to T_{H}. That is,

$$\begin{aligned}\Delta Q^{\text{rev}}_{\text{body}} &= m\int_{T_{\text{C}}}^{T_{\text{H}}} \mathrm{d}T\, C_{\text{V}}(T) = m\int_{T_{\text{C}}}^{T_{\text{H}}} \mathrm{d}T\left(A + BT + DT^2 + ET^3\right)\\ &= mA(T_{\text{H}} - T_{\text{C}}) + mB\left(\frac{T_{\text{H}}^2 - T_{\text{C}}^2}{2}\right)\\ &\quad + mD\left(\frac{T_{\text{H}}^3 - T_{\text{C}}^3}{3}\right) + mE\left(\frac{T_{\text{H}}^4 - T_{\text{C}}^4}{4}\right).\end{aligned} \tag{4.106}$$

The total change in the entropy of the universe, $\Delta S_{\text{universe}}$, that is the body plus the reservoir, is the sum of the two changes, i.e.,

$$\Delta S_{\text{universe}} = \Delta S_{\text{body}} + \Delta S_{\text{reservoir}}\,,$$

given in (4.104) and (4.105)–(4.106). Thus, $\Delta S_{\text{universe}}$ is equal to

$$\begin{aligned}\Delta S_{\text{universe}} = \Delta S_{\text{body}} + \Delta S_{\text{reservoir}} &= mA\left[\ln\left(\frac{T_{\text{H}}}{T_{\text{C}}}\right) - \left(\frac{T_{\text{H}} - T_{\text{C}}}{T_{\text{H}}}\right)\right]\\ &+ mB\left[T_{\text{H}} - T_{\text{C}} - \left(\frac{T_{\text{H}}^2 - T_{\text{C}}^2}{2T_{\text{H}}}\right)\right]\\ &+ mD\left[\left(\frac{T_{\text{H}}^2 - T_{\text{C}}^2}{2}\right) - \left(\frac{T_{\text{H}}^3 - T_{\text{C}}^3}{3T_{\text{H}}}\right)\right]\\ &+ mE\left[\left(\frac{T_{\text{H}}^3 - T_{\text{C}}^3}{3}\right) - \left(\frac{T_{\text{H}}^4 - T_{\text{C}}^4}{4T_{\text{H}}}\right)\right].\end{aligned} \tag{4.107}$$

It is readily checked that for $T_{\text{H}} = T_{\text{C}}$ the total entropy of the universe remains unchanged. Somewhat more challenging is the task of showing that, for $T_{\text{H}} > T_{\text{C}}$, the increase in the entropy is always positive. After some algebraic manipulation, contributions to the entropy from terms involving, B, D, and E can be recast into the following form:

$$\begin{aligned}&\left(\frac{B}{2T_{\text{H}}}\right)(T_{\text{H}} - T_{\text{C}})^2\,,\\ &\left(\frac{D}{6T_{\text{H}}}\right)(T_{\text{H}} - T_{\text{C}})^2(T_{\text{H}} + 2T_{\text{C}})\,,\\ &\left(\frac{E}{12T_{\text{H}}}\right)(T_{\text{H}} - T_{\text{C}})^2(T_{\text{H}}^2 + 2T_{\text{H}}T_{\text{C}} + 3T_{\text{C}}^2)\,.\end{aligned}$$

[22]This increase is actually negative because the reservoir loses positive amount of heat energy in the process.

All three of the above are manifestly positive. Recall that the remaining term proportional to the coefficient A is very similar to that treated earlier in this chapter where a detailed argument for its positivity was provided. Compare (4.62) and the discussion following it.

Problem 4.8 (Changes along Carnot paths)
Calculate the entropy changes along the four legs of the complete cycle—as shown in Fig. 4.1.a—that is followed by a Carnot engine. Remember, this engine is operating with perfect gas as its working substance.

Solution 4.8 The work done by n moles of a perfect gas in going from point 1 to 2 has been previously recorded in (4.7),

$$\Delta W^{\text{rev}}_{1\to 2} = nRT_{\text{H}} \ln\left(\frac{V_2}{V_1}\right). \tag{4.108}$$

Owing to the fact that this first leg of the travel occurs at constant temperature T_{H}, and we are dealing with a perfect gas whose internal energy depends only on the constants n, R, and the temperature, during this leg the internal energy must surely remain constant. That is,

$$\Delta U^{\text{rev}}_{1\to 2} = 0\,.$$

Therefore, according to the first law, the amount of heat energy reversibly added to the working substance at temperature T_{H} is

$$\Delta Q^{\text{rev}}(T_{\text{H}}) = \Delta U^{\text{rev}}_{1\to 2} + \Delta W^{\text{rev}}_{1\to 2} = nRT_{\text{H}} \ln\left(\frac{V_2}{V_1}\right). \tag{4.109}$$

Accordingly, the increase in the entropy of the perfect gas used as the working substance is

$$\Delta S_{1\to 2} = \left[\frac{\Delta Q^{\text{rev}}(T_{\text{H}})}{T_{\text{H}}}\right] = nR \ln\left(\frac{V_2}{V_1}\right). \tag{4.110}$$

Because the symbols 1 and 2 and H act as dummy indices referring to two points on an isothermal route at temperature T_{H}, without going through any additional algebra, we can use the above relationship to immediately write down the increase in the entropy of the working substance in going from point 3 to 4 at temperature T_{C}. In this manner we get

$$\Delta S_{3\to 4} = \left[\frac{\Delta Q^{\text{rev}}(T_{\text{C}})}{T_{\text{C}}}\right] = nR \ln\left(\frac{V_4}{V_3}\right). \tag{4.111}$$

Each of the two legs, $2 \to 3$ and $4 \to 1$, is traveled without any transfer of heat energy. Also both of these legs are traveled reversibly. Therefore, each of these two legs involves zero change in the entropy. That is,

$$\Delta S_{2\to 3} = \Delta S_{4\to 1} = 0\,. \tag{4.112}$$

Accordingly, the total change in the entropy—that is, ΔS_{total}—is

$$\begin{aligned}\Delta S_{\text{total}} &= \Delta S_{1\to 2} + \Delta S_{2\to 3} + \Delta S_{3\to 4} + \Delta S_{4\to 1} \\ &= \Delta S_{1\to 2} + \Delta S_{3\to 4} = nR\left[\ln\left(\frac{V_2}{V_1}\right) + \ln\left(\frac{V_4}{V_3}\right)\right].\end{aligned} \tag{4.113}$$

Equation (4.14) tells us that

$$\frac{V_2}{V_1} = \frac{V_3}{V_4},$$

or equivalently that

$$\ln\left(\frac{V_2}{V_1}\right) = -\ln\left(\frac{V_4}{V_3}\right).$$

Therefore, $\Delta S_{\text{total}} = 0$. That this is the case could have been anticipated from the fact that $\mathrm{d}S$ is an exact differential and its integral along any closed loop must be equal to zero.

Problem 4.9 (An object and a reservoir)
A Carnot engine withdraws heat energy from a finite object that is initially at temperature T_{H}. The mass of the object is m and its specific heat is C. The Carnot engine rejects heat energy to a very large reservoir at a lower temperature T_{R}.

(a) Calculate the total work output and therefore the efficiency, ε, of this Carnot engine. In particular, what is ε if $T_{\text{H}} = 2T_{\text{R}}$?
(b) Also calculate the efficiency, ϵ, of a Carnot engine that works between two very large reservoirs, again at temperatures $T_{\text{H}} = 2T_{\text{R}}$ and T_{R}.
(c) Describe the changes in entropy for the cases (a) and (b).

Solution 4.9 We are, in fact, given both the initial and final temperature of the finite object. Therefore we know the total amount of heat energy it must have yielded. This amount is, of course, equal to the heat energy received by the working substance.

Because of the use of a perfect Carnot engine, the total change in the entropy of the universe is vanishingly small. Therefore, all we need to do is calculate the change in the entropy of the finite object which then tells us what the change in the entropy of the reservoir is. Once we know that, we can readily calculate the amount of heat energy rejected by the working substance (and thereby added to the reservoir).

The Carnot engine operates cyclically. The difference between the heat energy input into the working substance and the heat energy it rejects to the reservoir is equal to the total work done.

The finite object, initially at temperature T_{H}, releases heat energy to the working substance of the Carnot engine. It continues releasing heat, little by little, until its temperature—as well as that of the working substance—eventually drops down to the lowest temperature that the working substance can possibly get to. Clearly, that lowest temperature must be equal to T_R, which is the temperature of the reservoir. This is so because the working substance gives away its unused heat energy to the reservoir. As a result its temperature drops. Indeed, it does not stop transferring heat energy out until its temperature drops down to that of the reservoir.

The "increase," ΔS_{object}, in the entropy of the finite object[23] is

$$\Delta S_{\text{object}} = mC \int_{T_H}^{T_R} \left(\frac{dT}{T}\right) = M \ln\left[\frac{T_R}{T_H}\right] = -M \ln\left[\frac{T_H}{T_R}\right], \tag{4.114}$$

where for convenience we have introduced the notation $mC = M$.

Next, we need to calculate the heat energy, H_{res}, that is added to the reservoir at its constant temperature T_R. This is best done by noting that the use of the Carnot engine ensures the constancy of the total entropy of the two sources of energy supply. That is,

$$\Delta S_{\text{object}} + \Delta S_{\text{res}} = -M \ln\left[\frac{T_H}{T_R}\right] + \Delta S_{\text{res}} = 0\,. \tag{4.115}$$

As a result the increase in the entropy of the reservoir, ΔS_{res}, is

$$\Delta S_{\text{res}} = M \ln\left[\frac{T_H}{T_R}\right]. \tag{4.116}$$

Because the temperature, T_R, of the reservoir stays constant, the heat energy, ΔH_{res}, added to the reservoir is

$$\Delta H_{\text{res}} = T_R \Delta S_{\text{res}} = M T_R \ln\left[\frac{T_H}{T_R}\right]. \tag{4.117}$$

The heat energy, ΔH_{obj}, contributed by the object to the working substance when its temperature falls from T_H to T_R is

$$\Delta H_{\text{obj}} = M(T_H - T_R)\,. \tag{4.118}$$

Therefore the total work, ΔW_{total}, done by the Carnot engine is

$$\Delta W_{\text{total}} = \Delta H_{\text{obj}} - \Delta H_{\text{res}} = M\left[(T_H - T_R) - T_R \ln\left(\frac{T_H}{T_R}\right)\right]. \tag{4.119}$$

The overall efficiency, ε, of the engine is the ratio of the total work done, ΔW_{total}, to the heat energy, ΔH_{obj}, withdrawn from the object which acts as the hot source. That is,

$$\varepsilon = \frac{\Delta W_{\text{total}}}{\Delta H_{\text{obj}}} = \frac{\Delta H_{\text{obj}} - \Delta H_{\text{res}}}{H_{\text{obj}}} = 1 - \frac{\Delta H_{\text{res}}}{\Delta H_{\text{obj}}} = 1 - \left(\frac{T_R}{T_H - T_R}\right) \ln\left(\frac{T_H}{T_R}\right). \tag{4.120}$$

[23] Of course, in reality, it is a decrease because the object has yielded heat energy in this process.

For $T_\mathrm{H} = 2T_\mathrm{R}$, this gives

$$\varepsilon = 1 - \ln(2) = 0.307\,, \tag{4.121}$$

or 30.7%.

Case (b) is equivalent to that of a Carnot engine operating between two fixed temperatures, T_H and T_R. Note that $T_\mathrm{H} > T_\mathrm{R}$. Therefore, its efficiency is

$$\varepsilon = 1 - \left(\frac{T_\mathrm{R}}{T_\mathrm{H}}\right), \tag{4.122}$$

which for $T_\mathrm{H} = 2T_\mathrm{R}$ is 50%.

(c) Because of the use of a Carnot engine, there is no net change in the entropy of the universe for either of the two cases, (a) or (b).

Problem 4.10 (Alternate solution for Problem 4.9(a))
The temperature of the object sometime during the course of this operation is T such that $T_\mathrm{H} \geq T \geq T_\mathrm{R}$.

Another method for solving this problem, which does not use the earlier result obtained in (4.71), is the following—see Fig. 4.7.

Solution 4.10 Consider a situation where the masses m_h and m_c are at temperatures, T'_h and T'_c, in-between their initial temperatures, T_h and T_c, and the final equilibrium temperature T_0, i.e.,

$$T_\mathrm{h} > T'_\mathrm{h} > T_0\,, \qquad T_\mathrm{c} < T'_\mathrm{c} < T_0\,.$$

Infinitesimal amount of heat energy $\mathrm{d}Q_\mathrm{h}^\mathrm{rev}$ is now reversibly extracted at temperature T'_h from the mass m_h and is introduced at the same temperature T'_h into the working substance of a Carnot engine. Similarly, heat energy $\mathrm{d}Q_\mathrm{c}^\mathrm{rev}$ is extracted from mass m_c at temperature T'_c and reversibly transferred to the working substance at the same temperature T'_c. These processes result in the engine performing work $\mathrm{d}W$ equal to the total amount of heat energy added to the working substance. That is,

$$\mathrm{d}W = \mathrm{d}Q_\mathrm{h}^\mathrm{rev} + \mathrm{d}Q_\mathrm{c}^\mathrm{rev}\,. \tag{4.123}$$

The extraction of heat energy $dQ_\mathrm{h}^\mathrm{rev}$ from the mass m_h causes a *decrease* $\mathrm{d}T'_\mathrm{h}$ in its temperature. Similarly, extraction of heat energy $\mathrm{d}Q_\mathrm{c}^\mathrm{rev}$ from mass m_c causes a decrease $\mathrm{d}T'_\mathrm{c}$ in its temperature. Thus (4.123) gives

$$\mathrm{d}W = -M_\mathrm{h}\mathrm{d}T'_\mathrm{h} - M_\mathrm{c}\mathrm{d}T'_\mathrm{c}\,. \tag{4.124}$$

Integration,

$$W = \int_\mathrm{initial}^\mathrm{final} \mathrm{d}W = -M_\mathrm{h}\int_{T_\mathrm{h}}^{T_0} \mathrm{d}T'_\mathrm{h} - M_\mathrm{c}\int_{T_\mathrm{c}}^{T_0} \mathrm{d}T'_\mathrm{c}\,, \tag{4.125}$$

yields

$$W = -M_\mathrm{h}(T_0 - T_\mathrm{h}) - M_\mathrm{c}(T_0 - T_\mathrm{c}) = (M_\mathrm{c}T_\mathrm{c} + M_\mathrm{h}T_\mathrm{h}) - (M_\mathrm{c} + M_\mathrm{h})T_0\,. \tag{4.126}$$

Fig. 4.6 Schematic plot for Problems 4.8 and 4.9. Heat energy is being withdrawn from an object M, initially at temperature T_{H}. As a result, its temperature decreases, eventually reaching the temperature, T_{R}, of the large reservoir. While cooling, it passes through an intermediate temperature T

This derivation has reproduced (4.103) whose existence had simply been asserted earlier.

In order to calculate T_0, we invoke the fact that the operation of the Carnot cycle requires

$$\frac{\mathrm{d}Q_{\mathrm{h}}^{\mathrm{rev}}}{T_{\mathrm{h}}'} + \frac{\mathrm{d}Q_{\mathrm{c}}^{\mathrm{rev}}}{T_{\mathrm{c}}'} = 0\,. \tag{4.127}$$

Therefore, we have

$$-M_{\mathrm{h}}\frac{\mathrm{d}T_{\mathrm{h}}'}{T_{\mathrm{h}}'} - M_{\mathrm{c}}\frac{\mathrm{d}T_{\mathrm{c}}'}{T_{\mathrm{c}}'} = 0\,. \tag{4.128}$$

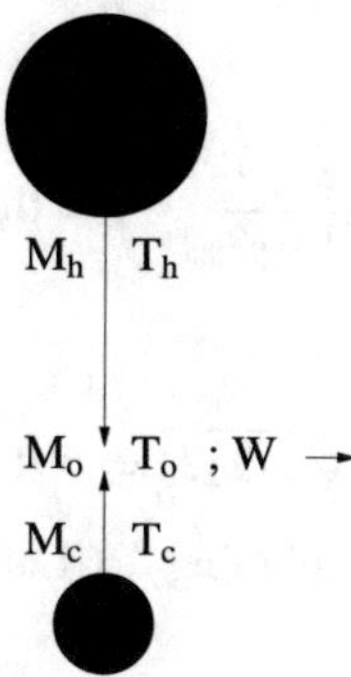

Fig. 4.7 Schematic plot for Problems 4.8 and 4.9. Two masses, m_h (specific heat C_h and at temperature T_h) and m_c (specific heat C_c and at a lower temperature T_c), are used as finite heat energy reservoirs for a perfect Carnot engine. When maximum work has been performed, the two masses reach a common temperature T_0. Note that this temperature is lower than the T_f that is obtained when the two masses are put directly into thermal contact. (Compare with T_f shown in Fig. 4.6.) The heat energy content of the system at temperatures T_f and T_0 differs by W, the work output of a perfect Carnot engine. For convenience of display, $m_h C_h$ and $m_c C_c$ are represented as M_h and M_c, and their sum $(M_h + M_c)$ as M_0

Integration—similar to that done in (4.125)—gives

$$-M_h \int_{T_h}^{T_0} \frac{dT'_h}{T'_h} - M_c \int_{T_c}^{T_0} \frac{dT'_c}{T'_c} = -M_h \ln\left(\frac{T_0}{T_h}\right) - M_c \ln\left(\frac{T_0}{T_c}\right) = 0 . \quad (4.129)$$

This immediately reproduces (4.101). Q.E.D.

Problem 4.11 (Contact between three finite masses)
Three identical objects, each of mass $m = 10$ mol and specific heat $C_P = 100$ J/(mol K), are contained in an adiabatic chamber maintained at constant pressure. Initially, the objects are at temperatures $T_1 = 500$ K, $T_2 = 400$ K, and $T_3 = 300$ K. Also available is a Carnot engine. Using these three objects appropriately as finite thermal sources, calculate: (1) the maximum work performed by the engine and (2) the change in the entropy of each of the objects.

Solution 4.11 When the Carnot engine has produced the maximum possible work, the three objects, 1, 2, and 3, reach a common temperature, T_f. The use of the Carnot engine ensures that the total increase in the entropy, ΔS(total), of the three objects is vanishing. That is,

$$\Delta S(\text{total}) = \Delta S_1 + \Delta S_2 + \Delta S_3 = mC_P \left[\int_{T_1}^{T_f} \frac{dT}{T} + \int_{T_2}^{T_f} \frac{dT}{T} + \int_{T_3}^{T_f} \frac{dT}{T} \right]$$

$$= mC_P \left[\ln \frac{T_f}{T_1} + \ln \frac{T_f}{T_2} + \ln \frac{T_f}{T_3} \right] = mC_P \ln\left(\frac{T_f^3}{T_1 T_2 T_3} \right) = 0 . \quad (4.130)$$

Thus

$$\ln\left(\frac{T_f^3}{T_1T_2T_3}\right) = 0\,,$$

which gives

$$T_F = \sqrt[3]{T_1T_2T_3} = 391.5\ \text{K}. \tag{4.131}$$

The work performed—which, according to Carnot, is for sure the maximum possible—is

$$\begin{aligned} W(\text{total}) &= \text{initial heat energy content} - \text{final heat energy content} \\ &= mC_P\big[(T_1 + T_2 + T_3) - 3T_f\big] \\ &= mC_P\big[(T_1 + T_2 + T_3) - 3\big(\sqrt[3]{T_1T_2T_3}\big)\big] \\ &= 1000(1200 - 1174.46) = 25.540\ \text{J}. \end{aligned} \tag{4.132}$$

Further, the increase in the entropy of each of the objects is:

$$\Delta S_1 = mC_P \ln\left(\frac{T_f}{T_1}\right) = -244.6\ \text{J/K}\,,$$

$$\Delta S_2 = mC_P \ln\left(\frac{T_f}{T_2}\right) = -21.5\ \text{J/K}\,,$$

$$\Delta S_3 = mC_P \ln\left(\frac{T_f}{T_3}\right) = 266.1\ \text{J/K}\,.$$

Problem 4.12 Provide an alternate solution for Problem 4.11.

Solution 4.12 Carnot engines that involve more than two objects can conveniently be handled by using a *cascade* procedure. This procedure consists in ordering the objects according to descending values of their initial temperatures and then treating first the top two objects as the hot and the cold energy sources for a Carnot engine.

After the Carnot engine has been run with these top two objects, and the maximum possible amount of work has been done, the temperatures of the two objects reach a common value. For the next process, these two objects together are used as the single composite energy source and the object that originally was the third on the descending temperature scale now acts as the second energy source for the Carnot engine.

This procedure can successively be followed until all the objects have been treated.

Let us calculate the result that follows from the first process, to be denoted with the index "I." Here objects 1 and 2 are the finite thermal sources at initial temperatures T_1 and T_2, respectively. Let the resultant temperature, after maximum possible amount of work has been produced by the Carnot engine, be T_f. When this temperature is reached, the total change, $\Delta S(\text{I})$, in the entropy of the two objects is calculated to be the following:

$$\Delta S(\text{I}) = \Delta S_1(\text{I}) + \Delta S_2(\text{I}) = mC_\text{P}\left[\int_{T_1}^{T_\text{f}} \frac{\text{d}T}{T} + \int_{T_2}^{T_\text{f}} \frac{\text{d}T}{T}\right]$$
$$= mC_\text{P}\left[\ln\frac{T_\text{f}}{T_1} + \ln\frac{T_\text{f}}{T_2}\right] = mC_\text{P}\ln\left(\frac{T_\text{f}^2}{T_1T_2}\right) = 2mC_\text{P}\ln\left(\frac{T_\text{f}}{\sqrt{T_1T_2}}\right). \tag{4.133}$$

Because the total entropy is conserved in each Carnot cycle that is completed, $\Delta S(\text{I}) = 0$, and as a result,

$$\ln\left(\frac{T_\text{f}}{\sqrt{T_1T_2}}\right) = 0 = \ln(1).$$

This gives

$$T_\text{f} = \sqrt{T_1T_2} = \sqrt{500 \times 400} = 447.2\ \text{K}. \tag{4.134}$$

The total work done, W_I, in the process I is equal to the total loss of heat energy in the two objects 1 and 2. That is,

$$W_\text{I} = mC_\text{P}\left[(T_1 + T_2) - 2T_\text{f}\right] = 5,573\ \text{J}. \tag{4.135}$$

The next process—i.e., process II—treats the composite body of mass $2m$ as the hot thermal source, at temperature T_f, and the third object as the cold thermal source at temperature T_3. The Carnot cycle, after doing the maximum possible amount of work, brings the system to the final temperature T_F. Again the counterpart of (4.133) is used and the total increase in the entropy in process II is found as follows:

$$\Delta S(\text{II}) = \Delta S_1(\text{II}) + \Delta S_2(\text{II})$$
$$= mC_\text{P}\left[2\int_{T_\text{f}}^{T_\text{F}} \frac{\text{d}T}{T} + \int_{T_3}^{T_\text{F}} \frac{\text{d}T}{T}\right] = mC_\text{P}\left[2\ln\frac{T_\text{F}}{T_\text{f}} + \ln\frac{T_\text{F}}{T_3}\right]$$
$$= mC_\text{P}\ln\left(\frac{{T_\text{F}}^3}{{T_\text{f}}^2T_3}\right) = mC_\text{P}\ln\left(\frac{{T_\text{F}}^3}{T_1T_2T_3}\right). \tag{4.136}$$

Now, demanding the conservation of entropy in the Carnot process II leads to the result

$$\Delta S(\text{II}) = mC_\text{P}\ln\left(\frac{{T_\text{F}}^3}{T_1T_2T_3}\right) = 0 = mC_\text{P}\ln(1),$$

which gives

$$T_\text{F} = \sqrt[3]{T_1T_2T_3} = 391.5\ \text{K}. \tag{4.137}$$

Similarly, much like (4.135), the amount of work done in stage II is

$$W_\text{II} = mC_\text{P}[T_3 + 2T_\text{f} - 3T_\text{F}] = 19,966\ \text{J}. \tag{4.138}$$

Note that T_F is the final common temperature of all the three objects and W_{II} is the work done during the second stage.

The total work done is the sum of the work done in stages I and II,

$$\begin{aligned} W(\text{total}) &= W_I + W_{II} \\ &= mC_P\big[(T_1 + T_2) - 2T_f\big] + mC_P[T_3 + 2T_f - 3T_F] \\ &= mC_P\big(T_1 + T_2 + T_3 - 3\sqrt[3]{T_1T_2T_3}\big) = 25,539\ \text{J}\,. \end{aligned} \tag{4.139}$$

The results for the entropy change are also clearly identical to those recorded in the first solution.

Instead of three, let the number of identical objects be n. By repeated application of (4.130), show that the final temperature, T_N, reached by all the objects is

$$T_N = \sqrt[n]{T_1T_2\cdots T_n}\,.$$

Also show that the maximum work performed by the Carnot engine is

$$W = mC_P(T_1 + T_2 + \cdots + T_n - nT_N)\,.$$

Problem 4.13 (Carnot engine and three reservoirs)
Three heat energy reservoirs, labeled I, II, and III, are at constant temperatures $T_1 = 600$ K, $T_2 = 500$ K, and $T_3 = 400$ K, respectively. A single Carnot engine exchanges heat energy with these reservoirs during complete cycles of operation. In the process it performs a total of 800 J of work. The total amount of heat energy withdrawn from reservoir I is 1800 J. Calculate the heat energy withdrawn from reservoirs II and III. Also calculate the changes in the entropy of the three reservoirs and the working substance.

Solution 4.13 Let the working substance of the Carnot engine receive heat energy Q_1, Q_2, and Q_3 from the three reservoirs I, II, and III that are at constant temperatures $T_1 = 600$ K, $T_2 = 500$ K, and $T_3 = 400$ K, respectively.

Because the derivative of the internal energy is an exact differential, the internal energy at the end of a—or a set off—complete cycle(s) is the same as it was at the beginning. Thus, conservation of total energy requires that the work done, i.e., 800 J, be equal to the total heat energy added to the working substance. That is,

$$800\ \text{J} = Q_1 + Q_2 + Q_3\,. \tag{4.140}$$

Also there is the fundamental requirement for a Carnot cycle, namely that the total gain—in fact, the total change—in the entropy of the working substance in any complete cycle be zero. Further, because all heat energy exchanges with the Carnot engine occur reversibly, the total entropy loss of the reservoirs is equal to the total entropy gain of the working substance—which we just stated is equal to zero. As a result, the total entropy loss of the three reservoirs is zero. That is,

$$0 = -\frac{Q_1}{T_1} - \frac{Q_2}{T_2} - \frac{Q_3}{T_3} = -\frac{Q_1}{600} - \frac{Q_2}{500} - \frac{Q_3}{400}\,. \tag{4.141}$$

The two equations, (4.140) and (4.141), appear to have three unknowns, Q_1, Q_2, and Q_3. But because we already know from the statement of the problem that heat energy Q_1 given away by the reservoir I at temperature 600 K—and added to the working substance—is

$$Q_1 = 1800\,\mathrm{J}\,,$$

there are, in fact, only two unknowns, Q_2 and Q_3.

It is convenient to rewrite (4.140.) That is,

$$800\,\mathrm{J} = 1800\,\mathrm{J} + Q_2 + Q_3\,.$$

Equivalently,

$$-1000\,\mathrm{J} = Q_2 + Q_3\,. \tag{4.142}$$

Similarly, let us also rewrite (4.141) as

$$0 = -\left(\frac{1800}{600}\right)\mathrm{J} - \frac{Q_2}{500} - \frac{Q_3}{400}\,.$$

Equivalently,

$$3\,\mathrm{J} = -\frac{Q_2}{500} - \frac{Q_3}{400}\,. \tag{4.143}$$

The solution of (4.142) and (4.143) is immediately found:

$$Q_2 = 1000\,\mathrm{J}\,, \qquad Q_3 = -2000\,\mathrm{J}\,. \tag{4.144}$$

Because the reservoirs give away heat energy, Q_1, Q_2, and Q_3, the change in their entropy is the following:

$$\Delta S_1 = -\frac{Q_1}{T_1} = -\frac{1800}{600} = -3\,\mathrm{J/K}\,,$$
$$\Delta S_2 = -\frac{Q_2}{T_2} = -\frac{1000}{500} = -2\,\mathrm{J/K}\,,$$
$$\Delta S_3 = -\frac{Q_3}{400} = \frac{2000}{400} = (+)5\,\mathrm{J/K}\,.$$

Problem 4.14 Provide an alternate solution for Problem 4.13.

Solution 4.14 *First Process.* Let the Carnot engine utilize first the warmer objects, I and II, as the hot and cold reservoirs.

We have already been told that the heat energy "added" to the working substance by the first reservoir, at temperature 600 K, is equal to 1800 J.

Let the heat energy "inserted into the working substance" by the second reservoir at temperature 500 K be Q_2^{A}.

Then the relevant Carnot requirement is as follows:

$$\left(\frac{1800}{600}\right)+\left(\frac{Q_2^{\mathrm{A}}}{500}\right)=0\,. \tag{4.145}$$

As a result, Q_2^{A} is

$$Q_2^{\mathrm{A}}=-1500\,\mathrm{J}\,. \tag{4.146}$$

In other words, at 500 K, by the end of the first process, the second reservoir will have "lost negative amount" of heat energy. This means that in actuality the second reservoir at temperature 500 K will have fattened its reserves by a positive amount of heat energy equal to +1500 J!

Second Process. At this juncture, we start the second process. We now remove Q_2^{B} from the second reservoir and insert it "into" the working substance at temperature $T_2 = 500$ K.

Additionally, we let the third reservoir also "add" Q_3 amount of heat energy to the working substance of the Carnot engine at temperature $T_3 = 400$ K.

The Carnot requirement is

$$\left(\frac{Q_2^{\mathrm{B}}}{500}\right)+\left(\frac{Q_3}{400}\right)=0\,.$$

This leads to the relationship

$$Q_2^{\mathrm{B}}+\left(\frac{5}{4}\right)Q_3=0\,. \tag{4.147}$$

Recall that we have been told in the statement of the problem that the total work done by the Carnot engine is $W = 800$ J.

We have also been told that 1800 J was extracted from reservoir I at temperature 600 K. As a result, all of that amount was added to the working substance at the same temperature.

Also, we know from (4.146) and (4.147) that first Q_2^{A} and later Q_2^{B} were added to the working substance by the reservoir II. And finally, an amount Q_3 was added to the working substance by the reservoir III. Therefore we have the equality

$$\text{total work done} = \text{total heat energy added to the working substance}\,.$$

Therefore, the total heat energy added by reservoirs I, II, and III is

$$\begin{aligned}\text{total work done} = 800\,\mathrm{J} &= \text{heat energy added to the working substance}\\ &\quad\ \text{by reservoirs I + II + III}\\ &= (1800\,\mathrm{J}) + \left(Q_2^{\mathrm{A}}+Q_2^{\mathrm{B}}\right)+Q_3\\ &= (1800\,\mathrm{J}) + \left(-1500\,\mathrm{J}+Q_2^{\mathrm{B}}\right)+Q_3\\ &= 300\,\mathrm{J}+Q_2^{\mathrm{B}}+Q_3\,.\end{aligned} \tag{4.148}$$

Rewrite the above as

$$Q_2^B + Q_3 = 500\,\text{J}\,. \tag{4.149}$$

Coupled (4.147) and (4.149) in the two unknowns Q_2^B and Q_3 are trivial to solve:

$$Q_2^B = 2500\,\text{J}\,, \qquad Q_3 = -2000\,\text{J}\,. \tag{4.150}$$

Our next task is to calculate the resultant change in the entropy of each of the reservoirs. This task is easily accomplished.

Reservoir I "gave away" 1800 J of heat energy at temperature 600 K. Therefore, its change in entropy, ΔS_I, is the following:

$$\Delta S_I = -\frac{1800}{600} = -3\ \text{J/K}\,.$$

Reservoir II "gave away" a total amount of $(Q_2^A + Q_2^B)$ of heat energy at temperature 500 K. Therefore, its change in entropy, ΔS_{II}, is

$$\Delta S_{II} = -\frac{(Q_2^A + Q_2^B)}{500} = -\left[\frac{-1500 + 2500}{500}\right] = -\left[\frac{1000}{500}\right] = -2\ \text{J/K}\,.$$

Finally, reservoir III "gave away" an amount $Q_3 = -2000$ J of heat energy at temperature 400 K. Therefore, its change in entropy is

$$S_{III} = -\left(\frac{Q_3}{400}\right) = \frac{2000}{400} = 5\ \text{J/K}\,.$$

Note that because $S_I + S_{II} + S_{III} = 0$, the Carnot requirement that the total change in the entropy be zero is satisfied.

Problem 4.15 (Two masses and reservoir)
A Carnot engine operates by exchanging heat energy with two finite objects and one very large heat energy reservoir. Each of the objects is of mass m and has heat capacity C_P.

The objects are initially at temperature T_1 and T_2, respectively. The large reservoir is at temperature T_0.

Calculate (1) the maximum amount of work, ΔW_{total}, performed by this Carnot engine, (2) its operating efficiency, ϵ, and (3) the change in the entropy of the two objects, $\Delta S_{objects}$, and that of the reservoir, $\Delta S_{reservoir}$.

Solution 4.15 The objects exchange heat energy with the reservoir. In the process their temperatures change from T_1 and T_2 to T_0, the temperature of the reservoir. As a result the total increase in their entropy, $\Delta S_{objects}$, is the following:

$$\Delta S_{objects} = mC_P\left[\int_{T_1}^{T_0}\frac{dT}{T} + \int_{T_1}^{T_0}\frac{dT}{T}\right] = mC_P\left[\ln\left(\frac{T_0}{T_1}\right) + \ln\left(\frac{T_0}{T_2}\right)\right]$$
$$= -2mC_P\ln\left(\frac{\sqrt{T_1T_2}}{T_0}\right). \tag{4.151}$$

In addition to the two objects, the Carnot engine also exchanges heat energy with the reservoir at temperature T_0. A simple method for determining the amount, $\Delta H_{\text{reservoir}}$, of such energy that actually gets added to the reservoir, is to calculate the change, $\Delta S_{\text{reservoir}}$, in the entropy of the reservoir.

To this purpose, we note two things: First, that the Carnot engine ensures that the total change, $\Delta S_{\text{universe}}$, in the entropy of the universe is vanishing. The universe here consists only of the reservoir and the two objects. Thus we get

$$\begin{aligned} \Delta S_{\text{universe}} &= \Delta S_{\text{reservoir}} + \Delta S_{\text{objects}} \\ &= \Delta S_{\text{reservoir}} - 2mC_P \ln\left(\frac{\sqrt{T_1 T_2}}{T_0}\right) = 0\,. \end{aligned} \tag{4.152}$$

Therefore,

$$\Delta S_{\text{reservoir}} = 2mC_P \ln\left(\frac{\sqrt{T_1 T_2}}{T_0}\right). \tag{4.153}$$

Accordingly, the heat energy, $\Delta H_{\text{reservoir}}$, added to the reservoir at temperature T_0 is

$$\Delta H_{\text{reservoir}} = T_0 \Delta S_{\text{reservoir}} = 2mC_P T_0 \ln\left(\frac{\sqrt{T_1 T_2}}{T_0}\right). \tag{4.154}$$

The second thing we note is that $\Delta H_{\text{reservoir}}$, the heat energy added to the reservoir, is equal to the difference between the heat energy, $\Delta H_{\text{objects}}$, introduced into the reservoir by the two objects,

$$\Delta H_{\text{objects}} = mC_P(T_1 + T_2 - 2T_0)\,,$$

and the total work, ΔW_{total}, done in the process, i.e.,

$$\Delta H_{\text{reservoir}} = \Delta H_{\text{objects}} - \Delta W_{\text{total}}\,.$$

Therefore,

$$\begin{aligned} \Delta W_{\text{total}} &= \Delta H_{\text{objects}} - \Delta H_{\text{reservoir}} \\ &= mC_P(T_1 + T_2 - 2T_0) - 2mC_P T_0 \ln\left(\frac{\sqrt{T_1 T_2}}{T_0}\right) \\ &= mC_P\left[(T_1 + T_2) - 2T_0 - 2T_0 \ln\left(\frac{\sqrt{T_1 T_2}}{T_0}\right)\right]. \end{aligned} \tag{4.155}$$

To calculate the operating efficiency, ϵ, of the engine, we need to divide ΔW_{total} by the amount of heat energy withdrawn from the two external sources. That is,

$$\epsilon = \frac{\Delta W_{\text{total}}}{mC_P[(T_1 + T_2) - 2T_0]} = 1 - \left[\frac{2T_0 \ln(\frac{\sqrt{T_1 T_2}}{T_0})}{T_1 + T_2 - 2T_0}\right]. \tag{4.156}$$

Problem 4.16 Alternate Solution of Problem 4.15.

Solution 4.16 Alternate solution of Problem 4.15 is given below.

Process I Let us treat first the two finite objects as the hot and cold energy sources for the Carnot engine.

After the Carnot engine has been run with these two objects, and the Carnot specified work has been done, the temperatures of the two objects reach a common value, T_c. A convenient procedure for calculating this common temperature, T_c, is via a calculation of the relevant entropy change, ΔS_c, of the two objects:

$$\Delta S_c = \Delta S_1 + \Delta S_2 = mC_P\left[\int_{T_1}^{T_c}\frac{dT}{T} + \int_{T_2}^{T_c}\frac{dT}{T}\right] = mC_P\left[\ln\frac{T_c}{T_1} + \ln\frac{T_c}{T_2}\right]$$
$$= mC_P\ln\left(\frac{T_c^2}{T_1T_2}\right) = 2mC_P\ln\left(\frac{T_c}{\sqrt{T_1T_2}}\right). \qquad (4.157)$$

Because the entropy is conserved in each Carnot cycle that is completed, $\Delta S_c = 0$. As a result,

$$\ln\left(\frac{T_c}{\sqrt{T_1T_2}}\right) = 0 = \ln(1).$$

This gives

$$T_c = \sqrt{T_1T_2}. \qquad (4.158)$$

The work done, ΔW_{first}, in the first process is equal to the loss of heat energy in the "composite body," which constitutes the two objects 1 and 2. That is,

$$\Delta W_{\text{first}} = mC_P\left[(T_1 + T_2) - 2T_c\right] = mC_P\left[(T_1 + T_2) - 2\sqrt{T_1T_2}\right]. \qquad (4.159)$$

Process II The second process treats the composite body of mass $2m$ as one thermal source, at temperature T_c, and the reservoir, at temperature T_0, as the other thermal source.

The temperature of the composite body, sometime during the course of this operation, is T.

At this temperature, T, let the composite body reversibly contribute an infinitesimal amount of heat energy, $dQ_{\text{body}}^{\text{rev}}$, to the working substance of the engine. Similarly, let $dQ_{\text{res}}^{\text{rev}}$ be the infinitesimal amount of heat energy reversibly contributed by the reservoir to the working substance. Note that because the reservoir is very large, its temperature, T_0, remains unchanged.

As a result of these two tiny additions of heat energy to the working substance, the infinitesimal work, dW_{second}, done by the Carnot engine in the second process is equal to the heat energy input into the working substance. That is,

$$dW_{\text{second}} = dQ_{\text{body}}^{\text{rev}} + dQ_{\text{res}}^{\text{rev}}. \qquad (4.160)$$

The Carnot requirement for this infinitesimal cycle is

$$\frac{dQ_{\text{body}}^{\text{rev}}}{T} + \frac{dQ_{\text{res}}^{\text{rev}}}{T_0} = 0. \qquad (4.161)$$

Using (4.160), it becomes

$$\frac{dQ_{\text{body}}^{\text{rev}}}{T} + \left[\frac{dW_{\text{second}} - dQ_{\text{body}}^{\text{rev}}}{T_0}\right] = 0\,. \tag{4.162}$$

Clearly, the withdrawal of heat energy $dQ_{\text{body}}^{\text{rev}}$ from the composite body causes its temperature to decrease by an infinitesimal amount dT.

For simplicity, let us introduce the notation $M = 2m\,C$. As a result, we can write

$$dQ_{\text{body}}^{\text{rev}} = -M dT\,. \tag{4.163}$$

Using (4.161) and (4.162), we can write

$$\begin{aligned}\frac{dW_{\text{second}}}{T_0} &= -\frac{dQ_{\text{body}}^{\text{rev}}}{T} + \frac{dQ_{\text{body}}^{\text{rev}}}{T_0} \\ &= -dQ_{\text{body}}^{\text{rev}}\left(\frac{1}{T} - \frac{1}{T_0}\right) = M\left(\frac{dT}{T} - \frac{dT}{T_0}\right).\end{aligned} \tag{4.164}$$

Because the path is reversible, we can integrate dW_{second} to calculate the work done during the second process:

$$\int_{\text{initial}}^{\text{final}} dW_{\text{second}} = W_{\text{second}} = M T_0 \int_{T_{\text{initial}}}^{T_{\text{final}}} \left(\frac{dT}{T}\right) - M \int_{T_{\text{initial}}}^{T_{\text{final}}} dT\,. \tag{4.165}$$

Recalling that the variable T denotes the temperature of the composite body of mass $2m$, for which $M = 2mC_{\text{P}}$, in the above equation $T_{\text{initial}} = T_{\text{c}}$ and $T_{\text{final}} =$ temperature of reservoir $= T_0$, the work done is

$$\begin{aligned}W_{\text{second}} &= M T_0 \int_{T_{\text{c}}}^{T_0} \left(\frac{dT}{T}\right) - M \int_{T_{\text{c}}}^{T_0} dT \\ &= M T_0 \ln\left(\frac{T_0}{T_{\text{c}}}\right) - M(T_0 - T_{\text{c}}) \\ &= M\left[(T_{\text{c}} - T_0) - T_0 \ln\left(\frac{T_{\text{c}}}{T_0}\right)\right].\end{aligned} \tag{4.166}$$

The total amount of work, ΔW_{total}, done is the sum of that performed during the two processes. That is,

$$\begin{aligned}\Delta W_{\text{total}} &= \Delta W_{\text{first}} + \Delta W_{\text{second}} \\ &= mC_{\text{P}}\left[(T_1 + T_2) - 2\sqrt{T_1 T_2}\right] + M\left[(T_{\text{c}} - T_0) - T_0 \ln\left(\frac{T_{\text{c}}}{T_0}\right)\right] \\ &= mC_{\text{P}}\left[(T_1 + T_2) - 2\sqrt{T_1 T_2}\right] \\ &\quad + 2mC_{\text{P}}\left[(\sqrt{T_1 T_2} - T_0) - T_0 \ln\left(\frac{\sqrt{T_1 T_2}}{T_0}\right)\right] \\ &= mC_{\text{P}}\left[(T_1 + T_2) - 2T_0 - 2T_0 \ln\left(\frac{\sqrt{T_1 T_2}}{T_0}\right)\right].\end{aligned} \tag{4.167}$$

To calculate the operating efficiency, ϵ, of the engine, we need to divide ΔW_{total} by the amount of heat energy withdrawn from the two external sources. That is,

$$\epsilon = \frac{\Delta W_{\text{total}}}{mC_P[(T_1 + T_2) - 2T_0]} = 1 - \left[\frac{2T_0 \ln(\frac{\sqrt{T_1 T_2}}{T_0})}{T_1 + T_2 - 2T_0}\right]. \tag{4.168}$$

Let us next calculate the change in the entropy of the reservoir. To this purpose, we need first to calculate the amount of heat energy, $\Delta H_{\text{reservoir}}$, that actually gets added to the reservoir at its fixed temperature, T_0.

Clearly, $\Delta H_{\text{reservoir}}$, is the difference between the total amount of heat energy added to the reservoir by the two objects and the work done in the process. That is,

$$\begin{aligned}\Delta H_{\text{reservoir}} &= mC_P(T_1 + T_2 - 2T_0) - \Delta W_{\text{total}} \\ &= 2mC_P T_0 \ln\left(\frac{\sqrt{T_1 T_2}}{T_0}\right).\end{aligned} \tag{4.169}$$

Therefore, the increase in the entropy of the reservoir is

$$\Delta S_{\text{reservoir}} = \frac{2mC_P T_0 \ln(\frac{\sqrt{T_1 T_2}}{T_0})}{T_0} = 2mC_P \ln\left(\frac{\sqrt{T_1 T_2}}{T_0}\right). \tag{4.170}$$

Knowing that the objects start off at temperatures T_1 and T_2 and eventually reach the temperature T_0 of the large reservoir, the increase in the entropy, $\Delta S_{\text{objects}}$, of the two objects is readily calculated as

$$\begin{aligned}\Delta S_{\text{objects}} &= mC_P\left[\int_{T_1}^{T_0} \frac{\mathrm{d}T}{T} + \int_{T_1}^{T_0} \frac{\mathrm{d}T}{T}\right] = mC_P\left[\ln\left(\frac{T_0}{T_1}\right) + \ln\left(\frac{T_0}{T_1}\right)\right] \\ &= -2mC_P \ln\left(\frac{\sqrt{T_1 T_2}}{T_0}\right).\end{aligned} \tag{4.171}$$

Carnot's requirement that the overall entropy be conserved would demand $\Delta S_{\text{universe}}$ to be equal to zero. We find this is indeed the case. That is,

$$\begin{aligned}\Delta S_{\text{universe}} &= \Delta S_{\text{reservoir}} + \Delta S_{\text{objects}} \\ &= 2mC_P \ln\left(\frac{\sqrt{T_1 T_2}}{T_0}\right) - 2mC_P \ln\left(\frac{\sqrt{T_1 T_2}}{T_0}\right) = 0.\end{aligned} \tag{4.172}$$

Problem 4.17 (Operating a Carnot engine)
Two identical objects, each of mass 0.5 mol, contained in an adiabatic enclosure which is maintained at constant pressure, are known to have molar specific heat

$$C_p(T) = a + bT\,, \quad a = 1\ \text{J/(mol K)}\,, \quad b = 0.002\ \text{J/(mol K}^2)\,,$$

in the temperature range 300–400 K. Given initial temperatures 300 K and 400 K, calculate their final temperature, T_f, after the bodies have been brought into thermal contact. If a perfect Carnot engine, which utilizes these objects as finite hot and cold thermal sources, is available for extracting maximum work, W, calculate W and the final temperature, T_f', achieved by the objects.

Solution 4.17 Upon thermal contact, the heat energy lost by the warmer object is equal to that gained by the colder one. Assume each of these objects has mass m mol. Then the heat energy loss is

$$m \int_{T_f}^{400} C_p dT = ma(400 - T_f) + \frac{mb}{2}\left[(400)^2 - (T_f)^2\right].$$

Similarly, the heat energy gain is

$$m \int_{300}^{T_f} C_p dT = ma(T_f - 300) + \frac{mb}{2}\left[(T_f)^2 - (300)^2\right].$$

Equating these two yields a quadratic equation

$$a(2T_f - 700) + b\left[(T_f)^2 - \frac{(300)^2 + (400)^2}{2}\right] = 0,$$

which is readily solved. For given values of a and b, it reduces to

$$(T_f)^2 + 1000T_f - 4.75 \times (10)^5 = 0, \tag{4.173}$$

yielding

$$T_f = 351.469 \text{ K}.$$

The calculation for the temperature $T_{f'}$—which is reached by the two bodies when the Carnot engine has produced the maximum possible work—proceeds as follows: Consider at some instant the temperatures of the two objects are T_h' and T_c'. Heat energy, dQ_h^{rev}, is extracted from the warmer body and is reversibly fed into the perfect Carnot engine at the same temperature T_h'. Similarly, heat energy dQ_c^{rev} is extracted from the colder body and reversibly added to the perfect Carnot engine at the same temperature T_c'. As always, the use of the perfect Carnot engine dictates that the total entropy change is vanishing. Consequently, we must have

$$\int_{400}^{T_f'} \left[\frac{dQ_h^{rev}}{T_h'}\right] dT_h' + \int_{300}^{T_f'} \left[\frac{dQ_c^{rev}}{T_c'}\right] dT_c' = 0, \tag{4.174}$$

which means

$$\int_{400}^{T_f'} \left[\frac{m(a + bT_h')}{T_h'}\right] dT_h' + \int_{300}^{T_f'} \left[\frac{m(a + bT_c')}{T_c'}\right] dT_c'$$
$$= m\left[a \ln\left(\frac{T_f'}{400}\right) + b(T_f' - 400)\right] m\left[a \ln\left(\frac{T_f'}{300}\right) + b(T_f' - 300)\right] = 0. \tag{4.175}$$

Simple manipulation leads to the expression

$$\ln\left(\frac{T_f'^2}{300 \times 400}\right) = \left(\frac{b}{a}\right)(300 + 400 - 2T_{f'}), \tag{4.176}$$

which can be recast as

$$\frac{T_f'}{\sqrt{300 \times 400}} = \exp\left[\left(\frac{b}{a}\right)(350 - T_f')\right]. \tag{4.177}$$

Outwardly, this transcendental expression looks quite fierce. But because for any reasonable choice for T_f',

$$\left[\left(\frac{b}{a}\right)(350 - T_f')\right] \ll 1,$$

the exponential can be expanded in powers of the exponent. That is, we can approximate $\exp(x)$ as $1 + x + O(x^2)$ when $x \ll 1$. This leads to a linear equation for T_f' with the result

$$T_f' = 100\sqrt{12}\left(\frac{1.7}{1 + 0.2\sqrt{12}}\right) = 347.879\ \text{K}. \tag{4.178}$$

Consideration of the quadratic term, $\frac{x^2}{2}$, in the expansion of $\exp(x)$ gives

$$T_f' = 347.881\ \text{K},$$

which hardly changes the earlier result. Clearly, inclusion of any additional terms in the expansion for $\exp(x)$ is unnecessary.

In conclusion, we note that as always $T_f > T_f'$. The difference between the heat energy content of the masses at the two temperatures is equal to the positive work, W, performed by the Carnot engine. That is,

$$W = 2m \int_{T_f'}^{T_f} dT\, C_p(T) = 2m \int_{T_f'}^{T_f} dT\,(a + bT)$$
$$= 2ma(T_f - T_f') + mb(T_f^2 - T_f'^2) = 6.10\ \text{J}. \tag{4.179}$$

It is interesting to also try another method for calculating W. Such a method is based on the direct integration of the energy conservation relationship.

For some temperature T', intermediate between 300 K and 400 K, elementary work done by the engine dW is equal to the difference between the positive amount of heat energy dQ_h, extracted from the warmer body, and the smaller positive amount of heat energy dQ_c, rejected to the colder body. Thus

$$W = \int dW = \int (dQ_h - dQ_c)$$
$$= m \int_{T_f'}^{400} dT' C_p(T') - m \int_{300}^{T_f'} dT' C_p(T')$$
$$= ma(700 - 2T_f') + \frac{mb}{2}\left[(300)^2 + (400)^2 - 2T_f'^2\right]. \tag{4.180}$$

A little algebra confirms the earlier result recorded in (4.179).

4.13 Carnot Refrigerator

A refrigerator, or equivalently an air-conditioner, is used for extracting positive amount of heat energy, $|Q_{\text{cold}}|$, from an object, or a room. Let us refer to such an object or a room as a cold reservoir[24] at temperature T_{cold}. Let us assume the positive amount of work required for this extraction is $|W|$.

Naturally, an efficient refrigerator/air-conditioner would produce a lot of cooling for a given amount of effort. That is, the ratio of heat energy extracted, $|Q_{\text{cold}}|$, to the work, $|W|$, required for its extraction, should be as large as possible. This ratio is the refrigerator's—or equivalently, an air-conditioner's—coefficient of performance, COP. That is,

$$COP = \frac{|Q_{\text{cold}}|}{|W|}. \tag{4.181}$$

Owing to the principle of conservation of energy enshrined in the statement of the first law, the total amount of positive heat energy, $|Q_{\text{hot}}|$, rejected to the hot reservoir at temperature T_{hot} is the sum of the corresponding positive amount of heat energy, $|Q_{\text{cold}}|$, extracted from the cold reservoir and the externally supplied positive work input, W. That is,

$$|Q_{\text{hot}}| = |W| + |Q_{\text{cold}}|, \quad \text{or}$$
$$|W| = |Q_{\text{hot}}| - |Q_{\text{cold}}|.$$

Thus

$$COP = \frac{|Q_{\text{cold}}|}{|Q_{\text{hot}}| - |Q_{\text{cold}}|}. \tag{4.182}$$

The foregoing applies to all cyclic refrigerators/air-conditioners. If all the processes are performed reversibly, we can employ a "perfect Carnot refrigerator" to provide the cooling. Accordingly, the perfect Carnot equality would hold. That is,[25]

$$\frac{|Q_{\text{cold}}|}{|Q_{\text{hot}}|} = \frac{T_{\text{cold}}}{T_{\text{hot}}}. \tag{4.184}$$

---

[24] Note that an object being cooled inside a refrigerator is a finite cold reservoir.

[25] Note that (4.184) below is exactly equivalent to the perfect Carnot equality recorded in (4.2), namely

$$\frac{Q^{\text{rev}}(T_{\text{C}})}{Q^{\text{rev}}(T_{\text{H}})} = -\left(\frac{T_{\text{C}}}{T_{\text{H}}}\right). \tag{4.183}$$

The appropriate translation of (4.184) in terms of the notation used in (4.183) is as follows: Because $|Q_{\text{hot}}|$ is rejected to the hot reservoir, an equal amount of heat energy is extracted from the working substance at temperature T_{H}. Thus the heat energy reversibly added to the working substance at temperature T_{H} is $Q^{\text{rev}}(T_{\text{H}}) = -|Q_{\text{hot}}|$. On the other hand, because $|Q_{\text{cold}}|$ is extracted from the cold object/reservoir, it is equal to $Q^{\text{rev}}(T_{\text{C}})$ added to the working substance at temperature T_{C}. Thus $Q^{\text{rev}}(T_{\text{C}}) = +|Q_{\text{cold}}|$.

Using (4.184) in (4.182) gives

$$(COP)_{\text{carnot}} = \left[\frac{T_{\text{cold}}}{T_{\text{hot}} - T_{\text{cold}}}\right]. \tag{4.185}$$

Note that unlike the efficiency of a work producing perfect Carnot engine—which is limited to being less than 100%—the coefficient of performance of a perfect Carnot refrigerator has no such limitation. If T_{cold} is not too cold and the difference ($T_{\text{hot}} - T_{\text{cold}}$) is relatively small, its $COP \times 100\%$ can be much larger than 100%.

Problem 4.18 (Work needed for cooling)
A mass m, with specific heat C_P, has been placed in a large laboratory room at temperature T_H.

(a) What is the minimum amount of work needed by a refrigerator to cool this mass to temperature T_C?
(b) Calculate the overall coefficient of performance of this refrigerator.

Note that the pressure is kept constant during the entire course of this operation and C_P is assumed to be independent of temperature.

Solution 4.18 (a) The most efficient machine for performing this task is, of course, the perfect Carnot refrigerator. As before, we consider a situation where the mass m is at some temperature T intermediate between the room temperature T_H and the desired final low temperature T_C. An infinitesimal amount of heat energy $|dQ_1|$ is extracted from the mass m at this temperature and is *introduced into* the working substance of the refrigerator. In order to help transfer this heat energy to the atmosphere in the room at the higher temperature T_H, it is also necessary, for some outside agency, to perform positive work $|dW|$. Such work gets added into the mix. As a result, the heat energy, $|dQ_2|$, actually transferred to the room is the sum of the two. That is,

$$|dQ_2| = |dW| + |dQ_1|. \tag{4.186}$$

The extraction of heat energy $|dQ_1|$ from the mass m causes a change dT in its temperature,

$$|dQ_1| = -mC_p dT. \tag{4.187}$$

Because of the reversible operation of the perfect Carnot machine, the total entropy of the universe—consisting of the mass and the room—remains unchanged,

$$\frac{|dQ_1|}{T} = \frac{|dQ_2|}{T_H}. \tag{4.188}$$

Using (4.188), $|dQ_2|$ is readily eliminated from (4.186). Next replacing $|dQ_1|$ by the right-hand side of (4.187) gives

$$-mC_P T_H\left(\frac{dT}{T}\right) = |dW| - mC_P dT. \tag{4.189}$$

Upon integrating both sides, we get

$$
\begin{aligned}
-mC_{\mathrm{P}}T_{\mathrm{H}}\int_{T_{\mathrm{H}}}^{T_{\mathrm{C}}}\left(\frac{\mathrm{d}T}{T}\right) &= \int_{T_{\mathrm{H}}}^{T_{\mathrm{C}}}|\mathrm{d}W| - mC_{\mathrm{P}}\int_{T_{\mathrm{H}}}^{T_{\mathrm{C}}}\mathrm{d}T \\
-mC_{\mathrm{P}}T_{\mathrm{H}}\ln(T_{\mathrm{C}}/T_{\mathrm{H}}) &= |W| - mC_{\mathrm{P}}(T_{\mathrm{C}} - T_{\mathrm{H}}),
\end{aligned} \tag{4.190}
$$

where we have used the fact that the initial temperature is T_{H} and the final temperature is T_{C}. This gives

$$
|W| = mC_{\mathrm{P}}\left[T_{\mathrm{H}}\ln\left(\frac{T_{\mathrm{H}}}{T_{\mathrm{C}}}\right) - (T_{\mathrm{H}} - T_{\mathrm{C}})\right], \tag{4.191}
$$

which is the minimum work needed for reducing the temperature of mass m from its initial value T_{H} to its final value T_{C}.

Problem 4.19 (Entropy change on cooling)
Consider an object of mass m with temperature dependent specific heat $C_{\mathrm{P}}(T) = A + BT + DT^2 + ET^3$, where A, B, D, and E are all positive constants. The object is initially at temperature T_{H} K. A perfect Carnot engine, acting as a refrigerator, is used to cool the object to a lower temperature T_{C} K. This process occurs at constant pressure. Calculate (I) the total energy input $|W|$ into the perfect Carnot engine for accomplishing this task and (II) the change of entropy of the working substance during this process.

Solution 4.19 Consider a small amount of heat energy, $|\mathrm{d}Q_1|$, which is extracted quasistatically from the object, at some temperature T K, intermediate between T_{H} and T_{C}, and is transferred to the reservoir by the use of a perfect Carnot refrigerator. This changes the entropy of the object by an amount

$$
\mathrm{d}S_{\mathrm{object}} = -\frac{|\mathrm{d}Q_1|}{T}. \tag{4.192}
$$

In order to extract this heat energy and eventually transfer it to the thermal reservoir at the higher temperature T_{H}, some external agent has to be called upon to do $|\mathrm{d}W|$ amount of work. Note that this work is usually provided by the electric power that drives the refrigerator. As a result, the total amount of energy equivalent that is quasistatically shifted—that is, added—to the reservoir is $|\mathrm{d}Q_2|$,

$$
|\mathrm{d}Q_2| = |\mathrm{d}Q_1| + |\mathrm{d}W|. \tag{4.193}
$$

The resultant change in the entropy of the reservoir, therefore, is

$$
\mathrm{d}S_{\mathrm{reservoir}} = \frac{|\mathrm{d}Q_2|}{T_{\mathrm{H}}} = \frac{|\mathrm{d}Q_1| + |\mathrm{d}W|}{T_{\mathrm{H}}}. \tag{4.194}
$$

Because the perfect Carnot refrigerator operates reversibly the total increase in the entropy of the universe, composed of the object and the reservoir, is zero. Therefore

$$\begin{aligned} \mathrm{d}S_{\text{object}} + \mathrm{d}S_{\text{reservoir}} &= -\frac{|\mathrm{d}Q_1|}{T} + \frac{|\mathrm{d}Q_1| + |\mathrm{d}W|}{T_{\mathrm{H}}} \\ &= |\mathrm{d}Q_1|\left[\frac{1}{T_{\mathrm{H}}} - \frac{1}{T}\right] + \frac{|\mathrm{d}W|}{T_{\mathrm{H}}} \\ &= 0\,. \end{aligned} \tag{4.195}$$

Equation (4.195) represents two relationships:

$$\mathrm{d}S_{\text{object}} = -\frac{|\mathrm{d}Q_1|}{T} \quad \text{and} \quad |\mathrm{d}W| = -|\mathrm{d}Q_1| + T_{\mathrm{H}}\left(\frac{|\mathrm{d}Q_1|}{T}\right). \tag{4.196}$$

The loss of heat energy $|\mathrm{d}Q_1|$ results in decreasing the temperature of the object, i.e.,

$$|\mathrm{d}Q_1| = -mC_{\mathrm{P}}(T)\mathrm{d}T\,.$$

Whence, from (4.196),

$$\begin{aligned} \mathrm{d}S_{\text{object}} &= \frac{mC_{\mathrm{P}}(T)\mathrm{d}T}{T} \quad \text{and} \\ \frac{|\mathrm{d}W|}{m} &= C_{\mathrm{P}}(T)\mathrm{d}T - T_{\mathrm{H}}C_{\mathrm{P}}(T)\left(\frac{\mathrm{d}T}{T}\right). \end{aligned} \tag{4.197}$$

The integration of both sides between the initial and the final temperatures, i.e.,

$$\begin{aligned} \int_{T_{\mathrm{H}}}^{T_{\mathrm{C}}} \mathrm{d}S_{\text{object}} &= m\int_{T_{\mathrm{H}}}^{T_{\mathrm{C}}} \frac{C_{\mathrm{P}}(T)\mathrm{d}T}{T} \\ \frac{1}{m}\int_{T_{\mathrm{H}}}^{T_{\mathrm{C}}} |\mathrm{d}W| &= \int_{T_{\mathrm{H}}}^{T_{\mathrm{C}}} C_{\mathrm{P}}(T)\mathrm{d}T - T_{\mathrm{H}}\int_{T_{\mathrm{H}}}^{T_{\mathrm{C}}} C_{\mathrm{P}}(T)\left(\frac{\mathrm{d}T}{T}\right), \end{aligned} \tag{4.198}$$

leads to

$$\begin{aligned} -\frac{S_{\text{object}}}{m} &= A\ln\left(\frac{T_{\mathrm{H}}}{T_{\mathrm{C}}}\right) + B(T_{\mathrm{H}} - T_{\mathrm{C}}) + D\frac{(T_{\mathrm{H}}^2 - T_{\mathrm{C}}^2)}{2} + E\frac{(T_{\mathrm{H}}^3 - T_{\mathrm{C}}^3)}{3} \\ \text{and} \quad \frac{|W|}{m} &= A\left[T_{\mathrm{H}}\ln\left(\frac{T_{\mathrm{H}}}{T_{\mathrm{C}}}\right) - (T_{\mathrm{H}} - T_{\mathrm{C}})\right] \\ &\quad + B\left[T_{\mathrm{H}}(T_{\mathrm{H}} - T_{\mathrm{C}}) - \left(\frac{T_{\mathrm{H}}^2 - T_{\mathrm{C}}^2}{2}\right)\right] \\ &\quad + D\left[T_{\mathrm{H}}\left(\frac{T_{\mathrm{H}}^2 - T_{\mathrm{C}}^2}{2}\right) - \left(\frac{T_{\mathrm{H}}^3 - T_{\mathrm{C}}^3}{3}\right)\right] \\ &\quad + E\left[T_{\mathrm{H}}\left(\frac{T_{\mathrm{H}}^3 - T_{\mathrm{C}}^3}{3}\right) - \left(\frac{T_{\mathrm{H}}^4 - T_{\mathrm{C}}^4}{4}\right)\right]. \end{aligned} \tag{4.199}$$

It is important to be assured that the increase in the entropy of the object that has been cooled is negative and the total work input for running the refrigerator is positive. This assurance is provided by the following facts:

Because $T_H \geq T_C$, the terms proportional to the coefficients B, D, and E are either equal to zero or they are readily seen to be manifestly positive. For the remaining terms that are proportional to A, clearly, $\ln(\frac{T_H}{T_C}) \geq 0$. Also, as demonstrated in (4.62)–(4.67), the following is true:

$$\left[T_H \ln\left(\frac{T_H}{T_C}\right) - (T_H - T_C)\right] \geq 0 .$$

(To check this, use the transliteration $T_{res} \rightarrow T_H$ and $T_{cold} \rightarrow T_C$ in (4.62–(4.67).)

Problem 4.20 (Effort needed for running Carnot heat pump)
For a Carnot heat energy pump show $\epsilon_{\text{heat energy pump}} = 1 + COP$.

Solution 4.20 A Carnot engine is like the proverbial Robin Hood. It robs the rich—that is, takes heat energy out of the hot body—and feeds the poor—i.e., puts part of it into a colder body. The remainder, of course, helps run the enterprize: that is, it produces useful work.

There are, of course, the anti-Robin Hoods—the Robber Barons—who do the opposite. Robbing even the poor requires some effort. But we need not shed any tears for the Barons because generally their loot is a lot greater than the effort expended! And that is how it is for a heat energy pump.

The objective of the exercise here is to keep a room warm during the winter. Normally, when a steady state is reached, the rate of loss of heat energy from the room through conduction, radiation, air leaks, etc., is equal to the rate at which the heat energy is being supplied to the room to maintain its temperature at the specified level.

A heat energy pump—being an air conditioner run in reverse—can be used to provide this heat energy for great savings in fuel bills. Let us see how this comes about. For simplicity, we employ a perfect Carnot machine.

The two temperature reservoirs are the room at temperature T_H and the outside atmosphere at temperature T_C. Let a positive amount of heat energy, $|Q_C|$, be extracted from the atmosphere and the energy used by the heat energy pump in the process be $|E|$. Then, according to the first law, heat energy $|Q_H|$ will be transferred to the room,

$$|Q_H| = |Q_C| + |E| .$$

Clearly, the relevant efficiency parameter is the ratio of the heat energy transferred to the room per unit input of energy. That is,

$$\epsilon_{\text{heat energy pump}} = \frac{|Q_H|}{|E|} = \frac{|Q_H|}{|Q_H| - |Q_C|} = \left[1 - \left(\frac{|Q_C|}{|Q_H|}\right)\right]^{-1} .$$

Therefore, for a perfect Carnot heat energy pump operating on the reversible perfect Carnot cycle,

$$\epsilon_{\text{heat energy pump–carnot}} = \left[1 - \left(\frac{T_C}{T_H}\right)\right]^{-1} .$$

In a temperate climate, during the winter $T_C \approx 50\,°F = 283$ K. Assuming the preferred room temperature is $T_H \approx 70\,°F = 294$ K, the idealized $\epsilon_{\text{heat energy pump–carnot}}$ would be $\approx 2,670\%$!

Of course, this estimate is far too optimistic for realistic systems. Perhaps a number that is 10 times smaller would be more realistic. The downside is the cost and the wear and tear of the equipment. Both the initial and maintenance costs may be high. But often the cycle direction of the heat energy pump can be reversed so that the equipment can also be used as an air conditioner during the summer.

In Philadelphia, winter temperatures can reach $\sim$ 253 K, and one often needs room temperature to be $\sim$ 295 K. Accordingly, even the idealized $\epsilon_{\text{heat energy pump–carnot}}$ would be only $\sim$ 700%. Realistically, if one could use a heat energy pump under these conditions, its operating efficiency would be about 150–200%. In practice, the very cold temperatures cause problems with the machinery and heat energy pumps are not a popular choice.

Problem 4.21 (Entropy increase on removing temperature gradient)
A metal rod of uniform density ρ, cross-section area A, and specific heat C has length L. Initially, its two ends are exposed to two very large heat energy reservoirs maintained at temperatures T_1 and T_2. (Note that either of these temperatures may be higher than the other.) The environment surrounding of the rod is so arranged that when steady state is reached the temperature gradient along the rod becomes constant. At this time the temperature reservoirs are removed and the entire rod is surrounded by adiabatic walls. In this initial state, the entropy of the rod is equal to S_0.

Calculate the entropy, S_{eq}, of the rod when an equilibrium state is reached. In this equilibrium state, assume the temperature along the rod is uniform.

Solution 4.21 Assume the rod is laid along the x-axis and extends from $x = 0$ to $x = L$. Further, the initial temperature at $x = 0$ is T_1. Therefore, the constancy of the temperature gradient ensures that initially the temperature at position x is

$$T_{\text{initial}} = T_1 + \left(\frac{T_2 - T_1}{L}\right)x . \tag{4.200}$$

At position x, choose a slice of length Δx. Its mass is

$$\Delta M = \rho A \Delta x . \tag{4.201}$$

Now, reversibly transfer an infinitesimal amount of heat energy $|\mathrm{d}Q|$ to the slice. This process will cause the temperature of the slice to rise by $\mathrm{d}T$. That is,

$$|\mathrm{d}Q| = (C \cdot \Delta M)\mathrm{d}T . \tag{4.202}$$

Accordingly, its entropy will increase by an amount dS, where

$$\mathrm{d}S = \frac{|\mathrm{d}Q|}{T} = (C \cdot \Delta M)\frac{\mathrm{d}T}{T}\,. \tag{4.203}$$

In reaching equilibrium, the total increase in the entropy of the slice,[26] (ΔS), would therefore be the following:

$$(\Delta S) = (C \cdot \Delta M)\int_{T_{\text{initial}}}^{T_{\text{final}}} \frac{\mathrm{d}T}{T} = (C \cdot \Delta M)\ln\left(\frac{T_{\text{final}}}{T_{\text{initial}}}\right). \tag{4.204}$$

Inserting the values of T_{initial} from (4.200), and ΔM from (4.201), into (4.204) leads to

$$\Delta S) = (C \cdot \rho A \Delta x)\int_{T_{\text{initial}}}^{T_{\text{final}}} \frac{\mathrm{d}T}{T} = (C \cdot \rho A \Delta x)\ln\left[\frac{T_{\text{final}}}{T_1 + (\frac{T_2 - T_1}{L})x}\right]. \tag{4.205}$$

In order to calculate the grand total of the entropy increase, $S_{\text{eq}} - S_0$, we need to sum, from $x = 0$ to $x = L$, both sides of (4.205). That is,

$$S_{\text{eq}} - S_0 \approx \sum_{x=0}^{L}(\Delta S) \approx \sum_{x=0}^{L}\left\{(C \cdot \rho A \Delta x)\ln\left[\frac{T_{\text{final}}}{T_1 + (\frac{T_2 - T_1}{L})x}\right]\right\}.$$

A more accurate way of calculating the entropy increase, $S_{\text{eq}} - S_0$, is to work with very small (ΔS)'s and convert all of these sums into the corresponding integrals. For convenience, we divide both sides by the factor $C\rho A$ and so

$$\begin{aligned}\frac{S_{\text{eq}} - S_0}{C\rho A} &= \int_0^L \mathrm{d}x \ln\left[\frac{T_{\text{final}}}{T_1 + (\frac{T_2 - T_1}{L})x}\right] \\ &= \int_0^L \mathrm{d}x \ln(T_{\text{final}}) - \int_0^L \mathrm{d}x \ln\left[T_1 + \left(\frac{T_2 - T_1}{L}\right)x\right] \\ &= L\ln(T_{\text{final}}) - L\ln T_1 - D\,, \end{aligned} \tag{4.206}$$

where

$$D = \int_0^L \mathrm{d}x \ln(1 + Bx) \tag{4.207}$$

and

$$B = \frac{T_2 - T_1}{LT_1}\,. \tag{4.208}$$

It is convenient to change the variable

$$y = 1 + Bx\,, \tag{4.209}$$

[26]Note that entropy is an extensive variable. So any increase in entropy is proportional to the system size. Here, a convenient measure of the size of the slice is its mass ΔM. Therefore, the increase in the entropy of the slice ΔS is proportional to the mass ΔM of the slice.

whereby

$$dx = \frac{dy}{B}$$

and

$$D = \frac{\int_1^{1+BL} dy \ln y}{B} = \frac{(y \ln y - y)}{B}\bigg|_1^{1+BL} = \left(\frac{LT_2}{T_2 - T_1}\right)\ln\left(\frac{T_2}{T_1}\right) - L\,. \quad (4.210)$$

Inserting the results of (4.210) into (4.206) yields

$$S_{\rm eq} - S_0 = C\rho AL\left[1 + \ln(T_{\rm final}) - \ln T_1 - \left(\frac{T_2}{T_2 - T_1}\right)\ln\left(\frac{T_2}{T_1}\right)\right]. \quad (4.211)$$

The entropy increase must not be affected by the interchange of the subscripts 1 and 2. This is dictated by the physics of the problem because it matters not whether at the start the right- or left-end is at temperature T_2 or T_1.

Before testing for this subscript-interchange invariance, we need to know $T_{\rm final}$. Equilibrium is reached when all the excess heat energy from the warmer parts of the rod has been transferred to the colder parts. This process continues until the temperature of the rod becomes uniform. Because of the reflection symmetry of the rod about its mid-point, the temperature of the mid-point remains constant during equilibration. And this temperature is equal to the average of the original temperatures of the two ends, i.e.,

$$T_{\rm final} = \frac{T_1 + T_2}{2}\,. \quad (4.212)$$

Clearly, $T_{\rm final}$ does satisfy the required invariance. But what about the rest of (4.211)?

The subscript interchange invariance becomes more transparent if we re-cast (4.211) into the following form:

$$\frac{S_{\rm eq} - S_0}{C\rho AL} = 1 + \ln\left(\frac{T_1 + T_2}{2}\right) - \ln T_1 - \left(\frac{T_2}{T_2 - T_1}\right)(\ln T_2 - \ln T_1)$$
$$= 1 + \ln\left(\frac{T_1 + T_2}{2}\right) + \left(\frac{1}{T_2 - T_1}\right)(T_1 \ln T_1 - T_2 \ln T_2)\,. \quad (4.213)$$

Clearly, interchanging subscript 1 with 2 leaves the result unaltered. In addition to the subscript-interchange invariance, the entropy increase must also satisfy the following requirements:

(i) It is imperative that $S_{\rm eq} - S_0$ be positive. In other words, the entropy must increase as a result of the irreversible process described in the problem under study.
(ii) Moreover, in the limit when T_2 tends to T_1, the entropy increase must tend to zero.

The satisfaction of both the above requirements can be confirmed by looking at the case

$$T_1/T_2 = [1-\epsilon]^{\nu},$$

where[27] $1 > \epsilon \geq 0$, $\nu = +1$ when $T_2 \geq T_1$ and[28] $\nu = -1$ for $T_1 \geq T_2$. A convenient check is provided by forming an expansion in powers of ϵ:

$$\frac{S_{eq} - S_0}{C\rho AL} = \frac{\epsilon^2}{24} + \frac{\epsilon^3}{24} + \frac{11\epsilon^4}{320} + \frac{13\epsilon^5}{480} + O\left(\epsilon^6\right). \tag{4.214}$$

Requirement (i) is obviously satisfied because when ϵ tends to zero so does $S_{eq} - S_0$. As for (ii), also because $\epsilon > 0$, the change in the entropy is positive.

Problem 4.22 (Maximum work available in Problem 4.21)
Calculate the maximum amount of work that can be extracted in Problem 4.21.

Solution 4.22 (See the preceding problem for some of the preliminary details)
The use of the perfect Carnot cycle ensures maximum extraction of work. Moreover, we are assured that after the extraction of all the available work has been completed the final temperature of the rod, T_f, will be consistent with zero total change in the entropy of the universe.[29]

Fortunately, (4.211) was derived for an arbitrary value of the final temperature, T_{final}. Therefore, to find the relevant final temperature, T_f, all we need to do is set the increase in the entropy $S_{eq} - S_0$ equal to zero in (4.211). That is,

$$0 = 1 + \ln(T_f) - \ln(T_1) - \left(\frac{T_2}{T_2 - T_1}\right) \ln\left(\frac{T_2}{T_1}\right), \tag{4.215}$$

which leads to

$$T_f = \left(\frac{T_1}{e}\right) \left[\frac{T_2}{T_1}\right]^{\left(\frac{T_2}{T_2 - T_1}\right)}. \tag{4.216}$$

The total work extracted, W_{max}, is equal to the difference between the initial and final heat energy content of the rod. That is,

$$W_{max} = C\rho AL \left[\left(\frac{T_1 + T_2}{2}\right) - T_f\right]. \tag{4.217}$$

It is helpful to present these results graphically. See Figs. 4.8.a–b where the entropy increase and the maximum available work are plotted as functions of the temperature ratio $R = T_2/T_1$. For ease of display, both the entropy and work are divided by $C\rho AL$ and T_1 is set equal to 1.

Exercise 4.1 Give a physical argument as to why T_f must approach T_1 when $T_2 \to T_1$. Also, prove that this holds true for (4.216).

[27] The ϵ used here is not to be confused with its earlier usage relating to engine efficiency.

[28] Note that whether we use +1 or −1, the expansion in powers of ϵ for the entropy change is the same.

[29] Note that the universe here consists only of the isolated rod!

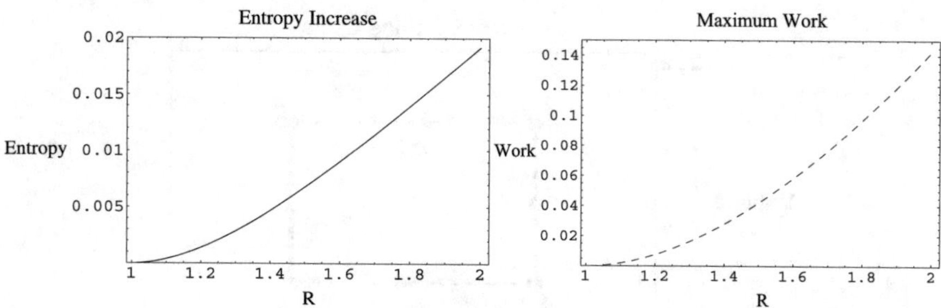

Fig. 4.8 (**a**) Setting $C\rho AL = 1$ and $T_1 = 1$, the entropy difference, $S_{eq} - S_0$, is plotted as a function of the temperature ratio $R = T_2/T_1$. (**b**) With $C\rho AL$ set equal to 1, the total work extracted, W_{max}, is plotted as a function of the temperature ratio $R = T_2/T_1$. For simplicity, T_1 is again set equal to 1

4.14 Idealized Version of Realistic Engine Cycles

A wide variety of engine types are used in practice. Below we describe four.

4.14.1 Stirling Cycle

Robert Stirling's conception of a cyclic gas engine occurred somewhat contemporaneously with that of the Carnot cycle. And, although less well known, in its idealized version the Stirling cycle has strikingly similar properties to those of a perfect Carnot engine.

Much as in the perfect Carnot cycle, the legs $1 \rightarrow 2$ and $3 \rightarrow 4$ are reversible isotherms (see Fig. 4.9).

But unlike the perfect Carnot cycle, the legs $2 \rightarrow 3$ and $4 \rightarrow 1$ are not adiabats. Rather, these two legs are traversed isochorically. Therefore, this part of the travel occurs both reversibly and at constant volume.

In going from $1 \rightarrow 2$, a positive amount of heat energy $|Q_H|$ is withdrawn reversibly (and is thereby reversibly added to the working substance) from the hot reservoir maintained at temperature T_H. Similarly, in traversing the link $3 \rightarrow 4$, positive heat energy $|Q_C|$ is rejected (after it has reversibly been extracted from the working substance) to the cold reservoir maintained at temperature T_C.

As in the perfect Carnot cycle, because of the constancy of the internal energy, the work done by the gas, $W_{1\rightarrow 2}$, in its isothermal expansion from volume $V_1 \rightarrow V_2$ is equal to the heat energy withdrawn from the hot reservoir,

$$|Q_H| = W_{1\rightarrow 2}\,, \tag{4.218}$$

where

$$W_{1\rightarrow 2} = \int_{V_1}^{V_2} P\mathrm{d}V = nRT_H \int_{V_1}^{V_2} \frac{\mathrm{d}V}{V} = nRT_H \ln\left(\frac{V_2}{V_1}\right). \tag{4.219}$$

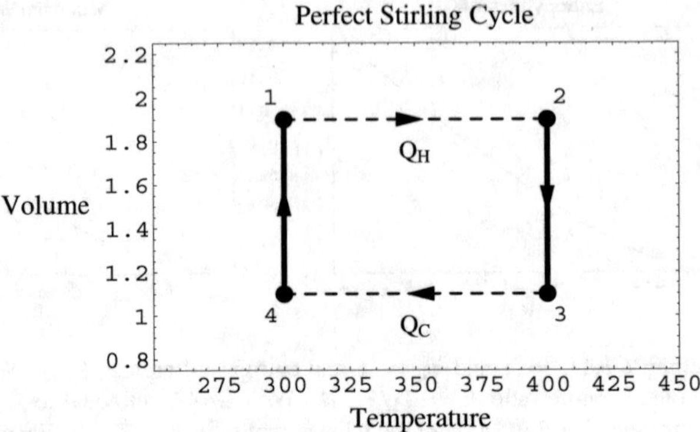

Fig. 4.9 Temperature versus volume. Here $Q_H = Q^{rev}(T_H)$ and $Q_C = Q^{rev}(T_C)$. Full lines are thermostats and broken lines isochors. Scale is arbitrary

Similarly, the work done by the gas in going along leg $3 \to 4$ is

$$W_{3\to4} = \int_{V_3}^{V_4} P\mathrm{d}V = nRT_H \int_{V_3}^{V_4} \frac{\mathrm{d}V}{V}$$
$$= nRT_H \ln\left(\frac{V_4}{V_3}\right) = -nRT_H \ln\left(\frac{V_3}{V_4}\right). \tag{4.220}$$

Therefore, the heat energy discarded to the cold reservoir during this leg is

$$|Q_C| = -W_{3\to4} = nRT_H \ln\left(\frac{V_3}{V_4}\right). \tag{4.221}$$

As a result, the ratio of the heat energy added to the working substance at T_H and that withdrawn from it at T_C is

$$\frac{|Q_H|}{|Q_C|} = \frac{T_H}{T_C}\left(\frac{\ln(\frac{V_2}{V_1})}{\ln(\frac{V_3}{V_4})}\right). \tag{4.222}$$

Owing to the fact that the traversal of legs $2 \to 3$ and $3 \to 4$ occurs at constant volume, the working substance does no work along them,

$$W_{2\to3} = 0 = W_{4\to1}.$$

Further, the algebraic value of increases in the internal energy of the working substance along these legs—equal to $nC_v(T_C - T_H)$ and $nC_v(T_H - T_C)$—cancel each other out. Thus, according to the first law, the sum of heat energy transfers along these two legs is zero.

Another important consequence of the constancy of volume along these two legs is the equality

$$\frac{V_2}{V_1} = \frac{V_3}{V_4}.$$

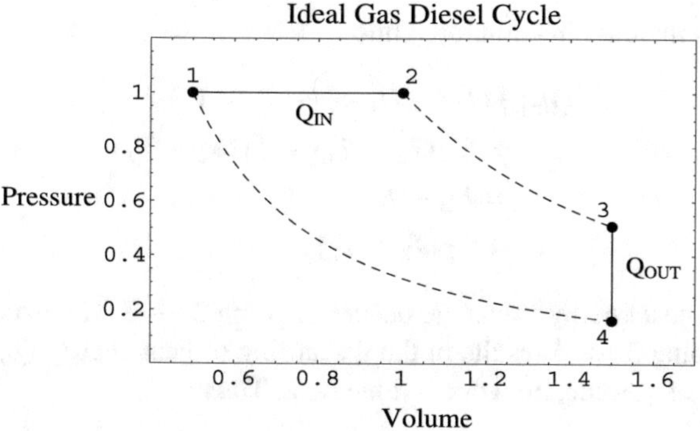

Fig. 4.10 Diesel cycle using ideal gas as the working substance

Therefore (4.222) yields

$$\frac{|Q_{\mathrm{H}}|}{|Q_{\mathrm{C}}|} = \frac{T_{\mathrm{H}}}{T_{\mathrm{C}}} .$$

The efficiency of the ideal Stirling cycle can now be calculated as

$$\begin{aligned}\epsilon_{\mathrm{stirling}} &= \frac{\text{total work output}}{\text{heat energy withdrawn from hot reservoir}} \\ &= \frac{W_{1\to 2} + W_{2\to 3} + W_{3\to 4} + W_{4\to 1}}{|Q_{\mathrm{H}}|} \\ &= \frac{W_{1\to 2} + W_{3\to 4}}{|Q_{\mathrm{H}}|} = \frac{|Q_{\mathrm{H}}| - |Q_{\mathrm{C}}|}{|Q_{\mathrm{H}}|} = 1 - \left(\frac{T_{\mathrm{C}}}{T_{\mathrm{H}}}\right) . \qquad (4.223)\end{aligned}$$

Thus, the hallmark of the perfect Carnot cycle is reproduced by the idealized Stirling cycle.

4.14.2 Diesel Cycle

A schematic plot of an idealized diesel cycle using a perfect gas as the working substance is given in Fig. 4.10.

The cycle consists of four links, all traversed reversibly. Expansion from volume $V_1 \to V_2$ occurs at constant pressure $P_1 = P_2$. The next link, $2 \to 3$, represents an adiabatic expansion $V_2 \to V_3$. It is followed by an isochoric pressure decrease, $P_3 \to P_4$. Note that here $V_3 = V_4$. The fourth and final link represents an adiabatic compression, $V_4 \to V_1$.

The engine absorbs heat energy, $|Q_{\mathrm{in}}|$, in going from position 1 to 2. According to the first law, it is equal to the resultant increase in internal energy plus the work done in the corresponding isobaric compression. Or, equivalently, it is equal to the corresponding increase in enthalpy. Thus

$$\begin{aligned} |Q_{\text{in}}| &= U_2 - U_1 + P_1(V_2 - V_1) \\ &= C_V(T_2 - T_1) + P_1(V_2 - V_1) \\ &= H_2 - H_1 \\ &= C_P(T_2 - T_1)\,. \end{aligned} \tag{4.224}$$

Of course, no heat energy exchange occurs in going $2 \to 3$. The isochoric pressure decrease in going $3 \to 4$ results in the discarding of heat energy Q_{out}. Because of the constancy of volume, no work is done here. Thus

$$|Q_{\text{out}}| = -(U_4 - U_3) = C_V(T_3 - T_4)\,. \tag{4.225}$$

Finally, the adiabat $4 \to 1$ entails no heat energy exchange. The total work output during the cycle is equal to the difference between heat energy absorbed and discarded, $|Q_{\text{in}}| - |Q_{\text{out}}|$. The efficiency ϵ_{diesel} of the engine therefore is

$$\epsilon_{\text{diesel}} = 1 - \frac{|Q_{\text{out}}|}{|Q_{\text{in}}|} = 1 - \left[\frac{C_V(T_3 - T_4)}{C_P(T_2 - T_1)}\right] = 1 - \frac{1}{\gamma}\left(\frac{T_3 - T_4}{T_2 - T_1}\right). \tag{4.226}$$

It is sometimes convenient to express the efficiency in terms of only the volumes instead of the temperatures. This is particularly so because while the cycle requires four temperature points for complete specification, only three volume points are needed. To this purpose, we invoke the ideal gas equation of state. As a result, we can write

$$\frac{T_1}{T_2} = \frac{P_1 V_1}{P_2 V_2}\,.$$

But $P_1 = P_2$. Therefore

$$T_1 = T_2\left(\frac{V_1}{V_2}\right). \tag{4.227}$$

Also because the links $2 \to 3$ and $4 \to 1$ are adiabats, we have

$$T_3 = T_2\left(\frac{V_2}{V_3}\right)^{\gamma-1} = T_2\left(\frac{V_2}{V_4}\right)^{\gamma-1} \tag{4.228}$$

and

$$T_4 = T_1\left(\frac{V_1}{V_4}\right)^{\gamma-1}. \tag{4.229}$$

Using (4.227) and replacing T_3 and T_4 by the expressions given in (4.228) and (4.229), (4.226) gives

$$\epsilon_{\text{diesel}} = 1 - \frac{1}{\gamma}\left[\frac{T_2(\frac{V_2}{V_4})^{\gamma-1} - T_1(\frac{V_1}{V_4})^{\gamma-1}}{T_2 - T_2(\frac{V_1}{V_2})}\right].$$

Multiplying both the numerator and denominator of the expression within square brackets by the factor $[V_2/(T_2 V_4)]$ leads to

$$\epsilon_{\text{diesel}} = 1 - \frac{1}{\gamma}\left[\frac{(\frac{V_2}{V_4})^{\gamma} - (\frac{T_1 V_2}{T_2 V_4})(\frac{V_1}{V_4})^{\gamma-1}}{\frac{V_2}{V_4} - \frac{V_1}{V_4}}\right]. \tag{4.230}$$

Division of both sides of (4.227) by V_4 allows the replacement

$$\frac{T_1 V_2}{T_2 V_4} = \frac{V_1}{V_4}.$$

As a result, (4.230) can be recast as

$$\epsilon_{\text{diesel}} = 1 - \frac{1}{\gamma}\left[\frac{(\frac{V_2}{V_4})^{\gamma} - (\frac{V_1}{V_4})^{\gamma}}{(\frac{V_2}{V_4}) - (\frac{V_1}{V_4})}\right]. \tag{4.231}$$

Because $V_4 = V_3$, according to one's visual preference, in (4.231) volume point V_4 may be replaced by V_3.

Problem 4.23 (Diesel engine)
Given an extremely efficient diesel cycle with compression ratio

$$\frac{V_4}{V_1} = 25$$

and the cut-off ratio

$$\frac{V_2}{V_1} = 2,$$

calculate its efficiency, ϵ_{diesel}. The working substance, air, may be treated as a diatomic ideal gas with $\gamma = \frac{7}{5} = 1.4$.

Solution 4.23 To use (4.231), we need

$$\frac{V_2}{V_4} = \frac{V_2}{V_1}\frac{V_1}{V_4} = \frac{2}{25}$$

and

$$\frac{V_1}{V_4} = \frac{1}{\frac{V_4}{V_1}} = \frac{1}{25}.$$

Then, according to (4.231), we have

$$\epsilon_{\text{diesel}} = 1 - \frac{5}{7}\left[\frac{(\frac{2}{25})^{1.4} - (\frac{1}{25})^{1.4}}{(\frac{2}{25}) - (\frac{1}{25})}\right] = 1 - 0.323 = 0.677. \tag{4.232}$$

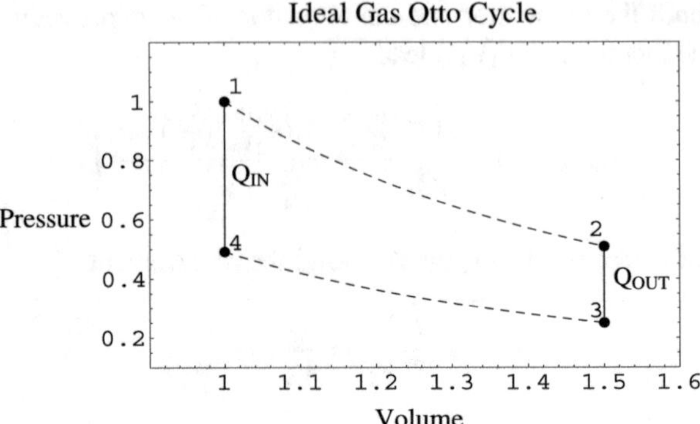

Fig. 4.11 Otto cycle using ideal gas as the working substance

4.14.3 Otto Cycle

In Fig. 4.11 we give a schematic plot of an idealized Otto cycle using a perfect gas as the working substance.

The cycle consists of four links, all traversed reversibly. Expansion of volume, $V_1 \rightarrow V_2$, occurs adiabatically. The next link, $2 \rightarrow 3$, represents an isochoric decrease in pressure, $P_2 \rightarrow P_3$. It is followed by an adiabatic compression of volume, $V_3 \rightarrow V_4$. During this process, the pressure increases, $P_3 \rightarrow P_4$. The fourth link represents an isochoric pressure increase, $P_4 \rightarrow P_1$.

The travel between 1 and 2 occurs without any exchange of heat energy. Between 2 and 3 the engine discards heat energy, $|Q_{\text{out}}|$. According to the first law, it is equal to the resultant decrease in internal energy of the working substance minus any work done by it during the travel. But because the volume remains unchanged, no work is done. Therefore, $|Q_{\text{out}}|$ is equal to the corresponding decrease in internal energy. That is,

$$|Q_{\text{out}}| = -(U_3 - U_2) = C_V(T_2 - T_3)\,. \tag{4.233}$$

Of course, no heat energy exchange occurs in traversing the next link $3 \rightarrow 4$ because it is an adiabat. The isochoric pressure increase in going $4 \rightarrow 1$ requires the adding of heat energy, $|Q_{\text{in}}|$. Because of the constancy of volume, no work is done here. Therefore, $|Q_{\text{in}}|$ is equal to the corresponding increase in the internal energy,

$$|Q_{\text{in}}| = C_V(T_1 - T_4)\,. \tag{4.234}$$

As usual, the total work output during the cycle is equal to the difference between the heat energy absorbed and discarded, that is, it is equal to $|Q_{\text{in}}| - |Q_{\text{out}}|$. And as a result, the efficiency ϵ_{otto} of the engine is

$$\epsilon_{\text{otto}} = 1 - \frac{|Q_{\text{out}}|}{|Q_{\text{in}}|} = 1 - \left[\frac{C_V(T_2 - T_3)}{C_V(T_1 - T_4)}\right] = 1 - \left[\frac{T_2 - T_3}{T_1 - T_4}\right]. \tag{4.235}$$

Outwardly the above result looks simple. But it makes reference to four temperatures. Can it be simplified?

To investigate this matter, we utilize the (T, V) equation for reversible adiabatic links. That is,

$$T(V)^{\gamma-1} = \text{const.}$$

For the links $1 \to 2$ and $3 \to 4$, we can write

$$\frac{T_1}{T_2} = \left(\frac{V_2}{V_1}\right)^{\gamma-1} \tag{4.236}$$

and

$$\frac{T_3}{T_4} = \left(\frac{V_4}{V_3}\right)^{\gamma-1}. \tag{4.237}$$

But $V_1 = V_4$ and $V_2 = V_3$. Thus (4.237) becomes

$$\frac{T_3}{T_4} = \left(\frac{V_1}{V_2}\right)^{\gamma-1}. \tag{4.238}$$

Equations (4.236)–(4.238) readily lead to the result

$$T_2 = T_1\left(\frac{T_3}{T_4}\right).$$

Upon insertion into the expression for ϵ_{otto} given in (4.235), we get

$$\epsilon_{\text{otto}} = 1 - \left[\frac{T_1(\frac{T_3}{T_4}) - T_3}{T_1 - T_4}\right] = 1 - \left(\frac{T_3}{T_4}\right)\left[\frac{T_1 - T_4}{T_1 - T_4}\right] = 1 - \left(\frac{T_3}{T_4}\right). \tag{4.239}$$

The ϵ_{otto} looks suspiciously similar to ϵ_{carnot} given in (4.16)! Let us compare the two. The highest temperature at which the Otto engine withdraws any heat energy is T_1. And the lowest such temperature in the cycle at which any heat energy is discarded is T_3. So, the relevant perfect Carnot cycle would have used heat energy reservoirs at temperatures $T_C = T_3$ and $T_H = T_1$, leading to

$$\epsilon_{\text{carnot}} = 1 - \left(\frac{T_3}{T_1}\right).$$

But $T_1 > T_4$. Therefore, ϵ_{otto} is less than the efficiency, ϵ_{carnot}, of a perfect Carnot engine that could possibly be operated here.

Often, ϵ_{otto} is expressed in terms of the compression ratio. Here that is readily done. Combining (4.238) and (4.239) gives

$$\epsilon_{\text{otto}} = 1 - \left(\frac{V_1}{V_2}\right)^{\gamma-1} = 1 - \left(\frac{1}{R_{\text{comp}}}\right)^{\gamma-1}, \tag{4.240}$$

where R_{comp} is the compression ratio.[30]

[30] R_{comp} is defined as the ratio of the largest to the smallest volumes achieved during the cycle. Here that would be equal to $\frac{V_2}{V_1} = \frac{V_3}{V_1} = \frac{V_2}{V_4} = \frac{V_3}{V_4}$.

Otto cycle can be mapped onto an idealized version of the familiar internal combustion gasoline engine used in most automobiles. Because we are dealing with a cyclic engine, we can describe the process beginning at any of the four points (nodes) displayed in Fig. 4.11.

Let us begin at point 3 with a cylinder full of what is called "the mixture," that is, air infused with gasoline vapor.[31] The adiabatic process $3 \to 4$ results in the compression of the mixture to a much smaller volume, V_4, equal to $1/R_{\rm comp}$ of the total working volume, V_3, of the uncompressed cylinder. At this point an electric spark plug—generally placed at the top or bottom of the compressed mixture, depending on the direction of compression—is actuated. The progress of the ignition process is approximately represented by an isochoric travel along the leg $4 \to 1$. Clearly, the corresponding heat energy input, $Q_{\rm IN}$, represents the energy generated by the firing of the mixture.

In actual practice the volume does not stay totally constant during this leg of the cycle. Rather, it increases some due to the tremendous rise in temperature. Also that, unless great care is exercised in the engineering, its combustion is often less than complete. Both these effects reduce the energy efficiency of the engine.

By the time the ignition phase is completed, the pressure of the gas has increased to P_1. This increased pressure exerts a large force that pushes the piston back, causing a rapid, nearly adiabatic expansion $V_1 \to V_2$. The expansion causes a crankshaft to turn, resulting in the production of torque that is transferred to the driving wheels.

As portrayed, the leg $2 \to 3$ is a relatively crude approximation to what actually takes place during this phase of the cycle. Here, the gas, now devoid of any—or almost any—live fuel and oxygen, is vented out through an outward opening valve in the cylinder while live mixture of gasoline and fresh air containing oxygen is sucked in from another inward opening valve. Again, the resulting process is not totally isochoric. The two additional "strokes" that are embedded (but not shown in the schematic diagram given in Fig. 4.11) in this leg—one involving the exhausting, and the other the sucking in, of the mixture—also use up some energy. Moreover, the exhausted gas is generally much hotter than the ambient atmospheric temperature. Thus the amount of heat energy being discarded is much greater than would be the case if the exhaust had occurred at the atmospheric temperature.

In addition to the caveats expressed above, there is also some energy loss due to friction[32] between the piston and the cylinder surfaces.

Let us next turn our attention to the compression ratio. According to (4.240), the larger the ratio $R_{\rm comp}$, the higher the efficiency. So, why do we not make this ratio very large? Perhaps this question is better rephrased as follows: How high is the compression ratio in practice?

A majority of the automobiles in operation today run on the so-called "regular" gasoline with octane value of 87. The compression ratio for these automobiles is usually ~ 9. High performance automobiles, on the other hand, require premium

[31] This mixture is the working substance and admittedly ideal gas is a rather crude approximation to it.

[32] Of course, the frictional loss is somewhat reduced by the presence of engine oil-lubricants.

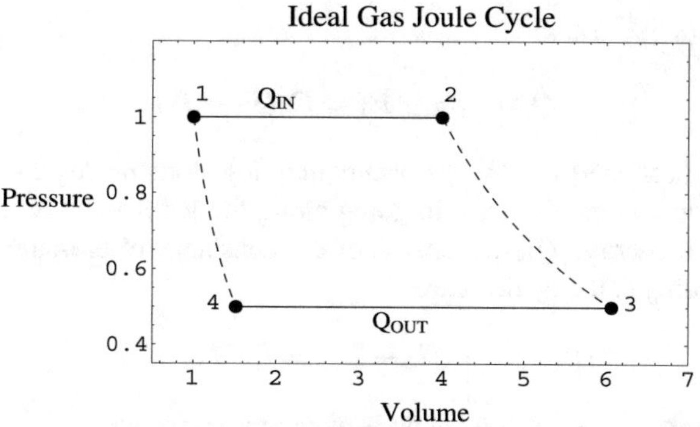

Fig. 4.12 Joule cycle using ideal gas as the working substance

fuel with octane values of 89–94. As a result, their engines can support higher compression ratios[33] $\sim$ 10.5. So, why the need for high octane fuel in high performance engines?

A typical automobile has a minimum of four cylinders. The sparking and the resultant firing of different cylinders occurs in a cleverly arranged sequence such that during any cycle the power producing stroke in one cylinder occurs as the other cylinders are progressing with their five non-power producing strokes. The result so achieved is that for a given rate of fuel injection—controlled by the driver through the fuel pedal—the torque transmitted to the driving wheels is constant in time. What puts a spanner—a monkey wrench—in the works is the "preignition." This means that the fuel mixture ignites before the compression stroke is fully completed. Such unscheduled preignition sends shock waves that cause knocking. This both reduces the power output and causes engine degradation. Increasing the octane value of the fuel helps avoid preignition by raising the combustion temperature of the fuel mixture.

4.14.4 Joule Cycle

In Fig. 4.12 we give a schematic plot of an idealized Joule cycle using a perfect gas as the working substance. The cycle consists of four links, all traversed reversibly.

Expansion of volume $V_1 \rightarrow V_2$ occurs at constant pressure $P_1 = P_2$. The next link, $2 \rightarrow 3$, represents an adiabatic expansion $V_2 \rightarrow V_3$. It is followed by an isobaric—at pressure $P_3 = P_4$—compression causing the volume to decrease from V_3 to V_4. The fourth and final link represents an adiabatic compression, $V_4 \rightarrow V_1$.

In going from position 1 to 2 the engine absorbs heat energy, $|Q_{\text{in}}|$. According to the first law, it is equal to the resultant increase in internal energy plus the work done during the isobaric compression. Or, equivalently, it is equal to the corresponding

[33] Mitsubishi, GDI engine, according to the manufacturer, achieves a compression ratio of 11.5. This requires clever engineering. For instance, part of the fuel is injected during the compression and in the process the mixture is cooled by the fuel spray.

increase in enthalpy. Thus

$$Q_{\text{in}} = H_2 - H_1 = C_P(T_2 - T_1)\,. \tag{4.241}$$

Of course, no heat energy exchange occurs in going along the leg $2 \rightarrow 3$.

The isobaric volume decrease in going along the leg $3 \rightarrow 4$ results in the discarding of heat energy, $|Q_{\text{out}}|$. Because of the constancy of pressure, it is equal to the corresponding enthalpy decrease

$$|Q_{\text{out}}| = -(H_4 - H_3) = C_P(T_3 - T_4)\,. \tag{4.242}$$

Again, the adiabat $4 \rightarrow 1$ results in no heat energy exchange.

As usual, the total work output during the cycle is equal to the difference between the heat energy absorbed and discarded, that is, $Q_{\text{in}} - Q_{\text{out}}$. And as a result, the efficiency ϵ_{joule} of the engine is

$$\epsilon_{\text{joule}} = 1 - \frac{|Q_{\text{out}}|}{|Q_{\text{in}}|} = 1 - \left[\frac{C_P(T_3 - T_4)}{C_P(T_2 - T_1)}\right] = 1 - \left[\frac{T_3 - T_4}{T_2 - T_1}\right]. \tag{4.243}$$

This result can be simplified. To this end, we use the (T, P) equation for adiabatic links. That is,

$$T \propto (P)^{\frac{\gamma-1}{\gamma}}\,.$$

For the links $4 \rightarrow 1$ and $2 \rightarrow 3$, we can write

$$\frac{T_1}{T_4} = \left(\frac{P_1}{P_4}\right)^{\frac{\gamma-1}{\gamma}}$$

and

$$\frac{T_2}{T_3} = \left(\frac{P_2}{P_3}\right)^{\frac{\gamma-1}{\gamma}}\,.$$

But $P_1 = P_2$ and $P_3 = P_4$. Thus the above become

$$\frac{T_1}{T_4} = \left(\frac{P_1}{P_3}\right)^{\frac{\gamma-1}{\gamma}}$$

and

$$\frac{T_2}{T_3} = \left(\frac{P_1}{P_3}\right)^{\frac{\gamma-1}{\gamma}}\,. \tag{4.244}$$

This leads to the result

$$T_4 = T_1\left(\frac{T_3}{T_2}\right),$$

and upon insertion into the expression for ϵ_{joule} given in (4.243), we get

$$\epsilon_{\text{joule}} = 1 - \left[\frac{T_3 - T_1(\frac{T_3}{T_2})}{T_2 - T_1}\right] = 1 - \left(\frac{T_3}{T_2}\right) = 1 - \left(\frac{T_4}{T_1}\right). \tag{4.245}$$

Outwardly, much like the Otto cycle, the efficiency of the Joule cycle, ϵ_{joule}, also looks similar to ϵ_{carnot} given in (4.16)! Again, let us compare the two. The highest temperature at which the Joule engine draws any heat energy is T_2. And the lowest temperature in the cycle where any heat energy is rejected is T_4. So, the relevant perfect Carnot cycle would have had $T_C = T_4$ and $T_H = T_2$ with efficiency

$$\epsilon_{\text{carnot}} = 1 - \left(\frac{T_4}{T_2}\right).$$

Because $T_3 > T_4$, ϵ_{carnot} is greater than the corresponding ϵ_{joule}. Note that, because $T_2 > T_1$, we are led to the same conclusion if we compare T_2 and T_1.

Often, the efficiency of the Joule cycle is expressed in terms of the pressure ratio. To do this, combine (4.244) and (4.245). Then

$$\epsilon_{\text{joule}} = 1 - \left(\frac{P_3}{P_1}\right)^{\frac{\gamma-1}{\gamma}} = 1 - \left(\frac{P_4}{P_1}\right)^{\frac{\gamma-1}{\gamma}} = 1 - \left(\frac{P_3}{P_2}\right)^{\frac{\gamma-1}{\gamma}} = 1 - \left(\frac{P_4}{P_2}\right)^{\frac{\gamma-1}{\gamma}}.$$

Problem 4.24 (Joule engine)
The pressure and temperature in a Joule engine at point 4 are P_4 = standard atmospheric pressure and $T_4 = 300$ K. Also given are $P_2 = 3.5 \times 10^5$ Pa and $T_2 = 500\,°\text{C}$. Calculate ϵ_{joule} and temperature T_3.

Solution 4.24

$$\epsilon_{\text{joule}} = 1 - \left(\frac{P_4}{P_2}\right)^{\frac{\gamma-1}{\gamma}} = 1 - \left(\frac{1.013 \times 10^5}{3.5 \times 10^5}\right)^{\frac{\gamma-1}{\gamma}} = 1 - 0.702 = 29.8\%,$$

$$T_3 = T_2\left(\frac{T_4}{T_1}\right) = T_2 \times \left(\frac{P_4}{P_2}\right)^{\frac{\gamma-1}{\gamma}} = T_2 \times 0.702 = 773 \times 0.702 = 542.6\ \text{K},$$

$$\text{and}\quad T_1 = \left[\frac{T_4}{(\frac{T_3}{T_2})}\right] = \frac{300}{0.702} = 429.4\ \text{K}.$$

4.15 Negative Temperature: Cursory Remark

By relating the average kinetic energy of a perfect gas to its absolute temperature T—see (2.31)—we have "psychologically" committed ourselves to treating T as a positive quantity. And this feeling has been further reinforced by Carnot's statement—see (4.5)—that the efficiency of a perfect Carnot engine, ϵ_{carnot}, is equal to $1 - (\frac{T_C}{T_H})$.

One may well ask: How has Carnot added to our distrust of negative temperatures? The answer is the following: Under no circumstances, the actual work produced may be greater than the energy used for its production. This means that the working efficiency of an engine may never be greater than 100%—if the opposite were ever true, the world would not have an "energy problem"! And, clearly, a negative value for T_C or T_H would lead to (100+)% efficiency! It turns out that the Kelvin–Planck formulation of the second law [2] is also uncomfortable with the concept of negative temperatures. The Clausius formulation[34] of the second law,[35] on the other hand, can be retained with qualifications, that is, it needs to be agreed that in the negative temperature regime, the "warmer" of the two bodies has the smaller absolute value for the temperature [3]. This means that just as +3 K is greater than +2 K, so is −2 K greater than −3 K. There is, however, an important "caveat." In increasing order, the relevant temperature in integer degrees Kelvin is

$$+0, +1, +2, \ldots, +\infty, -\infty, \ldots, -2, -1, -0\,.$$

Therefore, if we must use negative temperatures, we must also accept the fact that an object at any negative temperature is warmer than one at any positive temperature!

More seriously, the question to ask is how, using fundamental thermodynamic principles, a negative absolute temperature T must be defined. Following Ramsey's suggestion, T should be defined from the thermodynamic—see (7.101) above—identity

$$\left(\frac{\partial U}{\partial S}\right)_{\mathrm{V,N}} = T\,. \tag{4.246}$$

If this equation is used as the definition of the absolute temperature, then if and when the internal energy can be measured as a function of the entropy—big "if and when," considering it is very hard to precisely measure either of these quantities by any "direct" method—and if the derivative is negative, then we have a confirmed case of negative temperature. According to ter Haar, "For a system to be capable of negative temperatures, it is necessary for its energy to have an upper bound" [4]. Energy U is not bounded in normal systems. Rather, as a function of the entropy S, the energy is monotonically increasing. As a result, the derivative $(\frac{\partial U}{\partial S})_{\mathrm{V,N}}$ is positive. That, according to (7.102), results in the temperature being positive.

All this would work well unless the entropy is bounded from above. In that case when S reaches $S_{\max}$, then $[(\frac{\partial U}{\partial S})_{\mathrm{V,N}}]$ suddenly changes from $+\infty$ to $-\infty$ and the system starts its progress up the negative temperature scale! As noted above, the end point of the negative part of the temperature scale is $T = -0$, which surely must be just as unattainable as is $T = +0$.

[34]This is true despite the fact—as previously concluded—that "A violation of the Carnot version of the second law results in a violation of the Clausius statement of the second law."

[35]Namely, "Without assistance it is impossible to withdraw positive amount of heat energy from a colder object and transfer the same to a warmer object." In other words, heat energy does not spontaneously get transferred from a colder object to a warmer one.

An experiment with precisely these characteristics was first performed by Purcell and Pound [5]. Because the subject matter involves quantum statistical mechanics, for all relevant details the reader is referred to Chap. 11.

References

1. F.W. Sears, G.L. Salinger, *Thermodynamics, Kinetic Theory, and Statistical Thermodynamics*, 3rd edn. (Addison Wesley, Reading, 1975)
2. D. ter Haar, H. Wergeland, *Elements of Thermodynamics* (Addison-Wesley, Reading, 1966)
3. N.F. Ramsey, Thermodynamics and statistical mechanics at negative absolute temperature. Phys. Rev. **103**, 20 (1956)
4. H.B. Callen, See Callen's Rule in Chap. 8 (8.16), also Callen's scaling principle
5. E.M. Purcell, R.V. Pound, Phys. Rev. **81**, 279 (1951)

Introduction and the Zeroth Law

5

The first law reminds us of the well known fact that energy is conserved and all of it must be accounted for. Therefore when heat energy $\Delta Q'$ is added to a system, and none of it escapes, then all of it must still be there. And if there should exist a procedure to convert some of this heat energy to work, say $\Delta W'$, then after such conversion only $\Delta Q' - \Delta W'$ of it will still be present in the system. We call this increase in the system's internal energy the left-over amount, ΔU. Because the amount of heat energy input surely depends on what the temperature difference between the depositor and the depositee at any given time is, and because the total deposit of heat may have taken finite time, its magnitude will also depend on how long any particular instance, with some particular difference in temperature, lasted, etc. Similarly, the work that is done—namely, $\Delta W'$—will also crucially depend on where, what, how, and when the work was carried out. These facts are generally labeled as path dependencies. Clearly, therefore, the size of both $\Delta Q'$ and $\Delta W'$ must depend on the paths that are taken in carrying out these tasks.

All this is quite obvious. However, the first law also makes another very important, but less obvious statement, that is, "Despite the path dependence of the heat input and the work done, their difference, namely the internal energy, ΔU, is path-independent."

The centrality of the aforementioned ideas is recognized by the second law. Yet, following the ideas of Carnot, the second law makes additional qualifications to them. For instance, given the temperatures at which heat can be introduced and discarded, there is a maximum value for the efficiency with which work can be produced. Additionally, these ideas lead to the identification of a state function of universal[1] significance, the entropy.

In Sect. 5.1 we marry the first and second laws in a manner that achieves "a perfect union" of the two. And this marriage provides great insights into the workings of thermodynamics. Here this union is referred to as the "first–second" law. Thermodynamically this coupling is valid when all the processes being referred to occur quasistatically. The Clausius version of the first–second law is studied in

[1]For instance, as noted by Stephen Hawking with regard to black holes.

R. Tahir-Kheli, *General and Statistical Thermodynamics*,
https://doi.org/10.1007/978-3-030-20700-7_5

Sect. 5.1.1. Entropy is analyzed and its computation is studied as the first $t\mathrm{d}s$ equation in Sect. 5.1.2. Mixing of ideal gas and its effect on entropy change in two chambers is described in Sect. 5.2. The cases when the two variables p and v are independent is dealt with in Sect. 5.2.2. For the case when variables t and p are independent, the relationship $(\frac{\partial h}{\partial p})_{\mathrm{t}} - v = -t(\frac{\partial v}{\partial t})_{\mathrm{p}}$ is proven in Sect. 5.2.3. Newton's treatment of the velocity of sound is described in Sect. 5.3. The chapter is concluded in Sect. 5.4 with treatment of Van der Waal's gas, energy, and entropy changes and problems for the student. Fifteen different problems are worked out in this chapter.

5.1 The First and Second Laws

The first law of thermodynamics makes two distinct statements:

(i) Energy is conserved.[2]
(ii) In thermodynamic equilibrium, there exists a state function, U—that is referred to as the "internal energy"—such that the sum of the heat energy Q' that is *added to* and the work $-W'$ that is *done on* the system equal the *increase* U in the internal energy.

5.1.1 The First–Second Law: The Clausius Version

As stated above, the first law concludes that heat energy, $\mathrm{d}Q'$, added to an object plus the work, i.e., $-\mathrm{d}W'$, "done on it" are equal to the net increase, $\mathrm{d}U$, in the object's internal energy. That is,

$$\mathrm{d}Q' - \mathrm{d}W' = \mathrm{d}U \,. \tag{5.1}$$

It is important to note that $\mathrm{d}Q'$ and $\mathrm{d}W'$ may or may not be quasistatic.

The work $-\mathrm{d}W'$ "done on" a system means that work $+\mathrm{d}W'$ is "done by" the system. Combining (4.52) and (5.1) leads to—what we shall call—the Clausius version of the first–second law (compare with (4.58)):

$$T\mathrm{d}S \geq \mathrm{d}U + \mathrm{d}W' \,. \tag{5.2}$$

The equality sign in the statement (5.2) applies only when the work $\mathrm{d}W'$ is done quasistatically.

Assume that the transfer of heat energy and the work done both happen quasistatically. And an infinitesimal amount of heat energy input causes a single mole of such a system, originally in thermal equilibrium at temperature t, to move to a neighboring equilibrium state at temperature $t + \mathrm{d}t$. In the process, its thermodynamic state

[2] Occasionally, just the conservation of energy is referred to as the first law of thermodynamics. In fact, the conservation of energy has been known since the ancient Greek times and the real value of the first law lies in part (ii) which identifies an important state function, the internal energy.

functions,[3] (u, h, s), and state variables, (p, v), change to $(u + \mathrm{d}u, h + \mathrm{d}h, s + \mathrm{d}s)$ and $(p + \mathrm{d}p, v + \mathrm{d}v)$, respectively. These two neighboring equilibrium states have very interesting interrelationships.

As noted in (3.15), for what we call a simple system, when the infinitesimal work—equal to $\mathrm{d}w$—is done by an infinitesimal, quasistatic expansion equal to $\mathrm{d}v$, under pressure p, then $\mathrm{d}w = p\mathrm{d}v$. Then, according to the first law, the quasistatic increase, $\mathrm{d}u$, in the internal energy, and the relevant quasistatic heat energy input $\mathrm{d}q$, and the work done, $p\mathrm{d}v$, are related as follows:

$$\mathrm{d}u = \mathrm{d}q - \mathrm{d}w = \mathrm{d}q - p\mathrm{d}v\,. \tag{5.3}$$

Equivalently, we can write

$$\mathrm{d}q = \mathrm{d}u + p\mathrm{d}v\,. \tag{5.4}$$

In order fully to explore this subject, we need to focus on the fact that in (5.4) all the relevant processes occur quasistatically. Thus $\mathrm{d}q \to \mathrm{d}q^{\mathrm{rev}}$, and

$$\left(\frac{\mathrm{d}q^{\mathrm{rev}}}{t}\right) = \mathrm{d}s\,. \tag{5.5}$$

As a result, for quasistatic processes, the first–second law becomes

$$t\mathrm{d}s = \mathrm{d}u + p\mathrm{d}v\,. \tag{5.6}$$

While some authors call (5.6) the Gibbs relationship, here it will be referred to as the "first–second law," in which t and p are independent.

As before, expressing $u = u(t, v)$, the exact differential $\mathrm{d}u$ can be represented as

$$\mathrm{d}u = \left(\frac{\partial u}{\partial t}\right)_v \mathrm{d}t + \left(\frac{\partial u}{\partial v}\right)_t \mathrm{d}v\,.$$

Thus, the (t, v)-dependent form of the first–second law becomes

$$t\mathrm{d}s = \left(\frac{\partial u}{\partial t}\right)_v \mathrm{d}t + \left[p + \left(\frac{\partial u}{\partial v}\right)_t\right]\mathrm{d}v \equiv C_\mathrm{v}\mathrm{d}t + \Delta_v\mathrm{d}v\,. \tag{5.7}$$

This is the basic (t, v)-version of the first–second law. For ease of comparison, (5.7) is recast as

$$\mathrm{d}s = \frac{1}{t}\left[\left(\frac{\partial u}{\partial t}\right)_v \mathrm{d}t + \left(p + \left(\frac{\partial u}{\partial v}\right)_t\right)\mathrm{d}v\right]. \tag{5.8}$$

Given that here we are working in terms of the variables t and v, we can write

$$\mathrm{d}s = \left(\frac{\partial s}{\partial t}\right)_v \mathrm{d}t + \left(\frac{\partial s}{\partial v}\right)_t \mathrm{d}v\,. \tag{5.9}$$

[3]Recall that u, h, and s refer respectively to the internal energy, enthalpy, and entropy.

Further, because ds is an exact differential, we can also use the integrability relationship

$$\left(\frac{\partial(\frac{\partial s}{\partial t})_v}{\partial v}\right)_t = \left(\frac{\partial(\frac{\partial s}{\partial v})_t}{\partial t}\right)_v . \tag{5.10}$$

Proof of Relationship $(\frac{\partial u}{\partial v})_t = t(\frac{\partial p}{\partial t})_v - p$

We are now equipped to prove an important assertion that was made in (3.22), namely that

$$p + \left(\frac{\partial u}{\partial v}\right)_t = t\left(\frac{\partial p}{\partial t}\right)_v \equiv \Delta_{\mathrm{v}} . \tag{5.11}$$

To this end, we note that:

(i) The dependence of the perfect differential ds on the pair of differentials dt and dv must be the same whether we use (5.8) or (5.9). Equating the factors proportional to dt (in these equations) gives

$$\frac{1}{t}\left(\frac{\partial u}{\partial t}\right)_v = \left(\frac{\partial s}{\partial t}\right)_v . \tag{5.12}$$

Similarly, the factors proportional to dv can be equated. This gives

$$\frac{1}{t}\left[p + \left(\frac{\partial u}{\partial v}\right)_t\right] = \left(\frac{\partial s}{\partial v}\right)_t . \tag{5.13}$$

(ii) The integrability requirement for the perfect differential ds must also be satisfied. To this purpose, differentiate (both the right- and left-hand side) of (5.12) with respect to v while keeping t constant:

$$\frac{\partial^2 s}{\partial v \partial t} = \left(\frac{\partial(\frac{\partial s}{\partial t})_v}{\partial v}\right)_t = \frac{1}{t}\left[\frac{\partial^2 u}{\partial v \partial t}\right] . \tag{5.14}$$

Similarly, differentiate (5.13) with respect to t while keeping v constant:

$$\frac{\partial^2 s}{\partial t \partial v} = \left(\frac{\partial(\frac{\partial s}{\partial v})_t}{\partial t}\right)_v = -\frac{1}{t^2}\left[p + \left(\frac{\partial u}{\partial v}\right)_t\right] + \frac{1}{t}\left[\left(\frac{\partial p}{\partial t}\right)_v + \frac{\partial^2 u}{\partial t \partial v}\right] . \tag{5.15}$$

Because $\frac{\partial^2 s}{\partial v \partial t} = \frac{\partial^2 s}{\partial t \partial v}$, the left-hand sides of (5.14) and (5.15) are equal. Therefore we can equate their right-hand sides:

$$\frac{1}{t}\left[\frac{\partial^2 u}{\partial v \partial t}\right] = -\frac{1}{t^2}\left[p + \left(\frac{\partial u}{\partial v}\right)_t\right] + \frac{1}{t}\left(\frac{\partial p}{\partial t}\right)_v + \frac{1}{t}\left[\frac{\partial^2 u}{\partial t \partial v}\right] .$$

The terms, $\frac{1}{t}[\frac{\partial^2 u}{\partial v \partial t}]$ on the left and $\frac{1}{t}[\frac{\partial^2 u}{\partial t \partial v}]$ on the right, are equal and therefore can be eliminated from the equation. Multiplying the remainder by t^2 leads to the relationship

$$p + \left(\frac{\partial u}{\partial v}\right)_t = t\left(\frac{\partial p}{\partial t}\right)_v . \tag{5.16}$$

Upon invoking the cyclic identity, we arrive at the desired result:

$$\begin{aligned} p + \left(\frac{\partial u}{\partial v}\right)_t &= t\left(\frac{\partial p}{\partial t}\right)_v = -t \cdot \left[\left(\frac{\partial v}{\partial t}\right)_p \left(\frac{\partial p}{\partial v}\right)_t\right] \\ &= t \cdot \left[\frac{1}{v}\left(\frac{\partial v}{\partial t}\right)_p\right] \cdot \left[-v\left(\frac{\partial p}{\partial v}\right)_t\right] = t \cdot \alpha_{\mathrm{p}} \cdot \frac{1}{\chi_{\mathrm{t}}} = \Delta_{\mathrm{v}} . \end{aligned} \tag{5.17}$$

(Compare with (3.18) and (3.22).)

5.1.2 The First $t\mathrm{d}s$ Equation

To recapitulate: The first–second law in the (t, v)-representation can be displayed in four equivalent forms, namely

$$\begin{aligned} t\mathrm{d}s &= \left(\frac{\partial u}{\partial t}\right)_v \mathrm{d}t + \left[\left(\frac{\partial u}{\partial v}\right)_t + p\right]\mathrm{d}v = \left(\frac{\partial u}{\partial t}\right)_v \mathrm{d}t + t\left(\frac{\partial p}{\partial t}\right)_v \mathrm{d}v \\ &= C_{\mathrm{v}}\mathrm{d}t + t\left(\frac{\alpha_{\mathrm{p}}}{\chi_{\mathrm{t}}}\right)\mathrm{d}v = C_{\mathrm{v}}\mathrm{d}t + \Delta_{\mathrm{v}}\mathrm{d}v . \end{aligned} \tag{5.18}$$

Note that all processes described in (5.18) are to be carried out quasistatically. It is traditional to name (5.18) the *The First* $t\mathrm{d}s$ *Equation*. Also the positivity of both C_{v} and χ_{t} is required for the thermodynamic stability of states that are in thermal equilibrium.

Equation (5.18) yields

$$0 = C_{\mathrm{v}}(\mathrm{d}t)_{\mathrm{s}} + t\left(\frac{\alpha_{\mathrm{p}}}{\chi_{\mathrm{t}}}\right)(\mathrm{d}v)_{\mathrm{s}} = C_{\mathrm{v}}(\mathrm{d}t)_{\mathrm{s}} + \Delta_{\mathrm{v}}(\mathrm{d}v)_{\mathrm{s}} . \tag{5.19}$$

Dividing by $v(\mathrm{d}t)_{\mathrm{s}}\Delta_{\mathrm{v}}$ leads to

$$0 = \frac{C_{\mathrm{v}}}{v\Delta_{\mathrm{v}}} + \frac{1}{v}\left(\frac{\partial v}{\partial t}\right)_s . \tag{5.20}$$

This proves the equality

$$\alpha_{\mathrm{s}} = \frac{1}{v}\left(\frac{\partial v}{\partial t}\right)_s = -\left(\frac{C_{\mathrm{v}}}{v\Delta_{\mathrm{v}}}\right) = -\left(\frac{C_{\mathrm{v}}\chi_{\mathrm{t}}}{vt\alpha_{\mathrm{p}}}\right). \tag{5.21}$$

Analogous to the definition of an isobaric[4] volume expansion coefficient, i.e., $\alpha_\mathrm{p} = \frac{1}{v}(\frac{\partial v}{\partial t})_\mathrm{p}$, in (5.21) we have defined an isentropic[5] volume expansion coefficient $\alpha_\mathrm{s} = \frac{1}{v}(\frac{\partial v}{\partial t})_\mathrm{s}$. Looking at the first and last terms in (5.21), it is clear that α_p and α_s have opposite signs. This fact is ensured by the positivity of v, t, the specific heat C_v, and the isothermal compressibility χ_t.

It is important to note once more that the positivity of both C_v and χ_t is required for the thermodynamic stability of states that are in thermal equilibrium.

At constant pressure, normal systems expand with increase in temperature. This makes α_p positive. When such is the case, α_s is negative. The negativity of α_s, of course, implies that normal systems cool down during (quasistatic) adiabatic expansion. Equivalently, it can also be stated that the negativity of α_s implies that normal systems heat up during (quasistatic) adiabatic decrease of volume—a phenomenon that most of us have observed when air is rapidly let out of an inflated tyre. As we well know by now, the bizarre behavior of water below about 4 °C is not normal. Here, at constant pressure, water gets lighter—or equivalently, the volume per mole increases—with decrease in temperature, i.e.,

$$\alpha_\mathrm{p} = \frac{1}{v}\left(\frac{\partial v}{\partial t}\right)_p < 0\,. \tag{5.22}$$

As noted above, according to (5.21), if α_p is negative, α_s is positive. Therefore, below 4 °C, the temperature of water increases upon (quasistatic) adiabatic expansion—or, stated equivalently, upon (quasistatic) adiabatic decrease in volume, the temperature of water, below 4 °C, decreases still further.

Dependence of C_v on v

At constant volume, whenever the pressure is a linear function of the temperature, it leads to an important result for the system specific heat C_v. To see this, let us

(i) Write (5.16) as

$$\left(\frac{\partial u}{\partial v}\right)_t = t\left(\frac{\partial p}{\partial t}\right)_v - p\,,$$

and differentiate it with respect to t at constant volume v. We get

$$\frac{\partial^2 u}{\partial t \partial v} = \left(\frac{\partial(\frac{\partial u}{\partial v})_t}{\partial t}\right)_v = \left(\frac{\partial p}{\partial t}\right)_v + t\mathrm{d}ptv - \left(\frac{\partial p}{\partial t}\right)_v = t\mathrm{d}ptv\,. \tag{5.23}$$

(ii) Look at the second derivative $\frac{\partial^2 u}{\partial v \partial t}$, which is

$$\frac{\partial^2 u}{\partial v \partial t} = \left(\frac{\partial(\frac{\partial u}{\partial t})_v}{\partial v}\right)_t = \left(\frac{\partial C_v}{\partial v}\right)_t\,. \tag{5.24}$$

[4]Namely, that occurring at constant pressure.
[5]Namely, one that occurs at constant entropy.

(iii) Invoke the integrability condition and thereby equate the right-hand sides of (5.23) and (5.24). This leads to the relationship

$$\left(\frac{\partial C_\mathrm{V}}{\partial v}\right)_t = t\mathrm{d}ptv\,. \tag{5.25}$$

Now, if at constant volume the pressure p happens to be a linear function of the temperature, t, e.g.,

$$p(v,t) = a(v)t + b(v)\,, \tag{5.26}$$

where $a(v)$ and $b(v)$ are not dependent on the temperature t, we have

$$\left(\frac{\partial p}{\partial t}\right)_v = a(v)\,, \quad \mathrm{d}ptv = 0\,.$$

Therefore, (5.25) leads to an important result, namely that whenever at constant volume the system pressure is a linear function of the temperature[6]—as is the case specified in (5.26)—C_V is independent of the volume v for all isothermal processes,

$$\left(\frac{\partial C_\mathrm{V}}{\partial v}\right)_t = 0\,. \tag{5.27}$$

Alternate Proof of $(\frac{\partial u}{\partial v})_t = t(\frac{\partial p}{\partial t})_v - p$

In view of the importance of the relationship given in (5.16), i.e.,

$$\left(\frac{\partial u}{\partial v}\right)_t = t\left(\frac{\partial p}{\partial t}\right)_v - p\,,$$

below we rederive it by an alternative procedure.

Construct a state function[7] Φ such that

$$\Phi = s - (u/t)\,.$$

For neighboring equilibrium states, the difference in the value of Φ is

$$\mathrm{d}\Phi = \mathrm{d}s - \frac{\mathrm{d}u}{t} + \left(\frac{u}{t^2}\right)\mathrm{d}t\,.$$

According to the dictates of the first–second law, $t\mathrm{d}s = \mathrm{d}u + p\mathrm{d}v$. Therefore, $(\mathrm{d}s - \frac{\mathrm{d}u}{t})$ can be replaced by $(\frac{p}{t})\mathrm{d}v$. Hence, we have

$$\mathrm{d}\Phi = \left(\frac{p}{t}\right)\mathrm{d}v + \left(\frac{u}{t^2}\right)\mathrm{d}t\,.$$

[6]Note that this, for example, is the case for an ideal gas.

[7]Note that $\mathrm{d}\Phi$ is an exact differential because $\mathrm{d}s$, $\mathrm{d}u$, and $\mathrm{d}t$ are exact differentials. Note also that Φ, s, and u are state functions, and t is a state variable.

Because $d\Phi$ is an exact differential, the integrability requirement must be satisfied. That is,

$$\left(\frac{\partial(\frac{p}{t})}{\partial t}\right)_v = \left(\frac{\partial(\frac{u}{t^2})}{\partial v}\right)_t .$$

As a result, we have

$$-\frac{p}{t^2} + \frac{1}{t}\left(\frac{\partial p}{\partial t}\right)_v = \frac{1}{t^2}\left(\frac{\partial u}{\partial v}\right)_t .$$

Multiplying both sides by t^2 yields the desired identity

$$\left(\frac{\partial u}{\partial v}\right)_t = t\left(\frac{\partial p}{\partial t}\right)_v - p .$$

5.2 Two Chambers: Entropy Calculation

Assume an adiabatically isolated vessel having two chambers, to be referred to as 1 and 2.

5.2.1 Two Chambers: Mixing of Ideal Gases

The following part of this section contains a fully worked out problem.

Problem 5.1 (Isothermal process with different pressure and same number of atoms) The chambers are separated by a massless partition of zero volume. Both chambers contain the same monatomic ideal gas, are at the same temperature T, and the gas in them initially contains the same number, N, of atoms. However, their volumes, V_1 and V_2, and therefore their initial pressures, P_1 and P_2, are different. Calculate any resulting change in the entropy in terms of P_1 and P_2 if the partition is removed and the gas in the two chambers is allowed to mix homogeneously. Prove that the change in the entropy is positive. Also plot the result to show how the change in the entropy varies with change in the ratio of the two pressures.

Solution 5.1 Consider the same monatomic ideal gas placed in two different chambers of volume V_1 and V_2. The initial pressure of the ideal gas in the two chambers is P_1 and P_2. The temperature of the gas, T, and initially the number of atoms, N, in each of these chambers is the same. Therefore, initially the equation of state of the ideal gas in the two chambers is the following:

$$P_1V_1 = Nk_BT , \qquad P_2V_2 = Nk_BT .$$

Once the partition separating the chambers is opened, the gas achieves a joint status with total volume equal to $V_1 + V_2$. Because each portion has the same number of atoms, is at the same temperature T, and now is also at the same pressure, it is convenient, therefore, to treat the process as causing a change in volume of each of the two portions to a final volume V_{final} which is equal to half of the total volume. That is, $V_{\mathrm{final}} = (V_1 + V_2)/2$.

The thermodynamics of the process is specified by the first–second law in the form

$$T\mathrm{d}S = \mathrm{d}U + P\mathrm{d}V\,.$$

Because we are dealing with an ideal gas and its temperature, T, is constant, $\mathrm{d}U = 0$. Therefore, for either of the two portions, we have

$$\mathrm{d}S = \frac{P}{T}\mathrm{d}V\,, \tag{5.28}$$

where $PV = Nk_{\mathrm{B}}T$, or equivalently,

$$\frac{P}{T} = \frac{Nk_{\mathrm{B}}T}{V}\,. \tag{5.29}$$

Inserting this into (5.28) gives

$$\mathrm{d}S = Nk_{\mathrm{B}}\left(\frac{\mathrm{d}V}{V}\right). \tag{5.30}$$

We can now calculate the increase in the entropy of each of the two portions as follows:

$$\begin{aligned}
\Delta S_{\mathrm{portion\ 1}} &= \int_{\mathrm{initial\text{-}portion\ 1}}^{\mathrm{final\text{-}portion\ 1}} \mathrm{d}S = Nk_{\mathrm{B}} \int_{V_1}^{\frac{V_1+V_2}{2}} \left(\frac{\mathrm{d}V}{V}\right) \\
&= Nk_{\mathrm{B}} \ln\left[\frac{(\frac{V_1+V_2}{2})}{V_1}\right], \qquad (5.31) \\
\Delta S_{\mathrm{portion\ 2}} &= \int_{\mathrm{initial\text{-}portion\ 2}}^{\mathrm{final\text{-}portion\ 2}} \mathrm{d}S = Nk_{\mathrm{B}} \int_{V_2}^{\frac{V_1+V_2}{2}} \left(\frac{\mathrm{d}V}{V}\right) \\
&= Nk_{\mathrm{B}} \ln\left[\frac{(\frac{V_1+V_2}{2})}{V_2}\right]. \qquad (5.32)
\end{aligned}$$

The total increase in the entropy therefore is equal to

$$\Delta S_{\mathrm{total}} = \Delta S_{\mathrm{portion\ 1}} + \Delta S_{\mathrm{portion\ 2}} = Nk_{\mathrm{B}} \ln\left[\frac{(\frac{V_1+V_2}{2})^2}{V_1 V_2}\right]. \tag{5.33}$$

To change over to variables P_1 and P_2 we should rewrite the above equation in terms of the pressures. Using (5.29) for both these cases, we have

$$V_{\mathrm{i}} = \frac{Nk_{\mathrm{B}}T}{P_{\mathrm{i}}}\,, \qquad i = 1 \text{ or } 2\,.$$

Accordingly, we can write

$$\frac{(\frac{V_1+V_2}{2})^2}{V_1 V_2} = \frac{(Nk_B T)^2 \cdot (\frac{1}{P_1} + \frac{1}{P_2})^2 \cdot (\frac{1}{2})^2}{(Nk_B T)^2 \cdot (\frac{1}{P_1} \cdot \frac{1}{P_2})} = \left[\frac{(P_1 + P_2)^2}{4P_1 P_2}\right]. \tag{5.34}$$

Now using (5.33) and (5.34), we get

$$\Delta S_{\text{total}} = \Delta S_{\text{portion 1}} + \Delta S_{\text{portion 2}} = Nk_B \ln\left[\frac{(P_1 + P_2)^2}{4P_1 P_2}\right]. \tag{5.35}$$

Because

$$(P_1 + P_2)^2 - 4P_1 P_2 = (P_1 - P_2)^2 \geq 0\,,$$

or equivalently,

$$\frac{(P_1 + P_2)^2}{4P_1 P_2} \geq 1\,,$$

we have the following inequality:

$$\ln\left[\frac{(P_1 + P_2)^2}{4P_1 P_2}\right] \geq 0\,. \tag{5.36}$$

According to (5.35), (5.36) shows that

$$\Delta S_{\text{total}} \geq 0\,.$$

The equality, $\Delta S_{\text{total}} = 0$, is obtained only when $P_1 = P_2$. Why is this result not surprising? If we allowed two parts of the same (ideal) gas, at the same temperature and pressure, to mix together, we would have done nothing that makes the gas different from its unmixed version. For such a mixed gas to be tangibly different, the two parts should have had one or more of the following items different: either different pressures, or different temperatures, or different number of atoms, or indeed any combination of these differences.

It is helpful to see this result in a graphical form. In Fig. 5.1 the change in the entropy is plotted as a function of the ratio, P_2/P_1—or equivalently, P_1/P_2—of the two pressures. Notice that when the two pressures are equal, the change in the entropy is zero. And as the pressures begin to differ, the change in the entropy increases.

5.2.2 Two Chambers: Independent p and v

Problem 5.2 (Isothermal process with different pressure and different volumes)
An adiabatically isolated vessel has two chambers, to be referred to as 1 and 2. The chambers are separated by a massless partition of zero volume. Both chambers

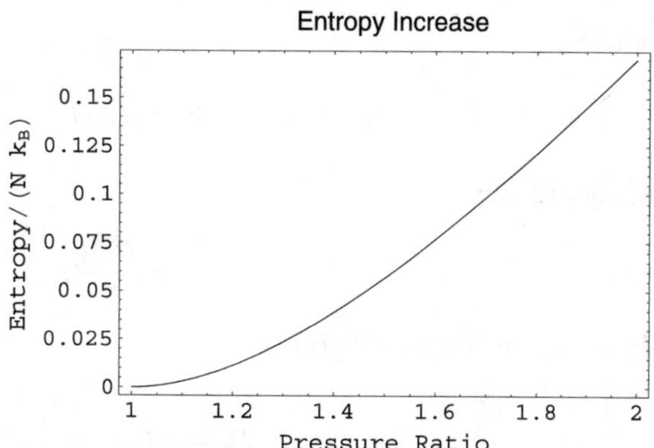

Fig. 5.1 Entropy increase/(Nk_B) versus the ratio of the two pressures

contain the same monatomic ideal gas, are at the same temperature T, but the gas in them initially contains different numbers, N_1 and N_2, of atoms. Also their volumes, V_1 and V_2, and therefore their initial pressures, P_1 and P_2, are different.

Let us calculate any resulting change in the entropy in terms of P_1 and P_2 when the partition is removed and the gas in the two chambers is allowed to mix homogeneously.

Solution 5.2 Two portions of the ideal gas with N_1 and N_2 molecules are initially contained in two different chambers of an adiabatically isolated vessel. These portions are titled 1 and 2. We are told that initially the pressure of the gas in these chambers is P_1 and P_2, respectively. Let us assume their volumes are V_1 and V_2. Because both these chambers are at the same temperature, T, initially the equation of state of the ideal gas is the following:

$$P_1 V_1 = N_1 k_B T \, , \qquad P_2 V_2 = N_2 k_B T \, . \tag{5.37}$$

Once the partition between the chambers is lifted, the gas achieves a joint status with total volume equal to $V_1 + V_2$.

While the fraction of the total number of molecules in portion 1 is $\frac{N_1}{N_1+N_2}$, in portion 2 it is equal to $\frac{N_2}{N_1+N_2}$. Therefore, it is convenient, to treat the isothermal process of mixing as causing a change in the volume of portion 1 from its initial value V_1 to its final value $(V_1 + V_2)(\frac{N_1}{N_1+N_2})$. And similarly, for portion 2, from the initial volume V_2 to its final volume $(V_1 + V_2)(\frac{N_2}{N_1+N_2})$.

The thermodynamics of the process is specified by the first–second law in the form

$$T\mathrm{d}S = \mathrm{d}U + P\mathrm{d}V \, .$$

There is no change in the internal energy of an ideal gas when its temperature, T, is constant. Therefore, $\mathrm{d}U = 0$. Accordingly, for either of the two portions i, where $i = 1$ or 2, we have

$$\mathrm{d}S = \frac{P}{T}\mathrm{d}V \, , \tag{5.38}$$

where

$$PV = Nk_B T\,,$$

or equivalently,

$$\frac{P}{T} = \frac{Nk_B}{V}\,. \tag{5.39}$$

Inserting this into (5.38) gives

$$dS = Nk_B\left(\frac{dV}{V}\right). \tag{5.40}$$

We can now calculate the increase in the entropy of each of the two portions as follows:

$$\begin{aligned}\Delta S_{\text{portion }1} &= \int_{\text{initial-portion }1}^{\text{final-portion }1} dS = N_1 k_B \int_{V_1}^{(V_1+V_2)(\frac{N_1}{N_1+N_2})}\left(\frac{dV}{V}\right)\\ &= N_1 k_B \ln\left[\frac{(V_1+V_2)(\frac{N_1}{N_1+N_2})}{V_1}\right]\end{aligned} \tag{5.41}$$

and

$$\begin{aligned}\Delta S_{\text{portion }2} &= \int_{\text{initial-portion }2}^{\text{final-portion }2} dS = N_2 k_B \int_{V_2}^{(V_1+V_2)(\frac{N_2}{N_1+N_2})}\left(\frac{dV}{V}\right)\\ &= N_2 k_B \ln\left[\frac{(V_1+V_2)(\frac{N_2}{N_1+N_2})}{V_2}\right].\end{aligned} \tag{5.42}$$

The total increase in the entropy therefore is equal to

$$\begin{aligned}\Delta S_{\text{total}} &= \Delta S_{\text{portion }1} + \Delta S_{\text{portion }2}\\ &= k_B \ln\left[\left(\frac{V_1+V_2}{N_1+N_2}\right)^{N_1+N_2}\left(\frac{N_1}{V_1}\right)^{N_1}\left(\frac{N_2}{V_2}\right)^{N_2}\right].\end{aligned} \tag{5.43}$$

Equation (5.43) tells us that if $\frac{N_1}{V_1}$ were equal to $\frac{N_2}{V_2}$, ΔS_{total} would be vanishing. Furthermore, when $\frac{N_1}{V_1} = \frac{N_2}{V_2}$, then P_1 would be equal to P_2. The upshot would be the mixing of two different amounts of the same gas, at the same temperature and pressure. Clearly, this process should not affect the total entropy of the gas.

Now using (5.39) for the portions 1 and 2 in the form

$$V_1 = k_B T\left(\frac{N_1}{P_1}\right) \quad\text{and}\quad V_2 = k_B T\left(\frac{N_2}{P_2}\right),$$

we can represent the above, in terms of the two pressures P_1 and P_2, as follows:

$$\begin{aligned}\Delta S_{\text{total}} &= \Delta S_{\text{portion }1} + \Delta S_{\text{portion }2} \\ &= k_B \ln\left[\left(\frac{\frac{N_1}{P_1}+\frac{N_2}{P_2}}{N_1+N_2}\right)^{N_1+N_2} P_1^{N_1} P_2^{N_2}\right].\end{aligned} \tag{5.44}$$

Again, when $P_1 = P_2$, the mixing would cause no change in the entropy. It is convenient here to rewrite the above in the following simple form:

$$\begin{aligned}\Delta S_{\text{total}} &= \Delta S_{\text{portion }1} + \Delta S_{\text{portion }2} \\ &= k_B(N_1+N_2)\cdot \ln\left[\left(\frac{p+n}{1+n}\right)\left(\frac{1}{p}\right)^{\frac{1}{1+n}}\right],\end{aligned} \tag{5.45}$$

where we have used the notation

$$\frac{P_2}{P_1} = p\,, \qquad \frac{N_2}{N_1} = n\,. \tag{5.46}$$

Of course, when $\frac{N_2}{N_1} = 1 = n$, the result given in (5.45) and (5.46) is identical to that obtained in the preceding problem—see (5.35).

To examine analytically how the entropy changes when the pressures P_1 and P_2 are nearly equal, let us work with the case

$$p = \frac{P_2}{P_1} = 1 + \epsilon\,,$$

where $\epsilon \ll 1$. After a bit of algebra, (5.45) yields the following result:

$$\frac{\Delta S_{\text{total}}}{k_B(N_1+N_2)} = \ln\left[\left(\frac{p+n}{1+n}\right)\left(\frac{1}{p}\right)^{\frac{1}{1+n}}\right]_{p=1+\epsilon} = \left[\frac{n}{2(1+n)^2}\right]\epsilon^2 + O(\epsilon)^3, \tag{5.47}$$

which shows that the entropy increases whenever the pressure difference between the two parts changes. Similarly, near the point where one of the two pressures is extremely small—that is, $p \ll 1$—the entropy increase, i.e., the following quantity,

$$\frac{\Delta S_{\text{total}}}{k_B(N_1+N_2)} \approx \ln\left[\left(\frac{n}{1+n}\right)\left(\frac{1}{p}\right)^{\frac{1}{1+n}}\right]_{p\ll 1}, \tag{5.48}$$

is large and positive.

All this is best shown on a three-dimensional plot given in Fig. 5.2. Notice that the entropy change is positive over the entire volume shown.

Problem 5.3 (Isothermal process with different pressure different number of atoms and different temperature)
An adiabatically isolated vessel has two chambers, to be referred to as 1 and 2. The chambers are separated by a massless partition of zero volume. Both chambers con-

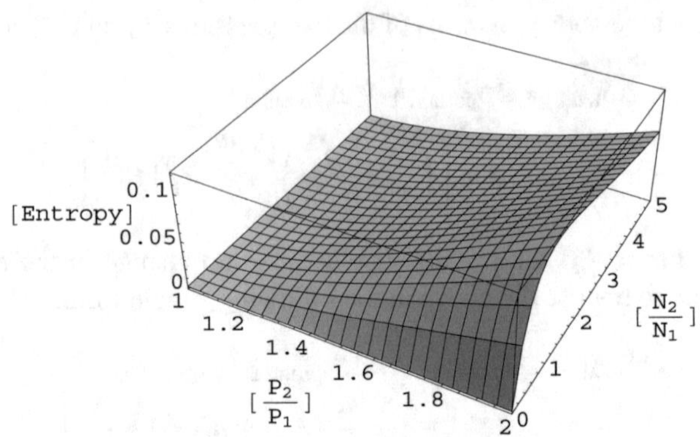

Fig. 5.2 For $1 \leq P_2/P_1 \leq 2$, and $0 \leq N_2/N_1 \leq 5$, twice the entropy increase/$[(N_1 + N_2)k_B]$ is plotted as a function of P_2/P_1 and N_2/N_1. Despite its changed appearance, notice that the curve for $N_2/N_1 = 1$ is identical to that given in Fig. 5.1

tain the same monatomic ideal gas but they are at different temperature, T_1 and T_2. Also the gas in these chambers initially contains different numbers, N_1 and N_2, of atoms. Further, their volumes, V_1 and V_2, and therefore their initial pressures, P_1 and P_2, are different. Calculate any resulting change in the entropy in terms of P_1 and P_2 when the partition is removed and the gas in the two chambers is allowed to mix homogeneously.

Solution 5.3 Two portions of the same ideal gas with N_1 and N_2 molecules are initially contained in two chambers of an adiabatically isolated vessel. The chambers are separated by a partition of zero weight. For convenience, the two portions of gas are titled 1 and 2. We are told that the number of molecules in these portions is N_1 and N_2, respectively. Also initially the temperature of the gas in the two chambers is T_1 and T_2, its pressure is P_1 and P_2, and we assume its volume is V_1 and V_2, respectively. Initially the equation of state of the ideal gas in these chambers is the following:

$$P_1 V_1 = N_1 k_B T_1 , \qquad P_2 V_2 = N_2 k_B T_2 . \tag{5.49}$$

Once the partition between the chambers is removed, the gas achieves a joint status. Because the specific heat of the ideal gas is temperature independent, and the atoms in the two vessels are identical, the only relevant parameter for calculating the final temperature is the number of atoms. Accordingly, the final temperature of the gas in the connected vessel is T_{final}, given by the relationship

$$T_{\text{final}} = (N_1 T_1 + N_2 T_2)/(N_1 + N_2) . \tag{5.50}$$

Clearly, the fraction of the total number of molecules in portion 1 is $\frac{N_1}{N_1+N_2}$, and in portion 2 it is equal to $\frac{N_2}{N_1+N_2}$. Therefore, it is convenient, to treat the physical process of mixing as causing a change in the volume of portion 1 from its initial value V_1 to its final equivalent value $(V_1 + V_2)(\frac{N_1}{N_1+N_2})$, and its initial temperature T_1 to the final temperature T_{final}. Similarly, for portion 2, the initial volume V_2 changes to its final equivalent volume $(V_1 + V_2)(\frac{N_2}{N_1+N_2})$, and its initial temperature T_2 changes to the final temperature T_{final}.

Assuming that the process is quasistatic, its thermodynamics is specified by the first–second law in the form

$$T\text{d}S = \text{d}U + P\text{d}V\,.$$

When the temperature of a monatomic ideal gas, with N_i atoms, increases by an amount $\text{d}T_\text{i}$, its internal energy increases by an amount (see (2.31))

$$\text{d}U_\text{i} = \frac{3}{2}N_\text{i}k_\text{B}\text{d}T_\text{i}\,. \tag{5.51}$$

Accordingly, for either of the two portions i, where $i = 1$ or 2, we have

$$T_\text{i}\text{d}S_\text{i} = \frac{3}{2}N_\text{i}k_\text{B}\text{d}T_\text{i} + P_\text{i}\text{d}V_\text{i}\,, \tag{5.52}$$

where $P_\text{i}V_\text{i} = N_\text{i}k_\text{B}T_\text{i}$, or equivalently,

$$\frac{P_\text{i}}{T_\text{i}} = \frac{N_\text{i}k_\text{B}}{V_\text{i}}\,. \tag{5.53}$$

Dividing (5.52) by T_i and using (5.53) gives

$$\text{d}S_\text{i} = \frac{3}{2}N_\text{i}k_\text{B}\left(\frac{\text{d}T_\text{i}}{T_\text{i}}\right) + N_\text{i}k_\text{B}\left(\frac{\text{d}V_\text{i}}{V_\text{i}}\right). \tag{5.54}$$

We can now calculate the increase in the entropy of each of the two portions as follows:

$$\begin{aligned}\Delta S_{\text{portion 1}} &= \int_{\text{portion 1 initial}}^{\text{portion 1 final}} \text{d}S_1 \\ &= \frac{3}{2}N_1k_\text{B}\int_{T_1}^{T_\text{final}}\left(\frac{\text{d}T_1}{T_1}\right) + N_1k_\text{B}\int_{V_1}^{(V_1+V_2)(\frac{N_1}{N_1+N_2})}\left(\frac{\text{d}V_1}{V_1}\right) \\ &= \frac{3}{2}N_1k_\text{B}\ln\left(\frac{T_\text{final}}{T_1}\right) + N_1k_\text{B}\,\ln\left[\frac{(V_1+V_2)(\frac{N_1}{N_1+N_2})}{V_1}\right]\end{aligned} \tag{5.55}$$

and

$$\begin{aligned}\Delta S_{\text{portion 2}} &= \int_{\text{portion 2 initial}}^{\text{portion 2 final}} \text{d}S_2 \\ &= \frac{3}{2}N_2k_B\int_{T_2}^{T_\text{final}}\left(\frac{\text{d}T_2}{T_2}\right) + N_2k_\text{B}\int_{V_2}^{(V_1+V_2)(\frac{N_2}{N_1+N_2})}\left(\frac{\text{d}V_2}{V_2}\right) \\ &= \frac{3}{2}N_2k_\text{B}\ln\left(\frac{T_\text{final}}{T_2}\right) + N_2k_\text{B}\ln\left[\frac{(V_1+V_2)(\frac{N_2}{N_1+N_2})}{V_2}\right].\end{aligned} \tag{5.56}$$

The total increase in the entropy therefore is equal to

$$\begin{aligned}
\Delta S_{\text{total}} &= \Delta S_{\text{portion 1}} + \Delta S_{\text{portion 2}} \\
&= \frac{3}{2} k_B \ln\left[\left(\frac{T_{\text{final}}}{T_1}\right)^{N_1} \cdot \left(\frac{T_{\text{final}}}{T_2}\right)^{N_2}\right] \\
&+ k_B \ln\left[\left(\frac{V_1 + V_2}{N_1 + N_2}\right)^{N_1+N_2} \left(\frac{N_1}{V_1}\right)^{N_1} \left(\frac{N_2}{V_2}\right)^{N_2}\right].
\end{aligned} \tag{5.57}$$

Note that when $\frac{N_1}{V_1} = \frac{N_2}{V_2}$ the second term on the right-hand side of (5.57) is vanishing. But the first term remains nonzero and positive as long as T_1 and T_2 are different. However, when the two temperatures are equal, this term is also vanishing. And then we are mixing the same gas with itself at the same temperature and pressure.

To change the variables from V_1 and V_2 to the pressures P_1 and P_2, we use (5.53) for the portions 1 and 2, that is, $V_1 = k_B(\frac{T_1 N_1}{P_1})$ and $V_2 = k_B(\frac{T_2 N_2}{P_2})$. Now the dependence upon V_1 and V_2 can be recast in terms of the pressures P_1 and P_2. That is,

$$\begin{aligned}
\Delta S_{\text{total}} &= \Delta S_{\text{portion 1}} + \Delta S_{\text{portion 2}} \\
&= \frac{3}{2} k_B \ln\left[\left(\frac{T_{\text{final}}}{T_1}\right)^{N_1} \cdot \left(\frac{T_{\text{final}}}{T_2}\right)^{N_2}\right] \\
&+ k_B \ln\left[\left(\frac{\frac{T_1 N_1}{P_1} + \frac{T_2 N_2}{P_2}}{N_1 + N_2}\right)^{N_1+N_2} \left(\frac{P_1}{T_1}\right)^{N_1} \left(\frac{P_2}{T_2}\right)^{N_2}\right].
\end{aligned} \tag{5.58}$$

Of course, when $T_2 = T_1 = T = T_{\text{final}}$, the result given in (5.58) is identical to that obtained in the preceding problem—see (5.44).

Problem 5.4 (If $(\frac{p}{t}) = (\frac{\alpha_p}{\chi_t})$ then $(\frac{\partial C_v}{\partial v})_t = 0$)
Given

$$\left(\frac{p}{t}\right) = \left(\frac{\alpha_p}{\chi_t}\right), \tag{5.59}$$

prove

$$\left(\frac{\partial C - v}{\partial v}\right)_t = 0\,, \tag{5.60}$$

and show that this problem is much the same as that investigated in the subsection above called "Dependence of C_v on v."

Solution 5.4 Expressing α_p and χ_t in terms of partial derivatives and using the cyclic identity,

$$\left(\frac{\alpha_p}{\chi_t}\right) = \frac{\frac{1}{v}(\frac{\partial v}{\partial t})_p}{-\frac{1}{v}(\frac{\partial v}{\partial p})_t} = -\left(\frac{\partial v}{\partial t}\right)_p \left(\frac{\partial p}{\partial v}\right)_t = \left(\frac{\partial p}{\partial t}\right)_v, \tag{5.61}$$

we can write (5.59) as

$$\frac{1}{t} = \frac{1}{p}\left(\frac{\alpha_p}{\chi_t}\right) = \frac{1}{p}\left(\frac{\partial p}{\partial t}\right)_v . \tag{5.62}$$

Differentiating with respect to t while keeping v constant, i.e.,

$$\left(\frac{\partial[\frac{1}{t}]}{\partial t}\right)_v = \left(\frac{\partial[\frac{1}{p}(\frac{\partial p}{\partial t})_v]}{\partial t}\right)_v ,$$

leads to the equality

$$-\frac{1}{t^2} = -\frac{1}{p^2}\left[\left(\frac{\partial p}{\partial t}\right)_v\right]^2 + \frac{1}{p}\mathrm{d}ptv .$$

Recast the above as

$$\mathrm{d}ptv = p\left(-\frac{1}{t^2} + \frac{1}{p^2}\left[\left(\frac{\partial p}{\partial t}\right)_v\right]^2\right). \tag{5.63}$$

Now, square each side in (5.62). This gives

$$\left(\frac{1}{t}\right)^2 = \left[\frac{1}{p}\left(\frac{\partial p}{\partial t}\right)_v\right]^2 .$$

Inserting the result into the right-hand side of (5.63) yields

$$\mathrm{d}ptv = 0 . \tag{5.64}$$

Finally, upon invoking the statement recorded in (5.25), i.e.,

$$t\mathrm{d}ptv = \left(\frac{\partial C_v}{\partial v}\right)_t , \tag{5.65}$$

we are led to the desired result

$$\left(\frac{\partial C_v}{\partial v}\right)_t = 0 . \tag{5.66}$$

To show that this problem is much the same as that investigated in the subsection above called "Dependence of C_v on v," let us reexamine the implications of (5.64). That is,

$$\mathrm{d}ptv = \left(\frac{\partial(\frac{\partial p}{\partial t})_v}{\partial t}\right)_v = \left(\frac{\partial X}{\partial t}\right)_v = 0 .$$

Therefore, X does not depend on t and is only a function of v. That is,

$$X = \left(\frac{\partial p}{\partial t}\right)_v = f(v) .$$

Therefore (5.62), i.e.,

$$\frac{1}{t} = \frac{1}{p}\left(\frac{\partial p}{\partial t}\right)_v ,$$

must lead to the result

$$p = f(v)t ,$$

which is much the same as was prescribed[8] in (5.26).

5.2.3 The First–Second Law: Independent *t* and *p*

In (3.83), through the use of the state function, h, called the "enthalpy," i.e., $h = u + pv$, the first law was expressed as a function of dh and dp. For quasistatic heat energy transfer, dq, it can be combined with the second law:

$$\begin{aligned} t\mathrm{d}s = \mathrm{d}q = \mathrm{d}u + p\mathrm{d}v &= \mathrm{d}u + p\mathrm{d}v + v\mathrm{d}p - v\mathrm{d}p \\ &= \mathrm{d}(u + pv) - v\mathrm{d}p = \mathrm{d}h - v\mathrm{d}p . \end{aligned} \tag{5.67}$$

Prove $(\frac{\partial h}{\partial p})_t - v = -t(\frac{\partial v}{\partial t})_p$.

Proof Much as was done for du, the perfect differential dh can be expressed in terms of the variables t and p. That is, one can represent $h = h(t, p)$. This leads to

$$\mathrm{d}h = \left(\frac{\partial h}{\partial t}\right)_p \mathrm{d}t + \left(\frac{\partial h}{\partial p}\right)_t \mathrm{d}p . \tag{5.68}$$

Inserting this into (5.67) yields

$$\mathrm{d}s = \frac{1}{t}\left(\frac{\partial h}{\partial t}\right)_p \mathrm{d}t + \frac{1}{t}\left[\left(\frac{\partial h}{\partial p}\right)_t - v\right]\mathrm{d}p . \tag{5.69}$$

Again, similar to what was done in the preceding section, we invoke the integrability requirement for the perfect differential ds in terms of its variables p and t, i.e., $\frac{\partial^2 s}{\partial p \partial t} = \frac{\partial^2 s}{\partial t \partial p}$. Note that (5.69) specifies the following:

$$\left(\frac{\partial s}{\partial t}\right)_p = \frac{1}{t}\left(\frac{\partial h}{\partial t}\right)_p \tag{5.70}$$

and

$$\left(\frac{\partial s}{\partial p}\right)_t = \frac{1}{t}\left[\left(\frac{\partial h}{\partial p}\right)_t - v\right] . \tag{5.71}$$

[8] The statement prescribed in (5.26) was as follows: $p(v, t) = a(v)t + b(v)$. Clearly, the term $b(v)$ can be transferred to the left-hand side and included in the general term $p(v, t)$.

Therefore, the integrability requirement—i.e., the equality of the two mixed derivatives—yields

$$\frac{\partial^2 s}{\partial p \partial t} = \left(\frac{\partial (\frac{\partial s}{\partial t})_p}{\partial p} \right)_t = \left(\frac{\partial [\frac{1}{t} (\frac{\partial h}{\partial t})_p]}{\partial p} \right)_t = \frac{1}{t} \left[\frac{\partial^2 h}{\partial p \partial t} \right] \tag{5.72}$$

and

$$\frac{\partial^2 s}{\partial t \partial p} = \left(\frac{\partial (\frac{\partial s}{\partial p})_t}{\partial t} \right)_p = -\frac{1}{t^2} \left[\left(\frac{\partial h}{\partial p} \right)_t - v \right] + \frac{1}{t} \left[\frac{\partial^2 h}{\partial t \partial p} - \left(\frac{\partial v}{\partial t} \right)_p \right]. \tag{5.73}$$

Because the left-hand sides of (5.72) and (5.73) are equal, their right-hand sides must also be equal:

$$\frac{1}{t} \left[\frac{\partial^2 h}{\partial p \partial t} \right] = -\frac{1}{t^2} \left[\left(\frac{\partial h}{\partial p} \right)_t - v \right] + \frac{1}{t} \left[\frac{\partial^2 h}{\partial t \partial p} - \left(\frac{\partial v}{\partial t} \right)_p \right].$$

The term $\frac{1}{t}[\frac{\partial^2 h}{\partial p \partial t}]$ on the left-hand side of the above relationship is equal to $\frac{1}{t}[\frac{\partial^2 h}{\partial t \partial p}]$ on the right-hand side. Canceling these and multiplying both sides of the remainder by t^2 gives

$$\left(\frac{\partial h}{\partial p} \right)_t - v = -t \left(\frac{\partial v}{\partial t} \right)_p = -t v \alpha_{\mathrm{p}} = \Delta_{\mathrm{p}}\,. \tag{5.74}$$

Second $t\mathrm{d}s$ Equation

Inserting the result that has been obtained in (5.74) into (5.69) allows us to display the first–second law in the (p, t)-representation

$$\begin{aligned} t\mathrm{d}s &= \left(\frac{\partial h}{\partial t} \right)_p \mathrm{d}t + \left[\left(\frac{\partial h}{\partial p} \right)_t - v \right] \mathrm{d}p = \left(\frac{\partial h}{\partial t} \right)_p \mathrm{d}t - t \left(\frac{\partial v}{\partial t} \right)_p \mathrm{d}p \\ &= C_{\mathrm{p}} \mathrm{d}t - t v \alpha_{\mathrm{p}} \mathrm{d}p = C_{\mathrm{p}} \mathrm{d}t + \Delta_{\mathrm{p}} \mathrm{d}p\,. \end{aligned} \tag{5.75}$$

(Compare and contrast with (3.57) and (3.59).) It is traditional to name (5.75) "The Second $t\mathrm{d}s$ Equation."

Isentropic Process

In order to discuss an isentropic process—i.e., a process in which the entropy of the system remains unchanged—we set $\mathrm{d}s = 0$,

$$0 = C_{\mathrm{p}} (\mathrm{d}t)_{\mathrm{s}} + \Delta_{\mathrm{p}} (\mathrm{d}p)_{\mathrm{s}} = C_{\mathrm{p}} (\mathrm{d}t)_{\mathrm{s}} - t v \alpha_{\mathrm{p}} (\mathrm{d}p)_{\mathrm{s}}\,.$$

This gives the pressure expansion coefficient at constant entropy,

$$\left(\frac{\partial p}{\partial t} \right)_s = -\frac{C_{\mathrm{p}}}{\Delta_{\mathrm{p}}} = \left(\frac{C_{\mathrm{p}}}{t v \alpha_{\mathrm{p}}} \right). \tag{5.76}$$

Compare and contrast this result with that of (5.17), namely

$$\left(\frac{\partial p}{\partial t}\right)_v = \frac{\Delta_\mathrm{v}}{t} = \left(\frac{\alpha_\mathrm{p}}{\chi_\mathrm{t}}\right). \tag{5.77}$$

Notice that three of the four quantities that appear in (5.76)—namely C_p, t, and v—are all > 0. Similarly in (5.77), thermodynamic stability requires χ_t also to be positive. Therefore, the sign of both types of pressure expansion coefficients—namely, $(\frac{\partial p}{\partial t})_\mathrm{v}$ and $(\frac{\partial p}{\partial t})_\mathrm{s}$—is determined only by α_p. Accordingly:

(a) Both $(\frac{\partial p}{\partial t})_\mathrm{v}$ and $(\frac{\partial p}{\partial t})_\mathrm{s}$ have the same sign. Let us contrast this fact with that noted earlier—this fact was recorded immediately following (5.21)—namely, that
(b) $(\frac{\partial v}{\partial t})_\mathrm{p}$ and $(\frac{\partial v}{\partial t})_\mathrm{s}$ have opposite signs.

For instance, at constant pressure, normal systems expand with increase in temperature. This makes $(\frac{\partial v}{\partial t})_\mathrm{p}$ positive. When such is the case, then, according to (5.21), normal systems would shrink in volume with increase in temperature if the system entropy were held constant. This is so because $(\frac{\partial v}{\partial t})_\mathrm{s}$ is negative, a fact that is insured by the positivity of the specific heat C_v and the isothermal compressibility χ_t—both required for the stability of thermal equilibrium. This requirement is noted in Sect. 9.4 and Sect. 9.4.1 of chapter titled "Le Châtelier Principle."

Dependence of C_p on p

Much like the dependence of C_v on volume, we can also explore the dependence of C_p on pressure. To this end, let us slightly rearrange (5.75) and write it as follows:

$$\mathrm{d}s = \frac{C_\mathrm{p}}{t}\mathrm{d}t - \left(\frac{\partial v}{\partial t}\right)_p \mathrm{d}p\,.$$

Again, invoking the exact differentiability of ds, we have

$$\left(\frac{\partial s}{\partial t}\right)_p = \frac{C_\mathrm{p}}{t}, \qquad \left(\frac{\partial s}{\partial p}\right)_t = -\left(\frac{\partial v}{\partial t}\right)_p.$$

The equality of the mixed second derivatives, $\frac{\partial^2 s}{\partial p \partial t}$ and $\frac{\partial^2 s}{\partial t \partial p}$, where

$$\frac{\partial^2 s}{\partial p \partial t} = \left(\frac{\partial(\frac{C_\mathrm{p}}{t})}{\partial p}\right)_t = \frac{1}{t}\left(\frac{\partial C_\mathrm{p}}{\partial p}\right)_t,$$

$$\frac{\partial^2 s}{\partial t \partial p} = \left(\frac{\partial(-(\frac{\partial v}{\partial t})_p)}{\partial t}\right)_p = -\mathrm{d}vtp\,,$$

yields the desired result,

$$\left(\frac{\partial C_\mathrm{p}}{\partial p}\right)_t = -t\mathrm{d}vtp\,. \tag{5.78}$$

If, for constant pressure, the volume of a thermodynamic system which can be represented as a linear function of the temperature, $\mathrm{d}vtp$, is vanishing. To see this, let

$$v(p,t) = a(p)t + b(p)\,. \tag{5.79}$$

Then

$$\left(\frac{\partial v}{\partial t}\right)_p = a(p)\,, \quad \mathrm{d}vtp = 0\,.$$

Thus (5.78) leads to an important result, namely that if for constant pressure, the volume v is linearly dependent on the temperature[9]—for example, this is the case in (5.79)—the specific heat C_p is independent of the pressure for all isothermal processes,

$$\left(\frac{\partial C_\mathrm{p}}{\partial p}\right)_t = -t\mathrm{d}vtp = 0\,. \tag{5.80}$$

Alternate Proof of $(\frac{\partial h}{\partial p})_t - v = -t(\frac{\partial v}{\partial t})_p$

The importance of the relationship given in (5.74),

$$\left(\frac{\partial h}{\partial p}\right)_t = v - t\left(\frac{\partial v}{\partial t}\right)_p\,,$$

warrants an alternate derivation.

To this purpose, it is helpful to construct a new state function Ψ,

$$\Psi = s - (h/t)\,,$$

whereby[10]

$$\mathrm{d}\Psi = \mathrm{d}s - \frac{\mathrm{d}h}{t} + \frac{h}{t^2}\mathrm{d}t\,.$$

The first–second law in the form $\mathrm{d}s - \frac{\mathrm{d}h}{t} = -(\frac{v}{t})\mathrm{d}p$ allows us to rewrite the above as

$$\mathrm{d}\Psi = -\left[\frac{v}{t}\right]\mathrm{d}p + \left[\frac{h}{t^2}\right]\mathrm{d}t\,.$$

Upon invoking the usual integrability requirement, i.e.,

$$-\left(\frac{\partial[\frac{v}{t}]}{\partial t}\right)_p = \left(\frac{\partial[\frac{h}{t^2}]}{\partial p}\right)_t\,,$$

[9]Note that this, for instance, is the case for an ideal gas.

[10]Note that because $\mathrm{d}s$, $\mathrm{d}t$, and $\mathrm{d}h$ are exact differentials, the same is true for $\mathrm{d}\Psi$.

we are led to the equality

$$\left[\frac{v}{t^2}\right]-\frac{1}{t}\left(\frac{\partial v}{\partial t}\right)_p=\left[\frac{1}{t^2}\right]\left(\frac{\partial h}{\partial p}\right)_t .$$

Multiplying both sides by t^2 and transferring v from the left side to the right yields the desired result

$$-t\left(\frac{\partial v}{\partial t}\right)_p=-v+\left(\frac{\partial h}{\partial p}\right)_t . \tag{5.81}$$

The First and Second $t\mathrm{d}s$ Equations Together

It is instructive to look at the first and second $t\mathrm{d}s$ equations together. To this end, we display, side by side and in seriatim, lines of (5.18) and (5.75) and note that $t\mathrm{d}s$ can be represented in any of the following forms:

$$\begin{aligned}
t ds &= t ds,\\
C_\mathrm{v}\mathrm{d}t+\left[\left(\frac{\partial u}{\partial v}\right)_t+p\right]\mathrm{d}v &= C_\mathrm{p}\mathrm{d}t+\left[\left(\frac{\partial h}{\partial p}\right)_t-v\right]\mathrm{d}p\,,\\
C_\mathrm{v}\mathrm{d}t+t\left(\frac{\partial p}{\partial t}\right)_v\mathrm{d}v &= C_\mathrm{p}\mathrm{d}t-t\left(\frac{\partial v}{\partial t}\right)_p\mathrm{d}p\,,\\
C_\mathrm{v}\mathrm{d}t+t\left(\frac{\alpha_\mathrm{p}}{\chi_\mathrm{t}}\right)\mathrm{d}v &= C_\mathrm{p}\mathrm{d}t-tv\alpha_\mathrm{p}\mathrm{d}p\,,\\
C_\mathrm{v}\mathrm{d}t+\Delta_\mathrm{v}\mathrm{d}v &= C_\mathrm{p}\mathrm{d}t+\Delta_\mathrm{p}\mathrm{d}p\,.
\end{aligned}$$

Thus we can write

$$\begin{aligned}
(C_\mathrm{p}-C_\mathrm{v})\mathrm{d}t &= \left[\left(\frac{\partial u}{\partial v}\right)_t+p\right]\mathrm{d}v-\left[\left(\frac{\partial h}{\partial p}\right)_t-v\right]\mathrm{d}p\\
&= t\left(\frac{\partial p}{\partial t}\right)_v\mathrm{d}v+t\left(\frac{\partial v}{\partial t}\right)_p\mathrm{d}p\\
&= t\left(\frac{\alpha_\mathrm{p}}{\chi_\mathrm{t}}\right)\mathrm{d}v+tv\alpha_\mathrm{p}\mathrm{d}p=\Delta_\mathrm{v}\mathrm{d}v-\Delta_\mathrm{p}\mathrm{d}p\,.
\end{aligned} \tag{5.82}$$

The difference in specific heats at constant pressure and constant volume is of great thermodynamic interest. For this reason, we derive it from two different processes. First, a process at constant pressure, that is, where $\mathrm{d}p=0$. We get

$$\begin{aligned}
C_\mathrm{p}-C_\mathrm{v} &= \left[\left(\frac{\partial u}{\partial v}\right)_t+p\right]\left(\frac{\partial v}{\partial t}\right)_p=t\left(\frac{\partial p}{\partial t}\right)_v\left(\frac{\partial v}{\partial t}\right)_p\\
&= t\left(\frac{\alpha_\mathrm{p}}{\chi_\mathrm{t}}\right)\left(\frac{\partial v}{\partial t}\right)_p=\Delta_\mathrm{v}\left(\frac{\partial v}{\partial t}\right)_p=tv\left(\frac{\alpha_\mathrm{p}^2}{\chi_\mathrm{t}}\right).
\end{aligned} \tag{5.83}$$

Second, a process at constant volume, that is, where $\mathrm{d}v = 0$, and (5.82) can be written in the following form:

$$C_\mathrm{p} - C_\mathrm{v} = -\left[\left(\frac{\partial h}{\partial p}\right)_t - v\right]\left(\frac{\partial p}{\partial t}\right)_v = t\left(\frac{\partial v}{\partial t}\right)_p\left(\frac{\partial p}{\partial t}\right)_v$$

$$= tv\alpha_\mathrm{p}\left(\frac{\partial p}{\partial t}\right)_v = -\Delta_\mathrm{p}\left(\frac{\partial p}{\partial t}\right)_v = tv\left(\frac{\alpha_\mathrm{p}^2}{\chi_\mathrm{t}}\right). \quad (5.84)$$

For this case, all the hard work has already been done in the chapter on the first law—see the work leading to (3.67), etc. And what remains is described below in the section on the third $t\mathrm{d}s$ equation.

The Third $t\mathrm{d}s$ Equation

Using the fact that quasistatic transfer of infinitesimal amount of heat energy, $\mathrm{d}q$, is equal to $t\mathrm{d}s$, (3.67) can be changed to the following:

$$t\mathrm{d}s = C_\mathrm{v}\left(\frac{\partial t}{\partial p}\right)_v \mathrm{d}p + C_\mathrm{p}\left(\frac{\partial t}{\partial v}\right)_p \mathrm{d}v$$

$$= C_\mathrm{v}\left(\frac{\chi_\mathrm{t}}{\alpha_\mathrm{p}}\right)\mathrm{d}p + \left(\frac{C_\mathrm{p}}{v\alpha_\mathrm{p}}\right)\mathrm{d}v. \quad (5.85)$$

Traditionally, (5.85) has been called "The Third $t\mathrm{d}s$ Equation."

Isentropic Processes

As before, an isentropic process can be examined by setting $\mathrm{d}s = 0$. Thus

$$0 = C_\mathrm{v}\left(\frac{\partial t}{\partial p}\right)_v (\mathrm{d}p)_\mathrm{s} + C_\mathrm{p}\left(\frac{\partial t}{\partial v}\right)_p (\mathrm{d}v)_\mathrm{s}.$$

Dividing both sides by $(\mathrm{d}p)_\mathrm{s}$ yields

$$0 = C_\mathrm{v}\left(\frac{\partial t}{\partial p}\right)_v + C_\mathrm{p}\cdot\left(\frac{\partial t}{\partial v}\right)_p\cdot\left(\frac{\partial v}{\partial p}\right)_s$$

$$= C_\mathrm{v}\left(\frac{\chi_\mathrm{t}}{\alpha_\mathrm{p}}\right) - C_\mathrm{p}\cdot v\left(\frac{\partial t}{\partial v}\right)_p\cdot\left(\frac{-1}{v}\left(\frac{\partial v}{\partial p}\right)_s\right)$$

$$= C_\mathrm{v}\left(\frac{\chi_\mathrm{t}}{\alpha_\mathrm{p}}\right) - C_\mathrm{p}\cdot\left(\frac{1}{\alpha_\mathrm{p}}\right)\cdot\chi_\mathrm{s}. \quad (5.86)$$

Analogous to the definition of isothermal compressibility $\chi_\mathrm{t} = \frac{-1}{v}(\frac{\partial v}{\partial p})_\mathrm{t}$, $\chi_\mathrm{s} = \frac{-1}{v}(\frac{\partial v}{\partial p})_\mathrm{s}$ is called the isentropic, or the adiabatic, compressibility. And we see from the above that

$$\gamma = \left(\frac{\chi_\mathrm{t}}{\chi_\mathrm{s}}\right) = \left(\frac{C_\mathrm{p}}{C_\mathrm{v}}\right). \quad (5.87)$$

Because C_p is always larger than C_v, χ_s is always smaller than its isothermal counterpart χ_t.

A technical explanation for this behavior can be found by examining (5.76), that is,

$$\left(\frac{\partial p}{\partial t}\right)_s = \left(\frac{C_p}{t v \alpha_p}\right).$$

Because in a normal system C_p, t, v, and α_p are all positive, an adiabatic increase in pressure is consistent with rise in temperature.[11] In turn, because of the positivity of α_p. That is,

$$\alpha_p = \frac{1}{v}\left(\frac{\partial v}{\partial t}\right)_p > 0,$$

this rise in temperature causes some increase in the volume. Clearly, such increase in the volume counteracts the effect—namely the decrease in volume—that the original increase in the pressure would normally have had.

Therefore, less decrease in volume is obtained than would be the case in a corresponding isothermal compression. The net result is that the adiabatic compressibility, $\chi_s = -\frac{1}{v}(\frac{\partial v}{\partial p})_s$, is smaller than the isothermal compressibility χ_t.

5.3 Velocity of Sound: Newton's Solution

There is an interesting historical anecdote, featuring Isaac Newton, with regard to (5.87). When the measured value of the density $\rho(t)$ of air at temperature t was used in the formula that was originally proposed for calculating the velocity of sound, i.e.,

$$c_{\text{original}} \sim \left[\rho(t)\chi_t\right]^{-0.5}, \tag{5.88}$$

it seemed to underestimate the result.

Newton noticed that typical wavelengths of ordinary sound are too long—being of the order of a meter or longer—to allow adequate thermalization during a typical oscillation time period $\sim 3 \times 10^{-3}$ s. Accordingly, he realized that compressions and rarefactions in the air occur without adequate heat energy exchange. Thus the energy transfer is not isothermal. Indeed, it is close to being adiabatic—meaning it occurs almost without any heat energy exchange. It was therefore suggested that rather than the isothermal compressibility χ_t, the formula for the velocity should involve adiabatic compressibility χ_s. That is,

$$c_{\text{Newton}} = \left[\rho(t)\chi_s\right]^{-0.5} = c_{\text{original}}\sqrt{\frac{\chi_t}{\chi_s}}. \tag{5.89}$$

[11]This is so because when $\frac{C_p}{p v \alpha_p} > 0$, positive $(\mathrm{d}p)_s$ implies positive $(\mathrm{d}t)_s$.

The net result of this interchange—from χ_t to χ_s—is that the original low estimate for the sound velocity given in (5.88) is increased when it is multiplied by the factor $\sqrt{\frac{\chi_t}{\chi_s} = \frac{C_p}{C_v} = \gamma = \frac{7}{5}}$.

It is amusing to carry out this exercise for a diatomic ideal gas of atomic weight 29 at $t = 293$ K. This, of course, is an approximation for air at room temperature.

The density of air is

$$\rho(t) = \frac{M}{v} = \left(\frac{Mp}{Rt}\right),$$

where M is the mass of one mole of air and v is its volume.

To calculate χ_t, let us differentiate with respect to p the equation of state for one mole, i.e.,

$$pv = Rt\,,$$

while t is kept constant. We get

$$\left(\frac{\partial v}{\partial p}\right)_t = -\frac{v}{p},$$

and therefore

$$\chi_t = -\frac{1}{v}\left(\frac{\partial v}{\partial p}\right)_t = \frac{1}{p}\,.$$

Accordingly, the original formula (5.88) leads to the result

$$c_{\text{original}} \sim \left[\rho(t)\chi_t\right]^{-0.5} = \left(\frac{Rt}{M}\right)^{0.5} = \left(\frac{8.31 \times 293}{29 \times 10^{-3}}\right)^{0.5} = 290 \text{ m/s}\,.$$

This should be contrasted with the experimental result

$$c_{\text{experiment}} = 343 \text{ m/s}\,.$$

However, when 290 is multiplied by $\sqrt{\gamma} = \sqrt{7/5}$ one gets

$$c_{\text{newton}} = 343 \text{ m/s}\,.$$

That is dead on. Bravo Newton!

Problems 5.5–5.15 are worked out below.

Problem 5.5 (Gas in contact with reservoir: change in u and s)
One mole of a gas which has an equation of state

$$p(v - b) = Rt\,, \tag{5.90}$$

where b is a constant, is in contact with an infinite thermal reservoir at temperature t_c. In an isothermal quasistatic expansion, the gas increases its volume from v_i to v_f and in the process does work equal to dw_c. Calculate dw_c, the resultant change in the internal energy du_c, the entropy of the gas and the reservoir.

Solution 5.5 Work done by the gas is

$$\mathrm{d}w_{\mathrm{c}} = \int_{\mathrm{v_i}}^{\mathrm{v_f}} p\mathrm{d}v = Rt_{\mathrm{c}} \int_{\mathrm{v_i}}^{\mathrm{v_f}} \frac{\mathrm{d}v}{v-b} = Rt_{\mathrm{c}} \ln\left[(v_{\mathrm{f}}-b)/(v_{\mathrm{i}}-b)\right]. \tag{5.91}$$

From (5.17) we have

$$\left(\frac{\partial u}{\partial v}\right)_t = -p + t\left(\frac{\partial p}{\partial t}\right)_v .$$

For the gas under review,

$$\left(\frac{\partial p}{\partial t}\right)_v = R/(v-b) . \tag{5.92}$$

Therefore,

$$\left(\frac{\partial u}{\partial v}\right)_t = -p + Rt/(v-b) = 0 . \tag{5.93}$$

Also, because

$$\mathrm{d}u = \left(\frac{\partial u}{\partial v}\right)_t \mathrm{d}v + \left(\frac{\partial u}{\partial t}\right)_v \mathrm{d}t ,$$

at $t = t_{\mathrm{c}}$, for the given isothermal process, i.e., where $\mathrm{d}t_{\mathrm{c}} = 0$, we have

$$\mathrm{d}u_{\mathrm{c}} = \left(\frac{\partial u}{\partial v}\right)_t \mathrm{d}v_{\mathrm{c}} + \left(\frac{\partial u}{\partial t}\right)_v \mathrm{d}t_{\mathrm{c}} = 0 \times (\mathrm{d}v_{\mathrm{c}}) + \left(\frac{\partial u}{\partial t}\right)_v \times 0 = 0 . \tag{5.94}$$

For reversible transfer of heat energy $\mathrm{d}q_{\mathrm{c}}^{\mathrm{rev}}$ to the gas, the first–second law dictates

$$\mathrm{d}q_{\mathrm{c}}^{\mathrm{rev}} = t_{\mathrm{c}}\mathrm{d}s_{\mathrm{c}} = \mathrm{d}u_{\mathrm{c}} + \mathrm{d}w_{\mathrm{c}} = 0 + \mathrm{d}w_{\mathrm{c}} = \mathrm{d}w_{\mathrm{c}} .$$

Thus

$$\mathrm{d}s_{\mathrm{c}} = \mathrm{d}w_{\mathrm{c}}/t_{\mathrm{c}} = t_{\mathrm{c}} \ln\left[(v_{\mathrm{f}}-b)/(v_{\mathrm{i}}-b)\right] \tag{5.95}$$

is the increase in the entropy of the gas. Because of reversible operation, the entropy of the universe remains unchanged. Therefore the entropy of the reservoir decreases by the same amount $\mathrm{d}s_{\mathrm{c}}$.

Problem 5.6 (ΔW, ΔU, ΔS for $P = (a/2b)T^2 + (1/bV)$)
A substance, which obeys the equation of state

$$P = (a/2b)T^2 + (1/bV) , \tag{5.96}$$

where a and b are constants, is in contact with an infinite thermal reservoir at temperature T_0. In an isothermal, quasistatic expansion, its volume is increased from V_0 to V and in the process it does work ΔW. Calculate ΔW, the resultant change in the internal energy ΔU, as well as the entropy change of the substance and the reservoir.

Solution 5.6 The work done is

$$\Delta W = \int_{V_0}^{V} P\mathrm{d}V = (a/2b)T_0^2 \int_{V_0}^{V} \mathrm{d}V + (1/b) \int_{V_0}^{V} (1/V)\mathrm{d}V$$
$$= (a/2b)T_0^2(V - V_0) + (1/b)\ln(V/V_0)\,. \tag{5.97}$$

From the identity

$$\left(\frac{\partial U}{\partial V}\right)_T = T\left(\frac{\partial P}{\partial T}\right)_V - P\,, \tag{5.98}$$

follows the result

$$\left(\frac{\partial U}{\partial V}\right)_{T=T_0} = (a/2b)T_0^2 - (1/bV)\,.$$

Integration leads to

$$\Delta U = (a/2b)T_0^2(V - V_0) - (1/b)\ln(V/V_0)\,. \tag{5.99}$$

The use of the first–second law in the form

$$\Delta Q = T_0\Delta S = \Delta U + \Delta W$$

yields

$$\Delta S = aT_0(V - V_0)/b\,. \tag{5.100}$$

Because the increase in the volume occurs reversibly, the gain in the entropy of the substance is compensated by an equal loss in the entropy of the reservoir.

Problem 5.7 ($\Delta W, \Delta U, \Delta S$ for $pv = Rt(1 + b_2/v + b_3/v^2 + \cdots + b_n/v^{n+1})$)
One mole of a gas that obeys an equation of state of the form

$$pv = Rt\big(1 + b_2/v + b_3/v^2 + \cdots + b_n/v^{n+1}\big)\,, \tag{5.101}$$

where b_n is the so-called nth virial coefficient which depends on the inter-molecular force, is in contact with an infinite thermal reservoir at temperature t_C. In an isothermal, quasistatic expansion the gas increases its volume from v_i to v_f and in the process does work Δw. Calculate Δw, the resultant change in the internal energy Δu, as well as the entropy Δs of the gas and the reservoir.

Solution 5.7 Work done by the gas is

$$\Delta w = \int_{v_i}^{v_f} p\mathrm{d}v = Rt_c\big\{\ln(v_f/v_i) + b_2(1/v_i - 1/v_f)$$
$$+ 2b_3\big(1/v_i^2 - 1/v_f^2\big) + \cdots + nb_{n+1}\big(1/v_i^n - 1/v_f^n\big)\big\}\,. \tag{5.102}$$

Because for constant volume, the equation of state specifies a linear dependence of p on t, for isothermal processes the internal energy u does not change with volume. Of course, this is also clear from (5.98) which states

$$\left(\frac{\partial u}{\partial v}\right)_t = t\left(\frac{\partial p}{\partial t}\right)_v - p = 0\,.$$

Accordingly, internal energy is unaffected by the isothermal volume expansion and the increase in the entropy of the gas is simply determined by the work done,

$$\Delta s = \frac{\Delta w}{t_c}\,.$$

Note that, due to the process being reversible, the entropy of the universe remains unchanged. But the entropy of the reservoir is reduced.

Problem 5.8 Show $\frac{\chi_t}{\chi_s} = 1 - \frac{\alpha_p}{\alpha_s}$.
The above is an important relationship.

Solution 5.8 Recall (3.92) and divide both sides by C_v to get

$$\frac{C_p - C_v}{C_v} = \frac{t v \alpha_p^2}{C_v \chi_t}\,. \tag{5.103}$$

On the left-hand side replace $\frac{C_p}{C_v}$ by $\frac{\chi_t}{\chi_s}$ and obtain

$$\frac{\chi_t}{\chi_s} = 1 + t\left(\frac{\alpha_p}{C_v \chi_t}\right) v \alpha_p\,.$$

From (5.19) we have

$$C_v(dt)_s = -t\left(\frac{\alpha_p}{\chi_t}\right)(dv)_s\,,$$

which can be written as

$$\frac{\alpha_p}{C_v \chi_t} = -\frac{1}{t}\left(\frac{\partial t}{\partial v}\right)_s\,.$$

Therefore

$$\frac{\chi_t}{\chi_s} = 1 + t\left[-\frac{1}{t}\left(\frac{\partial t}{\partial v}\right)_s\right] v \alpha_p = 1 - \left[v\left(\frac{\partial t}{\partial v}\right)_s\right]\alpha_p = 1 - \frac{\alpha_p}{\alpha_s}\,. \tag{5.104}$$

Q.E.D.

Problem 5.9 Show $C_p(\frac{\partial t}{\partial s})_h = t - \alpha_p t^2$.
Let us determine the validity of the above relationship.

Solution 5.9 Begin with

$$t\mathrm{d}s = \mathrm{d}h - v\mathrm{d}p\,.$$

For constant enthalpy, $\mathrm{d}h = 0$. Therefore

$$t(\mathrm{d}s)_\mathrm{h} = -v(\mathrm{d}p)_\mathrm{h}\,.$$

Next, divide both sides by $t(\mathrm{d}t)_\mathrm{h}$ and get

$$\left(\frac{\partial s}{\partial t}\right)_h = -\frac{v}{t}\left(\frac{\partial p}{\partial t}\right)_h,$$

then invert the result and multiply by C_p to obtain

$$C_\mathrm{p}\left(\frac{\partial t}{\partial s}\right)_h = C_\mathrm{p}\left[-\frac{t}{v}\left(\frac{\partial t}{\partial p}\right)_h\right]. \tag{5.105}$$

Next, we need to show the following equality:

$$C_\mathrm{p}\left[-\frac{t}{v}\left(\frac{\partial t}{\partial p}\right)_h\right] = t - \alpha_\mathrm{p} t^2\,.$$

To this end, let us recall that

$$C_\mathrm{p} = \left(\frac{\partial h}{\partial t}\right)_p.$$

As such the right-hand side of (5.105) is written as

$$C_\mathrm{p}\left[-\frac{t}{v}\left(\frac{\partial t}{\partial p}\right)_h\right] = \frac{t}{v}\left[-C_\mathrm{p}\left(\frac{\partial t}{\partial p}\right)_h\right] = \frac{t}{v}\left[-\left(\frac{\partial h}{\partial t}\right)_p\left(\frac{\partial t}{\partial p}\right)_h\right] = \frac{t}{v}\left(\frac{\partial h}{\partial p}\right)_t. \tag{5.106}$$

Note that the last term in (5.106) has been obtained by the use of the cyclic identity.

Recall the relationship (5.74), namely

$$\left(\frac{\partial h}{\partial p}\right)_t = v - \alpha_\mathrm{p} v t\,,$$

and insert it into the last term of (5.106). This leads to the desired result

$$C_\mathrm{p}\left(\frac{\partial t}{\partial s}\right)_h = C_\mathrm{p}\left[-\frac{t}{v}\left(\frac{\partial t}{\partial p}\right)_h\right] = \frac{t}{v}\left(\frac{\partial h}{\partial p}\right)_t = \frac{t}{v}(v - \alpha_\mathrm{p} v t) = t - \alpha_\mathrm{p} t^2\,. \tag{5.107}$$

Q.E.D.

Problem 5.10 Show $(\frac{v}{C_\mathrm{p}}) = (\frac{\partial t}{\partial p})_s - (\frac{\partial t}{\partial p})_h$.

In order to ascertain the validity of the above relationship proceed as follows.

Solution 5.10 As in the preceding problem, we use the cyclic identity

$$\left(\frac{\partial h}{\partial p}\right)_t = -\left(\frac{\partial h}{\partial t}\right)_p \left(\frac{\partial t}{\partial p}\right)_h = -C_\mathrm{P}\left(\frac{\partial t}{\partial p}\right)_h . \tag{5.108}$$

According to (5.74) the left-hand side of (5.108) is equal to $v - tv\alpha_\mathrm{p}$. The use of (5.76) allows us to replace $tv\alpha_\mathrm{p}$ by $C_\mathrm{P}(\frac{\partial t}{\partial p})_\mathrm{s}$. As a result the left-hand side of (5.108) can be recast as

$$\left(\frac{\partial h}{\partial p}\right)_t = v - tv\alpha_\mathrm{p} = v - C_\mathrm{P}\left(\frac{\partial t}{\partial p}\right)_s . \tag{5.109}$$

The left-hand sides of (5.108) and (5.109) are identical. As a result, their right-hand sides can also be equated. This leads to the equality

$$-C_\mathrm{P}\left(\frac{\partial t}{\partial p}\right)_h = v - C_\mathrm{P}\left(\frac{\partial t}{\partial p}\right)_s .$$

A alight rearrangement leads to the desired result

$$\left(\frac{v}{C_\mathrm{P}}\right) = \left(\frac{\partial t}{\partial p}\right)_s - \left(\frac{\partial t}{\partial p}\right)_h . \tag{5.110}$$

Problem 5.11 (Ideal rubber, isothermally stretched)
Ideal rubber in the form of a band of length l is found to have the following equation of state:

$$f = \beta t\left[\frac{l}{l_0} - \left(\frac{l_0}{l}\right)^2\right] . \tag{5.111}$$

Here f is the tension, l_0 is the unstretched length, β is a constant depending on the properties of given rubber, and t is the temperature.

(a) Show the internal energy is a function only of the temperature.
(b) Calculate the work w done on, and the heat energy q transferred to, the rubber band when it is isothermally stretched from l_0 to l. The temperature here is t_0.
(c) Given the specific heat at constant length is C_l, and the stretching is done adiabatically, how does the temperature of the band, and the corresponding work w' done on it change?

Solution 5.11 Under tension f, the work $\mathrm{d}w$ done on the rubber in a quasistatic extension $\mathrm{d}l$ is equal to $f\mathrm{d}l$. (Note that the work done by the rubber band is equal to $-f\mathrm{d}l$.) Hence the relevant statement of the first–second law is

$$t\mathrm{d}s = \mathrm{d}u - f\mathrm{d}l . \tag{5.112}$$

Clearly, this equation merely replaces p by $-f$ and v by l. Thus all the previous results can be readily transferred.

(a) For instance, (5.11) is recast as

$$-f + \left(\frac{\partial u}{\partial l}\right)_t = -t\left(\frac{\partial f}{\partial t}\right)_l . \tag{5.113}$$

The derivative $(\frac{\partial f}{\partial t})_l$ is easily found from the equation of state (5.111) as

$$\left(\frac{\partial f}{\partial t}\right)_l = \beta\left[\frac{l}{l_0} - \left(\frac{l_0}{l}\right)^2\right] = \frac{f}{t} . \tag{5.114}$$

Inserting this into (5.113) yields

$$\left(\frac{\partial u}{\partial l}\right)_t = 0 . \tag{5.115}$$

Thus, when the temperature is constant, u is not a function of the length l. The above can also be written as

$$\left(\frac{\partial u}{\partial l}\right)_t = \left(\frac{\partial u}{\partial f}\right)_t \left(\frac{\partial f}{\partial l}\right)_t = 0 .$$

Further, because of the equation of state (5.111), quite obviously $(\frac{\partial f}{\partial l})_t \neq 0$. Hence, the following must be the case:

$$\left(\frac{\partial u}{\partial f}\right)_t = 0 . \tag{5.116}$$

In other words, at constant temperature, the internal energy is also not a function of the tension f. Accordingly, u is a function only of t.

(b) When stretched isothermally, the work done *on the* band is

$$w = \int_{l_0}^{l} f \mathrm{d}l = \frac{\beta t}{2 l_0}(l^2 - l_0{}^2) + \beta t l_0^2\left(\frac{1}{l} - \frac{1}{l_0}\right) . \tag{5.117}$$

Noting the heat energy transferred *to* the band is q, and the fact that during the isothermal process the internal energy remains constant, the first law dictates

$$q = -w .$$

(c) Because the internal energy depends only on the temperature, the first–second law in (5.112) can be written as

$$t\mathrm{d}s = \left(\frac{\partial u}{\partial t}\right)_l \mathrm{d}t - f\mathrm{d}l = C_l \mathrm{d}t - f\mathrm{d}l . \tag{5.118}$$

At constant entropy, quasistatic stretching of the band by an infinitesimal amount $\mathrm{d}l$ requires the expenditure of work equal to $f(\mathrm{d}l)_{\mathrm{s}}$. Setting s constant in (5.118) gives

$$C_l(\mathrm{d}t)_{\mathrm{s}} = f(\mathrm{d}l)_{\mathrm{s}} = \beta t\left[\frac{l}{l_0} - \left(\frac{l_0}{l}\right)^2\right](\mathrm{d}l)_{\mathrm{s}}\,. \tag{5.119}$$

Thus total work done on the band during isentropic stretching from l_0 to l is

$$w' = \int_{l_0}^{l} f(\mathrm{d}l)_{\mathrm{s}} = C_{\mathrm{l}} \int_{t_0}^{t} (\mathrm{d}t)_{\mathrm{s}} = C_{\mathrm{l}}(t - t_0)\,. \tag{5.120}$$

Clearly, all we need to do now is find t.

To this purpose, transfer t to the left-hand side in (5.119) and integrate. This results in separating the variables l and t, and we get

$$\begin{aligned} C_{\mathrm{l}}\int_{t_0}^{T}\frac{(\mathrm{d}t)_{\mathrm{s}}}{t} &= C_{\mathrm{l}}\ln\left(\frac{t}{t_0}\right) = \int_{\mathrm{l_0}}^{l} ft(\mathrm{d}l)_{\mathrm{s}} = \beta\int_{l_0}^{l}\left[\frac{l}{l_0} - \left(\frac{l_0}{l}\right)^2\right](\mathrm{d}l)_{\mathrm{s}} \\ &= \frac{\beta}{2l_0}(l^2 - l_0{}^2) + \beta l_0^2\left(\frac{1}{l} - \frac{1}{l_0}\right). \end{aligned} \tag{5.121}$$

The temperature t can readily be found by first dividing both sides by C_{l} and then exponentiating them:

$$t = t_0 \exp\left\{\left(\frac{1}{C_{\mathrm{l}}}\right)\cdot\left[\left(\frac{\beta}{2l_0}\right)(l^2 - l_0{}^2) + \beta l_0^2\left(\frac{1}{l} - \frac{1}{l_0}\right)\right]\right\}.$$

5.4 Van der Waal's Gas: Energy and Entropy Change

Now we work out Problem 5.12 given below.

Problem 5.12 (Energy and entropy change in Van der Waal's gas)
One mole of a Van der Waal's gas,

$$\left(p + \frac{a}{v^2}\right)(v - b) = Rt\,,$$

where a and b are constants, is in contact with an infinite thermal reservoir at temperature t_{c}. In an isothermal, quasistatic expansion the gas increases its volume from v_{i} to v_{f} and in the process does work $(\Delta w)_{\mathrm{t_c}}$. Calculate $(\Delta w)_{\mathrm{t_c}}$ and the change in the internal energy $(\Delta u)_{\mathrm{t_c}}$. What is the change in the entropy of the gas, and the reservoir?

Solution 5.12 At temperature t_{c}, when the gas expands, $v_{\mathrm{i}} \to v_{\mathrm{f}}$, it does work Δw. Thus

$$(\Delta w)_{\mathrm{t_c}} = \int_{v_{\mathrm{i}}}^{v_{\mathrm{f}}} P\mathrm{d}v = Rt_{\mathrm{c}}\ln\left[\frac{v_{\mathrm{f}} - b}{v_{\mathrm{i}} - b}\right] - a\left(\frac{1}{v_{\mathrm{i}}} - \frac{1}{v_{\mathrm{f}}}\right).$$

From (5.17) we have

$$\left(\frac{\partial u}{\partial v}\right)_t = -p + t\left(\frac{\partial p}{\partial t}\right)_v,$$

where

$$\left(\frac{\partial p}{\partial t}\right)_v = \frac{R}{(v-b)}.$$

Therefore

$$\left(\frac{\partial u}{\partial v}\right)_t = -p + \frac{Rt}{(v-b)} = \left(\frac{a}{v^2}\right) = \left(\frac{\partial u}{\partial v}\right)_{t_c}.$$

Thus, for the isothermal process at t_c, net change in the internal energy is

$$(\Delta u)_{t_c} = \int_{v_i}^{v_f} \left(\frac{\partial u}{\partial v}\right)_{t_c} dv = \int_{v_i}^{v_f} \left(\frac{a}{v^2}\right) dv = a\left(\frac{1}{v_i} - \frac{1}{v_f}\right). \tag{5.122}$$

For quasistatic, reversible transfer of heat energy Δq to the gas, the first and the second laws dictate

$$(\Delta q)_{t_c} = t_c \Delta s = (\Delta u)_{t_c} + (\Delta w)_{t_c} = R t_c \ln\left[\frac{v_f - b}{v_i - b}\right].$$

As a result

$$\Delta s = \frac{(\Delta q)_{t_c}}{t_c} = R \ln\left[\frac{v_f - b}{v_i - b}\right]$$

is the increase in the entropy of the gas. Because of reversible operation, the entropy of the universe remains unchanged. Therefore the entropy of the reservoir decreases by the same amount Δs.

This shows that for isothermal processes volume dependence of the internal energy arises only out of the presence of the constant a: the constant b merely indicates an effective decrease of the volume. It is commonly assumed that b is an approximate measure of the volume that the molecules would occupy under strong compression. (See Chap. 6 on "Imperfect Gases" for further details.)

Problem 5.13 (Equation of state of a metal rod)
A metal rod at temperature T_0 has length L_0, cross-sectional area A_0, and temperature-independent Young's modulus Y. Construct its equation of state relating length L, tension Θ, and temperature T. The coefficient of linear expansion, β, is independent of the temperature and is small. Also, the extension due to tension is small compared to the original length of the wire.

Solution 5.13 Young's modulus is defined as the ratio of "stress" and "strain." That is,

$$Y = (\Theta/A)/\big[(\Delta L)_\theta/L\big].$$

Therefore

$$(\Delta L)_\theta = \frac{L\Theta}{YA},$$

where $(\Delta L)_\theta$ is the extension caused by the tension. Note that the total extension must also include the contribution, $(\Delta L)_\mathrm{T}$, caused by the temperature increase. Thus

$$L = L_0 + (\Delta L)_\mathrm{T} + (\Delta L)_\theta = L_0 + L_0\beta(T - T_0) + \left(\frac{L\Theta}{YA}\right). \tag{5.123}$$

Collecting L dependent terms on the left-hand side and dividing both sides by $L_0[1 - (\frac{\Theta}{YA})]$ yields

$$\begin{aligned}\frac{L}{L_0} &= \left[\frac{1 + \beta(T - T_0)}{1 - (\frac{\Theta}{YA})}\right] = \left[\frac{1 + \frac{(\Delta L)_\mathrm{T}}{L_0}}{1 - \frac{(\Delta L)_\theta}{L}}\right] \\ &= \left(1 + \frac{(\Delta L)_\mathrm{T}}{L_0}\right)\left[1 + \frac{(\Delta L)_\theta}{L} + O\left(\frac{(\Delta L)_\theta}{L}\right)^2\right] \\ &= 1 + \frac{(\Delta L)_\mathrm{T}}{L_0} + \frac{(\Delta L)_\theta}{L} + O\left(\frac{(\Delta L)_\theta}{L}\right)^2 + O\left[\left(\frac{(\Delta L)_\mathrm{T}}{L_0}\right)\left(\frac{(\Delta L)_\theta}{L}\right)\right].\end{aligned} \tag{5.124}$$

The second-order terms are negligibly small and can be ignored. Not all of the second-order terms are explicitly identified in (5.124). Such second-order terms as still remain are implicit in the ratio $\frac{(\Delta L)_\theta}{L} = \frac{\Theta}{YA}$. But they too can be ignored when A is replaced by A_0.

Thus, consistent with first order accuracy, the equation of state is

$$L = L_0\left[1 + \frac{(\Delta L)_\mathrm{T}}{L_0} + \frac{(\Delta L)_\theta}{L}\right] = L_0\left[1 + \beta(T - T_0) + \frac{\Theta}{YA_0}\right].$$

Problem 5.14 (Entropy change in extendable cord)
An extendable cord of length Λ obeys an equation of state

$$\Theta = T\left[a(\Lambda - \Lambda_0) + b(\Lambda - \Lambda_0)^2 + c(\Lambda - \Lambda_0)^3\right],$$

where Θ is the tension in the cord, T is its temperature, and the constants $a, b,$ and c are all > 0. The cord is in contact with a very large thermal reservoir at temperature T.

Calculate the change in the entropy of the cord when its length increases from Λ_0 to Λ.

Solution 5.14 As usual, change in the entropy is best calculated by devising an appropriate quasistatic process that takes the cord from its original length to the final length. Let the work done by the cord during such an increase in length be ΔW,

$$\Delta W = -\int_{\Lambda_0}^{\Lambda} \Theta \mathrm{d}\Lambda = -T\left[a\frac{(\Lambda-\Lambda_0)^2}{2} + b\frac{(\Lambda-\Lambda_0)^3}{3} + c\frac{(\Lambda-\Lambda_0)^4}{4}\right]. \tag{5.125}$$

The internal energy is unaffected by this process. This can be seen as follows:

Given $U = U(T, \Lambda)$, we have

$$\mathrm{d}U(T, \Lambda) = \left(\frac{\partial U}{\partial T}\right)_{\Lambda} \mathrm{d}T + \left(\frac{\partial U}{\partial \Lambda}\right)_{T} \mathrm{d}\Lambda . \tag{5.126}$$

Note that T is constant, i.e., $\mathrm{d}T = 0$. Further, in complete analogy with (5.17) which records the equality, $(\frac{\partial u}{\partial v})_{\mathrm{t}} = -p + t(\frac{\partial p}{\partial t})_{\mathrm{v}}$, we can write[12]

$$\left(\frac{\partial U}{\partial \Lambda}\right)_{T} = \Theta - T\left(\frac{\partial \Theta}{\partial T}\right)_{\Lambda} = \Theta - \Theta = 0 . \tag{5.127}$$

Since both $\mathrm{d}T$ and $(\frac{\partial U}{\partial \Lambda})_{\mathrm{T}}$ are equal to zero, due to (5.126), $\mathrm{d}U(T, \Lambda) = 0$. Accordingly,

$$\Delta U = \int_{\Lambda_0}^{\Lambda} \mathrm{d}U(T, \Lambda) = 0 .$$

Thus the change in the entropy, ΔS, of the cord is

$$\begin{aligned}\Delta S &= (\Delta U + \Delta W)/T = \Delta W/T \\ &= -\left[a\frac{(\Lambda-\Lambda_0)^2}{2} + b\frac{(\Lambda-\Lambda_0)^3}{3} + c\frac{(\Lambda-\Lambda_0)^4}{4}\right].\end{aligned} \tag{5.128}$$

Surprise! Surprise!

An extended cord has less entropy than an unextended one! Not to worry. All is well because given half a chance, the extended cord will revert to its original length. And this ‘spontaneous process" will, off course, result in increasing its entropy.

Problem 5.15 (Carnot engine: a trick question)
Noting the assertion that the efficiency of a Carnot engine is dictated by the second law without any regard to the working substance, calculate the efficiency of a Carnot engine slated to work with pure water as its working substance when the temperatures of the hot and cold reservoirs are set at 14 °C and 4 °C, respectively.

Solution 5.15 This is what, in the *lingo*, is called a *trick* question. Quite innocently one would calculate the efficiency, ϵ, as being equal to

$$\epsilon = 1 - \left(\frac{T_{\mathrm{C}}}{T_{\mathrm{H}}}\right) = 1 - \left(\frac{277}{287}\right).$$

[12]To transliterate (5.17) into (5.127), we need to use the analogy $-\Theta \to p$ and $\Lambda \to v$. To understand the transliteration, compare (5.125) and note that $p\mathrm{d}v$ is represented by $-\Theta \mathrm{d}\Lambda$.

Serious difficulties are encountered with the construction of the specified perfect Carnot cycle. Let us therefore examine in detail the four legs of a possible perfect Carnot cycle.

Proceeding isothermally along the leg $1 \rightarrow 2$ one reversibly withdraws—positive amount of—heat energy Q_H from the reservoir at temperature $T_H = 287$ °K and *adds* it to the working substance, which is water. This surely increases the system entropy from its initial value S_1 to some higher value that we shall call S_2,

$$S_2 = S_1 + \frac{Q_H}{287}.$$

Next, one proceeds reversibly along the adiabatic path $2 \rightarrow 3$ and hopefully reaches the lower temperature 277 °K. Accordingly, the entropy S_3 is equal to S_2. Therefore

$$S_3 = S_2 = S_1 + \frac{Q_H}{287}.$$

The third leg extends from $3 \rightarrow 4$ along a reversible path at constant temperature $T_C = 277$ °K. In the process, one hopes to *discard* a positive amount of heat energy Q_C to the reservoir at the lower temperature T_C. In this process one would reduce the system entropy and want it to become equal to the initial—lower—value S_1.

If all the above traveling does occur successfully, one would hopefully complete the cycle by returning along the reversible adiabat $4 \rightarrow 1$.

To check whether such a scenario can at all unfold, let us look at the isothermal trek $3 \rightarrow 4$ at the lower temperature 4 °C. Now, as is well known, fish do survive in relatively shallow lakes during severely cold weather. Of course, part of this miracle is owed to the fact that ice is not a very good conductor of "heat energy"—in this case, "heat energy" refers to the "severe cold" above the water. But the real savior here is the peculiar behavior of the bulk expansion coefficient of water,

$$\alpha_P = \left(\frac{\partial V}{\partial T}\right)_P / V.$$

It is usually positive for all normal substances. For water, α_P undergoes a crucial change as it crosses the temperature point ≈ 4 °C. Above this temperature, water behaves normally and α_P is positive. At ≈ 4 °C, α_P is zero; and below, it is negative.

Now, according to the second $T\mathrm{d}S$ equation,

$$T\mathrm{d}S = nC_v\mathrm{d}T + nT\left(\frac{\alpha_P}{\chi_T}\right).$$

Thus, along the isothermal path $3 \rightarrow 4$, where both $\mathrm{d}T$ and α_P are zero, we have

$$T\mathrm{d}S = 0,$$

which makes the entropies S_3 and S_4 equal. That is,

$$S_4 = S_3 = S_2 > S_1.$$

Clearly, the final leg of the cycle—the reversible adiabat $4 \to 1$—cannot exist because it would require S_4 to be equal to S_1. Thus the specified cycle is not a perfect Carnot cycle!

Given below are six exercises for the student. Occasionally, a little bit of help is also provided.

Exercise 5.1 Prove $(\frac{\partial t}{\partial p})_s(\frac{\partial v}{\partial t})_s = -v(\chi_t/\gamma)$.

Show $(\frac{\partial t}{\partial p})_s(\frac{\partial v}{\partial t})_s = -v(\chi_t/\gamma)$ and hence show that the relationship between the adiabatic compressibility χ_s and the isothermal compressibility χ_t is the following:

$$\frac{\chi_s}{\chi_s} = \frac{C_p}{C_v} = \gamma \,. \tag{5.129}$$

Exercise 5.2 Using the third $T\mathrm{d}S$ equation show that

$$\left(\frac{\partial p}{\partial v}\right)_s \left(\frac{\partial v}{\partial p}\right)_t = \frac{C_p}{C_v} \,. \tag{5.130}$$

Exercise 5.3 (Solution of Exercise 5.2 by Jacobians)

$$\begin{aligned}
\gamma &= \frac{C_p}{C_v} = \frac{tC_p}{tC_v} = \frac{t(\frac{\partial s}{\partial t})_p}{t(\frac{\partial s}{\partial t})_v} = \frac{\frac{\partial(s,p)}{\partial(t,p)}}{\frac{\partial(s,v)}{\partial(t,v)}} \\
&= \frac{\partial(s,p)}{\partial(s,v)} \cdot \frac{\partial(t,v)}{\partial(t,p)} = \frac{\partial(p,s)}{\partial(v,s)} \cdot \frac{\partial(v,t)}{\partial(p,t)} = \left(\frac{\partial p}{\partial v}\right)_s \left(\frac{\partial v}{\partial p}\right)_t \,.
\end{aligned} \tag{5.131}$$

Exercise 5.4 Show that the difference between the isothermal and adiabatic compressibility can be expressed as

$$\chi_t - \chi_s = tv\left(\frac{\alpha_p^2}{C_p}\right).$$

Help with the solution As shown in (5.84), the difference between the specific heats at constant pressure and constant volume can be represented as follows:

$$C_p - C_v = tv\left(\frac{\alpha_p^2}{\chi_t}\right).$$

Let us divide both sides by C_p, and multiply them both by χ_t, to get

$$\chi_t\left[1 - \left(\frac{C_v}{C_p}\right)\right] = tv\left(\frac{\alpha_p^2}{C_p}\right). \tag{5.132}$$

As proven earlier—see (5.87)—the ratio of the specific heats, at constant volume and pressure, is equal to the ratio of the isentropic and isothermal susceptibilities, i.e.,

$$\frac{C_v}{C_p} = \frac{\chi_s}{\chi_t} \,.$$

Therefore, (5.132) becomes

$$\chi_t - \chi_s = tv\left(\frac{\alpha_p^2}{C_p}\right). \tag{5.133}$$

Q.E.D.

Exercise 5.5 Show

$$\left(\frac{\partial t}{\partial v}\right)_s = -\left(\frac{\gamma}{v\chi_t}\right)\left(\frac{\partial t}{\partial p}\right)_s.$$

Solution to Exercise 5.5 Using the information given in (5.131), i.e., $\gamma = (\frac{\partial p}{\partial v})_s(\frac{\partial v}{\partial p})_t$, one readily finds

$$\begin{aligned} -\left(\frac{\gamma}{v\chi_t}\right)\left(\frac{\partial t}{\partial p}\right)_s &= -\left(\frac{(\frac{\partial p}{\partial v})_s(\frac{\partial v}{\partial p})_t}{-(\frac{\partial v}{\partial p})_t}\right)\left(\frac{\partial t}{\partial p}\right)_s \\ &= \left(\frac{\partial p}{\partial v}\right)_s\left(\frac{\partial t}{\partial p}\right)_s = \left(\frac{\partial t}{\partial v}\right)_s. \end{aligned} \tag{5.134}$$

Exercise 5.6 Show

$$C_p = -\frac{1}{\mu}\left(\frac{\partial h}{\partial p}\right)_t.$$

Van der Waals Theory of Imperfect Gases 6

The essential difference between molecules that are "real," and those that are postulated for the derivation of the ideal gas equation of state, is the following. Real molecules have finite size and they interact with each other. Ideal molecules are of zero size and have no interaction. Relevant description is presented in Sect. 6.1. In 1873 when Johannes Diderik Van der Waals presented his famous equation of state, little detail was available about interparticle forces. Both R. Laplace and C.-L. Berthollet suspected such forces to be short ranged. Also, it was clear that any interaction between microscopic constituents of a body must have two distinct features. Because these constituents congregate to form macroscopic entities, the overall interparticle potential must be negative. Yet, because matter does condense to finite densities, at small enough distances the potential must become large and repulsive and, as noted in Sect. 6.1.1, it must therefore have a hard core. The behavior of pressure change due to long range interaction is analyzed in Sect. 6.1.2.

Van der Waals offered separate treatment for the repulsive and attractive parts of the interparticle interaction. The standard thinking—which Van der Waals shared—was that both these interactions were "short ranged." It turns out, however, that his equation of state is somewhat more meaningful for a gas with both a hard core—much as he assumed—and an attractive interaction which is long ranged. Van der Waals virial expansion is discussed in Sect. 6.2. Section 6.3 deals with the critical point. Reduced equation of state is introduced in Sect. 6.4.

The critical region, the behavior below the critical point and Maxwell construction are discussed in Sects. 6.4.1 and 6.5.

Adding to the description in Sect. 6.5.1, a thermodynamic description of Maxwell construction is provided in Sect. 6.5.3. The metastable region is discussed in Sect. 6.5.4.

Molar specific volumes and densities, the Lever rule, smooth transition from liquid to gas, and vice versa are discussed in Sects. 6.6–6.8. The principle of corresponding states is described in Sect. 6.9. In Sect. 6.10, we discuss the Dieterici equation of state.

There are several figures. In particular, figures are shown and described in Sects. 6.5.2, 6.5.5, 6.5.6 and also in Sects. 6.9.1–6.9.6, as well as Sect. 6.10.1.

R. Tahir-Kheli, *General and Statistical Thermodynamics*,
https://doi.org/10.1007/978-3-030-20700-7_6

Critical constants P_c, V_c, T_c for a large number of gases—that could possibly be candidates for the Van der Waals equation of state—are recorded in Table 6.1. There are 11 solved problems.

6.1 Interacting Molecules

Accurate analysis of many-body systems with short range interaction is extremely difficult and certainly would have been impossible to carry out in Van der Waals time. Indeed, even today, exact solutions are available only for special cases. It is therefore a great testimony to Van der Waals genius that he came up with an approximation, albeit very crude, in which he could treat both the repulsive and attractive parts of the interaction.

6.1.1 Hard Core Volume Reduction

First, let us describe Van der Waals handling of the hard core repulsive potential.

In postulating an ideal gas, we have assumed that the molecules have zero size. Accordingly, at nonzero temperature, by increasing the pressure appropriately, the ideal gas volume could be reduced to any desired value.

Real molecules have finite size. Necessarily, therefore, the volume in which they can roam around must be less than that available in an ideal gas where all molecules have zero size. Furthermore, any such exclusion of volume (i.e., decrease by an amount V_{excluded} in the available volume) will be a function of the total number of all the other molecules present. Thus, if we are to use the ideal gas equation, we must replace V by something like $(V - V_{\text{excluded}})$. That is, instead of $P = Nk_{\text{B}}T/V$, we should use

$$P = \frac{Nk_{\text{B}}T}{V - V_{\text{excluded}}}. \tag{6.1}$$

It is best to consider V_{excluded} as an ad hoc phenomenological parameter approximately equal to the volume that would be occupied by the N molecules under large compression. An estimate for V_{excluded} may be obtained by treating molecules as incompressible, spherical hard-balls of radius r_0.

Consider, for instance, a pair of such molecules. The nearest distance that their centers may get to is $2r_0$. In other words, the centers of two identical molecules of radius r_0 are excluded from lying within a sphere of exclusion whose radius is $2r_0$.

For graphical representation, draw a sphere of radius $2r_0$ concentric with a "given" molecule.[1] Then the center of the nearest "other" molecule is excluded from falling within the sphere representing the "given" molecule. Clearly, the same holds true for all molecules in the system. Accordingly, for each molecule, the equivalent of half of the spherical volume $\frac{4\pi}{3}(2r_0)^3$ is excluded.[2] In this fashion, the total

[1] See Figs. 6.1.a–b where for generality we have shown two different types of molecules.

[2] Note that the factor $\frac{1}{2}$ is needed to avoid double counting.

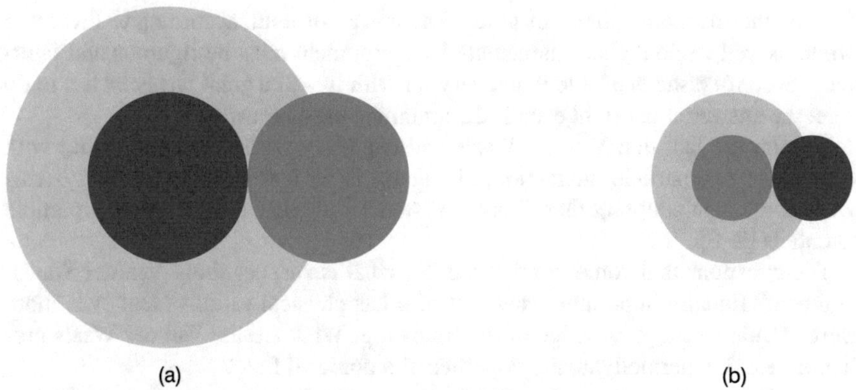

Fig. 6.1 Hard core sphere of exclusion for homogeneous pairs. For pairs of type 1 molecules—shown in (**a**)—the radius of the sphere of exclusion is $2r_1$. Similarly, for pairs of type 2 molecules—shown in (**b**)—the radius of the sphere of exclusion is $2r_2$. For simplicity the graph does not display the excluded volume for a nearest pair formed by one molecule of type 1 and another of type 2. In the description provided below, we initially treat only single type of molecules with radius of exclusion equal to $2r_0$. Later on, the case involving two different types of molecules is also considered. The shading for both types of molecules is dark and grey. The excluded volume is shaded very light-grey

excluded volume for N molecules is

$$V_{\text{excluded}} = \frac{N}{2}\left[\frac{4\pi}{3}(2r_0)^3\right] = 4N\left[\frac{4\pi}{3}(r_0)^3\right]. \tag{6.2}$$

Note that the total excluded volume is, in fact, four times the volume of N hard-balls of radius r_0.

6.1.2 Pressure Change due to Long Range Attraction

Van der Waals' treatment of the attractive part of the potential was similarly simple. According to science historian, M.J. Klein, Van der Waals argument went something like the following [1]:

"Effective force on a unit area of the surface due to these attractions comes only from a thin layer of molecules below the surface, because of the short range of the forces. The number of interacting pairs is proportional to the square of the density of the fluid, or inversely proportional to v^2 where v is the volume per mole. Introducing the proportionality factor a, the internal pressure P_{int} becomes just $(\frac{a}{v^2})$."

Note that the total pressure to be used in the ideal gas equation of state is thus $P + P_{\text{int}}$.

In regard to the above, I am reminded of a conversation I had with Professor Cyril Domb. Long ago when he was thinking of writing a historical review of some of Clerk Maxwell's work, he told me that like Van der Waals, who had used suspect argument to produce great science, Maxwell too had similar experience with his work on thermal conductivity of gases. The upshot of it all, according to Professor Domb, is that while a good mathematician's argument must be rigorous and fault-free—even if he/she is unable to get very far with it—all a great physicist has to do is get the answer about right even if the argument used is suspect!

Despite the fact that Van der Waals had expressly wished to treat a fluid with short range interparticle interaction, his theory is far less satisfactory for systems with short-range coupling than it is for systems with very long range interparticle potentials [2–6].

The questions that come to mind are: So what is suspect about Van der Waals' argument? Equally important, what part of it has physical validity? But even more useful would be the knowledge of the following: What are the Van der Waals predictions for the thermodynamic properties of a nonideal fluid?

In order to investigate these matters, let us assume the range of interaction to be practically infinite. Additionally, assume that the strength of such interaction is independent of the interparticle separation. As a result, the mutual potential energy of any pair of particles, separated by more than the hard core radius, is independent of their separation. Therefore, the total potential energy E of the N molecules is proportional to the total number of distinct pairs,

$$E \propto -\frac{N(N-1)}{2}\,. \tag{6.3}$$

That is, the overall attractive force resulting from this negative potential energy must lead to some reduction in the pressure that the gas would exert on its containing walls. Such reduction would be caused both from the slowing down and some decrease in the number of molecules actually hitting the walls. Clearly, therefore, the resulting reduction in pressure would be a function of the volume V within the containing walls and possibly also the pressure P itself and even the temperature. In other words, the equation of state would get changed to something like the following:

$$P = \frac{Nk_{\mathrm{B}}T}{V - V_{\mathrm{excluded}}} - Z(V, P, T)\left[\frac{N(N-1)}{2}\right], \tag{6.4}$$

where V_{excluded} is as given in (6.2) and $Z(V, P, T)$ is a function of V—and perhaps to a lesser extent also of P and T—and is positive.

Assume with Van der Waals that $Z(V)$ does not have any compelling dependence on P and T. That is,

$$Z(V, P, T) = Z(V)\,. \tag{6.5}$$

This assumption reduces the number of phenomenological parameters to only two as is explained below. Because when $N \gg 1$, to an accuracy of one part in N, $N(N-1) = N^2(1 - \frac{1}{N})$ can be replaced by N^2 in (6.4) above. Moreover, V being an extensive state variable, in the large N limit it scales linearly with N. Also, P is intensive. Therefore, for large N it scales as the zeroth power of N. In other words, P is independent of N. Clearly, therefore, V_{excluded} and $Z(V)$ in (6.4) must scale as

$$V_{\mathrm{excluded}} \propto N \tag{6.6}$$

and

$$Z(V) \propto \frac{1}{V^2} \,. \tag{6.7}$$

Accordingly, (6.4)—which may be called the equation of state—can be written as

$$P = \frac{N k_B T}{V - V_{\text{excluded}}} - z \frac{N^2}{V^2} \,, \tag{6.8}$$

where z is a constant independent of N. Note that both V_{excluded} and z are required to be positive.

It is usual to work in molal units and also use the notation

$$N = n N_A \,, \qquad z N_A^2 = a \,, \qquad N_A k_B = R \,. \tag{6.9}$$

Thus the equation of state takes on the convenient form:

$$P = \frac{nRT}{V - V_{\text{excluded}}} - a \left(\frac{n}{V} \right)^2 \,. \tag{6.10}$$

Traditionally, (6.10) is displayed as

$$\left(P + \frac{an^2}{V^2} \right) (V - V_{\text{excluded}}) = nRT \,. \tag{6.11}$$

Often one works with a single mole.[3] Because P is an intensive state variable, whenever we choose we may replace P by p where the latter refers to the pressure for a single mole. Other notation to be used for a single mole is the following:

$$V = vn \,, \qquad V_{\text{excluded}} = bnts, \qquad P = p \,. \tag{6.12}$$

Dividing both sides of (6.11) by n, the Van der Waals equation of state for a single mole of a fluid becomes

$$\left(p + \frac{a}{v^2} \right) (v - b) = Rt \,. \tag{6.13}$$

Differences between a and b should be noted. The parameter b relates to the physical size of the total number of molecules in one mole of the gas (fluid). On the other hand, a expresses the strength of the interparticle attractive force for the N_A molecules that are present in one mole.

---

[3]Van der Waals equation of state for 1 mole, gas or liquid, is $(p + \frac{a}{v^2})(v - v_{\text{excluded}}) = RT$, $p \equiv P$, $v = \frac{V}{n}$.

Van der Waals equation of state for n moles, gas or liquid, is $(P + \frac{an^2}{V^2})(V - V_{\text{excluded}}) = nRT$.

6.2 Van der Waals Virial Expansion

While a detailed analysis is deferred to an appendix, it is convenient here to record the fact that the equation of state for a mixture of Van der Waals gases (fluids) remains unchanged in form—see, for example, (G.20) and the discussion that leads to it. In other words, the equation of state can still be represented as

$$\left(p+\frac{a'}{v^2}\right)(v-b') = Rt\,. \tag{6.14}$$

Despite this similarity, Dalton's law of partial pressures is not necessarily valid for mixtures of Van der Waals gases.[4]

Note that a' and b' are the effective interaction and hard core size exclusion parameters of the mixture.

In the following we shall study[5] the so-called reduced form of the equation of state (see (6.30) below). Note that the above result, namely (6.14), which merely replaces the parameters a and b by a' and b', makes the following analysis for a single Van der Waals gas equally as valid for mixtures of Van der Waals gases.

The introduction of Van der Waals equation of state sparked a frenzy of theoretical and experimental activity. Kamerlingh Onnes' laboratory in the Netherlands was at the forefront of this effort. An important analytical tool that they devised for the study of real gases was the virial expansion. Experimentally observed values of the ratio $\frac{PV}{Nk_BT}$ were fitted to a series expansion in powers of $\frac{1}{V}$. For instance, (6.8) leads to the expansion

$$\begin{aligned}\frac{PV}{Nk_BT} &= 1+\frac{1}{V}\left(V_{\text{excluded}}-\frac{zN}{k_BT}\right)+O\left(\frac{V_{\text{excluded}}}{V}\right)^2\\ =\frac{PV}{nRT} &= 1+\frac{1}{V}\left(V_{\text{excluded}}-\frac{na}{RT}\right)+O\left(\frac{V_{\text{excluded}}}{V}\right)^2,\end{aligned} \tag{6.15}$$

where n is the mole number and a is as defined in (6.9).

Rather than the inverse powers of the actual volume V, a more convenient notation is arrived at by working with inverse powers of the specific volume, i.e., $v=\frac{V}{n}$ where n is the mole number. In this fashion, we can write (6.15) as follows:

$$\frac{pv}{Rt} = 1+\frac{1}{v}\left(v_{\text{excluded}}-\frac{a}{RT}\right)+O\left(\frac{v_{\text{excluded}}}{v}\right)^2 = b_1+\frac{b_2}{v}+\frac{b_3}{v^2}+\cdots\,. \tag{6.16}$$

As defined in (6.12), $v_{\text{excluded}} = \frac{V_{\text{excluded}}}{n} = b$.

The parameter b_{i} may be called the ith virial coefficient. In 1937 Mayer demonstrated how the virial coefficients may be related to the interparticle potential. Analytical calculations of b_{i} are usually carried out only for the gaseous phase. Even so, for realistic interparticle potential, the difficulty of calculation increases rapidly

[4]Dalton's law of partial pressures not necessarily valid for Van der Waals mixtures.
[5]Van der Waals virial expansion.

as we progress beyond $i = 2$. Returning to the Van der Waals gas, let us work out an inverse v expansion. To this end, multiply both sides of (6.13) by $\frac{v}{Rt(v-b)}$,

$$\frac{pv}{Rt} + \frac{a}{Rtv} = \frac{v}{(v-b)} = \left(1 - \frac{b}{v}\right)^{-1} = 1 + \frac{b}{v} + \left(\frac{b}{v}\right)^2 + \cdots, \tag{6.17}$$

and recast it as

$$\frac{pv}{Rt} = 1 + \frac{1}{v}\left(b - \frac{a}{Rt}\right) + \left(\frac{b}{v}\right)^2 + \cdots. \tag{6.18}$$

Comparison with (6.16) yields $b_1 = 1$, $b_3 = b^2$, and

$$b_2 = b - \frac{a}{Rt} = v_{\text{excluded}} - \frac{a}{Rt}. \tag{6.19}$$

For an ideal gas, $b_1 = 1$, and a and b are both zero. As a result, all virial coefficients, higher than 1, are vanishing.

Note that b_1 is equal to unity also for nonideal gases. Nonzero values of a and b_i, for $i > 1$, are reflective of the finite size of molecules and especially of the presence of interaction.

The b_2 for a Van der Waals gas, as given in (6.19), is correct in format. Because the second virial coefficient is straightforward to calculate, one can use realistic intermolecular potentials and thus obtain theoretical estimates for the parameters a and b for the actual gas under study.

An interesting feature of the second virial coefficient is that it changes sign at the so-called Boyle temperature t_b,[6]

$$t_\text{b} = a/(Rb). \tag{6.20}$$

Below t_b, b_2 is negative; at $t = t_\text{b}$, it is zero; and above t_b, it is positive. It is, of course, clear that when the second virial coefficient is vanishing, the gas is close to being *ideal*. Another contributing factor to the fame of the Boyle temperature is the prediction that cooling via the Joule–Kelvin effect—to be discussed in a later chapter—does not occur for a Van der Waals gas at temperatures higher than $2t_\text{b}$.

There is an interesting bit of folklore associated with the similarity in the formats of the Van der Waals result for the second virial coefficient and those that are obtained by the use of more realistic short-ranged intermolecular potentials. The folklore asserts—erroneously, of course—that the Van der Waals theory must indeed obtain for short range intermolecular potentials because it predicts a correct format for the second virial coefficient. The difficulty with this argument is that one can also get similar results from long-ranged interactions. And, indeed, all by themselves, the first and second virial coefficients do not make a theory.[7,8,9]

[6]Second virial coefficient changes sign at Boyle temperature.
[7]Critical point.
[8]Van der Waals critical point.
[9]Van der Waals two-phase region.

6.3 Critical Point

At constant temperature, the pressure p in an ideal gas is inversely proportional to the volume v. This relationship is obtained irrespective of how low the temperature may become.

Nonideal gases behave much like the ideal gas at high temperatures. As the temperature is lowered, real gases reach a critical temperature below which a process of saturation commences. That is, a gradual isothermal increase in pressure, resulting in slow decrease of volume v, eventually reaches a boundary—at some volume that we shall label v_G—of what turns out to be a two-phase region where liquid droplets begin to form within the vapor phase. Here, infinitesimal increase in pressure causes the volume to decrease while the proportion of liquid to vapor increases. This process continues all the way across to the opposite edge of the boundary, to some lower volume v_L, where all vapor condenses to a liquid. Any further increase in pressure carries the system into a phase that is all liquid and therefore has much smaller compressibility and a noticeably steeper isotherm, $(\frac{\partial p}{\partial v})_t$.

Let us investigate as to how well the Van der Waals gas mimics experimentally observed behavior of real gases. To this purpose, it is convenient to recast the equation of state (6.13) in the following manner. Multiply both sides by v^2/p, i.e., $(v^2 + \frac{a}{p})(v - b) = v^2 R \frac{t}{p}$, and rewrite it[10] as follows:

$$v^3 - \left(b + \frac{Rt}{p}\right)v^2 + \left(\frac{a}{p}\right)v - \frac{ab}{p} = 0\,. \tag{6.21}$$

Being a cubic in the variable v, for any p and t, it has three roots. For very high temperatures, there is a real root close to the ideal gas value $v \approx \frac{Rt}{p}$, while the other two roots are complex and mutually conjugate.

Let us consider next as to what is obtained when the temperature is reduced. On gradual lowering of the temperature, the imaginary parts of the complex roots become smaller and when a certain temperature t_c,

$$t_c = \frac{8a}{27Rb}\,, \tag{6.22}$$

is reached, the pressure approaches

$$p \to p_c = \frac{a}{27b^2}\,, \tag{6.23}$$

and the volume approaches

$$v \to v_c = 3b\,. \tag{6.24}$$

When this happens, the imaginary parts of the two complex roots become vanishingly small and all three roots of the resultant equation of state

$$v^3 - 9bv^2 + 27b^2v - 27b^3 = 0\,, \tag{6.25}$$

become real and indeed equal to v_c. The point (t_c, p_c, v_c) is called the *critical point.*

[10] Van der Waals equation as a cubic.

Usually, the parameters a and b are determined from the experimental results for p_c and t_c. Indeed, from the expressions given above, we readily get

$$a = \frac{27R^2t_c^2}{64p_c}, \qquad b = \frac{Rt_c}{8p_c}. \tag{6.26}$$

As an aside, we mention that experimental measurement of v_c is generally less accurate than that of p_c and t_c. Moreover—lest we should become sanguine about the validity of the Van der Waals approximation—measured values of v_c also do not fit well with the prediction $v_c = 3b$, when b is determined from the measurements of p_c and t_c.

Let us look at what is the case below the critical temperature t_c, especially when the temperature and pressure are raised towards the critical point. We find from the equation of state (6.21) that these roots move closer together. Eventually, as the temperature and pressure approach critical values, $t \to t_c$ and $p \to p_c$, from below, the equation of state approaches (6.25) and again the three real roots coalesce into one, $v = v_c$.

Thus there is something very interesting, indeed unique, about the critical point. Perhaps, some dramatic increase in the system response to external stimuli occurs there.

Because the parameters a and b are reflective of the finite size of atoms and any interparticle interaction, they are gas-specific. One notices that the critical value of the ratio

$$\Re = \frac{pv}{Rt}, \tag{6.27}$$

that is,

$$\Re_c = \frac{p_c v_c}{Rt_c} = \frac{\left(\frac{3ab}{27b^2}\right)}{\left(\frac{8aR}{27Rb}\right)} = \frac{3}{8}, \tag{6.28}$$

is independent of the gas-specific parameters. As such it applies equally well to all Van der Waals gases. Clearly, if the Van der Waals approximation is valid, the value of this ratio should be $3/8 = 0.375$. In this regard, we notice—see Table 6.1 below— that the agreement between the Van der Waals value for $\Re_c$ and the corresponding experimental results improves somewhat as interparticle interaction becomes less dominant. For instance, see helium, hydrogen, and neon critical constants P_c, V_c, T_c in Table 6.1.[11]

6.3.1 Critical Constants P_c, V_c, T_c

As mentioned above, corresponding Van der Waals parameters can be obtained as follows:

$$a = \frac{27R^2T_c^2}{64P_c}, \qquad b = \frac{RT_c}{8P_c}.$$

[11] Van der Waals parameters from critical constants.

Table 6.1 Critical constants, P_c, V_c, T_c, for a few real gasses, as quoted in [7]

Gas	T_c/K	P_c/MPa	$V_c/\mu m^3$	$\Re_c$
Water, H_2O	647.14	22.06	56	0.230
Nitric Oxide, NO	180	6.48	58	0.251
Bromine, Br_2	588	10.34	127	0.269
Carbon dioxide, CO_2	304.13	7.375	94	0.274
Ethane, C_2H_6	305.32	4.872	145.5	0.279
Ethylene, C_2H_4	282.34	5.041	131	0.281
Chlorine, Cl_2	416.9	7.991	123	0.284
Fluorine, Fl_2	144.13	5.172	66	0.285
Hydrogen sulfide, H_2S	373.2	8.94	99	0.285
Methane, CH_4	190.56	4.599	98.60	0.286
Oxygen, O_2	154.59	5.043	73	0.286
Xenon, Xe	289.77	5.841	118	0.286
Krypton, Kr	209.41	5.50	91	0.287
Nitrogen, N_2	126.21	3.39	90	0.291
Argon, Ar	150.87	4.898	75	0.293
Carbon monoxide	132.91	3.499	93	0.294
Helium, He	5.19	0.227	57	0.300
Hydrogen, H_2	32.97	1.293	65	0.307
Neon, Ne	44.4	2.76	42	0.314

6.4 Reduced Equation of State

Often it is the case that algebraic manipulation of the mathematical relations is simplified by transformation to an appropriately scaled system of variables. When these relations refer to physical phenomena, a convenient form of scaling is one where the variables are dimensionless. This transformation is especially useful if such a scaling can also help eliminate system-specific parameters.

To achieve such a scaling, let us transform to a set of the so-called *reduced* variables p_0, t_0, and v_0. Setting

$$\frac{p}{p_c} = p_0\,, \qquad \frac{t}{t_c} = t_0\,, \qquad \frac{v}{v_c} = v_0\,, \tag{6.29}$$

the equation of state (6.13) readily becomes

$$\left(p_0 + \frac{3}{v_0^2}\right)(3v_0 - 1) = 8t_0\,. \tag{6.30}$$

The interesting thing to note about this equation of state is that the variables are now dimensionless. Furthermore, both gas-specific parameters a and b have been eliminated. As a result, the equation has now become quite general in scope and should therefore apply to all Van der Waals gases.

6.4.1 Critical Region

At this juncture, the only explicit information we have about the system is contained in (6.30), which specifies a relationship among the reduced variables p_0, v_0, and t_0. Therefore, the simplest external stimulus that can be investigated is the isothermal compression that relates these three variables. Accordingly, we look at the reduced isothermal compressibility[12] $\chi_0 = -\frac{1}{v_0}(\frac{\partial v_0}{\partial p_0})_{t_0}$. Dramatic increase in χ_0 is signalled by the magnitude of the inverse compressibility,

$$|\chi_0^{-1}| = v_0 \left| \left(\frac{\partial p_0}{\partial v_0} \right)_{t_0} \right| , \tag{6.31}$$

tending to zero. Therefore we look for the satisfaction of the relation

$$\left(\frac{\partial p_0}{\partial v_0} \right)_{t_0} \to 0 . \tag{6.32}$$

Keeping t_0 constant, and differentiating the reduced equation of state (6.30) with respect to v_0, the above becomes

$$\left(\frac{\partial p_0}{\partial v_0} \right)_{t_0} = \left[\frac{6}{v_0^3} - \frac{24 t_0}{(3v_0 - 1)^2} \right] \to 0 . \tag{6.33}$$

Clearly, the choice $v_0 = 1$, $t_0 = 1$ satisfies the relationship. Thus, the $t_0 = 1$ isotherm becomes parallel to the volume axis at $v_0 = 1$ portraying the explosive increase in the isothermal volume compressibility.

To better understand the behavior of the isotherm at this point, we need also to examine the second derivative

$$\mathrm{d}p_0 v_0 t_0 = \frac{144 t_0}{(3v_0 - 1)^3} - \frac{18}{v_0^4} . \tag{6.34}$$

It is easily checked that the second derivative also vanishes at $v_0 = 1$, $t_0 = 1$. Indeed, as this *critical* point is approached from $v_0 > 1$ and is passed across toward $v_0 < 1$, the second derivative changes sign. As a result, the concavity of the isotherm changes from downwards to upwards, signalling the existence of an inflection point.

The third variable, p_0, at the critical point is also equal to unity, as is readily confirmed by inserting $t_0 = 1$, $v_0 = 1$ into the reduced equation of state (6.30).

Below, we examine the details of the critical region in the neighborhood of the critical point where the pressure, temperature, and volume are close to their critical values—namely, $p_0 \to 1$, $t_0 \to 1$, and $v_0 \to 1$. That is,

$$v_0 = 1 + \tilde{v} , \qquad p_0 = 1 + \tilde{p} , \qquad t_0 = 1 + \tilde{t} , \tag{6.35}$$

[12] Note that the reduced version of the isothermal compressibility is equal to p_c times the usual isothermal compressibility. That is, $\chi_0 = (\frac{a}{27b^2})\chi_t$.

where[13]

$$\tilde{v} = \frac{v - v_c}{v_c}\,, \qquad \tilde{p} = \frac{p - p_c}{p_c}\,, \qquad \tilde{t} = \frac{t - t_c}{t_c} \tag{6.36}$$

are all very small compared to unity, i.e.,

$$\tilde{v} \ll 1\,, \qquad \tilde{p} \ll 1\,, \qquad \tilde{t} \ll 1\,. \tag{6.37}$$

Multiplying both sides of equation of state (6.30) by v_0^2, introducing the notation suggested in (6.35) and (6.36), and doing slight rearranging leads to the result

$$\tilde{p} = \frac{\frac{-3\tilde{v}^3}{2} + 4\tilde{t}(1 + 2\tilde{v} + \tilde{v}^2)}{1 + \frac{7\tilde{v}}{2} + 4\tilde{v}^2 + \frac{3\tilde{v}^3}{2}}\,. \tag{6.38}$$

Problem 6.1 (Pressure versus volume for critical isotherm)
Calculate dependence of pressure on volume for the critical isotherm $t = t_c$.

Solution 6.1 At the critical temperature, $t_0 = 1$. Therefore, $\tilde{t} = 0$. Hence, according to (6.38), the relationship between the pressure and volume can be expressed as

$$\tilde{p} = -\frac{3\tilde{v}^3}{2} + O(\tilde{v}^4)\,,$$

which, according to (6.36), means

$$\frac{p - p_c}{p_c} = -\frac{3}{2}\left(\frac{v - v_c}{v_c}\right)^{\delta}\left[1 - O\left(\frac{v - v_c}{v_c}\right)\right], \tag{6.39}$$

where

$$\delta = 3\,. \tag{6.40}$$

(We have used the traditional notation for the exponent of the leading term on the right-hand side in (6.39).) Note that rather than the[14] result given by the Van der Waal's theory, i.e., $\delta = 3$, the experimental data suggest $\delta \approx 4.3$.

Problem 6.2 (Isothermal compressibility along critical isochore just above T_c)
Calculate the isothermal compressibility along the critical isochore,[15] i.e., for $v_0 = 1$, at just above the critical temperature t_c.

Solution 6.2 It is convenient first to represent the inverse isothermal compressibility χ_t^{-1} in terms of the reduced variables. That is,

$$\chi_t^{-1} = -v\left(\frac{\partial p}{\partial v}\right)_t = -p_c v_0 \left(\frac{\partial p_0}{\partial v_0}\right)_{t_0}\,. \tag{6.41}$$

[13] Note that like v_0, p_0, and t_0, all of $\tilde{v}$, $\tilde{p}$, and $\tilde{t}$ are dimensionless.
[14] Isothermal compressibility along critical isochore just above T_c.
[15] Isochore means occurring at constant volume. Critical isochore is the constant volume path such that the volume is equal to that which is obtained at the critical point.

Using the expression for $(\frac{\partial p_0}{\partial v_0})_{t_0}$ given in (6.33), we can write

$$\chi_{\mathrm{t}}^{-1} = -v_0 p_{\mathrm{c}} \left[\frac{6}{v_0^3} - \frac{24 t_0}{(3v_0 - 1)^2} \right]. \tag{6.42}$$

For the critical isochore, $v_0 = 1$. Expanding the right-hand side of (6.42) à la (6.35)—that is, setting $t_0 = 1 + \tilde{t}$—we readily get

$$\chi_{\mathrm{t}}^{-1} = -p_{\mathrm{c}} \left[6 - \frac{24(1+\tilde{t})}{4} \right] = 6 p_{\mathrm{c}} \tilde{t} \,. \tag{6.43}$$

Inverting the above gives

$$\chi_{\mathrm{t}} = \frac{1}{6 p_{\mathrm{c}} \tilde{t}} = \frac{1}{6 p_{\mathrm{c}}} \left(\frac{t_{\mathrm{c}}}{t - t_{\mathrm{c}}} \right) = \frac{1}{6 p_{\mathrm{c}}} \left(\frac{t_{\mathrm{c}}}{t - t_{\mathrm{c}}} \right)^{\gamma} . \tag{6.44}$$

Clearly,

$$\gamma = 1 \,. \tag{6.45}$$

Again, we have used the traditional notation γ for the exponent on the right-hand side of (6.44).

To recapitulate: For the critical isochore, as the temperature t approaches the critical temperature t_{c} from above, the Van der Waal's theory result for the isothermal compressibility diverges as $(\frac{t_{\mathrm{c}}}{t-t_{\mathrm{c}}})^{\gamma}$, where $\gamma = 1$. This result should be contrasted with the experimental data which suggest $\gamma \sim 1.3$.

6.5 Behavior Below T_{c}

As mentioned earlier, when $t \le t_{\mathrm{c}}$, or equivalently $t_0 \le 1$, for each isotherm there is a region in the (p_0, v_0)-space where the equation of state (6.30) has three real roots. Two such isotherms, for $t_0 = 1$ and 0.9, are shown in Fig. 6.2.a. While the three roots coalesce together at the critical point $t_0 = 1$, they separate as temperature falls, i.e., as t_0 decreases below 1. The flatness of the $t_0 = 1$ isotherm in the neighborhood of the critical point $v_0 = 1$ is clearly visible in Fig. 6.2.a.

6.5.1 Maxwell Construction

More interesting is the behavior of the subcritical, i.e., $t_0 = 0.9$, isotherm, which we examine in detail below.

6.5.2 Figures 6.2.a–b: (p_0, v_0) Isotherms

As mentioned earlier, the Van der Waals isotherms are plotted in Figs. 6.2.a–b. These are doubly curved, S-shaped, lines in the V–P plane as is shown by the route followed along $V_{0\mathrm{G}} \to V_{03} \to V_{04} \to V_{05} \to V_{0\mathrm{L}}$. However, quite unlike the prediction of the Van der Waals theory, the experimentally observed isotherms

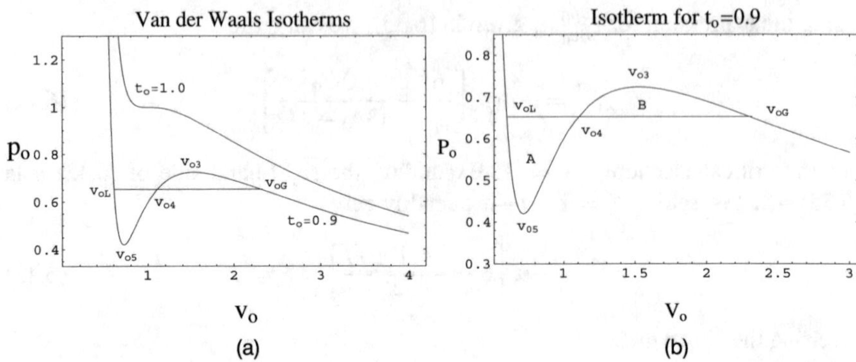

Fig. 6.2 (**a**) (p_0, v_0)-plot of the Van der Waals isotherms for $t_0 = 1.0$ and 0.9. (**b**) (p_0, v_0)-plot of the Van der Waals isotherm for $t_0 = 0.9$

that span the two phase region are isobars.[16] These isothermal-isobars are similar to the straight horizontal line $V_{0G} \rightarrow V_{04} \rightarrow V_{0L}$. This straight line is often referred to as the Maxwell isobar. According to Maxwell's prescription, the location of Maxwell isobar has to be so chosen as to make the algebraic sum of the areas enclosed, between the Van der Waals' S-shaped isotherm and Maxwell's isothermal-isobar, equal to zero. In other words, the magnitude of the area ($V_{04} \rightarrow V_{03} \rightarrow V_{0G} \rightarrow V_{04}$) should be equal to that of ($V_{0L} \rightarrow V_{04} \rightarrow V_{05} \rightarrow V_{0L}$). Based on the requirement that the Maxwell isothermal-isobar represent a state of thermal equilibrium, there is physical justification for following the Maxwell prescription.

6.5.3 Thermodynamic Justification for Maxwell Construction

It is instructive to provide a thermodynamic justification for the Maxwell construction via two different routes. Such double emphasis is in order because of the physical and historical importance of this issue.

Consider reversible travel along the closed loop

$$V_{0L} \rightarrow V_{04} \rightarrow V_{0G} \rightarrow V_{03} \rightarrow V_{04} \rightarrow V_{05} \rightarrow V_{0L}.$$

Then integrate, along this loop, the equation that represents the first law, namely the equation $\mathrm{d}u = t\mathrm{d}s - p\mathrm{d}v$. That is,

$$\oint \mathrm{d}u = \oint t\mathrm{d}s - \oint p\mathrm{d}v\,. \tag{6.46}$$

Because u is a state function, $\mathrm{d}u$ is an exact differential. Therefore, the left-hand side—which represents integration over the closed loop—is vanishing. Assume with Clerk Maxwell that the straight line portion of this path—i.e., the isobar $V_{0G} \rightarrow V_{04} \rightarrow V_{0L}$—is the proper isobar for the temperature t—that is, the temperature for which the S-shaped Van der Waals isotherm $V_{0G} \rightarrow V_{03} \rightarrow V_{04} \rightarrow V_{05} \rightarrow V_{0L}$ was

[16] Isobars are traversed at constant pressure.

drawn. Then the temperature t given in (6.46) is indeed the relevant temperature. Furthermore, this temperature t is constant along the whole loop. Thus t can be extracted out of the first integral on the right-hand side and the above equation can be rewritten as follows:

$$0 = t \oint \mathrm{d}s - \oint p \mathrm{d}dv\,. \tag{6.47}$$

Again, because the entropy s is a state function, the integral $\oint \mathrm{d}s$ is zero. Therefore, we are led to the relationship

$$0 = -\oint p\mathrm{d}v = -p_c v_c \oint p_0 \mathrm{d}v_0\,. \tag{6.48}$$

For ease of display, it is helpful to break the integral over the full loop up into several parts. Much like Figs. 6.2.a–b, these parts can also be displayed in the form of an equation:

$$0 = \oint p_0 \mathrm{d}v_0 = \int_{V_{0L}}^{V_{04}} p_0 \mathrm{d}v_0 + \int_{V_{04}}^{V_{0G}} p_0 \mathrm{d}v_0 + \int_{V_{0G}}^{V_{03}} p_0 \mathrm{d}v_0 + \int_{V_{03}}^{V_{04}} p_0 \mathrm{d}v_0 + \int_{V_{04}}^{V_{05}} p_0 \mathrm{d}v_0 + \int_{V_{05}}^{V_{0L}} p_0 \mathrm{d}v_0\,. \tag{6.49}$$

Therefore, the following relationship that represents the physical equality of the areas at the top and bottom of the Maxwell isothermal-isobar is thermodynamically valid:

$$\int_{V_{0L}}^{V_{04}} p_0 \mathrm{d}v_0 + \int_{V_{04}}^{V_{0G}} p_0 \mathrm{d}v_0 = \int_{V_{0L}}^{V_{05}} p_0 \mathrm{d}v_0 + \int_{V_{05}}^{V_{04}} p_0 \mathrm{d}v_0 + \int_{V_{04}}^{V_{03}} p_0 \mathrm{d}v_0 + \int_{V_{03}}^{V_{0G}} p_0 \mathrm{d}v_0\,. \tag{6.50}$$

Exercise 6.1 Show that Fig. 6.2.b agrees with the Maxwell prescription.

It is convenient to separate the six integrals given in (6.50) into two parts, each consisting of three integrals, such that their sums lead to equal areas underneath. This is done by shifting the second integral on the left-hand side, namely $\int_{V_{04}}^{V_{0G}} p_0 \mathrm{d}v_0$, to the right-hand side and moving the following two integrals, $\int_{V_{0L}}^{V_{05}} p_0 \mathrm{d}v_0 + \int_{V_{05}}^{V_{04}} p_0 \mathrm{d}v_0$, from the right- to the left-hand side. We get[17,18]

$$\int_{V_{0L}}^{V_{04}} p_0 \mathrm{d}v_0 - \int_{V_{0L}}^{V_{05}} p_0 \mathrm{d}v_0 - \int_{V_{05}}^{V_{04}} p_0 \mathrm{d}v_0 = \text{area A} = \int_{V_{04}}^{V_{03}} p_0 \mathrm{d}v_0 + \int_{V_{03}}^{V_{0G}} p_0 \mathrm{d}v_0 - \int_{V_{04}}^{V_{0G}} p_0 \mathrm{d}v_0 = \text{area B}\,. \tag{6.51}$$

Q.E.D.

[17] Maxwell prescription: alternate justification.

[18] Thermodynamic justification—Maxwell prescription. An alternate analysis.

An alternate physical argument for the thermodynamic justification of the Maxwell construction is the following. In thermal equilibrium the differential of the Gibbs[19] function can be written as

$$\mathrm{d}g = v\mathrm{d}p - s\mathrm{d}t\,. \tag{6.52}$$

But along an isotherm we have $\mathrm{d}t = 0$. Therefore,

$$(\mathrm{d}g)_\mathrm{t} = (v\mathrm{d}p)_\mathrm{t} = v_\mathrm{c}p_\mathrm{c}(v_\mathrm{O}\mathrm{d}p_\mathrm{O})_\mathrm{t}\,. \tag{6.53}$$

Dividing by $v_\mathrm{c}p_\mathrm{c}$, and integrating both sides along the path

$$V_\mathrm{OL} \to V_{05} \to V_{04} \to V_{03} \to V_\mathrm{OG}$$

gives

$$\left(\frac{1}{v_\mathrm{c}p_\mathrm{c}}\right)\left[\int_{V_\mathrm{OL}}^{V_{05}}(\mathrm{d}g)_\mathrm{t} + \int_{V_{05}}^{V_{04}}(\mathrm{d}g)_\mathrm{t} + \int_{V_{04}}^{V_{03}}(\mathrm{d}g)_\mathrm{t} + \int_{V_{03}}^{V_\mathrm{OG}}(\mathrm{d}g)_\mathrm{t}\right]$$
$$= \int_{V_\mathrm{OL}}^{V_{05}}(v_\mathrm{O}\mathrm{d}p_\mathrm{O})_\mathrm{t} + \int_{V_{05}}^{V_{04}}(v_\mathrm{O}\mathrm{d}p_\mathrm{O})_\mathrm{t} + \int_{V_{04}}^{V_{03}}(v_\mathrm{O}\mathrm{d}p_\mathrm{O})_\mathrm{t} + \int_{V_{03}}^{V_\mathrm{OG}}(v_\mathrm{O}\mathrm{d}p_\mathrm{O})_\mathrm{t}\,. \tag{6.54}$$

The integrals on the left-hand side are straightforward and we get:

$$\left(\frac{1}{p_\mathrm{c}v_\mathrm{c}}\right)[(g_{05} - g_\mathrm{OL}) + (g_{04} - g_{05}) + (g_{03} - g_{04}) + (g_\mathrm{OG} - g_{03})]$$
$$= \left(\frac{1}{p_\mathrm{c}v_\mathrm{c}}\right)[g_\mathrm{OG} - g_\mathrm{OL}]$$
$$= \int_{V_\mathrm{OL}}^{V_{05}}(v_\mathrm{O}\mathrm{d}p_\mathrm{O})_\mathrm{t} + \int_{V_{05}}^{V_{04}}(v_\mathrm{O}\mathrm{d}p_\mathrm{O})_\mathrm{t} + \int_{V_{04}}^{V_{03}}(v_\mathrm{O}\mathrm{d}p_\mathrm{O})_\mathrm{t} + \int_{V_{03}}^{V_\mathrm{OG}}(v_\mathrm{O}\mathrm{d}p_\mathrm{O})_\mathrm{t}\,. \tag{6.55}$$

In the above, g_OL and g_OG are the "specific"—meaning, for one mole only—Gibbs functions for the liquid and the gas phases, respectively. During phase transitions, thermodynamic stability requires the specific Gibbs function to be the same for all phases.[20] Thus g_OL is necessarily equal to g_OG. Accordingly, either of the top two terms—which of course are equal—in (6.55) are vanishing. The remainder can be written as

$$\int_{V_{05}}^{V_\mathrm{OL}}(v_\mathrm{O}\mathrm{d}p_\mathrm{O})_\mathrm{t} - \int_{V_{05}}^{V_{04}}(v_\mathrm{O}\mathrm{d}p_\mathrm{O})_\mathrm{t} = \text{area A}$$
$$= \int_{V_{04}}^{V_{03}}(v_\mathrm{O}\mathrm{d}p_\mathrm{O})_\mathrm{t} - \int_{V_\mathrm{OG}}^{V_{03}}(v_\mathrm{O}\mathrm{d}p_\mathrm{O})_\mathrm{t} = \text{area B}\,. \tag{6.56}$$

[19] Because the reader has not yet been introduced to the Gibbs free energy, upon first reading this subsection may be omitted. Equation (6.52) and the development of (6.55) will become more clear after the Gibbs free energy has been properly introduced and fully explained. See (10.31)–(10.44).

[20] Refer, for example, to the extremum principle for Gibbs free energy discussed in (10.39)–(10.40), etc.

Because the given drawings—that is, Figs. 6.2.a–b—were produced primarily for the purpose of displaying the Van der Waals results on a p versus v plot, the immediate feel for the v versus p plot may not be there. To assist those readers who do not readily see the equality of the areas A and B in the "volume versus pressure" picture, we add some additional information below which may be of help in this regard.

Notice that if we had constructed integrals like

$$\int_{V_{0G}}^{V_{04}} (v_0 \mathrm{d}p_0)_t \quad \text{and} \quad \int_{V_{04}}^{V_{0L}} (v_0 \mathrm{d}p_0)_t ,$$

they would both be equal to zero because their paths proceed along a straight line at constant value of the pressure p_0—for which $\mathrm{d}p_0$ is equal to zero. Therefore, without affecting any change in value, they can both be added to the right-hand side of (6.55). That is, we can write

$$0 = \int_{V_{0L}}^{V_{05}} (v_0 \mathrm{d}p_0)_t + \int_{V_{05}}^{V_{04}} (v_0 \mathrm{d}p_0)_t + \int_{V_{04}}^{V_{03}} (v_0 \mathrm{d}p_0)_t + \int_{V_{03}}^{V_{0G}} (v_0 \mathrm{d}p_0)_t + \int_{V_{0G}}^{V_{04}} (v_0 \mathrm{d}p_0)_t + \int_{V_{04}}^{V_{0L}} (v_0 \mathrm{d}p_0)_t . \tag{6.57}$$

Transferring the first, sixth, and second terms from the right-hand side to the left gives

$$\int_{V_{05}}^{V_{0L}} (v_0 \mathrm{d}p_0)_t + \int_{V_{0L}}^{V_{04}} (v_0 \mathrm{d}p_0)_t + \int_{V_{04}}^{V_{05}} (v_0 \mathrm{d}p_0)_t = \text{area A}$$
$$= \int_{V_{04}}^{V_{03}} (v_0 \mathrm{d}p_0)_t + \int_{V_{03}}^{V_{0G}} (v_0 \mathrm{d}p_0)_t + \int_{V_{0G}}^{V_{04}} (v_0 \mathrm{d}p_0)_t = \text{area B} . \tag{6.58}$$

Equations (6.56) and (6.58) form an equivalent, but alternate, theoretical form for the Maxwell construction. (Areas A and B are indicated graphically in Fig. 6.1.b.)

Exercise 6.2 Show that despite the fact the integral $\oint (v_0 \mathrm{d}p_0)_t$ over the above described loop is vanishing, $(v_0 \mathrm{d}p_0)_t$ is not an exact differential. This behavior reconfirms the fact that while the vanishing of the loop integral is necessary, it is not sufficient for a given differential to be exact.

6.5.4 Metastable Region

The Maxwell isobar, represented by the horizontal line $V_{0G} \rightarrow V_{04} \rightarrow V_{0L}$, is the thermodynamically appropriate isotherm within the two-phase region.

6.5.5 Figure 6.3: (p_0, v_0) Isotherms for Van der Waals Gas

On the other hand, as noted before,[21] rather than being a horizontal line, the Van der Waals isotherm has an S-shaped structure in this region. Can one make any sense of this odd behavior? Perhaps, one could say that the portion $V_{0G} \rightarrow V_{03}$, along

[21] Spinodal curve of Van der Waals.

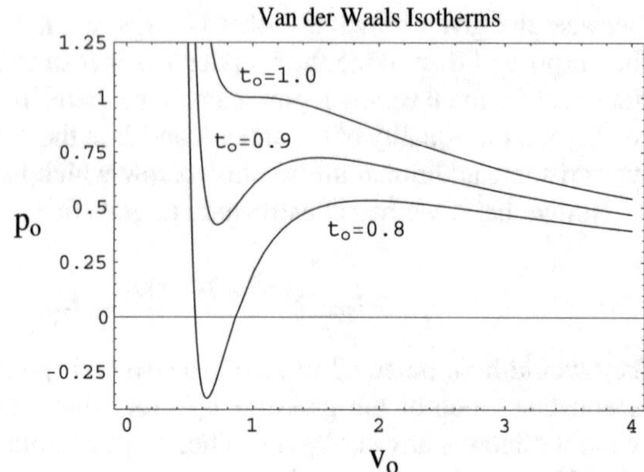

Fig. 6.3 (p_0, v_0)-plot of the Van der Waals isotherms for $t_0 = 1.0, 0.9$, and 0.8

which the pressure is above the saturation pressure, represents a state of metastable equilibrium where the vapor is supersaturated and, with scant inducement, would be ready to condense. Similarly, an argument could be advanced to suggest the portion $V_{05} \rightarrow V_{0L}$, which lies below the equilibrium isobar, represents a metastable superheated liquid, ready to evaporate.

Stranger still is the behavior of the isotherms for even lower values of temperature. In Fig. 6.3, an additional isotherm is included that refers to $t_0 = 0.8$. The negative pressure portion can surely not be related to a fluid state which cannot exist in a state of outwardly directed internal tension.

Could it then represent a state where the system has solidified? Of course, solids do support stresses that are the functional equivalent of negative pressure.

Despite these musings, no possible excuse can be offered for portions such as $V_{03} \rightarrow V_{04} \rightarrow V_{05}$ in Figs. 6.2.a–b along which the pressure and the volume decrease simultaneously, thus yielding a negative value for the isothermal compressibility, χ_t. We shall learn in a later chapter that in order for the system to be thermodynamically stable, χ_t must be positive. In other words, the Van der Waals equation has serious failings in its description of the two-phase state.

Yet, the very suggestion that the two-phase state could/should exist was a major accomplishment.

6.5.6 Figure 6.4: Spinodal Curve: Boundary of the Metastable–Unstable Region of Van der Waals

It is nevertheless interesting to outline the boundary between the unstable and what we have tentatively called the metastable region. This boundary is referred to as the *spinodal curve.*

To do this, we need to determine the locus of such points as V_{03} and V_{05} shown in Figs. 6.2.a–b. Fortunately, this is easily done because these are points of local maxima and minima of the subcritical isotherms. At both these points the tangents

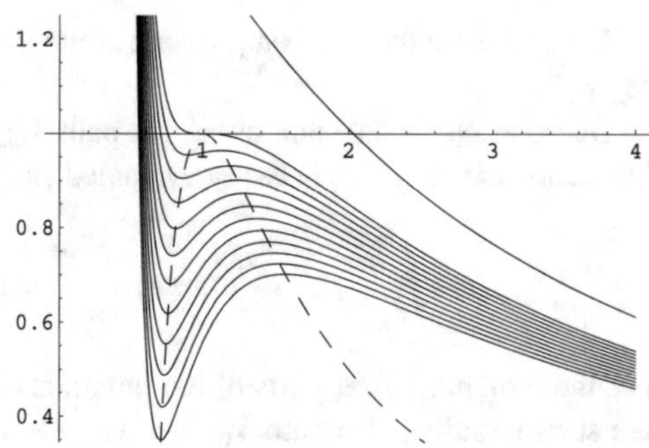

Fig. 6.4 The spinodal curve: Boundary of the metastable–unstable regions of Van der Waals. The spinodal curve—*dashed*—is shown against background of (p_0, v_0) isotherms. These are plotted for $t_0 = 0.9$–1.0 in increments of 0.01. Also given, at the top, is the curve for $t_0 = 1.1$

to the isotherms in the (p_0, v_0)-plane must be parallel to the v_0 axis. That is,

$$\left(\frac{\partial p_0}{\partial v_0}\right)_{t_0} = 0\,. \tag{6.59}$$

We proceed as was done in (6.33), where p_0 in the equation of state (6.30), namely

$$\left(p_0 + \frac{3}{v_0^2}\right)(3v_0 - 1) = 8t_0\,,$$

is differentiated at constant t_0. We get

$$\left(\frac{\partial p_0}{\partial v_0}\right)_{t_0} = \frac{6}{v_0^3} - \frac{24t_0}{(3v_0 - 1)^2} = 0\,. \tag{6.60}$$

Eliminating t_0 from these two equations leads to the desired locus

$$p_0 = \frac{(3v_0 - 2)}{v^3}\,. \tag{6.61}$$

In Fig. 6.4, the spinodal curve, that is, the locus—drawn as a dashed curve—is superimposed on a series of isotherms for $t_0 = 1.1, 1.0, 0.99, \ldots, 0.9$. Note that the region that lies within the spinodal curve is the thermodynamically unstable region.

6.6 Molar Specific Volume and Density

Below t_c, the difference in the reduced molar densities of the coexisting liquid and gaseous phases, i.e., ρ_{0L} and ρ_{0G}—which are the inverse of the corresponding specific volumes—that is, $\rho_{0L} = 1/V_{0L}$ and $\rho_{0G} = 1/V_{0G}$—is of considerable physical interest. In order to investigate this matter, first the dictates of the Maxwell rule have to be represented in an analytical form.

To this end, as mentioned earlier, we note that the construction of a Maxwell isobar, such as $V_{0L} \rightarrow V_{04} \rightarrow V_{0G}$ shown in Fig. 6.2.b, requires the satisfaction of (6.50), or equivalently (6.55).

Let us denote the reduced pressure along the Maxwell isobar $V_{0L} \rightarrow V_{04} \rightarrow V_{0G}$ as p_{00}.

Because p_{00} is constant along the path $V_{0L} \rightarrow V_{04} \rightarrow V_{0G}$, the integral on the left-hand side of (6.50) is easily evaluated as

$$\int_{V_{0L}}^{V_{0G}} p_0 \cdot \mathrm{d}v_0 = p_{00} \int_{V_{0L}}^{V_{0G}} \mathrm{d}v_0 = p_{00}(V_{0G} - V_{0L}) . \tag{6.62}$$

For the remaining four parts of the integrals on the right-hand side of (6.50) which are summed along the path $V_{0L} \rightarrow V_{05} \rightarrow V_{04} \rightarrow V_{03} \rightarrow V_{0G}$, the pressure p_0 is determined by the reduced equation of state (6.30) that meanders along the given path. That is,

$$p_0 = \frac{8t_0}{3v_0 - 1} - \frac{3}{v_0^2} . \tag{6.63}$$

Accordingly, they can be summed as follows:

$$\int_{V_{0L}}^{V_{0G}} \left(\frac{8t_0}{3v_0 - 1} - \frac{3}{v_0^2} \right) \mathrm{d}v_0 = \frac{8t_0}{3} \ln\left(\frac{3V_{0G} - 1}{3V_{0L} - 1} \right) - 3\left(\frac{1}{V_{0L}} - \frac{1}{V_{0G}} \right) . \tag{6.64}$$

Equating the right-hand sides of (6.62) and (6.64) gives

$$3\left(\frac{1}{V_{0L}} - \frac{1}{V_{0G}} \right) + p_{00}(V_{0G} - V_{0L}) = \frac{8t_0}{3} \ln\left(\frac{3V_{0G} - 1}{3V_{0L} - 1} \right) . \tag{6.65}$$

Our next task is to note that both the specific volumes V_{0G} and V_{0L} lie on the equation of state—see also, (6.30). Therefore, in addition to (6.65), they must also satisfy the relationships imposed by the equation of state (6.63). That is,

$$p_{00} = \frac{8t_0}{3V_{0G} - 1} - \frac{3}{V_{0G}^2} \tag{6.66}$$

and

$$p_{00} = \frac{8t_0}{3V_{0L} - 1} - \frac{3}{V_{0L}^2} . \tag{6.67}$$

6.6.1 Temperature Just Below the Critical Point

Below the critical point, for any given temperature t_0 there are three unknowns p_{00}, V_{0L}, and V_{0G} that can be determined by the simultaneous solution of (6.65), (6.66), and (6.67). Except for the immediate neighborhood of the critical point, where we can utilize series expansions in powers of small parameters, analytical solution of these equations is not possible. Nevertheless, if we should so desire, we can work out a numerical solution.

Problem 6.3 Calculate difference in the liquid and gas densities in the coexistence region just below the critical temperature.

Solution 6.3 Let us write

$$t_0 = \frac{t}{t_c} = 1 - \frac{(t_c - t)}{t_c} = 1 - \bar{t} \tag{6.68}$$

and

$$V_{0G} = 1 + \bar{v} . \tag{6.69}$$

Now what about V_{0L}? If we use a different expansion variable for V_{0L}, we would be forced to deal with the three simultaneous equations mentioned above. We note that the empirical principle of rectilinear diameters implies that when $\bar{t} \ll 1$, to a high degree of experimental accuracy,

$$V_{0L} = 1 - \bar{v} . \tag{6.70}$$

If we assume this principle to be true, we need to solve only two simultaneous equations, say (6.65) and (6.66), thus greatly simplifying the analysis. Most importantly, after working out a solution, we can double check it for accuracy by trying it out on the other pair of (6.65) and (6.67). The analysis given below will show that the principle of *rectilinear diameters* is correct to the leading order.

Eliminating p_{00} from (6.65) and (6.66) and using the expansions suggested in (6.68)–(6.70), we get the relationship

$$(1 - \bar{t})\left[\frac{8}{3} \ln\left(\frac{2 + 3\bar{v}}{2 - 3\bar{v}}\right) - \frac{16v}{2 + 3v}\right] = 3\left[\frac{1}{1 - \bar{v}} - \frac{1}{1 + \bar{v}}\right] - \frac{6\bar{v}}{(1 + \bar{v})^2} . \tag{6.71}$$

Small $\bar{v}$ expansions are readily obtained. For example,

$$\ln\left(\frac{2 + 3\bar{v}}{2 - 3\bar{v}}\right) = \ln\left(\frac{1 + \frac{3}{2}\bar{v}}{1 - \frac{3}{2}\bar{v}}\right) = 3\bar{v} + \frac{9\bar{v}^3}{4} + \frac{243\bar{v}^5}{80} + O(\bar{v}^7) \tag{6.72}$$

and

$$\frac{16v}{2 + 3v} = 8\bar{v} - 12\bar{v}^2 + 18\bar{v}^3 - 27\bar{v}^4 + \frac{81\bar{v}^5}{2} + O(\bar{v}^6) . \tag{6.73}$$

Using (6.72) and (6.73), the left-hand side of (6.71) becomes

$$(1 - \bar{t})\left[12\bar{v}^2 - 12\bar{v}^3 + 27\bar{v}^4 - \left(\frac{162\bar{v}^5}{5}\right) + O(\bar{v}^6)\right] . \tag{6.74}$$

The right-hand side is similarly calculated

$$3\left[\frac{1}{1 - \bar{v}} - \frac{1}{1 + \bar{v}}\right] - \frac{6\bar{v}}{(1 + \bar{v})^2} = 12\bar{v}^2 - 12\bar{v}^3 + 24\bar{v}^4 - 24\bar{v}^5 + O(\bar{v}^6) . \tag{6.75}$$

Dividing both by the factor

$$12\bar{v}^2 - 12\bar{v}^3 + 27\bar{v}^4 - \left(\frac{162\bar{v}^5}{5}\right) + O(\bar{v}^6)$$

gives

$$1 - \bar{t} = 1 - \frac{\bar{v}^2}{4} + \frac{9\bar{v}^3}{20} - \frac{21\bar{v}^4}{20} + O(\bar{v}^5).$$

That is,

$$\bar{t} = \frac{\bar{v}^2}{4} - \frac{9\bar{v}^3}{20} + \frac{21\bar{v}^4}{20} - O(\bar{v}^5). \tag{6.76}$$

To check on the consistency of the rectilinear diameters assumption contained in (6.69) and (6.70), we work next with the alternative pair of (6.65) and (6.67). Following a procedure similar to that used for deriving (6.76), we now get the result

$$\bar{t} = \frac{\bar{v}^2}{4} + \frac{9\bar{v}^3}{20} + \frac{21\bar{v}^4}{20} + O(\bar{v}^5). \tag{6.77}$$

Clearly, the violation of the principle of rectilinear diameters occurs first in the third order in $\bar{v}$. Thus we can confidently conclude that correct to the leading order $\bar{v}^2 = 4\bar{t}$.

The difference in the critical densities can be directly related to the difference in the corresponding specific volumes. As a result, just below the critical temperature,

$$\begin{aligned}\rho_{0L} - \rho_{0G} &= \frac{1}{V_{0L}} - \frac{1}{V_{0G}} = \frac{1}{1-\bar{v}} - \frac{1}{1+\bar{v}} = \frac{2\bar{v}}{1-\bar{v}^2} = 2\bar{v} + O(\bar{v})^3 \\ &= 2\sqrt{4\bar{t}} + O(\bar{t})^{\frac{3}{2}} = 4\left(\frac{t_c - t}{t_c}\right)^{\beta} + O\left(\frac{t_c - t}{t_c}\right)^{\beta+1},\end{aligned} \tag{6.78}$$

where

$$\beta = \frac{1}{2}. \tag{6.79}$$

Again, we have used the traditional notation whereby the leading term representing the difference between the specific densities of the coexisting liquid and gaseous phases just below the critical point is displayed as $(\frac{t_c-t}{t_c})^{\beta}$. Rather than being equal to $\frac{1}{2}$, as predicted here by the Van der Waals theory, experimental results for the exponent β are found to be ~ 0.3.

Exercise 6.3 Describe how the pressure of the coexistent liquid–gas phases—sometime called the saturation pressure—varies just below the critical point.

The behavior of the reduced saturation pressure[22] p_{00} just below the critical temperature can be examined by using either (6.66) or (6.67). This corresponds to setting the starting point in the gaseous or the liquid phase. Thus we can use either the

[22]Note that p_{00} is defined as the reduced pressure, that is, $\frac{p}{p_c}$, that is obtained along the Maxwell isobar in the coexistent region.

expansion indicated in (6.69) or (6.70). To the leading order, attesting to the consistency of the approximation used, both the equations give the same result. That is,

$$p_{00} = 1 - 4\left(\frac{t_c - t}{t_c}\right). \tag{6.80}$$

(Additional detail of p_{00} as a function of t_0 is provided in Figs. 6.8.a–b.)

Problem 6.4 Calculate the isothermal compressibility in the coexistence region just below the critical temperature t_c.

Solution 6.4 In order to examine the behavior of the isothermal compressibility just below the critical point, we need to start at the coexistence curve. We can begin either in the vapor phase where we set

$$v_0 = 1 + \bar{v}, \tag{6.81}$$

or the liquid phase where, in an equivalent approximation, we can write

$$v_0 = 1 - \bar{v}. \tag{6.82}$$

As before, the above prescription will be valid only if the results, for given value of temperature, are independent of the starting location on the relevant isotherm in the coexistence phase. Of course, for the present purposes, we should be satisfied if this is true to the leading order in the smallness parameter $\frac{t-t_c}{t_c}$. Fortunately, as before, the general expression for χ_t^{-1} given in (6.42) remains valid.

Thus depending on whether we begin in the gaseous—upper signs—or the liquid—lower signs—phase we get

$$\chi_t^{-1} = -(1 \pm \bar{v})p_c\left[\frac{6}{(1 \pm \bar{v})^3} - \frac{24(1-\bar{t})}{(2 \pm 3\bar{v})^2}\right], \tag{6.83}$$

which is readily expanded to yield

$$\chi_t^{-1} = -6\bar{t} + \frac{9}{2}\bar{v}^2 \pm 12\bar{v}\bar{t} + O\left(\bar{t}\bar{v}^2, \bar{v}^3\right). \tag{6.84}$$

Because $\bar{v}^2 = 4\bar{t}$, in retrospective validation of our approximation, to the leading order, we get the same result for both the starting positions:

$$\chi_t^{-1} = 12\bar{t}. \tag{6.85}$$

Accordingly, the compressibility just below t_c is given as follows:

$$\chi_t = \frac{1}{12p_c}\left(\frac{t_c}{t_c - t}\right)^{\gamma'}, \tag{6.86}$$

where

$$\gamma' = 1.$$

Note the coefficient $\frac{1}{12p_c}$, obtained for χ_t below t_c, is equal to one-half that obtained for χ_t above t_c. (Compare with (6.44).) Note also that rather than being equal to one, as predicted here by the Van der Waals theory, the experimentally observed value of the exponent γ' is ≈ 1.2.

6.7 Lever Rule

The validity of the principle of rectilinear diameters mentioned above is limited to the region in the immediate vicinity below the critical point. With decreasing temperature, the isothermal-isobar within the two phase region widens as does the difference in the specific molal volumes, V_{OG} and V_{OL}, of the coexisting phases. At any point in between the two ends of this isobar, the specific molal volume of the mixture, v_{OM}, is on average given by the relationship

$$n_G V_{OG} + n_L V_{OL} = v_{OM}(n_G + n_L) . \tag{6.87}$$

Because, the isothermal-isobar in question refers to a total of only one mole of the substance, the following two simultaneous equations determine n_G and n_L:

$$n_G + n_L - 1 = 0 , \tag{6.88}$$

$$n_G V_{OG} + n_L V_{OL} - v_{OM} = 0 . \tag{6.89}$$

The solution is

$$n_G = \frac{V_{OL} - v_{OM}}{V_{OL} - V_{OG}} , \qquad n_L = \frac{V_{OG} - v_{OM}}{V_{OG} - V_{OL}} . \tag{6.90}$$

6.8 Smooth Transition from Liquid to Gas and Vice Versa

Incorrectly, it may appear from the above discussion that any travel from the gaseous state to the liquid state of necessity has to pass through the coexistent two-phase region. A little reflection indicates that this is not always the case. A system starting in the gaseous phase can be made to travel above the transition point across to the other side of the coexistent region into the liquid phase. Also, the same process may be performed in reverse, whereby liquid state is smoothly converted into gaseous state without traversing across a two-phase region.

Question for Skeptics A sort of an experimental observation made while making coffee is of water boiling "directly" into steam at 100 °C. Additionally, the fact that the boiling point at the atmospheric pressure is much below the critical temperature which is $\sim$ 375 °C for H_2O,—as noted, for instance, in [8]—a reader might expect to "observe" the two phases occurring over a wide pressure–volume region. So an untutored observer is understandably skeptical about the foregoing description of the coexistent two-phase region. Indeed, as a result he/she may want to be skeptical about most of the theory presented in this chapter! If the coexistent two-phase region indeed occurs across a wide "region" [9], why doesn't the observer notice what should be a long Maxwell isobar [10]? Should he/she be skeptical?

6.9 Principle of Corresponding States

The advent of Van der Waals theory led to the expectation that perhaps the reduced equation of state (6.30) implied the existence of a more general rule called the Principle of Corresponding States (PCS). This rule conjectures that the true equation of state of a fluid should possess the form implicit in (6.30). That is, it should look like the following:

$$p_0 = f(v_0, t_0)\,. \tag{6.91}$$

Furthermore, according to this principle, even though the exact form of the function $f(v_0, t_0)$ is not known—and is certainly different from the Van der Waals approximation for it given in (6.30)—it should be the same for all physically similar fluids.

Clearly, the chemically inert gases such as neon, argon, krypton, and xenon are monatomic and physically similar. Therefore, they are the primary candidates for testing the hypothesis posited by the PCS. We should be aware that for the lighter inert gases, quantum effects—which are not included in the current discussion—can hold sway at very low temperatures. This is significantly the case for the lightest member of the group, helium. However, neon, the next heavier, is much less affected in this regard.

We can also try another group of fluids as possible candidates for testing the hypothesis. These have weak interparticle interaction—manifested by relatively low values for the critical temperature t_c. Additionally—except for hydrogen H_2, the very light member of this group—they are not plagued by quantum effects. Included in this category are nitrogen N_2, oxygen O_2, carbon monoxide CO, and methane CH_4, for which a variety of data are available.

6.9.1 Figure 6.5.a: X_0 as a Function of p_0 for Various Fluids

Returning to the PCS, the available data allow us to explore various relationships that are subsumed under (6.91).

We recall that the ratio $\Re_c = \frac{p_c v_c}{R t_c}$ is equal to $\frac{3}{8}$ for all Van der Waals fluids. Yet, experimentally observed values of $\Re_c$ for real fluids—see Table 6.1—are spread over a range, albeit a narrow one. It turns out that even though the parameter $\Re_c$ itself is not quite the same for different fluids, when we use it to scale the dimensionless ratio $\frac{p_0 v_0}{t_0}$, we find a nearly universal behavior—as is demonstrated in Fig. 6.5.a where such a scaled expression, X_0, i.e.,

$$X_0 = \frac{pv}{Rt} = \left(\frac{p_0 v_0}{t_0}\right)\Re_c\,, \tag{6.92}$$

is plotted as a function of p_0 for a number of fixed values of t_0 [12, 13].

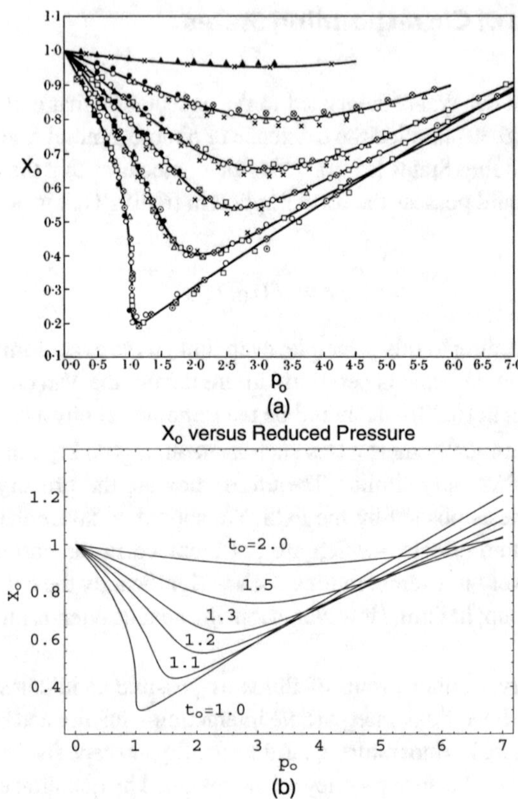

Fig. 6.5 (**a**) Experimental Results for X_0 as function of p_0. For a variety of fluids, X_0 is plotted as a function of the reduced pressure. The fluids represented are: carbon dioxide, ethane, ethylene, isopentane, methane, *n*-butane, *n*-heptane, nitrogen, and water. The experimental results are qualitatively similar to those predicted by the Van der Waals theory. Data fit remarkably well onto curves that are almost identical for different fluids. To avoid cluttering up the figure, the curves have not been labeled. Top curve refers to $t_0 = 2.0$. In descending order, t_0 for the next four lower curves is $= 1.5, 1.3, 1.2, 1.1, 1.0$. (Data from [11].) (**b**) X_0 versus p_0. Van der Waals theory results for X_0 versus p_0. The curves follow almost the same pattern as the experimental results shown in panel **a**

6.9.2 Figure 6.5.b: X_0 as a Function of p_0 for Van der Waals Gas

Corresponding results derived from the Van der Waals theory are also provided—see Fig. 6.5.b. It is noticed that there is qualitative agreement between the Van der Waals theory and the experiment.

6.9.3 Figure 6.6: The Reduced Second Virial Coefficient

Now we describe experimental data for the second virial coefficient for a number of real gases [13].

General form of the virial expansion within the context of the *PCS* would be

$$\frac{p_0 v_0}{t_0} = C_1 + \frac{C_2}{v_0} + \frac{C_3}{v_0^2} + \cdots, \tag{6.93}$$

where C_1, C_2, etc., depend on the reduced temperature t_0. In other words,

$$\frac{p_0 v_0}{t_0} = C_1 \left[1 + \frac{b_2}{v_0} + O\left(\frac{1}{v_0^2}\right) \right], \tag{6.94}$$

where $b_2 = \frac{C_2}{C_1}$. Note that when we work with the Van der Waals equation in reduced variables—see (6.30)—i.e.,

$$\left(p_0 + \frac{3}{v_0^2}\right)(3v_0 - 1) = 8t_0 ,$$

the expansion that corresponds to (6.94) is the following:

$$\frac{p_0 v_0}{t_0} = \frac{8}{3}\left[1 + \frac{1}{v_0}\left(\frac{1}{3} - \frac{9}{8t_0}\right) + O\left(\frac{1}{v_0^2}\right) \right]. \tag{6.95}$$

Hence, the Van der Waals theory counterpart of the reduced—that is, the PCS—second virial coefficient is

$$b_2 = \frac{1}{3} - \frac{9}{8t_0}. \tag{6.96}$$

Figure 6.6 shows experimental results for the reduced second virial coefficient b_2, obtained for a variety of fluids, as a function of the reduced temperature t_0. The corresponding results predicted by the Van der Waals theory are plotted as solid and dashed curves.

We note that while the Van der Waals estimates behave qualitatively similarly to the experimental results, quantitatively they are quite poor at temperatures well below the critical point.

For instance, the lower data curve records experimental results below the critical point. Agreement with the corresponding Van der Waals theory prediction, given as the dashed curve, is notably poor at $t = 0.5t_c$ but is seen to improve as the temperature rises towards t_c. Indeed, above the critical temperature—compare the upper data curve with the solid curve—the agreement is much improved.

It needs to be reiterated that Fig. 6.6 displays a parameter-free representation of both the experimental and theoretical results. The story would be a little different—meaning the fit between theory and experiment would be better than shown in Fig. 6.6—if we were to fit the data for a given fluid, by appropriately choosing the Van der Waals parameters a and b.

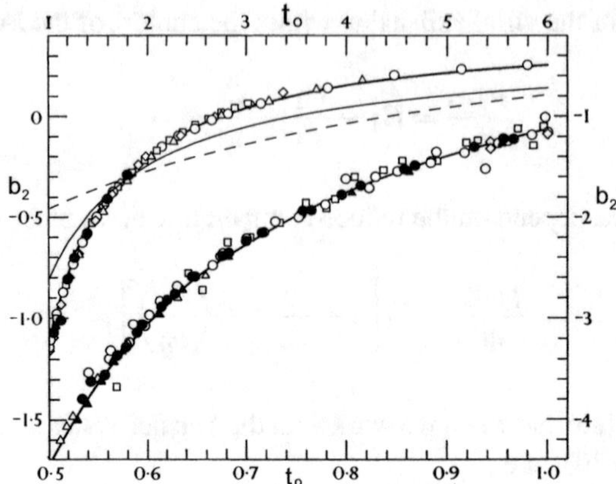

Fig. 6.6 Reduced Second Virial Coefficient: ○ ●Ar; △▲Kr; ◇ Xe; and □ CH_4. For ease of display, we have broken the data up into two parts. Experimental data shown on the lower curve refers to $0.5 \leq \frac{t}{t_c} \leq 1.0$. The right-hand and lower scales relate to the lower curve. The dashed curve represents the corresponding results of the Van der Waals theory. Data for $1.0 \leq \frac{t}{t_c} \leq 6.0$ is displayed in the upper curve. The left-hand and upper scales relate to the upper curve. Full curve records corresponding Van der Waals results

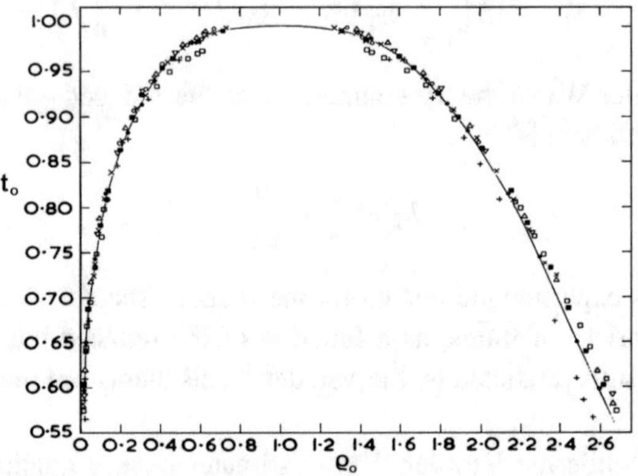

Fig. 6.7 Reduced Molar Densities in the Coexistent Regime. Experimental data—from [14]—for the reduced molar densities ρ_{0G} and ρ_{0L}—equal respectively to the inverse of the reduced molal volumes V_{0G} and V_{0L}—is displayed for a variety of fluids. Note, when $\rho_0 < 1$, the displayed results are for ρ_{0G}. Similarly, the data for ρ_{0L} is displayed in the region where $\rho_0 > 1$. The fluids studied are: argon, neon, krypton, xenon, nitrogen, oxygen, carbon monoxide, and methane. The ordinate is the reduced temperature t_0

6.9.4 Figure 6.7: Molar Densities of the Coexisting Phases and the PCS

As before we use the notation V_{0L} and V_{0G} for the reduced molal volumes of the mutually coexisting liquid and vapor phases that occur in the condensed region that obtains below the critical point. According to the dictates of the PCS, both V_{0L}

and V_{0G} should be common functions—that is, common within similar groups of fluids—of the reduced temperature t_0. Accordingly, results for similar fluids should all lie on the same curve. As is displayed in Fig. 6.7, this prediction is well supported by the experiment [15].

Problems 6.5–6.11

In the following problems extensive use will be made of (5.18) and (5.25). We also note the relationships

$$\left(\frac{\partial C_v}{\partial v}\right)_t = t\mathrm{d}ptv \tag{6.97}$$

and

$$\mathrm{d}u = \left(\frac{\partial u}{\partial t}\right)_{\mathrm{v}} \mathrm{d}t + \left(\frac{\partial u}{\partial v}\right)_{\mathrm{t}} \mathrm{d}v = C_{\mathrm{v}}\mathrm{d}t + \left[t\left(\frac{\partial p}{\partial t}\right)_{\mathrm{v}} - p\right]\mathrm{d}v\,. \tag{6.98}$$

Problem 6.5 (Internal energy and volume dependence of C_{v})
Describe the volume dependence of the specific heat C_{v} of a Van der Waals gas. Assuming the temperature is well above the critical point, give an expression for the internal energy, u.

Solution 6.5 For the Van der Waals gas,

$$p = \frac{Rt}{v-b} - \frac{a}{v^2}\,. \tag{6.99}$$

Therefore

$$\left(\frac{\partial p}{\partial t}\right)_{\mathrm{v}} = \left(\frac{R}{v-b}\right) \tag{6.100}$$

and

$$\mathrm{d}ptv = 0\,. \tag{6.101}$$

As such (6.97) gives

$$\left(\frac{\partial C_v}{\partial v}\right)_{\mathrm{t}} = t\mathrm{d}ptv = 0\,. \tag{6.102}$$

Consequently, for fixed temperature, C_{v} is independent of the volume. Indeed, this would be the case for any thermodynamic system for which the pressure is a linear function of the temperature because $\mathrm{d}ptv$ would then be zero. Keeping the temperature constant and integrating $(\frac{\partial C_{\mathrm{v}}}{\partial v})_{\mathrm{t}}$ with respect to v gives

$$\int \left(\frac{\partial C_{\mathrm{v}}}{\partial v}\right)_{\mathrm{t}} \mathrm{d}v = C_{\mathrm{v}} = f(t) = C_{\mathrm{v}}(t)\,, \tag{6.103}$$

where $f(t)$ is a constant independent of v, but possibly dependent on t.

Thus, C_v is independent of the volume. This is true even if we arrange the pressure so that the volume becomes large—i.e., the density becomes small making the gas dilute—while the temperature is kept constant. In the very dilute limit, the behavior of the Van der Waals gas must approach that of an ideal gas. Therefore, its specific heat C_v must equal that for an ideal gas. The latter, of course, is known to be independent of the temperature.

Combining (6.98) and (6.100), we can write

$$du = C_v dt + \left[\left(\frac{Rt}{v-b}\right) - p\right] dv = C_v dt + \frac{a}{v^2} dv\,. \tag{6.104}$$

Next we integrate this equation along the path $(t_0, v_0) \rightarrow (t, v_0) \rightarrow (t, v)$. This readily leads to the result

$$u - C_v t + (a/v) = u_0 - C_v t_0 + (a/v_0) = D\,, \tag{6.105}$$

where D is a constant.

6.9.5 Figure 6.8.a: Reduced Vapor Pressure in the Coexistent Regime and the PCS

In Fig. 6.8.a experimental data for the logarithm of the reduced vapor pressure[23] p_{00} within the coexisting vapor–liquid phases is plotted as a function of the inverse reduced temperature, $1/t_0$. Again, the dictates of the PCS are seemingly well followed, and the data for a wide variety of fluids appears to lie on the same curve.

Also, it is interesting to note that when plotted against the inverse of the reduced temperature, the experimental results for the logarithm of p_{00} seem to suggest a curve which is nearly a straight line. That is,

$$p_{00} \propto [t_0]^{\nu}\,, \tag{6.106}$$

with

$$\nu \sim 7.3\,. \tag{6.107}$$

6.9.6 Figure 6.8.b: Reduced Vapor Pressure for a Van der Waals Gas

Just below the critical point, the p_{00} for a Van der Waals gas can also be fitted to a similar result with $\nu = 4$. Accordingly, the saturation pressure for a Van der Waals gas—see Fig. 6.8.b—falls off less rapidly with decrease in temperature than is observed in experiment.[24] Moreover, for lower temperatures the Van der Waals

[23] Note that p_{00} was defined to be the reduced pressure, that is, $\frac{p}{p_c}$, that is obtained along the Maxwell isobar in the coexistent region.

[24] Note that $\nu = 4$ is consistent with the analytical result derived earlier. (See (6.80).)

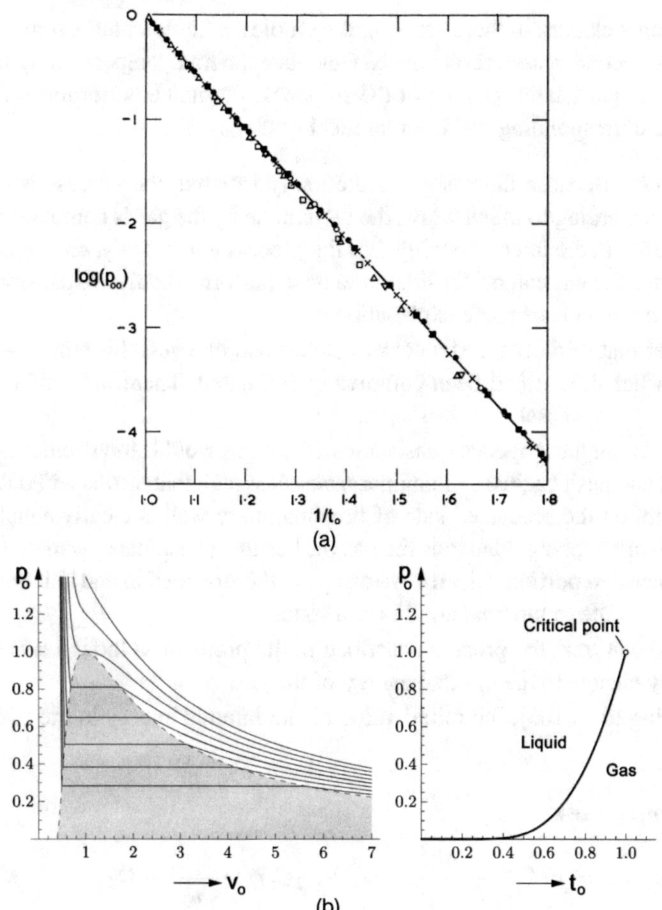

Fig. 6.8 (**a**) Reduced vapor pressure in the coexistent regime. The fluids studied are: argon, krypton, xenon, nitrogen, oxygen, carbon monoxide, and methane. Note that the curve appears to be a straight line, indicating a linear drop with $1/t_0$, which is equivalent to a functional dependence of the form $p_{00} \propto t_0^{\nu}$ with $\nu \sim 7.3$. (**b**) Reduced vapor pressure in the coexistent regime for a Van der Waals gas. The figure on the left-hand side shows Maxwell isobars on a P–V plot. Each such isobar has a corresponding temperature which is not shown in this figure. The figure on the right-hand side displays the pressure of such isobars against their corresponding temperature, both on the reduced scale. (This plot is reproduced courtesy of [16])

result for ν falls further below 4. This, of course, means that except possibly in the neighborhood of the critical point, for the Van der Waals gas the power relationship with a single ν does not obtain.

Problem 6.6 (Temperature change on mixing of Van der Waals gases)
A quantity of Van der Waals gas[25] is contained in two thermally isolated vessels, of volume v_1 and v_2, which are connected across a short tube with a stopcock. Initially,

[25] Temperature change on mixing Van der Waals gases.

with the stopcock closed, there are n_1 and n_2 moles, at temperatures t_1 and t_2, in the first and the second vessel, respectively. Calculate the final temperature t_f of the gas when, after a quasistatic opening of the stopcock, thermal equilibrium is achieved. What is the corresponding result for an ideal gas?

Solution 6.6 Because the system is thermally isolated, the process is adiabatic. Therefore, according to the first law, the work done by the gas is compensated by an equal decrease in the internal energy. For the process under study, no work has been done because the gas can be considered to have performed a free expansion. This is a subtle point and needs some explanation.

Consider that initially the stopcock is closed and one vessel is empty—meaning at some earlier time it had been completely evacuated. Therefore, initially all the gas is in the other vessel.

Upon opening the stopcock quasistatically, the gas would slowly enter the evacuated vessel and push against an imaginary, massless wall that moves without friction. The pressure on the evacuated side of this imaginary wall is clearly equal to zero. Hence no work is being done[26] as the gas pushes this (imaginary) wall and expands into the evacuated portion. Clearly, we can push the stopcock in and halt this process anywhere en route to the final equilibrium state.

Thus we can treat the process described in the problem as though it is executed without any change in the internal energy of the gas.

According to (6.105), the initial value of the internal energy in the two vessels is

$$\begin{aligned} u_{\text{initial}} &= u_1 + u_2 \\ &= n_1 C_v t_1 - \frac{a n_1^2}{v_1} + D_1 + n_2 C_v t_2 - \frac{a n_2^2}{v_2} + D_2 \,. \end{aligned} \tag{6.108}$$

When the stopcock is opened and the equilibrium has been reached, the number of moles in the two vessels become ν_1 and ν_2 where

$$\nu_1 = v_1 \left(\frac{n_1 + n_2}{v_1 + v_2} \right), \qquad \nu_2 = v_2 \left(\frac{n_1 + n_2}{v_1 + v_2} \right). \tag{6.109}$$

Note that

$$\nu_1 + \nu_2 = n_1 + n_2 \,. \tag{6.110}$$

Therefore, again according to (6.105), the expression for the total internal energy after the expansion is

$$u_{\text{final}} = \nu_1 C_v t_f - a\left(\nu_1^2/v_1\right) + D_1 + \nu_2 C_v t_f - a\left(\nu_2^2/v_2\right) + D_2 \,. \tag{6.111}$$

[26]This is ensured by the fact that the valve is opened quasistatically. As a result the gas enters the evacuated vessel with little kinetic energy.

Because there has been no change in the internal energy, we can equate the initial and final value of the internal energy. Using (6.108)–(6.111), we get

$$C_v(n_1t_1 + n_2t_2) - a\left(\frac{n_1^2}{v_1} + \frac{n_2^2}{v_2}\right) + D_1 + D_2$$
$$= C_v t_f(n_1 + n_2) - a\left(\frac{v_1^2}{v_1} + \frac{v_2^2}{v_2}\right) + D_1 + D_2 \tag{6.112}$$

and

$$t_f = \frac{(n_1t_1 + n_2t_2)}{(n_1 + n_2)} + \frac{a(n_1 + n_2)}{C_v(v_1 + v_2)} - \frac{a}{C_v(n_1 + n_2)} \cdot \left[\frac{n_1^2}{v_1} + \frac{n_2^2}{v_2}\right]. \tag{6.113}$$

The corresponding result for an ideal gas,

$$t_f = \left(\frac{n_1t_1 + n_2t_2}{n_1 + n_2}\right),$$

is obtained by setting $a = 0$.

Problem 6.7 (Van der Waals gas specific heat and enthalpy)
If the internal energy of one[27] mole of a Van der Waals gas is as given in (6.105), calculate:

(a) The difference in the specific heats, $C_p - C_v$,
(b) The enthalpy, h,
(c) The coefficients $\eta = (\frac{\partial t}{\partial v})_u$ and $\mu = (\frac{\partial t}{\partial p})_h$.
(d) What are the corresponding results for an ideal gas?

Solution 6.7 (a) Here it is convenient to use (5.83) and get

$$C_p - C_v = t\left(\frac{\partial p}{\partial t}\right)_v \left(\frac{\partial v}{\partial t}\right)_p. \tag{6.114}$$

The derivatives $(\frac{\partial p}{\partial t})_v$ and $(\frac{\partial v}{\partial t})_p$ are readily calculated from the equation of state

$$\left(p + \frac{a}{v^2}\right) \cdot (v - b) = Rt\,.$$

We have

$$\left(\frac{\partial p}{\partial t}\right)_v = \frac{R}{v - b}$$

[27]Van der Waals gas-specific heat, enthalpy, η, and μ.

and

$$\left(\frac{\partial v}{\partial t}\right)_{\mathrm{p}} = \frac{R(v-b)}{Rt - 2a(v-b)^2/v^3}\,.$$

Thus (6.114) gives

$$C_{\mathrm{p}} - C_{\mathrm{v}} = R\Big/\left[1 - \frac{2a(v-b)^2}{Rtv^3}\right]. \tag{6.115}$$

It is convenient to recast the above expression in terms of the reduced variables

$$v_0 = \frac{v}{v_{\mathrm{c}}} = \frac{v}{3b}$$

and

$$t_0 = \frac{t}{t_{\mathrm{c}}} = \left(\frac{27Rb}{8a}\right).$$

We get

$$C_{\mathrm{p}} - C_{\mathrm{v}} = R\Big/\left[1 - \frac{(3v_0-1)^2}{4v_0^3 t_0}\right]. \tag{6.116}$$

At very high temperatures, and/or very low densities—i.e., $t_0 \ll 1$ and/or $v_0 \ll 1$—the predicted difference in the specific heats equals R. This result is, of course, correct because here the gas closely approximates an ideal gas.

As the temperature and volume are decreased towards their critical values, $t_0 = 1$ and $v_0 = 1$, the difference between the specific heats increases, eventually approaching infinity at the critical point. Because above the critical point the specific heat C_{v} for the Van der Waals gas is finite—in fact, it is equal to that for an ideal gas—the divergence of C_{p} at the critical point has to be considered a positive feature of the theory.

(b) Equation (6.105) and the equation of state can also be used to write the expression for the enthalpy:

$$\begin{aligned} h = u + pv &= C_{\mathrm{v}}t - \frac{a}{v} + pv + D \\ &= C_{\mathrm{v}}t - \frac{a}{v} + \left[\frac{Rt}{v-b} - \frac{a}{v^2}\right]v + D \\ &= C_{\mathrm{v}}t - 2a/v + Rtv/(v-b) + D\,. \end{aligned} \tag{6.117}$$

(c) Next[28] we calculate η where

$$\eta = \left(\frac{\partial t}{\partial v}\right)_{\mathrm{u}}.$$

[28]Note that the physical relevance of the coefficients η and μ is explained in the chapter on Joule and Joule–Kelvin effects where we rederive the relevant expressions by a different procedure. Compare the results derived there and recorded in (7.9) and (7.19).

According to the cyclic identity, the right-hand side can be represented as

$$\left(\frac{\partial t}{\partial v}\right)_{\mathrm{u}} = -\left(\frac{\partial u}{\partial v}\right)_{\mathrm{t}}\left(\frac{\partial t}{\partial u}\right)_{\mathrm{v}} = -\left(\frac{\partial u}{\partial v}\right)_{\mathrm{t}}/\left(\frac{\partial u}{\partial t}\right)_{\mathrm{v}}.$$

As such for the Van der Waals gas we get

$$\eta = -\left(\frac{a}{v^2}\right)/C_{\mathrm{v}}. \tag{6.118}$$

Using a procedure similar to that followed above, we can also find the coefficient $(\frac{\partial t}{\partial p})_{\mathrm{h}}$,

$$\mu = -\left(\frac{\partial h}{\partial p}\right)_{\mathrm{t}}/\left(\frac{\partial h}{\partial t}\right)_{\mathrm{p}} = -\left(\frac{\partial h}{\partial p}\right)_{\mathrm{t}}/C_{\mathrm{p}}.$$

Next we write

$$\left(\frac{\partial h}{\partial p}\right)_{\mathrm{t}} = \left(\frac{\partial h}{\partial v}\right)_{\mathrm{t}}\left(\frac{\partial v}{\partial p}\right)_{\mathrm{t}} = \left(\frac{\partial h}{\partial v}\right)_{\mathrm{t}}/\left(\frac{\partial p}{\partial v}\right)_{\mathrm{t}}.$$

The partial differentials on the right-hand side are easy to calculate. From (6.117) we find

$$\left(\frac{\partial h}{\partial v}\right)_{\mathrm{t}} = \frac{2a}{v^2} + Rt\left[\frac{1}{v-b} - \frac{v}{(v-b)^2}\right].$$

The partial differential $(\frac{\partial p}{\partial v})_{\mathrm{t}}$ is found directly from the equation of state as

$$\left(\frac{\partial p}{\partial v}\right)_{\mathrm{t}} = -\frac{Rt}{(v-b)^2} + \frac{2a}{v^3}.$$

Thus for the Van der Waals gas,

$$\mu = \left(\frac{\partial t}{\partial p}\right)_{\mathrm{h}} = \frac{1}{C_{\mathrm{p}}}\left[\frac{2av(v-b)^2 - bRtv^3}{Rtv^3 - 2a(v-b)^2}\right]. \tag{6.119}$$

Corresponding results for an ideal gas can be obtained by setting $a = 0$ and $b = 0$ in the above equations. Accordingly, both η and μ are equal to zero and

$$h = C_{\mathrm{v}}t + pv = C_{\mathrm{v}}t + Rt.$$

Problem 6.8 (Work done and change in internal energy and entropy in Van der Waals gas)
One mole of a Van der Waals gas is in contact with an infinite thermal reservoir at temperature t_{c}. In an isothermal, reversible expansion the gas increases its volume from v_{i} to v_{f} and in the process does work w. Calculate w, the resultant change in the internal energy, u', and the entropy, s', of the gas and the reservoir.

Solution 6.8

$$w = \int_{v_i}^{v_f} p\mathrm{d}v = Rt_c \ln\left[(v_f - b)/(v_i - b)\right] - a(1/v_i - v_f)\,.$$

Since

$$\left(\frac{\partial p}{\partial t}\right)_v = R/(v-b)\,,$$

we get

$$\left(\frac{\partial u}{\partial v}\right)_t = -p + Rt/(v-b) = a/v^2\,.$$

Accordingly, for the isothermal process, the net change in the internal energy is

$$u' = \int_{v_i}^{v_f} \mathrm{d}v\left(a/v^2\right) = a(1/v_i - 1/v_f)\,. \tag{6.120}$$

For reversible, isothermal transfer of heat to the gas, the first–second law dictates

$$s' t_c = u' + w = Rt_c \ln\left[(v_f - b)/(v_i - b)\right].$$

Thus

$$s' = R\ln\left[(v_f - b)/(v_i - b)\right] \tag{6.121}$$

is the increase in the entropy of the gas. Because of reversible operation, the entropy of the universe remains unchanged. Therefore the entropy of the reservoir decreases by the same amount.

Note that for isothermal processes volume dependence of the internal energy arises only out of the presence of interparticle interaction, parameterized here by the constant a.

The occurrence of the constant b in the expression for the entropy change given above is quite interesting. Had we analyzed the statistical basis of entropy, we would have found that for isothermal processes the change in the entropy involves logarithmic dependence on the volume. This volume has to be the one that a molecule can actually roam around in. That means, rather than just v, the reduced volume $v - b$ must be involved.

Problem 6.9 (Adiabatic equation of state for Van der Waals gas in the region above the critical point)
Calculate the entropy of one mole of a Van der Waals gas above its critical point and use it to construct an adiabatic equation of state in that region.

Solution 6.9 Dividing both sides of the first $T\mathrm{d}S$ equation by t gives

$$\mathrm{d}s = C_v\frac{\mathrm{d}t}{t} + \left(\frac{\partial p}{\partial t}\right)_v \mathrm{d}v = C_v\frac{\mathrm{d}t}{t} + \frac{R}{(v-b)}\mathrm{d}v\,. \tag{6.122}$$

Because C_v is a constant in the region above the critical point, we can integrate the above along the path $(t_0, v_0) \to (t, v_0) \to (t, v)$ to get

$$s - s_0 = C_v \ln(t/t_0) + R \ln\big[(v-b)/(v_0-b)\big]. \tag{6.123}$$

For an adiabatic process,

$$s - s_0 = 0.$$

Consequently, (6.123) becomes

$$\ln(t/t_0)^{C_v} + \ln\big[(v-b)/(v_0-b)\big]^R = 0.$$

It can be recast as

$$\ln\big[t^{C_v}(v-b)^R\big] = \ln\big[t_0^{C_v}(v_0-b)^R\big].$$

A convenient form for the adiabatic equation of state is obtained by exponentiating both sides:

$$t^{C_v}(v-b)^R = \text{const.} \tag{6.124}$$

When $b = 0$, this result reduces to that for an ideal gas.

Problem 6.10 ($u, h, s, C_p - C_v$ and adiabatic equation of state)
One mole of[29] a certain gas has an equation of state

$$(p + a)v + (b/v) = Rt.$$

At a temperature above the point where the specific heat C_p peaks, calculate: (i) the internal energy u, (ii) the enthalpy h, and (iii) the entropy s and (iv) the difference in the specific heats C_p and C_v. Finally, (v) construct its adiabatic equation of state.

Solution 6.10 (i) Because $\mathrm{d}ptv = 0$, the specific heat C_v is independent of volume. Therefore, much like for the Van der Waals gas, C_v is independent of the temperature. Using the identity

$$\mathrm{d}u = \left(\frac{\partial u}{\partial t}\right)_v \mathrm{d}t + \left[t\left(\frac{\partial p}{\partial t}\right)_v - p\right]\mathrm{d}v,$$

we get

$$\mathrm{d}u = C_v \mathrm{d}t + \left(a + \frac{b}{v^2}\right)\mathrm{d}v.$$

Integrating along the path $(t_0, v_0) \to (t, v_0) \to (t, v)$ gives

$$u - u_0 = C_v(t - t_0) - \left(\frac{b}{v} - \frac{b}{v_0}\right) + a(v - v_0),$$

[29] Adiabatic state equation Van der Waals gas above critical point with $u, h, s, C_p - C_v$.

or

$$u = C_{\mathrm{v}}t + av - \left(\frac{b}{v}\right) + \text{const.} \tag{6.125}$$

(ii) The enthalpy is now readily evaluated as

$$h = u + pv = (C_{\mathrm{v}} + R)t - 2\left(\frac{b}{v}\right) + \text{const.}$$

(iii) To evaluate the entropy, we use the relationship

$$\mathrm{d}s = \frac{\mathrm{d}u}{t} + \frac{p}{t}\mathrm{d}v\,.$$

Next we use (6.125) to find du, i.e.,

$$\mathrm{d}u = C_{\mathrm{v}}\mathrm{d}t + \left[a + \frac{b}{v^2}\right]\mathrm{d}v\,.$$

Thus

$$\mathrm{d}s = C_{\mathrm{v}}\frac{\mathrm{d}t}{t} + R\frac{\mathrm{d}v}{v}\,.$$

Integrating along the path used for deriving (6.125), we get

$$s - s_0 = C_{\mathrm{v}}\ln(t/t_0) + R\ln(v/v_0)\,. \tag{6.126}$$

Whence

$$s = s_0 + \ln\left[\frac{T^{C_{\mathrm{v}}}v^{R}}{T_0^{C_{\mathrm{v}}}v_0^{R}}\right]. \tag{6.127}$$

(iv) As recorded earlier in (6.114),

$$C_{\mathrm{p}} - C_{\mathrm{v}} = t\left(\frac{\partial p}{\partial t}\right)_{\mathrm{v}}\left(\frac{\partial v}{\partial t}\right)_{\mathrm{p}}.$$

Here

$$\left(\frac{\partial p}{\partial t}\right)_{\mathrm{v}} = R/v$$

and

$$\left(\frac{\partial v}{\partial t}\right)_{\mathrm{P}} = \left[\frac{(v/t)}{1 - \frac{2b}{Rtv}}\right],$$

therefore

$$C_{\mathrm{p}} - C_{\mathrm{v}} = \left(\frac{R}{1 - \frac{2b}{Rtv}}\right). \tag{6.128}$$

As Rtv approaches $2b$ from above, the specific heat C_p becomes large and tends to ∞. While the first derivative $(\frac{\partial p}{\partial v})_t$ approaches zero, the second derivative $dpvt$ is nonvanishing. As such, there is no critical point.

Finally, for an adiabatic process, we set

$$s = s_0$$

in (6.127). The adiabatic equation of state, therefore, is

$$v^R t^{C_v} = \text{const.} \tag{6.129}$$

6.10 Dieterici's Equation of State

Much like the Van der Waals gas, the equation of state of a Dieterici gas employs two constants, a and b, that are supposed to represent interparticle interaction and molecular size effects. And, it leads to a coexistent region involving gaseous and liquid phases. Moreover, in the low density limit, both these equations yield identical results for the leading two virial coefficients.

For large v, expand the exponential in the small parameter $(-av^{-1}/RT)$ as

$$p(v-b) = Rt\left(1 - \frac{av^{-1}}{Rt}\right) + O\left(v^{-2}\right).$$

Problem 6.11 (Behavior of Dieterici gas)
Discuss the behavior of a Dieterici gas with the equation of state:

$$P(v-b) = Rt\exp\left(-av^{-1}/Rt\right). \tag{6.130}$$

Solution 6.11 Next, multiply both sides by $\frac{v}{Rt(v-b)}$ to obtain

$$\frac{pv}{Rt} = \frac{v}{v-b}\left(1 - \frac{av^{-1}}{Rt}\right) + O\left(v^{-2}\right),$$

and expand the term $\frac{v}{(v-b)}$ on the right-hand side in inverse powers of v to get

$$pv/Rt = 1 + (b_2)/v + O\left(v^{-2}\right).$$

Hence, identical with the Van der Waals gas, the second virial coefficient is

$$b_2 = b - a/(Rt).$$

Also, in principle, the constants a and b should be the same for both these equations. Despite this superficial similarity, we shall see below that the critical constants, p_c, v_c, and t_c, are quite different.

As noted for the Van der Waals gas, at the critical point two requirements have to be satisfied. First,

$$\left(\frac{\partial p}{\partial v}\right)_{\text{t}} = Rt(v-b)^{-1}\exp(-av^{-1}/Rt)\left(\frac{a}{Rtv^2} - \frac{1}{v-b}\right) = 0\,, \tag{6.131}$$

and second,

$$0 = \mathrm{d}pvt$$
$$= Rt\frac{\exp(-av^{-1}/Rt)}{v-b}\left[\frac{2}{(v-b)^2} + \left(\frac{av^{-2}}{RT}\right)^2 - 2\frac{av^{-3}}{RT} - \frac{2av^{-2}}{Rt(v-b)}\right].$$

After some algebra, it is determined that in order to satisfy these two requirements we must set

$$v_{\text{c}} = 2b\,, \qquad Rt_{\text{c}} = \frac{1}{4}(a/b)\,, \qquad p_{\text{c}} = a/(2eb)^2\,. \tag{6.132}$$

In particular, here the dimensionless ratio $p_{\text{c}}v_{\text{c}}/Rt_{\text{c}}$ is equal to $2/e^2 = 0.271$. Compared to the Van der Waals prediction, namely $3/8 = 0.375$, this result is seen to be—see Table 6.1 above—in somewhat better agreement with the experiment.

Using the above values for t_{c}, p_{c}, v_{c}, the Dieterici equation of state is transformed to its reduced variables t_0, p_0, v_0 as

$$p_0 = \frac{t_0 e^2}{2v_0 - 1}\exp\left(\frac{-2}{v_0 t_0}\right). \tag{6.133}$$

Similar to the Van der Waals gas, the reduced derivatives,

$$\left(\frac{\partial p_0}{\partial v_0}\right)_{\text{t}_0} = \left(\frac{2t_0e^2}{2v_0-1}\right)\left[\frac{-1}{2v_0-1} + \frac{1}{t_0 v_0^2}\right]\exp\left(\frac{-2}{v_0 t_0}\right) \tag{6.134}$$

and

$$\mathrm{d}p_0 v_0 t_0 = -\left(\frac{4}{2v_0-1}\right)\left(\frac{\partial p_0}{\partial v_0}\right)_{\text{t}_0} + \left[\frac{4e^2(1-t_0v_0)}{t_0v_0^4(2v_0-1)}\right], \tag{6.135}$$

are both vanishing at the critical point $t_0 = 1$, $p_0 = 1$, $v_0 = 1$.

6.10.1 Figure 6.9: Dieterici Isotherms

Shown in Fig. 6.9 below are a set of isotherms for Dieterici's equation of state. Close to $v_0 = 0.5$, the pressure increases rapidly, ultimately diverging at $v_0 = 0.5$. Therefore, much as Van der Waals would have intended, the smallest volume the Dieterici fluid can occupy is equal to b.

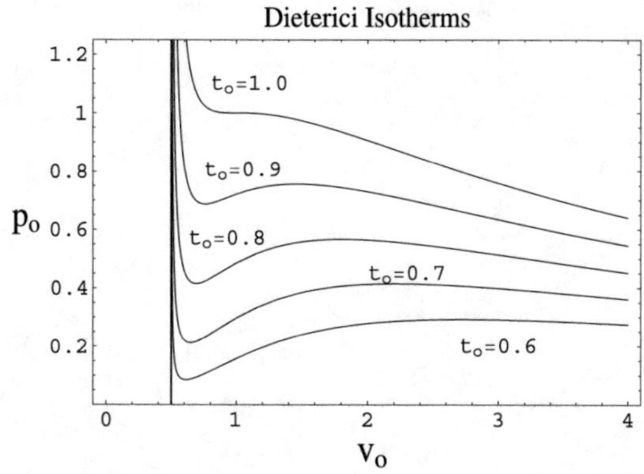

Fig. 6.9 Dieterici equation of state. Isotherms for $t_0 = 1.0, 0.9, 0.8, 0.7, 0.6$

References

1. M.J. Klein, in *The Proceedings of Van der Waals Centennial Conference on Statistical Mechanics* (North Holland, Amsterdam, 1974)
2. M. Kac, G.E. Uhlenbeck, P.C. Hemmer, J. Math. Phys. **4**, 216 (1963)
3. M. Kac, G.E. Uhlenbeck, P.C. Hemmer, J. Math. Phys. **5**, 60 (1964)
4. J.L. Lebovitz, in *Van der Waals Centennial Conference on Statistical Mechanics*, ed. by C. Prins (North Holland, Amsterdam, 1974)
5. P.C. Hemmer, M. Kac, G.E. Uhlenbeck, On the van der Waals theory to the vapor-liquid equilibrium
6. J.L. Lebovitz, Private communication. Rutgers University
7. CRC Handbook, 85th edition (2004–2005)
8. C.C. Dillio, E.P. Nye, *Thermal Physics* (International Textbook Company, Scranton, 1963)
9. E.P. Nye, C.C. Dillio
10. C.C. Dillio, E.P. Nye, *Thermal Engineering* (International Textbook Company, Scranton, 1963)
11. E.J. Su, Ind. Eng. Chem., Anal. Ed. **38**, 803 (1946)
12. H.E. Stanley, *Introduction to Phase Transitions and Critical Phenomena* (Oxford University Press, London, 1971)
13. G.J. Su, Ind. Eng. Chem., Anal. Ed. **38**, 303 (1946)
14. E.A. Guggenheim, J. Chem. Phys. **13**, 253 (1945)
15. E.A. Guggenheim, *Thermodynamics* (North Holland, Amsterdam, 1967)
16. D.V. Schroeder, An Introduction to Thermal Physics (Addison-Wesley-Longman-Pearson), p. 184, Fig. 5.23

Joule and Kelvin: Internal Energy and Enthalpy 7

Unlike the state variables, the volume, pressure, and temperature, other thermodynamic state functions such as the internal energy and enthalpy cannot easily be measured by a well-designed piece of apparatus. Rather, it is necessary to follow a circuitous route.

Measurement of the internal energy was first attempted in what is called the Gay-Lussac–Joule (GLJ) experiment. This experiment attempted to determine the amount of heat energy that is occasioned by a free-expansion of gas enclosed in vessels submerged in water. Unfortunately, the heat capacities of the water that was used, plus the enclosing tank, were vastly greater than the heat capacity of the gas being expanded. The resultant change in energy was immeasurably minuscule and could not reliably be measured. All that is discussed and three worked examples are provided in Gay-Lussac work described in Sect. 7.1.

To overcome these difficulties, Joule and Kelvin devised an experiment which shifted the focus from the internal energy to the enthalpy. Aspects of constant enthalpy and its ramifications are discussed in Sects. 7.2 and 7.3. A brief summary of some of the multivarious contributions that Joule and Kelvin made to thermodynamics theory is presented in Sect. 7.4. Enthalpy minimum for a gas with three virial coefficients is worked out in Sect. 7.5. Following the work of Joule–Gay-Lussac (JGL), various aspects of thermodynamic temperature are discussed in Sect. 7.6. The concept of negative temperature is mooted. There is a cursory remark regarding the issue in Sect. 7.7, the concluding section.

In addition, solutions to a total of 16 problems that are relevant to issues broached in Sects. 7.1–7.3 are presented in this chapter.

7.1 Gay-Lussac–Joule Coefficient

Let us deal first with the internal energy, $u \equiv u(t, v)$. Being a state function, it allows an exact differential

R. Tahir-Kheli, *General and Statistical Thermodynamics*,
https://doi.org/10.1007/978-3-030-20700-7_7

$$\begin{aligned} du &= \left(\frac{\partial u}{\partial t}\right)_{\mathrm{v}} dt + \left(\frac{\partial u}{\partial v}\right)_{\mathrm{t}} dv = c_{\mathrm{v}}(t)dt - \left(\frac{\partial u}{\partial t}\right)_{\mathrm{v}} \left(\frac{\partial t}{\partial v}\right)_{\mathrm{u}} dv \\ &= c_{\mathrm{v}}(t)dt - c_{\mathrm{v}}(t)\eta(t, v)dv\,. \end{aligned} \tag{7.1}$$

In the above, use was made of the cyclic identity (see (1.44) and (1.45) for a proof of the cyclic identity), namely

$$\left(\frac{\partial u}{\partial v}\right)_{\mathrm{t}} = -\left(\frac{\partial u}{\partial t}\right)_{\mathrm{v}} \left(\frac{\partial t}{\partial v}\right)_{\mathrm{u}}.$$

Also the Gay-Lussac–Joule (GLJ) coefficient, $\eta(t, v) = (\frac{\partial t}{\partial v})_{\mathrm{u}}$, was introduced.

According to (7.1), in order to fully calculate the internal energy, $u(t, v)$, detailed knowledge of the specific heat, $c_{\mathrm{v}}(t)$, as well as the GLJ coefficient, $\eta(t, v)$, is desired.[1]

7.1.1 Measurement of $\eta(t, v)$

Gay-Lussac, and separately Joule, were the first to attempt to measure $\eta(t, v)$—i.e., $(\frac{\partial t}{\partial v})_{\mathrm{u}}$—for a gas by using an apparatus that is displayed schematically in Fig. 7.1. The vessel "A" on the left-hand side is filled with a quantity of gas. It is connected to an evacuated vessel "B" via a pipe that can be opened and closed by the use of a stopcock. The vessels are immersed in water whose temperature is measured by an installed thermometer. The whole assembly is placed in a tank.

The opening of the stopcock causes the gas in the vessel "A" to expand so that both vessels are filled by gas at the same pressure. Because originally the vessel "B" is fully vacant, the expansion is free. As such, the expansion occurs without any work being done, i.e., $dw \approx 0$. According to the first law (see (3.7)),

$$dq = dw + du \approx 0 + du\,. \tag{7.2}$$

Therefore the heat energy, dq, introduced into the gas must be very close to equalling the increase, du, in its internal energy.

Clearly, the heat energy, dq, that has been added to the gas has come from the change, dt, in the temperature of the surrounding water. Both Gay-Lussac and Joule failed to observe any such change of temperature. In other words, they found $dt \approx 0$ and therefore concluded that $dq \approx 0$. Using (7.2), this leads to the conclusion that for the gas under consideration,

$$\begin{aligned} dt \approx 0\,, &\quad \text{therefore,} \quad dq \approx 0\,; \\ dq \approx 0 + du\,, &\quad \text{therefore,} \quad du \approx 0\,. \end{aligned} \tag{7.3}$$

[1] If, however, we needed to determine $u(t, v)$ at a fixed volume, say v_0, then we could set $dv = 0$ and use only the remainder of (7.1) in the form $du = c_{v_0}(t)dt$.

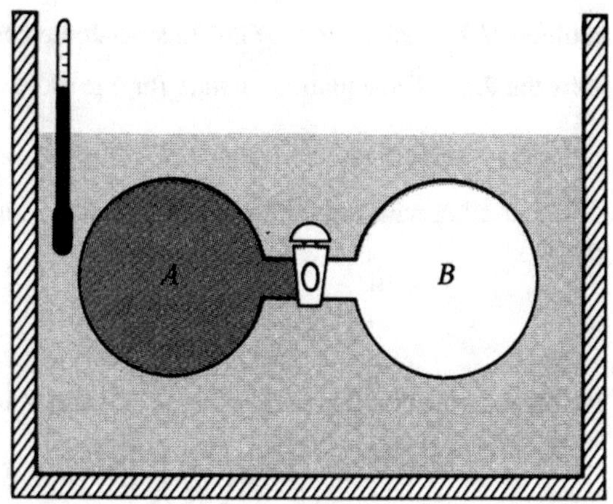

Fig. 7.1 Schematic view of the Gay-Lussac-Joule apparatus

Noting that both dt and du are close to zero, (7.1) yields

$$\begin{aligned} du &= c_{\mathrm{v}}(t)dt - c_{\mathrm{v}}(t)\eta(t, v)dv\,; \\ 0 &\approx -c_{\mathrm{v}}(t)\eta(t, v)ddv\,. \end{aligned} \tag{7.4}$$

In other words, Gay-Lussac–Joule found experimentally that $\eta(t, v)$ is "immeasurably" small at the given temperature. In accord with GLJ, below we show by a theoretical procedure that in a perfect gas the GLJ coefficient $\eta(t, v)$ is indeed vanishingly small.

7.1.2 Derivation of GLJ Coefficient

The GLJ coefficient, $\eta(t, v)$, can be expressed in terms of the specific heat C_{v}, isothermal compressibility χ_{t}, isobaric volume expansion coefficient α_{p}, pressure p, and temperature t. To this end, recall the relationship (see (5.17))

$$\left(\frac{\partial u}{\partial v}\right)_{\mathrm{t}} = t\left(\frac{\partial p}{\partial t}\right)_{\mathrm{v}} - p\,. \tag{7.5}$$

Accordingly, one can write

$$\begin{aligned} \eta(t, v) &= \left(\frac{\partial t}{\partial v}\right)_{\mathrm{u}} = -\left(\frac{\partial t}{\partial u}\right)_{\mathrm{v}}\left(\frac{\partial u}{\partial v}\right)_{\mathrm{t}} \\ &= \frac{-(\frac{\partial u}{\partial v})_{\mathrm{t}}}{(\frac{\partial u}{\partial t})_{\mathrm{v}}} = \frac{-(\frac{\partial u}{\partial v})_{\mathrm{t}}}{C_{\mathrm{v}}} = \frac{[p - t(\frac{\partial p}{\partial t})_{\mathrm{v}}]}{C_{\mathrm{v}}}\,. \end{aligned} \tag{7.6}$$

(Here the cyclic identity was used to derive the right-hand side of the second line from that of the first.)

A direct experimental measurement of $\eta(t, v)$ is not straightforward. Yet, if the usual type of equation of state is available, the quantity $(\frac{\partial p}{\partial t})_{\mathrm{v}}$ that occurs in (7.6) can readily be calculated.

Problem 7.1 Calculate the Gay-Lussac–Joule coefficient for a perfect gas.

Solution 7.1 The equation of state for one mole of a perfect gas is

$$pv = Rt\,.$$

Differentiating with respect to the temperature t and holding the volume v constant gives

$$\left(\frac{\partial p}{\partial t}\right)_{\mathrm{V}} v = R\,.$$

To calculate the coefficient η, we use (7.6) and obtain

$$\eta(t,v)C_{\mathrm{V}} = \left[p - t\left(\frac{\partial p}{\partial t}\right)_{\mathrm{V}}\right] = [p - Rt/v] = [p - p] = 0\,.$$

Thus the Gay-Lussac–Joule coefficient $\eta(t, v)$ for a perfect gas is zero.

Problem 7.2 (Nearly perfect gas)
If the gas does not greatly depart from being perfect, the addition of the so-called second virial coefficients, i.e.,

$$p = Rt\left[\frac{1}{v} + \frac{B_2(t)}{v^2}\right],$$

will improve its equation of state.

Solution 7.2 Again using (7.6), we have

$$\eta(t,v) = \frac{1}{C_{\mathrm{V}}}\left[p - t\left(\frac{\partial p}{\partial t}\right)_{\mathrm{V}}\right] = -\left(\frac{Rt^2}{v^2 C_{\mathrm{V}}}\right)\left(\frac{\partial B_2(t)}{\partial t}\right)_{\mathrm{V}}.$$

Because R, C_{V}, and $(\frac{\partial B_2(t)}{\partial t})_{\mathrm{V}} = \frac{\mathrm{d}B_2(t)}{\mathrm{d}t}$ are positive [1], the GLJ coefficient $\eta(t, v)$ for a slightly imperfect gas is nonzero and negative.

Problem 7.3 (Derive GLJ coefficient)
For the Van der Waals gas, one derivation of the GLJ coefficient, $\eta(t, v) = (\frac{\partial t}{\partial v})_{\mathrm{u}}$, is already available in a previous chapter. (See (6.118) and (6.119).) Work out another derivation.

Solution 7.3 As instructed, the present derivation is different. It utilizes the more compact expressions for these coefficients derived in the present chapter and also develops more fully their physical connotation.

Beginning with the equation of state for one mole of a Van der Waals gas,

$$p = \frac{Rt}{(v-b)} - \frac{a}{v^2}, \tag{7.7}$$

the partial derivatives $(\frac{\partial p}{\partial t})_{\mathrm{V}}$ and $(\frac{\partial v}{\partial t})_{\mathrm{P}}$ are readily determined. For instance, one gets

$$\left(\frac{\partial p}{\partial t}\right)_{\mathrm{V}} = \frac{R}{(v-b)}\,. \tag{7.8}$$

Accordingly, (7.6) yields

$$\eta(t, v) = \left(\frac{\partial t}{\partial v}\right)_{\mathrm{u}} = \frac{[p - t(\frac{\partial p}{\partial t})_{\mathrm{v}}]}{C_{\mathrm{v}}} = \frac{1}{C_{\mathrm{v}}}\left[p - \frac{Rt}{(v-b)}\right] = -\frac{a}{v^2 C_{\mathrm{v}}}. \tag{7.9}$$

Because both a and C_{v} are positive, similarly to the slightly imperfect gas, the GLJ coefficient $\eta(t, v)$ for a Van der Waals gas is negative.

Let us examine what happens if a Van der Waals gas undergoes a quasistatic, free expansion, in an adiabatic environment that allows no exchange of heat energy with the outside. Because the expansion is "free," the process does not do any work, i.e., $\mathrm{d}w = 0$. Further, because the expansion is quasistatic and occurs without exchange of heat energy, we have $t\mathrm{d}s = 0$. Accordingly, the first–second law equation, $t\mathrm{d}s = \mathrm{d}u + \mathrm{d}w$, leads to the result, $0 = \mathrm{d}u + 0$. This means the process occurs at constant u. Hence, (7.9) can be represented as

$$\left(\frac{\partial t}{\partial v}\right)_{\mathrm{u}} = \frac{dt}{dv} = -\frac{a}{v^2 C_{\mathrm{v}}}.$$

Thus the increase in temperature, $\mathrm{d}t$, as a result of the expansion (meaning, as a result of the increase, $\mathrm{d}v$, in its volume) is negative. That is, volume expansion cools the gas down. This, of course, is not surprising because the overall intermolecular force is attractive. Therefore, upon expansion of the volume, positive work is needed to overcome the attractive force. And such work has to be provided by the heat energy that is extracted from the gas. Hence the cooling.

7.2 Enthalpy: Description

Much like the internal energy, the enthalpy is a state function. Its exact differential can be represented in terms of the state variables t and p as

$$\begin{aligned}\mathrm{d}h(t, p) &= \left(\frac{\partial h}{\partial t}\right)_{\mathrm{p}} \mathrm{d}t + \left(\frac{\partial h}{\partial p}\right)_{\mathrm{t}} \mathrm{d}p = c_{\mathrm{p}}(t)\mathrm{d}dt - \left(\frac{\partial h}{\partial t}\right)_{\mathrm{p}}\left(\frac{\partial t}{\partial p}\right)_{\mathrm{h}} \mathrm{d}p \\ &= c_{\mathrm{p}}(t)\mathrm{d}t - c_{\mathrm{p}}(t)\mu(t, p)\mathrm{d}p\,. \end{aligned} \tag{7.10}$$

In the above use was made of the cyclic identity, namely

$$\left(\frac{\partial h}{\partial p}\right)_{\mathrm{t}} = -\left(\frac{\partial h}{\partial t}\right)_{\mathrm{p}}\left(\frac{\partial t}{\partial p}\right)_{\mathrm{h}}.$$

Further, the equality $(\frac{\partial h}{\partial t})_{\mathrm{p}} = c_{\mathrm{p}}$ was noted. Also, the so-called Joule–Kelvin (JK) coefficient,

$$\mu(t, p) = \left(\frac{\partial t}{\partial p}\right)_{\mathrm{h}}, \tag{7.11}$$

was introduced.

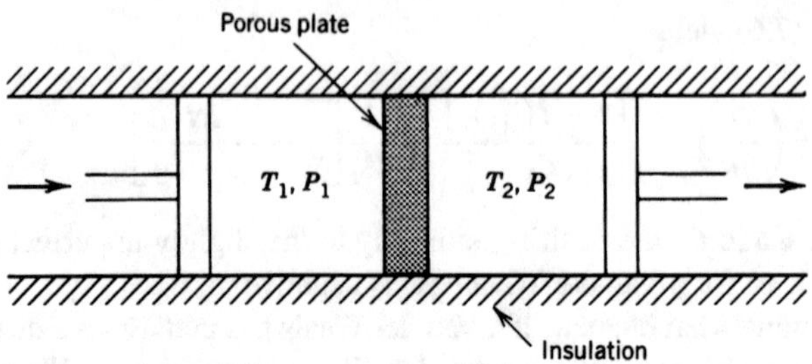

Fig. 7.2 Schematic view of the Joule–Kelvin apparatus

In order to fully calculate the enthalpy, $h(t, v)$, (7.10) recommends having detailed knowledge[2] of the specific heat, $c_{\mathrm{p}}(t)$, as well as the JK coefficient, $\mu(t, p)$.

7.3 Enthalpy Remaining Unchanged

During steady flow of a gas in the manner described below, its enthalpy (per mole) remains essentially unchanged.

Issues relating to the internal energy were the focus of the GLJ experiment. The attempt to measure an exceedingly small heat energy transfer that is occasioned by the free-expansion of gas enclosed in vessels submerged in water, is particularly handicapped by the enormous difference in the heat capacities of the gas and the surrounding water (including the enclosing tank).

To overcome these difficulties, Joule and Kelvin devised an experiment which did two things differently. One, it shifted the focus from the internal energy to the enthalpy. Two, rather than dealing with a fixed amount of gas that is stationary, it used a procedure that involved "steady-flow." Such a procedure circumvented the need to measure extremely small changes in energy which get masked by the very large heat capacities of the tank and its contents.

A schematic diagram of the apparatus used by Joule and Kelvin is displayed in Fig. 7.2. A stream of gas is steadily pushed in—at temperature T_1 and pressure P_1—from side 1 across a porous plate constriction. The stream is extracted at temperature T_2, and pressure P_2, by a piston on side 2. The flow continues until a steady, dynamic state is achieved.

After a steady state is reached, the temperature profile of the flowing gas, subject only to the vagaries of heat energy exchange with the environment across the insulation, remains essentially constant. To minimize any such exchange the tube is adiabatically shielded—meaning, it is extremely well insulated. Thus the exchange of heat energy with the environment is made essentially equal to zero, i.e., $\Delta Q \approx 0$. Moreover, because of the relative slowness of both the incoming and the outgoing portions of the gas, their kinetic energy is both small and not much different. Therefore, changes in the kinetic energy can also be assumed to be negligible.

[2]Note, however, that if the pressure is constant, say p_0, then $\mathrm{d}p = 0$. All we need then is the knowledge of $c_{\mathrm{p}_0}(t)$.

Even though one knows that very little or no heat energy is exchanged with the environment, i.e., that, $\Delta Q \approx 0$, in order to use the first law, $\Delta Q = \Delta U + \Delta W$, one also needs information about the increase in the internal energy, ΔU, and the work, ΔW, done during the expansion. All this, of course, refers to the process in which from side 1 a volume[3] V_1 of gas at temperature T_1 and pressure P_1, is pushed across to side 2. Side 2 is at temperature T_2 and some lower pressure P_2, and the outgoing volume is V_2. Therefore, during this traversal, the work done by the gas is

$$\Delta W = P_2 V_2 - P_1 V_1 , \tag{7.12}$$

and the increase in its internal energy is $U_2 - U_1$. Then, following the first law, one has

$$\Delta Q = \Delta U + \Delta W = (U_2 - U_1) + (P_2 V_2 - P_1 V_1) \approx 0 , \tag{7.13}$$

or equivalently,

$$H_1 = U_1 + P_1 V_1 \approx U_2 + P_2 V_2 = H_2 . \tag{7.14}$$

This is an important result. During steady flow of a gas in the manner described above, its enthalpy (per mole) remains essentially unchanged. And with appropriate planning, this behavior can be utilized for lowering the temperature of the gas.

Let us ask the question: Given that the starting point of the incoming gas is $(P_\mathrm{i}, T_\mathrm{i})$, what is obtained for the outgoing gas?

Because the value of the enthalpy H_i per mole for the incoming gas is fixed, the value of the enthalpy of the outgoing gas—which must also be equal to H_i per mole—is both fixed and predetermined. Now, two state variables completely determine the thermodynamic state of a simple system. As such, there remains only one other variable—say, for example, the pressure—of the outgoing gas that is subject to change. By adjusting the pumping rate, the pressure of the outgoing gas can be made to achieve a series of decreasing values, say, P_j, P_k, etc. As a result, the corresponding temperature[4] of the outgoing gas takes on the values T_j, T_k, etc., respectively.

Finally, the pressure P_i and the temperature T_i of the incoming, as well as the series of measured points of the outgoing gas—i.e., the points $(P_\mathrm{j}, T_\mathrm{j})$, $(P_\mathrm{k}, T_\mathrm{k})$, etc.—are all plotted. A smooth curve[5] is then drawn through these points, all of which refer to the same value of the enthalpy.

---

[3] A schematic plot of this is given in Fig. 7.2 above.

[4] Meaning, once the enthalpy and the pressure P_j of the outgoing gas have been chosen, its temperature T_j gets predetermined.

[5] Incidentally, this curve represents behavior that is quite likely nonquasistatic. Therefore, progress from one such point to the next does not necessarily occur through states that are in thermodynamic equilibrium.

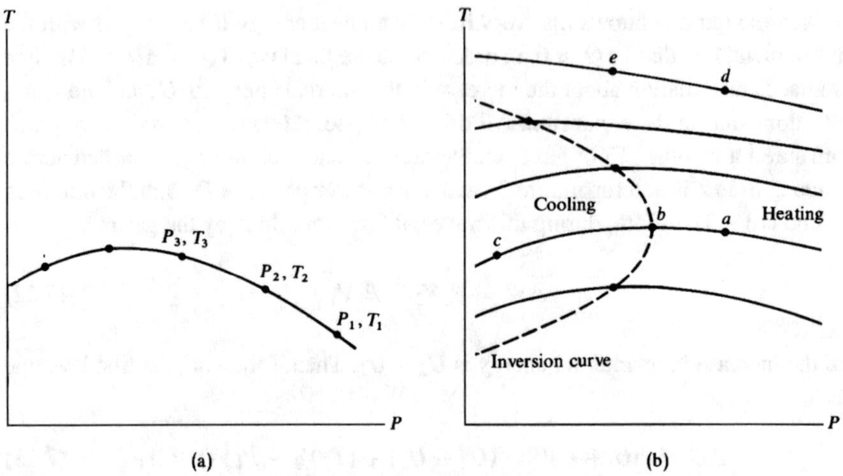

Fig. 7.3 Schematic view of the Joule–Kelvin experiment results. Each smooth curve refers to a constant value of the entropy while the dotted curve (**b**) represents the locus of the inversion points

7.3.1 Constant Enthalpy: Curves and Gaseous Cooling

A constant enthalpy curve[6] that starts at some appropriately chosen[7] incoming point (P_1, T_1) is schematically drawn in Fig. 7.3.a. Initially, the slope $(\frac{\partial T}{\partial P})_{\rm H}$ of this curve[8] is negative. As a result, the outgoing gas is warmer than the incoming gas. Continuing in the direction of lower values of the outgoing pressure, the curve rises toward its maximum, and the magnitude of the (negative) slope decreases. The position of the maximum, where the slope $(\frac{\partial T}{\partial P})_{\rm H}$ becomes zero, is called[9] the "inversion point." Further to the left,[10] the slope becomes positive, meaning here the outgoing gas cools down. By choosing different starting points and then, as was done above, performing a series of measurements of the outgoing gas, a family of such constant enthalpy curves is produced. Each member of such family has an inversion point—for example, see Fig. 7.3.b. Then, much like the dotted curves in Figs. 7.3.b and 7.4, a curve is drawn that passes through all these inversion points. This is the so-called "inversion curve." When extended to zero pressure, the inversion curve meets the temperature coordinate at two points, one higher than the other. These points are denoted as $T^{\rm max}$ and $T^{\rm min}$, respectively. From the comments in the present footnote[11] it is clear that when an isenthalpic curve starts at a point higher than

[6] To be called isenthalpic.

[7] See below for an explanation of what such appropriate choice should be.

[8] View Fig. 7.3.a and also observe the point a on the curve which is second from the bottom in Fig. 7.3.b.

[9] See point b on the noted curve in Fig. 7.3.b.

[10] As, for example, is the case at point c on the same curve in Fig. 7.3.b.

[11] When the constant enthalpy curves are extended toward zero pressure, the highest and the lowest such curves—assume they begin at the points $(P_{\rm upper}, T_{\rm upper})$ and $(P_{\rm lower}, T_{\rm lower})$, respectively—touch the temperature coordinate at $T^{\rm max}$ and $T^{\rm min}$. As shown in Fig. 7.4, in ni-

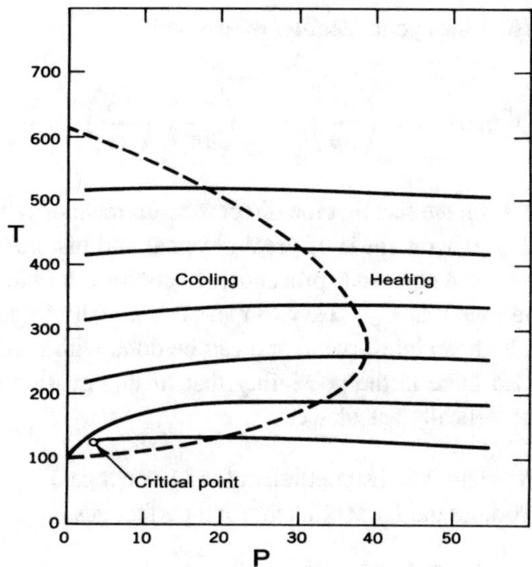

Fig. 7.4 Schematic view of the constant enthalpy curves for nitrogen. The *dotted line* represents the inversion curve. The pressure at the critical point should be ≈ 3.5 MPa and the temperature ≈ 125 K. (Compare with [2])

$(P_{\text{upper}}, T_{\text{upper}})$ it passes above the inversion curve—as, for example, does the top curve in Fig. 7.3.b—and no cooling results from it.

Initially, all the isenthalpic curves that start off at appropriately chosen locations[12] warm up as they begin heading toward lower pressure until they reach the inversion curve. After crossing this curve, they start to cool down.

Long ago gases were thought not to liquefy. The realization that the Joule–Kelvin process can lead to cooling, motivated its use in the liquefaction of gases. However, to reach really low temperatures, use is made of the reversible adiabatic demagnetization of paramagnetic salts—and indeed of nuclei themselves. This phenomenon will be investigated in a later chapter.

7.4 Joule–Kelvin Effect: Derivation

The Joule–Kelvin (JK) coefficient, $\mu = \mu(t, p) = (\frac{\partial t}{\partial p})_{\text{h}}$, can be expressed in terms of the specific heat C_{p}, isobaric volume expansion coefficient α_{p}, volume v, and temperature t.

Again we invoke a relationship that was obtained via the first–second law (see (5.74))

$$\left(\frac{\partial h}{\partial p}\right)_{\text{t}} = -t\left(\frac{\partial v}{\partial t}\right)_{\text{p}} + v\,. \tag{7.15}$$

trogen $T^{\max} \approx 610$ K and $T^{\min} \approx 100$ K. Regarding the magnitude of the starting pressures, P_{upper} and P_{lower}, note that they are also dependent on the relevant temperatures, T_{upper} and T_{lower}, respectively.

[12] An appropriately chosen location (P, T) is one that sits lower than the point $(P_{\text{upper}}, T_{\text{upper}})$ and higher than the point $(P_{\text{lower}}, T_{\text{lower}})$. See the preceding footnote for explanation.

Using the cyclic identity, we can write

$$\mu(t,p) = \left(\frac{\partial t}{\partial p}\right)_{\mathrm{h}} = -\left(\frac{\partial t}{\partial h}\right)_{\mathrm{p}}\left(\frac{\partial h}{\partial p}\right)_{\mathrm{t}} = -\frac{(\frac{\partial h}{\partial p})_{\mathrm{t}}}{(\frac{\partial h}{\partial t})_{\mathrm{p}}} = \frac{[t(\frac{\partial v}{\partial t})_{\mathrm{p}} - v]}{C_{\mathrm{P}}}. \tag{7.16}$$

Despite the fact that the direct measurement of $\mu(t, p)$ via the Joule–Kelvin porous plug experiment is of great physical and historical interest, making this measurement is a cumbrous procedure. In contrast, looking at the right-hand side of (7.16), we note that C_{P} is easy to measure. Equally important, the measurement of $(\frac{\partial v}{\partial t})_{\mathrm{P}}$ is both straightforward and can be done with much greater experimental accuracy. Also there is the possibility that, if an equation of state is known, $(\frac{\partial v}{\partial t})_{\mathrm{P}}$ can be theoretically calculated.

Problem 7.4 (JK coefficient for a perfect gas)
Workout the JK coefficient for a perfect gas.

Solution 7.4 The JK coefficient μ is readily found by using (7.16) as

$$\mu(t,p) = \left(\frac{\partial t}{\partial p}\right)_{\mathrm{h}} = \left[t\left(\frac{\partial v}{\partial t}\right)_{\mathrm{p}} - v\right]/C_{\mathrm{P}}.$$

Differentiating the equation of state, $pv = Rt$, with respect to t and keeping p constant gives

$$p\left(\frac{\partial v}{\partial t}\right)_{\mathrm{p}} = R,$$

which in turn yields

$$\mu(t,p)C_{\mathrm{P}} = t\left(\frac{\partial v}{\partial t}\right)_{\mathrm{p}} - v = (Rt/p) - v = v - v = 0.$$

Therefore, for a perfect gas, much like the GLJ coefficient $\eta(t, v)$, the JK coefficient $\mu(t, p)$ is also equal to zero.

Problem 7.5 (JK coefficient for a nearly perfect gas)
Calculate $\mu(t, p)$ for the nearly perfect gas with the equation of state

$$pv = Rt\left[1 + \frac{B_2(t)}{v}\right].$$

Solution 7.5 In order to calculate $\mu(t, p)$, first differentiate the equation of state with respect to temperature at constant pressure, then divide by pressure p and multiply by temperature t, and from the result subtract the volume v. Finally, multiply by $C_{\rm p}(\frac{pv}{Rt})$. This gives

$$\mu(t,p) = \frac{t(\frac{\partial v}{\partial t})_{\rm p} - v}{C_{\rm p}} = \left[\frac{t\frac{{\rm d}B_2(t)}{{\rm d}t} - (\frac{B_2(t)}{v})t(\frac{\partial v}{\partial t})_{\rm p}}{C_{\rm p}(1+\frac{B_2(t)}{v})}\right]$$
$$= \left[\frac{t\frac{{\rm d}B_2(t)}{{\rm d}t} - B_2(t)}{C_{\rm p}(1+\frac{B_2(t)}{v})}\right] - \left(\frac{B_2(t)}{v}\right)\mu(t,p), \qquad (7.17)$$

which can be written as

$$\mu(t,p) = \left[\frac{t\frac{{\rm d}B_2(t)}{{\rm d}t} - B_2(t)}{C_{\rm p}(1+\frac{B_2(t)}{v})^2}\right]. \qquad (7.18)$$

For a perfect gas, $B_2(t) \to 0$. In this manner, (7.18) again confirms the fact that for an ideal gas the Joule–Kelvin coefficient $\mu \to 0$.

7.4.1 JK Coefficient: Van der Waals Gas

To determine $\mu(t,p) = (\frac{\partial t}{\partial p})_{\rm h}$, for the Van der Waals gas, first we use (7.16), which equates $\mu(t,p)$ to $\frac{1}{C_{\rm p}}[t(\frac{\partial v}{\partial t})_{\rm p} - v]$. Next, we carry out direct differentiation of the equation of state—given in (7.7) above—to determine $(\frac{\partial v}{\partial t})_{\rm p}$. This gives

$$\left(\frac{\partial v}{\partial t}\right)_{\rm p} = \frac{R}{[(p+a/v^2) - 2a(v-b)/v^3]} = \frac{R(v-b)}{[Rt - 2a(v-b)^2/v^3]}.$$

Accordingly, for the Van der Waals gas, the JK coefficient is

$$\mu(t,p) = \left[t\left(\frac{\partial v}{\partial t}\right)_{\rm p} - v\right]/C_{\rm p} = \frac{1}{C_{\rm p}}\left[\frac{\frac{2a(v-b)^2}{v^2} - bRt}{Rt - \frac{2a(v-b)^2}{v^3}}\right] = \frac{\Gamma(t,p)}{\Delta(t,p)}. \qquad (7.19)$$

As noted before, for a perfect gas, the constants a and b are equal to zero and as a result the JK coefficient, $\mu(t,p)$, is vanishing.

7.4.2 JK Coefficient: Inversion Point

In the (t,p)-plane, any point at which the slope of the constant enthalpy curve for a Van der Waals gas is vanishing will be called an "inversion point." The temperature, volume, and pressure at which this happens will be denoted as $t_{\rm i}$, $v_{\rm i}$, and $p_{\rm i}$. Accordingly, at $(t_{\rm i}, p_{\rm i})$ the derivative $(\frac{\partial t}{\partial p})_{\rm h}$ tends to zero. When this happens, the numerator $\Gamma(t_{\rm i}, p_{\rm i})$, in (7.19) is vanishing. That is,

$$\frac{2a(v_{\rm i}-b)^2}{v_{\rm i}^2} - bRt_{\rm i} \to 0. \qquad (7.20)$$

We are assured of the accuracy of this statement by the fact that when Γ tends to zero the denominator in (7.19) is in general finite. This is easily demonstrated by multiplying the denominator by b. In general, b is not equal to v_i. As a result, the coefficient bRt_i cannot simultaneously be equal to $2ab(v_\mathrm{i} - b)^2/v_\mathrm{i}^3$—which would make the denominator vanish—and $2a(v_\mathrm{i} - b)^2/v_\mathrm{i}^2$—which would make the numerator vanish. Consequently, the numerator and denominator generally do not become zero at the same time.

7.4.3 JK Coefficient: Positive and Negative Regions

To investigate the other two possibilities, namely to find when $\mu(t, p)$ is positive or negative, we will also need to examine the behavior of the denominator. Let us therefore look again at $\mu(t, p)$, which is recorded in (7.19). As is shown in the equation below, the denominator Δ is directly proportional[13] to $C_\mathrm{p}/\chi_\mathrm{t}$.

$$\Delta = C_\mathrm{p} \cdot \left[Rt - 2a(v-b)^2/v^3\right] = C_\mathrm{p} \cdot (v-b)^2 \cdot \left[-\left(\frac{\partial p}{\partial v}\right)_\mathrm{t}\right]$$
$$= (C_\mathrm{p}/v) \cdot (v-b)^2 \cdot \left[-v\left(\frac{\partial p}{\partial v}\right)_\mathrm{t}\right] = (C_\mathrm{p}/v) \cdot (v-b)^2/\chi_\mathrm{t} \,. \tag{7.21}$$

Because in stable thermodynamic equilibrium C_p and χ_t are both positive, the denominator Δ is positive. Hence the sign of the Joule–Kelvin coefficient $\mu(t, p)$ is the same as that of the numerator Γ, i.e., $\mu \geq 0$ if $\Gamma \geq 0$. Similarly, $\mu \leq 0$ if $\Gamma \leq 0$.

As we pass through the inversion point, the Joule–Kelvin coefficient changes sign. Of course, as noted above, this means that at the inversion point, the Joule–Kelvin coefficient is equal to zero. In the (t, v)-plane, the inversion temperature is readily calculated from (7.20)[14] as

$$t_\mathrm{i} = \frac{2a}{bR} \cdot \left(1 - \frac{b}{v_\mathrm{i}}\right)^2 . \tag{7.24}$$

[13] To check this fact, use the equation of state (7.8) and determine that $(\frac{\partial p}{\partial v})_\mathrm{t} = (\frac{2a}{v^3}) - \frac{Rt}{(v-b)^2}$.

[14] As an aside, we mention a so-called maximum inversion temperature. Because the inversion volume $v_\mathrm{i} > b \geq 0$, such inversion temperature $(t_\mathrm{i})_\mathrm{max}$ would be obtained from (7.24) in the limit $(\frac{b}{v_\mathrm{i}}) \ll 1$,

$$(t_\mathrm{i})_\mathrm{max} = \frac{2a}{bR} = 2t_\mathrm{b} \,. \tag{7.22}$$

Therefore, for temperatures higher than twice the Boyle temperature, t_b, the throttling process for the Van der Waals gas does not result in cooling. Note that $(t_\mathrm{i})_\mathrm{max}$ is much higher than the critical temperature t_c,

$$(t_\mathrm{i})_\mathrm{max} = \frac{2a}{bR} = (27/4)(8a/27Rb) = (27/4)t_\mathrm{c} \,. \tag{7.23}$$

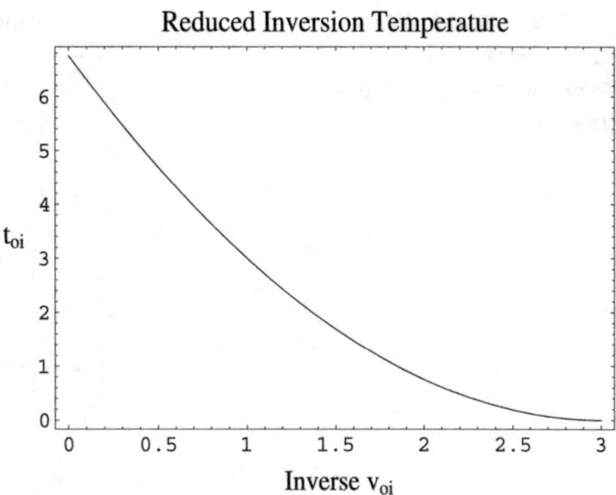

Fig. 7.5 Van der Waals reduced inversion temperature versus inverse of reduced volume

It is convenient to work in terms of the so-called reduced variables in which p, v, t, are divided by p_c, v_c, t_c, respectively. (See (6.22)–(6.24).) That is, we work in terms of p_0, v_0, t_0 instead of p, v, t:

$$
\begin{aligned}
p_0 &= p/p_c = p/\left(\frac{a}{27b^2}\right), \\
v_0 &= v/v_c = v/(3b), \\
t_0 &= t/t_c = t/\left(\frac{8a}{27Rb}\right).
\end{aligned}
\tag{7.25}
$$

In Fig. 7.5 above, we have plotted, in the temperature–volume plane, the dimensionless version of (7.24). Note, in this figure $t_{0i} = \frac{t_i}{t_c}$, and $v_{0i} = \frac{v_i}{v_c}$. It is important to note, however, that rather than the pair t and v, which is used in the above, for the discussion of the Joule–Kelvin effect, a more relevant pair of variables would have been t and p.

7.4.4 Upper and Lower Inversion Temperatures

Having recorded that, rather than v and t, the relevant state variables for the porous plug experiment are p and t, the equation of state is used to eliminate the variable v from the expression for the Joule–Kelvin coefficient μ given in (7.19). In this manner we obtain the desired p–t relationship. Again, this is best done by using the reduced variables given in (7.25).

The Van der Waals equation of state, that is,

$$
p_0 = \frac{8t_0}{3v_0 - 1} - \frac{3}{v_0^2}, \tag{7.26}
$$

is already given in terms of the reduced variables—see (6.30). Therefore, only the Joule–Kelvin coefficient μ given in (7.19) needs to be reexpressed in terms of the reduced variables. Fortunately, as noted before, in order to study the inversion curve,

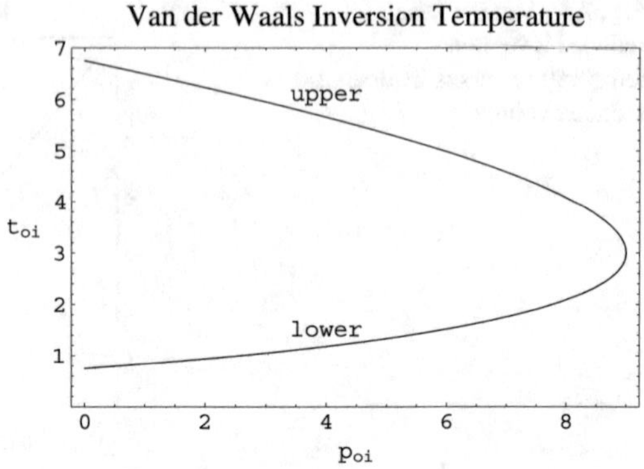

Fig. 7.6 Van der Waals reduced inversion temperature versus reduced pressure

all that needs to be analyzed is the numerator of this equation, i.e., Γ—as indeed was done in (7.20). In terms of the reduced variables, Γ is expressed as

$$\Gamma = \frac{2a(3v_0 - 1)^2}{9}\left[\frac{1}{v_0^2} - \frac{4}{3}\cdot\frac{t_0}{(3v_0-1)^2}\right]. \tag{7.27}$$

At the inversion points, (7.26) and (7.27) are written as follows:

$$p_{0i} = \frac{8t_{0i}}{3v_{0i} - 1} - \frac{3}{v_{0i}^2} \tag{7.28}$$

and

$$\Gamma = \frac{2a(3v_{0i} - 1)^2}{9}\left[\frac{1}{v_{0i}^2} - \frac{4}{3}\cdot\frac{t_{0i}}{(3v_{0i}-1)^2}\right] \to 0\,. \tag{7.29}$$

Next, in order to solve[15] the two simultaneous (7.28) and (7.29), we first eliminate v_{0i}. This leads us to the following quadratic equation in t_{0i}:

$$(-1080 + 24p_{0i})t_{0i} + 144t_{0i}^2 = -729 - 54p_{0i} - p_{0i}^2\,. \tag{7.30}$$

We shall call the two solutions of this quadratic equation t_{upper} and t_{lower}, where

$$t_{0i} = t_{\text{upper}} = \frac{1}{12}[45 - p_{0i} + 12\sqrt{9 - p_{0i}}]\,, \tag{7.31}$$

$$t_{0i} = t_{\text{lower}} = \frac{1}{12}[45 - p_{0i} - 12\sqrt{9 - p_{0i}}]\,. \tag{7.32}$$

Thus for a given value of the reduced pressure p_{0i}, there are two possible inversion temperatures, t_{upper} and t_{lower}. They are both plotted in Fig. 7.6. Note that at $p_{0i} = 9$ the upper and lower inversion temperatures are both equal to 3. (Compare with the measured values of the upper and lower inversion temperatures for nitrogen, shown as the dotted curve in Fig. 7.4.)

[15] Mathematica functions "eliminate" and "solve" have been employed here.

Problem 7.6 (Adiabatic-free expansion of Van der Waals gas)
An adiabatically enclosed vessel of negligibly small heat capacity is partitioned into two evacuated compartments of volume v_1 and v_2. One mole of a Van der Waals gas at temperature $t_\mathrm{i} = 300$ K is introduced into the first compartment with volume v_1. Upon removal of the partition, the gas expands into the second compartment. When thermal equilibrium is achieved, the temperature of the gas is found to be t_f.

Given $v_1 = 1$ l, $v_2 = 6$ l, and that the Van der Waals constants for the gas are $a = 0.37\ \mathrm{Jm}^3$, $b = 43 \times 10^{-6}\ \mathrm{m}^3$, $C_\mathrm{v} = 25\ \mathrm{J\,K}^{-1}$, calculate t_f.

Solution 7.6 The expansion occurs into an evacuated vessel of volume v_2. Thus no work is done during the expansion. Also, because of thermal isolation, no heat energy is exchanged with the environment. As a result, the internal energy remains unchanged during the expansion. This is the Joule–Gay-Lussac process. Clearly, therefore, the relevant equation to integrate is (7.6), i.e.,

$$t_\mathrm{f} - t_\mathrm{i} = \int_{v_1}^{v_1+v_2} \left(\frac{\partial t}{\partial v}\right)_\mathrm{u} \mathrm{d}v = \int_{v_1}^{v_1+v_2} \left(\frac{p - t(\frac{\partial p}{\partial t})_\mathrm{v}}{C_\mathrm{v}}\right)\mathrm{d}v\,. \tag{7.33}$$

Note that the final volume of the gas is $v_1 + v_2$.

For the Van der Waals gas, the integrand in (7.33) is as given in (7.9). Because we are told that C_v is constant, it can be pulled out of the integral. Further from the Van der Waals equation of state, namely $p + \frac{a}{v^2} = \frac{Rt}{v-b}$, we readily find $p - t(\frac{\partial p}{\partial t})_\mathrm{v} = -av^{-2}$. Therefore we have

$$\begin{aligned} t_\mathrm{f} - t_\mathrm{i} &= -\frac{a}{C_\mathrm{v}} \int_{v_1}^{v_1+v_2} v^{-2}\mathrm{d}v = -\frac{a}{C_\mathrm{v}}\left[\frac{1}{v_1} - \frac{1}{v_1+v_2}\right] \\ &= -\frac{0.37\ \mathrm{Jm}^3}{25\ \mathrm{J\,K}^{-1}}\left[\frac{1}{1 \times 10^{-3}\mathrm{m}^3} - \frac{1}{6 \times 10^{-3}\mathrm{m}^3}\right]. \end{aligned} \tag{7.34}$$

Thus $t_\mathrm{f} - t_\mathrm{i} = -12.3$ K. Finally,

$$t_\mathrm{f} = t_\mathrm{i} - 12.3\ \mathrm{K} = 300\ \mathrm{K} - 12.3\ \mathrm{K} = 287.7\ \mathrm{K}\,.$$

It is interesting to also do a "back of the envelope" calculation by assuming with Van der Waals that the gas is subject to an internal pressure equal to

$$\Delta P = -\frac{a}{v^2}\,.$$

And this pressure resists any further expansion of the gas. As a result, in expanding from v_1 to $v_1 + v_2$ the gas will have to "internally" do work equal to

$$\Delta w = -a \int_{v_1}^{v_1+v_2} v^{-2}\mathrm{d}v\,.$$

Because the "external expansion" is free, no external work results from it. Also the internal energy is constant. Thus the internal work done itself is equal to the change Δw in the heat energy content of the system. Because C_v is constant, ΔW can also be expressed as

$$\Delta w = C_{\mathrm{v}}(t_{\mathrm{f}} - t_{\mathrm{i}})\,.$$

Equating these two results for Δw leads to the expression given in (7.34)—a result obtained by the more formal procedure!

Problem 7.7 (From equation of state for hydrogen estimate of inversion temperature)
The equation of state for describing one mole of hydrogen in the limit of low pressure and moderate temperature—$t \sim 300$ K—is

$$pv = Rt\left[1 + \frac{B(t)}{v}\right]. \tag{7.35}$$

Assuming

$$B(t) = \text{const.} \times \left[3.15 - 2.9 \times \mathrm{e}^{\frac{10}{t}}\right], \tag{7.36}$$

estimate the inversion temperature for hydrogen.

Solution 7.7 Similar to (7.18), the Joule–Kelvin coefficient μ for the equation of state (7.35) is

$$\mu(t, p)C_{\mathrm{p}} = \frac{t\frac{\mathrm{d}B(t)}{\mathrm{d}t} - B(t)}{(1 + \frac{B(t)}{v})^2}\,. \tag{7.37}$$

At the inversion temperature, $t = t_i$,

$$\mu(t_{\mathrm{i}}, p) = 0\,.$$

Therefore,

$$t_i = \frac{B(t_{\mathrm{i}})}{[\frac{\mathrm{d}B}{\mathrm{d}t}]_{t=t_{\mathrm{i}}}}\,. \tag{7.38}$$

Using the expression for $B(t)$ given in (7.36), after a little algebra the above can be expressed as

$$3.15\,\exp(-z) - 2.9(1 + z) = 0\,, \tag{7.39}$$

where $z = (10/t_i)$. This equation is readily solved numerically using Mathematica, and we get $t_i = 239.4$ K. Surprisingly, this result is identical to the corresponding experimental value, 239.4 K!

Problem 7.8 Provide an alternate solution of Problem 7.7.

Solution 7.8 An alternate, more "physical"—but approximate—solution of Problem 7.7 is given below.

First, to get a feel for the size of z, we look for the Van der Waals estimate for the Boyle temperature t_b given in (7.22). An estimate for this can be found by setting t to be very large, or equivalently $z \ll 1$. Thus we would have

$$3.15 \sim 2.9 \times e^z(1+z) \approx 2.9(1+2z)\,. \tag{7.40}$$

Accordingly,

$$z \sim 0.04310 = (1/23.2) = \frac{10}{t_\mathrm{i}}\,,$$

which in turn gives

$$t_\mathrm{i} \sim \frac{10}{0.0413} \sim 242\ \mathrm{K}\,.$$

7.5 Enthalpy Minimum: Gas with Three Virial Coefficients

The equation of a state of one mole of a certain gas is given in terms of three leading virial coefficients in the form

$$pv = Rt + p\left(b - \frac{a}{Rt}\right) + ab\left(\frac{p}{Rt}\right)^2\,, \tag{7.41}$$

where a and b are gas-specific parameters that are > 0. The gas undergoes an isothermal compression at temperature t. If the temperature t is less than a certain t_0, the enthalpy has a minimum value at pressure p_0 such that

$$p_0 = \frac{R^2 t}{3a}(t_0 - t)\,. \tag{7.42}$$

Problem 7.9 Show $Rt_0 = 2(\frac{a}{b})$.

Solution 7.9 The relevant thermodynamic relationship to use is (7.15), which says $(\frac{\partial h}{\partial p})_\mathrm{t} = -t(\frac{\partial v}{\partial t})_\mathrm{p} + v$. With its help, the equation of state (7.41) readily leads to

$$\left(\frac{\partial h}{\partial p}\right)_\mathrm{t} = v - \frac{Rt}{p} - \frac{a}{Rt} + 2ab\frac{p}{R^2t^2} \tag{7.43}$$

and

$$\mathrm{d}hpt = 2\frac{ab}{R^2t^2}\,. \tag{7.44}$$

Clearly, because $ab > 0$, $\mathrm{d}hpt > 0$. Therefore, the remaining condition for determining the enthalpy minimum is the following:

$$\left(\frac{\partial h}{\partial p}\right)_\mathrm{t} = 0\,. \tag{7.45}$$

Multiplying (7.43) by p and combining the result with (7.41) and (7.45), and doing some simple algebra, yields the following result for the pressure p at the point where the enthalpy minimum occurs:

$$p = 2\frac{Rt}{3b} - \frac{R^2t^2}{3a}\,. \tag{7.46}$$

The statement of the problem tells us that the enthalpy minimum is obtained when the pressure p is equal to p_0,

$$p_0 = \frac{R^2t}{3a}(t_0 - t)\,. \tag{7.47}$$

The requirement that the pressure p becomes equal to p_0—see (7.46) and (7.47)—yields the desired result,

$$Rt_0 = 2\frac{a}{b}\,. \tag{7.48}$$

7.6 From Empirical to Thermodynamic Temperature

Let us assume that an appropriate physical property has been chosen as the basis for an empirical temperature scale. Then we examine how a relationship may be established between the empirical temperature θ and the true thermodynamic temperature t. To this end, we follow a suggestion of Lord Kelvin and choose thermodynamic relationships that explicitly involve the Kelvin temperature.

Equation (7.5) provides one such relationship. For convenience we rewrite it below:

$$\left(\frac{\partial u}{\partial v}\right)_t = t\left(\frac{\partial p}{\partial t}\right)_v - p\,. \tag{7.49}$$

In order to proceed further, we remind ourselves of the fact that $1 - \epsilon_{carnot}$, where ϵ_{carnot} is the efficiency of a perfect Carnot engine operating with an arbitrary working substance, is exactly equal to the ratio, $\frac{t_C}{t_H}$, of the Kelvin temperatures of the cold and hot reservoirs with which the working substance exchanges heat energy—see (4.2), (4.5), and (4.16). When the temperatures of the hot and cold reservoirs are measured in terms of an arbitrary thermometric property, this ratio is no longer directly that of the empirical temperatures, θ_C and θ_H. Rather, the needed ratio is that of an appropriate function, $f(\theta)$, of the relevant empirical temperatures, θ_C and θ_H, i.e., the needed ratio is $\frac{f(\theta_C)}{f(\theta_H)}$. Thus the Kelvin–Carnot temperature t can be represented as

$$t = f(\theta)\,. \tag{7.50}$$

Because, except for t, (7.50) does not depend on any other thermodynamic variables, say x and y, we can write

$$\left(\frac{\partial t}{\partial \theta}\right)_x = \frac{dt}{d\theta}\,, \quad \left(\frac{\partial \theta}{\partial t}\right)_y = \frac{d\theta}{dt}\,, \tag{7.51}$$

$$\left(\frac{\partial y}{\partial x}\right)_t = \left(\frac{\partial y}{\partial x}\right)_\theta\,, \tag{7.52}$$

and

$$\left(\frac{\partial y}{\partial t}\right)_{\mathrm{x}} = \left(\frac{\partial y}{\partial \theta}\right)_{\mathrm{x}}\left(\frac{\mathrm{d}\theta}{\mathrm{d}t}\right). \tag{7.53}$$

Similarly, (7.49) can be written as

$$\left(\frac{\partial u}{\partial v}\right)_{\mathrm{t}} = \left(\frac{\partial u}{\partial v}\right)_{\theta} = t\left(\frac{\partial p}{\partial \theta}\right)_{\mathrm{v}}\left(\frac{\mathrm{d}\theta}{\mathrm{d}t}\right) - p\,. \tag{7.54}$$

Slight rearrangement of (7.54) yields

$$\begin{gathered}\left(\frac{\partial u}{\partial v}\right)_{\theta} + p = t\left(\frac{\partial p}{\partial \theta}\right)_{\mathrm{v}} \cdot \frac{\mathrm{d}\theta}{\mathrm{d}t}\,, \qquad \frac{\mathrm{d}t}{\mathrm{d}\theta} = \frac{t(\frac{\partial p}{\partial \theta})_{\mathrm{v}}}{(\frac{\partial u}{\partial v})_{\theta} + p}\,, \\ \int_{t_1}^{t_2} \frac{\mathrm{d}t}{t} = \int_{\theta_1}^{\theta_2} \left(\frac{(\frac{\partial p}{\partial \theta})_{\mathrm{v}}}{(\frac{\partial u}{\partial v})_{\theta} + p}\right)\mathrm{d}\theta\,.\end{gathered} \tag{7.55}$$

First, we carry out indefinite integration of the last term in (7.55). Next we convert it into definite integration; we follow that by its exponentiation, and finally use the relationship between t and $f(\theta)$ that is given in (7.50). All this is indicated below:

$$\begin{gathered}\ln(t) = \left[\int \left(\frac{(\frac{\partial p}{\partial \theta})_{\mathrm{v}}}{(\frac{\partial u}{\partial v})_{\theta} + p}\right)\mathrm{d}\theta\right] + A = \ln\big[f(\theta)\big]\,, \\ t_2 - t_1 = \exp\left[\int_{\theta_1}^{\theta_2} \left(\frac{(\frac{\partial p}{\partial \theta})_{\mathrm{v}}}{(\frac{\partial u}{\partial v})_{\theta} + p}\right)\mathrm{d}\theta\right] = f(\theta_2) - f(\theta_1)\,.\end{gathered} \tag{7.56}$$

In so far as the above procedure allows for an experimental-numerical calculation of $f(\theta)$, in principle, it offers a method of determining the Kelvin temperature in terms of an empirical measurement. It is clear that only the difference between the two temperatures, t_2 and t_1, is provided. Therefore, in practise, a reference point is also needed for such a determination.

7.6.1 Thermodynamic Temperature Scale: Via JGL Coefficient

Rather than directly choosing a reference point, let us try an alternative procedure and see how far it takes us.

To this purpose, let us first measure the pressure p, the specific heat C_{v} and the Joule–Gay-Lussac (JGL) coefficient, η—all with respect to the empirical temperature scale θ. Next, let us rewrite the second line of (7.55) in the form[16]

$$\frac{\mathrm{d}t}{t} = \frac{(\frac{\partial p}{\partial \theta})_{\mathrm{v}}}{p - \eta(\theta)C_{\mathrm{v}}(\theta)}\,. \tag{7.57}$$

[16]Even though the temperatures θ and t are neither assumed to be equal, nor are their rates of change necessarily equal, when a statement refers to a specific value of t, it can legitimately also be expressed as referring to the appropriate specific value of θ. Therefore, in (7.57), we have used

Introducing the symbol

$$\beta_{\mathrm{p}} = \left(\frac{\partial p}{\partial \theta}\right)_{\mathrm{v}} / p\,, \tag{7.58}$$

the parameter, $(\frac{\partial p}{\partial \theta})_{\mathrm{v}}$, in the numerator can be replaced by $p\beta_{\mathrm{p}}(\theta)$. Dividing the numerator and denominator on the right-hand side of (7.57) by p, we get

$$\frac{\mathrm{d}t}{t} = \left[\frac{\beta_{\mathrm{p}}}{1 - Z(\theta)}\right] \mathrm{d}\theta\,, \tag{7.59}$$

where we have introduced the notation

$$Z(\theta) = \frac{C_{\mathrm{v}}(\theta)\eta(\theta)}{p}\,. \tag{7.60}$$

For helium and hydrogen near the ice point, $Z(\theta)$ is small compared to unity. Thus, to an accuracy of better than a couple of parts per thousand, we can ignore $Z(\theta)$. Accordingly,

$$\int_{t_0}^{t_0+\Delta} \left(\frac{\mathrm{d}t}{t}\right) = \int_0^{\Delta} \frac{\beta_{\mathrm{p}}}{1 - Z(\theta)} \mathrm{d}\theta = \int_0^{\Delta} \beta_{\mathrm{p}}[1 + Z(\theta) + \cdots] \mathrm{d}\theta \approx \int_0^{\Delta} \beta_{\mathrm{p}} \mathrm{d}\theta\,. \tag{7.61}$$

The meaning we have attached to the notation used in (7.61) given above is the following:

(i) The ice point on the true thermodynamic temperature scale—also called the absolute temperature scale or, as it is often the case, is called the Kelvin scale—is denoted as t_0.
(ii) The empirical scale is set so the ice point is at 0° empirical.
(iii) The thermodynamic temperature scale is such that any rise in temperature, Δ, beyond 0° empirical has the same numerical value on both the true thermodynamic and empirical scales.
(iv) Further, for convenience in the present calculation, work with a Δ such that $\Delta \ll t_0$.

Conditions (ii) and (iii) make the empirical scale identical to the Celsius scale.

Upon integration, (7.61) gives

$$\ln\left[\frac{t_0 + \Delta}{t_0}\right] \approx \int_0^{\Delta} \beta_{\mathrm{p}} \mathrm{d}\theta = \beta_{\mathrm{p}} \Delta = \ln\left[1 + \frac{\Delta}{t_0}\right] \approx \frac{\Delta}{t_0}\,. \tag{7.62}$$

---

the following relationships—compare with (7.6)—

$$\eta(\theta) = \left(\frac{\partial \theta}{\partial v}\right)_{\mathrm{u}} = -\frac{(\frac{\partial u}{\partial v})_{\theta}}{(\frac{\partial u}{\partial \theta})_{\mathrm{v}}} = -\frac{(\frac{\partial u}{\partial v})_{\theta}}{C_{\mathrm{v}}(\theta)}\,,$$

which yields $(\frac{\partial u}{\partial v})_{\theta} = -\eta(\theta) C_{\mathrm{v}}(\theta)$.

Comparing terms proportional to Δ leads to the sought after solution for the estimated value of the temperature, t_0, of the ice-point on the "true thermodynamic scale." The appropriate estimate for t_0 is thus found to be equal to the inverse of the "constant-volume, pressure expansion coefficient" β_p,

$$t_0 \approx \frac{1}{\beta_p} = \left[\frac{1}{p}\left(\frac{\partial p}{\partial \theta}\right)_v\right]^{-1} . \tag{7.63}$$

For hydrogen, the experimental result for β_p is 0.003663. Accordingly, the corresponding value for the temperature, t_0, of the ice-point on the true thermodynamic scale is ≈ 273.0 K.

7.6.2 Thermodynamic Scale: Via JK Coefficient

Continuing the study of convenient experimental procedures for relating an empirical measurement to the true thermodynamic temperature, we try next the use of the Joule–Kelvin (JK) relationship. In practice, this procedure is somewhat more convenient than those discussed above. And, to this purpose, it is helpful here to reproduce (7.16), namely

$$\mu(t,p) = \left(\frac{\partial t}{\partial p}\right)_h = \left[t\left(\frac{\partial v}{\partial t}\right)_p - v\right] \Big/ \left(\frac{\partial h}{\partial t}\right)_p . \tag{7.64}$$

The transformation to the empirical scale yields the following:[17]

$$\begin{aligned}\left(\frac{\partial t}{\partial p}\right)_h &= \left[t\left(\frac{\partial v}{\partial \theta}\right)_p \frac{d\theta}{dt} - v\right] \Big/ \left[\left(\frac{\partial h}{\partial \theta}\right)_p \frac{d\theta}{dt}\right] \\ = \left(\frac{\partial \theta}{\partial p}\right)_h \frac{dt}{d\theta} &= \left[t\left(\frac{\partial v}{\partial \theta}\right)_p \frac{d\theta}{dt} - v\right] \Big/ \left[C_p(\theta)\frac{d\theta}{dt}\right] .\end{aligned} \tag{7.67}$$

[17] Another route to deriving the same result is the following: Begin with (7.15). For convenience we rewrite it below,

$$\left(\frac{\partial h}{\partial p}\right)_t = -t\left(\frac{\partial v}{\partial t}\right)_p + v . \tag{7.65}$$

Following the prescription contained in (7.51) to (7.53), (7.65) can be represented as

$$\left(\frac{\partial h}{\partial p}\right)_t = \left(\frac{\partial h}{\partial p}\right)_\theta = -t\left(\frac{\partial v}{\partial \theta}\right)_p \frac{d\theta}{dt} + v .$$

Slight rearrangement leads to the following:

$$\frac{dt}{t} = \left[\frac{(\frac{\partial v}{\partial \theta})_p}{-(\frac{\partial h}{\partial p})_\theta + v}\right] d\theta = \left[\frac{(\frac{\partial v}{\partial \theta})_p}{v + C_p(\theta)(\frac{\partial \theta}{\partial p})_h}\right] d\theta . \tag{7.66}$$

Here we have noted that $-(\frac{\partial h}{\partial p})_\theta = C_p(\theta)(\frac{\partial \theta}{\partial p})_h$.

After multiplying both sides by $C_p(\theta)\frac{d\theta}{dt}$, the above is readily rearranged to give

$$\frac{dt}{t} = \frac{(\frac{\partial v}{\partial \theta})_p}{v + C_p(\theta)(\frac{\partial \theta}{\partial p})_h} d\theta = \left[\frac{(\frac{\partial v}{\partial \theta})_p}{v + C_p(\theta)\mu(\theta)}\right] d\theta \,. \tag{7.68}$$

First, let us look at the isobaric-volume expansion parameter, $(\frac{\partial v}{\partial \theta})_p$, in the numerator. Dividing the numerator and denominator on the right-hand side of (7.68) by v gives

$$\frac{dt}{t} = \left[\frac{\frac{1}{v}(\frac{\partial v}{\partial \theta})_p}{1+Y}\right] d\theta \,, \tag{7.69}$$

where

$$Y = \frac{C_p(\theta)\mu(\theta)}{v} \,. \tag{7.70}$$

Often, the empirical temperature is measured by a gas thermometer which utilizes the same gas that is also used for the measurement of $(\frac{\partial \theta}{\partial p})_h$. The optimal gas thermometer, of course, is an ideal gas thermometer. And, of all gases, helium is the closest to being an ideal gas. Hydrogen is the next closest, though it is dangerous to handle. In the following we shall work with both these gases.

For helium and hydrogen near the ice point, Y is small compared to unity. Thus, to an accuracy of better than a couple of parts per thousand, we can ignore Y. Accordingly,

$$\begin{aligned} \int_{t_0}^{t_0+\Delta} \left(\frac{dt}{t}\right) &= \int_0^{\Delta} \left[\frac{1}{v}\left(\frac{\partial v}{\partial \theta}\right)_p\right] (1 - Y + \cdots) d\theta \\ = \ln\left(\frac{t_0+\Delta}{t_0}\right) &\approx \int_0^{\Delta} \left[\frac{1}{v}\left(\frac{\partial v}{\partial \theta}\right)_p\right] d\theta \,. \end{aligned} \tag{7.71}$$

We use the same notation described in the preceding example. Thus:

(i) The ice point on the true thermodynamic temperature scale is again denoted as t_0.
(ii) The empirical scale is set so the ice point is at 0° empirical.
(iii) The thermodynamic temperature scale is such that any rise in temperature, Δ, beyond zero degree empirical has the same numerical value on both the absolute and the empirical scales.
(iv) Further, for convenience in the present calculation, we shall work with a Δ such that $\Delta \ll t_0$.

Conditions (ii) and (iii) make the empirical scale identical to the Celsius scale.

Expanding the logarithm, (7.71) leads to the following:

$$\Delta\left[\frac{1}{v}\left(\frac{\partial v}{\partial \theta}\right)_p\right] \approx \ln\left(\frac{t_0+\Delta}{t_0}\right) \approx \Delta/t_0 + O(\Delta/t_0)^2 \,.$$

Comparing terms proportional to Δ shows that the temperature of the ice-point on the true thermodynamic scale is equal to the inverse of the "constant-pressure, volume expansion coefficient." That is,

$$t_0 \approx \left[\frac{1}{v}\left(\frac{\partial v}{\partial \theta}\right)_{\mathrm{p}}\right]^{-1} = \frac{1}{\alpha_{\mathrm{p}}(\theta)} \,. \tag{7.72}$$

Note for the student: Quite remarkably this result looks similar to the earlier one that was obtained by following the Joule–Gay-Lussac procedure, namely $t_0 \approx [\frac{1}{p}(\frac{\partial p}{\partial \theta})_{\mathrm{v}}]^{-1}$. (See (7.63).) These two results merely interchange the variables v and p! Should it remind one of the equations of state of a nearly perfect gas?

For helium, and hydrogen, the measured values of $\alpha_{\mathrm{p}}(\theta)$ are as follows: 0.003659 for helium and 0.003660 for hydrogen. Accordingly, the corresponding values for the temperature, t_0, of the ice-point on the true thermodynamic scale are 273.3 K and 273.2 K, respectively.

7.6.3 Temperature of Ice-Point: An Estimate

The average value of the three estimates for t_0, the temperature of the ice-point, given in the two foregoing subsections is

$$t_0(\text{average}) \approx (273.0 + 273.3 + 273.2)/3 \approx 273.17\,\mathrm{K}\,. \tag{7.73}$$

In the above, for simplicity, the admittedly small contributions from the quantities Z and Y, given in (7.60)–(7.61) and (7.70)–(7.71), respectively, were ignored. Yet, a serious attempt at evaluating t_0 would surely have to include these contributions and, moreover, work with more accurate measurements of Z and Y.

We conclude with a word of caution. Despite the simplicity and attractiveness of the methods that we have proposed above, the modern practice is more elaborate and involved. Indeed, according to the dictates of the 1990 convention on the "International Temperature Scale" (see [3] for details), the Kelvin temperature scale is defined by separate processes that refer to different temperature ranges.

7.6.4 Temperature: Ideal Gas Thermodynamic Scale

Historically the most important thermodynamic system for establishing the perfect temperature scale has been the ideal gas—or the nearest approximation to it. Let us attempt to do the same below.

According to Joule's experiments, for an ideal gas or its closest equivalent, we have the result

$$\left(\frac{\partial u}{\partial v}\right)_{\theta} = 0\,. \tag{7.74}$$

Additionally, the Boyle–Charles observations indicate that

$$pv = a(t_0 + \theta)\,, \tag{7.75}$$

where a and t_0 are constants that can be determined by experiment and θ is the temperature on the Celsius scale. Equation (7.75) yields the following:

$$\frac{a}{pv} = \frac{1}{t_0 + \theta} \quad \text{and} \quad \left(\frac{\partial p}{\partial \theta}\right)_{\mathrm{v}} = \frac{a}{v}\,. \tag{7.76}$$

Therefore, from (7.74), (7.55), and (7.76), we get

$$\frac{\mathrm{d}t}{\mathrm{d}\theta} = \frac{t(\frac{\partial p}{\partial \theta})_{\mathrm{v}}}{(\frac{\partial u}{\partial v})_{\theta} + p} = \frac{t(\frac{a}{v})}{0 + p} = \frac{t}{t_0 + \theta}\,,$$

which is recast as

$$\frac{\mathrm{d}t}{t} = \frac{\mathrm{d}\theta}{t_0 + \theta}\,. \tag{7.77}$$

Integration and exponentiation lead to the result

$$t = b(t_0 + \theta)\,, \tag{7.78}$$

where b is a constant. The upshot of it all is that, by combining (7.75) and (7.78), the Boyle et al. observations can be codified into the relationship

$$pv = \left(\frac{a}{b}\right) t = Rt\,. \tag{7.79}$$

Note that $a/b = R$ is a constant and t, the so-called ideal gas temperature, is the true thermodynamic temperature.

To express the centigrade scale in terms of thermodynamic temperature, we determine the constant R by establishing a reference point. Taking the ice point of pure water as the reference point and using the Charles' experiments conducted in the early nineteenth century, we get $t_0 \approx 272$. Note that, by definition, the centigrade temperature for the ice-point is $\theta_{\mathrm{i}} = 0$.

Traditionally, it is the triple point of pure water, defined to be exactly at temperature $t_3 = 273.16$ K, that provides a highly reproducible reference point. To this end, let us for the ideal gas in question experimentally determine pv at the triple point, namely $(pv)_3$. Then

$$R = (pv)_3/273.16\,, \tag{7.80}$$

and as a result

$$pv = t(pv)_3/273.16\,. \tag{7.81}$$

Problem 7.10 Prove $(\frac{\partial t}{\partial v})_{\mathrm{u}} - (\frac{\partial t}{\partial v})_{\mathrm{s}} = \frac{p}{C_{\mathrm{v}}}$.

Solution 7.10 Using the cyclic identity, namely (1.44) and (1.45), we can write

$$\left(\frac{\partial u}{\partial v}\right)_{\mathrm{t}} = -\left(\frac{\partial u}{\partial t}\right)_{\mathrm{v}}\left(\frac{\partial t}{\partial v}\right)_{\mathrm{u}} = -C_{\mathrm{v}}\left(\frac{\partial t}{\partial v}\right)_{\mathrm{u}} .$$

Therefore

$$\left(\frac{\partial t}{\partial v}\right)_{\mathrm{u}} = -\left(\frac{\partial u}{\partial v}\right)_{\mathrm{t}} / C_{\mathrm{v}} .$$

Similarly, according to (7.5),

$$\left(\frac{\partial u}{\partial v}\right)_{\mathrm{t}} = t\left(\frac{\partial p}{\partial t}\right)_{\mathrm{v}} - p .$$

As a result,

$$\left(\frac{\partial t}{\partial v}\right)_{\mathrm{u}} = \left[-t\left(\frac{\partial p}{\partial t}\right)_{\mathrm{v}} + p\right] / C_{\mathrm{v}} . \tag{7.82}$$

Accordingly, what we are required to prove is the following relationship:

$$\left(\frac{\partial t}{\partial v}\right)_{\mathrm{u}} - \left(\frac{\partial t}{\partial v}\right)_{\mathrm{s}} = \left[-t\left(\frac{\partial p}{\partial t}\right)_{\mathrm{v}} + p\right] / C_{\mathrm{v}} - \left(\frac{\partial t}{\partial v}\right)_{\mathrm{s}} = \frac{p}{C_{\mathrm{v}}} . \tag{7.83}$$

Canceling $\frac{p}{C_{\mathrm{v}}}$ from the last two terms on the right-hand side of (7.83), what still remains to be proven is the equality

$$(t/C_{\mathrm{v}})\left(\frac{\partial p}{\partial t}\right)_{\mathrm{v}} + \left(\frac{\partial t}{\partial v}\right)_{\mathrm{s}} = 0 . \tag{7.84}$$

In order to prove this equality, let us revisit the first $T\mathrm{d}S$ equation given in (5.18), i.e.,

$$t\mathrm{d}s = C_{\mathrm{v}}\mathrm{d}t + t\left(\frac{\partial p}{\partial t}\right)_{\mathrm{v}} \mathrm{d}v ,$$

and keep s constant. This gives

$$0 = C_{\mathrm{v}}(\mathrm{d}t)_{\mathrm{s}} + t\left(\frac{\partial p}{\partial t}\right)_{\mathrm{v}} (\mathrm{d}v)_{\mathrm{s}} .$$

Now divide both sides by $(\mathrm{d}v)_{\mathrm{s}}$ to obtain

$$(t/C_{\mathrm{v}})\left(\frac{\partial p}{\partial t}\right)_{\mathrm{v}} + \left(\frac{\partial t}{\partial v}\right)_{\mathrm{s}} = 0 . \tag{7.85}$$

Q.E.D.

Problem 7.11 Prove $(\frac{\partial t}{\partial p})_{\mathrm{h}} - (\frac{\partial t}{\partial p})_{\mathrm{s}} = -\frac{v}{C_{\mathrm{p}}}$.

Solution 7.11 Using the cyclic identity (see (1.44) and (1.45)), we can write

$$\left(\frac{\partial h}{\partial p}\right)_{\mathrm{t}} = -\left(\frac{\partial h}{\partial t}\right)_{\mathrm{p}}\left(\frac{\partial t}{\partial p}\right)_{\mathrm{h}} = -C_{\mathrm{p}}\left(\frac{\partial t}{\partial p}\right)_{\mathrm{h}} .$$

Therefore

$$\left(\frac{\partial t}{\partial p}\right)_{\mathrm{h}} = -\left(\frac{\partial h}{\partial p}\right)_{\mathrm{t}} / C_{\mathrm{p}} .$$

Similarly, according to (7.15),

$$\left(\frac{\partial h}{\partial p}\right)_{\mathrm{t}} = -t\left(\frac{\partial v}{\partial t}\right)_{\mathrm{p}} + v .$$

As a result,

$$\left(\frac{\partial t}{\partial p}\right)_{\mathrm{h}} = -\left[-t\left(\frac{\partial v}{\partial t}\right)_{\mathrm{p}} + v\right] / C_{\mathrm{p}} . \tag{7.86}$$

Accordingly, what we are required to prove is

$$\left(\frac{\partial t}{\partial p}\right)_{\mathrm{h}} - \left(\frac{\partial t}{\partial p}\right)_{\mathrm{s}} = -\left[-t\left(\frac{\partial v}{\partial t}\right)_{\mathrm{p}} + v\right] / C_{\mathrm{p}} - \left(\frac{\partial t}{\partial p}\right)_{\mathrm{s}} = -\frac{v}{C_{\mathrm{p}}} . \tag{7.87}$$

Canceling $-\frac{v}{C_{\mathrm{p}}}$ from the last two terms on the right, what still remains to be proven is the equality

$$(t/C_{\mathrm{p}})\left(\frac{\partial v}{\partial t}\right)_{\mathrm{p}} - \left(\frac{\partial t}{\partial p}\right)_{\mathrm{s}} = 0 . \tag{7.88}$$

In order to prove this equality, let us revisit the second $T\mathrm{d}S$ equation, given in (5.75) in the form

$$t\mathrm{d}s = C_{\mathrm{p}}\mathrm{d}t - t\left(\frac{\partial v}{\partial t}\right)_{\mathrm{p}} \mathrm{d}p ,$$

and keep s constant. Then

$$0 = C_{\mathrm{p}}(\mathrm{d}t)_{\mathrm{s}} - t\left(\frac{\partial v}{\partial t}\right)_{\mathrm{p}} (\mathrm{d}p)_{\mathrm{s}} .$$

Therefore we have

$$(t/C_{\mathrm{p}})\left(\frac{\partial v}{\partial t}\right)_{\mathrm{p}} - \left(\frac{\partial t}{\partial p}\right)_{\mathrm{s}} = 0 , \tag{7.89}$$

which proves the desired equality given above in (7.88).

Problem 7.12 Prove $\frac{v\chi_\mathrm{t}}{\mu}(\frac{\partial h}{\partial v})_\mathrm{t} = C_\mathrm{p}$.
In other words, show that

$$(-v\chi_\mathrm{t}) \cdot \left(\frac{\partial h}{\partial v}\right)_\mathrm{t} = -\mu C_\mathrm{p} \,. \tag{7.90}$$

Solution 7.12 Let us look at the left-hand side. We have

$$(-v\chi_\mathrm{t}) \cdot \left(\frac{\partial h}{\partial v}\right)_\mathrm{t} = \left(\frac{\partial v}{\partial p}\right)_\mathrm{t} \cdot \left(\frac{\partial h}{\partial v}\right)_\mathrm{t} = \left(\frac{\partial h}{\partial p}\right)_\mathrm{t} . \tag{7.91}$$

Next, examine the right-hand side, namely

$$-\mu C_\mathrm{p} = -\left(\frac{\partial t}{\partial p}\right)_\mathrm{h} \cdot \left(\frac{\partial h}{\partial t}\right)_\mathrm{p} = \left(\frac{\partial h}{\partial p}\right)_\mathrm{t} . \tag{7.92}$$

(In the above we have used the cyclic identity given in (1.44) and (1.45).) Q.E.D.

Problem 7.13 Prove $\frac{1}{\eta v\chi_\mathrm{t}}(\frac{\partial u}{\partial p})_\mathrm{t} = C_\mathrm{v}$.
In other words, show

$$\eta \cdot v\chi_\mathrm{t} \cdot C_\mathrm{v} = \left(\frac{\partial u}{\partial p}\right)_\mathrm{t} . \tag{7.93}$$

Solution 7.13 The left-hand side of (7.93) is

$$\begin{aligned}\left(\frac{\partial t}{\partial v}\right)_\mathrm{u} \cdot \left[-\left(\frac{\partial v}{\partial p}\right)_\mathrm{t}\right] \cdot \left(\frac{\partial u}{\partial t}\right)_\mathrm{v} &= \left(\frac{\partial v}{\partial p}\right)_\mathrm{t} \cdot \left[-\left(\frac{\partial t}{\partial v}\right)_\mathrm{u} \left(\frac{\partial u}{\partial t}\right)_\mathrm{v}\right] \\ &= \left(\frac{\partial v}{\partial p}\right)_\mathrm{t} \cdot \left(\frac{\partial u}{\partial v}\right)_\mathrm{t} = \left(\frac{\partial u}{\partial p}\right)_\mathrm{t} .\end{aligned} \tag{7.94}$$

Q.E.D.

Problem 7.14 Prove $(\frac{\partial h}{\partial t})_\mathrm{v} = C_\mathrm{p}(1 - \mu\alpha_\mathrm{p}/\chi_\mathrm{t})$.

Solution 7.14 Recalling the identity given in (1.50),

$$\left(\frac{\partial A}{\partial x}\right)_\mathrm{y} = \left(\frac{\partial A}{\partial z}\right)_\mathrm{x} \left(\frac{\partial z}{\partial x}\right)_\mathrm{y} + \left(\frac{\partial A}{\partial x}\right)_\mathrm{z} , \tag{7.95}$$

and making the substitution

$$A \to h \,, \qquad x \to t \,, \qquad y \to v \,, \qquad z \to p \,,$$

we get

$$\left(\frac{\partial h}{\partial t}\right)_\mathrm{v} = \left(\frac{\partial h}{\partial p}\right)_\mathrm{t} \left(\frac{\partial p}{\partial t}\right)_\mathrm{v} + \left(\frac{\partial h}{\partial t}\right)_\mathrm{p} . \tag{7.96}$$

Noting

$$\left(\frac{\partial h}{\partial t}\right)_{\mathrm{p}} = C_{\mathrm{p}}$$

and

$$\left(\frac{\partial p}{\partial t}\right)_{\mathrm{v}} = -\left(\frac{\partial p}{\partial v}\right)_{\mathrm{t}}\left(\frac{\partial v}{\partial t}\right)_{\mathrm{p}} = \alpha_{\mathrm{p}}/\chi_{\mathrm{t}}\,,$$

the right-hand side of (7.96) can be written as

$$\begin{aligned} C_{\mathrm{p}} + \frac{\partial(h,t)}{\partial(p,t)}\cdot(\alpha_{\mathrm{p}}/\chi_{\mathrm{t}}) &= C_{\mathrm{p}} + (\alpha_{\mathrm{p}}/\chi_{\mathrm{t}})\left[\frac{\partial(h,t)}{\partial(p,h)}\cdot\frac{\partial(p,h)}{\partial(p,t)}\right] \\ &= C_{\mathrm{p}} + (\alpha_{\mathrm{p}}/\chi_{\mathrm{t}})\left[-\frac{\partial(t,h)}{\partial(p,h)}\cdot\frac{\partial(h,p)}{\partial(t,p)}\right] \\ &= C_{\mathrm{p}} + (\alpha_{\mathrm{p}}/\chi_{\mathrm{t}})\left[-\left(\frac{\partial t}{\partial p}\right)_{\mathrm{h}}\cdot\left(\frac{\partial h}{\partial t}\right)_{\mathrm{p}}\right] \\ &= C_{\mathrm{p}} + (\alpha_{\mathrm{p}}/\chi_{\mathrm{t}})\cdot[-\mu\cdot C_{\mathrm{p}}]\,. \end{aligned} \tag{7.97}$$

Q.E.D.

Problem 7.15 Prove $(\frac{\partial t}{\partial v})_{\mathrm{h}} = \frac{1}{v}(\alpha_{\mathrm{p}} - \frac{\chi_{\mathrm{t}}}{\mu})^{-1}$.
In other words, we are required to show

$$\frac{\partial(t,h)}{\partial(v,h)} = \frac{1}{v}\left[\alpha_{\mathrm{p}} - \frac{\chi_{\mathrm{t}}}{\mu}\right]^{-1}. \tag{7.98}$$

Solution 7.15

$$\frac{\partial(t,h)}{\partial(v,h)} = \frac{\partial(t,h)}{\partial(t,v)}\frac{\partial(t,v)}{\partial(v,h)} = \frac{\partial(t,h)}{\partial(t,v)}\Big/\frac{\partial(v,h)}{\partial(t,v)} = \frac{\partial(h,t)}{\partial(v,t)}\Big/\left[-\frac{\partial(h,v)}{\partial(t,v)}\right]. \tag{7.99}$$

Recasting (7.90) as

$$\frac{\partial(h,t)}{\partial(v,t)} = \frac{\mu C_{\mathrm{p}}}{v\chi_{\mathrm{t}}}\,,$$

and the combination of (7.96) and (7.97) as

$$\frac{\partial(h,v)}{\partial(t,v)} = \left(\frac{\partial h}{\partial t}\right)_{\mathrm{v}} = C_{\mathrm{p}}[1 - \mu\alpha_{\mathrm{p}}/\chi_{\mathrm{t}}]\,,$$

the right-hand sides of (7.98) and (7.99) are readily seen to be identical. Q.E.D.

Problem 7.16 Prove $(\frac{\partial u}{\partial s})_{\mathrm{v}} = t$ and $(\frac{\partial u}{\partial v})_{\mathrm{s}} = -p$.

Solution 7.16 Write

$$du(s, v) = \left(\frac{\partial u}{\partial s}\right)_v ds + \left(\frac{\partial u}{\partial v}\right)_s dv\,, \tag{7.100}$$

and compare with the first–second law written in the form

$$du = t ds + p dv\,.$$

Comparing coefficients yields:

$$\left(\frac{\partial u}{\partial s}\right)_v = t\,, \qquad \left(\frac{\partial u}{\partial v}\right)_s = -p\,. \tag{7.101}$$

7.7 Negative Temperature: Cursory Remark

By relating the average kinetic energy of a perfect gas to its absolute temperature T—see (2.31)—we have "psychologically" committed ourselves to treating T as a positive quantity. And this feeling has been further reinforced by Carnot's statement—see (4.5)—that the efficiency of a perfect Carnot engine, ϵ_{carnot}, is equal to $1 - \frac{T_C}{T_H}$.

One may well ask: How has Carnot added to our distrust of negative temperatures? The answer is the following: Under no circumstances, the actual work produced may be greater than the energy used for its production. This means that the working efficiency of an engine may never be greater than 100%—if the opposite were ever true, the world would not have an "energy problem"! And, clearly, a negative value for T_C or T_H would lead to (100+)% efficiency!

It turns out that the Kelvin–Planck formulation of the second law [4] is also uncomfortable with the concept of negative temperatures. The Clausius formulation[18] of the second law,[19] on the other hand, can be retained with qualifications, that is, it needs to be agreed that in the negative temperature regime, the "warmer" of the two bodies has the smaller absolute value for the temperature [5]. This means that just as +3 K is greater than +2 K, so is −2 K greater than −3 K. There is, however, an important "caveat." In increasing order, the relevant temperature in integer degrees Kelvin is

$$+0, +1, +2, \ldots, +\infty, -\infty, \ldots, -2, -1, -0\,.$$

Therefore, if we must use negative temperatures, we must also accept the fact that an object at any negative temperature is warmer than one at any positive temperature!

[18] This is true despite the fact—as previously concluded—that "A violation of the Carnot version of the second law results in a violation of the Clausius statement of the second law."

[19] Namely, "Without assistance it is impossible to withdraw positive amount of heat energy from a colder object and transfer the same to a warmer object." In other words, heat energy does not spontaneously get transferred from a colder object to a warmer one.

More seriously, the question to ask is: How, using fundamental thermodynamic principles, must a negative absolute temperature T be defined? Following Ramsey's suggestion, T should be defined from the thermodynamic—see (7.101) above—identity

$$\left(\frac{\partial U}{\partial S}\right)_{\mathrm{V,N}} = T . \tag{7.102}$$

If this equation is used as the definition of the absolute temperature, then if and when the internal energy can be measured as a function of the entropy—big "if and when," considering it is very hard to precisely measure either of these quantities by any "direct" method—and if the derivative is negative, then we have a confirmed case of negative temperature!

According to ter Haar [6]: "For a system to be capable of negative temperatures, it is necessary for its energy to have an upper bound." Energy U is not bounded in normal systems. Rather, as a function of the entropy S, the energy is monotonically increasing. As a result, the derivative $(\frac{\partial U}{\partial S})_{\mathrm{V,N}}$ is positive. That, according to (7.102), results in the temperature being positive.

All this would work well unless the entropy is bounded from above. In that case when S reaches S_{maximum}, then $(\frac{\partial U}{\partial S})_{\mathrm{V,N}}$ suddenly changes from $+\infty$ to $-\infty$ and the system starts its progress up the negative temperature scale! As noted above, the end point of the negative part of the temperature scale is $T = -0$, which surely must be just as unattainable as is $T = +0$.

An experiment with precisely these characteristics was first performed by Purcell and Pound [7]. Because the subject matter involves quantum statistical mechanics, for all relevant details the reader is referred to Chap. 11.

References

1. R.K. Pathria, Statistical Mechanics (Pergamon Press)
2. M.W. Zemanski, R.H. Dittman, *Heat and Thermodynamics* (McGraw-Hill, New York, 1981)
3. www.its-90.com
4. D. ter Haar, H. Wergeland, *Elements of Thermodynamics* (Addison-Wesley, Reading, 1966)
5. N.F. Ramsey, Thermodynamics and statistical mechanics at negative absolute temperature. Phys. Rev. **103**, 20 (1956)
6. D. ter Haar, *Elements of Statistical Mechanics*, 3rd edn. (1995)
7. E.M. Purcell, R.V. Pound, Phys. Rev. **81**, 279 (1951)

8 Euler Equation and Gibbs–Duhem Relation

In Chap. 5 we noted that in combination the first and second laws lead to an important relationship, that is, the difference, $\mathrm{d}U$, in the internal energy of two neighboring equilibrium states is linearly related to the corresponding difference, $\mathrm{d}S$, in their entropy. Of course, also included in this relationship is the heat energy quasistatically added to the system and the quasistatic work, $\mathrm{d}W_{\text{quasistatic}} = P\mathrm{d}V$, performed by the system when it transitions from a state with extensive variables (U, S, V) to one with $(U + dU, S + dS, V + dV)$. That is,

$$\mathrm{d}Q_{\text{quasistatic}} \to T\mathrm{d}S = \mathrm{d}U + \mathrm{d}W_{\text{quasistatic}} = \mathrm{d}U + P\mathrm{d}V\,. \tag{8.1}$$

Mostly thus far we have explicitly treated only single-phased, closed systems where the number of moles, n, is constant. Moreover, in addition to the internal energy[1] U, and the entropy S, the only extensive variable treated has been the volume V.

More general systems may possess properties such as magnetization, electric charge and polarization, surface tension, etc.; be composed of more than one variety of molecules; may separate into different phases; and, sometimes even undergo chemical decomposition, etc. In order to describe these phenomena, other extensive variables also need to be included in the expression for $\mathrm{d}W_{\text{quasistatic}}$. Thus, even while still considering a single-phase system, it is helpful to also include two additional terms to the quasistatic work that is performed by the system. To this end we write

$$\mathrm{d}W_{\text{quasistatic}} = P\mathrm{d}V - \sum_{\mathrm{j}=1}^{\ell_{\mathrm{c}}} \mu_{\mathrm{j}}\mathrm{d}n_{\mathrm{j}} + \sum_{\mathrm{i}} \mathcal{Y}_{\mathrm{i}}\mathrm{d}\mathcal{X}_{\mathrm{i}}\,. \tag{8.2}$$

The parameters, μ_{j}, are the intensive variables conjugate to the extensive variables n_{j}, the latter indicating the number of moles of the jth type of molecules

[1] And, of course, the enthalpy H—which is an extensive variable—that has also been considered before. Here the inclusion of H, however, is subsumed in that of U because the knowledge of the volume (and its conjugate variable, the pressure) relates U to H.

R. Tahir-Kheli, *General and Statistical Thermodynamics*,
https://doi.org/10.1007/978-3-030-20700-7_8

in the system.[2] The variable μ_j is generally called the molar chemical potential of the jth chemical constituent. We assume here a single phase constituted of ℓ_c different chemical constituents. Similarly, $\mathcal{X}_i$ are the other extensive variables that may be needed for a complete description of the system. As mentioned earlier, such variables may represent the magnetization $\mathcal{M}$, the electric polarization $\mathcal{P}$, the surface area $\mathcal{A}$ of a film, etc. The relevant conjugate variables, denoted as $\mathcal{Y}_i$, would then be the magnetic field $\mathcal{H}$, the electric filed $\mathcal{E}$, the surface tension $\mathcal{T}$, etc. Note the product of P and its conjugate V has dimensions of energy U.

The Euler equation is discussed in Sect. 8.1, chemical potential in Sect. 8.1.1, and multiple component systems in Sect. 8.1.2. Equations of state are introduced for a simple perfect gas. The three possible equations of state in the energy representation are identified. Two of these equations are well known. Several aspects of equation of state are discussed in Sects. 8.2–8.2.4. The third equation of state is mentioned in Sect. 8.2.5. Gibbs–Duhem relations in the energy and the entropy representations are worked out in Sect. 8.3. The fundamental equation for the ideal gas, both in the entropy and energy representations, is introduced in Sect. 8.5.

Three equations of state for a simple ideal gas are described in Sect. 8.5.3. The chapter concludes with a remark.

8.1 Euler Equation

Combining (8.1) with (8.2) gives

$$\sum_{j=1}^{\ell_c} \mu_j dn_j \equiv dG = dU - TdS + PdV + \sum_i \mathcal{Y}_i d\mathcal{X}_i . \tag{8.3}$$

Here G denotes the Gibbs potential—a thermodynamic potential that is put to much use later.

8.1.1 Chemical Potential

We are now more informatively able to identify the role that the chemical potential plays. First, when all other extensive variables—e.g., S, V, and $\mathcal{X}_i$—are held constant, an addition of a single mole of the jth chemical constituent increases the internal energy of the given thermodynamic system by an amount μ_j. Therefore, much like the intensive variable pressure, which provides the motive force for causing change in the extensive variable the volume, the intensive variable chemical potential provides the motive force for changing the extensive variable related to chemical composition. In open systems, the chemical potential is related to the rate

[2]Although, in the chapter on imperfect gases—see Chap. 6—we did describe the coexistence of liquid–vapor phases, it was done without an explicit treatment of the internal energy, U, and the entropy S, in the coexistent regime. Therefore, we were able to make do without having to introduce the concept of varying n.

at which a given chemical constituent is exchanged with the environment. Equally, in closed systems, the concept can be utilized for considering such phase transitions as affect changes of physical properties between coexisting phases, etc.[3]

It is important to note that when more than one phase is present, the chemical potential of any constituent is not dependent on the magnitude—i.e., the size—of the corresponding phase. Rather, it is a function of the intensive variables: the temperature, pressure, relative composition, and various $\mathcal{Y}_i$'s.

Having made the point that a complete description of a thermodynamic system may involve more than one extensive variable of the variety n_j and $\mathcal{X}_i$, for simplicity in the following we limit the description to a single chemical component system that involves only one n. Also, for further simplicity, only one additional extensive variable, $\mathcal{X}$, is included in the analysis. That is, the analysis is limited to a system where (8.3) reduces to the following:

$$\mathrm{d}U = T\mathrm{d}S - P\mathrm{d}V + \mu\mathrm{d}n - \mathcal{Y}\mathrm{d}\mathcal{X}\,. \tag{8.5}$$

Thus in the manner of (8.4), we have

$$\begin{aligned} \left(\frac{\partial U}{\partial S}\right)_{V,n,\mathcal{X}} &= T\,, & -\left(\frac{\partial U}{\partial V}\right)_{S,n,\mathcal{X}} &= P\,, \\ \left(\frac{\partial U}{\partial n}\right)_{S,V,\mathcal{X}} &= \mu\,, & -\left(\frac{\partial U}{\partial \mathcal{X}}\right)_{S,V,n} &= \mathcal{Y}\,. \end{aligned} \tag{8.6}$$

Because the internal energy is extensive—meaning its magnitude scales linearly with the size of the system—its dependence on the extensive variables, such as the entropy, number of moles n, volume V, and property $\mathcal{X}$, must be such that it constitutes a first-order homogeneous form. What this means is that if each of these variables is made bigger by a factor[4] equal to λ, then the resultant internal energy U must also get enlarged by the same factor λ. That is,

$$U' = U\big(S', V', n', \mathcal{X}'\big) = \lambda U(S, V, n, \mathcal{X})\,, \tag{8.7}$$

where

$$S' = \lambda\, S\,, \qquad V' = \lambda V\,, \qquad n' = \lambda n\,, \qquad \mathcal{X}' = \lambda \mathcal{X}\,. \tag{8.8}$$

[3]Note that according to (8.3), the rate of change of the extensive function U is completely described in terms of the rates of change of the extensive variables S, V, n_j's, and $\mathcal{X}_i$'s. Consequently, U is a function only of these extensive quantities. Moreover, because $\mathrm{d}U$ is an exact differential, its partial differentials with respect to any of the extensive variables are equal to, what we shall call, their conjugate intensive fields. For instance,

$$\left(\frac{\partial U}{\partial S}\right)_{V,n_j,\mathcal{X}_i} = T\,, \qquad \left(\frac{\partial U}{\partial V}\right)_{S,n_j,\mathcal{X}_i} = -P\,, \qquad \left(\frac{\partial U}{\partial n_l}\right)_{S,n_{j\neq l},\mathcal{X}_i} = \mu_l\,, \quad \text{etc.} \tag{8.4}$$

The subscripts in the above equation that include n_j need to be summed over all values of j. For notational convenience, the sum is not displayed.

[4]Usually, λ is called the "scaling parameter."

Using the chain rule for partial differentiation of an equation given in parametric form,[5] we have

$$\left(\frac{dU'}{d\lambda}\right) = \left(\frac{\partial U'}{\partial S'}\right)_{V',n',\mathcal{X}'}\left(\frac{dS'}{d\lambda}\right) + \left(\frac{\partial U'}{\partial V'}\right)_{S',n',\mathcal{X}'}\left(\frac{dV'}{d\lambda}\right)$$
$$+\left(\frac{\partial U'}{\partial n'}\right)_{S',V',\mathcal{X}'}\left(\frac{dn'}{d\lambda}\right) + \left(\frac{\partial U'}{\partial \mathcal{X}'}\right)_{S',V',n'}\left(\frac{d\mathcal{X}'}{d\lambda}\right). \quad (8.9)$$

If we look only at the display above, this equation looks fierce! In reality the differentials with respect to λ are easy to carry out—see (8.8)—and the result has a much tamer look:

$$U = \left(\frac{\partial U'}{\partial S'}\right)_{V',n',\mathcal{X}'} S + \left(\frac{\partial U'}{\partial V'}\right)_{S',n',\mathcal{X}'} V$$
$$+\left(\frac{\partial U'}{\partial n'}\right)_{S',V',\mathcal{X}'} n + \left(\frac{\partial U'}{\partial \mathcal{X}'}\right)_{S',V',n'} \mathcal{X}. \quad (8.10)$$

While this relationship is good for all finite values of the variable λ, it is particularly transparent for $\lambda = 1$. That, according to (8.8), means

$$U = \left(\frac{\partial U}{\partial S}\right)_{V,n,\mathcal{X}} S + \left(\frac{\partial U}{\partial V}\right)_{S,n,\mathcal{X}} V$$
$$+\left(\frac{\partial U}{\partial n}\right)_{S,V,\mathcal{X}} n + \left(\frac{\partial U}{\partial \mathcal{X}}\right)_{S,V,n} \mathcal{X}. \quad (8.11)$$

Introducing the results for $(\frac{\partial U}{\partial S})_{V,n,\mathcal{X}}$, etc., from (8.6), (8.11) yields, in the energy representation, the Euler equation[6]

$$U = TS - PV + \mu n - \mathcal{Y}\mathcal{X} = TS - PV + G - \mathcal{Y}\mathcal{X}. \quad (8.12)$$

8.1.2 Multiple-Component Systems

It is clear that if the thermodynamic system under discussion has multiple chemical constituents, then much like (8.3), instead of the single n in (8.12), molecular concentration of additional constituents would also need to be specified. This will result in the replacement of the Gibbs potential that has a single term, i.e., μn, by a Gibbs potential that is a sum over all the ℓ_c different chemical constituents, i.e.,

$$G = \sum_{j=1}^{\ell_c} \mu_j n_j. \quad (8.13)$$

[5]Note that here λ is the parameter.

[6]Occasionally, the Euler equation is also called the "complete fundamental equation."

Similarly, additional variables could also be included. As a result, the Euler equation will take the more general form:

$$\begin{aligned} U &= TS - PV + G - \sum_{\rm i} \mathcal{Y}_{\rm i}\mathcal{X}_{\rm i} \\ &= TS - PV + \sum_{\rm j=1}^{\ell_{\rm c}} \mu_{\rm j} n_{\rm j} - \sum_{\rm i} \mathcal{Y}_{\rm i}\mathcal{X}_{\rm i} \,. \end{aligned} \tag{8.14}$$

This then is the "Complete Euler Equation," or equivalently, the "Complete Fundamental Equation"—of the given thermodynamic system, in the energy representation. For a complex thermodynamic system, the knowledge of the complete fundamental equation is the holy grail of thermodynamics for it—according to Gibbs—contains all the thermodynamic information (about the given system).

Note that only a single component system will be treated in what follows in this chapter.

8.2 Equations of State

So far the only equation of state we have talked about is

$$PV = nRT \,.$$

8.2.1 Callen's Remarks

In formal terms, however, there is more than one equation of state. In fact, for a given thermodynamic system, the number of equations of state—in the energy or in the entropy representation—is equal to the number of all the intensive variables required for the description of thermodynamic states. In this context, Callen [1] refers to the intensive variables, T, P, and μ as follows: "The temperature, pressure, and the electrochemical potentials are partial derivatives of a function of $S,\ V, N_1, \ldots, N_{\rm r}$ and consequently are also functions of $S,\ V,\ N_1, \ldots, N_{\rm r}$. We thus have a set of functional relationships

$$\begin{aligned} T &= T(S, V, N_1, \ldots, N_{\rm r}) \,, \\ P &= P(S, V, N_1, \ldots, N_{\rm r}) \,, \\ \mu_{\rm j} &= \mu_{\rm j}(S, V, N_1, \ldots, N_r) \,. \end{aligned} \tag{8.15}$$

Such relationships, that express intensive parameters in terms of the independent extensive parameters, are called: "equations of state."

Callen further writes: "*... knowledge of all the equations of state of a system is equivalent to knowledge of the fundamental equation and consequently is thermodynamically complete.*"

This statement will be referred to as the Callen rule.

The fact that the fundamental equation of a system is homogeneous first-order in terms of the extensive variables[7] has direct implications for the functional form of the equations of state. It follows immediately that the equations of state are homogeneous zeroth-order. That is, multiplication of each of the independent extensive parameters by a scalar λ leaves the function unchanged:

$$\begin{aligned} T(\lambda S, \lambda V, \lambda N_{\mathrm{i}}) &= T(S, V, N_{\mathrm{i}})\,, \\ P(\lambda S, \lambda V, \lambda N_{\mathrm{i}}) &= P(S, V, N_{\mathrm{i}})\,, \\ \mu(\lambda S, \lambda V, \lambda N_{\mathrm{i}}) &= \mu(S, V, N_{\mathrm{i}})\,. \end{aligned} \tag{8.16}$$

It therefore follows that the temperature of a composite system, composed of two macroscopically sized subsystems, is equal to the temperature of either subsystem.

We shall refer to (8.16) as Callen's scaling principle.

8.2.2 Equation of State: Energy Representation

Consider a system with internal energy[8]

$$U = U(S, V, n, \mathcal{X})\,.$$

For such a system, in principle, up to four equations of state may be constructed. A formal display of these equations (compare with (8.6)) could be as follows:

$$\begin{aligned} T(S, V, n, \mathcal{X}) &= \left(\frac{\partial U}{\partial S}\right)_{\mathrm{V,n},\mathcal{X}}\,, & -P(S, V, n, \mathcal{X}) &= \left(\frac{\partial U}{\partial V}\right)_{\mathrm{S,n},\mathcal{X}}\,, \\ \mu(S, V, n, \mathcal{X}) &= \left(\frac{\partial U}{\partial n}\right)_{\mathrm{S,V},\mathcal{X}}\,, & -\mathcal{Y}(S, V, n, \mathcal{X}) &= \left(\frac{\partial U}{\partial \mathcal{X}}\right)_{\mathrm{S,V,n}}\,. \end{aligned} \tag{8.17}$$

Clearly, for a general system, involving more than one chemical potential μ_{j} and extensive variable $\mathcal{Y}$, there would be appropriate additional equations of state.

8.2.3 Equation of State: Entropy Representation

Equation (8.12) can readily be rearranged into

$$S = \left(\frac{1}{T}\right)U + \left(\frac{P}{T}\right)V - \left(\frac{\mu}{T}\right)n + \left(\frac{\mathcal{Y}}{T}\right)\mathcal{X}\,. \tag{8.18}$$

[7] For instance, multiply S, V, n_{j}, and $\mathcal{X}_{\mathrm{i}}$ each by λ and U changes to λU.

[8] Note that the phrase "energy representation" implies that the derivatives being considered here are those of the internal energy U.

As before, the equations of state are found as derivatives with respect to the extensive variables.[9] That is,

$$\begin{aligned}
\left(\frac{1}{T}\right) &= \left[\frac{1}{T(U,V,n,\mathcal{X})}\right] = \left(\frac{\partial S}{\partial U}\right)_{V,n,\mathcal{X}}, \\
\left(\frac{P}{T}\right) &= \left[\frac{P(U,V,n,\mathcal{X})}{T(U,V,n,\mathcal{X})}\right] = \left(\frac{\partial S}{\partial V}\right)_{U,n,\mathcal{X}}, \\
\left(\frac{\mu}{T}\right) &= \left[\frac{\mu(U,V,n,\mathcal{X})}{T(U,V,n,\mathcal{X})}\right] = -\left(\frac{\partial S}{\partial n}\right)_{U,V,\mathcal{X}}, \\
\left(\frac{\mathcal{Y}}{T}\right) &= \left[\frac{\mathcal{Y}(U,V,n,\mathcal{X})}{T(U,V,n,\mathcal{X})}\right] = \left(\frac{\partial S}{\partial \mathcal{X}}\right)_{U,V,n}.
\end{aligned} \tag{8.19}$$

8.2.4 Equations of State: Two Equations for an Ideal Gas

Of all thermodynamic systems, the ideal gases are the easiest to analyze. Indeed, for a simple ideal gas, with f degrees of freedom and no additional extensive variable $\mathcal{X}$, we already know (see, for example, Sect. 8.2.3) two of the possible three[10] equations of state in the entropy representation. That is,

$$\begin{aligned}
\left(\frac{1}{T}\right) &= \frac{fNk_B}{2U} = \frac{fnN_Ak_B}{2U} = \frac{fnR}{2U}, \\
\left(\frac{P}{T}\right) &= \frac{Nk_B}{V} = \frac{nN_Ak_B}{V} = \frac{nR}{V}.
\end{aligned} \tag{8.20}$$

A comparison with (8.19) indicates that these equations of state have been expressed in the entropy representation.

8.2.5 Where Is the Third Equation of State?

Clearly, the missing, third equation of state in the entropy representation has to be—see (8.19)—of the form

$$\left(\frac{\mu}{T}\right) = \left[\frac{\mu(U,n,V)}{T(U,n,V)}\right] = -\left(\frac{\partial S}{\partial n}\right)_{U,V}. \tag{8.21}$$

Because we have not yet worked out the functional details of the entropy, at this juncture it is not entirely clear how the above calculation is to be carried out. Therefore, to pursue the matter further, we need to take a different tack.

---

[9] Note that the phrase "entropy representation" implies that the derivatives being considered here are those of the system entropy S.

[10] The possible three equations of state in the entropy representation are specified in the first three equations in (8.19).

8.3 Gibbs–Duhem Relation: Energy Representation

In order to derive the Gibbs–Duhem relation in the energy representation, it is helpful first to examine the difference between the internal energy of two neighboring equilibrium states for the simple system being treated above, for which the fundamental, i.e., the Euler equation is given in (8.12). Further

$$\mathrm{d}U = (T\mathrm{d}S + S\mathrm{d}T) - (P\mathrm{d}V + V\mathrm{d}P) + (\mu\mathrm{d}n + n\mathrm{d}\mathrm{d}\mu) - (\mathcal{Y}\mathrm{d}\mathcal{X} + \mathcal{X}\mathrm{d}\mathcal{Y})\,. \quad (8.22)$$

Next, we compare this result with what would be the corresponding statement of the first–second law (compare with (8.3)), namely

$$\mathrm{d}U = T\mathrm{d}S - P\mathrm{d}V + \mu\mathrm{d}n - \mathcal{Y}\mathrm{d}\mathcal{X}\,. \quad (8.23)$$

Subtracting (8.23) from (8.22) yields the so-called Gibbs–Duhem relationship, for a simple one-component system, in the energy representation[11]

$$0 = S\mathrm{d}T - V\mathrm{d}P + n \Downarrow \mu - \mathcal{X}\mathrm{d}\mathcal{Y}\,. \quad (8.24)$$

8.3.1 Gibbs–Duhem Relation: Entropy Representation

Proceeding in an analogous fashion to that done above for the energy representation, we write first the Euler equation in a format best suited to the entropy representation. That is,

$$S = U\left(\frac{1}{T}\right) + V\left(\frac{P}{T}\right) - n\left(\frac{\mu}{T}\right) + \mathcal{X}\left(\frac{\mathcal{Y}}{T}\right). \quad (8.25)$$

Next, we find its derivative, which is

$$\begin{aligned}\mathrm{d}S = {} & U\mathrm{d}\left(\frac{1}{T}\right) + \left(\frac{1}{T}\right)\mathrm{d}U + V\mathrm{d}\left(\frac{P}{T}\right) + \left(\frac{P}{T}\right)\mathrm{d}V \\ & - n\mathrm{d}\left(\frac{\mu}{T}\right) - \left(\frac{\mu}{T}\right)\mathrm{d}n + \mathcal{X}\mathrm{d}\left(\frac{\mathcal{Y}}{T}\right) + \left(\frac{\mathcal{Y}}{T}\right)\mathrm{d}\mathcal{X}\,. \end{aligned} \quad (8.26)$$

Now we write the first–second law—compare to (8.23)—in the form

$$\mathrm{d}S = \left(\frac{1}{T}\right)\mathrm{d}U + \left(\frac{P}{T}\right)\mathrm{d}V - \left(\frac{\mu}{T}\right)\mathrm{d}n + \left(\frac{\mathcal{Y}}{T}\right)\mathrm{d}\mathcal{X}\,, \quad (8.27)$$

[11] Clearly, for the more general system the Gibbs–Duhem equation in the energy representation would be

$$0 = S\mathrm{d}T - V\mathrm{d}P + \sum_{\mathrm{j}} n_{\mathrm{j}}\mathrm{d}\mu_{\mathrm{j}} - \sum_{\mathrm{i}} \mathcal{X}_{\mathrm{i}}\mathrm{d}\mathcal{Y}_{\mathrm{i}}\,.$$

and subtract it from (8.26). The resultant relationship is the Gibbs–Duhem equation in the entropy representation[12]

$$0 = U\mathrm{d}\left(\frac{1}{T}\right) + V\mathrm{d}\left(\frac{P}{T}\right) - n\mathrm{d}\left(\frac{\mu}{T}\right) + \mathcal{X}\mathrm{d}\left(\frac{\mathcal{Y}}{T}\right). \tag{8.28}$$

8.4 Ideal Gas: Third Equation of State

We are now in a position to take a stab at finding the missing third equation of state in the entropy representation.[13] It is convenient to begin this effort by determining the fundamental equation. Also, for simplicity and convenience, we continue to consider only the simple ideal gas which does not involve extensive variables of the form $\mathcal{X}$.

Start by rearranging (8.28) as follows:

$$n\mathrm{d}\left(\frac{\mu}{T}\right) = U\mathrm{d}\left(\frac{1}{T}\right) + V\mathrm{d}\left(\frac{P}{T}\right). \tag{8.29}$$

In order to integrate (8.29), substitute the results of the first and second equations of state that were recorded in (8.20). That is,

$$\begin{aligned} n\mathrm{d}\left(\frac{\mu}{T}\right) &= U\mathrm{d}\left[\left(\frac{nfRT}{2U}\right)\frac{1}{T}\right] + V\mathrm{d}\left[\left(\frac{RV}{RV}\right)\left(\frac{P}{T}\right)\right] \\ &= \left[\frac{f}{2}R\right]U\mathrm{d}\left[\left(\frac{T}{U}\right)\frac{n}{T}\right] + RV\mathrm{d}\left[\left(\frac{PV}{RT}\right)\frac{1}{V}\right] \\ &= \left[\frac{f}{2}R\right]U\mathrm{d}\left(\frac{n}{U}\right) + RV\mathrm{d}\left(\frac{n}{V}\right) \\ &= -\left[\frac{f}{2}Rn\right]\frac{\mathrm{d}U}{U} - (Rn)\frac{\mathrm{d}V}{V} + \left(\frac{f+2}{2}\right)R\mathrm{d}n\,. \end{aligned} \tag{8.30}$$

In the top right-hand line of this equation, we first multiplied by two different factors each equal to unity, i.e., $\frac{nfRT}{2U}$ and $\frac{RV}{RV}$. We then extracted $\frac{f}{2}R$ from the left term and R from the right term. Finally, in the second term on the right, we used the fact that $PV = nRT$. (Note that for monatomic ideal gas there are only three degrees of freedom for each molecule. That is, $f = 3$. But for a diatomic ideal gas, at temperatures that are usually available in physics laboratories, $f = 5$.) Next we

[12]For the more general system, in the entropy representation, the Gibbs–Duhem equation would be

$$0 = U\mathrm{d}\left(\frac{1}{T}\right) + V\mathrm{d}\left(\frac{P}{T}\right) - \sum_{\mathrm{j}} n_{\mathrm{j}}\mathrm{d}\left(\frac{\mu_{\mathrm{j}}}{T}\right) + \sum_{\mathrm{i}} \mathcal{X}_{\mathrm{i}}\mathrm{d}\left(\frac{\mathcal{Y}_{\mathrm{i}}}{T}\right).$$

[13]For instance, look at (8.21) and note how it involves μ, T, S, n, U, and V.

divided both sides of (8.30) by n. Thus we have found

$$\mathrm{d}\left(\frac{\mu}{T}\right) = -\frac{f}{2}R\left(\frac{\mathrm{d}U}{U}\right) - R\left(\frac{\mathrm{d}V}{V}\right) + \frac{f+2}{2}R\left(\frac{\mathrm{d}n}{n}\right). \tag{8.31}$$

Integration is now easy to do and we get

$$\frac{\mu}{T} = -\left(\frac{f}{2}\right)R\ln(U) - R\ln(V) + \left(\frac{f+2}{2}\right)R\ln(n) - C_0\,, \tag{8.32}$$

where C_0 is a constant. Using (8.32), (8.25)[14] leads to the following:

$$\begin{aligned} S &= -n\left(\frac{\mu}{T}\right) + \left(\frac{U}{T}\right) + V\left(\frac{P}{T}\right) \\ &= \left(\frac{fRn}{2}\right)\ln(U) + Rn\ln(V) - \left(\frac{f+2}{2}\right)Rn\ln(n) + nC_0 \\ &\quad + \left(\frac{U}{T}\right) + V\left(\frac{P}{T}\right) \\ &= \left(\frac{fRn}{2}\right)\ln(U) + Rn\ln(V) - \left(\frac{f+2}{2}\right)Rn\ln(n) + nC_0 \\ &\quad + n\left(\frac{f+2}{2}\right)R\,. \end{aligned} \tag{8.33}$$

Note that the relationships $\frac{U}{T} = \frac{fRn}{2}$ and $V(\frac{P}{T}) = Rn$ have again been used here.

While the procedures of statistical mechanics lead to an analytical expression for the constant C_0, namely

$$C_0 = \left(\frac{3R}{2}\right)\ln\left[\frac{4\pi m}{3h^2 N_A^{\frac{5}{3}}}\right], \tag{8.34}$$

the discipline of thermodynamics does not lend itself to doing the same. Rather, the constant C_0 can be determined only if the entropy, S_0, for some reference state is known. At a certain reference state 0 let $S = S_0$, $U = U_0$, $V = V_0$, and $n = n_0$. Then (8.33) gives

$$\begin{aligned} n_0C_0 = S_0 &- \left(\frac{fRn_0}{2}\right)\ln(U_0) - Rn_0\ln(V_0) + \left(\frac{f+2}{2}\right)Rn_0\ln(n_0) \\ &- \left(\frac{f+2}{2}\right)Rn_0\,. \end{aligned} \tag{8.35}$$

[14] Remember, here we are considering the case where, in (8.25), $\mathcal{X}(\frac{\mathcal{Y}}{T}) = 0$.

8.5 Ideal Gas: Fundamental Equation

Multiplying the above by $\frac{n}{n_0}$ and inserting the result for nC_0 in (8.33) yields what is called the fundamental equation for the ideal gas in the entropy representation, in the following convenient form:

$$S=\left(\frac{n}{n_0}\right)S_0+Rn\left[\left(\frac{f}{2}\right)\ln\left(\frac{U}{U_0}\right)+\ln\left(\frac{V}{V_0}\right)\right]-\left(\frac{f+2}{2}\right)Rn\left[\ln\left(\frac{n}{n_0}\right)\right]. \quad (8.36)$$

Note that the above fundamental equation is still subject to a boundary condition.

8.5.1 Ideal Gas: Entropy Representation

Having derived the fundamental equation for an ideal gas in the entropy representation, the relevant three equations of state, that were defined in (8.21), are readily found to be

$$\left(\frac{1}{T}\right)=\left(\frac{\partial S}{\partial U}\right)_{\mathrm{V,n}}=\frac{Rnf}{2U}, \quad (8.37)$$

$$\left(\frac{P}{T}\right)=\left(\frac{\partial S}{\partial V}\right)_{\mathrm{U,n}}=\frac{Rn}{V}, \quad (8.38)$$

$$\frac{\mu}{T}=-\left(\frac{\partial S}{\partial n}\right)_{\mathrm{U,V}}=\left(\frac{f+2}{2}\right)R\left[1+\ln\left(\frac{n}{n_0}\right)\right]-\left(\frac{S_0}{n_0}\right)$$
$$-R\left[\left(\frac{f}{2}\right)\ln\left(\frac{U}{U_0}\right)+\ln\left(\frac{V}{V_0}\right)\right]. \quad (8.39)$$

Regarding Callen's scaling principle, note that in all of the three equations of state given above, multiplication of extensive parameters $(U, U_0, V, V_0, n, n_0, S_0)$ by λ leaves the equations unchanged.

8.5.2 Ideal Gas: Energy Representation

The fundamental equation for the ideal gas given in (8.36) is in the entropy representation. But it can readily be transformed into one in the energy representation. The algebra is straightforward and one gets

$$\frac{U}{U_0}=\left(\frac{n}{n_0}\right)^{\frac{f+2}{f}}\left(\frac{V}{V_0}\right)^{-\frac{2}{f}}\exp\left[\frac{2}{fR}\left\{\frac{S}{n}-\frac{S_0}{n_0}\right\}\right]. \quad (8.40)$$

8.5.3 Ideal Gas: Three Equations of State

The three equations of state in the energy representation were mentioned in (8.17). The first one is

$$T = \left(\frac{\partial U}{\partial S}\right)_{\mathrm{V,n}} = \left(\frac{2U_0}{fRn}\right)\left(\frac{n}{n_0}\right)^{\frac{f+2}{f}}\left(\frac{V}{V_0}\right)^{-\frac{2}{f}} \exp\left[\frac{2}{fR}\left\{\frac{S}{n} - \frac{S_0}{n_0}\right\}\right]. \quad (8.41)$$

The second and third equations in energy representation are:

$$\begin{aligned} P(S, V, n) &= -\left(\frac{\partial U}{\partial V}\right)_{\mathrm{S,n}} \\ &= \left(\frac{2U_0}{V_0 f}\right)\left(\frac{n}{n_0}\right)^{\frac{f+2}{f}}\left(\frac{V}{V_0}\right)^{-\frac{2+f}{f}} \exp\left[\frac{2}{fR}\left\{\frac{S}{n} - \frac{S_0}{n_0}\right\}\right], \quad (8.42) \\ \mu &= \left(\frac{\partial U}{\partial n}\right)_{\mathrm{S,V}} \\ &= \left(\frac{U_0}{n_0}\right)\left[\frac{f+2}{f} - \frac{2S}{fRn}\right]\left(\frac{nV_0}{n_0 V}\right)^{\frac{2}{f}} \exp\left[\frac{2}{fR}\left\{\frac{S}{n} - \frac{S_0}{n_0}\right\}\right]. \quad (8.43) \end{aligned}$$

It is interesting to check whether the Callen's rule regarding knowledge of all equations of state being equivalent to the knowledge of the complete fundamental equation itself is correct. This is especially interesting because an equation of state generally involves partial differentials whose integration introduces unknown constants.

In order to check the accuracy of Callen's rule (in the energy representation), let us introduce the results of the three equations of state, namely (8.41), (8.42), and (8.43), into the relevant version of the Euler equation (8.14). That is,

$$\begin{aligned} U &= TS - PV + \mu n \\ &= n\left(\frac{U_0}{n_0}\right)\left(\frac{nV_0}{n_0 V}\right)^{\frac{2}{f}} \exp\left[\frac{2}{fR}\left\{\frac{S}{n} - \frac{S_0}{n_0}\right\}\right] \\ &\quad \times \left\{\frac{2S}{fRn} - \frac{2}{f} + \left[\frac{f+2}{f} - \frac{2S}{fRn}\right]\right\} \\ &= n\left(\frac{U_0}{n_0}\right)\left(\frac{nV_0}{n_0 V}\right)^{\frac{2}{f}} \exp\left[\frac{2}{fR}\left\{\frac{S}{n} - \frac{S_0}{n_0}\right\}\right]. \quad (8.44) \end{aligned}$$

Note the last line on the right in (8.44) follows from (8.40).

Exercise 8.1 With the use of the Euler equation in the entropy representation given in (8.25)—remember the terms involving the extensive variable $\mathcal{X}$ are not being used here—and the relevant three equations of state given in (8.37), (8.38), and (8.39), show that Callen's rule also applies to the entropy representation.

Problem 8.1 Given the fundamental equation, $S = A \cdot (nVU)^{\frac{1}{3}}$, where A is a constant, determine the three equations of state in the entropy representation.

Solution 8.1 We have:

$$\frac{1}{T} = \left(\frac{\partial S}{\partial U}\right)_{\mathrm{V,n}} = \frac{A}{3} \cdot \left(\frac{nV}{U^2}\right)^{\frac{1}{3}},$$

$$\frac{P}{T} = \left(\frac{\partial S}{\partial V}\right)_{\mathrm{U,n}} = \frac{A}{3} \cdot \left(\frac{nU}{V^2}\right)^{\frac{1}{3}},$$

$$\frac{\mu}{T} = \left(\frac{\partial S}{\partial n}\right)_{\mathrm{V,U}} = \frac{A}{3} \cdot \left(\frac{UV}{n^2}\right)^{\frac{1}{3}}.$$

Callen's scaling clearly applies because when $n \to \lambda n$, $U \to \lambda U$, and $V \to \lambda V$, all the equations remain unchanged. Similarly, Callen's rule also applies because

$$\begin{aligned} S &= U \cdot \left(\frac{1}{T}\right) + V \cdot \left(\frac{P}{T}\right) - n \cdot \left(\frac{\mu}{T}\right) \\ &= U \cdot \frac{A}{3}\left(\frac{nV}{U^2}\right)^{\frac{1}{3}} + V \cdot \frac{A}{3}\left(\frac{nU}{V^2}\right)^{\frac{1}{3}} - n \cdot \frac{A}{3}\left(\frac{UV}{n^2}\right)^{\frac{1}{3}} \\ &= A \cdot (nVU)^{\frac{1}{3}}. \end{aligned} \tag{8.45}$$

Problem 8.2 Rework Problem 8.1, but this time in the energy representation.

Solution 8.2 In the energy representation, the relevant three equations of state are the following:

$$T = \left(\frac{\partial U}{\partial S}\right)_{\mathrm{V,n}} = \left(\frac{1}{A^3}\right)\left(\frac{3S^2}{nV}\right),$$

$$P = -\left(\frac{\partial U}{\partial V}\right)_{\mathrm{S,n}} = \left(\frac{1}{A^3}\right)\left(\frac{S^3}{nV^2}\right),$$

$$\mu = \left(\frac{\partial S}{\partial n}\right)_{\mathrm{V,U}} = -\left(\frac{1}{A^3}\right)\left(\frac{S^3}{n^2V}\right).$$

Again Callen's scaling applies because when $n \to \lambda n$, $S \to \lambda S$, and $V \to \lambda V$, all the equations remain unchanged.

Similarly, Callen's rule also applies because

$$\begin{aligned} U &= TS - PV + \mu n \\ &= \left(\frac{1}{A^3}\right)\left(\frac{S^3}{nV}\right)(3 - 1 - 1) = \left(\frac{1}{A^3}\right)\left(\frac{S^3}{nV}\right). \end{aligned} \tag{8.46}$$

Remark 8.1 Only for special cases can thermodynamic systems be exactly solved and a complete solution of the fundamental equation obtained. Nevertheless, the foregoing study of an extremely simple system has not been for naught. It has taught us that thermodynamics of a system in equilibrium can be formulated in two alternative but equivalent ways: one based on the representation of the internal energy as a function of the entropy S and extensive variables such as the volume V, the mole numbers n_j, etc., and the other, on the basis of the entropy as a function of U and the relevant extensive variables.

References

1. H.B. Callen, *Thermodynamics and an Introduction to Thermostatistics*, 2nd edn. (Wiley, New York, 1985)

Le Châtelier Principle

9

Detailed study and a second look at the zeroth law of thermodynamics is presented in Sect. 9.1, and the zeroth law is reconfirmed in the study presented in Sect. 9.1.1. The requirement that entropy must be maximum possible is analyzed in Sect. 9.2. The extremum in energy, namely that energy must be the minimum possible, is discussed in Sect. 9.2.4. The fact that the physical consequences of both these extrema are similar is noted and emphasized. Of course, as noted in a footnote, there are other extremum principles that are also equivalent to that for the entropy and internal energy. These refer to the system enthalpy, the Helmholtz free energy, and the Gibbs potential. Similarity of their predictions to the entropy and energy extrema is noted. Thermodynamic flows of heat energy and molecules are studied in Sects. 9.2.1 and 9.2.2. Isothermal compression is treated in Sect. 9.2.3. Thermodynamic motive forces and thermodynamic flows are described in detail in Sect. 9.3. Physical criteria for thermodynamic stability is studied in Sect. 9.4. Le Châtelier's principle and stable self-equilibrium are analyzed in Sects. 9.4.1 and 9.4.2. Two required inequalities, being called the first and second requirement, are given in (9.34) and (9.35). They are described in Sects. 9.5 and 9.5.1. Finally, the important subject of thermodynamic intrinsic stability is discussed in Sect. 9.6. The chapter concludes with a summary.

9.1 The Zeroth Law of Thermodynamics: Second Look

The zeroth law of thermodynamics asserts if a given macroscopic system is in thermal equilibrium simultaneously with two different systems, then those two (otherwise separate) systems are also in thermal equilibrium with each other. Further, whenever this happens, there is a property common to all three systems. This property is identified as the "temperature."

In the chapter on the second law, we learned that all spontaneous processes in an isolated thermodynamic system increase its total entropy. Spontaneous processes, of course, continue until the system achieves thermal equilibrium. In an isolated system, the state of thermal equilibrium—at least in theory—remains unchanged with

R. Tahir-Kheli, *General and Statistical Thermodynamics*,
https://doi.org/10.1007/978-3-030-20700-7_9

the passage of time. Accordingly, subject to the constraints under which the system has been maintained, and for the given value of its various extensive properties, its total entropy in thermodynamic equilibrium is the maximum possible.

Let us "mentally"—as distinct from "physically"—divide a given isolated macroscopic system that has achieved thermodynamic equilibrium into two subsystems, (a) and (b). Clearly, upon such imagined division, two conditions—named "First" and "Second"—have to be satisfied:

First: The total internal energy $U(0)$, total volume $V(0)$, total number of moles $n_j(0)$ for each chemical constituent j, and total magnetic moment, electric polarization, surface area, etc., cannot be affected by just imagining the stated division.[1] As a result,

$$\begin{aligned} U(0) &= u[a] + u[b]\,, \qquad & V(0) &= v(a) + v(b)\,, \\ n_j(0) &= n_j(a) + n_j(b)\,, \qquad & \mathcal{X}_i(0) &= \mathcal{X}_i(a) + \mathcal{X}_i(b)\,. \end{aligned} \tag{9.1}$$

Second: The following two facts—named (1), given in (9.2), and (2) given in (9.3), below—must also hold:

(1) If as a result of the "*gedanken experiment*," the extensive properties of the subsystem (a) were to change, say by infinitesimal amounts $du[a]$, etc., they would be compensated by corresponding changes in subsystem (b), i.e., $du[b]$, etc. Because the total amounts of the extensive properties, $U(0)$, $V(0)$, $n_j(0)$, and $\mathcal{X}_i(0)$, are conserved, their differentials are vanishing. In other words,

$$\begin{aligned} dU(0) &= du[a] + du[b] = 0\,, \qquad & dV(0) &= dv(a) + dv(b) = 0\,, \\ dn_j(0) &= dn_j(a) + dn_j(b) = 0\,, \qquad & d\mathcal{X}_i(0) &= d\mathcal{X}_i(a) + d\mathcal{X}_i(b) = 0\,. \end{aligned} \tag{9.2}$$

(2) Because the system is in thermodynamic equilibrium, its (extensive state function the) entropy is a maximum. As a result, the entropy is stationary with respect to any change occasioned by the *gedanken* experiment. That means the difference, ΔS (see below), between the entropy after and the entropy before the gedanken experiment is vanishing:

$$\begin{aligned} 0 = \Delta S = &\left[S(U, V, \mathcal{X}_i, n_j)\right]_{\text{after gedanken experiment}} \\ &- \left[S(U, V, \mathcal{X}_i, n_j)\right]_{\text{before gedanken experiment}}\,. \end{aligned} \tag{9.3}$$

The information recorded in (9.1)–(9.3) in hand, coupled with the concept of the fundamental equation, we are able to more fully examine the nature of the zeroth law.

The zeroth law is revisited in Sect. 9.1 and the resulting physical insights into determining the direction of thermodynamic motive forces are discussed at length in Sect. 9.1.1. Thermodynamic stability, as well as the Le Châtelier's principle, are discussed in Sect. 9.4 and Sect. 9.4.1. Two basic requirements for intrinsic thermodynamic stability are derived in the concluding Sect. 9.2.1.

[1] It is important to note that all these quantities are extensive—meaning they change linearly with the number of (relevant) particles in the system.

9.1.1 The Zeroth Law of Thermodynamics: Reconfirmed

Begin with the statement of the first–second law that is contained in (8.23). For brevity, again limit the discussion to a simple system with only n_j's and v as extensive variables (and, of course, s and u as extensive state functions). Equation (8.23) thus simplifies to

$$du(s, v, n_j) = t(ds) - p(dv) + \sum_j \mu_j(dn_j). \tag{9.4}$$

First, let us check on the status of the zeroth law. To this end, imagine the isolated thermodynamic system that has achieved equilibrium to be divided into two parts, (a) and (b). The specification that the system is isolated ensures that the total amount of the extensive variables, the entropy, mole numbers, and volume, remain constant. This fact is represented by the following equalities:

$$\begin{aligned} ds &= 0 = ds(a) + ds(b), \\ dv &= 0 = dv(a) + dv(b), \\ dn_j &= 0 = dn_j(a) + dn_j(b). \end{aligned} \tag{9.5}$$

Also, of course, because the internal energy is stationary, $du = 0 = du[a] + du[b]$. Let us now write (9.4) so that it refers to the duo (a) and (b).

$$\begin{aligned} du(s, v, n_j) &= \{du[a] + du[b]\}(s, v, n_j) = 0 \\ &= t(a)ds(a) + t(b)ds(b) - p(a)dv(a) - p(b)dv(b) \\ &\quad + \sum_j \{\mu_j(a)dn_j(a) + \mu_j(b)dn_j(b)\}. \end{aligned} \tag{9.6}$$

With the use of (9.5), (9.6) can be written as follows:

$$\begin{aligned} 0 &= [t(a) - t(b)]ds(a) - [p(a) - p(b)]dv(a) \\ &\quad + \sum_j [\mu_j(a) - \mu_j(b)]dn_j(a). \end{aligned} \tag{9.7}$$

Because in the energy representation the variables sn and v are linearly independent, in order for (9.7) to hold, each term on the right-hand side must be vanishing. Thus, much like the entropy stationarity principle, the stationarity of energy also correctly leads to the zeroth law:

$$t(a) = t(b), \qquad p(a) = p(b), \qquad \mu_j(a) = \mu_j(b). \tag{9.8}$$

Using the Euler equation in the entropy representation (that is recorded in (8.18)), the entropy before and after the gedanken experiment can be represented as follows:

$$\begin{aligned} &\{S[U, V, \mathcal{X}_i, n_j]\}_{\text{before gedanken experiment}} \\ &\quad = \{s[a]\}_{\text{before}} + \{s[b]\}_{\text{before}} \end{aligned}$$

$$
\begin{aligned}
&= \left\{\left[\frac{1}{t(a)}\right]u(a) + \left[\frac{p(a)}{t(a)}\right]v(a)\right\} + \left\{\left[\frac{1}{t(b)}\right]u(b) + \left[\frac{p(b)}{t(b)}\right]v(b)\right\} \\
&\quad + \left\{\sum_{\mathrm{i}}\left[\frac{\mathcal{Y}_{\mathrm{i}}(a)}{t(a)}\right]\mathcal{X}_{\mathrm{i}}(a) - \sum_{\mathrm{j}}\left[\frac{\mu_{\mathrm{j}}(a)}{t(a)}\right]n_{\mathrm{j}}(a)\right\} \\
&\quad + \left\{\sum_{\mathrm{i}}\left[\frac{\mathcal{Y}_{\mathrm{i}}(b)}{t(b)}\right]\mathcal{X}_{\mathrm{i}}(b) - \sum_{\mathrm{j}}\left[\frac{\mu_{\mathrm{j}}(b)}{t(b)}\right]n_{\mathrm{j}}(b)\right\}, \qquad (9.9)
\end{aligned}
$$

and

$$
\begin{aligned}
&\{S[U, V, \mathcal{X}_{\mathrm{i}}, n_{\mathrm{j}}]\}_{\text{after gedanken experiment}} \\
&\quad = \{s[a]\}_{\text{after}} + \{s[b]\}_{\text{after}} \\
&\quad = \left\{\left[\frac{1}{t(a)}\right]\{u[a] + \mathrm{d}u[a]\} + \left[\frac{p(a)}{t(a)}\right][v(a) + \mathrm{d}v(a)]\right\} \\
&\qquad + \left\{\left[\frac{1}{t(b)}\right]\{u[b] + \mathrm{d}u[b]\} + \left[\frac{p(b)}{t(b)}\right][v(b) + \mathrm{d}v(b)]\right\} \\
&\qquad + \left\{\sum_{\mathrm{i}}\left[\frac{\mathcal{Y}_{\mathrm{i}}(a)}{t(a)}\right][\mathcal{X}_{\mathrm{i}}(a) + \mathrm{d}\mathcal{X}_{\mathrm{i}}(a)] - \sum_{\mathrm{j}}\left[\frac{\mu_{\mathrm{j}}(a)}{t(a)}\right][n_{\mathrm{j}}(a) + \mathrm{d}n_{\mathrm{j}}(a)]\right\} \\
&\qquad + \left\{\sum_{\mathrm{i}}\left[\frac{\mathcal{Y}_{\mathrm{i}}(b)}{t(b)}\right][\mathcal{X}_{\mathrm{i}}(b) + \mathrm{d}\mathcal{X}_{\mathrm{i}}(b)] - \sum_{\mathrm{j}}\left[\frac{\mu_{\mathrm{j}}(b)}{t(b)}\right][n_{\mathrm{j}}(b) + \mathrm{d}n_{\mathrm{j}}(b)]\right\}. \qquad (9.10)
\end{aligned}
$$

Subtract (9.9) from (9.10) and make use of the fact that in equilibrium the entropy is a maximum. The latter fact is identified by the equality given in (9.3). That is,

$$
\begin{aligned}
0 &= \left[\frac{1}{t(a)} - \frac{1}{t(b)}\right]\mathrm{d}u[a] + \left[\frac{p(a)}{t(a)} - \frac{p(b)}{t(b)}\right]\mathrm{d}v(a) \\
&\quad + \sum_{\mathrm{i}}\left[\frac{\mathcal{Y}_{\mathrm{i}}(a)}{t(a)} - \frac{\mathcal{Y}_{\mathrm{i}}(b)}{t(b)}\right]\mathrm{d}\mathcal{X}_{\mathrm{i}}(a) - \sum_{\mathrm{j}}\left[\frac{\mu_{\mathrm{j}}(a)}{t(a)} - \frac{\mu_{\mathrm{j}}(b)}{t(b)}\right]\mathrm{d}n_{\mathrm{j}}(a). \qquad (9.11)
\end{aligned}
$$

Because the infinitesimal changes $\mathrm{d}u[a]$, $\mathrm{d}v(a)$, $\mathrm{d}\mathcal{X}_{\mathrm{i}}(a)$, and $\mathrm{d}n_{\mathrm{j}}(a)$ are linearly independent, the above equation can be satisfied only if each coefficient in the above equation is equal to zero.[2] Therefore to satisfy (9.11), we must have:

$$
t(a) = t(b)\,, \qquad p(a) = p(b)\,, \qquad \mathcal{Y}_{\mathrm{i}}(a) = \mathcal{Y}_{\mathrm{i}}(b)\,, \qquad \mu_{\mathrm{j}}(a) = \mu_{\mathrm{j}}(b)\,. \quad (9.12)
$$

Clearly, rather than just the two subsystems, a and b, this argument can be extended to any number of subsystems—$a, b, c, \ldots$, etc.

Similarly, the above discussion can immediately be applied also to real experiments on a macroscopic object composed of two separate thermodynamic systems, labeled (A) and (B). We arrange for these systems to be in complete thermal contact, and allow them to freely exchange all extensive properties. Note that jointly (A) and (B) are to be kept isolated from the rest of the universe.

[2]This can readily be checked, for example, by setting three of the four quantities $\mathrm{d}u(a)$, $\mathrm{d}v(a)$, $\mathrm{d}\mathcal{X}_{\mathrm{i}}(a)$, $\mathrm{d}n_{\mathrm{j}}(a)$ equal to zero.

Fortunately, all the needed work has already been done. It turns out that (9.11) is rather versatile. Above we used it to analyze two parts—(a) and (b)—of a single, isolated system in thermodynamic equilibrium where the entropy had achieved its maximum. Equation (9.11) can similarly be put to use to examine the adiabatically enclosed duo of two separate macroscopic systems (A) and (B) that are in thermodynamic contact.

Again, because these two systems will have reached mutual thermodynamic equilibrium, their total entropy will have attained its maximum value consistent with the given total value of the extensive properties U, V, n_j's, $\mathcal{X}_j$'s. Accordingly, much like the ΔS of the previous *gedanken experiment*, here too the change in the entropy between the "imagined" initial and final states of the experiment will be zero. (Compare with (9.3).) Therefore, the result of this process would be identical to that obtained in (9.12). The two separate systems, (A) and (B), therefore act as if they were part and parcel of one composite system. (In other words, (9.12) now holds with labels (A), (B) exchanged for (a), (b).)

Because merely the exchange in the indices from (a) and (b) to (A) and (B) recasts the result for the former case into that for the latter, in the following we shall assume the indices (a) and (b), etc., to be just generic indices. Thus they are applicable to all thermodynamic systems and/or their macroscopic parts. The requirement is that while being contained within adiabatic walls, they be in perfect physical contact.

Remark 9.1 Because (9.12) can be extended to an arbitrary number of thermodynamic systems, and/or to their macroscopic subsystems, it represents an important corollary of the zeroth law. It states unequivocally that when a set of systems in perfect physical contact are in thermodynamic equilibrium, at least theoretically speaking,[3] their temperature is the same all across their (macroscopic constituent) parts. And the same is true for the pressure, chemical potential for any constituent i, etc.[4] Remember, however, that allowance for gravity, or other forces that would cause changes in the pressure across the system, has not been made. If changes in pressure due to such forces were to be incorporated in this analysis, then their effect on the volume and some other variables conjugate to the relevant forces—e.g., the chemical potentials, μ_j, and other intensive variables such as $\mathcal{Y}_i$, etc.—would also need to be taken into account.

9.2 Entropy Extremum: Maximum Possible

As mentioned earlier, one of the important consequences of the second law is that any spontaneous process that occurs in an isolated thermodynamic system increases its total entropy. Therefore, when an isolated system with specified values of the

[3]Note that by saying "theoretically speaking" we are being appreciative of the fact that reality always militates against perfection.

In practise, somewhere a metastable fluctuation from what was thought to be perfect equilibrium may also appear.

[4]This, of course, proves Pascal's principle.

internal energy u, volume v, mole numbers n_j, and other extensive variables $\mathcal{X}_i$ is in thermodynamic equilibrium, its entropy s is the maximum possible.

Entropy, therefore, behaves as if it were, to coin a phrase, a thermodynamic "motive force" that drives an isolated system towards equilibrium which in the present context is defined as the state with maximum entropy. The question then is: How may one leverage this information both in regard to the two separate thermodynamic systems, (A) and (B), that are in mutual contact but are otherwise isolated from the rest of the universe, and equivalently, also in regard to two macroscopic subsystems[5] (a) and (b) which comprise a single thermodynamic system isolated from the universe?

9.2.1 Heat Energy Flow

The direction of heat energy flow is always from the warmer to the colder. Why is that the case? All macroscopic parts of a thermodynamic system in equilibrium are subject to undergoing spontaneous infinitesimal transformations that obey the standard conservation rules for the overall value of the property that is spontaneously being affected. That is, the system must follow the dictates of the first law.

However, in regard to following the second law, the spontaneous transformation is required to result either in increasing the overall entropy of the system or at best holding it unchanged. Because the entropy in the equilibrium state is the maximum possible, if the thermodynamic system being considered is not exactly at equilibrium, but is infinitesimally close to it, then the overall resultant, infinitesimal, entropy change, $(\mathrm{d}s)_{\text{spontaneous}}$, caused by the spontaneous transformation must be positive.

To study this matter, let us again consider macroscopic components (a), (b), etc., of a single thermodynamic system. Or equivalently, consider different thermodynamic systems (a), (b), etc., with or without other constituent parts. Again we use the Euler equation in the entropy representation.[6]

Often the spontaneous process will cause some infinitesimal change in each of the three extensive state variables, namely, the internal energy u, volume v, and/or the number of moles n_j of the jth chemical potential.[7] However, because the overall system is assumed to be adiabatically enclosed, the total value of all the extensive state variables is conserved. Thus, if we consider only two system, (a) and (b), we have:

$$\mathrm{d}u[a] + \mathrm{d}u[b] = 0\,, \qquad \mathrm{d}v(a) + \mathrm{d}v(b) = 0\,, \qquad \mathrm{d}n_j(a) + \mathrm{d}n_j(b) = 0\,. \quad (9.13)$$

The Euler equation tells us that the resultant change in the entropy is as given below. And the second law requires the corresponding spontaneous change in the entropy

[5]As mentioned earlier, it is convenient to call two separate thermodynamic systems by the same pair of indices, (a) and (b), that is used for two subsystems of a single thermodynamic object.

[6]Recall that such Euler equation is given in (8.18).

[7]We work with constant $\mathcal{X}_i$ so that $\mathrm{d}\mathcal{X}_i = 0$.

to be either greater than, or at best equal to, zero. Therefore, we can write

$$
\begin{aligned}
&\left[\frac{1}{t(a)}\right]\mathrm{d}u[a] + \left[\frac{p(a)}{t(a)}\right]\mathrm{d}(a) - \sum_{\mathrm{j}}\left[\frac{\mu_{\mathrm{j}}(a)}{t(a)}\right]\mathrm{d}n_{\mathrm{j}}(a) \\
&\quad + \left[\frac{1}{t(b)}\right]\mathrm{d}u[b] + \left[\frac{p(b)}{t(b)}\right]\mathrm{d}v(b) - \sum_{\mathrm{j}}\left[\frac{\mu_{\mathrm{j}}(b)}{t(b)}\right]\mathrm{d}n_{\mathrm{j}}(b) \\
&\quad = \left[\frac{1}{t(a)} - \frac{1}{t(b)}\right]\mathrm{d}u[a] + \left[\frac{p(a)}{t(a)} - \frac{p(b)}{t(b)}\right]\mathrm{d}v(a) \\
&\quad - \sum_{\mathrm{j}}\left[\frac{\mu_{\mathrm{j}}(a)}{t(a)} - \frac{\mu_{\mathrm{j}}(b)}{t(b)}\right]\mathrm{d}n_{\mathrm{j}}(a) \\
&\quad = (\mathrm{d}s)_{\text{spontaneous}} \geq 0\,.
\end{aligned}
\tag{9.14}
$$

Now, in order to answer the question posed regarding the direction of heat energy flow, let us proceed as follows.

Imagine the isolated composite system, composed of (a) and (b), is infinitesimally close to achieving equilibrium. Assume that during such a spontaneous process, the only contact allowed between (a) and (b) is across a fixed,[8] impermeable, diathermal wall,[9] meaning $\mathrm{d}v(a) = 0$ and $\mathrm{d}n_{\mathrm{j}}(a) = 0$. Therefore, (a) and (b) can exchange only the internal energy, i.e., $\mathrm{d}u(a)$.

Of the two options available—namely the spontaneous change in the entropy is either positive or it is equal to zero—consider first the case where the spontaneous change in the entropy is positive. Then according to (9.14),

$$
(\mathrm{d}s)_{\text{spontaneous}} = \left[\frac{1}{t(a)} - \frac{1}{t(b)}\right]du[a] > 0\,. \tag{9.15}
$$

As a result, if $t(b) > t(a)$, meaning if

$$
\left[\frac{1}{t(a)} - \frac{1}{t(b)}\right] > 0\,, \tag{9.16}
$$

then $\mathrm{d}u(a) > 0$.

Note that if we had used the equality option in (9.14), then nothing much would have happened. The temperatures on the two sides would be equal, $t(a) = t(b)$, and no heat energy would be transferred, namely $\mathrm{d}u(a) = \mathrm{d}u(b) = 0$.

In words, if the temperature on side (b) is higher than that on side (a), then heat energy will flow such that the resultant change in the internal energy[10] of the cooler side—i.e., side a—is positive. This means that positive heat energy is transferred from the warmer side (b) to the cooler side (a).

[8]Meaning the process is isochoric.

[9]Note that being diathermal ensures the flow of internal energy. The requirement that the diathermal wall be fixed ensures $\mathrm{d}v(a) = 0$. Similarly, the requirement of impermeability ensures that $\mathrm{d}n_{\mathrm{j}}(a) = 0$.

[10]In this case, the increase $\mathrm{d}u(a)$ in the internal energy is synonymous with the increase in the heat energy.

Simply put, the requirement that in an isolated system entropy must increase in any isochoric spontaneous process mandates that, across any fixed impermeable wall, heat energy can flow only from the "hot" side to the "cold."

9.2.2 Molecular Flow

At constant temperature and pressure, molecules tend to flow from regions of higher chemical potential to those of lower chemical potential. What is the relevant physics? To this end, let us examine the isothermal–isobaric behavior. Here, $t(a)$ and $t(b)$ are the same, say each is equal to t, and also $p(a) = p(b)$. Further, except for the exchange of the jth chemical component across an appropriately chosen membrane between the two systems, disallow the sharing of all remaining chemical components. Therefore, according to (9.14), the expression for the infinitesimal increase in the entropy is the following:[11]

$$(\mathrm{d}s)_{\text{spontaneous}} = -\left[\frac{\mu_{\mathrm{j}}(a)}{t} - \frac{\mu_{\mathrm{j}}(b)}{t}\right]\mathrm{d}n_{\mathrm{j}}(a) > 0\,. \tag{9.17}$$

Thus if $\mu_{\mathrm{j}}(a) > \mu_{\mathrm{j}}(b)$, then the change, $\mathrm{d}n_{\mathrm{j}}(a)$, in the jth mole number in system (a) is negative, or equivalently, the change in the jth mole number in system b is positive.

Simply put, the requirement that in an isolated system the entropy must increase in any spontaneous process mandates that at constant temperature and pressure, molecules always flow from a region of higher chemical potential towards a region of lower chemical potential.

A myriad of common experiences testify to this fact. Those of us with a sweet-tooth have surely noticed that, when added to a fluid, sweet taste flows away from a sugar cube to regions of lower sugar concentration. At constant temperature, the chemical potential for sugar would thus appear to be larger in regions where its density is higher.

9.2.3 Isothermal Compression

If the total volume is kept constant, in thermal contact two macroscopic parts of a system with a freely moveable but impermeable partition between them, adjust their volumes in such a way that the part with lower original pressure shrinks in volume. Why is that the case? With the total volume—that is, the sum of the volumes of a and b—kept constant, let the two systems at the same temperature t be allowed to exchange only the volume[12] across a freely moveable diaphragm. All

[11]Because nothing much happens if we use the equality sign, we shall not waste any energy looking at that option.

[12]This means that chemical potential of any given type, say of type j, in both the a and the b systems is the same. That is, $\mu_{\mathrm{j}}(a) = \mu_{\mathrm{j}}(b)$.

other exchanges are disallowed. Accordingly, the relevant relationship now is

$$(\mathrm{d}s)_{\text{spontaneous}} = \left[\frac{p(a)}{t} - \frac{p(b)}{t}\right]\mathrm{d}v(a) > 0\,. \tag{9.18}$$

Clearly, therefore, if the pressure on the side a is greater than the pressure on side b—i.e., $p(a) > p(b)$—then $dv(a)$ has to be positive. That is, the side a expands. As a result, the partitioning diaphragm moves towards system b, meaning the volume of the system[13] a increases and the system b shrinks in volume.[14] In other words, given two macroscopic systems in thermal contact are placed together in an adiabatically isolated chamber, if their total volume is constant, their chemical potentials are equal, they are at the same temperature, but they can freely affect each other's volume, then:

The requirement that the total entropy must increase in any isothermal spontaneous process, ensures that the side with originally lower pressure will shrink in volume. And if the process is allowed to continue, shrinking of the side with originally lower pressure will continue until the two pressures become equal.

Later we shall formally learn that this is an essential characteristic of thermodynamic systems in stable equilibrium: the isothermal compressibility in these systems is positive, which means the volume of a system at constant temperature decreases when it is subjected to increased compression.

9.2.4 Energy Extremum: Minimum Possible

In the foregoing, the entropy was treated as the all-important thermo-motive force that drove the flow towards thermodynamic equilibrium. Any rôle that the other basic state function, the internal energy, may play in nudging the system towards equilibrium was not investigated.

It turns out that the extremum principle for the entropy—that thermodynamic equilibrium requires the entropy of an isolated system to be the maximum consistent with any existing constraints that determine u, v, n_j, etc.—is equivalent to a corresponding extremum principle for the internal energy, namely that when an isolated thermodynamic system reaches equilibrium, the internal energy, u, achieves its minimum value, consistent with the existing constraints which specify the other extensive quantities s, v, n_j, etc.

Exploiting the internal energy extremum—which requires that in equilibrium energy is the minimum possible—provides an alternative formulation for the flow towards equilibrium.[15] The equivalence of the entropy and the energy extremum

[13]Note the system b is in such contact with a that the "volume" can flow from one to the other. Also that pressure in system b is lower than the pressure in a.

[14]Another way of looking at this is the fact that because $\mathrm{d}v(a) + \mathrm{d}v(b) = 0$, if $\mathrm{d}v(a)$ is positive, then $\mathrm{d}v(b)$ is negative.

[15]There are other extremum principles that are also equivalent to that for the entropy and the internal energy. These refer to the system enthalpy, the Helmholtz free energy, and the Gibbs potential.

principles can be demonstrated by showing in an isolated system unless the internal energy is the minimum possible, the entropy cannot be the maximum possible. Thus, for one to be true, the other must also hold. Consequently, the two extrema occur simultaneously.

To this end, consider an hypothetical circumstance when thermodynamic equilibrium is reached—that is, when the entropy is the maximum possible—without the internal energy having achieved its lowest possible value. As noted in the chapter on the second law, internal energy may be extracted out of a system even during reversible adiabatic processes. Clearly, the entropy of the isolated system under any such reversible adiabatic process must, by thermodynamic fiat, stay unchanged—at its maximum permissible value. On the other hand, the extraction of energy would decrease the system internal energy. Note that, under the above hypothesis, any such decrease would be permitted as long as the internal energy remained above its minimum possible level consistent with the specified entropy. Clearly, this course of action, would continue decreasing the internal energy of the system. Followed to its logical conclusion, these actions would lead the system to achieve the minimum possible internal energy.

Alternatively, and perhaps even more dramatically, another scenario can be made to unfold. The internal energy extracted as described above can in turn be transformed into equal amount of work. And, the work so produced converted into equal amount of heat energy that in turn can be added back into the system. This round-trip restores the original amount of internal energy. Because even a reversible introduction of such heat energy must result in increasing the system entropy, we are left with a situation where the system entropy has been increased from what was assumed to be its maximum possible value. The only situation in which this bootstrapping cannot be permitted is when the internal energy is already at its minimum permissible value.

To recapitulate, as has also been mentioned previously, when an isolated system with specified values of the internal energy u, volume v, mole numbers n_{j}, and other extensive variables $\mathcal{X}_{\mathrm{i}}$ is in thermodynamic equilibrium, its entropy s is the maximum possible. But the matter of interest in this section is the following:

Simultaneously with the entropy s being a maximum and with values of all other extensive variables that are obtained in the given state in equilibrium, the internal energy is the minimum possible.

Consequently, two requirements are obeyed in equilibrium. For any given value of the internal energy, u, and other unconstrained extensive parameters of a system in thermodynamic equilibrium, the corresponding value of the entropy is at its maximum. And similarly, for any given value of the entropy s, and other unconstrained extensive parameters of a system in thermodynamic equilibrium, the corresponding value of the internal energy, u, is a minimum. We display these statements as follows:

$$\mathrm{d}s = 0\,, \tag{9.19}$$

$$\mathrm{d}^2 s < 0\,, \tag{9.20}$$

and

$$\mathrm{d}u = 0\,, \tag{9.21}$$

$$\mathrm{d}^2u > 0\,. \tag{9.22}$$

Note that while (9.19) and (9.21) ensure the occurrence of thermodynamic equilibrium, (9.20) and (9.22) describe the bases for thermodynamic stability.

As we know from the work that led to (9.1)–(9.12), using the stationarity requirement for the entropy—a statement that is also embodied in (9.19)—leads to ("an extended version of") the zeroth law. Indeed, its implications have also been exploited to good effect for the study of thermodynamic motive forces—see (9.13)–(9.18). Therefore, we turn our attention to (9.21), which refers to the stationarity of the internal energy. We shall find that this equation also leads to the same physical conclusions as were derived from (9.19). It is instructive to see how this happens.

9.3 Motive Forces: Energy Formalism

Again, as was done for the entropy extremum analysis, imagine the composite system, composed of (a) and (b), to be infinitesimally close to achieving equilibrium. Further, assume the system undergoes an infinitesimal spontaneous transformation which actually brings it to thermal equilibrium. Because in equilibrium, for the given value of total entropy and other extensive variables, the total internal energy u is to be a minimum, such a spontaneous process must result in reducing the total internal energy. Accordingly, any change, $\mathrm{d}u$, must represent a decrease. That is,

$$\begin{aligned}(\mathrm{d}u)_{\text{spontaneous}} &= \mathrm{d}u[a] + \mathrm{d}u[b] \\ &= \big[t(a) - t(b)\big]\mathrm{d}s(a) - \big[p(a) - p(b)\big]dv(a) \\ &\quad + \sum_{\mathrm{j}}\big[\mu_{\mathrm{j}}(a) - \mu_{\mathrm{j}}(b)\big]\mathrm{d}n_{\mathrm{j}}(a) \\ &< 0\,. \end{aligned} \tag{9.23}$$

9.3.1 Isobaric Entropy Flow

Consider the case when the pressures p_{a} and p_{b} are equal and all molecular flows are disallowed. Then the inequality (9.23) simplifies to the following:

$$(\mathrm{d}u)_{\text{spontaneous}} = \mathrm{d}u[a] + \mathrm{d}u[b] = \big[t(a) - t(b)\big]\mathrm{d}s(a) < 0\,. \tag{9.24}$$

Let the temperature difference between a and b be positive, i.e.,

$$\big[t(a) - t(b)\big] > 0\,, \tag{9.25}$$

then (9.24) demands that $\mathrm{d}s(a) < 0$. In words, while the total entropy in this situation remains unchanged, positive entropy is transferred out of the warmer side "*a*" into the colder side "*b*."

Simply put, the requirement that in an isolated thermodynamic system in equilibrium, with no molecular flows occurring and a given amount of total entropy, the energy must decrease in an isobaric spontaneous process mandates that:

The entropy must flow outwards from a region of higher temperature towards a region of lower temperature.

9.3.2 Isothermal–Isobaric Molecular Flow

Next, examine the isothermal–isobaric behavior. Here, $t(a)$ and $t(b)$ are the same, say each is equal to t, and also $p(a) = p(b)$.

Further, except for the exchange of the lth chemical component across an appropriately chosen membrane between the two systems, disallow the sharing of all remaining extensive properties. Then (9.23) becomes

$$(\mathrm{d}u)_{\text{spontaneous}} = \big[\mu_l(a) - \mu_l(b)\big]\mathrm{d}n_l(a) < 0\,. \tag{9.26}$$

Thus in order to satisfy the inequality (9.26) for the case when the chemical potential $\mu_l(a)$ is greater than $\mu_l(b)$, the quantity $\mathrm{d}n_l(a)$ has to be negative. Consequently, the number of l-type molecules in system "*a*" must decrease. This is equivalent to saying more type l molecules leave system "*a*" than return to it. In other words, molecules tend to move away from regions with larger chemical potential, and go to regions with smaller chemical potential.

This behavior is summarized as follows: Assume an isolated system that is in thermal equilibrium. Its internal energy is the minimum possible for given values of the entropy and all other extensive variables. In such a system, the internal energy must decrease in any isothermal–isobaric spontaneous process that results in the interchange of, say, the l-type molecules. And for this to happen, the l-type molecules must flow from a region where the relevant chemical potential, μ_l, is higher to a region where that particular chemical potential is lower. This is exactly the result that was derived from the entropy maximum principle—see (9.17).

9.3.3 Isothermal Compression

Let the given two systems at the same temperature t be allowed to exchange only the volume across a freely moveable diaphragm. All other exchanges are to be disallowed. Accordingly, the relevant relationship now is

$$(\mathrm{d}u)_{\text{spontaneous}} = -\big[p(a) - p(b)\big]\mathrm{d}v(a) < 0\,. \tag{9.27}$$

Clearly, therefore, if the pressure $p(a)$ is less than the pressure $p(b)$, then $-[p(a) - p(b)]$ is positive. Therefore, $\mathrm{d}v(a)$ has to be negative. This means that the diaphragm must then move towards system a. As a result, system a will shrink in volume and system b will expand an equal amount.

Simply put, when the total volume of the two systems at the same temperature is constant but they can freely affect each other's volume, the requirement that total energy of the two systems must decrease in an isothermal spontaneous process mandates that the one with lower pressure shrink in volume.

This is exactly the same result that was derived from the entropy extremum principle. (See (9.18).)

9.4 Physical Criteria for Thermodynamic Stability

Physical criteria for thermodynamic stability are studied below.

9.4.1 Le Châtelier's Principle

Le Châtelier's principle asserts:

"Spontaneous processes caused by displacements from equilibrium help restore the system back to equilibrium."

In equilibrium, temperature of an isolated metal rod is uniform. If equilibrium were disturbed by—let us say, a virtual—fluctuation resulting in the rise of temperature at one of the ends and a compensating fall in the temperature at the other, heat energy would spontaneously flow from the hot end to the cold. In other words, displacement from equilibrium spawns spontaneous processes that help drive the system back towards equilibrium.

Moreover, as demonstrated in a solved example (see (4.211) and (4.213)), when this happens the entropy of the metal rod is higher in the equilibrium state.

A myriad of other familiar examples can be cited along these lines. So why does this happen?

Germane to this issue is the observation noted above, namely that the entropy of the equilibrium state is higher than the state displaced from equilibrium. This, of course, follows from the Clausius version of the second law: A spontaneous process causes the entropy of an isolated system to increase. Formally, this statement is embodied in the inequality (4.38), namely

$$\mathrm{d}S_{\mathrm{total}}(\text{spontaneous}) > 0\,. \tag{9.28}$$

Thus, spontaneous processes induced by displacements away from the equilibrium result in increasing the entropy. Clearly, this continues until the entropy, for specified constraints on, and given values of other extensive properties of, the system achieves its maximum, or in other words, the system achieves thermodynamic equilibrium.

Therefore, according to the above, when an isolated system—with given values of the internal energy U, volume V, mole numbers n_{j}, and other extensive variables $\mathcal{X}_{\mathrm{i}}$—is in thermodynamic equilibrium, its entropy S is the maximum possible.

Equally important, simultaneously with the above, another extremum principle is also obtained. That is, for a given amount of the entropy S and other extensive variables, the internal energy in an isolated system is the minimum possible. Therefore, according to the dictates of calculus, in such isolated systems, (9.19)–(9.22) must be satisfied.

The requirements related to only the first-order differential—namely those given in (9.19) and (9.21)—apply to all types of extrema.

For example, in the preceding part of this chapter—see (9.12)–(9.27)—they helped rederive the zeroth law—see (9.12) and (9.8)—and, more importantly, they helped determine the direction of thermodynamic motive forces.

The second set of requirements that use second-order differentials specified in (9.20) and (9.22) also have noteworthy—indeed, fundamental and far-reaching—physical consequences.

In particular, they are crucial to the understanding of thermodynamic stability, a topic that is addressed below.

The question that needs to be answered is the following: What are the essential criteria for a stable thermodynamic equilibrium? This question can refer either to a single system which is in equilibrium by itself, or to a composite system where its different subsystems, in addition to being in equilibrium by themselves, are also in equilibrium with each other. Because the results obtained are similar in both formulations, they can easily be transliterated from one case to the other. Therefore, in this chapter we shall treat only issues relating to self-equilibrium.

9.4.2 Stable Self-Equilibrium

As was discovered in the preceding sections, the entropy maximum formulation leads to exactly the same physical conclusions regarding thermodynamic properties as does the energy minimum formulation. Further, at the margins, the minimum energy formulation is easier to work with. Therefore, to simplify and reduce the amount of work needed, in the following we shall work only with the energy minimum formulation.

We do two things here.

First, we analyze as simple a system as possible. To this end, we shall limit extensive quantities to two state functions, namely the internal energy U and the entropy S, and one state variable, i.e., the volume.[16]

Second, we express the internal energy in terms of its characteristic variables, entropy S and volume V, i.e.,

$$U = U(S, V)\,. \tag{9.29}$$

According to the first–second law,

$$T\mathrm{d}S = \mathrm{d}U + P\mathrm{d}V\,,$$

[16] A system with an additional variable, namely the number of moles, requires somewhat more effort and therefore, for convenience, is deferred to Appendix H.

or equivalently,

$$dU(S, V) = TdS - PdV = \left(\frac{\partial U}{\partial S}\right)_V dS + \left(\frac{\partial U}{\partial V}\right)_S dV\,. \tag{9.30}$$

The stationarity of the energy—expressed in (9.21)—has already been fully exploited in the preceding work. Therefore, one needs now to examine only the second requirement—expressed in (9.22)—that comes into play because the energy is a minimum (for specified values of the extensive variables S and V).

Using (9.30), (9.22) can be written as[17]

$$\begin{aligned} d^2U(S, V) &= d\big[dU(S, V)\big] \\ &= \left(\frac{\partial[dU(S, V)]}{\partial S}\right)_V dS + \left(\frac{\partial[dU(S, V)]}{\partial V}\right)_S dV \\ &= dUSV(dS)^2 + \left(\frac{\partial^2 U}{\partial S \partial V}\right) dS \cdot dV \\ &\quad + \left(\frac{\partial^2 U}{\partial V \partial S}\right) dV \cdot dS + dUVS(dV)^2 \\ &> 0\,. \end{aligned} \tag{9.31}$$

It is convenient to display this inequality in the matrix form

$$d^2U(S, V) = \{dS, dV\} \cdot \hat{\mathbf{a}} \cdot \begin{Bmatrix} dS \\ dV \end{Bmatrix} > 0\,, \tag{9.32}$$

where

$$\hat{\mathbf{a}} = \begin{pmatrix} dUSV & \frac{\partial^2 U}{\partial S \partial V} \\ \frac{\partial^2 U}{\partial V \partial S} & dUVS \end{pmatrix}. \tag{9.33}$$

The right-hand side of (9.32) is a homogeneous quadratic form. The inequality > 0 stipulates that the given homogeneous form be positive definite. For this to be true, the two principal minors of the matrix $\hat{\mathbf{a}}$, that is, $|\mathbf{A_1}|$ and $|\mathbf{A_2}|$, must be positive.[18] In other words, the following two inequalities must hold:

$$|\mathbf{A_1}| = dUSV > 0\,, \tag{9.34}$$

and[19]

$$dUVS \cdot dUSV - \left(\frac{\partial^2 U}{\partial S \partial V}\right)^2 > 0\,.$$

---

[17]Note that according to the usual notation, the mixed derivative $\frac{\partial^2 U}{\partial S \partial V}$ is equivalent to $(\frac{\partial[(\frac{\partial U}{\partial V})_S]}{\partial S})_V$.

[18]See Appendix G, (H.10) and (H.11), which state that for the quadratic form to be positive-definite, the principal minors of the matrix $\hat{\mathbf{a}}$ must be positive.

[19]Note the inequality (9.35).

Furthermore, inequality (9.34),

$$dUSV > 0\,,$$

also implies the inequality

$$|\mathbf{A_1}| = dUVS > 0\,.$$

Indeed, the second inequality given above would itself have been mandated if, instead of the vector $\{S, V\}$, we had made an equally allowed choice. That is, working with the vector $\{S, V\}$,

$$|\mathbf{A_2}| = \begin{vmatrix} dUSV & \frac{\partial^2 U}{\partial S \partial V} \\ \frac{\partial^2 U}{\partial V \partial S} & dUVS \end{vmatrix} > 0\,. \tag{9.35}$$

9.5 The First and Second Requirement

Let us start with analysis of the first requirement.

9.5.1 Implications of the First Requirement

The first requirement is that given in (9.34). Because $(\frac{\partial U}{\partial S})_V = T$, we have

$$|\mathbf{A_1}| = dUSV = \left(\frac{\partial T}{\partial S}\right)_V = \frac{T}{C_V} > 0\,.$$

Because T is positive, the first requirement for thermodynamic stability is

$$C_V > 0\,. \tag{9.36}$$

In words,

For thermodynamic stability, the specific heat at constant volume must be positive.

The first requirement has an obvious physical basis: If heat energy is not allowed to be expended on expanding the volume, all any addition of (positive amount of) heat energy can do is increase the system temperature. Of course, one has complete confidence in this statement only when the given system is intrinsically stable against phase separation.

9.5.2 Implications of the Second Requirement

Next, let us now examine the second requirement which signifies thermodynamic stability. This is represented by the inequality (9.35). Expanding the determinant and writing the result in the Jacobian form gives

$$\begin{aligned}|\mathbf{A_2}| &= \mathrm{d}UVS \cdot \mathrm{d}USV - \left(\frac{\partial^2 U}{\partial V \partial S}\right) \cdot \left(\frac{\partial^2 U}{\partial S \partial V}\right)\\ &= \frac{\partial((\frac{\partial U}{\partial V})_\mathrm{S}, (\frac{\partial U}{\partial S})_\mathrm{V})}{\partial(V, S)}\\ &> 0\,.\end{aligned} \tag{9.37}$$

Because in thermodynamic equilibrium $(\frac{\partial U}{\partial V})_\mathrm{S} = -P$ and $(\frac{\partial U}{\partial S})_\mathrm{V} = T$, we can write (9.37) given above as follows:

$$\begin{aligned}|\mathbf{A_2}| &= \frac{\partial(-P, T)}{\partial(V, S)} = \frac{\partial(-P, T)}{\partial(V, T)} \frac{\partial(T, V)}{\partial(S, V)}\\ &= -\left(\frac{\partial P}{\partial V}\right)_\mathrm{T} \cdot \left(\frac{\partial T}{\partial S}\right)_\mathrm{V}\\ &= \left(\frac{1}{V \chi_\mathrm{T}}\right) \cdot \left(\frac{T}{C_\mathrm{V}}\right)\\ &> 0\,.\end{aligned} \tag{9.38}$$

And because both V and T are necessarily positive and, according to (9.36), for thermodynamic stability C_V is also positive, χ_T must be positive. This means

$$\chi_\mathrm{T} > 0\,. \tag{9.39}$$

In words,

For intrinsic thermodynamic stability, both the specific heat at constant volume, C_V, and the isothermal compressibility, χ_T, must be positive.

The positivity of the isothermal compressibility χ_T is testified to by observation. With the temperature maintained constant, an increase in compression shrinks the volume of the object being compressed. Again, this is necessarily true if the system is intrinsically stable against phase separation.

9.6 Intrinsic Thermodynamic Stability: Chemical Potential

The above adequately describes the physics of intrinsic—or, indeed, mutual—stability in simple thermodynamic systems. Because of the importance of the chemical potential, the variables used above must be extended to also include the mole number n. Understandably, such extension adds quite a little bit to the algebra. Therefore, the details are best deferred to an appendix—see Appendix H. It may, however, be helpful for a student to be apprised of the result.

In order to achieve and maintain intrinsic thermodynamic stability, the following three physical requirements must be met (see Appendix G, (I.19) for details):

$$C_V > 0\,, \qquad \chi_T > 0\,, \qquad \left(\frac{\partial \mu}{\partial n}\right)_{S,V} > 0\,. \tag{9.40}$$

Much like the first two requirements—already discussed above—the third also has an obvious physical basis: to maintain thermodynamic equilibrium when the total entropy and volume are held constant, the addition of a molecule (i.e., a tiny fraction) to an otherwise isolated system in equilibrium must increase its chemical potential. A hint of this phenomenon has already been noted in (9.17). There, an observation was made whose implications are similar to those of this requirement, namely that the chemical potential must be higher in a region of higher particle density.[20]

9.6.1 Intrinsic Stability: C_P and $\chi_S > 0$

The requirement that for intrinsic stability both C_V and χ_T must be > 0 implies that both C_P and χ_S are also > 0.

To see that C_P is > 0, let us refer to (5.83) and (5.84) where the exactness of the equality,

$$C_P = C_V + TV\left(\frac{\alpha_P^2}{\chi_T}\right), \tag{9.41}$$

was demonstrated. Because T, V, and the square of the real quantity α_P are all necessarily positive and in an intrinsically stable thermodynamic system both C_V and χ_T are also > 0, we have

$$C_P > 0\,. \tag{9.42}$$

To examine the positivity of χ_S, let us look at (5.87), which for convenience is reproduced in an equivalent form below as

$$\frac{\chi_S}{\chi_T} = \frac{C_V}{C_P}\,. \tag{9.43}$$

Because χ_T, C_V, and C_P are all positive,

$$\chi_S > 0\,. \tag{9.44}$$

Exercise 9.1 Show that in an intrinsically stable thermodynamic system χ_T is greater than χ_S.

[20]For instance, compare with the earlier statement. Simply put, the requirement that in an isolated system the entropy must increase in an isothermal spontaneous process mandates the molecular flow, at constant temperature, to occur outwards from a region of higher chemical potential towards a region of lower chemical potential.

Summary To recapitulate, positivity of the specific heat, C_V, and the isothermal compressibility, χ_T, in each of the two parts of a composite thermodynamic system ensures both the intrinsic and mutual thermodynamic stability of the two parts. Further, when changes in the particle density are also allowed, the rate of change of the chemical potential with respect to the particle density—for given values of the entropy and the volume—is also positive.

Clearly, the same must hold true even if the composite system has more than two parts. For instance, for a system with three parts, we can first lump two (of the three) parts together into a single system. This process can then be repeated to accommodate a system with arbitrary number of parts.

Gibbs, Helmholtz, and Clausius–Clapeyron 10

Section 10.1 begins with a description of the treatment of energy and entropy extrema. Two phases with a single variety of constituent atoms are studied in Sect. 10.2. Minimum energy in adiabatically isolated phases is considered in Sect. 10.3, while Sect. 10.4 deals with relative size of phases, equality of energy minimum and specific internal energy of the phases. Various aspects of entropy are examined in Sects. 10.5 and 10.6. Legendre transformations and Helmholtz free energy are discussed in Sects. 10.7 and 10.8. Some of Helmholtz' work relevant to this study is examined in Sects. 10.8.1 and 10.8.4–10.8.6. Clausius' inequality in differential form is described in Sect. 10.8.2. Validity of the suggestion that maximum possible work that a given system, in thermal contact with a constant temperature reservoir, can possibly perform is equal to the resulting decrease in its Helmholtz free energy is checked in Sect. 10.8.3. That specific Helmholtz free energy is equal for different phases in thermodynamic equilibrium is noted in Sect. 10.8.7.

Various aspects of Gibbs free energy are examined in Sect. 10.9 as well as in Sect. 10.10. Maximum available work when both temperature and pressure are kept constant is investigated in Sect. 10.9.1. For the simple thermodynamic system being considered here, there are only four possible choices for the pairs of independent characteristic variables. These are (v, s), (v, t), (p, t), and (p, s). Only the last of these four pairs containing pressure and entropy remains uninvestigated. Because both pressure and temperature are easy to measure, it is imperative that we learn to transform the Gibbs potential for constant pressure and temperature into one that only depends on pressure and entropy. To this end, one needs help from an appropriate Legendre transform that involves the fourth pair (p, s). This matter is studied in detail in Sect. 10.11. Heat of transformation, thermodynamic potentials, and characteristic equations are described in Sects. 10.12, 10.13, and 10.14. Thermodynamic paths from Helmholtz potential to internal energy, Gibbs potential to enthalpy are examined in Sects. 10.14.1 and 10.14.2.

Maxwell relations, metastable equilibrium, and Clausius–Clapeyron differential equation and its solution are studied in Sects. 10.15–10.17. Triple- and ice-points of water and Gibbs phase rule are described in Sects. 10.17.2 and 10.18. Section 10.18.1 deals with multiphase–multiconstituent systems. Phase equilibrium relations and phase rule are discussed in Sects. 10.19 and 10.21. Variance, as well as

R. Tahir-Kheli, *General and Statistical Thermodynamics*,
https://doi.org/10.1007/978-3-030-20700-7_10

invariant and monovariant systems, is studied in Sect. 10.20. Finally, a phase rule for systems with chemical reactions is analyzed in the concluding Sect. 10.21.

10.1 Energy and Entropy Extrema

In Chap. 9 some issues that pertain to intrinsic thermodynamic stability were noted. Also some matters relating to thermodynamic motive forces were investigated. These analyses were guided by the extremum principles for the internal energy and entropy. Here, in Chap. 10, other extremum principles are identified and some of their important consequences predicted.

We treat systems constituted of a single variety of molecules. The following sections deal with single variety constituent systems with two phases. In such systems, when more than one phase is present in thermodynamic equilibrium, the energy extremum principle requires that the relative size of the two phases must adjust itself to such value as minimizes the total internal energy. Further, the internal energy stipulates that the specific internal energy be the same in all phases. The same is also true of the entropy. Legendre transformations, which provide an essential tool for these studies, are discussed. Also, the extremum principles obeyed by the Helmholtz free energy, Gibbs potential, and enthalpy are analyzed and their consequences predicted. Characteristic equations for the four thermodynamic potentials—namely the internal energy, Helmholtz and Gibbs potentials, and enthalpy—are studied. Maxwell relations are described and the concept of metastable equilibrium is discussed. A detailed account of the Clausius–Clapeyron differential equation and its use in the study of thermodynamic phases is given. Finally, we discuss the Gibbs phase rule and variance, completing the description by referencing the application of the Gibbs phase rule to systems with internal chemical reactions.

10.2 Single Variety Constituent Systems: Two Phases

Assume the given isolated thermodynamic system is constituted of only a single variety of molecules but has two different thermodynamic phases.

10.3 Minimum Energy in an Adiabatically Isolated System

The requirement that in thermodynamic equilibrium the energy be the minimum possible for an adiabatically isolated system plays an important role in determining the physical viability of the thermodynamic state under review. This requirement is integral to achieving stable thermodynamics.[1] In particular, the requirement that "*At the equilibrium value of all the other extensive variables including the internal energy, the entropy of an adiabatically isolated thermodynamic system must be a maximum*", demands that under the appropriate conditions stated above:

[1] Refer to Chap. 9 where physical criteria for intrinsic thermodynamic stability are investigated.

(a) *In a completely isolated thermodynamic system, only those spontaneous processes occur that increase the total entropy of the system.*
(b) *And should an isolated, thermodynamic system exist in two or more states of stable equilibria, then the specific entropy is the same in each of those states.*

10.4 Relative Size of Phases and Energy Minimum

Consider a macroscopic system, constituted of two phases, placed within an adiabatically isolating chamber. The entropy extremum principle demands that in thermodynamic equilibrium the entropy summed over the two phases has to be a maximum. Therefore, during any spontaneous process the relative number of moles of the two phases will undergo appropriate thermodynamic changes so that in equilibrium the total entropy of the two phases is at its maximum.

Accordingly, given the equilibrium values of all the relevant extensive variables, including the internal energy, of the given biphase macroscopic system in thermodynamic equilibrium that has been placed in an adiabatically isolating chamber, the following holds true:

In an isolated, single constituent thermodynamic system, the relative number of moles of the two phases in the final equilibrium state are those that lead to the largest total entropy.

As discussed in Sect. 9.2.4, the energy extremum principle requires that the relative size[2] of the two phases must adjust itself to such value as to minimize the total internal energy, meaning, the relative size of the two phases adjusts itself so that the sum of the internal energy of the two phases is the minimum possible. Accordingly, this adjustment is such that for an isolated thermodynamic system, at the equilibrium values of all the relevant extensive variables, including the entropy, the following holds:

The relative numbers of moles of the two coexistent phases—of an isolated, macroscopic system in thermodynamic equilibrium—are those that lead to the lowest value for the total internal energy.

10.4.1 Specific Internal Energy of Two Phases Is Equal

Assume two different simultaneously present phases, A and B, constitute a given isolated thermodynamic system. Further, let their "specific" internal energy—meaning internal energy per mole—be u_A and u_B. For an equilibrium state titled 1, let the numbers of moles for the two phases be $n_{A,1}$ and $n_{B,1}$, respectively. Then the value of the internal energy, $u_{total;1}$, of the isolated thermodynamic system for the equilibrium state 1 is

$$u_{total;1} = n_{A,1}u_A + n_{B,1}u_B \,. \tag{10.1}$$

[2]Note, however, that while the relative size of the two phases does indeed depend upon the thermodynamic state of the system, their total mass does not, because it is necessarily conserved.

Now, if there is another equilibrium state titled 2, then with similar notation we can write

$$u_{\text{total};2} = n_{A,2}u_A + n_{B,2}u_B \,. \tag{10.2}$$

Thermodynamic equilibrium stipulates that the system be in the state with the lowest value of the internal energy. Therefore, the given isolated thermodynamic system cannot possibly be equally willing to be in either of the two equilibrium states, 1 and 2—unless, of course, the two states have the same total internal energy, $u(\text{total})$. Accordingly, we must have $u_{\text{total};1} = u_{\text{total};2} = u(\text{total})$. That is,

$$n_{A,1}u_A + n_{B,1}u_B = n_{A,2}u_A + n_{B,2}u_B = u(\text{total}) \,. \tag{10.3}$$

Also, let the total number of moles in the system be n_{total}. Then, because the total number of particles in a given thermodynamic system—not subject to any leakage of atoms—is conserved, the total number of moles in either of the two equilibrium states, 1 or 2, must be the same. That is,

$$n_{A,1} + n_{B,1} = n_{A,2} + n_{B,2} = n_{\text{total}} \,. \tag{10.4}$$

The only physically acceptable solution of the above two equations, i.e., (10.3) and (10.4), is

$$u_A = u_B \,. \tag{10.5}$$

To sum up: *When a biphase macroscopic system, placed in an adiabatically isolating chamber, is in stable thermodynamic equilibrium, the specific internal energy of the two phases is the same.*

10.5 Maximum Entropy in an Adiabatically Isolated System

The maximum in the entropy, for an adiabatically isolated macroscopic system, plays a similar role to the minimum in the internal energy. The latter was analyzed in Sect. 10.4. Either requirement leads to achieving stable thermodynamic equilibrium.[3] In particular, the requirement that "*At the equilibrium value of all the other extensive variables including the internal energy, the entropy of an adiabatically isolated thermodynamic system must be a maximum*", demands that under the appropriate conditions stated above:

(a) *In a completely isolated thermodynamic system, only those spontaneous processes occur that increase the total entropy of the system.*
(b) *And should an isolated, thermodynamic system exist in two or more states of stable equilibria, then the specific entropy is the same in each of those states.*

[3] Refer to Chap. 3 titled "The First Law" where issues of thermodynamic stability are investigated.

10.6 Relative Size of Phases and Entropy Maximum

Much like the internal energy, the entropy is a state function of fundamental importance.

Consider a macroscopic system, constituted of two phases, placed within an adiabatically isolating chamber. The entropy extremum principle demands that in thermodynamic equilibrium the entropy summed over the two phases has to be a maximum. Therefore, during any spontaneous process the relative number of moles of the two phases will undergo appropriate thermodynamic changes so that in equilibrium the total entropy of the two phases is at its maximum.

Accordingly, given the equilibrium values of all the relevant extensive variables, including the internal energy, of the given biphase macroscopic system in thermodynamic equilibrium that has been placed in an adiabatically isolating chamber, the following holds true:

In an isolated, single constituent thermodynamic system, the relative number of moles of the two phases in the final equilibrium state are those that lead to the largest total entropy.[4]

10.6.1 Specific Entropy of Two Phases Is Equal

Let the two different, simultaneously present, phases, A and B, constitute a given isolated single constituent thermodynamic system. Further, let their "specific" entropy—meaning entropy per mole—be s_{A} and s_{B}. For an equilibrium state titled 1, let the numbers of moles for the two phases be $n_{\text{A},1}$ and $n_{\text{B},1}$, respectively. Then the total value of the entropy, $s_{\text{total};1}$, of the isolated thermodynamic system for the equilibrium state 1 is

$$s_{\text{total};1} = n_{\text{A},1}s_{\text{A}} + n_{\text{B},1}s_{\text{B}} \,. \tag{10.6}$$

Now, if there is another equilibrium state titled 2, then with similar notation we can write

$$s_{\text{total};2} = n_{\text{A},2}s_{\text{A}} + n_{\text{B},2}s_{\text{B}} \,. \tag{10.7}$$

Thermodynamic equilibrium stipulates that the system be in the state with the largest value of the total entropy. Therefore, the given isolated thermodynamic system cannot possibly be in two different equilibrium states—unless, of course, the two states have the same amount of total entropy, $s(\text{total})$. Accordingly, we must have $s_{\text{total};1} = s_{\text{total};2} = s(\text{total})$. That is,

$$n_{\text{A},1}s_{\text{A}} + n_{\text{B},1}s_{\text{B}} = n_{\text{A},2}s_{\text{A}} + n_{\text{B},2}s_{\text{B}} = s(\text{total}) \,. \tag{10.8}$$

[4]Note, however, that while the relative size of the phases does indeed depend upon the thermodynamic state of the system, their total mass does not, because it is necessarily conserved.

Also, let the total number of moles in the system be n_{total}. Then, because the mole number in a given system—not subject to any leakage of atoms—is conserved, the total number of moles in either of the two equilibrium states must be the same. That is,

$$n_{A,1} + n_{B,1} = n_{A,2} + n_{B,2} = n_{\text{total}} \,. \tag{10.9}$$

The only physically acceptable solution of the above two equations, i.e., (10.8) and (10.9), is

$$s_A = s_B \,. \tag{10.10}$$

To sum up: *When a biphase, macroscopic system, placed in an adiabatically isolating chamber, is in stable thermodynamic equilibrium, the specific entropy of the two phases is the same.*

Remark 10.1 It turns out that, in addition to the internal energy, u, and the entropy, s, extremum principles are also obeyed by other thermodynamic potentials. And, much like u and s, these potentials also play compelling roles in thermal physics.

10.7 Legendre Transformations

As has been mentioned before, direct measurement of the entropy is not possible. This seriously affects the usefulness of those thermodynamic relationships where the entropy is an "independent variable." So one asks the question: Is there some way of transforming out a specific independent variable that is hard, or inconvenient, to measure? Further, can such a transformation be performed cleverly enough so that no loss of information occurs?

Happily, answers to both these questions are in the affirmative. And the methodology to use is that of Legendre transformations. Below we demonstrate, in the energy representation,[5] how to use Legendre transformations. For this purpose, let us choose first the most preeminent thermodynamic potential,[6] namely the internal energy, u.[7] In this regard, it is helpful to review the "Euler equation." A detailed description of the fundamental equation—namely the Euler equation which is (8.12)—is given in Chap. 8. For convenience, it is reproduced in an equivalent form below:

$$u = -pv + ts + \mu n - \mathcal{Y}\mathcal{X} \,. \tag{10.11}$$

This equation describes how the extensive variable of interest, i.e., the internal energy u, is related to other extensive variables: volume v, entropy s, mole number n, and parameters such as $\mathcal{X}$.

[5]Similar work in the entropy representation is also possible. The relevant potentials are then called Massieu functions. See Appendix I for details.

[6]What constitutes a "thermodynamic potential" is dealt with later in this chapter.

[7]Recall that, in the preceding two chapters, the internal energy was shown to play a central role in the description of thermodynamic equilibrium and stability.

Accordingly, the quasistatic increase in the potential energy of two neighboring equilibrium states is a function of the corresponding increases in the volume, entropy, mole numbers, and $\mathcal{X}$. This fact is represented in the form of the joint version of the first–second laws (compare with (8.5)). For convenience, that statement too is reproduced in an equivalent form below:

$$du = -p dv + t ds + \mu dn - \mathcal{Y} d\mathcal{X} . \tag{10.12}$$

The system that we treat first has only constant number of atoms—meaning $dn = 0$—and also does not have any dependence on the term $\mathcal{Y}d\mathcal{X}$. For such a simple system, a convenient appropriate relationship is provided by the statement of the first–second law (that was first given in (5.6)),

$$du = -p dv + t ds = \left(\frac{\partial u}{\partial v}\right)_s dv + \left(\frac{\partial u}{\partial s}\right)_v ds ,$$

$$\text{therefore,} \quad u \equiv u(v, s) . \tag{10.13}$$

Note that all the changes, i.e., du in the internal energy, dv in the volume, and ds in the entropy, refer to extensive variables, are infinitesimal in size, and occur quasistatically. Further note that the "canonical"—i.e., the characteristic—independent variables for the internal energy are the volume v and entropy s. And, while the volume is readily measured, precise measurement of the entropy s is not possible. Therefore, s is not the most desirable independent variable to have. Should we then attempt to transform out of the entropy as an independent variable? That is precisely what is achieved by the Helmholtz free energy. (For details, see the following section where the Helmholtz free energy is discussed.)

10.8 Helmholtz Free Energy

A potential to be introduced below is a function of two variables, namely volume and temperature. It is called the "Helmholtz free energy," or equivalently, the "Helmholtz thermodynamic potential."

10.8.1 Helmholtz Thermodynamic Potential

We often need to consider transformations in systems whose properties are available as functions of the two most accessible variables, volume and temperature. In (10.13) one of those two variables of interest—namely v through the occurrence of dv—is already present. The second variable of interest, namely t, can be included by transforming out of the entropy variable s—which occurs here via ds—into dt.

In other words, without engendering any loss in information, we need to transcribe the term $(+t ds)$ from (10.13), into a term that would involve dt. This can be done by using an appropriate Legendre transformation as follows:

First, recall that (10.13) represents the internal energy, u, as a function of its characteristic variables, volume v and entropy s. To transcribe ($+t\mathrm{d}s$) out of this equation, we need to add ($-ts$) to the primary[8] function u. Note that by this addition we have introduced an alternate function f—which we shall call the "Helmholtz thermodynamic potential,"

$$f = u - (ts) . \tag{10.14}$$

Second, we determine the differential $\mathrm{d}f$,

$$\mathrm{d}f = \mathrm{d}u - t\mathrm{d}s - s\mathrm{d}t . \tag{10.15}$$

Third, in (10.15) introduce the original expression for ($\mathrm{d}u$) that was given in the starting (10.13).

We get

$$\begin{aligned} \mathrm{d}f &= (\mathrm{d}u) - t\mathrm{d}s - s\mathrm{d}t \\ &= (-p\mathrm{d}v + t\mathrm{d}s) - t\mathrm{d}s - s\mathrm{d}t \\ &= -p\mathrm{d}v - s\mathrm{d}t \\ &= \left(\frac{\partial f}{\partial v}\right)_t \mathrm{d}v + \left(\frac{\partial f}{\partial t}\right)_v \mathrm{d}t . \end{aligned} \tag{10.16}$$

In this fashion, without any loss of information, the original function—that is, the internal energy u—has been replaced by a new function f. The important thing to note is that the new function f is a function only of the variables of v and t that are the most convenient for treating the thermodynamics of systems that are maintained at constant volume and temperature. That is,

$$f = f(v, t) . \tag{10.17}$$

Clearly, the new independent variable, temperature t, which is conjugate to the previous independent variable, entropy s, is one of the easiest thermodynamic parameters to measure.

The newly introduced potential, $f(v, t)$, is called the "Helmholtz free energy," or equivalently, the "Helmholtz thermodynamic potential"—and the variables v and t are often called its "characteristic, independent, variables."

10.8.2 Clausius Inequality in Differential Form

As we know from the chapter on the second law, heat engines need a minimum of two temperatures to successfully operate cyclically, namely the higher temperature, t_H, of the reservoir that supplies the needed heat energy and the lower temperature, t_C, of the dump into which the unused heat energy is discarded. The maximum work

[8] Reader, please note the procedure: To transform out of ($+t\mathrm{d}s$), we need to add ($-ts$) to the primary function.

efficiency, ϵ_{max}, is achieved by a perfect Carnot engine and is a direct function of the two temperatures, i.e., $\epsilon_{max} = 1 - \frac{t_C}{t_H}$.

Although it cannot operate cyclically—which a good heat energy engine must—a thermodynamic system in contact with a heat energy source at a single temperature can still produce work. The questions that arise are: How much work is produced? And what is its maximum possible value?

To answer these questions, let us refer to the differential form of the Clausius inequality (4.52). For convenience, it is reproduced below:

$$dS \geq \frac{dQ'(T)}{T} . \tag{10.18}$$

For the present purposes, it is helpful to multiply both sides by T and express dQ' according to the first law. In this fashion, the above inequality can be represented in the following equivalent form:

$$t\Delta s \geq \Delta u + \Delta w' . \tag{10.19}$$

Here, t is the temperature, and Δs the increase in the entropy of the system. Also, Δu is the increase in the internal energy of, and $\Delta w'$ the work done by, the system. Note the equality holds only when all these processes are fully quasistatic.

10.8.3 Maximum Possible Work

Consider a system that is maintained at constant temperature t by thermal contact with a temperature reservoir. Label the initial and final states as i and j.

Then $t\Delta s = t(s_j - s_i)$. Let the work done "by the system" in transforming from state i to state j be denoted as $\Delta w'_{i\to j}(t)$ and the relevant increase, Δu, in the internal energy be $u_j - u_i$. Then, using inequality (10.19), we can write

$$t(s_j - s_i) \geq (u_j - u_i) + \Delta w'_{i\to j}(t) ,$$

or equivalently,

$$t(s_j - s_i) - (u_j - u_i) = f_i(t) - f_j(t) \geq \Delta w'_{i\to j}(t) . \tag{10.20}$$

Note that $f_i(t) - f_j(t)$ is the decrease in the Helmholtz free energy—or equivalently stated, the decrease in the Helmholtz potential—when the system, in thermal contact with a heat energy reservoir at temperature t, undergoes a transformation from an initial equilibrium state i to a final equilibrium state j. Then according to inequality (10.19), the decrease, $f_i(t) - f_j(t)$, in the Helmholtz potential is almost always greater than—but, exceptionally, equal to—the work $\Delta w'_{i\to j}(t)$ done by the system in going from i to j.

In other words, consider an isolated macroscopic system in thermodynamic equilibrium. The system is in thermal contact with a single heat energy reservoir. In performing some work, it gets transformed from one equilibrium state, i, to another equilibrium state, j. In such a process:

The maximum possible work that a given system—in thermal contact with a constant temperature reservoir—can possibly perform is equal to the resulting decrease in its Helmholtz free energy.

And this maximum can only be achieved if, $\Delta w'_{i\to j}(t) \rightarrow \Delta w_{i\to j}(t)$. Because then, (10.20) becomes

$$f_i(t) - f_j(t) = \Delta w_{i\to j}(t)\,. \tag{10.21}$$

And this happens only when all the transformation processes occur quasistatically.

10.8.4 Helmholtz Free Energy Is Decreased

A general description of the quasistatic infinitesimal work that can be performed by a macroscopic system in thermodynamic equilibrium at temperature t is recorded in (8.2). For convenience it is reproduced in an equivalent form below:

$$\Delta w'(t) = p\Delta v' - \mu \Delta n' + \mathcal{Y}\Delta \mathcal{X}'\,. \tag{10.22}$$

If all the extensive variables, such as the volume v', mole numbers n', and the properties referred to by the general parameter $\mathcal{X}'$, remain constant and equal to their equilibrium values at temperature t then $\Delta v' = \Delta n' = \Delta \mathcal{X}' = 0$. Therefore, according to (10.22), $\Delta w'(t) = 0$.[9] A consequence of this is that in (10.20), $\Delta w'_{i\to j}(t) = 0$. And the decrease in the Helmholtz free energy in any spontaneous passage from a state i to a state j becomes

$$f_i(t) - f_j(t) \geq 0\,, \tag{10.23}$$

or equivalently,

$$f_i(t) \geq f_j(t)\,. \tag{10.24}$$

In words, for a macroscopic system that is in thermodynamic equilibrium, is in thermal contact with a temperature reservoir, and for which the volume, mole numbers, and quantity $\mathcal{X}$ are maintained constant at their equilibrium value:

The Helmholtz free energy decreases during any spontaneous thermodynamic process during which the equilibrium values of the volume, mole numbers, and other extensive variables such as $\mathcal{X}$ remain constant. Only in exceptional circumstances, when the process involved is fully quasistatic, does the Helmholtz free energy remain unchanged.

More briefly: *A spontaneous isothermal–isochoric process can proceed only if the Helmholtz free energy either decreases or holds steady.*

[9]Literally, this implies that its integral, or equivalently the sum, $\Delta w'(t)$, is constant. Hence w' in both the states i and j is the same.

Note that $\Delta w'$, etc., indicate that the changes being described may have occurred either wholly or partially nonquasistatically.

10.8.5 Helmholtz Free Energy: Extremum Principle

In systems where the equilibrium values of the volume, mole numbers, and other extensive variables such as $\mathcal{X}$ remain constant, spontaneous isothermal processes occur as long as they decrease the Helmholtz free energy. Clearly, such processes continue until the Helmholtz free energy cannot decrease any further. In other words:

At the equilibrium values of s and n, and at constant value of the volume v and temperature t—the latter is maintained constant by thermal contact with a heat energy reservoir—the inequality (10.24) demands that in thermodynamic equilibrium the Helmholtz free energy be a minimum.

Briefly stated: *The Helmholtz free energy of a system of the given volume, mole numbers and property* $\mathcal{X}$, *whose temperature is maintained by contact with a heat energy reservoir, is a minimum in thermodynamic equilibrium.*

10.8.6 Helmholtz Free Energy: Relative Size of Phases

While the relative sizes—i.e., the relative number of moles—of the two phases depend upon the equilibrium state, their total mass is necessarily conserved. In the foregoing analyses we have learned that for such a system, at the equilibrium values of all the unconstrained extensive variables and at constant temperature, t, the Helmholtz free energy, f, is a minimum.

In order that in equilibrium, the total Helmholtz free energy of such a biphase thermodynamic system be at its minimum, during any spontaneous process the relative masses—i.e., the relative numbers of moles—of the given two phases will adjust themselves to decrease the Helmholtz free energy by the largest amount possible. In other words:

For the amounts of all the relevant extensive variables, the relative numbers of moles of the two phases in the final equilibrium state are those that lead to the lowest total value for the Helmholtz free energy.

Note that the same ideas can also be extended to multiphase—i.e., more than two phases—and multiconstituent systems which for brevity are not explicitly treated here.

10.8.7 Specific Helmholtz Free Energy Is Equal for Different Phases

Let the two different simultaneously present phases, A and B, constitute a given thermodynamic system (at given fixed value of the total volume and given temperature). Further, let the "specific" Helmholtz free energies of the two phases—meaning Helmholtz free energy per mole of the two phases that are in mutual thermodynamic equilibrium—be f_{A} and f_{B}.

For an equilibrium state titled 1, let the numbers of moles for the two phases be $n_{\mathrm{A},1}$ and $n_{\mathrm{B},1}$, respectively. Then the value of the Helmholtz free energy, $f_{\mathrm{total};1}$, of the given thermodynamic system in state 1 is

$$f_{\mathrm{total};1} = n_{\mathrm{A},1} f_{\mathrm{A}} + n_{\mathrm{B},1} f_{\mathrm{B}} \,. \tag{10.25}$$

Now, if there is another equilibrium state titled 2, then with similar notation we can write

$$f_{\text{total};2} = n_{A,2} f_A + n_{B,2} f_B . \tag{10.26}$$

Thermodynamic equilibrium stipulates that the system must be in the state with the lowest value of the Helmholtz free energy. Therefore, the given thermodynamic system cannot possibly be in two different equilibrium states—unless, of course, the total Helmholtz free energy in each phase, i.e., $f_{\text{total};1}$ and $f_{\text{total};2}$, is the same for both the states 1 and 2. Accordingly, we have

$$f_{\text{total};1} = f_{\text{total};2} = f(\text{total}) , \tag{10.27}$$

where

$$\begin{aligned} n_{A,1} f_A + n_{B,1} f_B &= f_{\text{total};1} = f(\text{total}) , \\ n_{A,2} f_A + n_{B,2} f_B &= f_{\text{total};2} = f(\text{total}) . \end{aligned} \tag{10.28}$$

Also, let the total number of moles in the system be n_{total}. Then, because the total number of moles is conserved, the total number of moles in state 1 is the same as it is in state 2, i.e.,

$$n_{A,1} + n_{B,1} = n_{\text{total}} = n_{A,2} + n_{B,2} . \tag{10.29}$$

The only physically acceptable solution of the above two equations, i.e., (10.28) and (10.29), is

$$f_A = f_B . \tag{10.30}$$

To sum up: *In a biphase thermodynamic system, which is in thermal contact with a heat energy reservoir at a fixed temperature, the specific Helmholtz free energy of the two coexistent phases is the same.*

10.9 Gibbs Free Energy

Usually, the study of the Helmholtz free energy is the preferred option for theoretical calculations. But in experimental work, especially in chemistry, constant pressure is much easier to maintain than is constant volume. Thus, rather than v and t, it is even better to have the pressure p and temperature t as the two independent variables. To this purpose, in an analogous fashion to that followed before, we need to use an appropriate Legendre transformation.

Consider (10.16), namely $\mathrm{d}f = -p\mathrm{d}v - s\mathrm{d}t$. Because, while leaving $\mathrm{d}t$ alone, we wish to transform out of the Helmholtz free energy's dependence on $\mathrm{d}v$, namely from the form $-p\mathrm{d}v$, into a possible dependence on $\mathrm{d}p$, this process affects only the term $(-p\mathrm{d}v)$. Therefore, as in (8.12) we define an appropriate state[10] function

[10] Again to be called the Gibbs potential, or equivalently, the Gibbs free energy, but note that in this subsection the system being treated has both μ and $\mathcal{Y}$ absent.

g by adding $(+pv)$ to the Helmholtz potential f. That is,

$$g = pv + f \,. \tag{10.31}$$

Then $\mathrm{d}g = p\mathrm{d}v + v\mathrm{d}p + (\mathrm{d}f)$. But $(\mathrm{d}f)$ is equal to $(-p\mathrm{d}v - s\mathrm{d}t)$. Therefore, we can write

$$\mathrm{d}g = +p\mathrm{d}v + v\mathrm{d}p + (-p\mathrm{d}v - s\mathrm{d}t) = v\mathrm{d}p - s\mathrm{d}t \,. \tag{10.32}$$

Thus the characteristic independent variables for the Gibbs free energy, g, are the pressure and the temperature. That is,

$$\begin{aligned} g &\equiv g(p, t)\,, \\ \mathrm{d}g &= \left(\frac{\partial g}{\partial p}\right)_{\mathrm{t}} \mathrm{d}p + \left(\frac{\partial g}{\partial t}\right)_{\mathrm{p}} \mathrm{d}t \,. \end{aligned} \tag{10.33}$$

10.9.1 Maximum Available Work: Constant *t* and *p*

We previously learned that when all the transformation processes occur quasistatically, the amount of work any thermodynamic system in contact with a single heat energy reservoir—which maintains it at a fixed temperature—can possibly perform is equal to the resulting decrease in its Helmholtz free energy. It is, however, usually the case that in stable equilibrium the naturally provided reservoir—meaning the atmosphere around the heat energy engine—not only serves as an energy reservoir at constant temperature but also maintains its pressure constant. Then the question that arises is the following: What is the amount of work that can possibly be done by such an engine if, in addition to the constancy of the temperature, its pressure, p, were also kept constant?

General Analysis

Consider two equilibrium states, i and j, of a given thermodynamic system in thermal contact with a heat energy reservoir at temperature t—meaning both the temperatures, t_{i} and t_{j}, of the states i and j are equal to t. Additionally, let there also be a volume connection, across a freely movable piston, which ensures the constancy of pressure between the system and volume reservoir. The pressures, p_{i} and p_{j}, of the two states are therefore equal to the pressure, p, of the volume reservoir, i.e., $p_{\mathrm{i}} = p_{\mathrm{j}} = p$. Finally, let the internal energy of the two states be u_{i} and u_{j}, their volumes v_{i} and v_{j}, and their entropies s_{i} and s_{j}, respectively. Then, let us define the following equalities:

$$\begin{aligned} &\Delta u_{(\mathrm{i}\to\mathrm{j})} = u_{\mathrm{j}} - u_{\mathrm{i}}\,, \quad \Delta w'_{(\mathrm{i}\to\mathrm{j})} = p'\left(v'_{\mathrm{j}} - v'_{\mathrm{i}}\right) + \Delta\left(Z'_{\mathrm{i}\to\mathrm{j}}\right), \quad \Delta s_{(\mathrm{i}\to\mathrm{j})} = s_{\mathrm{j}} - s_{\mathrm{i}}\,, \\ &\Delta\left(Z'_{\mathrm{i}\to\mathrm{j}}\right) = \Delta\left[\left\{-\mu n' + \mathcal{Y}\mathcal{X}'\right\}_{\mathrm{j}} - \left\{-\mu n' + \mathcal{Y}\mathcal{X}'\right\}_{\mathrm{i}}\right]. \end{aligned} \tag{10.34}$$

Note that $\Delta(Z'_{i\rightarrow j})$ is the non-PdV "work done by the system" in going from the state i to state j. As a result, the inequality (10.20) becomes

$$t(s_j - s_i) - (u_j - u_i) \geq \Delta w'_{i\rightarrow j}(t)\,, \tag{10.35}$$

or equivalently,

$$t(s_j - s_i) - (u_j - u_i) - p'(v'_j - v'_i) \equiv (g'_i - g'_j) \geq \Delta(Z'_{i\rightarrow j})\,. \tag{10.36}$$

Thus the non-PdV work, $\Delta(Z'_{i\rightarrow j})$, "done by the system" in going from state i to state j, is almost always less than the corresponding decrease, $g'_i - g'_j$, in the Gibbs free energy. And only exceptionally is the non-PdV work done by the system equal to the decrease, $g_i - g_j$, in the Gibbs[11] free energy of the system. And that happens when all extensive thermodynamic variables undergo only quasistatic changes. If any of the intervening processes are nonquasistatic, then the non-PdV work done, $\Delta(Z'_{i\rightarrow j})$, is less than the corresponding decrease, $g'_i - g'_j$, in the Gibbs free energy. In other words, assuming that a given system is in thermal contact with a heat energy reservoir, which maintains its temperature at t, and it can also freely exchange volume with a volume reservoir, so that, in addition to the temperature, the pressure p is also equalized between the system and the reservoir, then:

The maximum possible non-PdV work that can be extracted from such a thermodynamic system, when it undergoes a transformation from one equilibrium state to another, is equal to the corresponding decrease in its Gibbs free energy. And the relevant maximum can only be achieved if all the transformation processes occur quasistatically.

10.10 Decrease in Gibbs Free Energy

In the following, we investigate how the Gibbs free energy decreases during spontaneous thermodynamic processes. To this purpose we examine the @PdV work.

10.10.1 The PdV Work

Here the mole numbers as well as all $\mathcal{X}$'s are constant. Therefore, in (10.36)

$$\Delta(Z'_{i\rightarrow j}) = 0\,.$$

This means, the non-dV work is vanishing. As a result, according to (10.36), the decrease in the Gibbs free energy in going from any initial equilibrium state i to any final equilibrium state j—or equivalently, during any spontaneous process—is either equal to, or greater than, zero. That is,

$$g_i(p,t) - g_j(p,t) \geq 0\,, \tag{10.37}$$

[11]Note the absence of the prime in the quantity $g_i - g_j$.

or equivalently,

$$g_{\rm i}(p,t) \geq g_{\rm j}(p,t)\,. \tag{10.38}$$

Therefore, in a system where the non-PdV work is vanishing, all isobaric–isothermal spontaneous processes decrease the Gibbs free energy. More completely stated:

In a system which does only the PdV *work, a spontaneous, isothermal-isobaric, transformation can proceed only if the Gibbs free energy either decreases, or holds steady. Note that, in order for the Gibbs free energy to hold steady, all the processes involved in the transformation must be quasistatic.*

10.10.2 Gibbs Free Energy: Extremum Principle

Clearly, these spontaneous processes continue until the Gibbs free energy cannot decrease any further. In other words:

The Gibbs free energy is a minimum in stable thermodynamic equilibrium for a system that does only PdV *work and whose temperature and pressure are maintained constant by contact with external reservoirs.*

10.10.3 Gibbs Potential Minimum: Relative Size of Phases

As mentioned before, while the relative number of moles of the two phases depends upon the equilibrium state, their total mass is conserved. Therefore, in order that in equilibrium the total value of the Gibbs free energy of such a biphase thermodynamic system be at its minimum, during any spontaneous process the relative number of moles of the given two phases must adjust itself to decrease the Gibbs free energy by the largest amount.

Thus, at the equilibrium values of s, v and n, and at constant temperature—maintained by thermal contact with a heat energy reservoir—and at constant pressure—maintained by contact with a volume reservoir across a freely compressible piston—a biphase, single constituent system in thermodynamic equilibrium behaves as follows:

The relative number of moles of the two phases in the final equilibrium state is such as leads to the lowest total value for the Gibbs free energy.

While the same ideas can also be extended to multiconstituent, multiphase systems, for brevity we continue to work with a single chemical constituent with up to two phases.

10.10.4 Specific Gibbs Free Energy: Equality for Different Phases

Let the two different simultaneously present phases, A and B, constitute a given thermodynamic system (at given fixed value of the pressure and temperature). Further, let the "specific" Gibbs free energy of the two phases—meaning Gibbs free energy per mole—be $g_{\rm A}$ and $g_{\rm B}$.

For an equilibrium state titled 1, let the numbers of moles for the two phases be $n_{A,1}$ and $n_{B,1}$, respectively. Then the value of the Gibbs free energy, $g_{total;1}$, of the given thermodynamic system in state 1 is

$$g_{total;1} = n_{A,1} g_A + n_{B,1} g_B . \tag{10.39}$$

Now, if there is another equilibrium state titled 2, then with similar notation we can write

$$g_{total;2} = n_{A,2} g_A + n_{B,2} g_B . \tag{10.40}$$

Thermodynamic equilibrium stipulates that the system be in the state with the lowest total value of the Gibbs free energy. Therefore, the given thermodynamic system cannot possibly be in two different equilibrium states—unless, of course, the total Gibbs free energy, $g_{total;1}$ and $g_{total;2}$, is the same for both the states 1 and 2. Accordingly, we have

$$g_{total;1} = g_{total;2} = g(\text{total}) , \tag{10.41}$$

where

$$\begin{aligned} n_{A,1} g_A + n_{B,1} g_B &= g_{total;1} = g(\text{total}) , \\ n_{A,2} g_A + n_{B,2} g_B &= g_{total;2} = g(\text{total}) . \end{aligned} \tag{10.42}$$

Also, let the total number of moles in the system be n_{total}. Then, because the total number of moles is conserved, the number of moles in state 1 is the same as it is in state 2, i.e.,

$$n_{A,1} + n_{B,1} = n_{total} = n_{A,2} + n_{B,2} . \tag{10.43}$$

The only physically acceptable solution of the above two equations, i.e., (10.42) and (10.43), is

$$g_A = g_B . \tag{10.44}$$

To sum up: At constant pressure, in a biphase thermodynamic system that is in thermal contact with a heat energy reservoir at a fixed temperature, the specific Gibbs free energy of the two coexistent phases is the same.

Clearly, the above argument is readily extended to three—and indeed to an arbitrary number of—phases.

It is interesting to recall that a system with two or, indeed, three possible thermodynamic phases—vapor–liquid or vapor–liquid–solid—was first treated in Chap. 6. There—see the statement presented between (6.55) and (6.56)—it was stated without proof that: "During phase transitions, thermodynamic stability requires the specific Gibbs free energy to be the same for all phases."

10.11 Enthalpy: Remark

As noted in (10.13), the characteristic independent variables for the internal energy u are the volume v and entropy s. That is, $u = u(v, s)$. The first time we used the Legendre transformation in this chapter, we exchanged the entropy (i.e., the independent variable s) for the temperature (i.e., the independent variable t) thereby transforming out of the internal energy $u = u(v, s)$ into the Helmholtz free energy $f = f(v, t)$. The Legendre transformation that was used next, exchanged volume (i.e., the independent variable v) for pressure (i.e., the independent variable p) thereby transforming the Helmholtz free energy $f(v, t)$ into the Gibbs free energy $g = g(p, t)$.

For the simple thermodynamic system being considered here, there are only four possible choices for the pairs of independent characteristic variables. These are (v, s), (v, t), (p, t), and (p, s). So far in this chapter, only three of these have been utilized, namely (v, s) in $u(v, s)$, (v, t) in $f(v, t)$, and (p, t) in $g(p, t)$. To make use of the last of these four possible pairs, namely (p, s), we need to transform the Gibbs potential $g(p, t)$ into an appropriate Legendre transform that involves the variable pair (p, s).

Consider (10.32). As stated, we wish to transform out of the term $(-s\mathrm{d}t)$ in the equation $\mathrm{d}g = v\mathrm{d}p - s\mathrm{d}t$. As such, we need to invent an appropriate potential[12] by adding $(+st)$ to g. That is,

$$h = g + st\,. \tag{10.45}$$

This leads to

$$\begin{aligned} \mathrm{d}h &= (\mathrm{d}g) + s\mathrm{d}t + t\mathrm{d}s \\ &= (v\mathrm{d}p - s\mathrm{d}t) + s\mathrm{d}t + t\mathrm{d}s = v\mathrm{d}p + t\mathrm{d}s\,. \end{aligned} \tag{10.46}$$

Thus the independent characteristic variables of the enthalpy h are the pressure p and the entropy s,

$$\mathrm{d}h(p, s) = \left(\frac{\partial h}{\partial p}\right)_{\mathrm{s}} \mathrm{d}p + \left(\frac{\partial h}{\partial s}\right)_{\mathrm{p}} \mathrm{d}s\,. \tag{10.47}$$

Note that despite this fact—namely that the characteristic independent variables for h are p and s—the enthalpy h can also be written as

$$h = g + st = u - ts + pv + st = u + pv\,. \tag{10.48}$$

Indeed, this is how the enthalpy is usually represented!

A word of caution is in order. To a beginner, the essential identity of (10.48) and (10.47)—both representing the enthalpy—may seem incongruous. Adding to any possible such incongruity may also be the fact that previously—meaning much earlier in this book, for example, in (3.89)—enthalpy was analyzed in terms of yet another pair of variables, namely p and t. Of course, while that analysis was perfectly legitimate for what was needed there, the more appropriate pair of variables—namely the characteristic variables—for the enthalpy are indeed p and s.

[12]Incidentally, such a potential was first introduced in (3.54) where it was equated to $u + pv$ and named the enthalpy.

10.12 Heat of Transformation

Liquids when sufficiently cooled usually solidify. Similarly, solids when heated tend to liquefy. Also, much as happens to frozen snow, occasionally, solids can evaporate without liquefying. And, at appropriately high temperatures, liquids boil, and in part vaporize. A convenient term for these occurrences is "phase transformations."

Enthalpy is an important state function. Because many issues that either relate to, or involve, the enthalpy have already been discussed in detail—see, (3.78) and related text—in what follows we shall only give a brief description of its use in the treatment of heats of transformation, where the enthalpy plays a central role.

Generally, during a phase transformation, the temperature, t, remains constant but the volume changes, say $v_{\mathrm{i}} \rightarrow v_{\mathrm{f}}$. There are exceptions to the volume change rule—as, for example, is the case for the so-called second-order phase transitions. In any event, because the transformation is isothermal, it is usually isobaric—so the initial, p_{i}, and the final, p_{f}, values of the pressure are the same, i.e., $p_{\mathrm{i}} = p_{\mathrm{f}} = p$.

Assuming the volume increase, $\Delta v = v_{\mathrm{f}} - v_{\mathrm{i}}$, occurs quasistatically, the work, Δw, done by the system during the isothermal–isobaric volume change is

$$\Delta w = p(v_{\mathrm{f}} - v_{\mathrm{i}}) \,.$$

If the heat energy used for the phase transformation—for one mole—is $\Delta q_{(\mathrm{PhTr})}$, then according to the first law,

$$\Delta q_{(\mathrm{PhTr})} = \Delta u + \Delta w \,,$$

where $\Delta u = (u_{\mathrm{f}} - u_{\mathrm{i}})$ is the relevant increase in the internal energy. Therefore

$$\begin{aligned}\Delta q_{(\mathrm{PhTr})} &= \Delta u + \Delta w \\ &= u_{\mathrm{f}} - u_{\mathrm{i}} + p v_{\mathrm{f}} - p v_{\mathrm{i}} \\ &= (u_{\mathrm{f}} + p v_{\mathrm{f}}) - (u_{\mathrm{i}} + p v_{\mathrm{i}}) \\ &= h_{\mathrm{f}} - h_{\mathrm{i}} = l_{\mathrm{i}\to\mathrm{f}} \,. \end{aligned} \tag{10.49}$$

The important fact to note here is the following. In stable thermodynamic equilibrium, and at the temperature and pressure at which the given system undergoes a phase transformation:

The heat energy, $l_{\mathrm{i}\to\mathrm{f}}$, needed for the phase transformation—from a state i to a state f—is equal to the corresponding increase, $h_{\mathrm{f}} - h_{\mathrm{i}}$, in the enthalpy.

Incidentally, for the so-called second order phase transitions, the heat of transformation is vanishingly small.

10.13 Thermodynamic Potentials: s, f, g, and h

In the foregoing we have studied the following important thermodynamic state functions: the internal energy u—for example, as in (10.13); the Helmholtz free energy f—as in (10.16); the Gibbs free energy g—as in (10.32); and the enthalpy h—as in (10.46). We notice that, much like the derivatives of electrostatic or mechanical potentials, the derivatives of these state functions also lead to appropriate "field functions." (See (10.50)–(10.53) below.) Therefore, in addition to the phrase thermodynamic state function, the other phrase to use for u, f, g, and h that suggests itself here is "thermodynamic potential."

For convenience, the relevant equations and their derivatives are represented below:

$$\big(\text{see (10.13)}\big) \quad \mathrm{d}u = -p\mathrm{d}v + t\mathrm{d}s = \left(\frac{\partial u}{\partial v}\right)_{\mathrm{s}} \mathrm{d}v + \left(\frac{\partial u}{\partial s}\right)_{\mathrm{v}} \mathrm{d}s\,,$$

$$\text{therefore,} \quad \left(\frac{\partial u}{\partial v}\right)_{\mathrm{s}} = -p\,, \quad \left(\frac{\partial u}{\partial s}\right)_{\mathrm{v}} = t\,, \tag{10.50}$$

$$\big(\text{see (10.16)}\big) \quad \mathrm{d}f = -p\mathrm{d}v - s\mathrm{d}t = \left(\frac{\partial f}{\partial v}\right)_{\mathrm{t}} \mathrm{d}v + \left(\frac{\partial f}{\partial t}\right)_{\mathrm{v}} \mathrm{d}t\,,$$

$$\text{therefore,} \quad \left(\frac{\partial f}{\partial v}\right)_{\mathrm{t}} = -p\,, \quad \left(\frac{\partial f}{\partial t}\right)_{\mathrm{v}} = -s\,, \tag{10.51}$$

$$\big(\text{see (10.32)}\big) \quad \mathrm{d}g = v\mathrm{d}p - s\mathrm{d}t = \left(\frac{\partial g}{\partial p}\right)_{\mathrm{t}} \mathrm{d}p + \left(\frac{\partial g}{\partial t}\right)_{\mathrm{p}} \mathrm{d}t\,,$$

$$\text{therefore,} \quad \left(\frac{\partial g}{\partial p}\right)_{\mathrm{t}} = v\,, \quad \left(\frac{\partial g}{\partial t}\right)_{\mathrm{p}} = -s\,. \tag{10.52}$$

$$\big(\text{see (10.46)}\big) \quad \mathrm{d}h = v\mathrm{d}p + t\mathrm{d}s = \left(\frac{\partial h}{\partial p}\right)_{\mathrm{s}} \mathrm{d}p + \left(\frac{\partial h}{\partial s}\right)_{\mathrm{p}} \mathrm{d}s\,,$$

$$\text{therefore,} \quad \left(\frac{\partial h}{\partial p}\right)_{\mathrm{s}} = v\,, \quad \left(\frac{\partial h}{\partial s}\right)_{\mathrm{p}} = t\,. \tag{10.53}$$

10.14 Characteristic Equations

The characteristic variables of two of the four thermodynamic potentials handled above—namely, the Helmholtz potential $f = f(v,t)$ and the Gibbs potential $g = g(p,t)$—are easy to measure.[13] If available, knowledge of these potentials—i.e., f and g—also helps determine the other two potentials, namely the internal energy $u = u(v,s)$ and the enthalpy $h = h(p,s)$.

10.14.1 Helmholtz Potential to Internal Energy

For instance, imagine that the functional form of the Helmholtz potential, $f(v,t)$, is known. Then, with the help of (10.51) and (10.14), the internal energy u may be obtained as follows:

$$\left(\frac{\partial f}{\partial t}\right)_{\mathrm{v}} = -s\,, \quad u = f + ts = f(v,t) - t\left(\frac{\partial f}{\partial t}\right)_{\mathrm{v}}. \tag{10.54}$$

Further, because $p = -(\frac{\partial f}{\partial v})_{\mathrm{t}}$, if we are interested we can also determine the pressure!

[13] These variables are, of course, (v,t) and (p,t), respectively.

10.14.2 Gibbs Potential to Enthalpy

Similarly, if the functional form of the Gibbs potential, $g(p, t)$, is known, then with the help of (10.31), (10.32), and (10.52) the enthalpy h can also be determined in the following manner:

$$\left(\frac{\partial g}{\partial t}\right)_{\mathrm{p}} = -s\,, \qquad h = g + ts = g(p,t) - t\left(\frac{\partial g}{\partial t}\right)_{\mathrm{p}} \,. \tag{10.55}$$

Because $v = (\frac{\partial g}{\partial p})_{\mathrm{t}}$, we can even determine the volume here!

10.15 Maxwell Relations

Recall that for an exact differential $\mathrm{d}Z$, where $Z = Z(x, y)$, one has the following relationships:

$$\mathrm{d}Z = \left(\frac{\partial Z}{\partial x}\right)_{\mathrm{y}} \mathrm{d}dx + \left(\frac{\partial Z}{\partial y}\right)_{\mathrm{x}} \mathrm{d}y\,, \tag{10.56}$$

and

$$\left(\frac{\partial(\frac{\partial Z}{\partial x})_{\mathrm{y}}}{\partial y}\right)_{\mathrm{x}} = \frac{\partial^2 Z}{\partial y \partial x} = \frac{\partial^2 Z}{\partial x \partial y} = \left(\frac{\partial(\frac{\partial Z}{\partial y})_{\mathrm{x}}}{\partial x}\right)_{\mathrm{y}} \,. \tag{10.57}$$

Because differentials of the state functions, u, f, g, and h are exact, (10.50)–(10.53) therefore lead to the following set of useful relationships:

$$\begin{aligned}
&(\text{see (10.50)}) \quad \left(\frac{\partial u}{\partial v}\right)_{\mathrm{s}} = -p\,, \quad \left(\frac{\partial u}{\partial s}\right)_{\mathrm{v}} = t\,,\\
&\quad\text{therefore,} \quad \frac{\partial^2 u}{\partial s \partial v} = -\left(\frac{\partial p}{\partial s}\right)_{\mathrm{v}} = \frac{\partial^2 u}{\partial v \partial s} = \left(\frac{\partial t}{\partial v}\right)_{\mathrm{s}}\,,\\
&(\text{see (10.51)}) \quad \left(\frac{\partial f}{\partial v}\right)_{\mathrm{t}} = -p\,, \quad \left(\frac{\partial f}{\partial t}\right)_{\mathrm{v}} = -s\,,\\
&\quad\text{therefore,} \quad \frac{\partial^2 f}{\partial t \partial v} = -\left(\frac{\partial p}{\partial t}\right)_{\mathrm{v}} = \frac{\partial^2 f}{\partial v \partial t} = -\left(\frac{\partial s}{\partial v}\right)_{\mathrm{t}}\,,\\
&(\text{see (10.52)}) \quad \left(\frac{\partial g}{\partial p}\right)_{\mathrm{t}} = v\,, \quad \left(\frac{\partial g}{\partial t}\right)_{\mathrm{p}} = -s\,,\\
&\quad\text{therefore,} \quad \frac{\partial^2 g}{\partial t \partial p} = \left(\frac{\partial v}{\partial t}\right)_{\mathrm{p}} = \frac{\partial^2 g}{\partial p \partial t} = -\left(\frac{\partial s}{\partial p}\right)_{\mathrm{t}}\,,\\
&(\text{see (10.53)}) \quad \left(\frac{\partial h}{\partial p}\right)_{\mathrm{s}} = v, \quad \left(\frac{\partial h}{\partial s}\right)_{\mathrm{p}} = t\,,\\
&\quad\text{therefore,} \quad \frac{\partial^2 h}{\partial s \partial p} = \left(\frac{\partial v}{\partial s}\right)_{\mathrm{p}} = \frac{\partial^2 h}{\partial p \partial s} = \left(\frac{\partial t}{\partial p}\right)_{\mathrm{s}}\,.
\end{aligned} \tag{10.58}$$

In (10.58) there are four equalities that are recorded with the symbol "=". Note that these equalities apply only to the so-called simple thermodynamic system.

Clearly, to the extent that the second and the third of these equalities, i.e.,

$$\left(\frac{\partial s}{\partial v}\right)_{\mathrm{t}} = \left(\frac{\partial p}{\partial t}\right)_{\mathrm{v}} \quad \text{and} \quad \left(\frac{\partial s}{\partial p}\right)_{\mathrm{t}} = -\left(\frac{\partial v}{\partial t}\right)_{\mathrm{p}}, \tag{10.59}$$

describe the rates of change of the entropy in terms of, and with respect to, the measurable parameters—the pressure, volume, and temperature—they are of great thermodynamic interest. Indeed, one expects these relations to be particularly useful for the calculation of the entropy.

Outwardly, that appears not to be the case for the relations 1 and 4. That view, however, changes if we look at the first and fourth relations when they are displayed upside-down. That is,

$$\left(\frac{\partial s}{\partial p}\right)_{\mathrm{v}} = -\left(\frac{\partial v}{\partial t}\right)_{\mathrm{s}} \quad \text{and} \quad \left(\frac{\partial s}{\partial v}\right)_{\mathrm{p}} = \left(\frac{\partial p}{\partial t}\right)_{\mathrm{s}}. \tag{10.60}$$

Because the differentials on the right-hand side of (10.60) are taken at constant value of the entropy, they outwardly do not appear to be as easy to handle experimentally as the two differentials on the right-hand side of (10.59), which do not involve the entropy.

Yet, in practice, keeping the entropy constant is much easier to achieve than finding the actual value of the entropy. Therefore, the relations given above in (10.60), where the entropy occurs only as a parameter that remains constant, are also useful.

For the record, we shall henceforth call the four relations that appear in (10.59) and (10.60) as the Maxwell relations for a simple thermodynamic system. Additional Maxwell relations for less simple thermodynamic systems, that involve changes in mole numbers and other extensive variables such as $\mathcal{X}$, can also be worked out. However, for brevity we shall not include that analysis here.

Exercise 10.1 A motivated student might want to work out the Maxwell relations for a less simple thermodynamic system, for example, one with different mole numbers, i.e., a multiconstituent system.

10.16 Metastable Equilibrium

The physical requirements that need to be satisfied for stable thermodynamic equilibrium have been examined in Sect. 9.2.1. Additional aspects of the issue are studied in Appendix G. When those requirements are not met, the equilibrium is unstable. There is, however, a possible third option.

At least in principle, a stable thermodynamic equilibrium is "almost infinitely long lived." This is so for two reasons (see Chap. 9).

The "Le Châtelier's principle" assures us that infinitesimal spontaneous processes that move the system out of equilibrium immediately get followed by ones that bring the system back to equilibrium.

More generally, finitely sized spontaneous processes are much less likely to occur. Why is that the case? The answer is as follows: For given values of all the constraints, the relevant values of all the unconstrained parameters, and the fixed values of the temperature and pressure, the Gibbs free energy is at its minimum in thermodynamic equilibrium. And any finitely sized spontaneous process, if it should occur, would of necessity want to additionally decrease the Gibbs free energy by a finite amount. And this decrease would need to happen even though the system Gibbs free energy is already at its minimum!

However, if some or all of the constraint are loosened, finitely sized spontaneous processes may occur. Furthermore, according to statistical mechanics—see Chap. 11—when left to its own devices, a thermodynamic system is subject to "tiny" spontaneous fluctuations that move fractional parts of the system away from the equilibrium. Occasionally, and in appropriate circumstances, these effects are additive and persist for a long enough period of time that the system can no longer be classified as being in a state of thermodynamic equilibrium. And when that happens, the system is in a state of "metastable" equilibrium.

Consider, for instance, a quantity of vapor, at temperature and pressure such that under normal circumstances—i.e., in the presence of some "condensation nuclei"—it would be ready to condense into the liquid form. Assume that there are no condensation nuclei—e.g., tiny dust particles or groups of ions—present. Further, assume that spontaneous processes that cause statistical fluctuations are few and far between.

Now, very, very slowly start reducing the temperature. Because the system is also completely free of any mechanical disturbance that might motivate the supercooled vapor to condense, the usual change to liquid phase does not occur. Indeed, here the vapor is being "supercooled," and often this process can continue for a while until the supercooled vapor undergoes a finite sized, spontaneous condensation to the thermodynamically stable liquid phase. Note, however, that the finite sized spontaneous condensation of the supercooled vapor can occur only if the Gibbs potential decreases in the process.

Liquids can also be supercooled. Instead of condensing into the solid phase, the liquid state continues to exist even when the temperature has fallen below the normal condensation point. Again the supercooled liquid eventually converts to the then thermodynamically stable solid phase.

Much like supercooling, superheating can also occur. Beginning with a system in the liquid phase, and while the pressure is kept constant, if the temperature is slowly and gently increased beyond the point where the liquid normally vaporizes, one can achieve a metastable state of superheating. Again, any small fluctuation here transforms the system out of this metastable state into a state of equilibrium involving the vapor phase. Again, any spontaneous process that brings the superheated liquid back to the equilibrium state would decrease the specific Gibbs potential.

Similar thoughts were expressed earlier in reference to the "Maxwell prescription" that was proposed for the Van der Waals isotherm. (See (6.52)–(6.58).) Note that after an appropriate recommendation—see below[14]—physical use of the Gibbs free energy was made, for the first time, there.

10.17 The Clausius–Clapeyron Differential Equation

In the foregoing we have learned that for the simultaneously present two phases—say, phases A and B—in a single constituent, biphase thermodynamic system in equilibrium, the specific Gibbs free energy u_{A} is equal to u_{B}. Of course, this equality holds at temperature t and pressure p—both of which are set by contact with an outside reservoir. The question that we ask here is the following: If the simultaneously present two phases should stay in equilibrium even when the reservoirs that control the pressure and temperature are reset at the values $p + \mathrm{d}p$ and $t + \mathrm{d}t$, how would that affect the results? In particular, would the rate of change of the pressure with respect to the temperature be relevant?

To this purpose let us recall (10.32)—which describes the relevant increase in the specific Gibbs free energy—and set it to refer separately to each of the two phases. That is,

$$\mathrm{d}g_{\mathrm{A}} = v_{\mathrm{A}}\mathrm{d}p - s_{\mathrm{A}}\mathrm{d}t \quad \text{and} \quad \mathrm{d}g_{\mathrm{B}} = v_{\mathrm{B}}\mathrm{d}p - s_{\mathrm{B}}\mathrm{d}t\,, \tag{10.61}$$

where $(g_{\mathrm{A}}, s_{\mathrm{A}}, v_{\mathrm{A}})$ and $(g_{\mathrm{B}}, s_{\mathrm{B}}, v_{\mathrm{B}})$ are the specific—i.e., the values per mole—Gibbs potential, entropy, and volume of the phases A and B, respectively. Because, as explained earlier, in equilibrium the specific Gibbs potential of the (two) coexisting phases is necessarily equal, i.e., $\mathrm{d}g_{\mathrm{A}} = \mathrm{d}g_{\mathrm{B}}$, (10.61) leads to the result

$$v_{\mathrm{A}}\mathrm{d}p - s_{\mathrm{A}}\mathrm{d}t = v_{\mathrm{B}}\mathrm{d}p - s_{\mathrm{B}}\mathrm{d}t\,, \tag{10.62}$$

or equivalently,

$$\frac{s_{\mathrm{B}} - s_{\mathrm{A}}}{v_{\mathrm{B}} - v_{\mathrm{A}}} = \frac{\mathrm{d}p}{\mathrm{d}t} = \left[\frac{\mathrm{d}p}{\mathrm{d}t}\right]_{\mathrm{A}\to\mathrm{B}} = \frac{t(s_{\mathrm{B}} - s_{\mathrm{A}})}{t(v_{\mathrm{B}} - v_{\mathrm{A}})}\,. \tag{10.63}$$

Recall that under these circumstances, the temperature multiplied by the relevant increase in the entropy per mole, i.e., $t(s_{\mathrm{B}} - s_{\mathrm{A}})$, is equal to the corresponding increase in the enthalpy, i.e., $h_{\mathrm{B}} - h_{\mathrm{A}}$. And the latter—according to (10.49) and the description that follows it—is equal to the heat energy, $l_{\mathrm{A}\to\mathrm{B}}$, that is needed for the phase transformation from A to B. (Recall that $l_{\mathrm{A}\to\mathrm{B}}$ was called the latent heat of transformation from A to B.)

[14]It was recommended there that "Because the reader has not yet been introduced to the Gibbs free energy, upon first reading a beginner might postpone the reading of that section until after the Gibbs potential had been introduced and fully discussed. Equation (6.52) and the development of (6.55) will become more clear after the Gibbs free energy has been properly introduced and fully explained. See (10.31)–(10.44)."

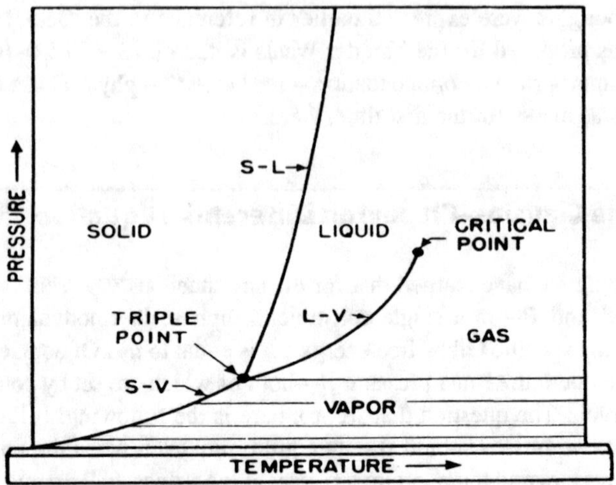

Fig. 10.1 Schematic plot of the Clausius–Clapeyron equations and the phase boundaries for real substances except H_2O

In other words—i.e., in the form of an equation—the following holds:

$$t(s_B - s_A) = h_B - h_A = l_{A\to B} \,. \tag{10.64}$$

On combining (10.63) and (10.64), we are led to the following result for the slope $[\frac{dp}{dt}]_{A\to B}$ of the equilibrium curve that is obtained—in the p–t plane—between the phases A and B:

$$\left[\frac{dp}{dt}\right]_{A\to B} = \frac{l_{A\to B}}{t(v_B - v_A)} \,. \tag{10.65}$$

This is the Clausius–Clapeyron differential equation. It describes the phase boundary between phases A and B—or equivalently stated, the slope of the equilibrium curve between the A and the B phases—that are simultaneously present in a multiphase thermodynamic system in equilibrium (at some given temperature[15] t and pressure[16] p).

In Fig. 10.1, a schematic plot is drawn of the phase boundaries that normally obtain for the three different phase-duos. These are:

(α) A = liquid and B = gas,
(β) A = solid and B = vapor,
(γ) A = solid and B = liquid.

Note that, while the numerator in (10.65)—i.e., the latent heat of phase transformation $l_{A\to B}$—can have a strong dependence on the temperature, and some dependence on the pressure, it is almost always[17] positive for all of the three cases,

[15] The temperature t is maintained by contact with a heat energy reservoir.

[16] The pressure p is maintained by contact with a so-called volume reservoir.

[17] An exception, usually cited, is liquid helium at ≈ 0.3 K.

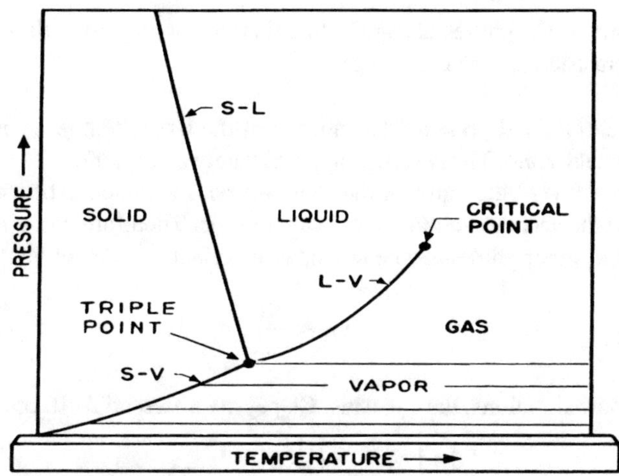

Fig. 10.2 Schematic plot of the Clausius–Clapeyron equations and the phase boundaries for H_2O

$\alpha \rightarrow \gamma$. Similarly, the denominator—i.e., $t(v_B - v_A)$—clearly depends on t, and usually also on p. Despite this fact, some general comments about the sign of the denominator can also be made.

With regard to the liquid–gas and solid–vapor phases, i.e., the cases referred to as α and β, we know that the specific volume of the gaseous—meaning the gas or the vapor—phases exceeds that of the liquid or the solid phases. Therefore, the difference $v_B - v_A > 0$. Also the latent heat, $l_{A\rightarrow B}$, for phase transformation is positive for both the "vapor pressure curve"—i.e., the case α—and the "sublimation pressure curve"—i.e., the case β. Thus, $[\frac{dp}{dt}]_{A\rightarrow B}$ is positive.

Next let us look at the solid–liquid phases, i.e., the case γ. For substances that shrink on freezing—which is usually the case—the specific volume v_A of the solid is less than v_B for the liquid. Thus $v_B - v_A > 0$. And because—as also stated earlier—$l_{A\rightarrow B}$ is positive for all the cases referred to in α, β, and γ, the slope $[\frac{dp}{dt}]_{A\rightarrow B}$ is also positive for the solid–liquid phases—see Fig. 10.1.

However, the opposite is the case for H_2O—i.e., pure water—which expands when it freezes to ordinary ice, namely Ice I—and $v_B - v_A < 0$. Therefore, for the solid–liquid phases, i.e., the case γ, the slope $[\frac{dp}{dt}]_{A\rightarrow B}$ is negative for H_2O (see Fig. 10.2).

Accordingly, while the equilibrium pressure-versus-temperature curves slope upwards to the "right" for all the three cases α, β, and γ—as in Fig. 10.1—for Ice I the solid–liquid curve—i.e., the case γ—slopes upwards to the "left"—as in Fig. 10.2.

10.17.1 Solution of Clausius–Clapeyron Differential Equation

Following S–S—that is, Sears and Salinger [1]—we describe below an approximate integration of the Clausius–Clapeyron differential equation that refers to the case where one of the phases is vapor. Note that this applies to both α and β. The approximations used are the following:

(a) All temperature and pressure dependence of the latent heat $l_{A\rightarrow B}$ is small and therefore negligible. Thus $l_{A\rightarrow B}$ is approximately a constant.
(b) Specific volume of the liquid or the solid phases is assumed to be small in comparison to the specific volume of the vapor phase. Therefore, $\upsilon_B - \upsilon_A \approx \upsilon_B$.
(c) Further, the vapor phase may be treated as a perfect gas. Accordingly,

$$\upsilon_B \approx \frac{Rt}{p}. \tag{10.66}$$

With these approximations, the Clausius–Clapeyron differential (10.65) becomes

$$\left[\frac{dp}{d}\right]_{A\rightarrow B} \approx \frac{l_{A\rightarrow B}}{t(\upsilon_B)} \approx \frac{l_{A\rightarrow B}}{t(\frac{Rt}{p})}. \tag{10.67}$$

This is a separable equation and can readily be integrated:

$$\int \frac{dp}{p} \approx \left(\frac{l_{A\rightarrow B}}{R}\right)\int \frac{dt}{t^2};$$
$$\ln(p) \approx \left(\frac{l_{A\rightarrow B}}{R}\right)\left[\frac{-1}{t}\right] + \ln(\text{constant}); \tag{10.68}$$
$$p \approx \text{constant} \times \exp\left(\frac{-l_{A\rightarrow B}}{Rt}\right).$$

Thus, instead of just the derivative of the phase-separation curve—for phases *A* and *B*—here we have an approximate representation of the phase separation curve itself for both of the phase-duos specified in α and β.

10.17.2 Triple- and Ice-Points of Water: Why the Separation?

Again, we follow S–S. This time to study how the Clausius–Clapeyron differential equation can be used to understand why the temperature of the ice-point—equal to 273.15 K—of pure water is slightly lower than that—i.e., 273.16 K—of the triple point. According to S–S, this fact: "... appears puzzling, since at both temperatures ice and water are in equilibrium."

At the triple point, which is at temperature $t_3 = 273.16$ K, the three phases—namely pure-water, i.e., the liquid, pure standard ice, i.e., the solid, and water vapor—are all in equilibrium. Here, the vapor pressure of water, the sublimation pressure of ice, and the system pressure are all equal to p_3 where

$$p_3 = 4.58 \text{ torr}, \qquad \text{atm} = 760 \text{ torr}.$$

(See the schematic plot of the triple point, and its neighborhood, in Fig. 10.2.)

On the other hand, at the ice point the following holds true: Pure ice, under total pressure $p_{\rm i} = 760$ torr of exactly one atm and at temperature $t_{\rm i}$, is in thermal equilibrium with air-saturated water.

Given that the pressures p_3 and $p_{\rm i}$ are noticeably different, it is no wonder that the triple point temperature t_3 should differ from the ice point temperature $t_{\rm i}$. Indeed, one can estimate the corresponding temperature difference by using the Clausius–Clapeyron differential equation.

To this end, let us first note that the expected difference, $t_{\rm i} - t_3$, in the temperature is very small. Accordingly, an acceptable approximation to the left-hand side of (10.65) is the following:

$$\left[\frac{dp}{d}\right]_{A\to B} = \frac{p_{\rm i} - p_3}{t_{\rm i} - t_3}\,. \tag{10.69}$$

Similarly, in the right-hand side of (10.65), to three significant figures, we may use the approximation:

$$t \approx \frac{t_{\rm i} + t_3}{2} \sim 273\ {\rm K}\,. \tag{10.70}$$

After first inverting (10.65) and then using the approximations indicated in (10.69) and (10.69), we get

$$\begin{aligned}\frac{t_{\rm i} - t_3}{p_{\rm i} - p_3} &\sim 273\ {\rm K} \times \frac{v_{\rm i} - v_3}{l_{\rm ice\to water}} \approx \frac{t_{\rm i} - t_3}{1.01 \times 10^5\ {\rm Nm}^{-2}}\\ &\approx 273\ {\rm K} \times \frac{(1.00 \times 10^{-3} - 1.09 \times 10^{-3})\ {\rm m^3\,kg^{-1}}}{3.34 \times 10^5\ {\rm J\,kg^{-1}}}\,,\end{aligned} \tag{10.71}$$

which leads to

$$\begin{aligned}t_{\rm i} - t_3 &\approx \left(1.01 \times 10^5\ {\rm N}\right) \times 273\ {\rm K} \times \frac{(1.00 \times 10^{-3} - 1.09 \times 10^{-3})\ {\rm Nm}}{3.34 \times 10^5\ {\rm J}}\\ &\approx -0.0075\ {\rm K}.\end{aligned} \tag{10.72}$$

Thus the difference in the pressures of the triple point and the equilibrium point of ice coupled with pure water lead to a small difference between the triple point temperature, t_3, and the equilibrium temperature, $t_{\rm i}$, of ice coupled with pure water. The triple point is very slightly warmer than the ice point.

In addition to the difference in pressures, the presence of dissolved air in water also affects the value of $t_3 - t_{\rm i}$. Dissolved air helps lower the temperature at which liquid air is in equilibrium with pure ice under atmospheric pressure by about 0.0023 K. Thus the triple point lies above the ice point by a total of approximately $0.0075 + 0.0023 = 0.0098$ K. Because by international agreement the triple point temperature is exactly 273.16 K, by this reckoning the ice point is very close to 273.15 K.

10.18 Gibbs Phase Rule

So far in this chapter single phased, uniconstituent thermodynamic systems—meaning systems that are in a single phase and composed of only one type of molecule—have been treated. Here we extend that analysis to multiphased, multiconstituent systems.

10.18.1 MultiPhase, Multiconstituent Systems

Consider a thermodynamic system with ℓ_p different phases, each composed of ℓ_c different chemical constituents. Let the ith constituent in the jth phase be characterized by specific chemical potential—i.e., chemical potential per mole—equal to $\mu_i^{(j)}$. Assume there are $n_i^{(j)}$ moles of the ith constituent in the jth phase where $i = 1, 2, \ldots, \ell_c$ and $j = 1, 2, \ldots, \ell_p$. In accord with the Euler equation (8.14), as noted in (8.13), for such multiphased, multiconstituent systems the total Gibbs free energy, G_{total}, should be a weighted sum of the specific chemical potentials over all of the ℓ_c constituents in each of the ℓ_p phases,

$$G_{\text{total}} = \sum_{i=1}^{i=\ell_c} \sum_{j=1}^{j=\ell_p} \mu_i^{(j)} n_i^{(j)} . \tag{10.73}$$

Note that of necessity the total number of molecules—or equivalently, total mole numbers—for each constituent is conserved (compare with (10.43)). That is, for any constituent i we have

$$\sum_{j=1}^{j=\ell_p} n_i^{(j)} = \text{const.}, \quad \text{therefore} \quad \sum_{j=1}^{j=\ell_p} dn_i^{(j)} = 0 . \tag{10.74}$$

Earlier in this chapter we learned two facts that are relevant here:

(1) The relative number of moles of different phases in the final equilibrium state of a multiphased thermodynamic system in thermodynamic equilibrium—at the equilibrium values of the entropy, volume, and at constant temperature and pressure—is such that the total Gibbs free energy, G_{total}, is the lowest possible. Accordingly, a system reaches thermodynamic equilibrium when its total Gibbs free energy is a minimum. Further, if the temperature and pressure are kept constant, the total Gibbs potential remains stationary with respect to infinitesimal changes in the equilibrium state. Because any such changes leave the specific chemical potentials unchanged, they result in infinitesimally changing only the mole numbers, $n_i^{(j)}$'s. And, of course, changes in the mole numbers happen subject to the requirement—described in (10.74)—that the total number of any given type of molecule is con-

served. In addition, the following must also hold:

$$\begin{aligned}
dG_{\text{total}} &= \sum_{i=1}^{i=\ell_c}\sum_{j=1}^{j=\ell_p} \mu_i^{(j)} dn_i^{(j)} \\
&= \sum_{i=1}^{i=\ell_c}\left[\mu_i^{(1)} dn_i^{(1)} + \mu_i^{(2)} dn_i^{(2)} + \mu_i^{(3)} dn_i^{(3)} + \cdots + \mu_i^{(\ell_p)} dn_i^{(\ell_p)}\right] \\
&= 0\,.
\end{aligned} \tag{10.75}$$

If every one of the $\ell_c \times \ell_p$ differentials $dn_i^{(j)}$ were independent—in the sense that their values could be chosen without any regard to the values of the others—then (10.75) would be satisfied only if all the chemical potentials $\mu_i^{(j)}$ were vanishing. Clearly, that solution would be unphysical.

But, as noted in (10.74), the sum of all the $\ell_c \times \ell_p$ differentials $dn_i^{(j)}$'s must be zero. Therefore, the differentials $dn_i^{(j)}$ are not all independent. As a result, there must exist a more physical solution than that where all chemical potentials are vanishing. Indeed, we have already reported such a solution for uniconstituent, dual phased systems in (10.44).

(2) Extending the previous result—i.e., that given in (10.44)—we assert that for any given chemical constituent i, the specific Gibbs free energy $\mu_i^{(j)}$ is the same for all the ℓ_p different phases j. In other words, we assert that the following is true:

$$\mu_i^{(1)} = \mu_i^{(2)} = \cdots = \mu_i^{(\ell_p)}, \quad i = 1, 2, \ldots, \ell_c\,. \tag{10.76}$$

For each of the ℓ_c constituents i, according to (10.76) above, there are a total of $\ell_p - 1$ different equalities that must be satisfied. As a result, for the whole thermodynamic system, this amounts to a total of $\ell_{\text{relationships}}$ different relationships where

$$\ell_{\text{relationships}} = \ell_c(\ell_p - 1)\,. \tag{10.77}$$

10.19 Phase Equilibrium Relationships

Remember that the changes, $dn_i^{(j)}$, in the mole numbers are not independent. Rather, they are constrained by the requirement imposed by (10.74).

Proof of (10.76) Write (10.74) and (10.75) in expanded form. Choose a constituent i—meaning, choose molecules of type i—from the total of ℓ_c different constituents. Because the total number of any given type of molecules is conserved, according to (10.74), a variation in the number of such molecules in phase 1 affects the number of the same type of molecules present in other $\ell_c - 1$ coexistent phases.

$$dn_i^{(1)} = -\left[dn_i^{(2)} + \cdots + dn_i^{(\ell_p)}\right]. \tag{10.78}$$

Next multiply both sides by $\mu_{\rm i}^{(1)}$ and insert the resulting expression into (10.75) on the right-hand side. In this fashion we readily get

$$\sum_{\rm i=1}^{\rm i=\ell_c}[(\mu_{\rm i}^{(2)}-\mu_{\rm i}^{(1)}){\rm d}n_{\rm i}^{(2)}+(\mu_{\rm i}^{(3)}-\mu_{\rm i}^{(1)}){\rm d}n_{\rm i}^{(3)}+\cdots+(\mu_{\rm i}^{(\ell_{\rm p})}-\mu_{\rm i}^{(1)}){\rm d}n_{\rm i}^{(\ell_{\rm p})}]=0\,. \tag{10.79}$$

Notice that the constraint for ${\rm d}n_{\rm i}^{(1)}$ prescribed by (10.78) has now been fully satisfied in (10.79). Therefore, ${\rm d}n_{\rm i}^{(j)}$'s for $j\geq 2$ can be handled independently. For instance, assume that ${\rm d}n_{\rm i}^{(2)}\neq 0$ but ${\rm d}n_{\rm i}^{(j)}=0$ for all $j\geq 3$. Then (10.79) becomes

$$\sum_{\rm i=1}^{\rm i=\ell_c}[\mu_{\rm i}^{(2)}-\mu_{\rm i}^{(1)}]{\rm d}n_{\rm i}^{(2)}=0\,. \tag{10.80}$$

The above equation can be satisfied only if

$$\mu_{\rm i}^{(1)}=\mu_{\rm i}^{(2)}\,. \tag{10.81}$$

Clearly, this argument can be repeated to obtain the equalities

$$\mu_{\rm i}^{(1)}=\mu_{\rm i}^{(3)}\,,\quad \ldots\,,\quad \mu_{\rm i}^{(1)}=\mu_{\rm i}^{(\ell_{\rm p})}\,. \tag{10.82}$$

As was also noted following (10.76), there are a total of $\ell_{\rm c}(\ell_{\rm p}-1)$ different relationships implicit in (10.81) and (10.82). These relationships are often called "phase equilibrium relationships." They generalize the previously derived result (see (10.44)) for a multiphased, single constituent thermodynamic system which was in thermal contact with a heat energy reservoir at a fixed temperature. The result there showed that chemical potentials in different coexistent phases are all equal.

10.19.1 Phase Rule

Clearly, if a multiphased thermodynamic system is not in equilibrium, then chemical potentials—for a given constituent in all the coexistent phases—are not necessarily equal. In an argument that was presented following (9.17), it was concluded that at constant temperature and pressure, molecules always flow out of a region of higher chemical potential into a region of lower chemical potential. When for any given constituent, differences in chemical potential exist between different phases, a motive force is created which causes the relevant molecules to escape from regions of higher chemical potential to those of lower chemical potential. This escaping process continues until the relevant chemical potential is the same in all phases.

Consider a homogeneous thermodynamic system in which molecules from $\ell_{\rm c}$ constituents are present in $\ell_{\rm p}$ different phases. Because by definition the sum of mole fractions of all the constituents in any given phase is equal to unity, only $\ell_{\rm c}-1$

mole fractions can be chosen independently in any given phase. Thus in ℓ_p different phases, the total number of possible choices of the mole fractions is equal to $\ell_p(\ell_c - 1)$. The temperature and the pressure are added to this mix making the total number of available variables, $\ell_{variables}$, equal to

$$\ell_{variables} = \ell_p(\ell_c - 1) + 2\,. \tag{10.83}$$

10.20 The Variance

The numerical difference between the total number of variables available, $\ell_{variables}$, and the number of relationships that need to be satisfied, $\ell_{relationships}$, is often called the "variance," $\ell_{varianc}$. It is equal to the number of variables that can arbitrarily be chosen,

$$\begin{aligned}\ell_{variance} &= \ell_{variables} - \ell_{relationships}\\ &= \ell_p(\ell_c - 1) + 2 - \left[\ell_c(\ell_p - 1)\right]\\ &= \ell_c - \ell_p + 2\,.\end{aligned} \tag{10.84}$$

This is the "Gibbs phase rule" for the so-called PdV systems that do not have any non-PdV interaction.

10.20.1 Invariant Systems

When the variance is vanishing, i.e., $\ell_{variance} = 0$, the system is labeled "invariant." And if there are no non-PdV forces, we have

$$\ell_p = \ell_c + 2\,. \tag{10.85}$$

Of course, being invariant does not guarantee that all the relationships—i.e., the $\ell_c(\ell_p - 1)$ different equations—can be solved and all the $\ell_p(\ell_c - 1) + 2$ variables—that is, the relative composition of all the phases as well as pressure and temperature—determined. What it does say, however, is that no arbitrary assignment can be made to the result for any of the variables.

As an example of this feature, let us consider the triple point of water. Clearly, there is only one chemical constituent,[18] namely H_2O molecules. So ℓ_c is equal to 1. Also, in equilibrium, the system has three phases: liquid, gas, and solid. Therefore, $\ell_p = 3$, $\ell_{variable} = \ell_p(\ell_c - 1) + 2 = 2$, $\ell_{relationships} = \ell_c(\ell_p - 1) = 2$, and $\ell_{variance} = \ell_{variables} - \ell_{relationships} = 0$. Consequently, the result that is obtained for the available two variables—meaning the pressure and temperature—is what it is and neither of them can be assigned an arbitrary value. Therefore, at the triple point, both the pressure and temperature have fixed values.

[18] Therefore, the relative composition in all the phases is exactly known. It is exactly one!

10.20.2 Monovariant Systems

Next consider two coexistent phases, liquid water and its vapor. Again, there is only a single chemical constituent, i.e., the H_2O molecules. Hence $\ell_c = 1$. But now there are two phases, thus $\ell_p = 2$. As a result, the number of variables is equal to two, i.e., $\ell_{\text{variable}} = 2$. However, there is only a single relationship, meaning $\ell_{\text{relationships}} = 1$. Clearly, the relationship here must refer to the equality of the two chemical potentials, i.e., $\mu_{\text{liquid}} = \mu_{\text{vapor}}$.

In addition, now the variance is unity, meaning $\ell_{\text{variance}} = 2 - 1 = 1$. Of the total of two available variables—clearly, the temperature T and pressure P—one can be assigned an arbitrary value. Everyday experience confirms this fact. Water can be heated to any desired temperature[19] and the vapor-pressure is determined accordingly. If the vapor pressure is increased beyond its appropriate value, the additional vapor will condense into liquid form, as water. On the other hand, if the vapor pressure is lower than its appropriate value for the temperature, more liquid will evaporate until the pressure regains its requisite value. Or indeed, while we do not often witness this in our daily lives, the vapor pressure can be set at any desired value, and that will determine the relevant, appropriate value of the system temperature.

10.21 Phase Rule for Systems with Chemical Reactions

In the above, all ℓ_c constituents were assumed to mix without any inter-reaction. If this should not be the case, and there should be a total of r different, independent, reversible chemical reactions that take place after the ℓ_c constituents have been placed within the system, then the total number of independent equations would be r more than was previously the case. In other words, now $\ell_{\text{relationships}} = \ell_c(\ell_p - 1) + r$. However, the number of variables is still the same, i.e., $\ell_{\text{variables}} = \ell_p(\ell_c - 1) + 2$. Therefore,

$$\begin{aligned} \ell_{\text{variance}} &= \ell_{\text{variables}} - \ell_{\text{relationships}} \\ &= \ell_p(\ell_c - 1) + 2 - \left[\ell_c(\ell_p - 1) + r\right] \\ &= \ell_c - \ell_p + 2 - r\,. \end{aligned} \qquad (10.86)$$

References

1. F.W. Sears, G.L. Salinger, *Thermodynamics, Kinetic Theory, and Statistical Thermodynamics*, 3rd edn. (Addison-Wesley, Reading, 1975)

[19] Physics for treating the boiling of water has not been included here. So this statement is valid only as long as the temperature is above the freezing point but below the boiling point of water.

Statistical Thermodynamics, the Third Law 11

So far, no effort has been made to carry out exact numerical calculation of state functions such as the entropy and internal energy.[1] Indeed, analyses of thermodynamic state functions have focused mainly on their rates of change and interrelationships. This has happened for a reason. Without considerable help from experimental observation, thermodynamics itself does not have any convenient method to precisely calculate the entropy or even the internal energy. Often, therefore, the good offices of statistical mechanics are needed for doing the same.

Using statistical mechanics is not unlike working with automobiles. There are important issues that relate to the physics and engineering of automobiles. While these issues are fundamental for, and integral to, achieving a full, proper, and complete understanding of automobiles, most of us get by without knowing anything about them. For an automobile to be any use to us, we must learn to drive it—and hopefully do that safely and well.

With a view to describing how statistical thermodynamics calculations are carried out, in the following we give first a brief and elementary description of the "standard route" that is often traversed for the calculation of state functions for a classical thermodynamic system which is in touch with a heat energy reservoir at some temperature T and where the number of particles is constant. Much as is the case for an automobile, there are important theoretical arguments in support of how and why the standard procedure leads us to where we want to go. But discussing them here would take us far afield. Therefore, we opt instead to learn only to drive! To carry out a statistical mechanical calculation, one needs to decide on the type of ensemble to use. Next one works out the relevant partition function. In Sects. 11.1 and 11.2 we provide a partition function for use in a canonical ensemble. Also reference is made to Boltzmann–Maxwell–Gibbs (BMG) ideas and the fact that they relate to Helmholtz free energy. The thermodynamic system being analyzed here is a classical monatomic perfect gas. The system partition function is described in Sect. 11.2.1. Using the statistical mechanical procedure, thermodynamics of mixed

[1]Except, of course, for the one attempt that was made by making rudimentary use of statistical mechanics in the calculation of the internal energy u—see (2.13)–(2.27). But the calculation of the entropy, and some other thermodynamic potentials, requires a more serious effort.

R. Tahir-Kheli, *General and Statistical Thermodynamics*,
https://doi.org/10.1007/978-3-030-20700-7_11

perfect gases in a variety of physical situations is studied in Sects. 11.2.2–11.4. Thermodynamic potentials are studied in Sect. 11.4.1 and the so-called Gibbs paradox in Sect. 11.4.2. Perfect gas of classical diatoms is analyzed in Sect. 11.5. Noninteracting free diatoms are treated in Sect. 11.5.1. Experimental observation of specific heat is noted in Sect. 11.5.2. Motion of center of mass, both around and about, is studied in Sects. 11.5.3 and 11.5.4. Several issues related to diatoms are mooted in Sects. 11.6–11.8.1. Thermodynamics of harmonic and anharmonic simple oscillators, classical dipole pairs, and Langevin paramagnets is discussed in Sects. 11.9–11.11. Additionally for the Langevin paramagnets, the classical picture, statistical average, the susceptibility, and behavior at high and low temperatures are looked at in Sects. 11.11.1–11.11.4. Extremely relativistic monatomic ideal gas is treated in Sect. 11.12. Relevant Hamiltonian, including that for gases with interaction, is in Sect. 11.13. Mayers cluster expansion, dilute classical gas with hard core interaction, Lennard-Jones potential, and the case of classical gas with interparticle attractive and repulsive interaction are studied in Sects. 11.13–11.16.2.

Next a study of quantum systems is begun: first, we make a cursory remark and introduce the relevant canonical partition function in Sects. 11.17 and 11.18. Various aspects of quantum particle systems are mentioned in Sects. 11.19–11.22. Also treated are quasiclassical quantum systems. Quasiclassical quantum—sometimes called classical coquantum—and quantum statistics are compared and contrasted. Rotational motion of heteronuclear diatoms is covered in Sect. 11.23, the quantum partition function and its analytical treatment are in Sects. 11.24 and 11.25. Thermodynamic potentials at low and high temperature are covered in Sects. 11.26 and 11.26.1. Sections 11.27, 11.28, and 11.27.1 respectively deal with homonuclear diatoms with rotational, vibrational motion, and rotational motion at very high temperature.

Langevin paramagnets are described both in quantum and classical statistical picture in Sects. 11.30 and 11.34.1. Helmholtz potential, partition function, and system entropy at high and low temperature are treated in Sects. 11.31–11.32.1. Internal energy at low and high temperature is described in Sects. 11.33 and 11.33.1. Specific heat at high and low temperature is studied in Sects. 11.33.2 and 11.34. As noted in Sect. 11.34.2, the attainment of very low temperatures usually requires the use of adiabatic demagnetization. Sections 11.34.3 and 11.35 deal with Nernst's heat theorem, the unattainability of absolute zero temperature, and the third law of thermodynamics. The concept of negative temperature is mooted in Sect. 11.36. Grand canonical ensembles play an important role in thermodynamic theory by helping with the calculation of the system partition function (see Sects. 11.36.1 and 11.37). Studied next in Sects. 11.38–11.54 are the multivarious aspects of quantum states such as Fermi–Dirac and Bose–Einstein. Black body radiation, and the fact that its chemical potential is zero, is described in Sects. 11.55 and 11.55.1. Phonons in a continuum or in lattices are explained in Sects. 11.56 and 11.56.1, respectively. We conclude with a discussion of the Debye temperature. A table of Debye temperatures and thermodynamic potentials is also provided in the concluding Sect. 11.57.

11.1 Boltzmann–Maxwell–Gibbs Ideas and Helmholtz Free Energy

The idea of calculating a partition function for classical systems is mooted in (b) that is given below.

Assume the temperature T of the given thermodynamic system, with a specified number of particles, is maintained at a fixed value by contact with a heat energy reservoir.

(a) Find a functional form for the system Hamiltonian. The Hamiltonian, $\mathcal{H} = \mathcal{H}(Q, p)$, is the sum of kinetic and potential energies of the N particles that constitute the thermodynamic system. Variables that are needed for specifying the Hamiltonian can often be expressed in terms of the number N and mutually conjugate variables, say Q and p. For instance, Q may represent the N δ-dimensional position vectors for the N particles, or equivalently, their $N\delta$ position variables $= (q_1, q_2, \ldots, q_{N\delta})$. Similarly, p may stand for the $N\delta$-dimensional momentum vectors for the N particles, or equivalently, their $N\delta$ momentum variables $= (p_1, p_2, \ldots, p_{N\delta})$. Here δ stands for the spatial dimensionality of the system, and if all of the N particles are identical, each will be assumed to have mass m.

(b) Calculate the so-called partition function, $\Xi(N, V, T)$, that is defined as follows:

$$\Xi(N, V, T) = (N!)^{-\xi} h^{-N\delta} \int_{-\infty}^{\infty} \cdots \int_{-\infty}^{\infty} \mathrm{d}Q \cdot \mathrm{d}p \cdot \exp\left[-\beta\mathcal{H}(Q, p)\right] . \quad (11.1)$$

The notation used here[2]—see also (2.13)–(2.27)—is the following: N is the total number of particles, each of mass m, V the volume, and T the "statistical-mechanical" temperature—called the Kelvin temperature (usually labeled as T_K but occasionally also as T_k). If all the N particles[3] are "indistinguishable," which is often the case, then $\xi = 1$; otherwise—that is, if they all are distinguishable—$\xi = 0$. Additionally, the following notation is used:

$$\begin{aligned} \mathrm{d}Q &= \mathrm{d}q_1 \mathrm{d}q_2 \cdots \mathrm{d}q_{N\delta} , \\ \mathrm{d}p &= \mathrm{d}p_1 \mathrm{d}p_2 \cdots \mathrm{d}p_{N\delta} , \\ \beta &= \frac{1}{k_B T} = \frac{N_A}{RT} = \frac{N}{nRT} , \end{aligned} \quad (11.2)$$

n and N_A refer to the number of moles and the Avogadro's number, respectively; R is called the molar gas constant; k_B is the Boltzmann constant; h is the Planck constant. In addition, the velocity of light, c, in vacuum is a useful number to have:[4]

$$N_A = 6.02214179(30) \times 10^{23} \text{ mol}^{-1},$$

$$R = 8.3144\,72(15) \text{ J mol}^{-1} \text{ K}^{-1},$$

[2]Note that there are situations where the appropriate variables to use are different than Q and P. Such is often the case for systems involving angular motion. See (11.45)–(11.55) for examples.

[3]Be careful when dealing with mixtures; see, for example, (11.22) and compare with (11.29)–(11.33) and the argument following (11.29).

[4]Also, see (2.12)—and the statement that follows it—for details.

$$
\begin{aligned}
k_B &= 1.38065\,04(24) \times 10^{-23}\,\mathrm{J\,K^{-1}}, \\
h &= 6.62606896(33) \times 10^{-34}\,\mathrm{J\,s}, \\
c &= 2.99792458 \times 10^{8}\ \mathrm{m\,s^{-1}}\,.
\end{aligned}
\tag{11.3}
$$

In (11.1) the integrations over the Q variables are carried out over the maximum volume available to each of the N particles. The integration over the momentum variables is over the—in principle—infinite range available to the momenta from $-\infty$ to $+\infty$.

It is helpful to deal directly with the factorial of N that occurs in (11.1). Because N is a very large number, one may use the Stirling asymptotic expansion [1–8], namely

$$
N! = (2\pi N)^{1/2}\left(\frac{N}{e}\right)^{\mathrm{N}}\left[1 + \frac{1}{12N} + \frac{1}{288N^2} - O\left(\frac{1}{N^3}\right)\right]. \tag{11.4}
$$

This looks fierce! It turns out, however, that statistical mechanics generally works with the natural logarithm of $N!$ Therefore (11.4) can be simplified to

$$
\begin{aligned}
\ln(N!) &= \ln\left[(2\pi N)^{1/2}\left(\frac{N}{e}\right)^{N} O\left\{\exp\left(\frac{1}{N}\right)\right\}\right] \\
&= \frac{1}{2}\ln(2\pi N) + \ln\left(\frac{N}{e}\right)^{N} + O\left(\frac{1}{N}\right) \\
&= \left\{\ln\left(\frac{N}{e}\right)^{N}\right\}\left[1 + O\left(\frac{(1/2)\ln(2\pi N)}{N\ln(N/e)}\right)\right] \\
&= \ln\left(\frac{N}{e}\right)^{N}\left[1 + O\left(\frac{1}{N}\right)\right] \approx \ln\left(\frac{N}{e}\right)^{N}.
\end{aligned}
\tag{11.5}
$$

Thus, to the leading order, we can use the approximation

$$
N! \approx \left(\frac{N}{e}\right)^{N}. \tag{11.6}
$$

Because N, the number of particles, is $\gg 1$, the above equality is almost exact.

(c) Following Boltzmann–Maxwell's ideas—as extended by Gibbs and to be referred to as BMG—the partition function, $\Xi(N, V, T)$, determines the Helmholtz free energy, F. Once F is known, other thermodynamic potentials—as well as the pressure—are readily evaluated. For convenience these relationships are listed below:

$$
\begin{aligned}
F(N, V, T) &= -k_B T \ln \Xi(N, V, T)\,, \\
P(N, V, T) &= -\left(\frac{\partial F}{\partial V}\right)_{\mathrm{N,T}}, \\
S(N, V, T) &= -\left(\frac{\partial F}{\partial T}\right)_{\mathrm{N,V}},
\end{aligned}
\tag{11.7}
$$

$$G(N, V, T) = N\left(\frac{\partial F}{\partial N}\right)_{\mathrm{T,V}} = n\mu = F + PV\,,$$

$$U(N, V, T) = -\left(\frac{\partial \ln \Xi(N, V, T)}{\partial \beta}\right)_{\mathrm{N,V}} = F + TS\,,$$

$$H(N, V, T) = U + PV\,.$$

We have used the standard notation here. That is, F is the Helmholtz free energy; P is the pressure; S is the entropy; G is the Gibbs potential; μ is the chemical potential per mole; n is the number of moles; U is the internal energy; and H is the enthalpy.

11.2 Noninteracting Classical Systems: Monatomic Perfect Gas in Three Dimensions

The Hamiltonian $\mathcal{H}(Q, p)$ for a perfect gas is assumed not to include the effects of any external forces. Therefore, (11.1) is isotropic—which implies that both the Hamiltonian and all its physical consequences are the same in all directions. Also, because interatomic coupling is completely absent, and there are no single-body potentials, the system Hamiltonian has no potential energy terms—which would normally be position-dependent. Therefore, $\mathcal{H}(Q, p)$ is independent of the position vectors Q. Indeed, although an atom has finite mass, here we shall assume that it is so tiny that it does not rotate around its own center. With these assumptions, the Hamiltonian is just the total translational kinetic energy of the N atoms. Accordingly, the Hamiltonian depends only on the sum of the squares of the $3N$ components $(p_1, p_2, \ldots, p_{3\mathrm{N}})$ of the N, three-dimensional, momentum vectors,

$$\mathcal{H}(Q, p) = \mathcal{H} = \sum_{\mathrm{i}=1}^{3\mathrm{N}} \frac{p_{\mathrm{i}}^2}{2m}\,. \tag{11.8}$$

Recall that all atoms have the same mass m.

11.2.1 Partition Function: Classical Monatomic Perfect Gas in Three Dimensions

We are now in a position to calculate the thermodynamics of N noninteracting classical single-atoms in three dimensions. While the subject of indistinguishability will be treated later when we get to study quantum manybody systems, we assume that these atoms are "indistinguishable for the present purposes." In the partition function

$$\Xi(N, V, T) = (N!)^{-1} h^{-3\mathrm{N}} \int_Q \mathrm{d}Q \int_{-\infty}^{\infty} \ldots \int_{-\infty}^{\infty} \mathrm{d}p \exp[-\beta\mathcal{H}]\,, \tag{11.9}$$

the integral over Q is trivial because (here) $\exp[-\beta\mathcal{H}]$ does not depend on Q. Also each of the N atoms contributes a factor V that is equal to the maximum volume available to it. Therefore

$$\int_Q \mathrm{d}Q = \prod_{\mathrm{i=1}}^{\mathrm{i=N}} \left[\int_{q_{\mathrm{i,x}}} \int_{q_{\mathrm{i,y}}} \int_{q_{\mathrm{i,z}}} \mathrm{d}q_{\mathrm{i,x}} \mathrm{d}q_{\mathrm{i,y}} \mathrm{d}q_{\mathrm{i,z}} \right] = V^N , \tag{11.10}$$

and as a result, (11.9) becomes

$$\begin{aligned} \Xi(N,V,T) &= (N!)^{-1} h^{-3N} \int_Q \mathrm{d}Q \int_{-\infty}^{\infty} \cdots \int_{-\infty}^{\infty} \mathrm{d}p \exp[-\beta\mathcal{H}] \\ &= \frac{1}{N!}\left(\frac{V}{h^3}\right)^N \int_{-\infty}^{\infty} \cdots \int_{-\infty}^{\infty} \mathrm{d}p_1 \ldots \mathrm{d}p_{3\mathrm{N}} \exp\left(-\beta \sum_{\mathrm{i=1}}^{3\mathrm{N}} \frac{p_{\mathrm{i}}^2}{2m}\right). \end{aligned} \tag{11.11}$$

The remaining integrals in (11.11) are of a standard form and are worked out in detail in (B.1)–(B.5). Therefore, only a brief description is provided below:

$$\begin{aligned} &\int_{-\infty}^{\infty} \cdots \int_{-\infty}^{\infty} \exp\left(\frac{-\beta \sum_{\mathrm{i=1}}^{3\mathrm{N}} p_{\mathrm{i}}^2}{2m}\right) \mathrm{d}p_1 \mathrm{d}p_2 \cdots \mathrm{d}p_{3\mathrm{N}} \\ &= \prod_{\mathrm{i=1}}^{3\mathrm{N}} \left[\int_{-\infty}^{\infty} \exp\left(\frac{-\beta p_{\mathrm{i}}^2}{2m}\right) \mathrm{d}p_{\mathrm{i}} \right] = \prod_{\mathrm{i=1}}^{3\mathrm{N}} \left(\frac{2m\pi}{\beta}\right)^{\frac{1}{2}} = \left(\frac{2m\pi}{\beta}\right)^{\frac{3N}{2}} . \end{aligned} \tag{11.12}$$

Inserting the result of (11.12) into (11.11) leads to the partition function for the perfect gas:

$$\begin{aligned} \Xi(N,V,T) &= \frac{1}{N!}\left(\frac{V}{h^3}\right)^N \left(\frac{2m\pi}{\beta}\right)^{\frac{3N}{2}} \\ &\approx \left(\frac{eV}{Nh^3}\right)^N (2\pi m k_{\mathrm{B}} T)^{\frac{3\mathrm{N}}{2}} = \Xi(N,V,T)_{\mathrm{IG}} . \end{aligned} \tag{11.13}$$

Because the partition function for the perfect gas—sometimes called the ideal gas—will be needed again later, it is helpful to introduce a convenient notation, e.g., $\Xi(N,V,T)_{\mathrm{IG}}$, for it. Note, the subscript IG stands for "ideal gas."

11.2.2 Monatomic Perfect Gas: Thermodynamic Potentials

Once the partition function, $\Xi(N,V,T;m)_{\mathrm{IG}}$, has been calculated, then with the use of (11.7) the Helmholtz free energy,

$$F_{\mathrm{IG}}(N,V,T) = -Nk_{\mathrm{B}}T \ln\left[\left(\frac{eV}{Nh^3}\right)(2\pi m k_{\mathrm{B}} T)^{\frac{3}{2}}\right], \tag{11.14}$$

the pressure, P_{IG}, and thermodynamic potentials—such as the entropy S_{IG}, Gibbs potential G_{IG}, internal energy U_{IG}, and enthalpy H_{IG}—are readily found. Note that the suffix IG is specific to the monatomic ideal—i.e., perfect—gas.

Using (11.7) and (11.14), we find:

$$P_{IG}(N,V,T) = -\left(\frac{\partial F_{IG}}{\partial V}\right)_{N,T} = \frac{Nk_BT}{V}, \tag{11.15}$$

$$\begin{aligned} S_{IG}(N,V,T) &= -\left(\frac{\partial F_{IG}}{\partial T}\right)_{N,V} \\ &= Nk_B \ln\left(\frac{V}{N}\right) + \frac{3}{2}Nk_B \ln\left(\frac{2\pi m k_B T}{h^2}\right) + \frac{5}{2}Nk_B, \end{aligned} \tag{11.16}$$

and
$$\begin{aligned} G_{I,G}(N,V,T) &= F_{I,G}(N,V,T) + PV = n\mu \\ &= -k_BTN \ln\left\{\left(\frac{V}{Nh^3}\right)(2\pi m k_B T)^{\frac{3}{2}}\right\}. \end{aligned} \tag{11.17}$$

Next, the internal energy can be evaluated from (11.14) and (11.16) as

$$\begin{aligned} U_{I,G}(N,V,T) &= F_{I,G}(N,V,T;m) + TS_{I,G}(N,V,T) \\ &= -\left(\frac{\partial(\ln \Xi)}{\partial \beta}\right)_{N,V} = \frac{3}{2}Nk_BT. \end{aligned} \tag{11.18}$$

Because in an ideal gas, at temperature T, each degree of freedom contributes to the free energy an amount equal to $\frac{k_BT}{2}$ per particle, the above proves that the number of degrees of freedom is equal to three. Also, since $PV = Nk_BT$ the enthalpy $H_{I,G}(N,V,T)$, being equal to $U_{I,G}(N,V,T) + PV$, is $\frac{5}{2}Nk_BT$.

11.3 Same Monatomic Perfect Gas at Different Pressure: Changes Due to Isothermal Mixing

Problem 11.1 (Isothermal mixing of ideal gas: different pressures but same gas and number of atoms)
An adiabatically isolated vessel has two chambers, to be referred to as 1 and 2. The chambers are separated by a massless, friction free, partition of zero volume. Both chambers contain the same ideal gas; are maintained at the same temperature T; and the gas in them initially contains the same number, N, of atoms. However, the volumes, V_1 and V_2, and therefore the initial pressures, P_1 and P_2, of the gas in the two chambers are different. Calculate the resulting change in thermodynamic potentials if the partition is removed and the gas in the two chambers is allowed to mix homogeneously?

Solution 11.1 We are told that two portions of the ideal gas—to be labeled 1 and 2—are placed in two different chambers of volume, V_1 and V_2. The initial pressure of the ideal gas in the two chambers is P_1 and P_2. The temperature of the gas,

T, and initially the number of particles, N, in each of these chambers is the same. Once the partition separating the chambers is removed, the gas from the two sides mixes together and achieves a total volume equal to $V_1 + V_2$.

The resulting change in thermodynamic potential was evaluated in an earlier chapter—see (5.33) and (5.35) where use was made of only the standard thermodynamic techniques. Furthermore, the earlier calculation for change was limited to only one thermodynamic potential, namely the entropy. In the following we use statistical mechanics instead. It enables us to determine change in all the relevant thermodynamic potentials. Let us begin first with the entropy.

11.3.1 Change in the Entropy Due to Mixing

The expression for the entropy, both before the mixing of portions 1 and 2—namely S_1 and S_2—and after their mixing—namely S_{1+2}—is available in (11.16) in the form $S_{IG}(N, V, T)$. All we need to do here is transcribe the appropriate variables. Thus we have:[5]

$$\begin{aligned} S_1 &= S_{IG}(N, V_1, T) \\ &= Nk_B \ln\left(\frac{V_1}{N}\right) + \frac{3}{2}Nk_B \ln\left(\frac{2\pi m k_B T}{h^2}\right) + \frac{5}{2}Nk_B\,, \\ S_2 &= S_{IG}(N, V_2, T) \\ &= Nk_B \ln\left(\frac{V_2}{N}\right) + \frac{3}{2}Nk_B \ln\left(\frac{2\pi m k_B T}{h^2}\right) + \frac{5}{2}Nk_B\,, \\ S_{1+2} &= S_{IG}(2N, V_1 + V_2, T) \\ &= 2Nk_B \ln\left(\frac{V_1 + V_2}{2N}\right) + 3Nk_B \ln\left(\frac{2\pi m k_B T}{h^2}\right) + 5Nk_B\,. \end{aligned} \tag{11.19}$$

Therefore,

$$\begin{aligned} S_{1+2} - S_1 - S_2 &= 2Nk_B \ln\left(\frac{V_1 + V_2}{2N}\right) - Nk_B \ln\left(\frac{V_1}{N}\right) - Nk_B \ln\left(\frac{V_2}{N}\right) \\ &= Nk_B \ln\left[\frac{(V_1 + V_2)^2}{4V_1 V_2}\right]. \end{aligned} \tag{11.20}$$

The result presented in (11.20) is identical to that obtained by employing standard thermodynamics procedures (see (5.33)). Using the simple equation of state for the ideal gas in its initial state, that is, $P_i V_i = Nk_B T$ for $i = 1$ and 2, (11.20) can also readily be expressed in terms of the pressures P_1 and P_2. That is,

$$S_{1+2} - S_1 - S_2 = Nk_B \ln\left[\frac{(P_1 + P_2)^2}{4P_1 P_2}\right]. \tag{11.21}$$

Again, the result given in (11.21) is identical—as it ought to be—to that found earlier—see (5.35).

[5]Note that after mixing there are $2N$ particles and their volume is $V_1 + V_2$.

Exercise 11.1 Using the above procedure, calculate the following differences: $F_{1+2} - F_1 - F_2$, in the Helmholtz potential; $G_{1+2} - G_1 - G_2$, in the Gibbs potential; $U_{1+2} - U_1 - U_2$, in the internal energy; and $H_{1+2} - H_1 - H_2$, in the enthalpy. The subscripts 1 and 2 refer to the portions of the gas labeled 1 and 2, respectively, while the subscript $1 + 2$ describes their mixture.

Problem 11.2 (Isothermal mixing of monatomic ideal gas: different pressures and number of atoms)
Two different vessels are placed together in an isolating chamber. Both vessels contain the same ideal gas, at the same temperature T, but their volumes, V_1 and V_2, are different. Initially these vessels have different numbers, N_1 and N_2, of particles and have different pressures, P_1 and P_2, respectively. Calculate any resulting changes in thermodynamic potentials if the vessels are connected and the gas in the two vessels is allowed to mix homogeneously.

Solution 11.2 By using standard thermodynamics techniques, the resulting change in the entropy was evaluated in an earlier chapter—see (5.43) and (5.44). However, by employing statistical mechanics, the entropy before the mixing, namely, S_1 and S_2, and after the mixing, namely S_{1+2}, can be determined directly from (11.16). As was done in the preceding subsection, all we need to do is transliterate (11.16) to the variables relevant to the present case. We get

$$\begin{aligned} S_1 &= S_{\mathrm{IG}}(N_1, V_1, T) \\ &= N_1 k_{\mathrm{B}} \ln\left(\frac{V_1}{N_1}\right) + \frac{3}{2} N_1 k_{\mathrm{B}} \ln\left(\frac{2\pi m k_{\mathrm{B}} T}{h^2}\right) + \frac{5}{2} N_1 k_{\mathrm{B}}\,, \\ S_2 &= S_{\mathrm{IG}}(N_2, V_2, T) \\ &= N_2 k_{\mathrm{B}} \ln\left(\frac{V_2}{N_2}\right) + \frac{3}{2} N_2 k_{\mathrm{B}} \ln\left(\frac{2\pi m k_{\mathrm{B}} T}{h^2}\right) + \frac{5}{2} N_2 k_{\mathrm{B}}\,, \\ S_{1+2} &= S_{\mathrm{IG}}(N_1 + N_2, V_1 + V_2, T) \\ &= (N_1 + N_2) k_{\mathrm{B}} \ln\left(\frac{V_1 + V_2}{N_1 + N_2}\right) + \frac{3}{2}(N_1 + N_2) k_{\mathrm{B}} \ln\left(\frac{2\pi m k_{\mathrm{B}} T}{h^2}\right) \\ &\quad + \frac{5}{2}(N_1 + N_2) k_{\mathrm{B}}\,. \end{aligned} \tag{11.22}$$

Note that after mixing there are $N_1 + N_2$ particles—that is, $N \to N_1 + N_2$—and their volume is $V_1 + V_2$—that is, $V \to V_1 + V_2$.

Therefore,

$$\begin{aligned} S_{1+2} - S_1 - S_2 = (N_1 + N_2) k_{\mathrm{B}} \ln\left(\frac{V_1 + V_2}{N_1 + N_2}\right) \\ - N_1 K_{\mathrm{B}} \ln\left(\frac{V_1}{N_1}\right) - N_2 K_{\mathrm{B}} \ln\left(\frac{V_2}{N_2}\right). \end{aligned} \tag{11.23}$$

This result is, of course, identical to that obtained by using standard thermodynamics procedures (see (5.43)). Utilizing the equation of state, $P_i V_i = N_i k_{\mathrm{B}} T$ for $i = 1$ or 2, we can rewrite the result given in (11.23) as follows:

$$S_{1+2} - S_1 - S_2 = k_B \ln\left[\frac{(\frac{N_1}{P_1} + \frac{N_2}{P_2})^{N_1+N_2}}{(N_1 + N_2)^{N_1+N_2}} P_1^{N_1} P_2^{N_2}\right]. \tag{11.24}$$

Again this result is the same as previously obtained for the mixing of identical perfect gases that are at the same temperature but have different pressure and number of atoms—see (5.44).

Exercise 11.2 Using the above procedure, calculate the following differences: $F_{1+2} - F_1 - F_2$, in the Helmholtz potential; $G_{1+2} - G_1 - G_2$, in the Gibbs potential; $U_{1+2} - U_1 - U_2$, in the internal energy; and $H_{1+2} - H_1 - H_2$, in the enthalpy. As before, the subscripts 1 and 2 refer to the portions of the gas labeled 1 and 2, respectively, while the subscript $1 + 2$ describes their mixture.

Problem 11.3 (Mixing of ideal gas: different temperature, pressure and number of atoms)
Two different vessels are placed together in an isolating chamber. While both vessels contain the same ideal gas, their volumes, V_1 and V_2, and their initial temperatures, T_1 and T_2, are different. Also, initially these vessels have different numbers, N_1 and N_2, of particles and have different pressures, P_1 and P_2, respectively. Calculate any resulting changes in thermodynamic potentials if the vessels are connected and the gas in the two vessels is allowed to mix homogeneously?

Solution 11.3 We need to remind ourselves of two things: First, before mixing, the equation of state of each of the two parts of the ideal gas is $P_i V_i = N_i k_B T_i$ where $i = 1$ or 2. Second, while before mixing, the temperatures of the two parts of the given ideal gas are T_1 and T_2, after mixing the joint system has only one temperature. We shall call it T_{final},

$$T_{\text{final}} = (N_1 T_1 + N_2 T_2)/(N_1 + N_2)\,. \tag{11.25}$$

The entropy before the mixing—namely S_1 and S_2—and after the mixing—namely S_{1+2}—can be determined[6] directly from (11.16) as

$$\begin{aligned}
S_1 &= S_{\text{IG}}(N_1, V_1, T_1)\\
&= N_1 k_B \ln\left(\frac{V_1}{N_1}\right) + \frac{3}{2} N_1 k_B \ln\left(\frac{2\pi m k_B T_1}{h^2}\right) + \frac{5}{2} N_1 k_B\,,\\
S_2 &= S_{\text{IG}}(N_2, V_2, T_2)\\
&= N_2 k_B \ln\left(\frac{V_2}{N_2}\right) + \frac{3}{2} N_2 k_B \ln\left(\frac{2\pi m k_B T_2}{h^2}\right) + \frac{5}{2} N_2 k_B\,,\\
S_{1+2} &= S_{\text{IG}}(N_1 + N_2, V_1 + V_2, T_{\text{final}})\\
&= (N_1 + N_2) k_B \ln\left(\frac{V_1 + V_2}{N_1 + N_2}\right) + \frac{3}{2}(N_1 + N_2) k_B \ln\left(\frac{2\pi m k_B T_{\text{final}}}{h^2}\right)\\
&\quad + \frac{5}{2}(N_1 + N_2) k_B\,.
\end{aligned} \tag{11.26}$$

[6] Note that after mixing there are $N_1 + N_2$ particles, their volume is $V_1 + V_2$ and their temperature is T_{final}.

Therefore we have

$$S_{1+2} - S_1 - S_2 = \frac{3}{2} k_B \ln\left[\left(\frac{T_{\text{final}}}{T_1}\right)^{N_1} \cdot \left(\frac{T_{\text{final}}}{T_2}\right)^{N_2}\right]$$
$$+ k_B \ln\left[\left(\frac{V_1 + V_2}{N_1 + N_2}\right)^{N_1+N_2} \left(\frac{N_1}{V_1}\right)^{N_1} \left(\frac{N_2}{V_2}\right)^{N_2}\right]. \quad (11.27)$$

Now using (5.53) for the portions 1 and 2,

$$V_1 = k_B \left(\frac{T_1 N_1}{P_1}\right) \quad \text{and} \quad V_2 = k_B \left(\frac{T_2 N_2}{P_2}\right),$$

the dependence upon V_1 and V_2 can be recast in terms of the pressures P_1 and P_2. That is,

$$S_{1+2} - S_1 - S_2 = \frac{3}{2} k_B \ln\left[\left(\frac{T_{\text{final}}}{T_1}\right)^{N_1} \cdot \left(\frac{T_{\text{final}}}{T_2}\right)^{N_2}\right]$$
$$+ k_B \ln\left[\left(\frac{\frac{T_1 N_1}{P_1} + \frac{T_2 N_2}{P_2}}{N_1 + N_2}\right)^{N_1+N_2} \left(\frac{P_1}{T_1}\right)^{N_1} \left(\frac{P_2}{T_2}\right)^{N_2}\right]. \quad (11.28)$$

Of course, when $T_2 = T_1 = T_{\text{final}}$, the result given in (11.28) is identical to that obtained in (5.58).

Exercise 11.3 Using the above procedure, calculate the following differences: $F_{1+2} - F_1 - F_2$, in the Helmholtz potential; $G_{1+2} - G_1 - G_2$, in the Gibbs potential; $U_{1+2} - U_1 - U_2$, in the internal energy; and $H_{1+2} - H_1 - H_2$, in the enthalpy. As in the above, the subscripts 1 and 2 refer to the portions of the gas labeled 1 and 2, respectively, while the subscript $1 + 2$ describes their mixture.

11.4 Different Monatomic Ideal Gases: Mixed at Same Temperature

Assume that two different monatomic ideal gases are both at the same temperature T. They consist of N_1 and N_2 atoms of mass[7] m_1 and m_2, and are placed in two different chambers of an adiabatically isolated vessel. The volume of these chambers is V_1 and V_2, respectively. When there is gas present, it can be exchanged via an interconnecting stop-cock of zero heat capacity. Initially, the gas in these chambers is at pressures P_1 and P_2, respectively.

Two things need to be noted.

First, before mixing, the equation of state of the ideal gas in either of the two chambers is $P_i V_i = N_i k_B T$ where $i = 1$ or 2.

[7]Examples of such mixing could be the mixing of helium and neon, or argon, etc., or indeed, $^3He_2$ and $^4He_2$.

Second, while before mixing, the different monatomic ideal gases occupy different volumes, V_1 and V_2, after mixing both the gases occupy the same joint volume $V_1 + V_2$.

We shall begin by calculating the various thermodynamic potentials of the gas initially present in each of the two vessels. To this end, we need their partition functions.

Clearly, either of the ideal gases in their single, unmixed state, can be represented by the partition function recorded earlier. All what needs to be done here is to make appropriate changes in the variables N, V, and m. Thus (11.13) specifies the following partition functions for the unmixed, monatomic but different ideal gases 1 and 2:

$$\begin{aligned} \Xi(N_1, V_1, T; m_1)_{\mathrm{IG}} &= \left(\frac{eV_1}{N_1 h^3}\right)^{N_1} (2\pi m_1 k_{\mathrm{B}} T)^{\frac{3N_1}{2}} , \\ \Xi(N_2, V_2, T; m_2)_{\mathrm{IG}} &= \left(\frac{eV_2}{N_2 h^3}\right)^{N_2} (2\pi m_2 k_{\mathrm{B}} T)^{\frac{3N_2}{2}} . \end{aligned} \tag{11.29}$$

Regarding the mixture of "different" ideal gases, the partition function is the product of their individual partition functions. It is, however, not the product of exactly the partition functions as given in (11.29) above. The reason is that any atom in either of the gases in their combined status has access to the full volume $V_1 + V_2$. Therefore, as anticipated from (11.13), the partition function of the mixture of the two different ideal gases is the following:

$$\begin{aligned} \Xi(\text{mixture}) &= \Xi\big(N_1, (V_1 + V_2), T; m_1\big)_{\mathrm{IG}} \times \Xi\big(N_2, (V_1 + V_2), T; m_2\big)_{\mathrm{IG}} \\ &= \left(\frac{e(V_1 + V_2)}{N_1 h^3}\right)^{N_1} (2\pi m_1 k_{\mathrm{B}} T)^{\frac{3N_1}{2}} \\ &\times \left(\frac{e(V_1 + V_2)}{N_2 h^3}\right)^{N_2} (2\pi m_2 k_{\mathrm{B}} T)^{\frac{3N_2}{2}} . \end{aligned} \tag{11.30}$$

11.4.1 Thermodynamic Potentials

As before, we first work out the Helmholtz potential energy:

$$\begin{aligned} F_1 &= F_{\mathrm{IG}}(N_1, V_1, T; m_1) = -k_{\mathrm{B}} T \ln \Xi(N_1, V_1, T; m_1)_{\mathrm{IG}} , \\ F_2 &= F_{\mathrm{IG}}(N_2, V_2, T; m2) = -k_{\mathrm{B}} T \ln \Xi(N_2, V_2, T; m_2)_{\mathrm{IG}} , \\ F_{1+2} &= -k_{\mathrm{B}} T \ln\big\{\Xi(\text{mixture})\big\} . \end{aligned} \tag{11.31}$$

The entropy is calculated next. We have $S = -(\frac{\partial F}{\partial T})_{\mathrm{N,V}}$. Accordingly,

$$\begin{aligned} S_1 &= S_{\mathrm{IG}}(N_1, V_1, T; m_1) , \\ S_2 &= S_{\mathrm{IG}}(N_2, V_2, T; m_2) , \\ S_{1+2} &= S_{\mathrm{IG}}(N_1, V_1 + V_2, T; m_1) + S_{\mathrm{IG}}(N_2, V_1 + V_2, T; m_2) . \end{aligned} \tag{11.32}$$

Therefore the increase in the entropy as a result of the mixing of two different ideal gases at the same temperature T is equal to

$$
\begin{aligned}
S_{1+2} - S_1 - S_2 &= N_1 k_B \ln\left(\frac{V_1 + V_2}{N_1}\right) + N_2 k_B \ln\left(\frac{V_1 + V_2}{N_2}\right) \\
&\quad - N_1 k_B \ln\left(\frac{V_1}{N_1}\right) - N_2 k_B \ln\left(\frac{V_2}{N_2}\right) \\
&= N_1 k_B \ln\left(\frac{V_1 + V_2}{V_1}\right) + N_2 k_B \ln\left(\frac{V_1 + V_2}{V_2}\right) \\
&= k_B \ln\left[\frac{(\frac{N_1}{P_1} + \frac{N_2}{P_2})^{N_1+N_2}}{N_1{}^{N_1} \cdot N_2{}^{N_2}} P_1^{N_1} P_2^{N_2}\right] .
\end{aligned}
\tag{11.33}
$$

11.4.2 Gibbs Paradox

G.W. Gibbs was arguably the most important contributor to what may be called early-modern thermodynamics. Not having the space—or the time—to dwell on the details of his multifaceted contributions, here we shall refer only to his work on thermodynamics of mixtures—and, in particular, to what is known as the Gibbs paradox. Actually, in the following we discuss only an intimately related but more accessible version of the Gibbs paradox.

The increase in the entropy caused by the mixing of two different gases[8] is recorded in (11.29)–(11.33). Although the masses of the atoms in the two different gases discussed there are defined to be m_1 and m_2 and are not necessarily equal, the final result for $S_{1+2} - S_1 - S_2$ does not show any presence of that information! Outwardly, therefore, this result could just as well have applied to the mixing of identical gases for which m_1 is necessarily equal to m_2. These thoughts are clearly disconcerting, and might possibly imply that our analysis given in the foregoing is not quite right!

Gibbs noticed this conundrum and managed to explain why both (11.24) and (11.33) should in fact be correct. He argued that even though the increase in the entropy for the mixing of identical gases—as recorded in (11.24)—and that for the similar mixing of different gases—as in (11.33)—are both seemingly independent of the two masses m_1 and m_2, they are in fact quite different in detail (for instance, compare the denominators within the logarithm). And the reason for the difference is the following:

The partition function, for a mixture of two identical gases with total number of particles equal to $N_1 + N_2$, should be divided by $(N_1 + N_2)!$ (Notice that we have followed the Gibbs suggestion in our work here—see, for instance, (11.22)–(11.24) and the statement that follows (11.22).)

In contrast, when the gases being mixed are different, irrespective of whether their atomic masses are the same, the partition functions of the type 1 and type 2 atoms should be dealt with separately. This way, the relevant dividing factor would

[8] See the comment following (11.29).

equal $N_1!$ for type 1 atoms and $N_2!$ for type 2 atoms. Of course, then the total partition function will be the product of these two partition functions. And, indeed, this is exactly what we have done here.

Exercise 11.4 Extending the above work and that presented in (11.29)–(11.33), calculate thermodynamic potentials for a mixture of different ideal gases with different initial volume, number of particles, initial temperature, and, of course, particles with different masses. However, rather than calculating the actual value of the final temperature—which in addition to the masses and initial temperatures of the two gases will also involve their specific heats—for simplicity assume that the final temperature of the mixture is defined just by the symbol T_{final}.

11.5 Perfect Gas of Classical Diatoms

It is easy to predict thermodynamics of a perfect gas of such "completely open, free diatomic molecules" by using the results obtained for a perfect gas of monatomic molecules. As mentioned in Chap. 2, a "monatomic molecule"—i.e., a single, zero-sized atom—by definition, cannot have any vibrational or rotational motion of its own. Therefore, it has only three translational degrees of freedom. A diatomic molecule with zero interatomic interaction—and no associated electronic or nuclear dynamics whatsoever—consists of two completely free monatoms with no inter-atom interaction. Each such molecule, of course, has six translational degrees of freedom.

11.5.1 Noninteracting Free Diatoms

Denote the time as "t" and the two noninteracting atoms, that make up a single diatom, by suffices "1" and "2." In three dimensions, at time t, the position of the ith atom—$i = 1$ or 2—can be described by its three Cartesian coordinates, $x_{\text{i}}(t)$, $y_{\text{i}}(t)$, and $z_{\text{i}}(t)$. Because an atom is assumed to be of size zero so that any notion of it rotating around its own center is meaningless, and interatomic interaction is also assumed to be absent, the Hamiltonian, $\mathcal{H}$(one free diatom), contains only the sum of the kinetic energy of the two single atoms. That is,

$$\begin{aligned}\mathcal{H}(\text{one free diatom}) &= \frac{m_1}{2}\left[\left(\frac{\partial x_1(t)}{\partial t}\right)^2 + \left(\frac{\partial y_1(t)}{\partial t}\right)^2 + \left(\frac{\partial z_1(t)}{\partial t}\right)^2\right] \\ &\quad + \frac{m_2}{2}\left[\left(\frac{\partial x_2(t)}{\partial t}\right)^2 + \left(\frac{\partial y_2(t)}{\partial t}\right)^2 + \left(\frac{\partial z_2(t)}{\partial t}\right)^2\right] \\ &= \sum_{i=1}^{2}\left(\frac{p_{\text{i,x}}^2(t)}{2m_i} + \frac{p_{\text{i,y}}^2(t)}{2m_{\text{i}}} + \frac{p_{\text{i,z}}^2(t)}{2m_{\text{i}}}\right), \qquad (11.34)\end{aligned}$$

where m_{i} denotes the mass and $p_{\text{i,x}}(t)$ the x component of the momentum vector of the ith atom at time t.

Thermodynamics of N_d Free-Diatoms

Given the fact that the Hamiltonian for one classical free-diatom is the same as that for two noninteracting free classical monatoms, the thermodynamics of N_d such free-diatoms is identical to that of N_d noninteracting free monatoms of mass m_1 plus another N_d noninteracting free monatoms of mass m_2. And, as per (11.14)–(11.18), this means that each of the free-diatoms contributes to the internal energy the amount provided by two single free monatoms, one of mass m_1 and the other of mass m_2. That is, at temperature T, the contribution to the internal energy per dipole is $= 2(\frac{3}{2})k_B T$. Accordingly, a single free-diatom has six degrees of freedom.

Also, using (11.14)–(11.18), at temperature T and volume V_d, we can predict that such a system would exert pressure

$$P = \frac{2N_d k_B T}{V_d}, \tag{11.35}$$

and would have internal energy

$$U(N_d \text{ noninteracting free-diatoms}) = N_d(3k_B T), \tag{11.36}$$

which would lead to specific heat equal to $3k_B$ per (diatomic) molecule. This is equivalent to c_v being equal to $3R$ per (diatomic) mole.

11.5.2 Experimental Observation of Specific Heat

But the specific heat, c_v, of real diatomic gases at low temperatures is observed to be $\frac{3R}{2}$ per (diatomic) mole, or equivalently, $\frac{3k_B}{2}$ per diatom. Experimental result, therefore, is only half of that predicted for a noninteracting free diatomic gas. Why is that the case? It is clearly not a result of diatoms dissociating into two monatoms because the dissociation energies are of the order of electron volts. And that translates into $\approx 10^4 \rightarrow 10^5$ K, namely temperatures that are much, much higher than where laboratory experiments are done.

In a real diatomic molecule, the two constituent atoms are coupled by attractive interaction causing the two to get bound together. And at low temperature a "bound" diatom acts much like a single particle of mass $m_1 + m_2$. Because the specific heat per free particle—of whatever mass—is $\frac{3k_B}{2}$, the experimental result is entirely as expected.

As the temperature rises, in addition to the translational motion, a bound diatom also begins to experience molecular rotation around its center of mass. Such rotation has two components that are mutually transverse and are transverse to the line joining the two atoms. Thus the rotational motion adds two degrees of freedom to the system thermodynamics.

With further rise in the temperature, other physical features of the bound diatom also become relevant. For example, depending on the molecular weight of the diatom, at appropriately high temperature a diatomic molecule may experience—in addition to translational motion of the center of mass and the two rotational motions around it—intradiatom vibration.[9]

To study these behaviors, in addition to doing purely algebraic manipulation, it is helpful also to be aware of the experimental results over a wide range of temperature. In particular, the fact that while at low temperatures the specific heat c_v of most light-diatomic gases starts off being equal to that of noninteracting monatomic gases—that is, it is equal to $\frac{3R}{2}$ per mole[10]—at higher temperatures the specific heat increases to become $\frac{5R}{2}$ per mole. And it generally stays at that level for a range of temperature—the range itself depending on the particularity of the diatom. But with even further increase in the temperature, the bond stiffness begins to weaken. The specific heat breaks through the rotational threshold of $\frac{5R}{2}$ per mole, and starts to rise toward a value equal to $\frac{7R}{2}$ per mole. Let us work through the algebraic manipulation first.

11.5.3 Center of Mass: Motion of and Around

The total kinetic energy of a system composed of $\tilde{N}$ particles has two parts that are mutually independent. The first part consists of the kinetic energy of the center of mass. This is equivalent to the motion of a single particle moving with the velocity of the center of mass. The mass of such a single particle is equal to that of all the $\tilde{N}$ particles of the system. The second part represents motion with respect to the center of mass of all the $\tilde{N}$ particles.

When measured from the center of mass, let the velocity of the ith atom be denoted as $\acute{V}_i$. Also, let the velocity of the center of mass itself—as measured from the origin of the Cartesian coordinates—be V_0. Then, clearly, the Cartesian representation of the velocity, V_i, of the ith particle is the vectorial sum of the velocity of the center of mass and the velocity measured with respect to the center of mass. That is,

$$V_i = V_0 + \acute{V}_i \,. \tag{11.37}$$

The kinetic energy of the $\tilde{N}$ particles of masses m_i, where $i = 1, \ldots, \tilde{N}$, is

$$\begin{aligned} \frac{1}{2}\sum_{i=1}^{\tilde{N}}(m_i * V_i^2) &= \frac{1}{2}\sum_{i=1}^{\tilde{N}} m_i * (V_0 + \acute{V}_i)^2 \\ &= \frac{V_0^2}{2}\sum_{i=1}^{\tilde{N}} m_i + \frac{1}{2}\sum_{i=1}^{\tilde{N}}(m_i * \acute{V}_i^2) + V_0 * \sum_{i=1}^{\tilde{N}}(m_i \acute{V}_i) \,. \end{aligned} \tag{11.38}$$

[9]Indeed, in principle, at even higher temperatures, electronic and nuclear excitations may also be obtained.

[10]See Fig. 11.1, for example, where the experimental values of $\frac{C_v}{R}$ for (diatomic) hydrogen are plotted as a function of the temperature.

The second row contains the kinetic energy of the center of mass and the kinetic energy of the $\tilde{N}$ particles due to their motion relative to the center of mass. The last term in the above equation—i.e., in (11.38)—can be demonstrated to be equal to zero. To this end, we proceed as follows. First look at

$$\sum_{\mathrm{i}=1}^{\tilde{\mathrm{N}}} m_{\mathrm{i}} \acute{V}_{\mathrm{i}} = \sum_{\mathrm{i}=1}^{\tilde{\mathrm{N}}} \left(m_{\mathrm{i}} \times \frac{\mathrm{d}\acute{r}_{\mathrm{i}}}{\mathrm{d}t} \right) = \frac{\mathrm{d}[\sum_{\mathrm{i}=1}^{\tilde{\mathrm{N}}} (m_{\mathrm{i}} \acute{r}_{\mathrm{i}})]}{\mathrm{d}t}, \tag{11.39}$$

where $\acute{r}_{\mathrm{i}}$ is the (time-dependent) position vector of the ith atom with respect to the center of mass. Next, notice that an expression like $\frac{\sum_{\mathrm{i}=1}^{\tilde{\mathrm{N}}} m_{\mathrm{i}} \acute{r}_{\mathrm{i}}}{\sum_{\mathrm{i}=1}^{\tilde{\mathrm{N}}} m_{\mathrm{i}}}$ would represent the position vector of the center of mass when measured from the center of mass itself. In other words, it would be vanishing by definition. Because the denominator of such an expression equals the total mass of the system, it is necessarily positive. Therefore, only the numerator would be vanishing. But the differential of the numerator of such an expression—being equivalent to differential of zero— represents the right-hand side of (11.39). Therefore, as initially stated, the left-hand side of (11.39)—or equivalently, the last term in (11.38)—is equal to zero. Therefore we have

$$\text{kinetic energy of } \tilde{N} \text{ particles} = \frac{V_0^2}{2} \sum_{\mathrm{i}=1}^{\tilde{\mathrm{N}}} m_{\mathrm{i}} + \frac{1}{2} \sum_{\mathrm{i}=1}^{\tilde{\mathrm{N}}} \left(m_{\mathrm{i}} * \acute{V}_{\mathrm{i}}^{2} \right). \tag{11.40}$$

Thus the total kinetic energy of the system indeed consists only of two separate parts, (a) and (b). Part (a) represents motion much like that of a single particle—whose mass is equal to the total mass of the system that is translating at the velocity, V_0, of the center of mass. We shall call it the "kinetic energy of the center of mass." And part (b) is the kinetic energy of motion (of all the given $\tilde{N}$ particles) with respect to the center of mass. We shall call it the motion "about the center of mass." The important thing to note is that the variables for these two motions—namely their respective velocities V_0 and $\acute{V}_i$ for $i = 1, \ldots, \tilde{N}$—are completely unrelated. Therefore whether one wishes to calculate the partition function, or determine the degrees of freedom, parts (a) and (b) can be treated independently.

11.5.4 Center of Mass: Translational Motion

The treatment of the translational motion of the center of mass of a given diatom is identical to that of the translational motion of the perfect gas monatoms discussed in the preceding section. Indeed, all the previous results can be transferred here with the following trivial caveat: the number N—which was the number of free monatoms—is now to be changed to N_{d}—which is the number of diatoms. Also, wherever the mass m of a single monatom occurs in (11.13)–(11.18), it has to be changed to $m_1 + m_2$ which is the mass of a single diatom. In particular, following (11.13), the partition function for the translational motion of the center of mass coordinates for N_d diatoms is

$$\begin{aligned}
&\Xi\,(\text{center of mass motion of } N_\mathrm{d} \text{ diatoms}) \\
&= \frac{1}{N_\mathrm{d}!}\left(\frac{V_\mathrm{d}}{h^3}\right)^{N_d}\left(\frac{2(m_1+m_2)\pi}{\beta}\right)^{\frac{3N_d}{2}} \\
&\approx \left(\frac{eV_\mathrm{d}}{N_\mathrm{d}h^3}\right)^{N_d}\left(\frac{2(m_1+m_2)\pi}{\beta}\right)^{\frac{3N_d}{2}}\,,
\end{aligned} \tag{11.41}$$

where V_d is the maximum volume in which a single diatom may roam around.

Next, by using (11.14)–(11.18), the contribution to the system thermodynamics from the center of mass motion is readily determined. The relevant Helmholtz potential is

$$F_{(\mathrm{c.of.m})} = -N_\mathrm{d}k_\mathrm{B}T\ln\left[\left(\frac{eV_\mathrm{d}}{N_\mathrm{d}h^3}\right)\left(\frac{2(m_1+m_2)\pi}{\beta}\right)^{\frac{3}{2}}\right]. \tag{11.42}$$

Therefore, the pressure, $P_{(\mathrm{c.of.m})}$, and thermodynamic potentials—such as the entropy, $S_{(\mathrm{c.of.m})}$, the Gibbs potential $G_{(\mathrm{c.of.m})}$, the internal energy $U_{(\mathrm{c.of.m})}$, and the enthalpy, $H_{(\mathrm{c.of.m})}$—are as follows:

$$\begin{aligned}
P_{(\mathrm{c.of.m})} &= \frac{N_\mathrm{d}k_\mathrm{B}T}{V_\mathrm{d}}\,, \\
S_{(\mathrm{c.of.m})} &= N_\mathrm{d}k_\mathrm{B}\ln\left(\frac{V_\mathrm{d}}{N_\mathrm{d}}\right) + \frac{3}{2}N_\mathrm{d}k_\mathrm{B}\ln\left(\frac{2(m_1+m_2)\pi}{\beta h^2}\right) + \frac{5}{2}N_\mathrm{d}k_\mathrm{B}\,, \\
G_{(\mathrm{c.of.m})} &= -k_BTN_\mathrm{d}\ln\left[\left(\frac{V_\mathrm{d}}{N_\mathrm{d}}\right)\left(\frac{2(m_1+m_2)\pi}{\beta h^2}\right)^{\frac{3}{2}}\right], \\
U_{(\mathrm{c.of.m})} &= -\left(\frac{\partial(\ln\Xi\,(\text{center of mass motion of } N_\mathrm{d} \text{ diatoms}))}{\partial\beta}\right)_{\mathrm{N_d,V_d}} \\
&= \left(\frac{3}{2}\right)N_\mathrm{d}k_\mathrm{B}T\,, \\
H_{(\mathrm{c.of.m})} &= U_{(\mathrm{c.of.m})} + P_{(\mathrm{c.of.m})}V_\mathrm{d} = \frac{5}{2}N_\mathrm{d}k_\mathrm{B}T\,.
\end{aligned} \tag{11.43}$$

According to the equipartition theorem, each degree of freedom contributes to the internal energy, U, an amount equal to $\frac{k_\mathrm{B}T}{2}$ per particle. Therefore the number of degrees of freedom for the center of mass translational motion is equal to three. Also consider motion about the center of mass.

11.6 Transformation to Spherical Coordinates: Classical Diatom with Stationary Center of Mass

Let its two monatoms—both of which are of infinitesimal size—be separated from each other by distance $|r_0|$. Using the notation $r_0 \equiv |r_0|$, $r_1 \equiv |r_1|$, $r_2 \equiv |r_2|$, the center of mass of the diatom is at distance r_i from the monatom of mass m_i so that

$r_1 + r_2 = r_0$ and $m_1 r_1 = m_2 r_2$. Accordingly,

$$r_1 = \frac{m_2 r_0}{m_1 + m_2}, \qquad r_2 = \frac{m_1 r_0}{m_1 + m_2}. \tag{11.44}$$

It is convenient to transform the Cartesian representation—with variables (x_i, y_i, z_i) that were used in (11.34)—to one using spherical coordinates whose variables are r_i, θ_i, and ϕ_i. That is,

$$\begin{aligned} x_i(t) &= r_i(t)\sin(\theta_i)\cos(\phi_i), \\ y_i(t) &= r_i(t)\sin(\theta_i)\sin(\phi_i), \\ z_i(t) &= r_i(t)\cos(\theta_i). \end{aligned} \tag{11.45}$$

For additional convenience, the origin of both the spherical and Cartesian coordinate systems is set at the position of the center of mass so that it remains stationary. The index "i" stands for one of the two monatoms—i.e., $i = 1$ or $i = 2$—that make up the diatom; $r_i(t)$ represents the distance of the ith monatom from the center of mass. Both θ_i and ϕ_i are also time-dependent. And, as usual, "t" represents the time. To avoid cluttering, in the above we have not explicitly displayed the time dependence of θ_i and ϕ_i. For the diatom described above, the angles θ_1 and θ_2—and similarly, ϕ_1 and ϕ_2—are related to each other. That is,

$$\theta_2 = \pi - \theta_1, \qquad \phi_2 = \phi_1 - \pi. \tag{11.46}$$

Making use of this identity helps transform the angles for atom 2, i.e., θ_2, ϕ_2, into angles for atom 1, i.e., θ_1 and ϕ_1. Because

$$\begin{aligned} \sin(\pi - \theta_1) &= \sin(\theta_1), \qquad & \cos(\pi - \theta_1) &= -\cos(\theta_1), \\ \sin(\phi_1 - \pi) &= -\sin(\phi_1), \qquad & \cos(\phi_1 - \pi) &= -\cos(\phi_1), \end{aligned} \tag{11.47}$$

(11.45) can be written as

$$\begin{aligned} x_1(t) &= r_1(t)\sin(\theta_1)\cos(\phi_1), \\ y_1(t) &= r_1(t)\sin(\theta_1)\sin(\phi_1), \\ z_1(t) &= r_1(t)\cos(\theta_1), \\ x_2(t) &= -r_2(t)\sin(\theta_1)\cos(\phi_1), \\ y_2(t) &= -r_2(t)\sin(\theta_1)\sin(\phi_1), \\ z_2(t) &= -r_2(t)\cos(\theta_1). \end{aligned} \tag{11.48}$$

In order to calculate the kinetic energy, we need the time derivative of the six terms given in (11.48) above. Now that all angles carry the suffix 1, no confusion is caused if in what follows we use the notation $\theta \equiv \theta_1$, $\phi \equiv \phi_1$. Further, to avoid writing something like $\frac{\partial a}{\partial t}$ many times, we shall use the simpler notation $a^{\cdot} \equiv \frac{\partial a}{\partial t}$,

$$
\begin{aligned}
x_1^{\cdot} &= r_1^{\cdot} \sin(\theta)\cos(\phi) + r_1\left[\theta^{\cdot} \cos(\theta)\cos(\phi) - \phi^{\cdot} \sin(\theta)\sin(\phi)\right], \\
y_1^{\cdot} &= r_1^{\cdot} \sin(\theta)\sin(\phi) + r_1\left[\theta^{\cdot} \cos(\theta)\sin(\phi) + \phi^{\cdot} \sin(\theta)\cos(\phi)\right], \\
z_1^{\cdot} &= r_1^{\cdot} \cos(\theta) - r_1\theta^{\cdot} \sin(\theta), \\
x_2^{\cdot} &= -r_2^{\cdot} \sin(\theta)\cos(\phi) - r_2\left[\theta^{\cdot} \cos(\theta)\cos(\phi) - \phi^{\cdot} \sin(\theta)\sin(\phi)\right], \\
y_2^{\cdot} &= -r_2^{\cdot} \sin(\theta)\sin(\phi) - r_2\left[\theta^{\cdot} \cos(\theta)\sin(\phi) + \phi^{\cdot} \sin(\theta)\cos(\phi)\right], \\
z_2^{\cdot} &= -r_2^{\cdot} \cos(\phi) + r_2\theta^{\cdot} \sin(\theta).
\end{aligned}
\tag{11.49}
$$

It should be noted that none of our diatoms is assumed to have any intra-diatom potential energy. Furthermore, the system of diatoms being treated here is also considered not to have any interdiatom coupling. This means that the total potential energy of the system is equal to zero and the Hamiltonian consists only of the sum of the kinetic energy of $N_{\rm d}$ diatoms. Accordingly, the Hamiltonian for one diatom[11] is[12]

$$
\begin{aligned}
\mathcal{H}_{\text{one diatom}} &= \frac{m_1}{2}\left[(x_1^{\cdot})^2 + (y_1^{\cdot})^2 + (z_1^{\cdot})^2\right] + \frac{m_2}{2}\left[(x_2^{\cdot})^2 + (y_2^{\cdot})^2 + (z_2^{\cdot})^2\right] \\
&= \frac{1}{2}\left[m_1(r_1^{\cdot})^2 + m_2(r_2^{\cdot})^2\right] + \frac{M}{2}\left[(\theta^{\cdot})^2 + (\phi^{\cdot})^2 \sin^2(\theta)\right].
\end{aligned}
\tag{11.53}
$$

11.7 Diatom with Stiff Bond: Rotational Kinetic Energy

At room temperature, in light diatomic ideal gases, the intradiatom bond is quite "stiff." This effectively stops all intradiatomic vibration at such ordinary temperatures. As a result, the distance separating the two monatoms in a given light diatom

[11] Remember that here we have assumed that the center of mass of the diatom is stationary. So the kinetic energy refers only to motion with respect to—that is, "to and from" and "around"—the center of mass.

[12] While the algebra needed for deriving the result given in (11.53) is trivial, it is a hassle doing it by hand. Therefore, we recommend using Mathematica, which of course is both easy to use and takes much less time. Also it should be mentioned that a more elegant, but more abstruse, alternative description of the above result is available in [9]. The authors show that setting the origin of the spherical coordinates at the "center of mass" of the diatom, the relevant transformation for an infinitesimal increase, ds, in the position $\boldsymbol{s}$ of a particle is as given below:

$$
\mathrm{d}s = \mathrm{d}r\boldsymbol{e}_{\rm r} + r\mathrm{d}\theta\boldsymbol{e}_\theta + r\sin(\theta)\mathrm{d}\phi\boldsymbol{e}_\phi. \tag{11.50}
$$

Here $\boldsymbol{e}_{\rm r}$, $\boldsymbol{e}_\theta$, and $\boldsymbol{e}_\phi$ are unit vectors along the direction of $\boldsymbol{r}$, and angles θ and ϕ, respectively. These vectors are orthogonal. Accordingly, the velocity vector $\boldsymbol{v}$ and its square are

$$
\begin{aligned}
\boldsymbol{v} &= \dot{r}\boldsymbol{e}_{\rm r} + r\dot{\theta}\boldsymbol{e}_\theta + r\sin(\theta)\dot{\phi}\boldsymbol{e}_\phi, \\
\boldsymbol{v}.\boldsymbol{v} &= v^2 = \dot{r}^2 + r^2\dot{\theta}^2 + r^2\sin(\theta)^2\dot{\phi}^2,
\end{aligned}
\tag{11.51}
$$

M is the moment of inertia of the diatom,

$$
M = m_1 r_1^2 + m_2 r_2^2. \tag{11.52}
$$

remains essentially constant—meaning it does not depend on time. So, for such a stiff-diatom,

$$r_{\rm i}(t) = r_{\rm i}\,, \qquad \dot{r}_{\rm i} = 0\,. \tag{11.54}$$

Using (11.54), we can write (11.53) as follows:

$$\mathcal{H}_{\text{rotation one stiff diatom}} = \frac{M}{2}\big[(\theta^{\cdot})^2 + (\phi^{\cdot})^2 \sin^2(\theta)\big]. \tag{11.55}$$

In order to find the contribution that the rotational motion of the stiff diatom makes to the partition function, we need first to find the momenta that are conjugate to the two rotational angles θ and ϕ. To this purpose, we need its Lagrangian, Γ. Now, Γ is equal to the difference between the kinetic and the potential energies. The given diatom has a stiff bond, its center of mass is stationary, and it does not have any intradiatom potential energy. Therefore, the Γ here is equal just to its rotational kinetic energy. That is,

$$\Gamma = \frac{1}{2}M\big[(\theta^{\cdot})^2 + (\phi^{\cdot})^2 \sin^2(\theta)\big], \tag{11.56}$$

where, as noted in (11.52), M is the moment of inertia. According to the theory of Lagrange, the momenta L_θ and L_ϕ that are conjugate to the angles θ and ϕ, are calculated as follows:

$$L_\theta = \left(\frac{\partial \Gamma}{\partial \theta^{\cdot}}\right)_{\theta,\phi,\phi^{\cdot}} = M\theta^{\cdot}\,, \qquad L_\phi = \left(\frac{\partial \Gamma}{\partial \phi^{\cdot}}\right)_{\theta,\phi,\theta^{\cdot}} = M\sin^2(\theta)\phi^{\cdot}\,. \tag{11.57}$$

Therefore

$$\Gamma = \frac{1}{2}M\left[\left(\frac{L_\theta}{M}\right)^2 + \left(\frac{L_\phi}{M\sin(\theta)}\right)^2\right] = \frac{L_\theta^2}{2M} + \frac{L_\phi^2}{2M\sin^2(\theta)}\,. \tag{11.58}$$

We can now set up the expression for rotational contribution to the partition function for one classical—homonuclear or heteronuclear—diatom with stiff-bond:

$$\begin{aligned}
&\Xi\,(\text{rotation one stiff diatom})\\
&= \left(\frac{1}{h^2}\right)\int_0^\pi d\theta \int_{-\infty}^{\infty} dL_\theta \int_{-\infty}^{\infty} dL_\phi \exp\left(\frac{-\Gamma}{k_{\rm B}T}\right)\int_0^{\phi_{\max}} d\phi\\
&= \left(\frac{\phi_{\max}}{h^2}\right)\int_0^\pi d\theta \int_{-\infty}^{\infty} dL_\theta \exp\left(\frac{-\frac{(L_\theta)^2}{2M}}{k_{\rm B}T}\right)\int_{-\infty}^{\infty} dL_\phi \exp\left(\frac{-\frac{(L_\phi)^2}{2M\sin^2(\theta)}}{k_{\rm B}T}\right)\\
&= \left(\frac{\phi_{\max}}{h^2}\right)\int_0^\pi d\theta \int_{-\infty}^{\infty} dL_\theta \exp\left(\frac{-\frac{(L_\theta)^2}{2M}}{k_{\rm B}T}\right)\big[\sin(\theta)\sqrt{2\pi M k_{\rm B}T}\big]\\
&= \left(\frac{2\pi\phi_{\max}}{h^2}Mk_{\rm B}T\right)\int_0^\pi d\theta\big[\sin(\theta)\big] = \frac{4\pi\phi_{\max}Mk_{\rm B}T}{h^2}\,,
\end{aligned} \tag{11.59}$$

where $\phi_{\max} = 2\pi$ for heteronuclear diatoms. For homonuclear diatoms, on the other hand, because the azimuthal angles ϕ and $\phi+\pi$ differ only in the interchange of two identical nuclei, they correspond to only one distinct state of the system. Therefore, for the calculation of the classical partition function for homonuclear diatoms, the relevant range for the angle ϕ is just $0 \to \pi$ rather than $0 \to 2\pi$. Accordingly, for homonuclear diatoms $\phi_{\max} = \pi$.

Because the interdiatom interaction is absent, each diatom operates independently. Therefore, the total partition function is the product of N_{d} partition functions, each referring to a single diatom.

Now the thermodynamics of rotational motion is readily evaluated. We have:

$$\begin{aligned}
\Xi\,(\text{rotation of } N_d \text{ stiff diatoms}) &= \left[\frac{4\pi\,\phi_{\max} M k_{\mathrm{B}} T}{h^2}\right]^{N_d}, \\
F_{\text{rotation}} &= -k_{\mathrm{B}} T \ln\left[\frac{4\pi\,\phi_{\max} M k_{\mathrm{B}} T}{h^2}\right]^{N_d}, \\
P_{\text{rotation}} &= 0\,, \\
S_{\text{rotation}} &= k_{\mathrm{B}} N_{\mathrm{d}} \ln\left[\frac{4\pi\,\phi_{\max} M k_{\mathrm{B}} T}{h^2}\right] + k_{\mathrm{B}} N_{\mathrm{d}}\,, \\
U_{\text{rotation}} &= H_{\text{rotation}} = N_{\mathrm{d}} k_{\mathrm{B}} T\,.
\end{aligned} \tag{11.60}$$

Clearly, the internal energy is the same for both homonuclear and heteronuclear classical diatoms. This, of course, is not strictly true for quantum diatoms except at very high temperature. We are now in a position to calculate the thermodynamics of N_{d} classical diatoms each with stiff intradiatom bond. Because the contribution to the partition function from different parts of the Hamiltonian is multiplicative, therefore the total value of the thermodynamic potentials is the sum of that obtained for the motion of the center of mass—that is, the result given in (11.43)—and the rotational motion of the diatoms—i.e., that given in (11.60). In particular, we have for the pressure, the entropy and the internal energy the result:

$$\begin{aligned}
&P(N_{\mathrm{d}} \text{ diatoms with stiff bond}) = P_{(\text{c.of.m})} + P_{\text{rotation}} = \frac{N_{\mathrm{d}} k_{\mathrm{B}} T}{V_{\mathrm{d}}}\,, \\
&S(N_{\mathrm{d}} \text{ diatoms with stiff bond}) \\
&\quad = S_{(\text{c.of.m})} + S_{\text{rotation}} \\
&\quad = N_{\mathrm{d}} k_{\mathrm{B}} \ln\Big[\Big(\frac{V_{\mathrm{d}}}{N_{\mathrm{d}} h^5}\Big) 4\pi\,\phi_{\max} M [2\pi(m_1+m_2)]^{\frac{3}{2}} (k_{\mathrm{B}} T)^{\frac{5}{2}}\Big] + \frac{7}{2} N_{\mathrm{d}} k_{\mathrm{B}}\,, \\
&U(N_{\mathrm{d}} \text{ diatoms with stiff bond}) = U_{(\text{c.of.m})} + U_{\text{rotation}} = \frac{5}{2} N_{\mathrm{d}} k_{\mathrm{B}} T\,.
\end{aligned} \tag{11.61}$$

This means that each of the N_{d} diatoms with a stiff bond contributes an amount equal to $\frac{5}{2} k_{\mathrm{B}} T$ to the internal energy. Therefore, for classical diatoms with stiff intradiatom bonds, the number of available degrees of freedom is equal to 5.

11.7.1 Diatoms with Free Bonds

If the intradiatom bond is completely free, then at all times (11.45) can be reversed back to (11.34). As a result, each diatom becomes identical to two free monatoms and the thermodynamics of N_d diatoms is identical to that of $2N_d$ free monatoms.

Remark 11.1 So far diatoms have been treated classically. Also, nowhere has it been necessary to introduce the fallacious notion, namely that the sixth degree of freedom consists of the self-rotation of the infinitesimally sized two atoms—each around its own center. However, in order fully to appreciate the temperature dependence of the system thermodynamics, it is helpful to also use some quantum mechanical ideas. (See, for instance, (11.144)–(11.151).) At very low temperature only the translational motion is excited. But soon thereafter, rotational motion begins getting tweaked. As a result the internal energy per diatom starts to increase from its initial value of $\frac{3}{2}k_B T$. And, at the temperature where the rotational motion has become fully available, but intradiatom bond still remains essentially stiff, the internal energy per diatom becomes $\frac{5}{2}k_B T$. In other words, a diatom with stiff intradiatom bond will, with appropriate rise in the temperature,[13] access up to 5 degrees of freedom. And when this process is completed, the specific heat c_v per mole of such diatoms would have risen to become $\frac{5}{2}R$. In a "real" diatomic gas—as distinct from an "ideal" diatomic gas—when the system temperature continues to rise beyond this point, the sixth degree of freedom—i.e., the vibrational—and the seventh degree of freedom[14] would also get excited. Note that the seventh degree is a theoretical construct. It arises because of the occurrence of the two-body, intradiatom, potential.

It is instructive to examine experimental results—over a wide range of temperature—for the specific heat of diatomic gases.[15] Here the relevant quantities are the characteristic temperatures for rotation and vibration of diatomic molecules.

For hydrogen (i.e., H_2) the "characteristic rotational" temperature—which is 85.5 K—is almost as low as the temperature of liquid air (i.e., $\approx$ 77 K). The rotational degrees of freedom become fully excited long before the "characteristic vibrational" temperature is reached. This is so because generally for the light diatomic molecules the characteristic vibrational temperature is much higher than the rotational temperature. Figure 11.1 indicates that in diatomic hydrogen, while the bond stiffness very slowly begins to relax as early as $\approx$ 750 K, it persists until a fantastically high temperature $\approx$ 6140 K.

[13]Accessing the five degrees of freedom at this stage is true both in the classical and quantum pictures.

[14]We have ignored electronic and nuclear excitations, and also rotation of single atoms around their center. These excitations may, in principle, occur and thereby add to the number of degrees of freedom. However, in practice, for small diatoms these excitations are often irrelevant because the temperatures needed for them are inordinately high.

[15]See the attached table for characteristic temperatures at which diatomic rotations and intradiatomic vibration in various diatoms begin to take effect. Note that the characteristic temperature for rotation is defined as the temperature where, in addition to the three translational degrees of freedom of the center of mass, two degrees of freedom for rotation also begin to be excited—and clearly where the bond is still effectively "stiff."

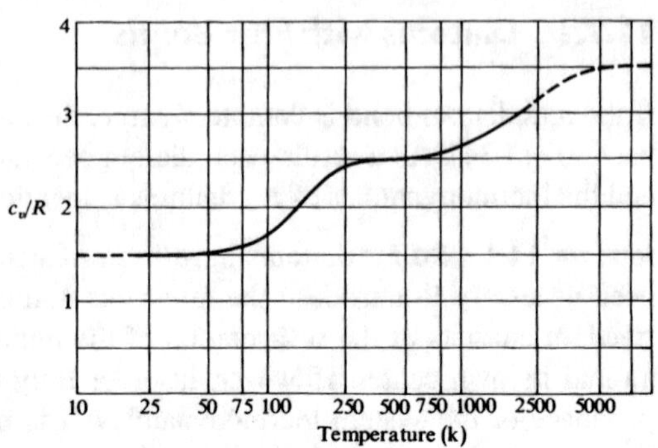

Fig. 11.1 Schematic plot of $\frac{c_V}{R}$ versus temperature for H_2. (Compare [10, Figs. 12–16]) For diatomic hydrogen, some experimental results for the specific heat c_V, in units of R, are schematically shown as a function of temperature. The scale used is logarithmic

The next diatom listed in the given table is the heteronuclear OH. Here the relevant two numbers are 27.5 K for the rotational motion and 5360 K for the vibrational motion. Again, the intradiatom bond appears mostly to remain stiff at laboratory temperatures. HCl, CH, and CO appear to behave in similar fashion.

Like H_2, one of the other homonuclear diatoms listed in the table is potassium (K_2). Here the rotational motion occurs at very low temperature; indeed, not far above absolute zero. And, in contrast with the light, heavy diatoms allow vibrations to set in much sooner. For instance, here bond stiffness begins decreasing at relatively low temperatures and the vibrational modes are fully available at temperature as low as $\approx$ 140 K.

To recapitulate: It is clear that the bond-stiffness in most "light" diatoms—which one would normally apply the ideal gas theory to—is a real, genuine, physical phenomenon. This stiffness holds back both the sixth and, of course, the seventh degrees of freedom. Often in light diatoms, while the rotational degrees of freedom are available at moderate temperatures, substantial increase in temperature is needed to overcome the bond stiffness and excite vibrational motion. (For instance, see Fig. 11.1.)

11.8 Classical Diatoms: High Temperature

With rise in temperature beyond the region where the bond stiffness holds sway, interaction between the two atoms in the diatomic molecule leads to simple harmonic like vibration of the two atoms. As a result, the bond length begins to vibrate at a rate specified by the interaction (see Sect. 11.8.1). When this behavior is fully established, it adds two more degrees of freedom to the diatom. Therefore, eventually, the specific heat per mole achieves a value equal to $\frac{7}{2}k_B T$. Clearly, with even further rise in temperature the interatomic bond vibration will become more and more enharmonic. Also eventually other issues such as electronic and nuclear excitations—and even possibly self-rotation of the single atoms around the line joining them—will begin to contribute. And last, and perhaps the most, the present treatment—that employs classical statistics—will need to be revamped with the use of quantum statistics.

Table 11.1 T_{rotation} and $T_{\text{vibration}}$ for various atoms

Diatom	T_{rotation} (K)	$T_{\text{vibration}}$ (K)
H_2	85.5	6140
OH	27.5	5360
HCl	15.3	4300
CH	20.7	4100
CO	2.77	3120
NO	2.47	2740
O_2	2.09	2260
Cl_2	0.347	810
Br_2	0.117	470
Na_2	0.224	230
K_2	0.081	140

11.8.1 Classical Diatoms: End of Stiffness

For a diatom, in addition to the Hamiltonian that have already been considered—meaning that referring to the translational motion of the center of mass, and that given in (11.55), namely $\mathcal{H}_{\text{rotation one stiff diatom}}$—there remains the part that refers to the intradiatom potential energy, and the part that refers to the kinetic energy of the two atoms when the bond unstiffens. Let us deal first with the intradiatom potential energy (see Table 11.1).

Assume that when the bond unstiffens, the Hooke's like force of attraction between the two atoms—in a given classical diatom—will cause the atoms to start vibrating in a simple harmonic-like motion. Because the diatom is not subject to any external forces, its total momentum is conserved. Assuming it starts with zero total momentum, its momentum will stay zero throughout.

The Hooke's law equations of motion of the two atoms are:

$$
\begin{aligned}
m_1 \frac{d^2 r_1(t)}{dt^2} &= -K r_0(t)\,, \\
m_2 \frac{d^2 r_2(t)}{dt^2} &= -K r_0(t)\,, \\
r_0(t) &\equiv r_1(t) + r_2(t)\,.
\end{aligned}
\tag{11.62}
$$

Here $r_0(t)$ is the bond length at time t; $r_1(t)$ and $r_2(t)$ are distances from the center of mass to atoms 1 and 2, respectively. Rather than focusing on $r_1(t)$ and $r_2(t)$, it is convenient instead to study the time dependence of $r_0(t)$. This can be done by adding the two equations as follows:

$$
\frac{d^2 r_1(t)}{dt^2} + \frac{d^2 r_2(t)}{dt^2} = \frac{d^2 r_0(t)}{dt^2} = -\left(\frac{K}{m_1} + \frac{K}{m_2}\right) r_0(t) = -\omega^2 r_0(t)\,, \tag{11.63}
$$

where

$$
K = \left(\frac{m_1 m_2}{m_1 + m_2}\right)\omega^2\,, \tag{11.64}
$$

and ω is the angular frequency of the simple harmonic motion. Clearly, the part of the Hamiltonian that contains the intradiatom coupling which gives rise to the potential energy—and describes the Hooke's law given above—is $\frac{K}{2}r_0^2$.

Next we look at the kinetic energy part of the Hamiltonian. This kinetic energy—namely $\frac{1}{2}[m_1(\dot{r_1})^2 + m_2(\dot{r_2})^2]$—represents the motion of the two atoms after the diatomic bond has unstiffened. Such kinetic energy was, of course, present in (11.53). But it was then dropped from that equation because of the effect of bond stiffness which is described in (11.54).

Therefore, the total (remaining part of the) Hamiltonian for a single diatom—to be called H(vibration single diatom)—is the following:

$$\begin{aligned} H(\text{vibration single diatom}) &= \frac{K}{2}r_0^2 + \frac{1}{2}\left[m_1(\dot{r_1})^2 + m_2(\dot{r_2})^2\right] \\ &= \frac{K}{2}r_0^2 + \left(\frac{p_1^2}{2m_1} + \frac{p_2^2}{2m_2}\right). \end{aligned} \tag{11.65}$$

Clearly, because there are no external forces present, the total momentum is conserved and the momenta of the two atoms must add up to zero. That is,

$$\mathbf{p_1}(t) = -\mathbf{p_2}(t)\,, \qquad p_1(t)^2 = p_2(t)^2 \equiv p^2(t)\,. \tag{11.66}$$

This fact and the use of (11.64) help write the remaining Hamiltonian in terms of only two variables:

$$H(\text{vibration single diatom}) = \left(\frac{m_1 m_2 \omega^2}{2(m_1+m_2)}\right) r_0^2 + p^2 \frac{m_1+m_2}{2m_1 m_2}\,. \tag{11.67}$$

Note that this Hamiltonian is identical to that for one-dimensional harmonic oscillator.

The contribution to the partition function from the vibrational motion of a single diatom is now readily calculated as

$$\begin{aligned} \Xi(\text{vibration single diatom}) &= h^{-1}\int_{-\infty}^{\infty} \exp\left[-\beta\left(\frac{m_1 m_2\omega^2}{2(m_1+m_2)}\right) r_0^2\right] \mathrm{d}r_0 \\ &\quad \times \int_{-\infty}^{\infty} \exp\left[-\beta\left(\frac{(m_1+m_2)}{2m_1 m_2}\right) p^2\right] \mathrm{d}p \\ &= h^{-1}\left(\frac{2\pi(m_1+m_2)}{\beta m_1 m_2 \omega^2}\right)^{\frac{1}{2}} \times \left(\frac{2\pi m_1 m_2}{\beta(m_1+m_2)}\right)^{\frac{1}{2}} \\ &= \frac{2\pi}{h\beta\omega}\,. \end{aligned} \tag{11.68}$$

The resultant vibrational partition function for N_d diatoms is simply the N_dth power of the above. The vibrational motion of N_d diatoms adds to the Helmholtz potential the amount:

$$\begin{aligned} F_{(\text{vibration})} &= -k_B T N_d \ln \Xi(\text{vibration single diatom}) \\ &= N_d k_B T \ln\left(\frac{\beta h \omega}{2\pi}\right). \end{aligned} \tag{11.69}$$

Because the Helmholtz potential is independent of the (one-dimensional) volume, the corresponding change in the pressure is vanishing. The contributions to other thermodynamic potentials—due to the vibrational motion—are the following:

$$\begin{aligned}
S_{(\text{vibration})} &= -\left(\frac{\partial F_{(\text{vibration})}}{\partial T}\right)_{N_\text{d}} = N_\text{d}k_\text{B}\left[1+\ln\left(\frac{2\pi k_\text{B}T}{h\omega}\right)\right],\\
G_{(\text{vibration})} &= N_\text{d}\left(\frac{\partial F_{(\text{vibration})}}{\partial N_\text{d}}\right)_\text{T} = N_\text{d}k_\text{B}T\ln\left(\frac{h\omega}{2\pi k_\text{B}T}\right),\\
U_{(\text{vibration})} &= F_{(\text{vibration})} + TS_{(\text{vibration})} = N_\text{d}k_\text{B}T\,,\\
H_{(\text{vibration})} &= U_{(\text{vibration})}\,.
\end{aligned} \tag{11.70}$$

Adding $U_{(\text{vibration})}$ to $U_{(\text{c.of.m})}$ and U_{rotation} leads to the result

$$U_{\text{total}} = \frac{3}{2}N_\text{d}k_\text{B}T + N_\text{d}k_\text{B}T + N_\text{d}k_\text{B}T = \frac{7}{2}N_\text{d}k_\text{B}T\,. \tag{11.71}$$

This means that a diatom, when raised to an appropriately high temperature, in addition to undergoing translational and rotational motion, also experiences vibrational motion of its intradiatom bond. And when that happens, it contributes an amount equal to $\frac{7}{2}k_\text{B}T$ to the internal energy, making the number of degrees of freedom equal to 7. (A quantum-statistical treatment of these issues is provided in Sect. 11.5.3; see (11.144)–(11.166).)

11.9 Simple Oscillators: Anharmonic

Elementary theoretical physics owes much to the idea of simple harmonic oscillators. Indeed, the bond length vibration studied in the previous section—see (11.67)–(11.70)—was treated as a simple harmonic motion. In real systems, however, such simple oscillators must often be assumed to possess some anharmonicity.

Consider a collection of N distinguishable, one-dimensional simple oscillators, each with a tiny bit of anharmonic potential. Assume that each oscillator is independent of the other because there is no interaction between different oscillators. Further assume that at time t the ith such oscillator has momentum $p_\text{i}(t)$ and separation length equal to $q_\text{i}(t)$, and its Hamiltonian $\mathcal{H}_\text{i}(t)$ is

$$\mathcal{H}_\text{i}(t) = \frac{p_\text{i}^2}{2m} + m\omega^2\left(\frac{q_\text{i}^2}{2} - aq_\text{i}^3 - bq_\text{i}^4\right), \tag{11.72}$$

where $b \approx O(a^2)$ and $a \ll 1$. Then the partition function is

$$\begin{aligned}
\Xi(N,V,T) &= \left(\frac{2\pi k_\text{B}T}{h\omega}\right)^N\\
&\times\left[1+\left(3b+\frac{15}{2}a^2\right)\left(\frac{k_\text{B}T}{m\omega^2}\right)+o\left(b^2\left(\frac{k_\text{B}T}{m\omega^2}\right)^2\right)\right]^N.
\end{aligned} \tag{11.73}$$

Note that to see the algebra that takes us from (11.72) to (11.73) refer to the long footnote below.[16] Therefore we have:

$$\begin{aligned} F &= -k_B T \ln \Xi(N, V, T) \\ &= -Nk_B T \ln\left(\frac{2\pi k_B T}{h\omega}\right) \\ &\quad - Nk_B T\left[\left(3b + \frac{15}{2}a^2\right)\left(\frac{k_B T}{m\omega^2}\right) - O\left(\left(3b + \frac{15}{2}a^2\right)^2\left(\frac{k_B T}{m\omega^2}\right)^2\right)\right], \end{aligned}$$

[16]We need to work out the following:

$$\begin{aligned} \Xi(N, V, T) &= \left[\left(\frac{1}{h}\right)^N \int_{-\infty}^{\infty} \cdots \int_{-\infty}^{\infty} \mathrm{d}p_1 \cdots \mathrm{d}p_\mathrm{N} \exp\left(-\beta \sum_{i=1}^{\mathrm{N}} \frac{p_\mathrm{i}^2}{2m}\right)\right] \\ &\quad \times \int_{-\infty}^{\infty} \cdots \int_{-\infty}^{\infty} \mathrm{d}q_1 \cdots \mathrm{d}q_\mathrm{N} \\ &\quad \times \exp\left\{-\beta \sum_{i=1}^{\mathrm{N}} m\omega^2\left(\frac{q_\mathrm{i}^2}{2} - aq_\mathrm{i}{}^3 - bq_\mathrm{i}{}^4\right)\right\} \\ &= \left(\frac{1}{h}\right)^N \left(\frac{2m\pi}{\beta}\right)^{\frac{N}{2}} \\ &\quad \times \int_{-\infty}^{\infty} \cdots \int_{-\infty}^{\infty} \mathrm{d}q_1 \cdots \mathrm{d}q_\mathrm{N} \\ &\quad \times \exp\left\{-\beta \sum_{\mathrm{i}=1}^{\mathrm{N}} m\omega^2\left(\frac{q_\mathrm{i}^2}{2} - aq_\mathrm{i}{}^3 - bq_\mathrm{i}{}^4\right)\right\}. \end{aligned} \tag{11.74}$$

To this purpose, we introduce the simplifying notation, $\alpha = \frac{\beta m\omega^2}{2}$, $\lambda_1 = \beta m\omega^2 a$, and $\lambda_2 = \beta m\omega^2 b$, and make a Taylor expansion of the exponential in powers of the small quantities λ_1 and λ_2. That is,

$$\begin{aligned} \exp\left\{-\beta m\omega^2\left(\frac{q^2}{2} - aq^3 - bq^4\right)\right\} &= \exp(-aq^2 + \lambda_1 q^3 + \lambda_2 q^4) \\ = \exp\left(-\alpha q^2\right)\left[1 + \lambda_1 q^3 + \lambda_2 q^4 + \frac{\lambda_1^2}{2}q^6 + \lambda_1\lambda_2 q^7 + \frac{\lambda_2^2}{2}q^8 + O(\lambda_1^3 q^9)\right]. \end{aligned} \tag{11.75}$$

Now using (B.5) and (B.6), we can write

$$\begin{aligned} &\int_{-\infty}^{\infty} \mathrm{d}q \exp\left(-\alpha q^2 + \lambda_1 q^3 + \lambda_2 q^4\right) \\ &= \int_{-\infty}^{\infty} \mathrm{d}q\ \exp\left(-\alpha q^2\right)\left[1 + \lambda_1 q^3 + \lambda_2 q^4 + \frac{\lambda_1^2 q^6}{2} + \lambda_1\lambda_2 q^7 + \frac{\lambda_2^2 q^8}{2} + \cdots\right] \\ &= \sqrt{\left(\frac{\pi}{\alpha}\right)}\left[1 + \lambda_2\left(\frac{3}{4\alpha^2}\right) + \lambda_1^2\left(\frac{15}{16\alpha^3}\right) + \lambda_2^2\left(\frac{105}{32\alpha^4}\right)\right] \\ &= \left(\frac{2\pi k_B T}{m\omega^2}\right)^{\frac{1}{2}}\left[1 + \left(3b + \frac{15}{2}a^2\right)\left(\frac{k_B T}{m\omega^2}\right) + O\left(b^2\left(\frac{k_B T}{m\omega^2}\right)^2\right)\right]. \end{aligned} \tag{11.76}$$

$$
\begin{aligned}
S &= -\left(\frac{\partial F}{\partial T}\right)_{\mathrm{N,V}} \\
&= Nk_{\mathrm{B}}\left[1+\ln\left(\frac{2\pi k_{\mathrm{B}}T}{h\omega}\right)\right] + 2Nk_{\mathrm{B}}\left(\frac{k_{\mathrm{B}}T}{m\omega^2}\right)\left(3b+\frac{15}{2}a^2\right) \\
&\quad - O\left(Nk_{\mathrm{B}}\left(3b+\frac{15}{2}a^2\right)^2\left(\frac{k_{\mathrm{B}}T}{m\omega^2}\right)^2\right), \\
U &= F+TS = Nk_{\mathrm{B}}T\left[1+\left(\frac{k_{\mathrm{B}}T}{m\omega^2}\right)\left(3b+\frac{15}{2}a^2\right)\right], \\
C_{\mathrm{v}} &= \left(\frac{\partial U}{\partial T}\right)_{\mathrm{V,N}} = Nk_{\mathrm{B}}\left[1+2\left(\frac{k_{\mathrm{B}}T}{m\omega^2}\right)\left(3b+\frac{15}{2}a^2\right)\right].
\end{aligned}
\tag{11.77}
$$

The results in (11.77) are recorded to the same accuracy as in (11.76).

Exercise 11.5 Assume spatial symmetry and work out the foregoing problem in three dimensions.

11.10 Classical Dipole Pairs: Average Energy and Force

A thermodynamic system, consisting of N pairs of—magnetic or electrostatic—classical dipoles of zero mass, is at temperature T. The intrapair separation (of each pair of dipoles) is R. Interaction between different pairs of dipoles—namely the interpair interaction—is assumed to be zero.

The potential energy of the ith pair of dipoles, $\Delta\mathcal{H}_{\mathrm{i}}$, is

$$
\begin{aligned}
\Delta\mathcal{H}_{\mathrm{i}} &= C\left[\frac{\overline{\mu_{\mathrm{i;1}}}\cdot\overline{\mu_{\mathrm{i;2}}} - 3\mu_{\mathrm{iz;1}}\mu_{\mathrm{iz;2}}}{R^3}\right] \\
&= C\left[\frac{\mu_{\mathrm{ix;1}}\mu_{\mathrm{ix;2}} + \mu_{\mathrm{iy;1}}\mu_{\mathrm{iy;2}} - 2\,\mu_{\mathrm{iz;1}}\mu_{\mathrm{iz;2}}}{R^3}\right].
\end{aligned}
\tag{11.78}
$$

Here $\overline{\mu_{\mathrm{i;1}}}$ and $\overline{\mu_{\mathrm{i;2}}}$ are the vector moments of the ith pair of dipoles. Because there is no interpair interaction, the dipoles do not translate spatially, and are assumed to be of zero mass, the sum of (only) the potential energies of all the N pairs of dipoles is the system Hamiltonian $\mathcal{H}$. That is,

$$
\mathcal{H} = \sum_{\mathrm{i=1}}^{\mathrm{N}} \Delta\mathcal{H}_{\mathrm{i}}\,. \tag{11.79}
$$

The notation used in (11.78) is such that $\mu_{\mathrm{ix;1}}$, $\mu_{\mathrm{iy;1}}$ and $\mu_{\mathrm{iz;1}}$ are the x-, y-, and the z-components of the dipole moment vector, $\overline{\mu_{\mathrm{i;1}}}$. For convenience, the z-direction is chosen to be along the line joining the centers of the two dipoles. Let us first calculate the observed—i.e., the thermodynamic average—value of the energy.

11.10.1 Distribution Factor and Thermal Average

Whenever we have available the system Hamiltonian, $\mathcal{H}(\Theta, \Phi)$, then following (2.13), the BMG distribution factor $f(\Theta, \Phi)$ can be written as

$$f(\Theta, \Phi) = \frac{\exp[-\beta\mathcal{H}(\Theta, \Phi)]}{\int\int d\Theta \cdot d\Phi \exp[-\beta\mathcal{H}(\Theta, \Phi)]} . \tag{11.80}$$

Accordingly, the observed value of the total energy—namely the thermal average of the total Hamiltonian, $\mathcal{H}(\Theta, \Phi)$—is the following:

$$\langle\![\mathcal{H}(\Theta, \Phi)]\!\rangle = \frac{\int\int d\Theta \cdot d\Phi[\mathcal{H}(\Theta, \Phi)] \exp[-\beta\mathcal{H}(\Theta, \Phi)]}{\int\int d\Theta \cdot d\Phi \exp[-\beta\mathcal{H}(\Theta, \Phi)]} . \tag{11.81}$$

Using (11.79), we can write the above as

$$U = \langle\mathcal{H}\rangle = \left\langle\sum_{j=1}^{N} \Delta\mathcal{H}_j\right\rangle = \frac{\int\int d\Theta \cdot d\Phi[\sum_{j=1}^{N} \Delta\mathcal{H}_j] \exp(-\beta\mathcal{H})}{\int\int d\Theta \cdot d\Phi \exp(-\beta\mathcal{H})} . \tag{11.82}$$

The differentials $d\Theta$ and $d\Phi$ are defined in (11.84) below.

For the problem in hand, (11.82) can be greatly simplified (compare with (2.16)–(2.21)). While any given pair of dipoles do have angular interaction with each other, they remain unaware of the presence of the other $N-1$ dipole-pairs. Also, the temperature T experienced by any given pair of dipoles is that established by the very large number, $N \gg 1$, of dipole pairs present. Further, because of spherical symmetry, rather than Cartesian, spherical are the best coordinates to use. To this end, we chose the angular orientations of the ith pair of dipole to be (θ_{i1}, ϕ_{i1}) and (θ_{i2}, ϕ_{i2}).

In light of the above, all $N-1$ terms for which the subscript j is different from some given subscript i are the same in both the numerator and denominator of (11.82). Therefore they cancel each other out. As a result, (11.82) simplifies to the following:

$$u_i = \langle\Delta\mathcal{H}_i\rangle = \frac{\int\int d\theta_i \cdot d\phi_i(\Delta\mathcal{H}_i) \exp(-\beta\Delta\mathcal{H}_i)}{\int\int d\theta_i \cdot d\phi_i \exp(-\beta\Delta\mathcal{H}_i)} . \tag{11.83}$$

The differentials in (11.83) and (11.82) are denoted as given below:

$$d\theta_i \cdot d\phi_i = \sin\theta_{i1} d\theta_{i1} \sin\theta_{i2} d\theta_{i2} d\phi_{i1} d\phi_{i2}$$
$$\text{and} \quad d\Theta \cdot d\Phi = \prod_{i=1}^{N} [d\theta_i \cdot d\phi_i] . \tag{11.84}$$

In the spherical representation, the dipole vectors that occur in $\Delta\mathcal{H}_i$—see (11.78)—are the following:

$$\begin{aligned}\mu_{ix;1} &= \mu_1 \sin(\theta_{i1})\cos(\phi_{i1}),\\ \mu_{iy;1} &= \mu_1 \sin(\theta_{i1})\sin(\phi_{i1}),\\ \mu_{iz;1} &= \mu_1 \cos(\theta_{i1}).\end{aligned} \tag{11.85}$$

The corresponding vectors for the second dipole of the ith pair—namely number 2—are similarly expressed. (Just change 1 to 2!)

Because all N pairs of dipoles are identical, the observed value of the potential energy u_i of the ith pair of dipoles must be the same as the observed corresponding value for any one of the other $N-1$ dipole pairs. Therefore, the thermodynamic average $u_i = \langle \Delta\mathcal{H}_i\rangle = u$ is independent of the index i. Considerable notational simplification is achieved by ignoring any future mention of this index. Indeed, we can represent (11.78) more simply as follows:

$$\begin{aligned}\Delta\mathcal{H}_i &= -\left(\frac{C}{R^3}\right)[2\mu_{1z}\mu_{2z} - \mu_{1x}\mu_{2x} - \mu_{1y}\mu_{2y}]\\ &= -\left(\frac{C\mu_1\mu_2}{R^3}\right)\big[2\cos\theta_1\cos\theta_2 - \sin\theta_1\sin\theta_2\cos(\phi_1-\phi_2)\big]\\ &= F(1,2),\end{aligned} \tag{11.86}$$

and (11.83) as

$$\begin{aligned}u_i = u &= \langle F(1,2)\rangle\\ &= \frac{\int_0^\pi \sin(\theta_1)d\theta_1 \int_0^{2\pi} d\phi_1 \int_0^\pi \sin(\theta_2)d\theta_2 \int_0^{2\pi} d\phi_2 F(1,2)\exp[-\beta F(1,2)]}{\int_0^\pi \sin(\theta_1)d\theta_1 \int_0^{2\pi} d\phi_1 \int_0^\pi \sin(\theta_2)d\theta_2 \int_0^{2\pi} d\phi_2 \exp[-\beta F(1,2)]},\end{aligned} \tag{11.87}$$

where

$$\exp\big[-\beta F(1,2)\big] = 1 - \beta F(1,2) + \frac{1}{2}\big[\beta F(1,2)\big]^2 + \cdots. \tag{11.88}$$

The integral (11.87) above can only be done by numerical methods. If the system is at high enough temperature such that $\frac{\mu_1\mu_2}{R^3} \ll T$, the exponential can be expanded in powers of the exponent. The resultant integrals are straightforward, but somewhat tedious to evaluate. Therefore, it is convenient to append the relevant details in an appendix. (See (K.1)–(K.20).)

The average value, U, of the energy of N dipole pairs is N times the average energy, u, of a single dipole pair. Note that u is the ratio of the results given in (K.17) and (K.16). That is,

$$\begin{aligned}U = N\langle\Delta\mathcal{H}_i\rangle = Nu &= \frac{-N\beta A_3}{(4\pi)^2[1+\frac{\beta^2}{3}(\frac{C\mu_1\mu_2}{R^3})^2]} + \cdots\\ &= -N\beta\frac{2}{3}\left(\frac{C\mu_1\mu_2}{R^3}\right)^2\left[1 - \frac{\beta^2}{3}\left(\frac{C\mu_1\mu_2}{R^3}\right)^2\right] + \cdots\\ &\approx -N\beta\frac{2}{3}\left(\frac{C\mu_1\mu_2}{R^3}\right)^2.\end{aligned} \tag{11.89}$$

Compare development of this equation in (K.18).

11.10.2 Average Force Between a Pair

The force Λ between any pair of dipoles can be determined by differentiating their intrapair potential energy $\Delta\mathcal{H}_{\rm i}$ that was specified in (11.86) as

$$\Lambda = -\left(\frac{\partial \Delta\mathcal{H}_{\rm i}}{\partial R}\right)_{\mu_1,\mu_2,\theta_1,\theta_2,\phi_1,\phi_2} = \left(\frac{3}{R}\right)\Delta\mathcal{H}_{\rm i}\,. \tag{11.90}$$

Similar to the average energy, $\langle\Delta\mathcal{H}_{\rm i}\rangle$, of a pair—see (11.83) and (11.84)—we can also represent the average value of the intradipolar force, Λ, as $\langle\Lambda\rangle$, where

$$\langle\Lambda\rangle = \left(\frac{3}{R}\right)\langle\Delta\mathcal{H}_{\rm i}\rangle = \left(\frac{3}{RN}\right)U = -\beta\frac{2}{R}\left(\frac{C\mu_1\mu_2}{R^3}\right)^2. \tag{11.91}$$

Compare development of this equation in (K.20). Note that U is as given in (K.18). Clearly, the thermodynamic average of the force between any single pair of dipoles is attractive.

11.11 Langevin Paramagnetism: Classical Picture

Magnetic dipoles arise as a result of the presence of electronic and/or nuclear angular momenta. Consider a uniform, constant magnetic field $\boldsymbol{B}$ applied to a system of N such—assumed to be classical—noninteracting magnetic dipoles. We further assume that the spatial location of all dipoles and the size, $\mu_{\rm c}$, of their dipole moments is fixed. Yet, every dipole is assumed to be free to rotate around its midpoint. The Hamiltonian consists only of the potential energy,

$$\mathcal{H} = -\boldsymbol{B}\cdot\sum_{\rm i=1}^{\rm N}\boldsymbol{\mu}_{\rm i} = -B_0\mu_{\rm c}\sum_{\rm i=1}^{\rm N}\cos(\theta_{\rm i})\,, \tag{11.92}$$

where $\theta_{\rm i}$ is the angle between the dipole moment vector $\boldsymbol{\mu}_{\rm i}$ and the applied magnetic field $\boldsymbol{B}$. (For convenience the z-axis is chosen parallel to the applied field. The z-components of the applied field and the dipole moment vectors are B_0 and μ_c, respectively.)

11.11.1 Langevin Paramagnetism: Statistical Average

Because there is no interdipole interaction, all thermodynamic averages—e.g., $\langle\boldsymbol{B}\cdot\mu_{\rm i}\rangle$—for a given dipole i are independent of their location indices i, etc. Much as in (11.83)–(11.85), the spherical coordinate representation of the statistical average of the magnetic moment—which will, of course, equal its thermodynamic average—M can be written as follows:

$$\begin{aligned} M &= N\frac{\langle\boldsymbol{B}\cdot\mu_{\rm i}\rangle}{B_0} = -\frac{\langle\mathcal{H}\rangle}{B_0} = \left(\frac{-U}{B_0}\right) = N\mu_{\rm c}\big\langle\cos(\theta_{\rm i})\big\rangle \\ &= N\mu_{\rm c}\frac{\int_0^{\pi}[\cos(\theta_{\rm i})]\cdot\sin(\theta_{\rm i})\exp[\beta B_0\mu_{\rm c}\cos(\theta_{\rm i})]{\rm d}\theta_i\int_0^{2\pi}{\rm d}\phi_{\rm i}}{\int_0^{\pi}\sin(\theta_{\rm i})\exp[\beta B_0\mu_{\rm c}\cos(\theta_{\rm i})]{\rm d}\theta_{\rm i}\int_0^{2\pi}{\rm d}\phi_{\rm i}}\,. \end{aligned} \tag{11.93}$$

To compute the integral in the numerator of the last term on the right-hand side, we employ standard integration techniques. Namely,

(1) Introduce the substitution:

$$\alpha = \beta B_0 \mu_c , \qquad y = \cos(\theta_i) . \tag{11.94}$$

(2) Integrate the numerator by parts:

$$\begin{aligned}
&\int_0^{\pi} \cos(\theta_i) \exp\left[\beta B_0 \mu_c \cos(\theta_i)\right] \sin(\theta_i) d\theta_i \cdot \int_0^{2\pi} d\phi_i \\
&= \int_1^{-1} y \exp(\alpha y)(-dy) \cdot (2\pi) = 2\pi \int_{-1}^{1} \exp(\alpha y) y dy \\
&= 2\pi \left\{ \left[\frac{\exp(\alpha y)}{\alpha} y \right]_{-1}^{1} - \frac{1}{\alpha^2} \left[\exp(\alpha) - \exp(-\alpha)\right] \right\} \\
&= 2\pi \left[\exp(\alpha) \left(\frac{1}{\alpha} - \frac{1}{\alpha^2} \right) + \exp(-\alpha) \left(\frac{1}{\alpha} + \frac{1}{\alpha^2} \right) \right] .
\end{aligned} \tag{11.95}$$

Using the same notation, the integral in the denominator of (11.93) is

$$2\pi \int_{-1}^{1} \exp(\alpha y) dy = \frac{2\pi}{\alpha} \left[\exp(\alpha) - \exp(-\alpha)\right] . \tag{11.96}$$

The ratio of the numerator given in (11.95) and the denominator given in (11.96) yields $\langle \cos(\theta_i) \rangle$. And as noted in (11.93), $\langle \cos(\theta_i) \rangle$ leads to both the total energy $\langle \mathcal{H} \rangle$ and the statistical average of the magnetic moment M,

$$\begin{aligned}
U = \langle \mathcal{H} \rangle &= -N B_0 \mu_c \big\langle \cos(\theta_i) \big\rangle = -B_0 M \\
&= -N B_0 \mu_c \left[\frac{2\pi \left[\exp(\alpha)(\frac{1}{\alpha} - \frac{1}{\alpha^2}) + \exp(-\alpha)(\frac{1}{\alpha} + \frac{1}{\alpha^2})\right]}{\frac{2\pi}{\alpha} \left[\exp(\alpha) - \exp(-\alpha)\right]} \right] \\
&= -N B_0 \mu_c \left[\coth(\alpha) - \frac{1}{\alpha} \right] = -N B_0 \mu_c L(\alpha) .
\end{aligned} \tag{11.97}$$

11.11.2 Langevin Paramagnetism: High Temperature

In the limit of high temperature—that is, when α is small—we can expand the Langevin function $L(\alpha)$ in powers of α. Recall that α, i.e., $\beta B_0 \mu_c$, is the ratio of the magnetic potential energy (per dipole)—which is $B_0 \mu_c$—and the thermal kinetic energy (per dipole)—which is $\approx k_B T$. We get

$$\begin{aligned}
L(\alpha) = \coth(\alpha) - \frac{1}{\alpha} &= \left[\frac{\alpha}{3} - \frac{\alpha^3}{45} + \cdots \right] \\
&= \frac{B_0 \mu_c}{3 k_B T} \left[1 - \frac{(\beta B_0 \mu_c)^2}{15} + \cdots \right] ,
\end{aligned}$$

$$
\begin{aligned}
(\chi)_{\text{langevin-classical}} &= \left(\frac{\partial M}{\partial B_0}\right)_T = N\mu_c\left(\frac{\partial L(\alpha)}{\partial B_0}\right)_T \\
&\approx \frac{N\mu_c{}^2}{3k_B T}\left[1-\frac{1}{5}\left(\frac{B_0\mu_C}{k_B T}\right)^2\right], \qquad (11.98)\\
[C_{\text{specific heat}}]_{\text{langevin-classical}} &= \left(\frac{\partial U}{\partial T}\right)_{N,B_0} \\
&= \frac{Nk_B}{3}\left(\frac{B_0\mu_c}{k_B T}\right)^2\left[1-\frac{1}{5}\left(\frac{B_0\mu_C}{k_B T}\right)^2\right].
\end{aligned}
$$

Note that the above results have been obtained by the use of classical statistics and are applicable at high temperatures. Here $(\frac{\partial M}{\partial B_0})_T$ is the magnetic susceptibility of the Langevin paramagnet. It will be referred to as $(\chi)_{\text{langevin-classical}}$. At what we have called high temperature, it varies as the inverse power of the temperature. This result is known as Curie's law of paramagnetism. The quantity, $\frac{N\mu_c{}^2}{3k_B}$, is called the Curie constant for a system consisting of N Langevin classical paramagnetic dipoles.

11.11.3 Langevin Susceptibility

In paramagnetic salts, such as gadolinium ethylsulphate, the high temperature Curie's law is found to hold down to about 10 K. However, the high temperature law for the Langevin susceptibility must surely break down—and the system undergo a phase transition—at some finite low temperature because in its present form—compare (11.98)—the law can predict an unphysically large alignment of dipoles that exceeds even 100%!

Note that at high temperature, the specific heat decreases as the square of the inverse power of the temperature, i.e., $[C_{\text{specific heat}}]_{\text{langevin-classical}} \propto \frac{1}{T^2}$.

11.11.4 Langevin Paramagnetism: Low Temperature

On the other hand, at low temperature where $\alpha \gg 1$, the Langevin factor $L(\alpha)$ can be approximated as[17]

$$
\coth(\alpha) - \frac{1}{\alpha} = \left(1-\frac{1}{\alpha}\right) + O\{\exp(-4\alpha)\}. \qquad (11.99)
$$

Therefore, according to (11.97), at very low temperature $\langle\cos(\theta_i)\rangle$ is equal to unity and all the Langevin classical dipoles are completely aligned with the applied field; and quite understandably, the magnetic susceptibility is extremely small.

[17]See Fig. 11.2 for a plot of the Langevin function and a demonstration of its low and high temperature behavior.

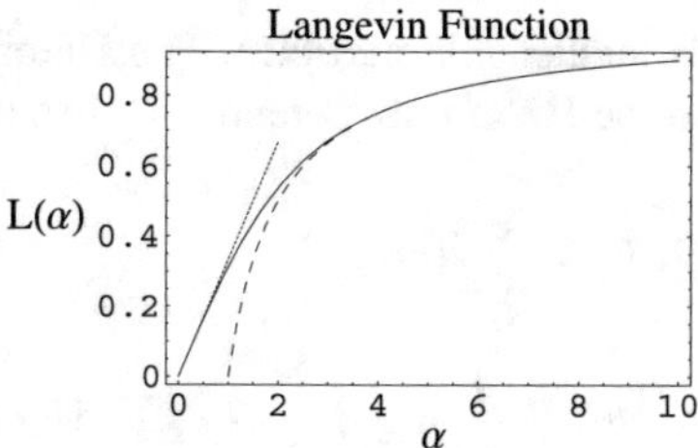

Fig. 11.2 The Langevin function $L(\alpha)$. Recall that the average energy, U, of a Langevin paramagnet is related to $L(\alpha)$. That is, $U = -NB_0\mu_c L(\alpha)$. Plotted above is $L(\alpha)$ versus $\alpha = \frac{B_0\mu_c}{k_B T}$. The thinly dashed curve represents the high temperature approximation, i.e., $L(\alpha) \approx \frac{\alpha}{3}$, that is valid for $\alpha \ll 1$. The curve with wider dashes represents the low temperature approximation valid for $\alpha \gg 1$, that is, $L(\alpha) \approx 1 - \frac{1}{\alpha}$

11.12 Extremely Relativistic Monatomic Ideal Gas

As noted in (2.83), the special theory of relativity predicts that particles with rest mass m_0 and velocity $\boldsymbol{v}_i$ have energy

$$E_i = m_i c^2 = \left(\frac{m_0}{\sqrt{1 - \frac{\mathbf{v}_i^2}{c^2}}}\right) c^2 , \tag{11.100}$$

and momentum

$$\boldsymbol{p}_i = m_i \boldsymbol{v}_i = \left(\frac{m_0}{\sqrt{1 - \frac{\boldsymbol{v}_i^2}{c^2}}}\right) \boldsymbol{v}_i , \tag{11.101}$$

where c is the velocity of light. Combining these equations leads to

$$E_i = \sqrt{c^2 \boldsymbol{p}_i^2 + m_0^2 c^4} . \tag{11.102}$$

In the extreme relativistic limit

$$c^2 \boldsymbol{p}_i^2 \gg m_0^2 c^4 . \tag{11.103}$$

Accordingly, in (11.102) we can use the approximation

$$E_i \approx c p_i . \tag{11.104}$$

Note that here $p_i = \sqrt{\boldsymbol{p}_i^2}$.

Consider N such indistinguishable particles. The Hamiltonian $\mathcal{H}$, the partition function $\Xi(N, V, T)$, and the Helmholtz potential F then are

$$
\begin{aligned}
\mathcal{H} &= \mathcal{H}(Q, P) = \sum_{\mathrm{i}=1}^{\mathrm{N}} (cp_{\mathrm{i}}) , \\
\Xi(N, V, T) &= (N!)^{-1} h^{-3N} \int_Q \mathrm{d}Q \int_0^{\infty} \cdots \int_0^{\infty} \mathrm{d}P \exp[-\beta\mathcal{H}] \\
&= \frac{1}{N!}\left(\frac{V}{h^3}\right)^N \left[\int_0^{\infty} \exp(-\beta cp) 4\pi p^2 \mathrm{d}p\right]^N \\
&= \left(\frac{Ve}{Nh^3}\right)^N \left[\frac{8\pi}{(\beta c)^3}\right]^N , \\
F &= -k_{\mathrm{B}} T \ln \Xi(N, V, T) \\
&= -k_{\mathrm{B}} T N \left[\ln\left(\frac{8\pi e V}{N}\right) + 3 \ln\left(\frac{k_{\mathrm{B}} T}{hc}\right)\right] .
\end{aligned}
\tag{11.105}
$$

Other thermodynamic potentials and the pressure can now readily be determined from the Helmholtz potential as

$$
\begin{aligned}
P &= -\left(\frac{\partial F}{\partial V}\right)_{\mathrm{N,T}} = \frac{N k_{\mathrm{B}} T}{V} , \\
S &= -\left(\frac{\partial F}{\partial T}\right)_{\mathrm{N,V}} = k_{\mathrm{B}} N \left[\ln\left(\frac{8\pi e V}{N}\right) + 3 \ln\left(\frac{k_{\mathrm{B}} T}{hc}\right)\right] + 3 k_{\mathrm{B}} N , \\
U &= F + TS = 3 N k_{\mathrm{B}} T .
\end{aligned}
\tag{11.106}
$$

Equation (11.106) shows that

$$
(PV)_{\text{relativistic}} = N k_{\mathrm{B}} T = \frac{U}{3} . \tag{11.107}
$$

This extreme-relativistic result contrasts with that for nonrelativistic, monatomic ideal gases recorded in (2.31). Indeed, all nonrelativistic monatomic ideal gases in three dimensions, whether they be classical or quantum, obey the following relationship:

$$
(PV)_{\text{nonrelativistic classical or quantum ideal gas in 3 dim}} = \frac{2U}{3} . \tag{11.108}
$$

Noninteracting classical gases considered in the preceding subsections of this chapter have been easy to treat. The treatment of systems with interactions, however, is more involved.

11.13 Hamiltonian: Gas with Interaction

Hamiltonian for a gas of N particles, with no ability for self-rotation but with the presence of two-body interparticle potential, can be written as follows:

$$\mathcal{H} = \mathcal{H}(Q, P) = \sum_{\mathrm{i=1}}^{\mathrm{N}} \frac{\boldsymbol{p}_{\mathrm{i}}^2}{2m} + \sum_{\mathrm{j=2;\, j>i}}^{\mathrm{N}} \sum_{\mathrm{i=1}}^{\mathrm{N-1}} U_{\mathrm{i,j}} . \tag{11.109}$$

Note that by definition the two-body potential $U_{\mathrm{i,i}} = U_{\mathrm{j,j}} = 0$ and by symmetry $U_{\mathrm{i,j}} = U_{\mathrm{j,i}} = U(|\boldsymbol{r}_{\mathrm{i}} - \boldsymbol{r}_{\mathrm{j}}|)$ where $\boldsymbol{r}_{\mathrm{i}}$ and $\boldsymbol{r}_{\mathrm{j}}$ are the three-dimensional position vectors of the ith and jth particles.[18] Note also that $\boldsymbol{p}_{\mathrm{i}}$ is the three-dimensional momentum vector of the ith particle and the form of the summation over the pairs (i, j) ensures that the total number of pairs summed is equal to $\frac{N(N-1)}{2}$ and that as a result no double counting occurs.[19]

11.14 Mayer's Cluster Expansion: Partition Function

Using the Hamiltonian specified in (11.109), the partition function for fixed number of particles—that is, equal to N—is

$$\begin{aligned}
\Xi(N, V, T) &= (N!)^{-1} h^{-3N} \int_{-\infty}^{\infty} \cdots \int_{-\infty}^{\infty} \mathrm{d}P \int_{-\infty}^{\infty} \cdots \int_{-\infty}^{\infty} \mathrm{d}Q \exp\left[-\beta \mathcal{H}\right] \\
&= \frac{V^N}{N!} \left(\frac{1}{h^3}\right)^N \left[\int_{-\infty}^{\infty} \exp\left(-\frac{\beta p^2}{2m}\right) \mathrm{d}p\right]^{3N} \\
&\quad \times V^{-N} \int_{-\infty}^{\infty} \cdots \int_{-\infty}^{\infty} \prod_{\mathrm{i<j}} \exp(-\beta U_{\mathrm{i,j}}) \mathrm{d}Q \\
&= \left[\frac{V^N}{N!} \left(\frac{2\pi m k_{\mathrm{B}} T}{h^2}\right)^{\frac{3N}{2}}\right] \int_{-\infty}^{\infty} \cdots \int_{-\infty}^{\infty} V^{-N} \prod_{\mathrm{i<j}} (1 + f_{\mathrm{i,j}}) \mathrm{d}Q \\
&= \left[\frac{V^N}{N!} \left(\frac{2\pi m k_{\mathrm{B}} T}{h^2}\right)^{\frac{3N}{2}}\right] \exp\{A\} .
\end{aligned} \tag{11.110}$$

In (11.110) we have made use of Mayer's notation for his two particle function $f_{\mathrm{i,j}}$. Also, introduced a new notation, $\exp\{A\}$ That is,

$$\begin{aligned}
\exp(-\beta U_{\mathrm{i,j}}) &= 1 + f_{\mathrm{i,j}} , \\
\int_{-\infty}^{\infty} \cdots \int_{-\infty}^{\infty} V^{-N} \prod_{\mathrm{i<j}} (1 + f_{\mathrm{i,j}}) \mathrm{d}Q &= \exp\{A\} .
\end{aligned} \tag{11.111}$$

[18] Here $|\boldsymbol{r}_{\mathrm{i}} - \boldsymbol{r}_{\mathrm{j}}| = r$ is the separation between the ith and jth atoms.

[19] A beginner might benefit from a numerical example. For simplicity let us consider $N = 3$. Then the summation of $\sum_{\mathrm{j>i}} U_{\mathrm{i,j}}$ will consist of the following possible choices: Start with $i = 1$. As a result, j can be either 2 or 3. These two possible choices lead to two possible potentials, $U_{1,2}$ and $U_{1,3}$. Next we choose $i = 2$. Then j can only be equal to 3 leading to the third possible choice for the two-body potential, namely $U_{2,3}$. Note that $i = 2$ was the maximum possible allowed value for i, that is, $i = N - 1 = 3 - 1 = 2$. Therefore the total of the three choices for the U's, that have already been made, exhaust the possibility of making any further choices. And, of course, $3 = \frac{N(N-1)}{2} = \frac{3\times 2}{2} = 3$.

In ideal gases there is no interparticle interaction, i.e., $U_{i,j} = 0$. As a result, $\exp(-\beta U_{i,j}) = 1$ and therefore—see (11.111)—$f_{i,j}$ is also equal to zero.

Similarly, when either the system is weakly interacting, i.e., $U_{i,j}$ is very small, or the temperature is high, i.e., $\beta U_{i,j} \ll 1$, we have an equally simple result because then

$$f_{i,j} = \exp(-\beta U_{i,j}) - 1 = -\beta U_{i,j} + O(\beta U_{i,j})^2 \ll 1 . \tag{11.112}$$

Thus $f_{i,j}$ is a small number compared to unity for: (a) the weekly interacting systems; (b) at high temperatures; and (c) when both (a) and (b) are true.

The integral on the right-hand side of (11.110) is extraordinarily difficult to evaluate. Indeed, it cannot be done exactly for any but the most trivial cases. Therefore approximations have to be made. Central to these approximations is the fact that $f_{i,j}$ is small. For more details, see [11–13].

Because the needed integrals increase in complexity very rapidly as the number of f-functions being multiplied increases, we need to look for a power expansion of the product integrals in $\exp\{A\}$ that occur in the second line of (11.111). Note that because eventually we shall need the logarithm of the partition function, it is convenient to express the product integrals as an exponential. There is also another important benefit that this form of expression shares with other studies that involve interacting manybody systems. The relevant series expansion for $\exp\{A\}$ is restricted only to a class of multiply-connected f-functions. In fact, we get

$$\begin{aligned}
\{A\} &= V^{-2}\left(\int_{-\infty}^{\infty}\int_{-\infty}^{\infty}\sum_{i<j} f_{i,j}\mathrm{d}^3\boldsymbol{r}_i\mathrm{d}^3\boldsymbol{r}_j\right) \\
&\quad + V^{-3}\left(\int_{-\infty}^{\infty}\int_{-\infty}^{\infty}\int_{-\infty}^{\infty}\sum_{i<j<k} f_{i,j}f_{j,k}f_{k,i}\mathrm{d}^3\boldsymbol{r}_i\mathrm{d}^3\boldsymbol{r}_j\mathrm{d}^3 r_k\right) + \cdots \\
&= V^{-2}\frac{N(N-1)}{2!}\left(\int_{-\infty}^{\infty}\int_{-\infty}^{\infty} f_{i,j}\mathrm{d}^3\boldsymbol{r}_i\mathrm{d}^3\boldsymbol{r}_j\right) \\
&\quad + V^{-3}\frac{N(N-1)(N-2)}{3!}\left(\int_{-\infty}^{\infty}\int_{-\infty}^{\infty}\int_{-\infty}^{\infty} f_{i,j}f_{j,k}f_{k,i}\mathrm{d}^3\boldsymbol{r}_i\mathrm{d}^3\boldsymbol{r}_j\mathrm{d}^3 r_k\right) \\
&\quad + \cdots .
\end{aligned} \tag{11.113}$$

In (11.113) we have recognized that in an N particle system in toto there are $\frac{N(N-1)}{2!}$ pairs and $\frac{N(N-1)(N-2)}{3!}$—multiply connected—triplets available. Also that each particle can be represented by any dummy position index i, j, or k. And further that the volume available to each particle is equal to V.

As is well known, for two identical particles i and j, $\mathrm{d}^3\boldsymbol{r}_i\mathrm{d}^3\boldsymbol{r}_j$ can be represented as $\mathrm{d}^3\boldsymbol{R}\mathrm{d}^3\boldsymbol{r}$ where $\boldsymbol{R}$ is the position vector of the center of mass, i.e., $\boldsymbol{R} = (\boldsymbol{r}_j + \boldsymbol{r}_i)/2$, and $\boldsymbol{r}$ is their vector separation, that is, $\boldsymbol{r} = \boldsymbol{r}_i - \boldsymbol{r}_j$. Because $f_{i,j} = f(|\boldsymbol{r}_i - \boldsymbol{r}_j|) = f(r)$, we can write (for dummy indices i, j)

$$\int_{-\infty}^{\infty}\cdots\int_{-\infty}^{\infty} f_{i,j}\mathrm{d}^3\boldsymbol{r}_i\mathrm{d}^3\boldsymbol{r}_j = \int \mathrm{d}^3 R\int_0^{\infty} f(r)\mathrm{d}^3 r = V\int_0^{\infty} f(r)4\pi r^2\mathrm{d}r .$$

Therefore, the Helmholtz potential, calculated from the partition function specified in (11.110) and (11.113), becomes

$$\begin{aligned} F &= -k_B T \ln\left[\Xi(N,V,T)\right] = -k_B T \ln\left\{\Xi(N,V,T)_{IG}\right\} \\ &- k_B T\left(\frac{N^2}{2V}\right)\int_0^\infty f(r)4\pi r^2 \mathrm{d}r \\ &- k_B T\left(\frac{N^3}{6V^3}\right)\left(\int_{-\infty}^{\infty}\int_{-\infty}^{\infty}\int_{-\infty}^{\infty} f_{i,j} f_{j,k} f_{k,i} \mathrm{d}^3\boldsymbol{r}_i \mathrm{d}^3\boldsymbol{r}_j \mathrm{d}^3\boldsymbol{r}_k\right) \\ &- \cdots . \end{aligned} \tag{11.114}$$

Note that $\Xi(N, V, T)_{IG}$ is as defined in (11.13). Also note that in the above we have used an approximation that is valid for large N, i.e., $N(N-1) = N^2(1-\frac{1}{N}) \approx N^2$ and $N(N-1)(N-2) \approx N^3$.

The third term on the right-hand side of (11.114), involves three- and higher-body integrals. Even the simplest of these—namely the three-body integral with only the hard-core potential—requires considerable effort to evaluate. However, because in the limit of weak-interaction and/or high temperature, the physical contribution of these terms is much smaller than that of the second term, in the following we shall work with an approximation where they can be ignored.

With this approximation the Helmholtz potential energy is

$$F \approx -k_B T \ln\left\{\Xi(N,V,T)_{IG}\right\} - k_B T\left(\frac{N^2}{2V}\right)\int_0^\infty f(r)4\pi r^2 \mathrm{d}r . \tag{11.115}$$

Therefore the pressure, P, is as follows:

$$\begin{aligned} P = -\left(\frac{\partial F}{\partial V}\right)_T &= \frac{N k_B T}{V} - \left(\frac{N^2 k_B T}{2V^2}\right)\int_0^\infty f(r)4\pi r^2 \mathrm{d}r + \cdots \\ &= \frac{N k_B T}{V}\left[1 - \left(\frac{N}{2V}\right)\int_0^\infty f(r)4\pi r^2 \mathrm{d}r + \cdots\right]. \end{aligned} \tag{11.116}$$

Clearly, therefore, the second virial coefficient, b_2, defined in (6.16), is the following:

$$b_2 = -\frac{N_A}{2}\int_0^\infty f(r)4\pi r^2 \mathrm{d}r . \tag{11.117}$$

In the following we shall attempt to evaluate it.

11.15 Hard-Core Interaction

For simplicity we consider first a dilute classical gas where the interparticle interaction is of the "hard-core" variety. In a hard-core gas, each particle offers infinite repulsion to any other particle that attempts to come closer than a specified distance. If the hard-core radius is r_0 then the nearest the centers of any two particles can get to is $2r_0$. Therefore, once the centers of the given two particles are separated by a distance greater than $2r_0$, interaction between them vanishes. In other words,

$$U(r) = \infty\,, \quad \text{for} \quad r \le 2r_0, \\ = 0\,, \quad \text{otherwise}\,. \tag{11.118}$$

Thus

$$f(r) = -1\,, \quad \text{for} \quad r \le 2r_0, \\ = 0\,, \quad \text{otherwise}\,. \tag{11.119}$$

Accordingly,

$$-\frac{1}{2}\int_0^\infty f(r)4\pi r^2 \mathrm{d}r = -\frac{1}{2}\int_0^{2r_0}(-1)4\pi r^2 \mathrm{d}r \\ = \frac{16\pi r_0{}^3}{3} = \frac{1}{N}\left\{4N\left[\frac{4\pi}{3}\left(r_0{}^3\right)\right]\right\} = \frac{V_{\text{excluded}}}{N}\,. \tag{11.120}$$

The notation V_{excluded} is the same as first introduced in (6.2).

Inserting the result given in (11.120) into (11.116) yields

$$P = \frac{Nk_B T}{V}\left[1 + \frac{V_{\text{excluded}}}{V} + \cdots\right]. \tag{11.121}$$

It is convenient—as was done in (6.16)—to reexpress (11.121) in terms of the inverse powers of the specific volume $v = \frac{V}{n}$. We get

$$\frac{Pv}{Rt} = 1 + \frac{v_{\text{excluded}}}{v} + \cdots = 1 + \frac{b}{v} + \cdots\,. \tag{11.122}$$

As is apparent from (6.19), when the (long-range) interparticle interaction is absent, i.e., when $a = 0$, the first two terms of the virial expansion recorded above in (11.122) are identical to those obtained for the Van der Waals gas.

Exercise 11.6 Using (11.115) and (11.120) work out thermodynamic potentials for the hard-core gas.

11.16 Lennard-Jones Potential

The total electric charge of an atom is zero. Yet, because of the presence of a (quantum-mechanical) cloud of negatively charged electrons, a pair of atoms experiences interaction. Lennard-Jones, in 1924, proposed what appears to have been a formalization of the classical Van der Waals ideas of a weak attractive interatomic force in the limit of large separation and a strong repulsive force for short separation. The interatomic potential $U(r)$ expressed below is phenomenologically similar to the original version of the Lennard-Jones (L-J) potential, yet is somewhat more convenient for computational purposes:

$$U(r) = U_0\left[\left(\frac{r_0}{r}\right)^{12} - 2\left(\frac{r_0}{r}\right)^6\right]. \tag{11.123}$$

The parameters U_0 and r_0 can be fitted to reproduce experimental data or results of accurate quantum chemistry calculations.

11.16.1 Attractive Potential

The attractive long-range interatomic potential is related to (what is known as) the "London dispersion interaction" [14].

Clearly, there exists a nonzero probability that the dynamic behavior of electron occupancy (variables) will result in temporary occurrences of (electric) dipoles in each of the given pair of atoms. To see how this might come about, consider one of the atoms—to be labeled as the first atom—spontaneously achieving an instantaneous asymmetry of charge distribution. Such asymmetry results in the first atom developing a temporary electric dipole. This dipole then induces an oppositely directed charge distribution in the second atom, which amounts to the development of a dipole in the second atom.

The above "bootstrap" process, spontaneously results in the temporary occurrence of two dipoles. As a result, a temporary attractive force is produced. We know from classical electrostatics that the potential (between two dipoles) falls off as the third power of their separation r.

Quantum mechanical calculations of the resultant energy require the use of perturbation theory. Because of symmetry, first order perturbation is vanishing. The second order perturbation clearly must depend on the square of the interparticle potential. Therefore, the overall result for the energy falls off as $\propto (\frac{1}{r})^6$.

Amusingly, as noted in the preceding section, the thermodynamics of classical electric dipoles also ordains an interparticle potential that falls off as $\propto (\frac{1}{r})^6$.

The L-J potential is a relatively good approximation and due to its simplicity is often used to describe the properties of gases, and to model dispersion and overlap interactions in molecular models. It is particularly accurate for noble gas atoms and is a good approximation at long and short distances for neutral atoms and molecules.

At very long distances the interparticle potential $U(r)$ approaches zero from below. As the interparticle separation decreases, $U(r)$ also (slowly) decreases; $U(r)$ approaches its minimum at $r = r_0$. For shorter distances, it starts to increase rapidly. Indeed, when r is only about $0.705\, r_0$, $U(r)$ has risen[20] more than 50 times as high as the lowest value it attained at $r = r_0$.

[20] See Fig. 11.3.

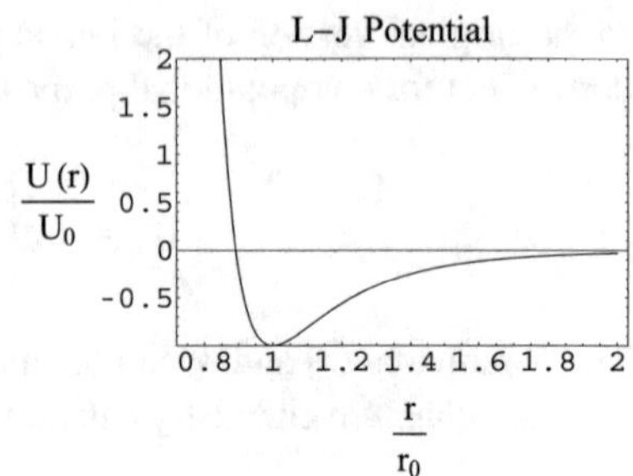

Fig. 11.3 The Lennard-Jones interatomic potential

11.16.2 Repulsive Potential

Clearly, the Lennard-Jones potential (also referred to as the L-J potential, 6–12 potential, or, less commonly, 12–6 potential) is an approximation. The exponent, 12, continues to be chosen because it is equal to the square of the more reliable attractive potential of power 6. This is done largely for the resultant ease in computation. Indeed, the precise form of the repulsive term has scant theoretical justification. Given that its physical origin is quite possibly related to the Pauli principle—meaning, when the electronic clouds surrounding the atoms start to overlap, the energy of the system increases appropriately. Therefore, one might have expected the repulsive force to have had an exotic dependence on the interparticle separation. Still, as long as the repulsion is steep, the precise nature of the repulsion is not critical to the physics of the problem.

Most undergraduates reading physics take one or two courses in quantum mechanics. Therefore, a long winded introduction is not necessary. And we limit ourselves to a few cursory remarks.

11.17 Quantum Mechanics: Cursory Remark

After a Hamiltonian, $\mathcal{H}$, for the quantum system is assembled as a function of ϱ—which represents translational and other relevant variables—the so-called Schrödinger equation, i.e.,

$$\mathcal{H}(\varrho)\psi_{\rm n}(\varrho) = E_{\rm n}\psi_{\rm n}(\varrho)\,, \tag{11.124}$$

is put together. Any function of ϱ that satisfies the above equation, $\psi_{\rm n}(\varrho)$, is known as an eigenfunction and the corresponding $E_{\rm n}$ as the eigenvalue.

The Hamiltonian is always Hermitian. Therefore, its eigenvalues are real. Physically, they represent the observable value of energy that the system is allowed to have. It turns out, in stark contrast with classical physics, that the system is allowed to have only "quantized levels" of energy. These levels may be "discrete" or "continuous."

Often it is convenient to use Dirac's notation. The square-integrable version of wave-function $\psi_{\rm n_1}(\varrho)$ is represented as an eigenvector in the form of a "ket," i.e., as $|n_1\rangle$. If needed, in addition to n_1 other indices, say n_2, etc., that identify the wave-function $\psi_{\mathrm{n}_1,\mathrm{n}_2}(\varrho)$ may also be added to n_1, e.g., $|n_1; n_2\rangle$. The complex conjugate of $\psi_{\mathrm{n}_1;\mathrm{n}_2}(\varrho)$ is written as a "bra" eigenvector, i.e., as $\langle n_2; n_1|$. Then the following notation applies:

$$\begin{aligned}\langle n_2'; n_1'|n_1; n_2\rangle &= \int \psi^*_{\mathrm{n}_1';\mathrm{n}_2'}(\varrho)\psi_{\mathrm{n}_1;\mathrm{n}_2}(\varrho)\mathrm{d}\varrho \\ &= \delta_{\mathrm{n}_1',\mathrm{n}_1}\delta_{\mathrm{n}_2',\mathrm{n}_2}\,, \\ \langle n_2'; n_1'|V|n_1; n_2\rangle &= \int \psi^*_{\mathrm{n}_1';\mathrm{n}_2'}(\varrho)V(\varrho)\psi_{n_1;n_2}(\varrho)\mathrm{d}\varrho \\ &= V^{\mathrm{n}_1';\mathrm{n}_2'}_{\mathrm{n}_1;\mathrm{n}_2}\,.\end{aligned} \tag{11.125}$$

11.18 Canonical Partition Function

Newton's laws that describe classical mechanics do not demand quantization of energy, angular momentum, or indeed of any observable quantity. The rules of quantum mechanics, on the other hand, specify quantization. As noted above, such quantization may lead to discrete, or continuous values.

Equally important in quantum mechanics, the Hamiltonian and other "operators" for observable quantities are required to be Hermitian. The eigenvalues of such operators represent physically allowed—and therefore, measurable—values. Being eigenvalues of Hermitian operators, these values are necessarily "real."

Let us indicate the eigenvectors and eigenvalues by energy and other relevant indices, for instance, γ and ν. In accord with the BMG theories, a convenient partition function is the following:

$$\Xi = \mathrm{Tr}\big[\exp(-\beta\mathcal{H})\big] = \sum_{\gamma,\nu}\big[\langle\nu,\gamma|\exp(-\beta\mathcal{H})|\gamma,\nu\rangle\big]\,, \tag{11.126}$$

where $|\gamma;\nu\rangle$ is the $\gamma - \nu$th eigenvector of the Hamiltonian and Tr stands for the "trace." Also we have the BMG distribution factor, $f(\mathcal{H})$,

$$f(\mathcal{H}) = \frac{\exp(-\beta\mathcal{H})}{\Xi}\,, \tag{11.127}$$

which can be used to determine the thermal average, $\langle\hat{O}\rangle$, of any observable whose operator is $\hat{O}$,

$$\langle\hat{O}\rangle = \frac{\mathrm{Tr}[\hat{O}\exp(-\beta\mathcal{H})]}{\mathrm{Tr}[\exp(-\beta\mathcal{H})]} = \sum_{\gamma,\nu}\big[\langle\nu,\gamma|\hat{O}f(\mathcal{H})|\gamma,\nu\rangle\big]\,. \tag{11.128}$$

Note that by definition, and common sense, $\langle 1\rangle = 1$.

The thermodynamic potentials are related to the partition function in the usual manner:

$$F = -k_B T \ln\{\Xi\}, \quad P = -\left(\frac{\partial F}{\partial V}\right)_T, \quad S = -\left(\frac{\partial F}{\partial T}\right)_V, \quad G = F + PV,$$
$$U = F + TS = -\left(\frac{\partial \ln\{\Xi\}}{\partial \beta}\right)_V, \quad H = U + PV. \tag{11.129}$$

Again, standard notation has been used. That is, F is the Helmholtz potential; P is the pressure; S is the entropy; G is the Gibbs potential; U is the internal energy; and H is the[21] enthalpy.

11.19 Quantum Particles

Let us imagine a gas of noninteracting—i.e., ideal—quantum particles, each of mass m. This sounds much like a quantum perfect gas!

On second thought, we have to, alas, admit that we are not yet ready to reliably study a truly quantum, manybody thermodynamic system. This is in particular true of a system consisting of light particles (such as electrons) and especially if the system is gaseous or is in liquid state. A reliable study of these can be done only after the grand canonical ensemble has been introduced. And then, we shall need also to treat the effects of fundamental symmetry rules that the particles that make up such a manybody system have to follow.

So while we wait to possibly sketch how to do some of that, we shall amuse ourselves by playing around with a single quantum particle. The simplest example is a particle without any external field that is free to move within a cuboid of lengths, l_1, l_2, and l_3.

11.19.1 Quantum Particles: Motion in One Dimension

To begin with, let us consider motion only in one dimension, say, from $x = 0$ to $x = l$. Let us assume the end points of this range are infinitely thick so the particle cannot cross out of them.

In quantum mechanics, the space and momentum variables, x and p_x, are transformed into Hermitian operators $\hat{X}$ and $\hat{P}_x$, respectively. These operators are inter-related:

$$\hat{P}_x = -i\hbar\left(\frac{\partial}{\partial \hat{X}}\right), \qquad [\hat{X}, \hat{P}_x]_- = \hat{X}\hat{P}_x - \hat{P}_x\hat{X} = i\hbar. \tag{11.130}$$

[21] Partition function determines F, P, S, G, U, and H.

The Hamiltonian $\mathcal{H}$, dependent only on the kinetic energy of the single particle, is

$$\mathcal{H} = \frac{\hat{P}_x^{\;2}}{2m} = -\frac{\hbar^2}{2m}\left(\frac{\partial(\frac{\partial}{\partial \hat{X}})}{\partial \hat{X}}\right). \tag{11.131}$$

The Schrödinger expression is a differential equation,

$$\mathcal{H}\psi_{n_1}(X) = E_{n_1}\psi_{n_1}(X) = -\frac{\hbar^2}{2m}\mathrm{d}\psi_{n_1}(X)X\,, \tag{11.132}$$

which is readily solved. Its solutions are the eigenfunctions,

$$\psi_{n_1}(X) = \left(\frac{2}{l}\right)^{\frac{1}{2}}\sin\left(\frac{n_1\pi X}{l}\right). \tag{11.133}$$

The relevant eigenvalues are

$$E_{n_1} = \frac{\hbar^2}{2m}\left(\frac{n_1\pi}{l}\right)^2. \tag{11.134}$$

Here n_1 is a positive integer $1, 2, \ldots$ Notice that the eigenfunctions, $\psi_{n_1}(X)$, are orthonormal, i.e.,

$$\left(\sqrt{\frac{2}{l}}\right)^2\int_0^l \sin\left(\frac{n_1\pi X}{l}\right)\sin\left(\frac{n_2\pi X}{l}\right)\mathrm{d}X = \delta_{n_1,n_2}\,. \tag{11.135}$$

The choice of the solution given in (11.133) has been motivated by the Dirichlet boundary condition whereby $\psi_{n_1}(X)$ is required to be zero at the hard-core boundaries, $X = 0$ and $X = l$.

Quantum Particles: Partition Function

For a free quantum particle moving in one dimension, between $x = 0 and x = l$, the partition function is

$$\begin{aligned}\Xi(l,T) = \mathrm{Tr}\big[\exp(-\beta\mathcal{H})\big] &= \sum_{n_1=1}^{\infty}\big[\langle n_1|\exp(-\beta E_{n_1})|n_1\rangle\big] \\ &= \sum_{n_1=1}^{\infty}\left[\langle n_1|\exp\left(-\frac{\pi^2\hbar^2 n_1^{\;2}}{2mk_B T l^2}\right)|n_1\rangle\right].\end{aligned} \tag{11.136}$$

For a hydrogen atom,

$$\frac{\pi^2\hbar^2}{2mk_B} \approx 2.3\times 10^{-18}\,\mathrm{m^2K}\,. \tag{11.137}$$

For distance $l \geq 10^{-6}$ m and temperature $T \geq 1$ K, the exponent $-\frac{\pi^2\hbar^2 n^2}{2mk_\mathrm{B}Tl^2}$ changes slowly with change in n. Therefore much as in (11.142) the discrete sum in (11.136) may be approximated by an integral, i.e.,

$$\begin{aligned}
\Xi(l,T) &= \int_1^\infty \exp\left(-\beta\frac{\pi^2\hbar^2 n^2}{2ml^2}\right)\mathrm{d}n = \int_0^\infty \exp\left(-\beta\frac{\pi^2\hbar^2 n^2}{2ml^2}\right)\mathrm{d}n \\
&\quad - \int_0^1 \exp\left(-\beta\frac{\pi^2\hbar^2 n^2}{2ml^2}\right)\mathrm{d}n \\
&= l\sqrt{\frac{2\pi mk_\mathrm{B}T}{h^2}} - \int_0^1 \exp\left(-\frac{\pi^2\hbar^2 n^2}{2ml^2 k_\mathrm{B}T}\right)\mathrm{d}n \\
&\approx l\sqrt{\frac{2\pi mk_\mathrm{B}T}{h^2}}\,.
\end{aligned} \tag{11.138}$$

Note that, at moderate-to-high temperature, in the fourth row on the right-hand side, the first part is much larger than the integral over dn—the latter being ≤ 1. Therefore, the integral has been ignored in the last row.

Quantum Particle: Motion in Three Dimensions

Briefly, the relevant Schrödinger equation is

$$\mathcal{H}\Psi = E\Psi = -\frac{\hbar^2}{2m}[\mathrm{d}\Psi x + \mathrm{d}\Psi y + \mathrm{d}\Psi z]\,, \tag{11.139}$$

whose solution—appropriate to the Dirichlet boundary condition—is as follows:

$$\Psi = \left(\frac{2^3}{l_1 l_2 l_3}\right)^{\frac{1}{2}} \sin\left(\frac{n_1\pi x}{l_1}\right)\sin\left(\frac{n_2\pi x}{l_2}\right)\sin\left(\frac{n_3\pi x}{l_3}\right), \tag{11.140}$$

where n_1, n_2, and n_3 are positive integers, 1, 2, 3, ... Accordingly, the allowed energy levels, $E = E(n_1, n_2, n_3)$, are quantized, namely

$$E(n_1,n_2,n_3) = \left(\frac{\pi^2\hbar^2}{2m}\right)\left\{\left(\frac{n_1}{l_1}\right)^2 + \left(\frac{n_2}{l_2}\right)^2 + \left(\frac{n_3}{l_3}\right)^2\right\}. \tag{11.141}$$

Therefore, the partition function for a single quantum particle in three dimensions is the following:

$$\begin{aligned}
\Xi(V,T) &= \mathrm{Tr}\big[\exp(-\beta\mathcal{H})\big] \\
&= \sum_{(\mathrm{n_1 n_2, n_3})=1}^{\infty} \exp\{-\beta E(n_1,n_2,n_3)\} \\
&\approx l_1 l_2 l_3\left(\sqrt{\frac{2\pi mk_\mathrm{B}T}{h^2}}\right)^3 = V\left(\frac{2\pi mk_\mathrm{B}T}{h^2}\right)^{\frac{3}{2}},
\end{aligned} \tag{11.142}$$

where $V = (l_1 l_2 l_3)$ is the available volume. Note that here we have made use of the results of the integration described in (11.138).

As is clear from (11.13), the above result—i.e., (11.142) for a free quantum particle at moderate-to-high temperature—is identical to the corresponding result for a classical single particle.

11.20 Classical Coquantum Gas

Having now studied the partition function of a single quantum particle with only translational degrees of freedom, the temptation is to make a simple extension to a system of similar manyparticles. We shall succumb to this temptation but be honest in admitting that the resultant gas is not a quantum ideal gas, but rather one that—when described most euphemistically—is a "classical coquantum" gas.

The partition function of N "classical coquantum" particles—assumed here to be distinguishable—of the type analyzed above is equal simply to the Nth power of the single particle partition function given in (11.142). That is,

$$\Xi(N, V, T) = \left[\Xi(V, T)\right]^N = \left[V\left(\frac{2\pi m k_B T}{h^2}\right)^{\frac{3}{2}}\right]^N. \tag{11.143}$$

It is no surprise that this result, and therefore all thermodynamic potentials that follow from it, are identical to those that can be found from (11.13) for a classical ideal gas of N distinguishable particles.

11.21 Noninteracting Particles: Classical Coquantum Versus Quantum Statistics

Whether labeled distinguishable or indistinguishable, any given group of noninteracting classical coquantum particles behaves as though it is a collection of distinct particles. Each particle in the group is "physically" recognizable as a distinct individual and two particles cannot be interchanged without the interchange being noticeable. In addition, there are no fundamental symmetry rules that the (quantum mechanical) wave-functions of such particles have to obey.

In contrast, any two indistinguishable quantum particles can be interchanged without the interchange being physically noticeable. Indeed, in a system consisting of identical quantum particles, a single quantum particle can no longer be considered as occupying only one single-particle constituent state. Rather, each particle may be considered as occupying different fractions of all single particle constituent states. Yet, for those indistinguishable quantum particles that are called fermions, no single-particle state may have more than $2S + 1$ particles present. Note that for electrons the spin S is equal to $\frac{1}{2}$, so $2S + 1 = 2$. Thus any single particle state may have either one electron with spin pointing in any given direction, or two electrons with spins pointing in mutually opposite directions. Therefore, for electrons at any specific time, each state can accommodate a maximum of only two particles.

On the other hand, if we are dealing with quantum particles that are called bosons, there is no limit to the occupancy level of any single-particle state. Both for fermions and bosons, the relevant occupancy prescriptions are a product of the fundamental symmetry rules that these indistinguishable quantum particles are required to obey.

In view of the supremacy of the single particle state occupancy rules, a system of indistinguishable noninteracting quantum particles is often most conveniently treated in terms of the properties of single particle state functions.

We shall have occasion to study these phenomena later in this chapter.

11.22 Quasiclassical Statistical Thermodynamics of Rigid Quantum Diatoms

Statistical treatment of a classical gas of N_D diatoms with stiff bonds and no interdiatom coupling was given in (11.34)–(11.71). The Hamiltonian of such a gas was the sum of N_D Hamiltonians of single diatoms. And the dynamics of any diatom could be described in terms of the translational motion of its center of mass and the rotational motion of the two monatoms around it. These motions were shown to be uncorrelated. Moreover, the center of mass motion was identical to that of free monatoms of equivalent mass. Classical thermodynamics of free monatoms was extensively studied earlier in this chapter. Thus the interesting part of the motion that needed to be treated was the rotational motion of the rigid diatom.

Earlier in this chapter when classical diatoms were analyzed it was noted that for molecular dissociation the temperatures needed were far higher than those available in the laboratory. Mostly the same is true for achieving—by thermal excitation—quantum states that are higher than the ground state. Therefore for calculating thermodynamics at usual laboratory temperature, only the lowest electronic state of the diatom—which is usually nondegenerate—needs to be considered.

At temperatures lower than those where rotation sets in, we can consider a diatom as a single particle in its ground atomic state. Its motion then is much like that of a classical free particle—with mass equal to that of the two atoms—moving at the velocity of the center of mass. Such motion has already been treated in an early part of this chapter (see Sect. 11.5.4, "Translational Motion of Center of Mass", and (11.41)–(11.43)). Therefore we consider first the quantum statistics at slightly higher temperature where rotational motion begins to occur. We assume that the neighboring diatoms are sufficiently far apart so that only intradiatom—and not interdiatom—quantum rules of symmetry are relevant. Such diatoms will be called "quasiclassical." And using quasiclassical version of quantum statistical mechanics, their rotational motion is analyzed below.

While the effects of mutual interaction between the nuclear and the rotational states can be relevant for homonuclear diatoms, they are much less important for heteronuclear diatoms. Therefore, for convenience, we treat first the latter case.

11.23 Heteronuclear Diatoms: Rotational Motion

At temperatures lower than those where the bond stiffness weakens and vibrational modes get excited, the dipoles have stiff bonds. And when rotational modes get excited, these dipoles behave as rigid rotators, each with moment of inertia equal to M. The classical Hamiltonian for the rotational motion of a rigid diatom is the quantity labeled Γ that was given in (11.56) and (11.58). For quantum use, the same Hamiltonian for the ith diatom with angular momentum $\hat{L}$ is, say, $\mathcal{H}_\mathrm{i}$. Then the Schrödinger equation, its eigenfunctions, $\psi_\ell(\theta,\phi)$, and eigenvalues, E_ℓ, are as follows (see [15, 16]):

$$\begin{aligned}
\mathcal{H}_\mathrm{i} &= \frac{\hat{L}^2}{2M} = \frac{\hbar^2}{2M}\left[\left(\frac{\partial^2}{\partial^2\theta}\right) + \left(\frac{1}{\tan\theta}\right)\left(\frac{\partial}{\partial\theta}\right) + \left(\frac{1}{\sin^2(\theta)}\right)\left(\frac{\partial^2}{\partial^2\phi}\right)\right], \\
\mathcal{H}_\mathrm{i}\psi_\ell(\theta,\phi) &= E_\ell\psi_1(\theta,\phi) = \left(\frac{\ell(\ell+1)\hbar^2}{2M}\right)\psi_\ell(\theta,\phi), \qquad (11.144)\\
\psi_\ell(\theta,\phi) &\equiv Y_\ell^{\bar{\mathrm{m}}}(\theta,\phi),
\end{aligned}$$

where $\hat{L}$ is an operator that represents the angular momentum vector of a rigid diatom and, as before, M is its moment of inertia[22]—equal to $\frac{m_1 m_2 r_0^2}{m1+m_2}$, where m_1 and m_2 are the masses of the two monatoms that comprise the diatom and r_0 is the length of the stiff bond separating the two. Also, $\ell = 0, 1, 2, \ldots, \infty$. Because all diatoms being considered are alike, ℓ is independent of whichever of the N_d diatoms—such as the ith—it may refer to. The relevant eigenfunctions are the well known spherical harmonics, $Y_\ell^{\bar{\mathrm{m}}}(\theta,\phi)$, where $\bar{m}$ can take on a total of $2\ell+1$ different values, e.g., $\ell, \ell-1, \ldots, -\ell+1, -\ell$. An important point to note is that as a result, the eigenvalue E_ℓ is $(2\ell+1)$-fold degenerate. This has significant consequences for the partition function.

11.24 Partition Function: Quantum

In a previous calculation—see (11.136)—each state being treated was only singly degenerate. That, however, is not the case here. Therefore, the diagonal matrix elements, i.e., $\langle\ell|\exp(-\beta\mathcal{H}_\mathrm{i})|\ell\rangle$, to be summed for calculating the partition function must also include their degeneracy factor, which is equal to $2\ell+1$. Accordingly, the partition function of the ith rigid diatom is

$$\begin{aligned}
(\Xi_\mathrm{i})_{\text{quantum rotation}} &= \sum_{\ell=0}^{\infty}(2\ell+1)\langle\ell|\exp(-\beta\mathcal{H}_\mathrm{i})|\ell\rangle \\
&= \sum_{\ell=0}^{\infty}(2\ell+1)\exp\{-\beta E_\ell\}
\end{aligned}$$

[22] See (11.52).

$$= \sum_{\ell=0}^{\infty}(2\ell+1)\exp\left\{-\beta\left(\frac{\hbar^2\ell(\ell+1)}{2M}\right)\right\}$$

$$\equiv \sum_{\ell=0}^{\infty} f(\ell) = \sum_{\ell=0}^{\infty}(2\ell+1)\exp\{-\varrho\ell(\ell+1)\},$$

$$\text{where} \quad \varrho = \frac{\beta\hbar^2}{2M} = \frac{h^2}{8\pi^2 M k_B T}, \tag{11.145}$$

and the degeneracy factor, $2l+1$, identifies the number of times the particular state with eigenvalue E_ℓ may occur.

11.25 Partition Function: Analytical Treatment

Numerically, (11.145) is straightforward to compute. Its analytical evaluation, however, can adequately be done only for very low or very high temperature.

For low temperature, i.e., where $\varrho \gg 1$ and as a result $\exp(-\varrho) \ll 1$, the successive terms in (11.145) rapidly decrease in value. Therefore, only the first few terms of the expansion suffice.

$$\left[(\Xi_i)_{\text{quantum rotation}}\right]_{(\text{low T})} \approx 1 + 3\exp(-2\varrho) + 5\exp(-6\varrho) + \cdots. \tag{11.146}$$

For high temperature, where $\varrho \ll 1$, many, many terms contribute to the sum in (11.145). Therefore, much as was done in (11.138), we might conveniently approximate the convergent sum by an integral.[23] In this fashion, by using the Euler–Maclaurin formula for $\lim \rho \ll 1$—that is, given

$$(\Xi_i)_{\text{quantum rotation}} = \sum_{\ell=0}^{\infty} f(\ell), \quad \text{we can write}$$

$$\left[(\Xi_i)_{\text{quantum rotation}}\right]_{\text{high T}}$$
$$=_{\lim \rho \ll 1} \int_0^{\infty} f(x)\mathrm{d}x + \frac{1}{2}f(0) - \frac{1}{12}f'(0) + \frac{1}{720}f'''(0) + \cdots \tag{11.147}$$
$$= \frac{1}{\varrho} + \frac{1}{3} + \frac{\varrho}{15} + \frac{4\varrho^2}{315} + \cdots.$$

[23] A beginner might benefit by noting that for large N the leading term in the sum, $\sum_{\ell=0}^{N}(2\ell+1) = N^2+2N$, is $= N^2$, which is also the case for the corresponding integral, namely $\int_0^N (2x+1)\mathrm{d}x$. The integral approximation for (11.145) is even more satisfactory because successive terms, after ℓ has become large, begin to decrease in value and go rapidly to zero.

Here

$$\begin{aligned} f(\ell) &= (2\ell+1)\exp\{-\varrho\ell(\ell+1)\}, \quad \text{and} \\ \int_0^\infty f(x)\mathrm{d}x &= \int_0^\infty (2x+1)\exp\{-\varrho x(x+1)\}\mathrm{d}x \\ &= \exp\left(\frac{\varrho}{4}\right)\int_0^\infty 2\left(x+\frac{1}{2}\right)\exp\left\{-\varrho\left(x+\frac{1}{2}\right)^2\right\}\mathrm{d}x \\ &= \exp\left(\frac{\varrho}{4}\right)\int_{\frac{1}{2}}^\infty 2y\exp\{-\varrho y^2\}\mathrm{d}y = \frac{1}{\varrho}. \end{aligned} \tag{11.148}$$

We observe that at very high temperature, quantum-statistical partition function for the rotational motion of an heteronuclear rigid diatom with stiff bond is identical to its classical counterpart that was recorded in (11.59). That is,

$$\begin{aligned} \left[(\Xi_{\mathrm{i}})_{\text{quantum rotation}}\right]_{(\text{high T})} &\rightarrow \left[\Xi\,(\text{rotation one stiff diatom})\right]_{\text{classical}} \\ &= \frac{1}{\varrho}. \end{aligned} \tag{11.149}$$

Thermodynamic potential for both low and high temperature are studied next.

11.26 Thermodynamic Potential: Low Temperature

First, let us examine the quantum results for thermodynamic potential at low temperature where $\varrho \gg 1$. The Helmholtz free energy, entropy, internal energy, and specific heat energy are as follows:

$$\begin{aligned} F_{\text{lowTrot}} &= -k_{\mathrm{B}}T\log\left\{\left[(\Xi_i)_{\text{quantum rotation}}\right]_{(\text{lowT})}\right\}^{N_{\mathrm{d}}} \\ &= -3k_{\mathrm{B}}TN_{\mathrm{d}}\exp(-2\varrho)\left[1-\frac{3}{2}\exp(-2\varrho)+O\big(\exp(-4\varrho)\big)\right], \\ S_{\text{lowTrot}} &= 3k_{\mathrm{B}}N_{\mathrm{d}}\exp(-2\varrho)\left[1-\frac{3}{2}\exp(-2\varrho)+O\big(\exp(-4\varrho)\big)\right] \\ &\quad + 6\varrho k_{\mathrm{B}}N_{\mathrm{d}}\exp(-2\varrho)\left[1-3\exp(-2\varrho)+O\big(\exp(-4\varrho)\big)\right], \\ U_{\text{lowTrot}} &= 6\varrho k_{\mathrm{B}}TN_{\mathrm{d}}\exp(-2\varrho)\left[1-3\exp(-2\varrho)+O\big(\exp(-4\varrho)\big)\right], \\ C_{\text{lowTrot}} &= 12k_{\mathrm{B}}N_{\mathrm{d}}\varrho^2\exp(-2\varrho)\left[1-O\big(\exp(-2\varrho)\big)\right]. \end{aligned} \tag{11.150}$$

Note that while the classical specific heat for the rotational motion of a stiff-bonded diatom is constant at all temperature, and is equal to that provided by the two rotational degrees of freedom, at low temperature the quantum specific heat for rotational motion is exponentially small. Thus according to the quantum picture, for the specific heat, only the translational motion of the center of mass with its three degrees of freedom is relevant at low temperatures. And that is exactly what the experiment observes. (See Fig. 11.1.)

11.26.1 Thermodynamic Potential: High Temperature

For one diatom at high temperature, the quantum partition function is given in (11.149). Helmholtz potential for a system of $N_{\rm d}$ diatoms is the following:

$$\begin{aligned} F_{\rm highTrot} &= -k_{\rm B}T\ \ln\left\{\left[(\Xi_{\rm i})_{\rm quantum\ rotation}\right]_{\rm (highT)}\right\}^{N_{\rm d}} \\ &= -k_{\rm B}T N_{\rm d}\left[\ln\left(\frac{1}{\varrho}\right)+\frac{\varrho}{3}+\frac{\varrho^2}{90}+\frac{8\varrho^3}{2835}+\cdots\right]. \end{aligned} \tag{11.151}$$

By using (11.7), the quantum statistical results for thermodynamic potential in the limit of very high temperature can be found from the Helmholtz potential given in (11.151) above. In particular, the high temperature quantum result for the specific heat for rotational motion is

$$C_{\rm highTrot} = N_{\rm d}k_{\rm B}\left[1+\frac{\varrho^2}{45}+\frac{16\varrho^3}{945}+\cdots\right]. \tag{11.152}$$

Again, in the limit of very large temperature, this approaches the classical result.

11.27 Homonuclear Diatoms: Rotational Motion

At very high temperatures, the specific heat of homonuclear and heteronuclear diatoms is the same, approaching as it does their classical value. (See (11.60).) At lower temperatures, the use of quantum statistics results in important difference between the two types of diatoms. When the two atoms in a diatom are identical, respectively depending on whether they obey the Bose[24] or Fermi[25] statistics, the total wave function is symmetric or antisymmetric with respect to their interchange. Equally important is the symmetry characteristic of the rotational state, which is symmetric or antisymmetric depending on whether the quantum number ℓ is even or odd.[26]

As for the nuclear wave-function, it consists of an appropriate linear combination of the spin functions of the two nuclei. Given that the nuclear spin is S, there are a total of $(2S+1)^2$ different combinations, out of which $(S+1)(2S+1)$ are symmetric with respect to the nuclear interchange, and $S(2S+1)$ are antisymmetric. For instance, in H_2 where the nuclei are fermions with $S=\frac{1}{2}$, three[27] nuclear states are symmetric and one[28] state is antisymmetric. We can figuratively represent the symmetric states as (up × up), (down × down), $\sqrt{\frac{1}{2}(\text{up}\times\text{down}+\text{down}\times\text{up})}$, and

[24] Boson spins are equal to 0, 1, 2, . . . , etc.

[25] Fermions have half-odd-integral spin, namely $\frac{1}{2}, \frac{3}{2}, \ldots$, etc.

[26] Note that in contrast, in heteronuclear diatoms, ℓ takes on both even and odd values—see, for example, (11.145).

[27] Note that $(\frac{1}{2}+1)(2\times\frac{1}{2}+1)=3$.

[28] Again note that $\frac{1}{2}(2\times\frac{1}{2}+1)=1$.

the antisymmetric state as $\sqrt{\frac{1}{2}}$(up × down − down × up). Note that both (up) and (down) states are normalized.

When the two nuclei in a diatom are fermions, or bosons, the total wave function is required to be antisymmetric, or symmetric, respectively. The total wave function, of course, represents the product of the rotational and the nuclear states. According to Dennison [17], as noted by Pathria [18], rather than working directly with the partition function, the correct procedure is to use the appropriate fraction of the contribution to the internal energy that arises from the antisymmetric and the symmetric states. In other words, the Dennison suggestion is that one should proceed as follows: Consider a single spin state. Then, by using (11.145), calculate $[(\Xi_{\mathrm{i}})_{\text{quantum rotation}}]$ for both odd and even values of ℓ. That is,

$$\left[(\Xi_{\mathrm{i}})_{\text{quantum rotation}}\right]_{\text{odd}} = \sum_{\ell=1,3,5,\ldots}^{\infty} f(\ell)\,,$$
$$\left[(\Xi_{\mathrm{i}})_{\text{quantum rotation}}\right]_{\text{even}} = \sum_{\ell=0,2,4,\ldots}^{\infty} f(\ell)\,. \tag{11.153}$$

Next by using (11.129), calculate the corresponding value for the internal energy for a single spin state:

$$\left[U(T)\right]_{\text{odd}} = -\left(\frac{\partial \ln\left[(\Xi_{\mathrm{i}})_{\text{quantum rotation}}\right]_{\text{odd}}}{\partial\beta}\right)_{\mathrm{M}},$$
$$\left[U(T)\right]_{\text{even}} = -\left(\frac{\partial \ln\left[(\Xi_{\mathrm{i}})_{\text{quantum rotation}}\right]_{\text{even}}}{\partial\beta}\right)_{\mathrm{M}}. \tag{11.154}$$

Finally, multiply the above by the corresponding ratio of the odd and even spin states so that while the total result for a diatom with fermion-like atoms is antisymmetric, for boson-like atoms it is symmetric. This gives

$$\left[U(T)\right]_{\text{fermi}} = \frac{S(2S+1)}{(2S+1)^2}\left[U(T)\right]_{\text{even}} + \frac{(S+1)(2S+1)}{(2S+1)^2}\left[U(T)\right]_{\text{odd}},$$
$$\left[U(T)\right]_{\text{bose}} = \frac{S(2S+1)}{(2S+1)^2}\left[U(T)\right]_{\text{odd}} + \frac{(S+1)(2S+1)}{(2S+1)^2}\left[U(T)\right]_{\text{even}}. \tag{11.155}$$

11.27.1 Homonuclear Diatoms: Very High Temperature

As is clear from (11.153), at very high temperature only the very large values of ℓ contribute significantly to the partition functions $[(\Xi_{\mathrm{i}})_{\text{quantum rotation}}]_{\text{odd}}$ and $[(\Xi_{\mathrm{i}})_{\text{quantum rotation}}]_{\text{even}}$. Therefore, either of these partition functions for a single spin state approaches the limiting value equal to $\frac{1}{2}\sum_{\ell=0}^{\infty} f(\ell)$, which according to (11.151), is $\approx \frac{1}{2\varrho}$. Multiplying this result by the total number of spin states yields the high temperature limiting value for the partition function.

The high temperature partition function—and therefore the internal energy—is independent of the total number of spin states and the result for the internal energy approaches the classical result—see (11.60)—namely, $U_{\mathrm{rotation}} = H_{\mathrm{rotation}} = N_{\mathrm{d}} k_{\mathrm{B}} T$.

11.27.2 Homnuclear Diatoms: Very Low and Intermediate Temperature

At very low temperature, because the decreasing exponential decreases rapidly in value, only the leading term is relevant. Indeed, we can write:

$$\begin{aligned} \left[(\Xi_{\mathrm{i}})_{\mathrm{quantum\ rotation}}\right]_{\mathrm{odd}} &\approx 3\exp(-2\varrho)\,, \\ \left[(\Xi_{\mathrm{i}})_{\mathrm{quantum\ rotation}}\right]_{\mathrm{even}} &\approx 1\,. \end{aligned} \tag{11.156}$$

Accordingly, only the odd term, namely

$$\left[U(T)\right]_{\mathrm{fermi}} \approx \frac{(S+1)(2S+1)}{(2S+1)^2}\left[U(T)\right]_{\mathrm{odd}}\,, \tag{11.157}$$

contributes to the internal energy (and therefore the specific heat). For H_2—which is the Fermi case with $S = \frac{1}{2}$—the spin dependent factor, $\frac{(S+1)(2S+1)}{(2S+1)^2}$, is equal to $\frac{3}{4}$. In any case, the specific heat at very low temperature decreases extremely rapidly, i.e., exponentially, as it tends to zero when $T \to 0$.

For intermediate temperature the $[U(T)]_{\mathrm{even}}$ and $[U(T)]_{\mathrm{odd}}$, that appear in (11.155), are best calculated numerically. According to Pathria, the calculated result for the temperature derivative of $[U(T)]_{\mathrm{fermi}}$, for $S = \frac{1}{2}$, agrees well with the experimental result for the specific heat of diatomic hydrogen.

11.28 Diatoms with Vibrational Motion

As mentioned earlier, vibrational states of diatoms—particularly those with small masses—come into play at temperatures much higher than the rotational states. The temperature range where vibrations begin contributing to the system thermodynamics is of the order 10^3 K. And, it is estimated that the classical equipartition predictions would begin holding only at temperatures of the order 10^4 K or higher.

The Hamiltonian that describes the vibrational motion of a single diatom—let us say the ith such diatom—has been given in (11.67). It is convenient to change the earlier notation, namely

$$\left[H(\text{vibration single diatom})\right] = \left(\frac{m_1 m_2 \omega^2}{2(m_1+m_2)}\right) r_0^2 + p^2 \frac{m_1+m_2}{2m_1 m_2}\,,$$

to the following:

$$\left[H(\text{vibration single diatom})\right] \to [\mathcal{H}_{\mathrm{i}}] \equiv \frac{1}{2} m\omega^2 {q_{\mathrm{i}}}^2 + \frac{p_{\mathrm{i}}^2}{2m}\,, \tag{11.158}$$

where we have set $r_0 \to q_i$, $p \to p_i$, and $\frac{m_1 m_2}{m_1+m_2} \to m$. Clearly, this Hamiltonian is identical to that for a classical one-dimensional harmonic oscillator. So when we are focusing exclusively on the thermodynamics of vibrational modes of diatoms, it is convenient to refer to it as "the thermodynamics of one-dimensional harmonic oscillators."

For quantum-mechanical use of the Hamiltonian, both q_i and p_i are to be treated as operators—denoted as $\hat{q}_i$ and $\hat{p}_i$—that obey the relationship $\hat{q}_i\hat{p}_i - \hat{p}_i\hat{q}_i = i\hbar$. However, instead of $\hat{q}_i$ and $\hat{p}_i$, the diagonal nature of the above Hamiltonian is best utilized by working with the operator $\hat{a}_i$ and its Hermitian conjugate $\hat{a}_i{}^+$. That is,

$$\begin{aligned} \hat{a}_i &= \sqrt{\frac{m\omega}{2\hbar}}\left(\hat{q}_i + i\frac{\hat{p}}{m\omega}\right), \\ \hat{a}_i{}^+ &= \sqrt{\frac{m\omega}{2\hbar}}\left(\hat{q}_i - i\frac{\hat{p}}{m\omega}\right), \\ \hat{\mathcal{H}}_i &= \hbar\omega\left(\hat{a}^+\hat{a} + \frac{1}{2}\right). \end{aligned} \tag{11.159}$$

11.29 Quantum-Statistical Treatment: Quasiclassical

Assume that the system can be described by a quasiclassical quantum picture whereby individual Harmonic oscillators—that is, diatoms—are treated quantum mechanically but the collection of N oscillators is not required to obey the quantum manybody symmetry rules. The Schrödinger equation and the matrix elements of the quantum Hamiltonian, $\mathcal{H}_i$, for the ith harmonic oscillator can be expressed in[29] very compact form. That is,

$$\begin{aligned} \mathcal{H}_i|\nu\rangle &= \hbar\omega\left(\nu + \frac{1}{2}\right)|\nu\rangle, \\ \langle\nu|\mathcal{H}_i|\nu'\rangle &= \hbar\omega\left(\nu + \frac{1}{2}\right)\delta_{\nu,\nu'}, \end{aligned} \tag{11.160}$$

where ν and ν' are $= 0, 1, 2, 3, \ldots, \infty$. Equation (11.160) readily leads to

$$\langle\nu|\exp(-\beta\mathcal{H}_i)|\nu\rangle = \exp\left\{-\beta\hbar\omega\left(\nu + \frac{1}{2}\right)\right\} = \exp\left\{-\gamma\left(\nu + \frac{1}{2}\right)\right\},$$

where, for convenience, we have used the notation $\gamma = \beta\hbar\omega$.

The quantum partition function for the ith simple harmonic oscillator is

$$\Xi_i = \sum_{\nu=0}^{\infty}\langle\nu|\exp(-\beta\mathcal{H}_i)|\nu\rangle = \sum_{\nu=0}^{\infty}\exp\left[-\gamma\left(\nu + \frac{1}{2}\right)\right]$$

[29] Schrödinger equation matrix element of quantum Hamiltonian.

$$= \exp\left(-\frac{\gamma}{2}\right)\sum_{\nu=0}^{\infty}\exp(-\gamma\nu) = \frac{\exp\left(-\frac{\gamma}{2}\right)}{1-\exp(-\gamma)}$$

$$= \left[2\sinh\left(\frac{\gamma}{2}\right)\right]^{-1}, \quad \text{where} \quad \gamma = (\beta\hbar\omega). \tag{11.161}$$

Because there is no inter-harmonic-oscillator potential, any given oscillator is independent of the other $N-1$ oscillators. Further, because the total partition function of a quasiclassical composite system is the product of the partition functions of its components, the partition function of N quasiclassical quantum simple harmonic oscillators is as follows:

$$\Xi(N,T) = \prod_{\mathrm{i}=1}^{N}\Xi_{\mathrm{i}} = \left[2\sinh\left(\frac{\gamma}{2}\right)\right]^{-N}. \tag{11.162}$$

Using (11.129) and (11.162), we get

$$\begin{aligned}
F(N,T) &= Nk_{\mathrm{B}}T\ln\left[2\sinh\left(\frac{\gamma}{2}\right)\right],\\
U(N,T) &= -\left(\frac{\partial\{\ln\Xi(N,T)\}}{\partial\beta}\right)_{\mathrm{N,V}} = \frac{N\hbar\omega}{2}\coth\left(\frac{\gamma}{2}\right),\\
S(N,T) &= \frac{U(N,T)-F(N,T)}{T}\\
&= \left(\frac{N\hbar\omega}{2T}\right)\coth\left(\frac{\gamma}{2}\right) - Nk_{\mathrm{B}}\ln\left[2\sinh\left(\frac{\gamma}{2}\right)\right].
\end{aligned} \tag{11.163}$$

Because $\Xi(N,T)$ does not involve V and P, $H(N,T) = U(N,T)$. As a result,

$$\left(\frac{\partial U}{\partial T}\right)_{\mathrm{N,p,v}} = C_{\mathrm{v}} = C_{\mathrm{p}} = C = Nk_{\mathrm{B}}\gamma^2\frac{\exp(\gamma)}{[\exp(\gamma)-1]^2}. \tag{11.164}$$

Let us refer to temperature T as being "high" when $\frac{k_{\mathrm{B}}T}{\hbar\omega}$ is of the order of, or larger than, 2. At high temperatures, it is instructive to compare the quantum results—see (11.163) and (11.164)—with classical ones—see (11.72)–(11.77) and set $a \equiv b = 0$.

As noted, the parameter $\gamma = \frac{\hbar\omega}{k_{\mathrm{B}}T}$ is $< \frac{1}{2}$ at such a "high" temperature. Therefore we have

$$\begin{aligned}
U &= U_{\mathrm{classical}}\left[1+\frac{\gamma^2}{12}-\cdots\right],\\
C &= C_{\mathrm{classical}}\left[1-\frac{\gamma^2}{12}+\cdots\right],\\
S &= S_{\mathrm{classical}} + Nk_{\mathrm{B}}\gamma^2\left(\frac{11}{24}\right)-\cdots.
\end{aligned} \tag{11.165}$$

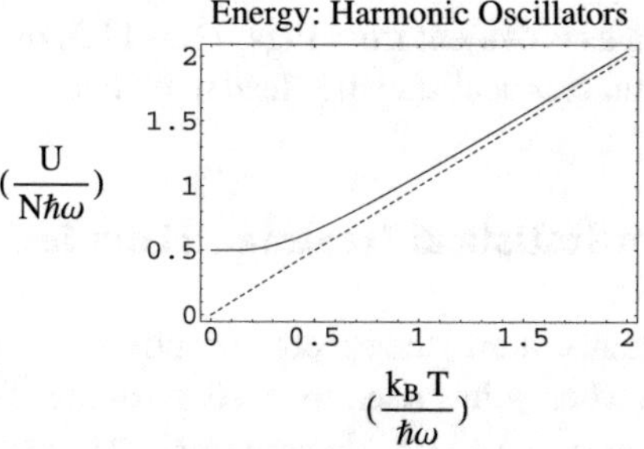

Fig. 11.4 Energy of a quantum and a classical simple harmonic oscillator. The solid curve represents the average energy of a quantum simple harmonic oscillator as a function of the dimensionless variable $\gamma^{-1} = \frac{k_B T}{\hbar\omega}$. The dashed curve represents the corresponding result for a classical simple harmonic oscillator

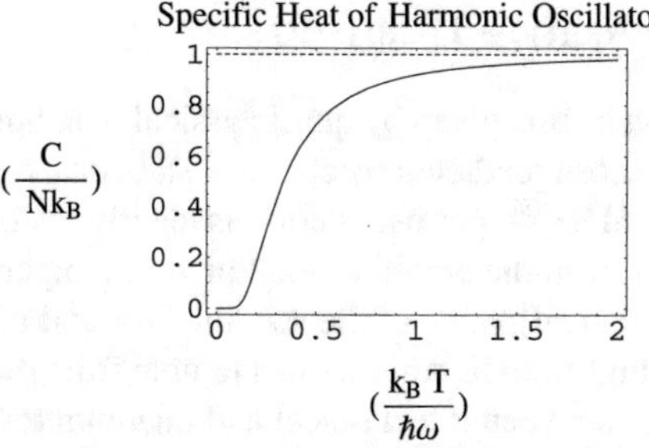

Fig. 11.5 Specific heat of a quantum and a classical simple harmonic oscillator. The solid curve represents $\frac{C}{Nk_B}$ of a quantum simple harmonic oscillator as a function of the dimensionless variable $\gamma^{-1} = \frac{k_B T}{\hbar\omega}$. The dashed curve represents the corresponding result for a classical simple harmonic oscillator. Remember $C_p = C_v = C$

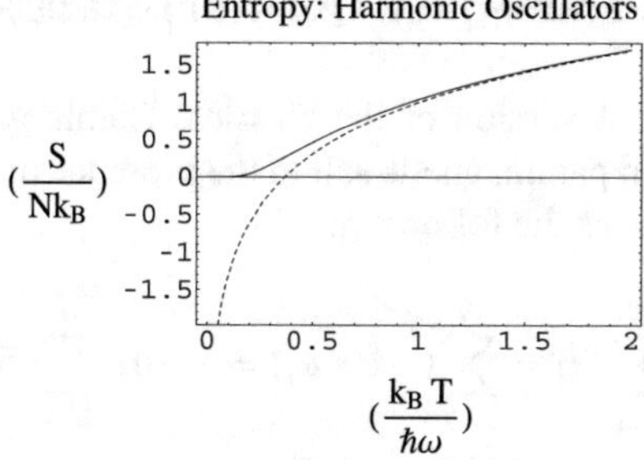

Fig. 11.6 Entropy of a quantum and a classical simple harmonic oscillator. The solid curve represents the entropy of a quantum simple harmonic oscillator as a function of the dimensionless variable $\gamma^{-1} = \frac{k_B T}{\hbar\omega}$. The entropy is given in units of k_B. The dashed curve represents the corresponding result for a classical simple harmonic oscillator

In Figs. 11.4, 11.5, and 11.6, the energy, U, specific heat, $C = (\frac{\partial U}{\partial T})_{p,v}$, and entropy, S, are plotted for both the classical and quantum one-dimensional harmonic oscillators. With rise in temperature, all these quantities are seen to approach their classical value. Indeed, as is evident from Figs. 11.3–11.5, by the time $\frac{k_B T}{\hbar\omega}$ has risen to ≈ 2, the quantum and classical statistics lead to essentially the same result.

11.29.1 Quantum Statistical Treatment: Low Temperature

With the lowering of temperature, the specific heat decreases rapidly. In the limit of very low temperature, when γ has risen to ≈ 10, meaning $\frac{k_B T}{\hbar\omega}$ has fallen to ≈ 0.1, $\exp(-\gamma)$—indeed, even $\gamma^2 \exp(-\gamma)$—is very small. Then we have

$$
\begin{aligned}
U &= N\left(\frac{\hbar\omega}{2}\right)\left[1 + 2\exp(-\gamma) + \cdots\right], \\
C &= N k_B \gamma^2 \exp(-\gamma)\left[1 + 2\exp(-\gamma) + \cdots\right], \\
S &= N k_B (\gamma + 1) \exp(-\gamma) .
\end{aligned}
\tag{11.166}
$$

These results—meaning results given by quasiclassical quantum statistic—are qualitatively different from those predicted by classical statistics. Note that the quantum, zero-point energy—equal to $\frac{\hbar\omega}{2}$ per oscillator—is clearly visible in Fig. 11.4. As is demonstrated in Fig. 11.5, in the neighborhood of zero temperature, both the (quasiclassical quantum-statistical) value of the specific heat and its rate of change with temperature are vanishingly small. Another fact to note from these figures is that not only does the difference between the classical and quantum results become large as the temperature falls below (about) $\frac{\hbar\omega}{k_B}$, even the qualitative behavior of the classical results begins to be suspect—for instance, see the classical result that the system entropy becomes negative as the temperature heads further below $\approx 0.4(\frac{\hbar\omega}{k_B})$.

11.30 Langevin Paramagnet: Quantum-Statistical Picture

The quantum-mechanical version of the classical Hamiltonian that was given in (11.92)—for a Langevin paramagnetic salt in the presence of applied field $\boldsymbol{B}$ which in the z-direction is B_0—is the following:

$$
\begin{aligned}
\mathcal{H} &= \sum_{i=1}^{N} \mathcal{H}_i = \sum_{i=1}^{N} (-\boldsymbol{B} \cdot \boldsymbol{\mu}_i) = -g\mu_B \sum_{i=1}^{N} (\boldsymbol{B} \cdot \boldsymbol{J}_i) \\
&= -g\mu_B B_0 \sum_{i=1}^{N} J_{i,z} .
\end{aligned}
\tag{11.167}
$$

The vector parameter, $\boldsymbol{\mu}_i$, represents the magnetic moment of the ith dipole. Quantum mechanics stipulates that $\boldsymbol{\mu}_i$ be related to its quantized angular momentum vector variable $\boldsymbol{J}_i$. That is,

$$
\boldsymbol{\mu}_i = g\mu_B \boldsymbol{J}_i , \qquad \langle \nu | J_{i,z} | \nu' \rangle = \nu \delta_{\nu,\nu'} . \tag{11.168}
$$

Here g is the Landé g-factor, and $\mu_{\rm B}$ is the Bohr magneton. Note that a Bohr magneton is the magnetic dipole moment of an electron. In CGS units, it is equal to

$$\mu_{\rm B} = \frac{e\hbar}{2mc} = 9.27400915(23) \times 10^{-21}\ \mathrm{Erg\ Oe^{-1}}\,, \tag{11.169}$$

where e is the charge of an electron, m is its rest mass, and c is the speed of light in vacuum.

The eigenvalues of $J_{\rm i,z}$ are denoted as ν such that ν can take on—indeed, it may take on only—the values:

$$\nu = J, J-1, \ldots, -J+1, -J\,. \tag{11.170}$$

Clearly, the total number of allowed eigenvalues of $J_{\rm i,z}$ is equal to $2J+1$. However, depending on the paramagnetic salt being studied, the value of J may be integral or half-odd integral. That is,

$$J = 0, 1, 2, \ldots, \text{ or } \frac{1}{2}, \frac{3}{2}, \ldots \tag{11.171}$$

11.31 Helmholtz Potential and Partition Function

Much like the collection of quasiclassical quantum simple harmonic oscillators, the Hamiltonian here is completely separable. Therefore, following the procedure outlined in (11.126)–(11.163), we can write the relevant single-particle partition function as follows:

$$\begin{aligned}
\Xi_{\rm i} &= \mathrm{Tr}\big[\exp(-\beta\mathcal{H})\big] = \sum_{\nu=-J}^{+J} \langle\nu|\exp(-\beta\mathcal{H}_{\rm i})|\nu\rangle \\
&= \sum_{\nu=-J}^{+J} \langle\nu|\exp(\beta B_0 g\mu_{\rm B} J_{\rm i,z})|\nu\rangle = \sum_{\nu=-J}^{+J} \exp(\beta B_0 g\mu_{\rm B}\nu) \\
&= \sum_{\nu=-J}^{+J} \exp\left(\frac{x\nu}{J}\right) = \exp(-x)\left[1 + \exp\left(\frac{x}{J}\right) + \cdots + \exp\left(2J\frac{x}{J}\right)\right] \\
&= \exp(-x)\left[\frac{1-\exp\{\frac{x}{J}(2J+1)\}}{1-\exp(\frac{x}{J})}\right] = \frac{\sinh\{x(1+\frac{1}{2J})\}}{\sinh(\frac{x}{2J})}\,.
\end{aligned} \tag{11.172}$$

In the above and henceforth, $x = \frac{B_0 g\mu_{\rm B} J}{k_{\rm B}T}$. For convenience, the applied magnetic field has been chosen to be along the z-axis and the z-component of $\boldsymbol{B}$ is B_0. The full partition function, $\Xi(N, T, B_0)$, and the Helmholtz free energy are

$$\begin{aligned}
\Xi(N, T, B_0) &= \prod_{\rm i=1}^{N} \Xi_{\rm i} = \left[\frac{\sinh\{x(1+\frac{1}{2J})\}}{\sinh(\frac{x}{2J})}\right]^N, \\
F &= -k_{\rm B}T\ln\big[\Xi(N, T, B_0)\big] \\
&= -Nk_{\rm B}T\ln\left[\frac{\sinh\{x(1+\frac{1}{2J})\}}{\sinh(\frac{x}{2J})}\right].
\end{aligned} \tag{11.173}$$

Entropy

As noted in (11.129), the entropy can now be determined from the temperature derivative of the Helmholtz free energy F given in (11.173).

$$S = -\left(\frac{\partial F}{\partial T}\right)_{\mathrm{N,J}} = Nk_{\mathrm{B}} \ln\left[\frac{\sinh\{x(1+\frac{1}{2J})\}}{\sinh(\frac{x}{2J})}\right]$$
$$- xNk_{\mathrm{B}}\left\{\left(1+\frac{1}{2J}\right)\coth\left[x\left(1+\frac{1}{2J}\right)\right] - \left(\frac{1}{2J}\right)\coth\left(\frac{x}{2J}\right)\right\}. \quad (11.174)$$

For given N and J, the entropy S is a function only of x. But because $x = \frac{B_0 g \mu_{\mathrm{B}} J}{k_{\mathrm{B}} T}$, the relevant variable in x is the ratio $\frac{B_0}{T}$. Therefore, for constant S, $\frac{B_0}{T}$ is constant. That is,

$$\left(\frac{B_0}{T}\right)_{\mathrm{S}} = \text{const.} \quad (11.175)$$

11.32 System Entropy: High Temperature

At high temperatures, that is, when $x \ll 1$, the system entropy S approaches the limit[30] $Nk_{\mathrm{B}} \ln(2J+1)$,

$$\frac{S}{Nk_{\mathrm{B}}} = \ln(2J+1) - \left(\frac{J+1}{6J}\right)x^2 + \left(\frac{2J^3+4J^2+3J+1}{120J^3}\right)x^4 + O\left(x^6\right). \quad (11.176)$$

This result is in agreement with the Boltzmann prescription which gives the high temperature limiting value as being equal to $S = Nk_{\mathrm{B}} \ln(\Omega)$ where Ω is the multiplicity per magnetic dipole equal to $2J+1$. Indeed, even when $\frac{k_{\mathrm{B}} T}{B_0 g \mu_{\mathrm{B}} J}$ is only as large as 5—this means, when x is only as small as $\frac{1}{5}$—it is noticed—see Fig. 11.7—that $\frac{S}{Nk_{\mathrm{B}}}$ has already approached a value close to its limiting values of $\ln(2J+1)$.

11.32.1 System Entropy: Low Temperature

As the temperature is lowered and x becomes > 1, the entropy begins to decrease rapidly. Indeed, according to (11.174), the entropy reduces exponentially at low temperatures. That is,

$$S = Nk_{\mathrm{B}}(2x+1)\exp(-2x) + O\left[x\exp(-4x)\right]. \quad (11.177)$$

Thus, the entropy is vanishingly small at zero temperature (where $x \to \infty$). Again this fact is visible in Fig. 11.7.

[30]In particular, when $J = \frac{1}{2}$, the small x, high temperature expansion of the entropy yields $(\frac{S}{Nk_{\mathrm{B}}})_{\mathrm{J}=\frac{1}{2}} = \ln(2) - \frac{x^2}{2} + \frac{x^4}{4} - \frac{x^6}{9} + O(x^8)$.

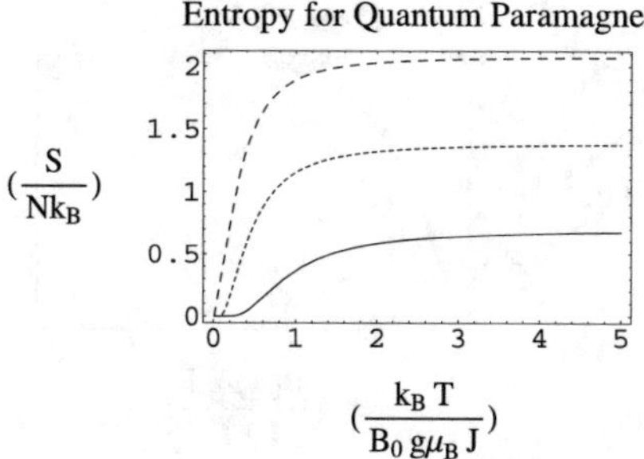

Fig. 11.7 Entropy of a quasiclassical quantum Langevin paramagnet as a function of $\frac{1}{x} = \frac{k_B T}{B_0 g \mu_B J}$. The solid curve refers to $J = \frac{1}{2}$; the short-dashed curve relates to $J = \frac{3}{2}$, and the curve with sightly longer dashes is for $J = \frac{7}{2}$. Notice that at high end of the temperature scale, the curves are clearly heading toward their limiting value of $\ln(2J + 1)$. The plots are in dimensionless units

11.33 Internal Energy

As noted in (11.164), the internal energy U is equal to the negative of the derivative with respect to β of the logarithm of the partition function,

$$\begin{aligned} -U &= \left(\frac{\partial(\ln \Xi)}{\partial \beta}\right)_{\mathrm{N,V,B_0}} \\ &= B_0 N g \mu_B J \left[\left(1+\frac{1}{2J}\right)\coth\left[\left(1+\frac{1}{2J}\right)x\right] - \left(\frac{1}{2J}\right)\coth\left(\frac{x}{2J}\right)\right] \\ &= B_0 N g \mu_B J \left[B_{\mathrm{J}}(x)\right] = B_0 M_{\mathrm{sat}} B_{\mathrm{J}}(x) = B_0 M\,, \end{aligned} \tag{11.178}$$

where

$$M_{\mathrm{sat}} = N g \mu_B J\,, \qquad B_{\mathrm{J}}(x) = \frac{M}{M_{\mathrm{sat}}}\,, \qquad x = \frac{B_0 g \mu_B J}{k_B T}\,. \tag{11.179}$$

In (11.179), M is the—temperature and the magnetic field dependent—quasi-classical quantum statistical average of the magnetic moment of the paramagnetic salt. And M_{sat} is its saturation value, while $B_{\mathrm{J}}(x)$, which is often called Brillouin function of order J, is equal to the ratio of the two. Also note that the internal energy U in terms of x—the latter being related to the ratio of two state variables B_0 and T—represents an equation of state.

The ratio $\frac{M}{M_{\mathrm{sat}}}$ is plotted in Fig. 11.8 as a function of $\frac{B_0 g \mu_B J}{k_B T}$ for four different values of J, namely $J = \frac{1}{2}, \frac{3}{2}, \frac{5}{2}$ and $J \to \infty$. Note that for specified values of N and B_0, the ratio $\frac{M}{M_{\mathrm{sat}}}$ is a function only of x, or more particularly, of the ratio $\frac{B_0}{T}$. Therefore for constant M, we have

$$\left(\frac{B_0}{T}\right)_{\mathrm{M}} = \text{const.} \tag{11.180}$$

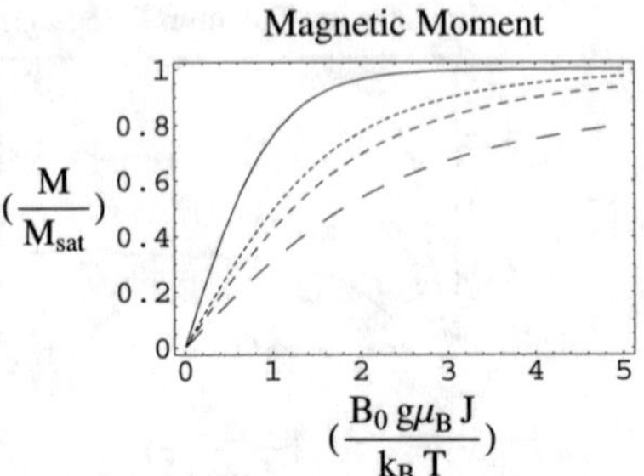

Fig. 11.8 The ratio $\frac{M}{M_{sat}}$ is plotted as a function of $\frac{B_0 g\mu_B J}{k_B T}$. Here M represents the quasi-classical quantum statistical average of the system magnetic moment and M_{sat} its saturation value. The temperature is T and the magnetic field is B_0. The solid curve refers to $J = \frac{1}{2}$; the thinnest, dashed curve is for $J = \frac{3}{2}$; the next lower curve refers to $J = \frac{5}{2}$; while the lowest curve—with long dashes—represents the classical limit, i.e., $J \to \infty$. Notice that the solid curve has nearly reached its maximum possible value of unity when $\frac{B_0 g\mu_B J}{k_B T} \approx 3$. For a $J = \frac{1}{2}$ system, this means that the magnetic moment is close to reaching saturation when $k_B T$ falls below $B_0 \mu_B/3$, or equivalently, when the ratio $\frac{B_0}{T}$ rises above the value $\frac{3k_B}{\mu_B} = \frac{3\times 8.617343(15)\times 10^{-5}}{5.7883817555(79)\times 10^{-5}} (\text{eV K}^{-1}/\text{eV T}^{-1}) \approx 4.4662 \frac{\text{Tesla}}{\text{Kelvin}}$

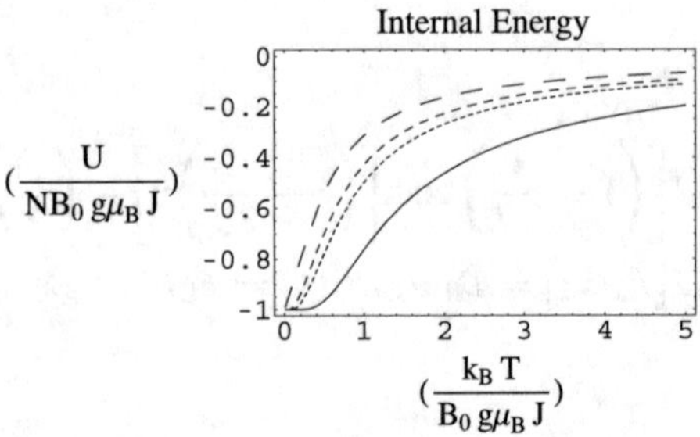

Fig. 11.9 Temperature dependence of the internal energy of the paramagnetic salt is plotted in dimensionless units. The solid curve refers to $J = \frac{1}{2}$, the thinnest, dashed curve is for $J = \frac{3}{2}$, the next higher curve refers to $J = \frac{5}{2}$, while the highest curve—with long dashes—represents the classical limit, i.e., $J \to \infty$

An interesting consequence of (11.180) and (11.175) is that for given B_0 and T, in those reversible processes where the entropy, S, is constant, the magnetization, M, is also constant.

In Fig. 11.9, the internal energy U—as described by (11.178)—is displayed as a function of the temperature. When plotted as dimensionless variables, namely $\frac{U}{N B_0 g\mu_B J}$ versus $\frac{k_B T}{B_0 g\mu_B J}$, the curves for $J = \frac{1}{2}$ and $J = \frac{3}{2}$ are well separated from each other. In contrast, the separation between the curves for $J = \frac{3}{2}$ and $J = \frac{5}{2}$ is much smaller. This fact plays a role in the relative spacing between the location of the Schottky anomaly.

11.33.1 Internal Energy: High Temperature

For high temperatures, where the energy spacing between different allowed energy levels–i.e., the different eigenvalues of the Hamiltonian—is *small* compared with $k_B T$, all the levels are amply occupied. For such cases, the specific heat—as to be shown in (11.186)—is inversely proportional to the square of the temperature. Because of copious occupation of the energy levels, this result also holds for the relevant classical system. Also, here β is small and x is $\ll 1$. Then

$$\left[B_J(x)\right]_{\text{hightemp}} \to \left(\frac{J+1}{3J}\right)x\,. \tag{11.181}$$

Therefore, according to (11.178) and (11.181), the internal energy and the magnetization at high temperatures are

$$\begin{aligned}
-U_{\text{hightemp}} &= B_0 M_{\text{hightemp}} = B_0 M_{\text{sat}}\left[B_J(x)\right]_{\text{hightemp}} \\
&= B_0 M_{\text{sat}}\left[\left(\frac{J+1}{3J}\right)x - \left(\frac{2J^3+4J^2+3J+1}{90J^3}\right)x^3 + O\left(x^5\right)\right] \\
&= NJ(J+1)\frac{(B_0 g\mu_B)^2}{3k_B T}\left[1 - \left(\frac{2J^3+4J^2+3J+1}{30J^2(J+1)}\right)x^2 + O\left(x^4\right)\right].
\end{aligned} \tag{11.182}$$

11.33.2 Internal Energy: Low Temperature

At low temperatures, where the energy spacing between different allowed energy levels–i.e., the different eigenvalues of the Hamiltonian—is large compared with $k_B T$, the specific heat has a very different character. As noted in (11.187), the specific heat decreases exponentially as the temperature is lowered toward zero—see Fig. 11.10.

An interesting feature of the heat capacity in the paramagnetic salts (with applied magnetic field) is the "bump" in the specific heat recorded in Fig. 11.9. This is known as the Schottky anomaly.

The occurrence of a bump is anomalous because in solids the heat capacity generally shows steady change with temperature. It usually increases, or stays constant. Schottky anomaly signifies the presence of a limited number of energy states—e.g., the system being treated here has $2J+1$ states—and because its location shifts rapidly for small J values—see the graphical representation of the change between $J = \frac{3}{2}$ and $J = \frac{1}{2}$—and quite slowly for large J values, it can be helpful in determining the relevant J value and therefore the number of available energy levels.

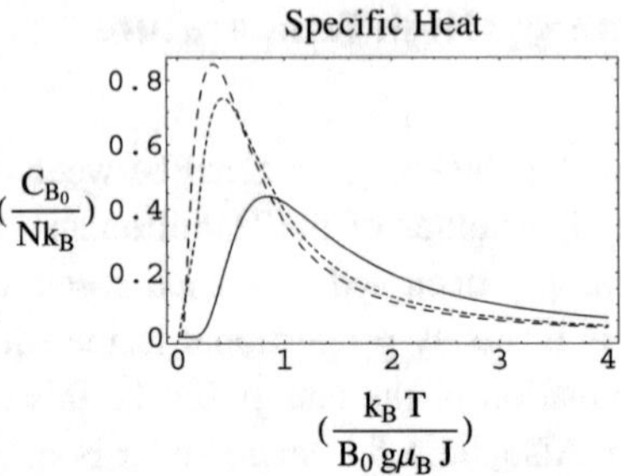

Fig. 11.10 Specific heat of paramagnetic salt at constant applied field, B_0, is plotted as function of the temperature. The plot is in dimensionless units. The solid curve refers to $J = \frac{1}{2}$; the short-dashed curve relates to $J = \frac{3}{2}$, while the curve with slightly longer dashes represents the case $J = \frac{7}{2}$. Notice that even though the difference in the J values is half as large, the separation between the bumps for $J = \frac{3}{2}$ and $J = \frac{1}{2}$ is noticeably greater than the corresponding separation between the bumps for $J = \frac{7}{2}$ and $J = \frac{3}{2}$

At low temperatures β is large and, when the magnetic field B_0 is not too week, $x \gg 1$ for all allowed values[31] of J. And for large x the internal energy, U, and the magnetization, M_{lowtemp}, are as given below in (11.183), namely[32]

$$-U_{\text{lowtemp}} = B_0 M_{\text{lowtemp}} = B_0 M_{\text{sat}}\left[1 - \frac{1}{J}\exp\left(\frac{-x}{J}\right)\right]. \tag{11.183}$$

Under these conditions magnetic saturation begins to be reached, that is, $M_{\text{lowtemp}} \to M_{\text{sat}}$. And essentially all the magnetic dipoles in the paramagnet begin to get aligned in the same direction. Here the magnetic susceptibility for the paramagnet is as given below

$$[\chi]_{\text{lowtemp quasiclassical}} = \left(\frac{\partial M_{\text{lowtemp}}}{\partial B_0}\right)_{\text{T}} = N\frac{(g\mu_B)^2}{k_B T}\exp\left(\frac{-x}{J}\right). \tag{11.184}$$

11.34 Specific Heat: General Temperature

At general temperature, the specific heat, i.e.,

$$\frac{C_{B_0}}{Nk_B} = \left(\frac{1}{Nk_B}\right)\left(\frac{\partial U}{\partial T}\right)_{B_0,N} = \left(\frac{-B_0}{Nk_B}\right)\left(\frac{\partial M}{\partial T}\right)_{B_0,N}$$
$$= x^2\left\{\left(\frac{1}{2J}\right)\frac{1}{\sinh(\frac{x}{2J})}\right\}^2 - x^2\left\{\left(1+\frac{1}{2J}\right)\frac{1}{\sinh[(1+\frac{1}{2J})x]}\right\}^2, \tag{11.185}$$

is plotted in Fig. 11.10.

[31] Note that this means J is not allowed to be equal to zero.

[32] Exercise for the student: Show that for $J = \frac{1}{2}$, $B_J(x) = \tanh(x)$.

11.34.1 Specific Heat: High Temperature

At high-temperatures—that is, for $x \ll 1$—and constant applied field—that is when B_0 is constant—the specific heat and the magnetic susceptibility are readily found as

$$[C_{B_0}]_{\text{hightemp quasiclassical}} = \left(\frac{\partial U_{\text{hightemp}}}{\partial T}\right)_{N,B_0} = -B_0\left(\frac{\partial M_{\text{hightemp}}}{\partial T}\right)_{B_0,N}$$
$$= \frac{Nk_B}{3}J(J+1)\left(\frac{B_0 g\mu_B}{k_B T}\right)^2\left[1-\left(\frac{2J^3+4J^2+3J+1}{10J^2(J+1)}\right)x^2+O(x^4)\right],$$
$$[\chi]_{\text{hightemp quasiclassical}} = \left(\frac{\partial M_{\text{hightemp}}}{\partial B_0}\right)_T$$
$$= NJ(J+1)\frac{(g\mu_B)^2}{3k_B T}\left[1-\left(\frac{2J^3+4J^2+3J+1}{10J^2(J+1)}\right)x^2+O(x^4)\right]. \tag{11.186}$$

11.34.2 Specific Heat: Low Temperature

Similarly, for week but constant applied field and low temperature—that is, when $x \gg 1$—the specific heat for the paramagnet is

$$[C_{B_0}]_{\text{lowtemp quasiclassical}} = \left(\frac{\partial U_{\text{lowtemp}}}{\partial T}\right)_{N,B_0} = -B_0\left(\frac{\partial M_{\text{lowtemp}}}{\partial T}\right)_{B_0,N}$$
$$= Nk_B\left(\frac{B_0 g\mu_B}{k_B T}\right)^2 \exp\left(\frac{-x}{J}\right). \tag{11.187}$$

Notice that at low temperatures the specific heat, and therefore also $(\frac{\partial M}{\partial T})_{B_0,N}$, decrease with temperature at an extremely rapid rate.

11.34.3 Langevin Paramagnet: Classical Statistical Picture

Recall that the dipole moment—expressed as $g\mu_B J$ in the quantum version given above where μ_B denotes a Bohr magneton—is represented as μ_c in the classical version of the Hamiltonian. Consequently, we can relate x that appears in the quantum mechanical version—see its description immediately following (11.172) above—to the α that was defined in (11.94) and used in the classical version. That is,

$$g\mu_B J = \left(\frac{x}{\beta B_0}\right)_{\text{quantum version}} \rightarrow \mu_c = \left(\frac{\alpha}{\beta B_0}\right)_{\text{classical version}}. \tag{11.188}$$

In fact, the quantum-mechanical result that actually corresponds to $(\mu_c)^2$ is not $(Jg\mu_B)^2$, instead, it is $J(J+1)(g\mu_B)^2$. This behavior results from the fact that the eigenvalues of $(\boldsymbol{J}.\boldsymbol{J})$ are not (J^2). Rather, they are $J(J+1)$. The classical limit is achieved when $(\frac{1}{J}) \ll 1$. Then $J(J+1) = J^2(1+\frac{1}{J}) \approx J^2$, and $B_J(x) \approx L(\alpha)$,

$$\begin{aligned} B_J(x) &\approx -\left(\frac{1}{2J}\right)\left[\left(\frac{2J}{x}\right)+O\left(\frac{x}{J}\right)+\cdots\right]+\coth(x)\left[1+O\left(\frac{1}{J}\right)\right] \\ &= \coth(x)-\left(\frac{1}{x}\right)+O\left(\frac{1}{J}\right)=L(x)+O\left(\frac{1}{J}\right). \end{aligned} \tag{11.189}$$

That is,

$$\left[B_J(x)\right]_{\text{quasiclassical quantum}} \to \left[L(\alpha)\right]_{\text{classical}}. \tag{11.190}$$

Using the aforementioned procedure, the quasiclassical quantum results for the specific heat and the magnetic susceptibility, that are given in (11.186), are exactly transformed into the corresponding results for the classical Langevin paramagnet. The latter were recorded in (11.98).

11.35 Adiabatic Demagnetization: Very Low Temperatures

Long ago no satisfactory method existed for reducing the temperature much below the freezing point of water. It was, of course, understood that if there were to be any hope of liquefying gases (other than water vapor), temperature must be lowered a great deal below zero degree centigrade.

The realization that the Joule–Kelvin process could lead to cooling, motivated its use in liquefaction of gases. But to liquefy some of the rare gases, temperatures of the order of a few degrees Kelvin would be needed. And such low temperatures were well beyond the Joule–Kelvin reach.

In order to achieve really low temperatures, use must be made of isothermal magnetization of paramagnetic salts followed by reversible adiabatic decrease of the applied magnetic field.[33] One such process is schematically displayed in Fig. 11.10 where the entropy of the salt is plotted against its temperature for two very different strengths of the applied magnetic field.

The upper curve relates to the paramagnetic salt being under the influence of a weak applied magnetic field. For the lower curve, in contrast, the applied magnetic field is stronger than that for the upper curve. Notice that, as per (11.174), at any finite, fixed temperature, the entropy decreases with increase in the applied field. Accordingly, at constant temperature the derivative of the entropy with respect to the applied field—that is, $(\frac{\partial S}{\partial B_0})_T$—is negative.[34] (See also Fig. 11.12.)

[33] And indeed, for attaining even lower temperatures, reversible adiabatic demagnetization of nuclei themselves is needed.

[34] When, at any fixed finite temperature T, the derivative of the entropy is expanded in powers of the strength of the applied field B_0 we get

$$\left(\frac{\partial S}{\partial B_0}\right)_T = \left(\frac{N g\mu_B J}{T}\right)\left[-\left(\frac{J+1}{3J}\right)x+\left(\frac{2J^3+4J^2+3J+1}{30j^3}\right)x^3-O(x^5)\right],$$

Before the experiment is begun, the temperature of the paramagnetic salt is reduced to the lowest value achievable[35] and the salt is subjected to a weak applied magnetic field. At this instant the entropy versus temperature graph of the salt is represented by the upper curve in Fig. 11.10 and the salt is assumed to be positioned at the point a.

Next, reversibly and isothermally,[36] the applied magnetic field B_0 is increased—meaning the salt is magnetized. In accordance with the prediction that isothermal reversible increase in the strength of the applied field must decrease the entropy of the salt, an isothermal reversible increase in the applied field is shown in Fig. 11.10 as transferring the system from the upper curve to the lower one. In particular, if the system starts at the point a then the given isothermal increase in the applied field will drop the system down to the point b in the lower curve. This exchange of locations—i.e., the process of isothermal magnetization—results in the system having lower energy. Accordingly, during the travel from a to b, the system releases positive amount of heat energy to the liquid helium reservoir.

The salt is now at position b. Its temperature still has its original value. It is magnetized and has less energy. This is so because it is under the influence of a larger magnetic field than was originally the case. Also, because of the increase in the applied field, it has less disorder. Therefore, it has lower entropy.

At this juncture, contact with liquid helium reservoir is broken and the system is thermally isolated.

Next, a reversible, adiabatic reduction of the applied field[37] to its original low value is engineered. In order to be reversible, the reduction occurs very slowly. And being adiabatic, no heat energy is transferred to, or out of, the system.

Because reversible adiabatic processes ensure constancy of the entropy, this part of the travel is isentropic. It takes the system smoothly from point b on the lower curve, to point c on the upper curve. Additionally, any energy needed for the work to be done by the system during its travel from b to c necessarily comes out of its own internal energy.

The final temperature—at the position c—can be seen to be much lower than the original temperature at points a and b.

In principle the above two-step process may be repeated and the system taken from the position c, via a point d, to a position e, which would be at even lower temperature than point c. In practise, however, this additional two-step process requires the use of a new cold-temperature bath at temperature defined by the new

---

where $x = (\frac{g\mu_B J}{k_B T})B_0$. In particular, for $J = \frac{1}{2}$, we have

$$\left(\frac{\partial S}{\partial B_0}\right)_T = \left(\frac{Ng\mu_B J}{T}\right)\left[-x + x^3 - \frac{2}{3}x^5 + O(x^7)\right].$$

[35] Typically this would involve making contact with liquid helium.

[36] Meaning the system stays in contact with the liquid helium bath which is maintained at constant temperature.

[37] This part of the process is often incorrectly called reversible, adiabatic demagnetization of the paramagnetic salt.

starting point c. Clearly, therefore, to reach ever closer to absolute zero we would need cold temperature reservoirs at ever lower temperatures.

11.36 Third Law: Nernst's Heat Theorem

Results of experiments on heats of reaction in chemical processes conducted by Pierre Eugène Marcellin Berthelot and H. Julius Thomsen, and W.H. Nernst's experiments with galvanic cells, indicated that with decreasing temperature, changes in the Gibbs function become ever closer to the corresponding changes in the enthalpy. Let us denote these changes as ΔG and ΔH, respectively, and start with the set of equations numbered (11.7). That is,

$$G = F + PV\,, \qquad U = F + TS\,, \qquad H = U + PV\,. \tag{11.191}$$

As usual, F, G, and H are the Helmholtz potential, Gibbs potential, and enthalpy. The first two relationships in (11.191) give $G = U - TS + PV$. Combining that with the third relationship yields $H = G + TS$. And using the identity given in (10.52) leads to the expression:

$$H = G + TS = G - T\left(\frac{\partial G}{\partial T}\right)_{\mathrm{V}}. \tag{11.192}$$

Thus, in any experiment, changes in the Gibbs function can be related to the enthalpy. Assuming the initial and the final values of the enthalpy and the Gibbs potential are H_{i}, G_{i} and H_{f}, G_{f}, we can write (11.192) as follows:

$$\begin{aligned} H_{\mathrm{f}} - H_{\mathrm{i}} &= (G_{\mathrm{f}} - G_{\mathrm{i}}) + T(S_{\mathrm{f}} - S_{\mathrm{i}}) \\ &= (G_{\mathrm{f}} - G_{\mathrm{i}}) - T\left(\frac{\partial(G_{\mathrm{f}} - G_{\mathrm{i}})}{\partial T}\right)_{\mathrm{V}}, \\ \text{or equivalently,} \qquad \Delta H &= \Delta G + T\Delta S \\ &= \Delta G - T\left(\frac{\partial(\Delta G)}{\partial T}\right)_{\mathrm{V}}. \end{aligned} \tag{11.193}$$

The results of the experiments mentioned above indicate that not only do ΔH and ΔG become ever closer as the temperature is reduced even at relatively high temperatures, but the rate at which their equality is achieved is faster than the first power of the temperature. Equation (11.193) therefore suggests that ΔS, or equivalently, the differential $(\frac{\partial(\Delta G)}{\partial T})_{\mathrm{V}}$, is also decreasing as T heads toward zero. (See Fig. 11.13, which is a schematic drawing of the experimental results.)

These observations led Nernst to postulate[38] that in thermodynamic equilibrium the change ΔH in the enthalpy and the corresponding change ΔG in the Gibbs potential must achieve equality at absolute zero. Additionally, the upshot [22] of the

[38] This postulate is well supported by experiment; see, for example, [20, 21].

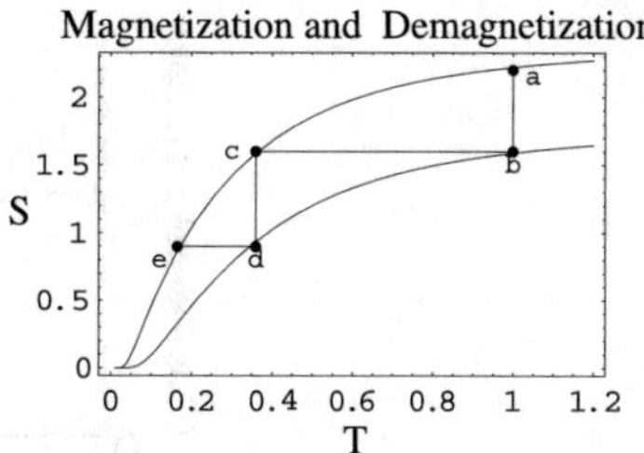

Fig. 11.11 Shown above are instances of a paramagnetic crystal under isothermal increase in the applied field followed by reversible adiabatic decrease of the same. This is a schematic plot and arbitrary scale is used for both the entropy S and the temperature T. While the upper curve refers to the system being under a week applied field, the magnetic field for the lower curve is relatively strong. The journey from point a to b represents isothermal magnetization of the paramagnetic salt. Here heat energy is released by the salt to the helium reservoir. During the final part of the journey, which takes the system from b to c, the magnetic field is reduced back to its original strength. This travel occurs reversibly and adiabatically. Hence it is isentropic. A consequence of (11.180) and (11.175) is that in any of the given processes, when the entropy, S, is constant, the magnetization, M, is also constant. Therefore, during the travel from b to c, the system magnetization also remains constant. *Accordingly, in addition to being isentropic, this process is also "isomagnetic." Clearly, calling it "adiabatic demagnetization" is a misnomer!* More importantly, the process from b to c lowers the system temperature from that of the liquid helium reservoir—recall that both points a and b are at the same temperature that is set by contact with the liquid helium reservoir—to that of point c. If another reservoir at the lower temperature represented by the point c were also available, then a similar two-step process could be repeated. In this fashion, the system would move from c to a point shown as d and then onto a point e at an even lower temperature

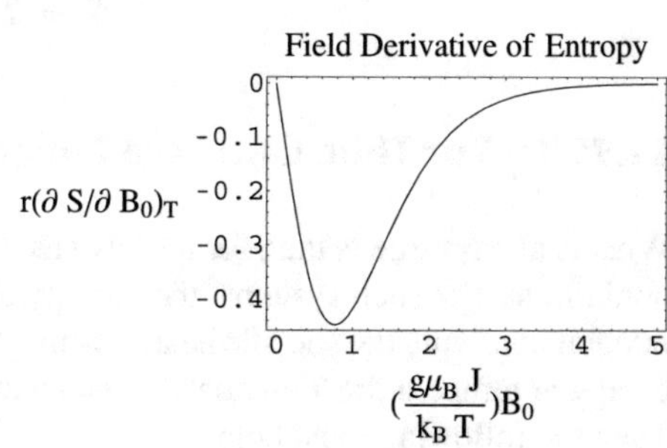

Fig. 11.12 As per (11.174), at some arbitrary, finite temperature T that is held constant, the derivative $r(\frac{\partial S}{\partial B_0})_\mathrm{T}$ is plotted as a function of $(\frac{g\mu_\mathrm{B} J}{k_\mathrm{B} T})B_0$. Here $r = \frac{T}{Ng\mu_\mathrm{B} J}$

equality:

$$\lim_{\mathrm{T}\to 0}\left[-\left(\frac{\partial(\Delta G)}{\partial T}\right)_\mathrm{P} = \Delta S\right] = 0\,, \tag{11.194}$$

in this limit is the Nernst's heat theorem, which can be expressed as follows:

"As absolute zero is approached, all chemical and/or physical transformations in thermodynamic systems—that are in internal equilibrium—occur with zero change in entropy."

Following the enunciation of the Nernst's heat theorem, Max Planck hypothesized that:

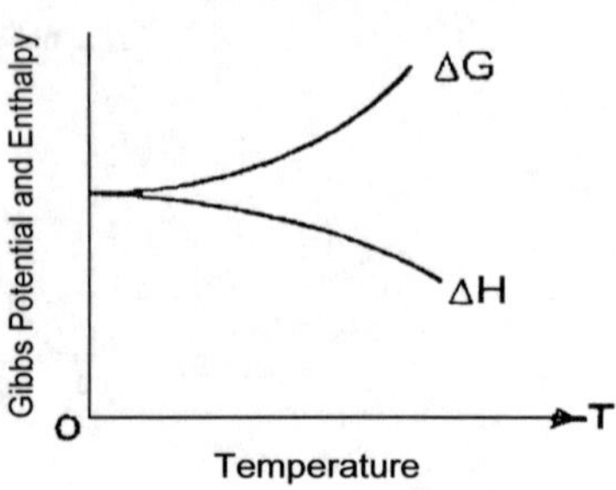

Fig. 11.13 Schematic drawing—similar to that given in [19, Fig. 7-5]—of ΔH and ΔG versus (very low) temperature

"The entropy of all thermodynamic systems in equilibrium is vanishing at absolute zero."

The above is often called the Max Planck hypothesis of a "third law of thermodynamics." Unfortunately, while this hypothesis has physical validity much of the time, unlike the zeroth, first, and second laws of thermodynamics, it is not always true. In particular, for some noncrystalline, geometrically frustrated systems, the entropy approaches a set of nonzero minima. An interesting read on the subject is: Goldstein, M. and Inge, F. (1993). "The Refrigerator and the Universe", Harvard University Press, Cambridge, MA. Therefore, a revamped version of the hypothesis would be the following:

"The entropy of all perfectly crystalline thermodynamic systems in equilibrium is vanishing at absolute zero" [23].

That is, as $T \rightarrow 0$,

$$S \rightarrow T^{\alpha}\,. \tag{11.195}$$

11.36.1 The Third Law: Zero Temperature Unattainable

What is always true is that, for finitely sized thermodynamic systems, the entropy is noninfinite. For such systems, the entropy at temperature T can be expressed as an integral involving the specific heat. Assume that the latter, say $C_{\mathrm{y}}(T)$, is specific heat energy at temperature T at constant value of some thermodynamic state variable y. Then the following must hold:

$$\begin{aligned} S(T) &= \int_0^T \frac{\mathrm{d}Q}{T} = \int_0^T C_{\mathrm{y}}(T)\frac{dT}{T} \\ &\neq \infty\,. \end{aligned} \tag{11.196}$$

An important implication—see below—of the Planck hypothesis is that absolute zero cannot be reached with finite number of (any type of temperature lowering) operations. In other words, reaching absolute zero temperature will require an infinite amount of effort.

The most efficient procedure for cooling a system is to thermally isolate it at the lowest temperature available and adiabatically conduct the system through a process where it does positive work. As demanded by the first law and because the process proceeds adiabatically, such work will necessarily have to be done at the expense of the system's own internal energy. And reduction in the internal energy will lower its temperature.

In a "gedanken" experiment, the given system is reversibly taken along two different paths, one which stretches from absolute zero to some point i at temperature T_i K; and the other, starting at absolute zero but stretching to a different point j at some other temperature T_j. These two travels occur while certain thermodynamic properties remain at their fixed values. Let these properties be y_i and y_j for the two paths. Then according to the first–second law, we may calculate the entropy at the two end points as follows:

$$S_\mathrm{i}(T_\mathrm{i}, y_\mathrm{i}) = \int_0^{T_\mathrm{i}} \frac{C_{y_\mathrm{i}}(T)}{T} \mathrm{d}T\,,$$
$$S_\mathrm{j}(T_\mathrm{j}, y_\mathrm{j}) = \int_0^{T_\mathrm{j}} \frac{C_{y_\mathrm{j}}(T)}{T} \mathrm{d}T\,. \qquad (11.197)$$

Let us assume that during this gedanken experiment, through superhuman effort totally committed to attaining absolute zero, the temperature T_i that we have reached is finite, but is very, very close to absolute zero. Then, hopefully an appropriate reversible adiabatic process would be available which will help us travel from point i to point j—the latter being at temperature T_j exactly equal to absolute zero. Clearly, this will be the last leg of a very, very long journey!

Because the travel from i to the end point j—the latter being exactly at absolute zero—is reversible and adiabatic, the entropy is the same at points i and j.[39] This means, $S_\mathrm{i}(T_\mathrm{i}, y_\mathrm{i})$ would be equal to $S_\mathrm{j}(T_\mathrm{j} = 0, y_\mathrm{j})$. That is,

$$\int_0^{T_\mathrm{i}} \frac{C_{y_\mathrm{i}}(T)}{T} \mathrm{d}T =_{\mathrm{T_j}=0} \int_0^{T_\mathrm{j}} \frac{C_{y_\mathrm{j}}(T)}{T} \mathrm{d}T$$
$$= \int_0^{0} \frac{C_{y_\mathrm{j}}(T)}{T} \mathrm{d}T = 0\,. \qquad (11.198)$$

Therefore, according to the above equation—namely (11.198)—the following must be true:

$$\int_0^{T_\mathrm{i}} \frac{C_{y_\mathrm{i}}(T)}{T} \mathrm{d}T = 0\,. \qquad (11.199)$$

And this happens while the temperature T_i, albeit very, very low, is still finite and nonzero.

Recall that, according to Max Planck, at this very low temperature the specific heat $C_{y_\mathrm{i}}(T)$ depends on temperature as $\propto T^\alpha$ where $\alpha > 0$—see (11.195). In other words, we must have

$$\int_0^{T_\mathrm{i}} T^{\alpha-1} \mathrm{d}T = \frac{T_\mathrm{i}^\alpha}{\alpha} = 0\,, \qquad (11.200)$$

with $\alpha > 0$. But this cannot be unless T_i itself is equal to zero!

[39]For instance, compare j with point e and i with point d in Fig. 11.11.

Thus, we are informed by the Max Planck hypothesis of a possible third law that, in the

"Never-Never Land, Zero Temperature Can Easily Be Reached—But Only After It Has Already Been Reached!"

11.37 Negative Temperatures

Negative temperatures were cursorily referred to in Chap. 7 ("Internal Energy and Enthalpy").[40] In this regard, it is helpful to remind ourselves of some features of the experiment that first led to the idea of negative temperature.

A quantum spin-a-half particle, in the presence of applied magnetic field, can present itself in two possible modes: either it is aligned in parallel, or it is antiparallel, to the field. Let the two modes differ in energy by an amount $+\hbar\omega$. According to the BMG prescription,[41] in a thermodynamically large system, the ratio of the number, N_{high}, of the high-energy modes and, N_{low}, of the low-energy modes that occur is as follows: $\frac{N_{\text{high}}}{N_{\text{low}}} = \exp(-\beta\hbar\omega)$. Thus a majority of spins are at the lower energy level (and are oriented parallel to the field). Purcell and Pound [24] were able—through experimental trickery—to reverse the relative orientation of the applied field and the spins. This is equivalent to arranging a majority of spins, in a thermodynamically large system, to be at the higher energy level—an occurrence that the BMG prescription would ascribe to a state of negative temperature!

Regarding the negative temperature, following D. ter Haar, it was mentioned in Chap. 7 that—

Firstly, we must consider the increasing order of the temperature—in integer degrees Kelvin—to be the following:

$$+0, +1, +2, \ldots, +\infty, -\infty, \ldots, -2, -1, -0 .$$

Therefore, an object at any negative temperature—even as low as $-\infty$—must be treated as being warmer than one at any positive temperature—including $+\infty$!

Secondly, following Ramsey's suggestion, the temperature T should be treated as though it is specified by the thermodynamic identity given in (7.101). That is,

$$\left(\frac{\partial U}{\partial S}\right)_{\text{V,N}} = T . \tag{11.201}$$

Thirdly, "For a system to be capable of having negative temperature, it is necessary for its energy to possess an upper bound." The system treated by Purcell and Pound had precisely this characteristic.

[40] See Sect. 4.8. Also see the discussion associated with (11.201)–(11.204).

[41] See (2.13) and (11.80), (11.81), (11.82).

In order to simplify both the notation and algebra, while still preserving the essentials of the physics, let us proceed as follows:

In complete accord with the $J = \frac{1}{2}$ magnetic spin case studied earlier, and the N, spin-$\frac{1}{2}$ nuclear spins in a magnetized crystal referred to above, we assume these spins to be noninteracting. Further, for notational simplicity, we assume that when a spin is aligned parallel with the applied field it has energy equal to zero. And we again assume that when the spin is antiparallel to the field its energy is $\hbar\omega$. Thus there are only two states.

Accordingly, the relevant partition function for a single pair of spins is

$$\begin{aligned} \Xi(1,T) &= \mathrm{Tr}\big[\exp(-\beta\mathcal{H})\big] \\ &= \left[1+\exp\left(\frac{-\varpi}{T}\right)\right], \\ \text{where} \quad \varpi &= \frac{\hbar\omega}{k_{\mathrm{B}}}. \end{aligned} \tag{11.202}$$

And for N pairs we have

$$\Xi(N,T) = \big\{\mathrm{Tr}\big[\exp(-\beta\mathcal{H})\big]\big\}^N = \left\{1+\exp\left(\frac{-\varpi}{T}\right)\right\}^N. \tag{11.203}$$

As described in (11.129), the partition function is related to various thermodynamic potentials:

$$\begin{aligned} F &= -Nk_{\mathrm{B}}T\ln\left[1+\exp\left(\frac{-\varpi}{T}\right)\right], \\ \frac{S}{Nk_{\mathrm{B}}} &= \frac{-1}{Nk_{\mathrm{B}}}\left(\frac{\partial F}{\partial T}\right)_{\mathrm{N},\epsilon} = \ln\left[1+\exp\left(\frac{-\varpi}{T}\right)\right] + \left[\frac{\frac{\varpi}{T}\exp(\frac{-\varpi}{T})}{1+\exp(\frac{-\varpi}{T})}\right] \\ &= \ln\left[1+\exp\left(\frac{\varpi}{T}\right)\right] - \left[\frac{\frac{\varpi}{T}}{1+\exp(\frac{-\varpi}{T})}\right], \\ U &= F+TS = \frac{Nk_{\mathrm{B}}\varpi}{1+\exp(\frac{\varpi}{T})}, \\ C &= \left(\frac{\partial U}{\partial T}\right)_{\mathrm{N}} = Nk_{\mathrm{B}}\left(\frac{\varpi}{T}\right)^2\left[\frac{\exp(\frac{\varpi}{T})}{\{1+\exp(\frac{\varpi}{T})\}^2}\right]. \end{aligned} \tag{11.204}$$

For such a two state nuclear paramagnet, in Fig. 11.14 we have plotted $\frac{U}{Nk_{\mathrm{B}}\varpi}$ versus $\frac{T}{\varpi}$. At $T = +0$, the energy U is seen to be equal to zero—which is its minimum value. The energy rises with the increase in temperature and levels off at $+\frac{Nk_{\mathrm{B}}\varpi}{2}$, a value that is attained at $T \to +\infty$. As the temperature moves infinitesimally "upwards" beyond $T = +\infty$ to $T = -\infty$, the energy remains constant at $\frac{Nk_{\mathrm{B}}\varpi}{2}$. However, during the final leg of the travel, as the temperature increases from $-\infty$ to its "highest" value -0, the energy slowly rises beyond $+\frac{Nk_{\mathrm{B}}\varpi}{2}$ to its maximum value of $+Nk_{\mathrm{B}}\varpi$.

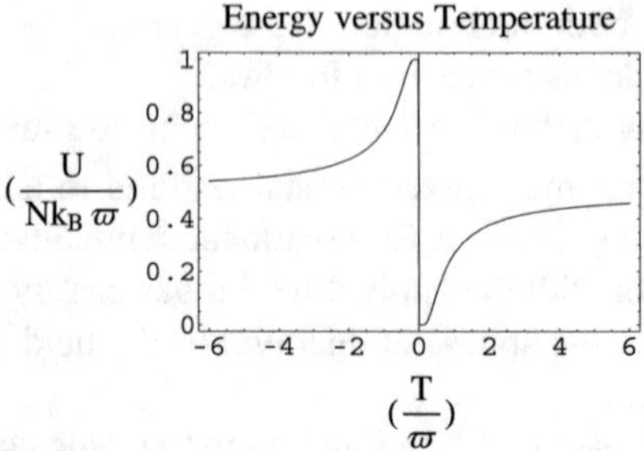

Fig. 11.14 Energy U, in units of $Nk_B\varpi$, as a function of temperature T, in units of ϖ, for a two state nuclear paramagnet. Notice that the energy is equal to zero—which is its minimum value—at $T = +0$. It rises with the increase in temperature and levels off at $\frac{Nk_B\varpi}{2}$ as the temperature heads upward in the direction of $T \to +\infty$. Just beyond $T = +\infty$ is $T = -\infty$. (Note, in theory, all negative temperatures are higher than $+\infty$.) From here, as the temperature moves "upwards" toward $T = -0$, the energy slowly rises above $\frac{Nk_B\varpi}{2}$ and reaches its maximum value of $Nk_B\varpi$ at the end of its journey at $T = -0$

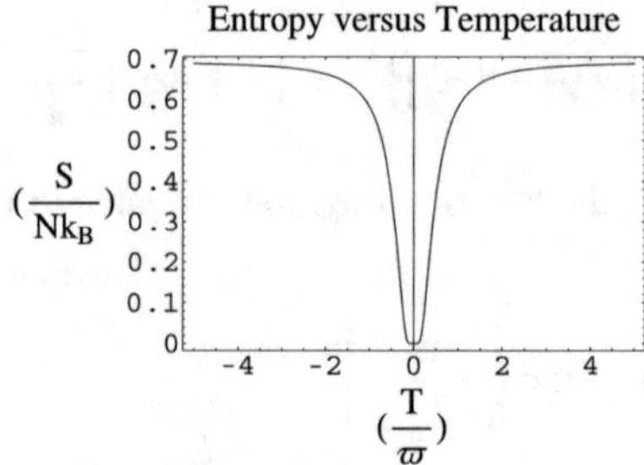

Fig. 11.15 Entropy S, in units of Nk_B, as a function of the temperature T, in units of ϖ, for a two state nuclear paramagnet. Notice that the entropy is equal to zero—which is its minimum here—at $T = +0$. It rises with the increase in temperature and, as the temperature heads to $T \to +\infty$, it seems to level off at its maximum value of $Nk_B \ln\{2(\frac{1}{2}) + 1\} = Nk_B \ln\{2\}$. Also notice that the entropy is symmetric with respect to the temperature, that is, $S(+T) = S(-T)$

As a function of the temperature, $\frac{T}{\varpi}$, the entropy, $\frac{S}{Nk_B}$, is plotted in Fig. 11.15. In contrast with the behavior of the energy, the entropy, however, is symmetric with respect to the temperature, that is, $S(+T) = S(-T)$. It is equal to zero—which is its minimum—at $T = +0$, it rises with the increase in temperature and, as the temperature heads to $T \to +\infty$, it seems to level off at its maximum value of $Nk_B \ln\{2(\frac{1}{2}) + 1\} = Nk_B \ln\{2\}$.

Beyond $T = +\infty$—see the left half of the figure—as the temperature rises toward[42] $T = -0$, the entropy begins returning to its erstwhile minimum value of zero.

[42]Note that all negative temperatures are supposed to be higher than $+\infty$.

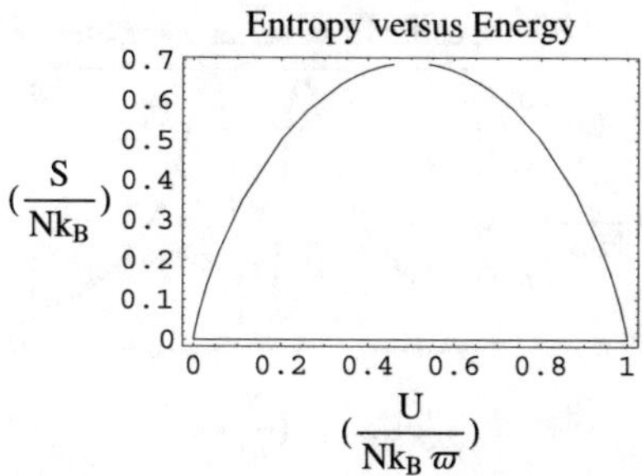

Fig. 11.16 Entropy S, in units of Nk_B, as a function of the energy U, in units of $Nk_B\varpi$, for a two state nuclear paramagnet. The shown value of the entropy is equal to zero at both the minimum value of the energy—i.e., at $U = 0$—and the maximum value—i.e., at $U = Nk_B\varpi$. The entropy rises with the increase in temperature and seems to level off at its maximum value of $Nk_B \ln\{2(\frac{1}{2}) + 1\} = Nk_B \ln\{2\}$. Recall that $S(T)$ is symmetric with respect to the temperature, that is, $S(+T) = S(-T)$. At the position marked by a small empty space in the otherwise continuous looking curve, the derivative $(\frac{\partial S}{\partial U})_{V,N}$ changes its value from $+0$ to -0, which indicates that at this point the temperature instantly changes from $+\infty$ to $-\infty$. Note that the derivative $(\frac{\partial(\frac{S}{Nk_B})}{\partial(\frac{U}{Nk_B\varpi})})_{N,V}$ is equal to $\frac{\varpi}{T}$—and its positivity from the position $U = 0$ to $\frac{U}{Nk_B\varpi} = 0.5$ indicates positive values of the temperature while its negativity in the range $\frac{U}{Nk_B\varpi} = 0.5 \to 1$ indicates negative values for the temperature. From left to right the temperature increases according to $(T = 0 \to +\infty, -\infty \to -0)$

The entropy (as in the expression) $\frac{S}{Nk_B}$ is plotted in Fig. 11.16 as a function of the energy (in the form) $\frac{U}{Nk_B\varpi}$. Note that its derivative $(\frac{\partial(\frac{S}{Nk_B})}{\partial(\frac{U}{Nk_B\varpi})})_{N,V}$ is equal to $\frac{\varpi}{T}$—and its positivity from the point $U = 0$ to $\frac{U}{Nk_B\varpi} = 0.5$ indicates positive values for the temperature while its negativity in the range $\frac{U}{Nk_B\varpi} = 0.5 \to 1$ indicates negative values for the temperature.

Reader's attention is recommended to the point marked by a small empty spacing in the continuous curve. Here, the derivative $(\frac{\partial S}{\partial U})_{V,N}$ alters its course from a positive value to a negative one. Indeed, immediately to the left of this point, the derivative is equal to $+0$ and it is equal to -0 immediately to its right. The inverse of this derivative is, of course, the temperature, which therefore instantly changes from $+\infty$ to $-\infty$. The reason for this behavior is that while the energy continues increasing, at this point the entropy starts to decrease. Beyond this point to the right, the temperature continues to "increase" until it reaches -0. And at this "highest" temperature, the entropy is vanishing but the energy is at its maximum value of $\frac{U}{Nk_B\varpi} = 1$.

The specific heat is plotted in Fig. 11.17. At $T = +0$ the specific heat, as also noted in Fig. 11.10, is equal to zero—which is its minimum value. It rises rapidly with the increase in temperature and, as expected, undergoes a Schottky anomaly before trailing off toward its minimum value of zero as the temperature heads toward $+\infty$. Its behavior, however, is symmetrical with respect to the interchange of temperature, that is, $C(+T) = C(-T)$.

In the following we study the grand canonical ensemble and some of its usage.

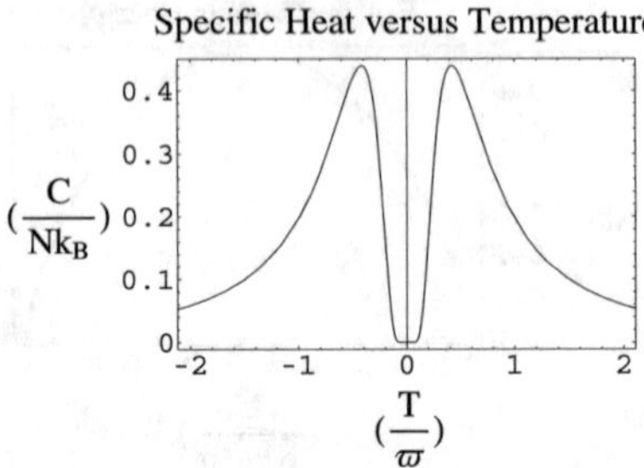

Fig. 11.17 The specific heat of the two-state nuclear paramagnet. At $T = +0$ the specific heat, as also shown in Fig. 11.10, is equal to zero—which is its minimum value. It rises—first slowly and then rapidly—with the increase in temperature and, as expected, undergoes a Schottky anomaly before trailing off toward its minimum value of zero as the temperature heads toward $+\infty$. Note that $C(T)$ is symmetric with respect to the interchange of temperature, that is, $C(+T) = C(-T)$

11.38 Grand Canonical Ensemble: Classical Systems

The canonical partition function for a thermodynamic system that obeys classical statistics and is at temperature T, has volume V, and a fixed number, N, of particles was described in (11.1). Its relationship to the various thermodynamic potentials was recorded in (11.7).

By freely exchanging the heat energy back and forth, the system temperature could be maintained at a constant value equal to that of the outside reservoir. The result, however, was that the system energy was not conserved. Indeed, it is the rules of thermodynamics that actually established the observed, average value of the system energy.

The thermodynamic systems being treated were assumed to be "closed" in that they consisted of fixed numbers of particles. In practice, however, this assumption is not well founded. With the particle numbers in trillion-trillions, there is no practical way of ensuring that they do not leak back and forth. And, of course, there is no possible way of measuring their exact number.

Therefore, an appropriate thermodynamic format is one where the system is in contact with an outside, heat energy and particle, reservoir: a reservoir with which it can freely exchange both particles and the heat energy. The result is that both the temperature—which is conjugate to the energy—and the chemical potential—which is conjugate to the particle number—are conserved. As a result of firm contact between the system and reservoir, the values of temperature and chemical potential are the same in both systems.

Yet, even though the particle number is not conserved, its statistical average must be consistent with the measured value of its thermodynamic average $\bar{N}$. Thus appropriate changes in the formulation of the canonical ensemble are needed.

Accordingly, our basic task in the construction of grand canonical ensemble is to ensure that in addition to it obeying the usual requirements of thermodynamics—which are assured by it having the Boltzmann–Maxwell–Gibbs (BMG)[43] form—its

[43] BMG: Boltzmann–Maxwell–Gibbs.

temperature and the chemical potential must be equal to that of the reservoir with which it is in "energy and number contact." In addition, the statical average of the particle number must be equal to its measured value $\bar{N}$.

A convenient procedure to achieve these objectives is to introduce a Lagrange multiplier, say μ, and use Calculus. This is done as follows:

(1) Replace the Hamiltonian, $\mathcal{H}_{\mathrm{n}}$, by its Lagrange version $\mathcal{H}_{\mathrm{Ln}}$. That is,

$$\mathcal{H}_{\mathrm{Ln}} = \mathcal{H}_{\mathrm{n}} - \mu\hat{n}\,. \tag{11.205}$$

This changes $\mathrm{Tr}[\exp(-\beta\mathcal{H}_{\mathrm{n}})]$ to $\mathrm{Tr}[\exp(-\beta\mathcal{H}_{\mathrm{Ln}})]$. Then we can write:

$$\mathrm{Tr}\big[\exp(-\beta\mathcal{H}_{\mathrm{Ln}})\big] = z^{\mathrm{n}}\mathrm{Tr}\big[\exp(-\beta\mathcal{H}_{\mathrm{n}})\big]\,, \quad \text{where} \quad z = \exp(\beta\mu)\,. \tag{11.206}$$

Note that here n stands for the number of moles.

The grand canonical partition function, to be denoted as $\Psi(z, V, T)$, can now be defined as

$$\Psi(z, V, T) = \sum_{\mathrm{n}=0}^{\infty} \mathrm{Tr}\big[\exp(-\beta\mathcal{H}_{\mathrm{Ln}})\big]\,. \tag{11.207}$$

Note that the logarithm of the grand canonical partition function is often needed. We shall call it the "grand potential" and denote it as $Q(z, V, T, \text{etc.})$, where

$$\begin{aligned} Q(z, V, T) = \ln\big[\Psi(z, V, T)\big] &= \ln\sum_{\mathrm{n}=0}^{\infty} \mathrm{Tr}\big[\exp\{-\beta(\mathcal{H}_{\mathrm{n}} - \mu n)\}\big] \\ &= \ln\sum_{\mathrm{n}=0}^{\infty} z^{\mathrm{n}}\mathrm{Tr}\big[\exp\{-\beta\mathcal{H}_{\mathrm{n}}\}\big]\,. \end{aligned} \tag{11.208}$$

We are required to choose the Lagrange multiplier μ to be equal to the chemical potential such that the calculated value of the total number of moles, $\langle\hat{n}\rangle$, is equal to their measured average, $\bar{n}$.

The BMG relationship that determines $\langle\hat{n}\rangle$ and $\langle\mathcal{H}_{\mathrm{Ln}}\rangle$ is the following:

$$\begin{aligned} \langle\hat{n}\rangle &= \bar{n} \\ &= \left[\frac{\sum_{\mathrm{n}=0}^{\infty} z^{n} \cdot n \cdot \mathrm{Tr}[\exp(-\beta\mathcal{H}_{\mathrm{n}})]}{\Psi(z, V, T)}\right] \\ &= z\left(\frac{\partial Q(z, V, T)}{\partial z}\right)_{\mathrm{V,T}} \\ &= k_{\mathrm{B}}T\left(\frac{\partial Q(z, V, T)}{\partial \mu}\right)_{\mathrm{V,T}}, \\ U(z, V, T) &= \langle\mathcal{H}_{\mathrm{Ln}}\rangle = \big\langle\{\mathcal{H}_{\mathrm{n}} - \mu\hat{n}\}\big\rangle \\ &= -\left(\frac{\partial Q(z, V, T)}{\partial \beta}\right)_{\mathrm{z,V}} \\ &= k_{\mathrm{B}}T^{2}\left(\frac{\partial Q(z, V, T)}{\partial T}\right)_{\mathrm{z,V}}. \end{aligned} \tag{11.209}$$

In addition to (11.209), there is a fundamental relationship that is also needed for successful use of the grand canonical partition function. According to Pathria, this relationship "represents the essential link between the thermodynamics of the given system and the grand canonical ensemble." Much as was done before, this relationship is recorded below without proof:[44]

$$\frac{V\,P(z,V,T)}{k_{\mathrm{B}}T} = Q(z,V,T)\,. \tag{11.210}$$

11.38.1 Grand Canonical Ensemble: Partition Function

Because the grand canonical partition function appears to require[45] knowledge of the canonical partition function, $\mathrm{Tr}[\exp(-\beta\mathcal{H})]$, it is redundant for studying systems where the particle number is fully conserved. Quite obviously, therefore, one would rather use just the canonical partition function there. Still it may be helpful for a beginner to see how the grand canonical partition function $\Psi_{\mathrm{classical}}(z,V,T)$ actually works. To this end, for convenience, we treat only a very simple system, namely the monatomic perfect gas—see (11.8).

The partition function, worked out in (11.13) for the perfect gas, can be written as

$$\Xi(N,V,T;m) = \frac{1}{N!}\left(\frac{V}{h^3}\right)^N\left(\frac{2m\pi}{\beta}\right)^{\frac{3N}{2}} = \frac{1}{N!}\left(\frac{V}{\lambda^3}\right)^N . \tag{11.211}$$

(If we were working with only with the canonical partition function, we would not need any further analysis.)

Reminiscent of the de Broglie length in quantum mechanics, λ in (11.211) above is the quantum-statistical, temperature-dependent unit of length, i.e.,

$$\lambda = \left(\frac{2\pi m k_{\mathrm{B}}T}{h^2}\right)^{-\frac{1}{2}} . \tag{11.212}$$

The grand canonical partition function is the following:

$$\Psi(z,V,T) = \sum_{n=0}^{\infty}\frac{z^n}{n!}\left(\frac{V}{\Lambda^3}\right)^n = \exp\left(\frac{zV}{\lambda^3}\right). \tag{11.213}$$

According to (11.210),

$$\frac{PV}{k_{\mathrm{B}}T} = Q(z,V,T) = \ln\Psi(z,V,T) = \frac{zV}{\lambda^3}\,, \tag{11.214}$$

which gives

$$P = \frac{zk_{\mathrm{B}}T}{\lambda^3}\,. \tag{11.215}$$

[44] The relevant proof is available in [25].
[45] See (11.207).

Additionally, following (11.209) and (11.214), we have:

$$\begin{aligned}\bar{N} &= k_{\rm B}T\left(\frac{\partial Q(z, V, T)}{\partial \mu}\right)_{\rm V,T} = k_{\rm B}T\left(\frac{\partial(\frac{zV}{\lambda^3})}{\partial \mu}\right)_{\rm V,T} \\ &= k_{\rm B}T\left(\frac{V}{\lambda^3}\right)\left(\frac{\partial(z)}{\partial \mu}\right)_{\rm V,T} = k_{\rm B}T\left(\frac{V}{\lambda^3}\right)\beta z = \frac{zV}{\lambda^3}\,. \end{aligned} \tag{11.216}$$

By combining (11.215) and (11.216), we can find the equation of state for the ideal gas. That is,

$$\begin{aligned} z &= \frac{\lambda^3 P}{k_{\rm B}T}\,, \\ z &= \frac{\lambda^3 \bar{N}}{V}\,, \\ \text{therefore,} \qquad PV &= \bar{N}k_{\rm B}T\,. \end{aligned} \tag{11.217}$$

The Gibbs and Helmholtz potentials are also readily found as

$$\begin{aligned} G &= \bar{N}\mu = \frac{\bar{N}}{\beta}\ln(z) = \bar{N}k_{\rm B}T\ln(\beta P\lambda^3) \\ &= -\bar{N}k_{\rm B}T\ln\left\{\left(\frac{V}{\bar{N}h^3}\right)(2\pi m k_{\rm B}T)^{\frac{3}{2}}\right\}, \\ F &= G - PV \\ &= -\bar{N}k_B T\ln\left\{\left(\frac{V}{\bar{N}h^3}\right)(2\pi m k_{\rm B}T)^{\frac{3}{2}}\right\} - Nk_{\rm B}T\,. \end{aligned} \tag{11.218}$$

To arrive at (11.218), we first used (11.206) to connect μ to $\ln(z)$; then, according to (11.217), P was replaced by $\bar{N}k_B T/V$, and finally, the logarithm was inverted.

Notice that the equations in (11.218) exactly reproduce the corresponding results that were derived earlier (see (11.14), (11.16), and (11.17)).

The internal energy U is evaluated next. According to (11.209), (11.213), and (11.214), we have

$$\langle \mathcal{H}_{\rm L}\rangle = k_{\rm B}T^2\left(\frac{\partial Q(z, V, T)}{\partial T}\right)_{\rm z,V} = k_{\rm B}T^2 zV\left(\frac{\partial(\frac{1}{\lambda^3})}{\partial T}\right)_{\rm z,V} = \frac{3}{2}k_{\rm B}T\left(\frac{zV}{\lambda^3}\right). \tag{11.219}$$

As derived in (11.217), $\frac{zV}{\lambda^3} = \bar{N}$. Therefore the above leads to the well-known result

$$U = \frac{3}{2}\bar{N}k_{\rm B}T\,. \tag{11.220}$$

The entropy, S, is now easily found. And the result is identical with that found earlier—see (11.16):

$$S = \frac{U - F}{T} = Nk_{\rm B}\ln\left(\frac{V}{N}\right) + \frac{3}{2}Nk_{\rm B}\ln\left(\frac{2\pi m k_{\rm B}T}{h^2}\right) + \frac{5}{2}Nk_{\rm B}\,. \tag{11.221}$$

Remark 11.2 While for systems with conserved numbers of particles both the canonical and the grand canonical partition functions give the same result, the former is clearly much more convenient to use. Therefore, in practise, the latter—i.e., the grand canonical partition function—is used for systems where particle exchanges are a normal occurrence, as, for example, is the case in chemical processes.

The need for employing grand canonical procedures is even more acute in quantum statistical systems. Except for relatively trivial cases, the number of elementary excitations in these systems is not conserved. Rather, it is determined by the thermodynamic state of the system.

11.39 Quantum States: Statistics

While a formulation based on grand canonical statistical mechanics of indistinguishable, noninteracting, quantum particles is best deferred to an appendix—see (L.1)–(L.9)—it is convenient to present here a less demanding, alternate formulation. We consider, rather than the particles themselves, only their possible single-particle energy states.

Consider one such state with momentum $\boldsymbol{p}$, energy ε_{p}, and eigenvalue ℓ for the spin operator S. When singly occupied, we shall assume its contribution to the Hamiltonian to be just its kinetic energy, ε_{p}, equal to $\frac{p^2}{2m}$. (Note that we are assuming no spin dependence of the energy levels.) On the other hand, if the occupation number of this state were n_{p}, its contribution to the Hamiltonian would be $\mathcal{H}_{\mathrm{n_p}} = n_{\mathrm{p}}\varepsilon_{\mathrm{p}}$. However, with a view to ensuring that the actual particle number of the given state is equal to its statistical average, we need—as explained in (11.205)—to work with, rather than $\mathcal{H}_{\mathrm{n_p}}$, the Lagrange multiplier version of the Hamiltonian, namely $\mathcal{H}_{\mathrm{Ln_p}} = \mathcal{H}_{\mathrm{n_p}} - n_{\mathrm{p}}\mu$. Therefore, in analogy with the BMG prescription,[46] an appropriate choice for the grand probability distribution factor, $f(n_{\mathrm{p}})$, and the partition function, $\Xi_{(\varepsilon_{\mathrm{p}}-\mu)}$, for this state is the following:

$$
\begin{aligned}
f(n_{\mathrm{p}}) &= \frac{\exp[-\beta(\mathcal{H}_{\mathrm{Ln_p}})]}{\Xi_{(\varepsilon_{\mathrm{p}}-\mu)}} = \frac{\exp[-\beta(\mathcal{H}_{\mathrm{n_p}} - \mu n_{\mathrm{p}})]}{\Xi_{(\varepsilon_{\mathrm{p}}-\mu)}} \\
&= \frac{\exp[-\beta n_{\mathrm{p}}(\varepsilon_{\mathrm{p}} - \mu)]}{\Xi_{(\varepsilon_{\mathrm{p}}-\mu)}}, \\
\text{where} \quad \Xi_{(\varepsilon_{\mathrm{p}}-\mu)} &= \sum_{\mathrm{n_p}} \exp\big[-\beta n_{\mathrm{p}}(\varepsilon_{\mathrm{p}} - \mu)\big]. \qquad (11.222)
\end{aligned}
$$

Reminder: The extent of the summation over n_{p} is yet to be decided—see (11.223) below—and the Lagrange multiplier μ must be so chosen that the statistical average of the relevant particle number is equal to its observed (i.e., thermal) average.

[46] Versions of the BMG prescription have been used in this chapter—see (11.80)–(11.82) and (11.126)–(11.128)—as well as in an earlier chapter in equation numbered (2.13).

11.40 Fermi–Dirac: Noninteracting System

A noninteracting Fermi–Dirac (F–D) system is studied below.

11.40.1 Fermi–Dirac: Single-State Partition Function

Electrons have spin $S = \frac{1}{2}$. As such, a system of electrons is described by wave-functions which are antisymmetric with regard to the interchange of any given pair of particles. The system, therefore, is required—by the dictates of quantum mechanics—to obey the Pauli principle. Accordingly, at any given time, no more than one electron may be in any one quantum state. In other words, for the spin vector component equal to, say, $+\frac{1}{2}$ and time t, a given state with kinetic energy ε_{p} has only two options. It is either empty at that moment—meaning it is unoccupied and $n_{\mathrm{p}}(t) = 0$—or it has only one electron in it, i.e., $n_{\mathrm{p}}(t) = 1$. Of course, the same requirement applies to the state with energy ε_{p} when it has spin vector component equal to $-\frac{1}{2}$.

Therefore, according to the definition given in (11.222), the partition function for this state—either with spin-up or with spin-down—is

$$\begin{aligned} \Xi_{(\varepsilon_{\mathrm{p}}-\mu)} &= \sum_{\mathrm{n_p}=0}^{\mathrm{n_p}=1} \exp\left[-\beta n_{\mathrm{p}}(\varepsilon_{\mathrm{p}} - \mu)\right] \\ &= 1 + \exp\left[-\beta(\varepsilon_{\mathrm{p}} - \mu)\right] \\ &= 1 + z \exp(-\beta\varepsilon_{\mathrm{p}}) , \end{aligned} \tag{11.223}$$

where $z = \exp(\beta\mu)$.

Because partition functions for different parts of a system are multiplicative, in order to calculate the Fermi–Dirac partition function, $\Psi_{\mathrm{F-D}}(z, V, T)$, for the whole system, we need to multiply $\Xi_{(\varepsilon_{\mathrm{p}}-\mu)}$ over all the possible single-particle states. (Remember that if our particles have spin S, there would be a total of $2S+1$ spin vector components. And because the system Hamiltonian does not depend on S, each of these components would contribute overall the same amount to the partition function. Also remember that different contributions affect the partition function multiplicatively.)

Thus, for spin S particles, we would have

$$\begin{aligned} \Psi_{\mathrm{F-D}}(z, V, T) &= \prod_{\mathrm{p}} \left[\Xi_{(\varepsilon_{\mathrm{p}}-\mu)}\right]^{2S+1} \\ &= \prod_{\mathrm{p}} \left[1 + z \exp(-\beta\varepsilon_{\mathrm{p}})\right]^{2\mathrm{S}+1} . \end{aligned} \tag{11.224}$$

It is convenient to introduce a spin-degeneracy factor, $\Delta_{\mathrm{F-D}} = 2S + 1$, and also record the grand potential $[Q(z, V, T)]_{\mathrm{F-D}}$ which we recall is equal to the logarithm of the $F - D$ partition function,

$$\begin{aligned}\left[Q(z, V, T)\right]_{\mathrm{F-D}} &= \ln \Psi_{\mathrm{F-D}}(z, V, T)\\ &= \Delta_{\mathrm{F-D}} \sum_{\mathrm{p}} \ln\left[1 + z \exp(-\beta\varepsilon_{\mathrm{p}})\right]. \end{aligned} \tag{11.225}$$

For electrons, $S = \frac{1}{2}$. Therefore, the spin-degeneracy factor $\Delta_{\mathrm{F-D}} = 2(\frac{1}{2}) + 1 = 2$.

The grand potential $[Q(z, V, T)]_{\mathrm{F-D}}$ will be put to important use in analyses to be presented later.

11.41 Bose–Einstein: Noninteracting System

Partition function for single-particle state of a noninteracting Bose–Einstein (B–E) gas is analyzed below.

Here all choices are allowed for possible occupancy of any given single-particle state. We describe this by saying that the state with energy ε_{p} is occupied n_{p} number of times where, unlike for the F–D partition function, n_{p} can range between zero and ∞. Therefore the partition function for this state is

$$\begin{aligned}\Xi_{(\varepsilon_{\mathrm{p}}-\mu)} &= \sum_{\mathrm{n_p}=0}^{\mathrm{n_p}=\infty} \exp\left\{\beta n_{\mathrm{p}}(-\varepsilon_{\mathrm{p}} + \mu)\right\}\\ &= \sum_{\mathrm{n_p}=0}^{\mathrm{n_p}=\infty} \left[z \exp(-\beta\varepsilon_{\mathrm{p}})\right]^{\mathrm{n_p}} = \sum_{\mathrm{n_p}=0}^{\mathrm{n_p}=\infty} [\alpha]^{\mathrm{n_p}}, \end{aligned} \tag{11.226}$$

where

$$z = \exp(\beta\mu) \quad \text{and} \quad \alpha = z \exp(-\beta\varepsilon_{\mathrm{p}}). \tag{11.227}$$

11.41.1 Bose–Einstein: Chemical Potential Always Negative

Both β and ε_{p} are necessarily nonnegative. Therefore

$$1 \geq \exp(-\beta\varepsilon_{\mathrm{p}}) \geq 0. \tag{11.228}$$

For chemical potential, the limits unity, or zero, are reached for a perfect Bose–Einstein gas when its temperature is very high, i.e., is infinity, or is exceedingly low, i.e., is zero. Also, of course, the same effect can be brought about by ε_{p} being either zero, or infinity.

The infinite series—see (11.226)—representing the partition function is divergent[47] if $|\alpha| \geq 1$. Again, this contrasts with a perfect-gas of spin $S = \frac{1}{2}$, F–D quantum particles, because there—see (11.223)—the corresponding partition function consisted of only two finite terms: therefore, the series there was absolutely convergent.

Consequently, in order for the partition function, for any given single state of a noninteracting B–E gas to remain finite, the magnitude of α must be less than unity. In other words, the following inequality must hold:

$$|\alpha| = \left|z \exp(-\beta\varepsilon_p)\right| < 1\,. \tag{11.229}$$

Given the two inequalities (11.228) and (11.229), the following inequality also must hold:

$$|z| < 1\,. \tag{11.230}$$

We are reminded, however, that z is equal to $\exp(\beta\mu)$, and β can range between 0 and ∞. Clearly, therefore, the molar chemical potential μ—which is an important thermodynamic function—cannot take on any positive value. That is,

$$0 \geq \mu \geq -\infty\,. \tag{11.231}$$

In other words, for a perfect B–E gas, the chemical potential is always negative or at best it approaches zero. This contrasts with the perfect F–D gas where the chemical potential can have either sign.

11.41.2 Bose–Einstein: Partition Function

Because—as stated above—partition functions for different parts of a system are multiplicative, in order to calculate the Bose–Einstein partition function, $\Psi_{\mathrm{B-E}}(z, V, T)$, for the whole system, we need to multiply $\Xi_{(\varepsilon_{\mathrm{p}}-\mu)}$ over all the possible single-particle states. That is,

$$\begin{aligned}\Psi_{\mathrm{B-E}}(z, V, T) &= \prod_{\mathrm{p}}\left[\Xi_{(\varepsilon_{\mathrm{p}}-\mu)}\right]\\ &= \prod_{\mathrm{p}}\left[1 - z\exp(-\beta\varepsilon_{\mathrm{p}})\right]^{-1}.\end{aligned} \tag{11.232}$$

We introduce a spin-degeneracy factor, $\Delta_{\mathrm{B-E}} = 1$, and as before also define the grand potential $[Q(z, V, T)]_{\mathrm{B-E}}$ that is equal to the logarithm of the B–E partition function given above,

$$\left[Q(z, V, T)\right]_{\mathrm{B-E}} = \Delta_{\mathrm{B-E}} \ln \Psi_{\mathrm{B-E}}(z, V, T)$$

[47] This is obviously the case if $\alpha \geq 1$. But even if the same series has alternating signs and $\alpha \leq -1$, the series is divergent. See, for example, [26].

$$= \Delta_{\text{B-E}} \sum_{\text{p}} \ln\left[1 - z\exp(-\beta\varepsilon_{\text{p}})\right]^{-1}$$
$$= -\Delta_{\text{B-E}} \sum_{\text{p}} \ln\left[1 - z\exp(-\beta\varepsilon_{\text{p}})\right]. \tag{11.233}$$

While traditionally the spin degeneracy weight factor for the Bose–Einstein system is equal to unity, the case of bosons with spin degeneracy factor > 1 is now of at least formal interest in connection with ultracold Bose gases such as Rb_{87} (see, for example, [27]). It is, therefore, helpful for future work to have it included as $\Delta_{\text{B-E}}$. The grand potential $[Q(z, V, T)]_{\text{B-E}}$, like that for the F–D system, will be used in analyses to be presented later.

11.42 Fermi–Dirac and Bose–Einstein Systems

Following a suggestion by Pathria, we can combine (11.225) and (11.233) and thus describe in a single equation both the quantum noninteracting Bose–Einstein and Fermi–Dirac systems. Also, and while we do not necessarily need it, such description simultaneously contains the corresponding representation for a noninteracting classical gas.

Let us write (11.225) and (11.233) as follows:

$$Q(z, V, T) = \frac{\Delta}{a} \sum_{\text{p}} \ln\left[1 + az\exp(-\beta\varepsilon_{\text{p}})\right],$$

$$\begin{aligned}
\text{where} \quad a &= +1\,, \quad \text{for F–D system,}\\
&= -1\,, \quad \text{for B–E system,}\\
&\to 0\,, \quad \text{for classical system,}\\
\text{and} \quad \Delta &= (2S + 1)\,, \quad \text{for F–D system,}\\
&= 1\,, \quad \text{otherwise.}
\end{aligned} \tag{11.234}$$

Note that for electrons, $S = \frac{1}{2}$.

For a Fermi–Dirac and Bose–Einstein gas, in the following we look at the pressure, internal energy, and chemical potential.

11.42.1 Pressure, Internal Energy, and Chemical Potential

Having determined the grand potential $Q(z, V, T)$ for the perfect F–D and B–E gases—see (11.234)—we can use (11.209) to calculate their internal energy, $U(z, V, T)$. Also we can find the necessary relationship that the chemical potential μ has to obey so that the statistical average of the number operator $\bar{N}$ is equal to its actual value N. Finally, we can use (11.210) to calculate the pressure, $P(z, V, T)$.

To this end, we replace, in (11.234), the energy ε_{p} by $\frac{\bar{p}.\bar{p}}{2m} = \frac{p^2}{2m}$, which is the free particle translational energy for momentum $\bar{p}$, and replace the sum, $\sum_{\mathrm{p}}$, by an integral over the relevant volume V and the momentum $\boldsymbol{p}$, that is, $\sum_{\mathrm{p}} = \frac{1}{h^3}\int \mathrm{d}\boldsymbol{x} \int \mathrm{d}\boldsymbol{p} \cdots$ where $\int \mathrm{d}\boldsymbol{x} = V$ and $\int \mathrm{d}\boldsymbol{p} \cdots = \int (4\pi p^2)\mathrm{d}p \cdots$.

As noted before, an important requirement of the grand canonical ensemble is the following relationship between the grand potential $Q(z, V, T)$ and the ratio $\frac{V\,P(z,V,T)}{k_{\mathrm{B}}T}$.[48]

$$\frac{VP(z, V, T)}{k_{\mathrm{B}}T} = Q(z, V, T) = \frac{\Delta}{a}\sum_{\mathrm{p}} \ln[1 + az\exp(-\beta\varepsilon_{\mathrm{p}})]$$
$$= \frac{\Delta}{a}\left(\frac{V}{h^3}\right)\int_0^\infty \ln\left[1 + az\exp\left(-\frac{\beta p^2}{2m}\right)\right]4\pi p^2\mathrm{d}p\,. \tag{11.235}$$

Also we have

$$U(z, V, T) = -\left(\frac{\partial Q(z, V, T)}{\partial \beta}\right)_{\mathrm{z,V}} = \frac{\Delta}{a}\sum_{\mathrm{p}} \varepsilon_{\mathrm{p}}\left[a^{-1}z^{-1}\exp(\beta\varepsilon_{\mathrm{p}}) + 1\right]^{-1}$$
$$= \frac{\Delta}{a}\left(\frac{V}{h^3}\right)\int_0^\infty \left(\frac{p^2}{2m}\right)\left[a^{-1}z^{-1}\exp\left(\frac{\beta p^2}{2m}\right) + 1\right]^{-1} 4\pi p^2 dp \tag{11.236}$$

and

$$\bar{N} = z\left(\frac{\partial Q(z, V, T)}{\partial z}\right)_{\mathrm{V,T}} = \frac{\Delta}{a}\sum_{\mathrm{p}}\left[a^{-1}z^{-1}\exp(\beta\varepsilon_p) + 1\right]^{-1} = \sum_{\mathrm{p}} N_{\mathrm{p}}$$
$$= \frac{\Delta}{a}\left(\frac{V}{h^3}\right)\int_0^\infty \left[a^{-1}z^{-1}\exp\left(\frac{\beta p^2}{2m}\right) + 1\right]^{-1} 4\pi p^2\mathrm{d}p\,. \tag{11.237}$$

In (11.236) and (11.237), the quantity $N_{\mathrm{p}'}$ stands for the mean occupation number of the state with energy $\varepsilon_{\mathrm{p}'}$. That is,

$$N_{\mathrm{p}'} = -\frac{1}{\beta}\left(\frac{\partial Q(z, V, T)}{\partial \varepsilon_{\mathrm{p}'}}\right)_{\mathrm{z,T}} = \frac{\Delta}{a}\left[a^{-1}z^{-1}\exp(\beta\varepsilon_{\mathrm{p}'}) + 1\right]^{-1}. \tag{11.238}$$

It is convenient to integrate (11.235) by parts. We get

$$\frac{VP(z, V, T)}{k_{\mathrm{B}}T} = \left(\frac{V4\pi\Delta}{ah^3}\right)\left\{\frac{p^3}{3}\ln\left[1 + az\exp\left(-\frac{\beta p^2}{2m}\right)\right]\right]_0^\infty$$

[48] As in (11.235)–(11.237), we shall use the notation

$$\sum_{\mathrm{p}}[\hat{O}(p)] = \left(\frac{V}{h^3}\right)\int_0^\infty [\hat{O}(p)]4\pi p^2 dp\,.$$

$$+\left(\frac{V4\pi\beta\Delta}{3mh^3}\right)\int_0^\infty p^4\left[z^{-1}\exp\left(\frac{\beta p^2}{2m}\right)+a\right]^{-1}dp$$
$$=0+\left(\frac{V4\pi\beta\Delta}{3mh^3}\right)\int_0^\infty p^4\left[z^{-1}\exp\left(\frac{\beta p^2}{2m}\right)+a\right]^{-1}dp\,. \quad (11.239)$$

Remark 11.3 Whether they be quantum or classical, the above four equations—namely, (11.236)–(11.239)—along with the prescription given in (11.234), can be used to study perfect gases.

11.43 Perfect Fermi–Dirac System

For a perfect Fermi–Dirac system, with spin $S = \frac{1}{2}$, (11.236)–(11.239) are to be used with $\Delta = 2$ and $a = +1$.

Let us introduce a variable $x = \frac{\beta p^2}{2m}$. Then $p\mathrm{d}p = \frac{m\mathrm{d}x}{\beta}$. And according to (11.237), we have

$$\begin{aligned}\bar{N} &= \left(\frac{V\Delta}{h^3}\right)\int_0^\infty\left[z^{-1}\exp\left(\frac{\beta p^2}{2m}\right)+1\right]^{-1}4\pi p^2\mathrm{d}p\\ &= \left(\frac{2V\Delta}{\lambda^3\sqrt{\pi}}\right)\int_0^\infty x^{\frac{1}{2}}\left[z^{-1}\exp(x)+1\right]^{-1}\mathrm{d}x\\ &= \left(\frac{V\Delta}{\lambda^3}\right)f_{\frac{3}{2}}(z)\,,\end{aligned} \quad (11.240)$$

where

$$f_\mathrm{n}(z) = \frac{1}{\Gamma(n)}\int_0^\infty x^{\mathrm{n}-1}\left[z^{-1}\exp(x)+1\right]^{-1}\mathrm{d}x\,. \quad (11.241)$$

Note that z is to be chosen such that $\bar{N} = N$.

Similarly, we can also write (11.236) as

$$\begin{aligned}U(z,V,T) &= \left(\frac{V\Delta}{h^3}\right)\int_0^\infty\left(\frac{4\pi p^4}{2m}\right)\left[a^{-1}z^{-1}\exp\left(\frac{\beta p^2}{2m}\right)+1\right]^{-1}\mathrm{d}p\\ &= \left(\frac{2V\Delta k_\mathrm{B}T}{\lambda^3\sqrt{\pi}}\right)\int_0^\infty x^{\frac{3}{2}}\left[z^{-1}\exp(x)+1\right]^{-1}\mathrm{d}x\\ &= \left(\frac{3V\Delta k_\mathrm{B}T}{2\lambda^3}\right)f_{\frac{5}{2}}(z)\,.\end{aligned} \quad (11.242)$$

It is convenient to eliminate $\frac{V}{\lambda^3}$ from (11.242) and (11.240). This can be done by dividing the two equations. We get

$$\frac{U(z,V,T)}{N} = \frac{3}{2}k_\mathrm{B}T\left(\frac{f_{\frac{5}{2}}(z)}{f_{\frac{3}{2}}(z)}\right). \quad (11.243)$$

Finally, we deal with (11.239). It can be written as

$$P(z, V, T) = \left(\frac{4\Delta k_B T \pi \beta}{3mh^3}\right) \int_0^\infty p^4 \left[z^{-1} \exp\left(\frac{\beta p^2}{2m}\right) + 1\right]^{-1} dp .$$
$$= \left(\frac{\Delta k_B T}{\lambda^3}\right) f_{\frac{5}{2}}(z) . \tag{11.244}$$

Note that in (11.240)–(11.244) we have needed $\Gamma(n)$. It is available from the relationship

$$\Gamma(n+1) = n\Gamma(n) = n! \qquad \text{Therefore,}$$
$$\Gamma\left(\frac{5}{2}\right) = \frac{3}{2}\Gamma\left(\frac{3}{2}\right) = \left(\frac{3}{2}\right)! = \left(\frac{3}{2}\right)\left(\frac{1}{2}\right)! = \left(\frac{3}{2}\right)\left(\frac{1}{2}\right)\sqrt{\pi} . \tag{11.245}$$

Dividing (11.244) by (11.242) leads to the well-known relationship

$$PV = \frac{2}{3}U . \tag{11.246}$$

This result, namely $PV = \frac{2}{3}U$, applies to all non-relativistic ideal gases in three dimensions whether they be classical—see, for example, (11.108)—or quantum. And it contrasts with the corresponding result for extremely relativistic ideal gases, which according to (11.107) is $PV = \frac{1}{3}U$.

Finally, we write the mean occupation number, N_p, for a single particle state with energy $\varepsilon_p = \frac{p^2}{2m}$. To this end, we use (11.238) and set $\Delta = 2$ and $a = 1$ to get

$$N_p = 2\left[z^{-1} \exp(\beta\varepsilon_p) +\right]^{-1} = 2\left[\exp\{\beta(\varepsilon_p - \mu)\} +\right]^{-1} . \tag{11.247}$$

It is important to note that (11.247) includes both spin states, namely those with—what may be called—spin-up and others with spin-down. As the temperature is lowered, the electrons in the system begin to place themselves in those states that lower the total energy. At zero temperature Kelvin, the system energy is the lowest possible—subject, of course, to the dictates of the Pauli principle, which demands that for a prescribed value of the spin component and at a given time no more than a single electron may occupy any given state.[49]

Because the Hamiltonian does not involve electron spin, any single particle state with energy ε_p can accommodate a total of two electrons with the same energy, one with spin-up and the other with spin-down. Accordingly, at zero temperature, the electrons are packed, two—that have the higher energy—on top of other two that have the lower energy. This packing continues up the energy chain to some maximum value E_F which is such that all N electrons have been accommodated.

Therefore, at zero temperature, all single particle states at energy ε_p less than or equal to $\mu = \mu_0 = E_F$ are occupied. (In what follows we shall use the notation μ_0 for the chemical potential μ at zero temperature—see also (11.275).)

[49]Note that single particle states for noninteracting electrons are identified by their kinetic energy as well as the direction of their spins.

11.43.1 Weakly-Degenerate Fermi–Dirac System

For low particle density and high temperature, an F–D system is in a state of weak degeneracy. Here z is less than unity. Therefore, (11.241) can be expanded as follows:

$$\begin{aligned} f_{\mathrm{n}}(z) &= \frac{1}{\Gamma(n)} \int_0^{\infty} x^{n-1}\left[z^{-1}\exp(x)+1\right]^{-1}\mathrm{d}x \\ &= \frac{1}{\Gamma(n)} \int_0^{\infty} x^{n-1} \sum_{l=1}^{\infty}(-1)^{l-1}\left[z\exp(-x)\right]^{l}\mathrm{d}x \\ &= \sum_{l=1}^{\infty}(-1)^{l-1}\frac{z^l}{\Gamma(n)}\int_0^{\infty} x^{n-1}\exp(-lx)\mathrm{d}x \\ &= \sum_{l=1}^{\infty}(-1)^{l-1}\frac{z^l}{l^n}\left[\frac{\int_0^{\infty}\eta^{n-1}\exp(-\eta)\mathrm{d}\eta}{\Gamma(n)}\right] \\ &= \sum_{l=1}^{\infty}(-1)^{l-1}\frac{z^l}{l^n}\,. \end{aligned} \tag{11.248}$$

Indeed, for a highly nondegenerate Fermi gas, z is very small and only a few terms in power expansion for the function $f_{\mathrm{n}}(z)$ suffice. Therefore, (11.248) gives

$$f_{\mathrm{n}}(z) = z - \frac{z^2}{2^n} + \frac{z^3}{3^n} - O\left(z^4\right). \tag{11.249}$$

By using (11.249), equations (11.240), (11.242), and (11.244) are readily represented in terms of the leading three powers of z. However, rather than z, we are interested in expressing the results in terms of the temperature. To this end, we need to invert the series in powers of z that is implicit in (11.240). Such inversions are commonplace and are easily done using mathematical software. It may, however, be of interest to a beginner to see how this inversion might be done analytically.

Let us rewrite the equation to be inverted as follows:

$$\rho = \frac{\bar{N}\lambda^3}{2V} = f_{\frac{3}{2}}(z) = z - \frac{z^2}{2^{\frac{3}{2}}} + \frac{z^3}{3^{\frac{3}{2}}} - O\left(z^4\right). \tag{11.250}$$

The first step is to notice that for small z we get $z \approx \rho$. An appropriate trial representation to work with is of the form

$$z = \rho + a_2\rho^2 + a_3\rho^3 + O\left(\rho^4\right), \tag{11.251}$$

where self-consistently z and ρ can both be expected to be $\ll 1$. The second step is to use the defining (11.250), then plug (11.251) into it, and work exactly to the second order. We get

$$\rho = z - \frac{z^2}{2^{\frac{3}{2}}} + O\left(z^3\right)$$

$$= (\rho + a_2\rho^2) - \frac{(\rho + a_2\rho^2)^2}{2^{\frac{3}{2}}} + O(\rho^3)$$

$$= \rho + \rho^2\left(a_2 - \frac{1}{2^{\frac{3}{2}}}\right) + O(\rho^3). \tag{11.252}$$

Comparing similar terms on the left- and right-hand sides leads to the result

$$a_2 = \frac{1}{2^{\frac{3}{2}}}. \tag{11.253}$$

And continuing this procedure to the third order readily yields $a_3 = \frac{1}{4} - \frac{1}{3^{\frac{3}{2}}}$. Therefore (11.251) becomes

$$z = \rho + \left(\frac{1}{2^{\frac{3}{2}}}\right)\rho^2 + \left(\frac{1}{4} - \frac{1}{3^{\frac{3}{2}}}\right)\rho^3 + O(\rho^4). \tag{11.254}$$

11.44 Virial Expansion for a Perfect Fermi–Dirac Gas

As shown in (11.242) and (11.244), both U and P depend on $f_{\frac{5}{2}}(z)$ which—with the use of (11.254)—becomes

$$f_{\frac{5}{2}}(z) = z - \frac{z^2}{2^{\frac{5}{2}}} + \frac{z^3}{3^{\frac{5}{2}}} - O(z^4)$$

$$= \rho + \left(\frac{1}{2^{\frac{3}{2}}}\right)\rho^2 + \left(\frac{1}{4} - \frac{1}{3^{\frac{3}{2}}}\right)\rho^3 - \frac{(\rho + (\frac{1}{2^{\frac{3}{2}}})\rho^2)^2}{2^{\frac{5}{2}}} + \frac{\rho^3}{3^{\frac{5}{2}}} + O(\rho^4)$$

$$= \rho + \left(\frac{1}{2^{\frac{5}{2}}}\right)\rho^2 + \left(\frac{1}{8} - \frac{2}{3^{\frac{5}{2}}}\right)\rho^3 + O(\rho^4). \tag{11.255}$$

Therefore, using (11.244) and (11.255), we can construct a kind of virial expansion[50] for the noninteracting perfect F–D gas. Such an expansion relates the ratio $\frac{PV}{\bar{N}k_BT}$ to a power expansion in ρ (or equivalently, an inverse power of the temperature T). We have

$$PV = V\left(\frac{2k_BT}{\lambda^3}\right) f_{\frac{5}{2}}(z)$$

$$= V\left(\frac{2k_BT}{\lambda^3}\right)\left\{\rho + \left(\frac{1}{2^{\frac{5}{2}}}\right)\rho^2 + \left(\frac{1}{8} - \frac{2}{3^{\frac{5}{2}}}\right)\rho^3 + O(\rho^4)\right\}$$

$$= k_BT\bar{N}\left\{1 + \left(\frac{1}{2^{\frac{5}{2}}}\right)\rho + \left(\frac{1}{8} - \frac{2}{3^{\frac{5}{2}}}\right)\rho^2 + O(\rho^3)\right\}. \tag{11.256}$$

[50] The classical virial expansion, of course, is in the inverse powers of T rather than in inverse powers of $T^{3/2}$ as is the case here.

Here, as in (11.250), $\rho = \frac{\bar{N}\lambda^3}{V\Delta} = (\frac{\bar{N}h^3}{V\Delta})(2\pi m k_B T)^{\frac{-3}{2}}$, with $\Delta = 2$. Therefore, the above equation approaches the corresponding result for a perfect classical gas when the temperature is high and ρ small. Comparing this result with that of a classical gas with nonzero interparticle interaction, the above equation appears to represent the behavior of a gas with "repulsive" interparticle force. *Notice that the "appearance" of nonzero repulsive interaction is being produced by a completely noninteracting gas! But, of course, the gas is made of quantum particles. Perhaps it should also be mentioned that if the quantum particles had obeyed, rather than the Fermi–Dirac, the Bose–Einstein statistics, the apparent interparticle interaction would have been "attractive."* (See, for instance, the corresponding result for the B–E gas. Refer to (11.365).)

The internal energy, specific heat, Gibbs potential, and Helmholtz free energy are given below.

$$
\begin{aligned}
U &= \frac{3k_B T\bar{N}}{2}\left\{1+\left(\frac{1}{2^{\frac{5}{2}}}\right)\rho+\left(\frac{1}{8}-\frac{2}{3^{\frac{5}{2}}}\right)\rho^2+O(\rho^3)\right\}=\frac{3}{2}PV\,,\\
C_v &= \left(\frac{\partial U}{\partial T}\right)_{\bar{N},V}=\frac{3}{2}\bar{N}k_B\left\{1-\left(\frac{1}{2^{\frac{7}{2}}}\right)\rho+\left(\frac{4}{3^{\frac{5}{2}}}-\frac{1}{4}\right)\rho^2+O(\rho^3)\right\},\\
G &= \bar{N}\mu = k_B T\bar{N}\ln(z)\\
&= k_B T\bar{N}\ln\left[\rho\left\{1+\left(\frac{1}{2^{\frac{3}{2}}}\right)\rho+\left(\frac{1}{4}-\frac{1}{3^{\frac{3}{2}}}\right)\rho^2+O(\rho^3)\right\}\right],
\end{aligned}
\tag{11.257}
$$

$$
\begin{aligned}
F &= G - PV = G - \frac{2}{3}U\\
&= k_B T\bar{N}\ln\left[\rho\left\{1+\left(\frac{1}{2^{\frac{3}{2}}}\right)\rho+\left(\frac{1}{4}-\frac{1}{3^{\frac{3}{2}}}\right)\rho^2+O(\rho^3)\right\}\right]\\
&\quad - k_B T\bar{N}\left\{1+\left(\frac{1}{2^{\frac{5}{2}}}\right)\rho+\left(\frac{1}{8}-\frac{2}{3^{\frac{5}{2}}}\right)\rho^2-O(\rho^3)\right\}.
\end{aligned}
$$

Notice that unless the temperature is truly infinite, the specific heat for the F–D gas is less than that for a corresponding classical ideal gas. Indeed, unlike the classical ideal gas—for which the specific heat is constant—we shall learn in the following that the specific heat for the F–D gas continues to decrease monotonically (all the way to zero) when the temperature decreases.

Finally, let us look at the entropy

$$
\begin{aligned}
S = \frac{U-F}{T} &= \frac{5}{2}\bar{N}k_B\left\{1+\left(\frac{1}{2^{\frac{5}{2}}}\right)\rho+\left(\frac{1}{8}-\frac{2}{3^{\frac{5}{2}}}\right)\rho^2\right\}\\
&\quad - \bar{N}k_B\ln\left[\rho\left\{1+\left(\frac{1}{2^{\frac{3}{2}}}\right)\rho+\left(\frac{1}{4}-\frac{1}{3^{\frac{3}{2}}}\right)\rho^2+O(\rho^3)\right\}\right].
\end{aligned}
\tag{11.258}
$$

Remark 11.4 The canonical ensembles are sometimes hard put to treat quantum systems. Such is the case as much due to the intrinsic quantum character of these systems as it is due to their particle indistinguishability. To relieve these difficulties, the grand canonical ensemble is used.

An important feature of its predictions is that even when the quantum particles are totally noninteracting—except in the limit of very low density and very high temperature—their thermodynamics is qualitatively different from that of similar classical particles.

For instance, a low density noninteracting quantum system at very high temperature is essentially classical and has little or no spatial correlation.

In contrast, at high particle density and low temperature, spatial correlation exists even in a noninteracting quantum system. This is especially true for particle densities and temperature such that the mean thermal wave-length, λ, is comparable to or longer than the average interparticle separation.

Numerically, the above[51] statements can be summarized in terms of the following rule: Given N, V, Δ, and $\lambda = (\frac{h^2}{2\pi m k_B T})^{\frac{1}{2}}$, if

$$\rho = \frac{N(\frac{h^2}{2\pi m k_B T})^{\frac{3}{2}}}{V\Delta} \geq 1\,, \quad \text{thermodynamics is quantum, but if} \qquad (11.259)$$

$$\rho \ll 1\,, \quad \text{thermodynamics is classical.}$$

The above rule is physically significant and ρ—which is known as the degeneracy discriminant—acts as a useful expansion parameter. The rule reads:

When the[52] system is relatively dense, is at low temperature, and ρ is of the order of unity or larger, the system begins to be degenerate and displays truly quantum effects.

On the other hand, in the limit when the particle number density, $\frac{N}{V}$, is very low and the temperature, T, is very high—that is, when ρ is $\ll 1$—the system tends toward complete nondegeneracy, and all physical quantities approach their classical limit.

Exercise 11.7 Confirm the above statement by observing that if in all expressions with the appearance $\{1+\text{terms of order } \rho\}$—that occur in (11.256)–(11.258)—terms of order ρ are ignored compared with 1, then the results reduce to those obtained for a classical noninteracting gas. This affirms the fact that in the limit of extreme nondegeneracy, a quantum gas approaches its classical limit.

11.45 Highly or Partially-Degenerate Fermi–Dirac

In the above we have learned that at high temperatures and low density, the parameters ρ and therefore z are both very small compared to unity. At very low temperatures, in complete contrast to what the case is at high temperatures, z becomes exponentially large (and indeed tends to ∞ at zero temperature). Such is the case because even at room temperature in a typical system of conduction electrons, $\frac{\mu}{k_B T}$ is of order 25 to 150. Accordingly, $z = \exp(\frac{\mu}{k_B T}) \gg 1$.

[51] Thermal wave-length $\lambda = (\frac{h^2}{2\pi m k_B T})^{\frac{1}{2}}$.

[52] Degeneracy discriminant $\rho = \frac{N(\frac{h^2}{2\pi m k_B T})^{\frac{3}{2}}}{2V}$.

Thus at low to normal temperatures, $\ln z$ is large compared to unity and $(\ln z)^{-1}$ is correspondingly small compared to 1.

Therefore, in order to proceed further, we need to determine an expansion for the function $f_n(z)$—note that $f_n(z)$ was defined in (11.241). Such an expansion should be appropriate for both the partially or completely degenerate cases. Recall that the expansion for $f_n(z)$ that was given earlier in (11.248) is appropriate only for the highly nondegenerate case.

Evaluation of the integral $f_n(z)$ For notational convenience, in order to evaluate the integral $f_n(z)$, we introduce a related integral $A_n(z)$ which is given below:

$$\begin{aligned}
\text{Let}\quad & B_n(\epsilon) = \epsilon^{n-1}, \\
\text{then}\quad & A_n(z) = \int_0^\infty \left[\frac{B_n(\epsilon)}{z^{-1}\exp(\frac{\epsilon}{k_B T}) + 1}\right] d\epsilon \\
&= \int_0^\infty \left[\frac{B_n(\epsilon)}{\exp(\frac{\epsilon-\mu}{k_B T}) + 1}\right] d\epsilon \\
&= (k_B T)^n \int_0^\infty x^{n-1}\big[z^{-1}\exp(x) + 1\big]^{-1} dx . \\
&= \Gamma(n)(k_B T)^n f_n(z) .
\end{aligned} \tag{11.260}$$

In order to calculate $A_n(z)$, let us begin by introducing the substitution

$$\tau = k_B T , \qquad x = \frac{\epsilon - \mu}{\tau} . \tag{11.261}$$

Then $\epsilon = \mu + \tau x$ and $d\epsilon = \tau dx$. And

$$\begin{aligned}
A_n(z) &= \tau \int_{-\frac{\mu}{\tau}}^\infty \left[\frac{B_n(\mu + \tau x)}{\exp(x) + 1}\right] dx \\
&= \tau \int_0^{\frac{\mu}{\tau}} \left[\frac{B_n(\mu - \tau y)}{\exp(-y) + 1}\right] dy + \tau \int_0^\infty \left[\frac{B_n(\mu + \tau x)}{\exp(x) + 1}\right] dx .
\end{aligned} \tag{11.262}$$

To arrive at the second row, we separated the integral in the first row into two parts, one ranging from $-\frac{\mu}{\tau}$ to 0 and the other from 0 to ∞. Then we introduced the substitution $y = -x$ into the first part.

Noting the fact that $[\exp(-y) + 1]^{-1} = 1 - [\exp(y) + 1]^{-1}$ and introducing a new substitution $x = \tau y$, which gives $dx = \tau dy$, into the first term in the second row in (11.262), gives

$$\begin{aligned}
A_n(z) &= \tau \int_0^{\frac{\mu}{\tau}} B_n(\mu - \tau y) dy - \tau \int_0^{\frac{\mu}{\tau}} \left[\frac{B_n(\mu - \tau y)}{\exp(y) + 1}\right] dy \\
&\quad + \tau \int_0^\infty \left[\frac{B_n(\mu + \tau x)}{\exp(x) + 1}\right] dx .
\end{aligned} \tag{11.263}$$

The first integral on the right-hand side looks better if we introduce yet another change in the variables. To do that, set $\mu - \tau y = x$. Then $\mathrm{d}y = (\frac{-1}{T})\mathrm{d}x$ and we can write

$$A_{\mathrm{n}}(z) = \int_0^{\mu} B_{\mathrm{n}}(x)\mathrm{d}x - \tau \int_0^{\frac{\mu}{\tau}} \left[\frac{B_{\mathrm{n}}(\mu - \tau x)}{\exp(x) + 1}\right]\mathrm{d}x$$
$$+ \tau \int_0^{\infty} \left[\frac{B_{\mathrm{n}}(\mu + \tau x)}{\exp(x) + 1}\right]\mathrm{d}x \,. \quad (11.264)$$

Note that, in order to improve its looks, surreptitiously we also changed the dummy variable y to another dummy variable x in the second term in (11.264).

We have now reached a point beyond which approximations are necessary. Fortunately, however, for a typical electron gas—for example, the conduction electrons in sodium—at room temperature, $\frac{\mu}{\tau}$ is of order 100, or higher. At these values—namely $x \approx \frac{\mu}{T}$—the presence of the exponential in the denominator makes the integrand become essentially equal to zero. Therefore the upper limit in the last two integrals in (11.264) can safely be extended beyond $\frac{\mu}{\tau}$ to ∞. That is,

$$A_{\mathrm{n}}(z) \approx \int_0^{\mu} B_{\mathrm{n}}(x)\mathrm{d}x - \tau \int_0^{\infty} \left[\frac{B_{\mathrm{n}}(\mu - \tau x)}{\exp(x) + 1}\right]\mathrm{d}x$$
$$+ \tau \int_0^{\infty} \left[\frac{B_{\mathrm{n}}(\mu + \tau x)}{\exp(x) + 1}\right]\mathrm{d}x \,. \quad (11.265)$$

The presence of the exponential in the denominator also has another salutary effect. The bulk of the integral necessarily comes from those values of x for which $\exp(x)$ is not large compared to unity. Also, for essentially all laboratory temperatures, τx is $\ll \mu$. Accordingly, we can expand $B(\mu \pm \tau x)$ in powers of $\frac{\tau x}{\mu}$. That is,

$$B_{\mathrm{n}}(\mu \pm \tau x) = B(\mu) \pm (\tau x) B'_{\mathrm{n}}(\mu)$$
$$+ \frac{(\tau x)^2}{2!} B''_{\mathrm{n}}(\mu) \pm \frac{(\tau x)^3}{3!} B'''_{\mathrm{n}}(\mu) + \cdots \,, \quad (11.266)$$
$$B_{\mathrm{n}}(\mu + \tau x) - B_{\mathrm{n}}(\mu - \tau x) = 2(\tau x) B'_{\mathrm{n}}(\mu) + 2\frac{(\tau x)^3}{3!} B'''_{\mathrm{n}}(\mu) + \cdots \,.$$

Therefore

$$A_{\mathrm{n}}(z) \approx \int_0^{\mu} B_{\mathrm{n}}(x)\mathrm{d}x + \left(2\tau^2\right) B'_{\mathrm{n}}(\mu) \int_0^{\infty} \left[\frac{x}{\exp(x) + 1}\right]\mathrm{d}x$$
$$+ \left(\frac{2\tau^4}{3!}\right) B'''_{\mathrm{n}}(\mu) \int_0^{\infty} \left[\frac{x^3}{\exp(x) + 1}\right]\mathrm{d}x + \cdots \,. \quad (11.267)$$

The integrals in (11.267) are well known. That is,

$$\int_0^\infty \left[\frac{x}{\exp(x)+1}\right]\mathrm{d}x = \frac{\pi^2}{12},$$
$$\int_0^\infty \left[\frac{x^3}{\exp(x)+1}\right]\mathrm{d}x = \frac{7\pi^4}{120}, \tag{11.268}$$
$$\int_0^\infty \left[\frac{x^5}{\exp(x)+1}\right]\mathrm{d}x = \frac{31\pi^6}{252}, \quad \text{etc.}$$

Students of mathematics will recognize that these integrals are related to the well known—and extensively available—Riemann zeta function, $\varsigma(n)$. That is,

$$\begin{aligned}
\int_0^\infty \left[\frac{x^n}{\exp(x)+1}\right]\mathrm{d}x &= \sum_{l=1}^{\infty}(-1)^{l-1}\left[\int_0^\infty x^n\left[\exp(-x)\right]^l \mathrm{d}x\right] \\
&= \Gamma(n+1)\sum_{l=1}^{\infty}(-1)^{l-1}\left(\frac{1}{l}\right)^{n+1} \\
&= \Gamma(n+1)\left[1-\frac{1}{2^{n+1}}+\frac{1}{3^{n+1}}-\cdots\right] \\
&= \Gamma(n+1)\left[\left(1-\frac{1}{2^n}\right)\varsigma(n+1)\right].
\end{aligned} \tag{11.269}$$

And, of course, $\varsigma(2)=\frac{\pi^2}{6}$, $\varsigma(4)=\frac{\pi^4}{90}$, $\varsigma(6)=\frac{\pi^6}{945}$, etc.

Having thus calculated $A_{\mathrm{n}}(z)$—see (11.267)–(11.269)—we in effect have an expansion, in ascending powers of the variable $(\frac{\tau}{\mu})^2=\zeta^{-2}$, for the functions $f_{\mathrm{n}}(z)$. (In this regard, see (11.260) which specifies a direct relationship between $A_{\mathrm{n}}(z)$ and $f_{\mathrm{n}}(z)$. Also see (11.241) for the definition of $f_{\mathrm{n}}(z)$.) We get

$$f_{\frac{3}{2}}(z)=\frac{4\zeta^{\frac{3}{2}}}{3\sqrt{\pi}}\left[1+\frac{\pi^2}{8}\zeta^{-2}+\frac{7\pi^4}{640}\zeta^{-4}+O\left(\zeta^{-6}\right)\right] \tag{11.270}$$

and

$$f_{\frac{5}{2}}(z)=\frac{8\zeta^{\frac{5}{2}}}{15\sqrt{\pi}}\left[1+\frac{5\pi^2}{8}\zeta^{-2}-\frac{7\pi^4}{384}\zeta^{-4}+O\left(\zeta^{-6}\right)\right], \tag{11.271}$$

where[53]

$$\zeta^{-1}=(\ln z)^{-1}=\frac{\tau}{\mu}=\frac{k_{\mathrm{B}}T}{\mu}\ll 1. \tag{11.272}$$

[53] It is important to distinguish the nth power of this variable, namely ζ^n, from the somewhat similar looking Riemann zeta function, $\varsigma(n)$.

Now let us combine (11.240) and (11.270). We get

$$\frac{N}{V} = \frac{2}{\lambda^3} f_{\frac{3}{2}}(z) = \left(\frac{8\zeta^{\frac{3}{2}}}{3\sqrt{\pi}\lambda^3}\right)\left[1 + \frac{\pi^2}{8}\zeta^{-2} + \frac{7\pi^4}{640}\zeta^{-4} + o\left(\zeta^{-6}\right)\right]. \tag{11.273}$$

Similarly, combine (11.270) and (11.271). Then (11.243) yields

$$\begin{aligned} U(z, V, T) &= \left(\frac{3k_B T N}{2}\right)\frac{f_{\frac{5}{2}}(z)}{f_{\frac{3}{2}}(z)} \\ &= \left(\frac{3k_B T N\zeta}{5}\right)\left[\left(\frac{1 + \frac{5\pi^2}{8}\zeta^{-2} - \frac{7\pi^4}{384}\zeta^{-4}}{1 + \frac{\pi^2}{8}\zeta^{-2} + \frac{7\pi^4}{640}\zeta^{-4}}\right) + o\left(\zeta^{-6}\right)\right] \\ &= \left(\frac{3k_B T N\zeta}{5}\right)\left[1 + \frac{\pi^2}{2}\zeta^{-2} - \frac{11\pi^4}{120}\zeta^{-4} + o\left(\zeta^{-6}\right)\right]. \end{aligned} \tag{11.274}$$

While we should have liked to have had both $\frac{N}{V}$ and U given in terms of just the temperature, the above two equations involve ζ which, in addition to the temperature, also depends on the chemical potential μ. So we still have some work to do! See (11.283)–(11.295).

11.45.1 Complete Degeneracy: Zero Temperature

Simple Analysis

As stated earlier, at zero degrees absolute, the chemical potential μ is denoted as μ_0. Similarly, at zero temperature, we shall denote the occupation number N_p of the state with momentum p as $(N_p)_0$. Of course, at this temperature, $\beta = \frac{1}{k_B T} \to \infty$. Then, (11.247), i.e., $N_p = 2[\exp\{\beta(\varepsilon_p - \mu_0)\} + 1]^{-1}$, readily leads to

$$\begin{aligned} N_p \to_{T=0} (N_p)_0 &=_{\beta\to\infty} 2\,, \quad \text{if } \varepsilon_p \text{ is} < \mu_0, \\ &= 0\,, \quad \text{if } \varepsilon_p \text{ is} > \mu_0, \\ \text{and exceptionally,} \quad (N_p)_0 &= 1\,, \quad \text{if } \varepsilon_p = \mu_0\,. \end{aligned} \tag{11.275}$$

Electrons in the system occupy states with momentum p—which ranges from zero to (the so-called) Fermi momentum, p_F. Note that p_F is the maximum allowed value of momentum that any single-electron (in the noninteracting system of N electrons) can possibly have at zero temperature. Also note that p_F is defined by the relationship $(p_F)^2/2m = E_F = \mu_0$. The energy E_F is known as the "Fermi energy."

Recall that N denotes the total number of electrons in the system. That is, $N = \sum_p N_p = \sum_{p=0}^{p_F}(N_p)_0$. Therefore, using the value of $(N_p)_0$ as given in (11.275), at zero temperature we can write the sum, $\sum_p N_p = (\frac{V}{h^3})\int_0^\infty (N_p)4\pi p^2 dp$, as follows:

$$\begin{aligned} N = \sum_{p=0}^{p_F}(N_p)_0 &= \left(\frac{V}{h^3}\right)\int_0^{p_F}[2]4\pi p^2 dp \\ &= \int_0^{p_F}\nu(p)dp = \frac{8\pi V {p_F}^3}{3h^3}\,. \end{aligned} \tag{11.276}$$

The notation $\nu(p)\mathrm{d}p$ indicates the number of single particles associated with states that lie in momentum between p and $p+\mathrm{d}p$. It is often convenient to replace $\nu(p)\mathrm{d}p$ by $f(\varepsilon)\mathrm{d}\varepsilon$ such that $p^2 = 2m\varepsilon$ and $\mathrm{d}p = \sqrt{\frac{m}{2\varepsilon}}\mathrm{d}\varepsilon$. As a result we have

$$\begin{aligned}\nu(p)\mathrm{d}p &= \frac{2V}{h^3}(4\pi p^2)dp \\ &= \left[\frac{V}{2}\sqrt{\frac{\varepsilon}{\pi}}\left(\frac{8\pi m}{h^2}\right)^{\frac{3}{2}}\right]\mathrm{d}\varepsilon = f(\varepsilon)\mathrm{d}\varepsilon\,. \end{aligned} \tag{11.277}$$

Like $\nu(p)\mathrm{d}p$, $f(\varepsilon)\mathrm{d}\varepsilon$ is equal to the number of single particles associated with single particle states with energy between ε and $\varepsilon+\mathrm{d}\varepsilon$. Therefore $f(\varepsilon)$ can legitimately be called the density of states because it is equal to the number of single particles per unit of energy—or, equivalently, because the spin can be up or down—twice the number of single particle states that have energy between ε and $\varepsilon + \mathrm{d}\varepsilon$.

We shall use the following notation to express the result of (11.276):

$$\begin{aligned} p_{\mathrm{F}} &= (2m\mu_0)^{\frac{1}{2}} = \left(\frac{3N}{8\pi V}\right)^{\frac{1}{3}} h \quad \text{and} \\ E_{\mathrm{F}} = \mu_0 &= \frac{(p_{\mathrm{F}})^2}{2m} = \left(\frac{3N}{8\pi V}\right)^{\frac{2}{3}}\left(\frac{h^2}{2m}\right). \end{aligned} \tag{11.278}$$

For conduction electrons in sodium, the Fermi energy E_{F} is ≈ 3.15 electron volts. In terms of the standard thermal energy unit divided by k_{B}, i.e., the temperature T, the sodium Fermi energy is equivalent to approximately 37 thousand degrees Kelvin—a temperature much higher than even the boiling-evaporation temperature of sodium!

Similarly, by using (11.236), we can calculate the total energy U_0 of the system at zero temperature. Note that U_0 is the so-called zero point, or the ground state, energy:

$$\begin{aligned} U = \sum_{\mathrm{p}} \varepsilon_{\mathrm{p}} N_{\mathrm{p}}\,, \quad \lim_{(\mathrm{T}=0)} U &= \sum_{\mathrm{p}=0}^{\mathrm{p}=\mathrm{p_F}} \varepsilon_{\mathrm{p}}(N_{\mathrm{p}})_0 = U_0 \\ &= \left(\frac{V}{h^3}\right)\int_0^{p_{\mathrm{F}}}\left(\frac{p^2}{2m}\right)[2]4\pi p^2 \mathrm{d}p = \left(\frac{4\pi V}{mh^3}\right)\left(\frac{p_{\mathrm{F}}{}^5}{5}\right). \end{aligned} \tag{11.279}$$

Equation (11.278) allows us to replace p_{F}^3 by its equivalent, $(\frac{3N}{8\pi V})h^3$, and represent the above result in the following better-known form:

$$\begin{aligned} U_0 &= \left(\frac{4\pi V}{mh^3}\right)\left(\frac{p_{\mathrm{F}}{}^2}{5}\right)(p_{\mathrm{F}})^3 = \left(\frac{4\pi V}{mh^3}\right)\left(\frac{p_{\mathrm{F}}{}^2}{5}\right)\left(\frac{3N}{8\pi V}\right)h^3 \\ &= \frac{3}{5}N\left(\frac{p_{\mathrm{F}}{}^2}{2m}\right) = \frac{3}{5}N\mu_0 = \frac{3}{5}NE_{\mathrm{F}} \\ &= \left(\frac{3V}{2}\right)\left(\frac{3}{8\pi}\right)^{\frac{2}{3}}\left(\frac{h^2}{5m}\right)\left(\frac{N}{V}\right)^{\frac{5}{3}}. \end{aligned} \tag{11.280}$$

The average, quantum statistical energy per particle is $\frac{U_0}{N} = \frac{3}{5}E_F$. Considering that a reasonable approximation to such a particle is a conduction electron in a low alkali metal, this energy would appear to be much larger than the typical thermal energy at room temperature. So why do these electrons, with all this "fantastic, hot" energy "in their pocket," still stay around within these metals? The answer is twofold:

An appropriate description of the first part of the answer has to await the discussion—see a few paragraphs below—of the results given in (11.282).

The second part of the answer is the following: Figuratively speaking, at zero temperature, all the electrons are sunk within the so-called Fermi sea. The two electrons at the bottom have little momentum. The next higher two have slightly more. The process continues until all the N electrons are placed atop each other so that the final Nth electron is placed with momentum p_F. However, when the temperature rises above zero by, say, an amount ΔT, then some electrons from the top of the Fermi sea increase their energy by an amount $k_B \Delta T$. As a result, while normally their momentum is close to p_F, the added bit of thermal energy raises their momentum a bit and, as a result, they begin to evaporate out. We shall have occasion to amplify on this statement when we examine the Richardson effect which describes the number of electrons that leave the metal surface as a result of the increase in temperature.

Let us next deal with the pressure. The so-called ground state, degeneracy pressure, P_0, is clearly related to U_0 through the usual relationship

$$P_0 = \frac{2U_0}{3V} = \frac{2}{5}\left(\frac{N}{V}\right)E_F = \left(\frac{3}{8\pi}\right)^{\frac{2}{3}}\left(\frac{h^2}{5m}\right)\left(\frac{N}{V}\right)^{\frac{5}{3}}. \tag{11.281}$$

The bulk modulus, $-P_0(\frac{\partial P_0}{\partial V})_{N,T}$, can now be calculated. We get

$$-V\left(\frac{\partial P_0}{\partial V}\right)_{N,T} = \frac{5}{3}\left(\frac{3}{8\pi}\right)^{\frac{2}{3}}\left(\frac{h^2}{5m}\right)\left(\frac{N}{V}\right)^{\frac{5}{3}} = \frac{2}{3}\left(\frac{N}{V}\right)E_F. \tag{11.282}$$

(In (11.280) and (11.281) we have used (11.278). This was to equate E_F to $(\frac{3N}{8\pi V})^{\frac{2}{3}}(\frac{h^2}{2m})$.)

The bulk modulus in sodium is approximately equal to 7×10^9 N/m^2. This is of the same order of magnitude as the number predicted by the above relationship.

Notice that the pressure P_0 in (11.281) is nonzero even though the system is at zero temperature. Indeed, the Fermi–Dirac degeneracy pressure P_0 is in fact very large, being typically of the order of a hundred thousand times the atmospheric pressure. With all this pressure wanting to expand the electrons in a metal, why do they not fly off? Typically, the attractive electrostatic interaction—i.e., the Coulomb force—between the positive metal ions and the free electrons is sufficient to hold the electrons back. While the number of free electrons in metals is large compared to the number of atoms in a similar volume of free air, the chemical potential in a "white dwarf" is even larger than that in metals. That causes an enormous degeneracy pressure to build up within the dwarf which holds the gravitational pull of many of the average sized dwarfs in check and stops them from collapsing into "black holes."

Further notice the fact that the pressure increases rapidly with decrease in volume. While for classical systems this behavior would indicate the presence of interparticle interaction, the given system is completely noninteracting. All it has are the laws of quantum mechanics, whereby even a noninteracting, single Fermi–Dirac particle expresses its distaste for being confined to a finite volume; additionally, the electrons follow the laws of "natural philosophy" which demand that they obey the Fermi–Dirac symmetry rules.

Remark 11.5 When $k_{\rm B}T \to 0$, by classical description all thermal activity ceases. This should make both the pressure and the kinetic energy of the system go to zero. Yet, (11.280) and (11.281) indicate otherwise. Why is that the case, one might ask.

Clearly, it is a quantum effect, unknown in classical mechanics. Yet, as we shall discover when we consider the Bose–Einstein statistics—where, at zero temperature, all the Bose quantum particles may gather together in a single energy state—the details of the given results are crucially dependent upon the statistics of the quantum states being treated here.

In stark contrast with bosons, a maximum of only two spin a-half fermions, i.e., electrons—one with spin-up and the other with spin-down—are allowed to settle down in one single (spin-less) state. So, the best they can do is to follow the general dictates of thermodynamics, which enjoin that at zero temperature only the lowest available energy state has a finite probability of being occupied. Accordingly, as the particles are accommodated two by two, the lowest momentum state is occupied first. Higher momentum states get occupied at the rate of two particles each until the Nth particle is settled in at momentum $p_{\rm F}$. At this momentum, its—spin-less—single particle kinetic energy is $\frac{p_{\rm F}^2}{2m} = \mu_0 = E_{\rm F}$.

Formal Treatment

If we were at low temperature, then the inverse powers of ζ—such as ζ^{-2} and ζ^{-4}—would be very small. Indeed, they would be equal to zero if $T \to 0$. (See (11.272) which states $\zeta^{-1} = \frac{k_{\rm B}T}{\mu}$.)

Let us now set the temperature equal to zero and rewrite (11.273) as

$$\zeta^{\frac{3}{2}} =_{\rm T\to 0} \frac{3\sqrt{\pi}\lambda^3 N}{8V}, \tag{11.283}$$

and convert it to the following more convenient form:

$$\begin{aligned} k_{\rm B}T\zeta = \mu =_{\rm T\to 0} k_{\rm B}T\left(\frac{3\sqrt{\pi}\lambda^3 N}{8V}\right)^{\frac{2}{3}} = \mu_0 \\ = \left(\frac{3N}{8\pi V}\right)^{\frac{2}{3}}\left(\frac{h^2}{2m}\right) = E_{\rm F}. \end{aligned} \tag{11.284}$$

Similarly, (11.274) can be written as

$$U =_{\mathrm{T}\to 0} \frac{3k_{\mathrm{B}}TN\zeta}{5} = U_0 = \frac{3}{5}NE_{\mathrm{F}}\,. \tag{11.285}$$

Why is the average value of the energy—namely U_0—not equal to $\frac{1}{2}$, but instead it is three-fifths of the maximum value E_{F}? This has to do with the fact that the number of states per unit volume of the momentum space increase with rise in momentum—meaning $\nu(p)$ increases as p^2. So a larger number of states are found in the upper-half than the lower-half and, as a result, the average energy is higher than one-half, because the latter assumes uniform spacing.[54]

In (11.284) we have used the equality, $\lambda = (\frac{2\pi mk_{\mathrm{B}}T}{h^2})^{-\frac{1}{2}}$, that was recorded in (11.212). Also we have continued the use of the notation E_{F} which is equal to the chemical potential μ_0 at zero temperature. Recall that E_{F} is best known as the system Fermi energy. Also, and once again, because (11.246) tells us that the equality $PV = \frac{2}{3}U$ holds at all temperatures and applies to all noninteracting systems whether they be quantum or classical. Therefore we can relate the system pressure, P_0, at zero temperature to E_{F} or U_0 as

$$P = \frac{2}{3}\frac{U}{V} =_{\mathrm{T}\to 0} P_0 = \frac{2}{3}\frac{U_0}{V} = \frac{2N}{5V}E_{\mathrm{F}}\,. \tag{11.286}$$

All the above results for zero temperature, which show that for given N, m, and V, the internal energy U_0 is a constant, are the same as those previously obtained through the use of a simpler physical procedure—compare (11.280) and (11.281).

11.45.2 Partial Degeneracy: Finite but Low Temperature

It is of interest to know how the internal energy changes with T for finite but low temperature. To this end, we need to look at higher order approximations to (11.273) and (11.274) than the zeroth-order approximation used in (11.284)–(11.286).

Therefore, let us first notice that according to (11.284), $\zeta_{\mathrm{T}=0} \to \frac{E_{\mathrm{F}}}{k_{\mathrm{B}}T}$. Second, rearrange (11.273) in the following form:

$$\zeta^{\frac{3}{2}} = \left(\frac{3\sqrt{\pi}\lambda^3 N}{8V}\right)\left[1 + \frac{\pi^2}{8}\zeta^{-2} + \frac{7\pi^4}{640}\zeta^{-4} + O(\zeta^{-6})\right]^{-1}, \tag{11.287}$$

and raise both sides to the power $\frac{2}{3}$. This gives

$$\begin{aligned}\zeta &= \left(\frac{3\sqrt{\pi}\lambda^3 N}{8V}\right)^{\frac{2}{3}}\left[1 + \frac{\pi^2}{8}\zeta^{-2} + \frac{7\pi^4}{640}\zeta^{-4} + O(\zeta^{-6})\right]^{-\frac{2}{3}} \\ &= \left(\frac{E_{\mathrm{F}}}{k_{\mathrm{B}}T}\right)\left[1 - \frac{\pi^2}{12}\zeta^{-2} + \frac{\pi^4}{720}\zeta^{-4} + O(\zeta^{-6})\right].\end{aligned} \tag{11.288}$$

[54]Fermi energy $E_{\mathrm{F}} = (\frac{3N}{8\pi V})^{\frac{2}{3}}(\frac{h^2}{2m})$.

As indicated before, the above relationship would be fine except that we need to know ζ as a function of the temperature. To this end, we proceed as follows:

First, replace ζ on the right-hand side of the last row—in (11.288)—by its zeroth-order value $\frac{E_F}{k_B T}$. As a result, (11.288) becomes

$$\zeta = \frac{E_F}{k_B T}\left[1 - \frac{\pi^2}{12}\left(\frac{k_B T}{E_F}\right)^2 + o\left(\frac{k_B T}{E_F}\right)^4\right]. \tag{11.289}$$

The second correction can now be made by inserting the result of ζ—given in (11.289) above—into (11.288). This gives

$$\zeta = \frac{E_F}{k_B T}\left[1 - \frac{\pi^2}{12}\left(\frac{k_B T}{E_F}\right)^2 - \left(\frac{\pi^4}{80}\right)\left(\frac{k_B T}{E_F}\right)^4 + o\left(\frac{k_B T}{E_F}\right)^6\right]. \tag{11.290}$$

11.46 Thermodynamic Potentials

We are now able to determine the various thermodynamic potentials. First, let us find the temperature dependence of the Gibbs potential G,

$$\begin{aligned} G &= N\mu = N k_B T \zeta \\ &= N E_F\left[1 - \frac{\pi^2}{12}\left(\frac{k_B T}{E_F}\right)^2 - \left(\frac{\pi^4}{80}\right)\left(\frac{k_B T}{E_F}\right)^4 + o\left(\frac{k_B T}{E_F}\right)^6\right]. \end{aligned} \tag{11.291}$$

Similarly, using the value of ζ given in (11.290) and a little bit of algebra, the internal energy according to (11.274) is found to be the following:

$$\begin{aligned} U &= \frac{3}{5} N E_F\left[1 + \frac{5\pi^2}{12}\left(\frac{k_B T}{E_F}\right)^2 - \frac{\pi^4}{16}\left(\frac{k_B T}{E_F}\right)^4 + o\left(\frac{k_B T}{E_F}\right)^6\right] \\ &= \frac{3}{2} P V . \end{aligned} \tag{11.292}$$

Note that the last row above is a statement of the well-known fact that for a three-dimensional, nonrelativistic, ideal system, whether it be classical or quantum, the pressure $P = \frac{2U}{3V}$.

Because both the Gibbs free energy G and the internal energy U are now known, we can calculate the Helmholtz free energy F as follows:

$$\begin{aligned} F &= G - PV = G - \frac{2}{3} U \\ &= N E_F\left(\frac{3}{5}\right)\left[1 - \frac{5}{12}\left(\frac{k_B T \pi}{E_F}\right)^2 + \frac{1}{48}\left(\frac{k_B T \pi}{E_F}\right)^4 + o\left(\frac{k_B T \pi}{E_F}\right)^6\right]. \end{aligned} \tag{11.293}$$

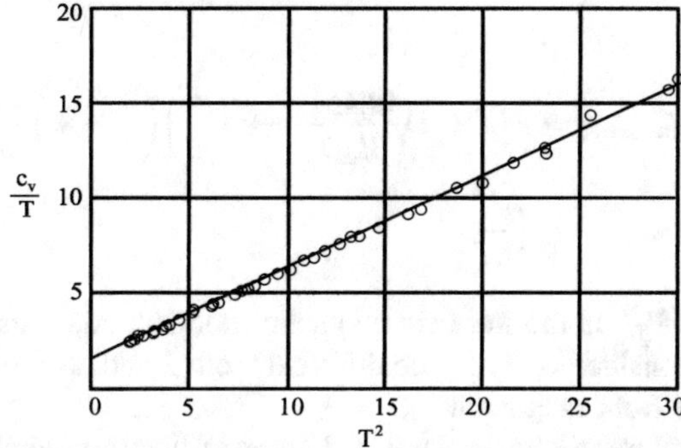

Fig. 11.18 $\frac{C_V}{T}$ for sodium versus T^2. Following the work of [28], the specific heat, C_V, of Na divided by the temperature, T, is plotted as a function of T^2. Notice that the plot has an intercept at $T \to 0$ indicating that at very low temperature C_V has a linear dependence on T. For slightly higher temperature, the specific heat is seen to follow the T^3 law which is a result of the lattice vibrations. (See Fig. 11.19 and the related text for details of the lattice contribution)

Finally, we deal with the entropy S. We have

$$S = \frac{U - F}{T} = \frac{N k_B \pi}{2} \left(\frac{k_B T \pi}{E_F} \right) \left[1 - \frac{1}{10} \left(\frac{k_B T \pi}{E_F} \right)^2 + O\left(\frac{k_B T \pi}{E_F} \right)^4 \right]. \quad (11.294)$$

The specific heat C_V is also readily found as

$$C_V = \left(\frac{\partial U}{\partial T} \right)_{N,V} = N k_B \left[\frac{\pi^2}{2} \left(\frac{k_B T}{E_F} \right) - \frac{3\pi^4}{20} \left(\frac{k_B T}{E_F} \right)^3 + O\left(\frac{k_B T}{E_F} \right)^5 \right]. \quad (11.295)$$

Remark 11.6 Notice that for the electron gas the specific heat tends toward zero linearly with temperature (see Fig. 11.18). This fact is in dramatic contrast with the classical result which asserts that the specific heat is (both large and) constant at all temperatures. Indeed, thereby hangs a long tale!

Drude and Lorentz' theory based on the BMG—that is, the Boltzmann–Maxwell–Gibbs—classical statistics, predicted that the free conduction electrons would add to the specific heat, C_V, an amount equal to $\frac{3}{2} k_B$ per electron. Experimentally, however, it was observed that—except for very low temperatures where the specific heat seemed to vary linearly with the temperature—the cubic temperature dependence of the lattice vibrations provided a satisfactory description of the system specific heat. So what happened to the effect of classical motion of the conduction electrons? Why and where did that disappear?

A similar such occurrence also afflicted the quasiclassical Langevin prediction for the electronic paramagnetic susceptibility. For electrons, $J = \frac{1}{2}$, and the Landé g-factor is (very close to) $= 2$. Therefore, at high temperature and for weak magnetic field, the quasiclassical value for the magnetic susceptibility for one dipole should be

$$(\chi)_{\text{quasiclassical}} = J(J+1)\frac{(g\mu_B)^2}{3k_BT} \rightarrow \left(\frac{1}{2}\right)\left(\frac{1}{2}+1\right)\frac{(2\mu^*)^2}{3k_BT}$$

$$= \frac{\mu^{*2}}{k_BT} . \tag{11.296}$$

Note that $\mu^* \rightarrow \frac{g\mu_B}{2}$ is the intrinsic magnetic moment. Also, that depending on the object being considered, g may be different from 2 and, as a result, μ^* may be different from one Bohr magneton μ_B.

At low temperature, the quasiclassical Langevin theory predicted that the system will approach magnetic saturation. The experimental results, in contrast, told a different story. At low temperatures, the susceptibility, although dependent on the density, became independent of the temperature and tended towards a small, but constant value.

Wolfgang Pauli was the first to recognize that the system constituted a highly degenerate electron gas that should be treated by using Fermi–Dirac statistics. His results for paramagnetism are often referred to as Pauli paramagnetism.

11.47 Pauli Paramagnetism

For a system of N quantum particles with spin S, where all spin directions are symmetrical, the spin degeneracy factor, Δ, is $2S+1$. This is equal to the total number of spin states available to each particle. The fact that each electron has spin $S = \frac{1}{2}$ means that $\Delta = 2$ and an electron can lie either parallel or anti-parallel to any chosen direction. And unless an external field is applied—which would break the symmetry between the up- and the down-directions—roughly a half of a large number N of electrons, in a system which is in thermodynamic equilibrium, would on the average point either up or down. More precisely, each direction would have approximately $\frac{N}{2}[1 \pm O(\frac{1}{\sqrt{N}})]$ electrons.

With the application of an external magnetic field $\boldsymbol{B_0}$, at least three new things happen. First, the energy of a given electron changes—that is, it increases, if it is antiparallel to the field. Or the energy decreases, if the electron is parallel to the field. In other words, the applied field causes a change in the energy by an amount equal to $-\boldsymbol{B}_0.\boldsymbol{\mu}_B = \pm B_0\mu_B$. (Recall that μ_B is the Bohr magneton.) Second, the symmetry is broken and the number of electrons pointing in the up- or the down-direction is no longer the same. Third, the spin degeneracy factor for each of the $2S+1=2$ directions—namely, either up or down—is now equal to just 1.

An important principle of thermodynamics—see (10.44)—is that: "At constant pressure, in a biphase thermodynamic system that is in thermal contact with a heat energy reservoir at a fixed temperature, the specific Gibbs free energy of the two coexistent phases is the same."

Here the two coexistent phases are the up- and the down-spin parts of the electron gas and the equality of the specific Gibbs function translates into the chemical potential μ being the same for both the up- and the down-spin parts. Be warned, however, that the equality of the chemical potential does not mean that the number of electrons with different spins have to be the same. Indeed, as we shall see shortly, they are different.

Armed with this information, we can use (11.247) to write

$$N_{\mathrm{p}}^{+} = \left[\exp\{\beta(\varepsilon_{\mathrm{p}}^{+} - \mu)\} + 1\right]^{-1},$$
$$N_{\mathrm{p}}^{-} = \left[\exp\{\beta(\varepsilon_{\mathrm{p}}^{-} - \mu)\} + 1\right]^{-1}, \tag{11.297}$$

where $\varepsilon_{\mathrm{p}}^{+}$ and $\varepsilon_{\mathrm{p}}^{-}$ are the energies of single electron states with momentum p that refer to an electron pointing up—meaning parallel to the applied magnetic field—or down, respectively. In view of the description given two paragraphs above, we have

$$\varepsilon_{\mathrm{p}}^{+} = \frac{p^2}{2m} - B_0\mu_{\mathrm{B}}\,, \qquad \varepsilon_{\mathrm{p}}^{-} = \frac{p^2}{2m} + B_0\mu_{\mathrm{B}}\,. \tag{11.298}$$

Similar to the statements made for (11.247), N_{p}^{+} and N_{p}^{-} are the mean occupation numbers for a state with energy $\varepsilon_{\mathrm{p}}^{+}$ and $\varepsilon_{\mathrm{p}}^{-}$, respectively.

11.47.1 Pauli Paramagnetism: Zero Temperature

Simple Treatment

At zero temperature, $\beta \to \infty$ and $\mu \to E_{\mathrm{F}}$. With these changes, (11.297) tells us that the positive spin electrons have occupancy N_{p}^{+} equal to unity for all those momenta for which $(\varepsilon_{\mathrm{p}}^{+}) \le E_{\mathrm{F}}$. Similarly, the negative spin electrons are occupied, i.e., $N_{\mathrm{p}}^{-} = 1$, for those momenta for which $(\varepsilon_{\mathrm{p}}^{-}) \le E_{\mathrm{F}}$.

For $(\varepsilon_{\mathrm{p}}^{+}) > E_{\mathrm{F}}$ and $(\varepsilon_{\mathrm{p}}^{-}) > E_{\mathrm{F}}$, both N_{p}^{+} and N_{p}^{-} are zero. Therefore, there is no occupancy of the electrons parallel to the field beyond the momentum $p_{+}^{\mathrm{F}} = \sqrt{2m(E_{\mathrm{F}} + B_0\mu_{\mathrm{B}})}$. Similarly, electrons antiparallel to the field are unoccupied beyond the momentum $p_{-}^{\mathrm{F}} = \sqrt{2m(E_{\mathrm{F}} - B_0\mu_{\mathrm{B}})}$.

This means that for spins that are parallel to the field, the momentum ranges between 0 and p_{+}^{F} where the occupancy, $[\exp\{\beta(\varepsilon_{\mathrm{p}}^{+} - \mu_0)\} + 1]^{-1}$, stays equal to 1. And similarly for spins that are antiparallel to the field, the momentum ranges between 0 and p_{-}^{F} where the occupancy, $[\exp\{\beta(\varepsilon_{\mathrm{p}}^{-} - \mu_0)\} + 1]^{-1}$, stays equal to 1.

As a result, by using (11.276), and noting the $\mu_0 = E_{\mathrm{F}}$, the numbers of electrons with spins $+\frac{1}{2}$ and $-\frac{1}{2}$ are the following:

$$\begin{aligned} N_{+} &= \frac{V}{h^3}\int_0^{p_{+}^{\mathrm{F}}} [N_{\mathrm{p}}^{+}] 4\pi p^2 \mathrm{d}p \\ &= \frac{V}{h^3}\int_0^{p_{+}^{\mathrm{F}}} \left[\left[\exp\{\beta(\varepsilon_{\mathrm{p}}^{+} - \mu_0)\} + 1\right]^{-1}\right] 4\pi p^2 \mathrm{d}p \end{aligned}$$

$$
\begin{aligned}
&= \frac{V}{h^3}\int_0^{p_+^F}[1]4\pi p^2 \mathrm{d}p \\
&= \frac{V}{h^3}\frac{4\pi}{3}\left(p_+^F\right)^3 = \left(\frac{4\pi V}{3h^3}\right)\{2m(E_F + B_0\mu_B)\}^{\frac{3}{2}}
\end{aligned} \tag{11.299}
$$

and

$$
\begin{aligned}
N_- &= \frac{V}{h^3}\int_0^{p_-^F}[N_p^-]4\pi p^2 \mathrm{d}p \\
&= \frac{V}{h^3}\int_0^{p_-^F}\left[[\exp\{\beta(\varepsilon_p^- - \mu_0)\} + 1]^{-1}\right]4\pi p^2 \mathrm{d}p \\
&= \frac{V}{h^3}\int_0^{p_-^F}[1]4\pi p^2 \mathrm{d}p \\
&= \frac{V}{h^3}\frac{4\pi}{3}\left(p_-^F\right)^3 = \left(\frac{4\pi V}{3h^3}\right)\{2m(E_F - B_0\mu_B)\}^{\frac{3}{2}}.
\end{aligned} \tag{11.300}
$$

Without the applied field, the number of up and down spins is the same, that is, for $B_0 = 0$, $N_+ = N_- = \frac{N}{2}$. So there is no net magnetic moment of the electron gas. However, in the presence of an applied field, the magnetic moment, NM, of the system is nonzero. Therefore, its small field magnetic susceptibility per electron at temperature T, i.e., $(\chi(T))_{\text{pauli}}$, is also nonzero:

$$
\begin{aligned}
NM &= \mu_B(N_+ - N_-), \\
(\chi(T))_{\text{pauli}} &=_{B_0\to 0} \frac{M}{B_0}.
\end{aligned} \tag{11.301}
$$

Because the Fermi energy is so much larger than the usual size of the magnetic field factor $B_0\mu_B$, we can write

$$
\begin{aligned}
NM &=_{T\to 0} \mu_B\left(\frac{4\pi V}{3h^3}\right)(2m)^{\frac{3}{2}}\left[(E_F + \mu_B B_0)^{\frac{3}{2}} - (E_F - \mu_B B_0)^{\frac{3}{2}}\right] \\
&= B_0 V\left(\frac{4\pi \mu_B{}^2}{h^3}\right)(2m)^{\frac{3}{2}}\sqrt{E_F}\left[1 - O\left(\frac{\mu_B B_0}{E_F}\right)^2\right].
\end{aligned} \tag{11.302}
$$

The small field, single electron, Pauli susceptibility at zero temperature—to be called $(\chi_0)_{\text{pauli}}$—therefore is

$$
\begin{aligned}
\lim_{(B_0\to 0;T\to 0)}(\chi(T))_{\text{pauli}} &= \frac{M}{B_0} = (\chi_0)_{\text{pauli}}, \\
(\chi_0)_{\text{pauli}} &= \left(\frac{4V\pi\mu_B{}^2}{Nh^3}\right)(2m)^{\frac{3}{2}}\sqrt{E_F} \\
&= \left(\frac{3}{2E_F}\right)\mu_B{}^2.
\end{aligned} \tag{11.303}
$$

Note that the result for $(\chi_0)_{\text{pauli}}$ given above is exact. To derive the third row from the second, make the change suggested in the last part of (11.278), that is, from $(2m)^{\frac{3}{2}}$ to $(\frac{3N}{8\pi V})(\frac{h^2}{E_F})^{\frac{3}{2}}$.

11.47.2 Pauli Paramagnetism: Finite Temperature

At finite temperature the occupancy of electrons extends beyond the Fermi surface. Therefore, the integration over the momentum—in contrast to that specified in both (11.299) and (11.300)—needs to be extended beyond the Fermi sea to some appropriate, higher value. However, because the integrand falls off very rapidly with the rise in momentum—indeed, as fast as $\exp(-\frac{\beta p^2}{2m})$—the integration may be extended, without any noticeable amount of error, to ∞. This gives

$$\begin{aligned}
N_+ &= \frac{V}{h^3}\int_0^\infty \left[\exp\{\beta(\varepsilon_p^+ - \mu)\} + 1\right]^{-1} 4\pi p^2 \mathrm{d}p \\
&= \frac{V}{h^3}\int_0^\infty \left[\exp\left\{\beta\left(\frac{p^2}{2m} - B_0\mu_B - \mu\right)\right\} + 1\right]^{-1} 4\pi p^2 \mathrm{d}p \\
&= \frac{V}{h^3}\int_0^\infty \left[(z_+)^{-1}\exp\left\{\beta\left(\frac{p^2}{2m}\right)\right\} + 1\right]^{-1} 4\pi p^2 \mathrm{d}p \\
&= \frac{V}{\lambda^3} f_{\frac{3}{2}}(z_+)
\end{aligned} \tag{11.304}$$

and

$$\begin{aligned}
N_- &= \frac{V}{h^3}\int_0^\infty \left[\exp\{\beta(\varepsilon_p^- - \mu)\} + 1\right]^{-1} 4\pi p^2 \mathrm{d}p \\
&= \frac{V}{h^3}\int_0^\infty \left[\exp\left\{\beta\left(\frac{p^2}{2m} + B_0\mu_B - \mu\right)\right\} + 1\right]^{-1} 4\pi p^2 \mathrm{d}p \\
&= \frac{V}{h^3}\int_0^\infty \left[(z_-)^{-1}\exp\left\{\beta\left(\frac{p^2}{2m}\right)\right\} + 1\right]^{-1} 4\pi p^2 \mathrm{d}p \\
&= \frac{V}{\lambda^3} f_{\frac{3}{2}}(z_-),
\end{aligned} \tag{11.305}$$

where we have introduced the notation

$$z_\pm = \exp\{\beta(\mu \pm B_0\mu_B)\} = \exp(\beta\mu_\pm) = \exp(\zeta_\pm). \tag{11.306}$$

It is important here to notice that (11.240) can immediately be transformed into (11.304) by making the following trivial changes: The degeneracy factor 2 changes to 1 and z changes to z_+. The same transformation with $z \to z_-$ reproduces (11.305).

As a result of this transformation, (11.273)–(11.270), (11.272), (11.290), and (11.291), are readily generalized to accommodate changes from N to N_+ and N_-, from z to z_+ and z_-, from μ to μ_+ and μ_-, and, from ζ to ζ_+ and ζ_-. Recall that because the up and down spins are being treated separately, here the spin degeneracy factor is 1 for each of them. We get

$$N_+ = \frac{V}{\lambda^3} f_{\frac{3}{2}}(z_+) = \frac{4V\zeta_+^{\frac{3}{2}}}{3\sqrt{\pi}\lambda^3}\left[1 + \frac{\pi^2}{8}\zeta_+^{-2} + \frac{7\pi^4}{640}\zeta_+^{-4} + O\left(\zeta_+^{-6}\right)\right],$$
$$N_- = \frac{V}{\lambda^3} f_{\frac{3}{2}}(z_-) = \frac{4V\zeta_-^{\frac{3}{2}}}{3\sqrt{\pi}\lambda^3}\left[1 + \frac{\pi^2}{8}\zeta_-^{-2} + \frac{7\pi^4}{640}\zeta_-^{-4} + O\left(\zeta_-^{-6}\right)\right], \tag{11.307}$$

where

$$\mu_\pm = k_B T\zeta_\pm = (E_F \pm B_0\mu_B) \times \left[1 - \frac{\pi^2}{12}\left(\frac{k_B T}{E_F \pm B_0\mu_B}\right)^2 - \left(\frac{\pi^4}{80}\right)\left(\frac{k_B T}{E_F \pm B_0\mu_B}\right)^4 - \cdots\right] \tag{11.308}$$

and

$$\zeta_\pm = \ln z_\pm = \frac{\mu_\pm}{k_B T} = \frac{\mu \pm B_0\mu_B}{k_B T} \ll 1,$$
$$\omega = \frac{B_0\mu_B}{\mu} \ll 1. \tag{11.309}$$

In order to determine the Pauli paramagnetic susceptibility, we need to calculate $N_+ - N_-$ as a function of $B_0\mu_B$ and the temperature T. This requires combining (11.307) and (11.308), with help from (11.309). We get

$$\begin{aligned} N_+ - N_- &= \frac{V}{\lambda^3} f_{\frac{3}{2}}(z_+) - \frac{V}{\lambda^3} f_{\frac{3}{2}}(z_-) \\ &= \frac{4V}{3\sqrt{\pi}\lambda^3}\left[\zeta_+^{\frac{3}{2}} + \frac{\pi^2}{8}\zeta_+^{-\frac{1}{2}} + \frac{7\pi^4}{640}\zeta_+^{-\frac{5}{2}}\right] \\ &\quad - \frac{4V}{3\sqrt{\pi}\lambda^3}\left[\zeta_-^{\frac{3}{2}} + \frac{\pi^2}{8}\zeta_-^{-\frac{1}{2}} + \frac{7\pi^4}{640}\zeta_-^{-\frac{5}{2}}\right] \\ &\quad + O\left[\frac{4V}{3\sqrt{\pi}\lambda^3}\zeta_+^{-\frac{9}{2}}\right]. \end{aligned} \tag{11.310}$$

We need next to expand $\zeta_\pm = \frac{\mu \pm B_0\mu_B}{k_B T}$ as a power expansion in terms of the variable $\omega = \frac{B_0\mu_B}{\mu}$, which is exceedingly small for most available magnetic fields. Indeed, because we are interested in only the small field magnetic susceptibility, we shall retain only the leading term in ω, i.e.,

$$\zeta_\pm^n = \left(\frac{\mu \pm B_0\mu_B}{k_B T}\right)^n = \left(\frac{\mu}{k_B T}\right)^n \left[1 \pm n\omega + O(\omega^2)\right]. \tag{11.311}$$

Accordingly, (11.310) leads to the following:

$$\begin{aligned}
\frac{N_+ - N_-}{\left(\frac{4V}{3\sqrt{\pi}\lambda^3}\right)} &= \left(\zeta_+^{\frac{3}{2}} - \zeta_-^{\frac{3}{2}}\right) + \frac{\pi^2}{8}\left(\zeta_+^{-\frac{1}{2}} - \zeta_-^{-\frac{1}{2}}\right) \\
&+ \frac{7\pi^4}{640}\left(\zeta_+^{-\frac{5}{2}} - \zeta_-^{-\frac{5}{2}}\right) + O\left[\zeta_+^{-\frac{9}{2}}\right] \\
&= \left(\frac{\mu}{k_B T}\right)^{\frac{3}{2}}(3\omega) - \frac{\pi^2}{8}\left(\frac{k_B T}{\mu}\right)^{\frac{1}{2}}(\omega) \\
&- \frac{7\pi^4}{640}\left(\frac{k_B T}{\mu}\right)^{\frac{5}{2}}(5\omega) + O\left[\left(\frac{k_B T}{\mu}\right)^{\frac{9}{2}}\omega; \omega^2\right] \\
&= \left(\frac{\mu}{k_B T}\right)^{\frac{3}{2}}(3\omega)\left[1 - \frac{\pi^2}{24}\left(\frac{k_B T}{\mu}\right)^2 - \frac{7\pi^4}{384}\left(\frac{k_B T}{\mu}\right)^4\right] \\
&+ O\left[\left(\frac{k_B T}{\mu}\right)^{\frac{9}{2}}\left(\frac{\mu_B B_0}{\mu}\right); \omega^2\right].
\end{aligned} \tag{11.312}$$

Reverting the notation ω back to its original form, $\frac{B_0 \mu_B}{\mu}$, (11.312) is re-arranged as follows:

$$\begin{aligned}
&\frac{(N_+ - N_-)(k_B T)^{\frac{3}{2}}}{\left(\frac{4V}{3\sqrt{\pi}\lambda^3}\right)(3 B_0 \mu_B \sqrt{E_F})} \\
&= \left(\frac{\mu}{E_F}\right)^{\frac{1}{2}} - \frac{\pi^2}{24}\left(\frac{k_B T}{E_F}\right)^2\left(\frac{E_F}{\mu}\right)^{\frac{3}{2}} - \frac{7\pi^4}{384}\left(\frac{k_B T}{E_F}\right)^4\left(\frac{E_F}{\mu}\right)^{\frac{7}{2}} \\
&+ O\left[\left(\frac{k_B T}{\mu}\right)^{\frac{9}{2}}\left(\frac{B_0 \mu_B}{\mu}\right); \left(\frac{B_0 \mu_B}{\mu}\right)^2\right].
\end{aligned} \tag{11.313}$$

Our next task is to replace μ by its temperature dependent value given in (11.291)—or, equivalently, in (11.308). For the reader's convenience, we reproduce it below:

$$\frac{\mu}{E_F} = 1 - \frac{\pi^2}{12}\left(\frac{k_B T}{E_F}\right)^2 - \left(\frac{\pi^4}{80}\right)\left(\frac{k_B T}{E_F}\right)^4 + O\left(\frac{k_B T}{E_F}\right)^6. \tag{11.314}$$

Therefore we have:

$$\begin{aligned}
\left(\frac{\mu}{E_F}\right)^{\frac{1}{2}} &= 1 - \frac{\pi^2}{24}\left(\frac{k_B T}{E_F}\right)^2 - \left(\frac{41\pi^4}{180 \times 32}\right)\left(\frac{k_B T}{E_F}\right)^4 - O\left(\frac{k_B T}{E_F}\right)^6, \\
\left(\frac{E_F}{\mu}\right)^{\frac{3}{2}} &= 1 + \frac{\pi^2}{8}\left(\frac{k_B T}{E_F}\right)^2 - O\left(\frac{k_B T}{E_F}\right)^4, \\
\left(\frac{E_F}{\mu}\right)^{\frac{7}{2}} &= 1 + O\left(\frac{k_B T}{E_F}\right)^2.
\end{aligned} \tag{11.315}$$

Inserting (11.315) into (11.313), we finally get to the desired result:

$$\frac{(N_+ - N_-)(k_B T)^{\frac{3}{2}}}{(\frac{4V}{3\sqrt{\pi}\lambda^3})(3B_0\mu_B\sqrt{E_F})} = \frac{\frac{\mu_B}{B_0}(N_+ - N_-)}{(\frac{4V}{3\sqrt{\pi}\lambda^3})(3B_0\mu_B\sqrt{E_F})(k_B T)^{\frac{-3}{2}}(\frac{\mu_B}{B_0})}$$
$$=_{B_0\to 0}\left[\frac{(\chi(T))_{\text{pauli}}}{(\chi_0)_{\text{pauli}}}\right]_{\text{exact}}$$
$$= 1 - \frac{\pi^2}{12}\left(\frac{k_B T}{E_F}\right)^2 - \frac{11\pi^4}{360}\left(\frac{k_B T}{E_F}\right)^4 - O\left(\frac{k_B T}{E_F}\right)^6. \tag{11.316}$$

Note that in the above we have used the definition of the Pauli susceptibility $(\chi(T))_{\text{pauli}}$, and the corresponding zero temperature susceptibility $(\chi_0)_{\text{pauli}}$, as they were specified in (11.301), (11.302), and (11.303).

11.47.3 Pauli Paramagnetism: Very High Temperature

Rather than directly solving (11.299), (11.305) and (11.306), we can take advantage of the fact that the behavior of the quantum electrons at very high temperatures is essentially the same as that of quasiclassical electrons. Therefore, to the leading order we can expect the result for the Pauli susceptibility, $(\chi_\infty)_{\text{pauli}}$, at very high temperature to be very much the same as that given by the quasiclassical Langevin theory for what we have called the "high temperature magnetic susceptibility for paramagnetic salts." (For the case $J = \frac{1}{2}$, that result is for convenience reproduced in (11.317) below.)

$$\left[(\chi)_{\text{quasiclassical}}\right]_{\text{at very high temp}} = \frac{(g\mu_B)^2}{4k_B T} \approx (\chi_\infty)_{\text{pauli}}\,. \tag{11.317}$$

Remark 11.7 Apropos of an earlier issue, we recall that the result for Langevin paramagnetic susceptibility at high temperature—reproduced in (11.317) above—is generally also quite accurate for temperatures as low as $\approx$ 100 K $\to$ 20 K. At these temperatures the "zero temperature" result for the magnetic susceptibility of quantum electrons is also applicable because here the neglected higher order temperature-dependent terms are very small. Because these two different results are valid at roughly the same range of moderate temperatures, it is interesting to compare the proper statistical result for the magnetic susceptibility of a system of noninteracting quantum electrons, as given in (11.303)—that is, $(\chi_0)_{\text{pauli}} = \frac{3\mu_B{}^2}{2E_F}$—with the corresponding quasi-classical result, $(\chi)_{\text{quasiclassical}}$, for paramagnetic salts that is reproduced in (11.317) above. The essential difference between these two results lies in their denominators. Because $E_F \gg k_B T$, the magnetic susceptibility per particle, $(\chi_0)_{\text{pauli}}$, of the quantum gas is much smaller than $(\chi)_{\text{quasiclassical}}$ of the paramagnetic salts. Why should that be the case?

In a paramagnetic salt—or equivalently, in the noninteracting, quantum electron gas at very, very high temperature—essentially all the spins participate in the magnetization process. In contrast, at temperature below the Fermi temperature, the Fermi–Dirac spins are forced to settle down in the Fermi sea. And the imposition of the magnetic field can affect the energy of only a small fraction of such spins that lie near the top. In the following, this effect is analyzed by a hand-waving-quantitative procedure.

Using a so-called hand-waving argument, we study the behavior of the system specific heat, Pauli paramagnetism at zero temperature, and Pauli paramagnetism at finite temperature.

11.48 Hand-Waving Argument: Specific Heat

The low temperature specific heat, calculated by a formal quantum-statistical procedure—see (11.295)—tended toward zero linearly with the temperature and was noted to be rather small. That result is in dramatic contrast with one given by classical theories that predict the specific heat—to be both quite large and—to remain constant at all temperatures.

In the following we attempt to understand the quantum-statistical result through the use of simple-minded physical arguments.

As outlined in (11.247), the mean occupation number N_p—for a state with momentum p, energy ε_p, in a spin S perfect F–D gas that is at temperature $T = \frac{1}{\beta k_B}$ and has chemical potential μ—is

$$N_p = (2S + 1)\left[\exp\{\beta(\varepsilon_p - \mu)\} + 1\right]^{-1} . \tag{11.318}$$

When the temperature tends to zero, $\mu \to \mu_0 = E_F$ and $\beta \to \infty$. Accordingly, N_p is equal to $2S + 1$ for all values of ε_p that range between zero and the top of the Fermi sea which is at energy E_F. (As stated before, $2S + 1$ different quantum particles—with, say, the z-component of spin ranging between $-S$ and $+S$—may occupy any given single particle state. For electrons, however, $S = \frac{1}{2}$. Therefore, the plot of N_p versus $\frac{\varepsilon_p}{E_F}$ would be a rectangle of height $2S + 1 = 2$ and width 1.)

As the temperature is increased slightly from 0 to T, the electrons begin to rise in energy—meaning they begin to spread out of the extreme right hand side of the rectangle. The electrons so "moved" are clearly those that are seated near the top of the Fermi sea with energy close to E_F. Indeed, thermodynamics suggests that such electrons are mostly those that lie within energy levels that range approximately between $E_F - k_B T$ and $E_F + k_B T$. This represents a width in energy of approximately $2k_B T$ out of the total width which is $\approx E_F$. Accordingly, the number of electrons that are likely to move out of the sea due to the rise in temperature is approximately $(\delta N)_{\text{temp rise}}$. This number should be of order

$$(\delta N)_{\text{temp rise}} \approx \left(\frac{2k_B T}{E_F}\right) N . \tag{11.319}$$

Approximately, each of these $(\delta N)_{\text{temp rise}}$ electrons is shifted up in energy by an amount equal to $k_B T$. Therefore, our estimate for the increase, $(\delta E)_{\text{estimate}}$, in energy of N electrons due to the rise in temperature is the following:

$$(\delta E)_{\text{estimate}} \approx (\delta N)_{\text{temp rise}}(k_B T) = \left(\frac{2Nk_B T}{E_F}\right)(k_B T) = \frac{2N(k_B T)^2}{E_F}. \tag{11.320}$$

The estimated total specific heat is therefore equal to

$$(C_v)_{\text{estimate}} \approx \left(\frac{\partial(\delta E)_{\text{estimate}}}{\partial T}\right)_V = 4\left(\frac{N(k_B)^2 T}{E_F}\right). \tag{11.321}$$

The above estimate for the total specific heat compares well with the leading term of the exact result given in (11.295) (which, for convenience, is reproduced below),

$$(C_v)_{\text{exact}} = \frac{\pi^2}{2}\left(\frac{N(k_B)^2 T}{E_F}\right) - \left(\frac{3\pi^4}{20}\right)Nk_B\left(\frac{k_B T}{E_F}\right)^3. \tag{11.322}$$

11.48.1 Hand-Waving Argument: Zero Temperature Pauli Paramagnetism

Upon application of the magnetic field B_0, the energies of the down- and up-spin electrons are shifted by $\pm B_0\mu_F$. At zero temperature, the energy shift is $2B_0\mu_F$. As in (11.319), we estimate that the electrons that are affected most are those that reside at the top of the Fermi sea. This amounts to a fraction $\frac{2B_0\mu_F}{E_F}$ of the total number of electrons. Therefore, we estimate—much as was done above in (11.319)—their number to be approximately the following:

$$(\delta N)_{\text{magnetic field}} \approx \left(\frac{2B_0\mu_F}{E_F}\right)N. \tag{11.323}$$

From $(\delta N)_{\text{magnetic field}}$, we can generate an estimate for, $(M)_{\text{estimate}}$, the magnetic moment per electron to be

$$(M)_{\text{estimate}} \approx \mu_F(\delta N)_{\text{magnetic field}}/N = \frac{2B_0\mu_F^2}{E_F}. \tag{11.324}$$

Therefore, an estimate for the single electron, zero temperature zero-field magnetic susceptibility is

$$(\chi_0)_{\text{estimate}} \approx \frac{(M)_{\text{estimate}}}{B_0} = 2\left(\frac{{\mu_F}^2}{E_F}\right). \tag{11.325}$$

Qualitatively, the above estimate is similar to the corresponding exact result given in (11.303), namely

$$(\chi_0)_{\text{exact}} = (\chi_0)_{\text{pauli}} = \frac{3}{2}\left(\frac{{\mu_F}^2}{E_F}\right). \tag{11.326}$$

11.48.2 Hand-Waving Argument: Pauli Paramagnetism at Finite Temperature

At zero temperature, the relevant Pauli paramagnetic electrons sit contiguously to the top of the Fermi surface. As was shown in (11.323), their number is

$$(\delta N)_{\text{magnetic field}} \approx 2B_0\left(\frac{\mu_F}{E_F}\right)N . \tag{11.327}$$

With slight increase in temperature, each of these electrons is shifted up in energy from approximately E_F to $E_F + \Delta E$. According to (11.320), the increase in energy, ΔE, per electron is of order $\frac{2(k_B T)^2}{E_F}$. Therefore, their total number should shift from its value $(\delta N)_{\text{magnetic field}}$—that is given in (11.327) above—to approximately the following value:

$$\begin{aligned}
&(\delta N)_{\text{magnetic field and finite temperature}} \\
&\approx 2B_0\left(\frac{\mu_F}{E_F + \Delta E}\right) \approx 2B_0\left(\frac{\mu_F}{E_F + \frac{2(k_B T)^2}{E_F}}\right)N \\
&\approx 2B_0\left(\frac{\mu_F}{E_F}\right)N\left[1 - \frac{2(k_B T)^2}{E_F^2}\right].
\end{aligned} \tag{11.328}$$

The estimate for the temperature dependent susceptibility is therefore the following:

$$\begin{aligned}
(\chi)_{\text{temperature dependent estimate}} &\approx \frac{(M)_{\text{magnetic field at finite temperature}}}{B_0} \\
&\approx \left(\frac{\mu_F}{B_0}\right)(\delta N)_{\text{magnetic field at finite temperature}}/N \\
&\approx \left(\frac{2\mu_F^2}{\mu_0}\right)\left[1 - 2\left(\frac{k_B T}{E_F}\right)^2\right] \\
&= (\chi_0)_{\text{estimate}}\left[1 - 2\left(\frac{k_B T}{E_F}\right)^2\right].
\end{aligned} \tag{11.329}$$

The above estimate for the single particle temperature dependent susceptibility at low temperatures is also qualitatively similar to the corresponding exact result that was given in (11.316), namely

$$(\chi)_{\text{exact}} = \big(\chi(T)\big)_{\text{pauli}} = (\chi_0)_{\text{exact}}\left[1 - \frac{\pi^2}{12}\left(\frac{k_B T}{E_F}\right)^2\right]. \tag{11.330}$$

11.49 Landau Diamagnetism

In addition to the paramagnetism studied above, conduction electrons are also subject to a sort of negative paramagnetism whereby the effect of applied field in a given direction produces a magnetic moment in the opposite direction. Such behavior, resulting in negative magnetic susceptibility, is termed diamagnetic.

So far in our study of the conduction electrons, we have not considered any effects of the orbital motion. In the presence of an applied magnetic field—say, B_0 along the z-axis—an electron follows a helical path around the z-axis. The rotational motion is caused by the so called Lorentz force. Its projection on the $x-y$ plane is completely circular. While a single rotating electron may be diamagnetic, when a large number N of electrons are present, and they are subject to reflection from the enclosing walls, classical statistics predicts that there is "complete"—i.e., to the leading order in N—cancelation of diamagnetic effect. Therefore, as first noted by Bohr, van Leeuwen, and others, classical statistical description of this motion cannot lead to diamagnetism.

Unlike classical statistics, L.D. Landau has shown that quantum statistics do lead to nonzero diamagnetism. The details of Landau's work are somewhat involved, but a relatively simple general comment can still be made. Due to reflection from the bounding walls, the quantum electrons near the boundary—unlike the classical, Maxwell–Boltzmann electrons—have on the average different quantized velocities from the electrons that have not been reflected. Therefore, complete compensation of the diamagnetic effect that occurs for perfectly reflecting classical electrons does not take place for quantum electrons.

As described in an appendix—see (M.1) to (M.13)—at relatively high temperature, the thermodynamic average of $(M)_{\text{landau}}$, the diamagnetic moment per particle, and the resultant Landau susceptibility, $(\chi)_{\text{landau}}$, are as follows:

$$\begin{aligned}(M)_{\text{landau}} &= -\left(\frac{B_0\mu_B^2}{3k_BT}\right) + O\left[\mu_B(\beta B_0\mu_B)^3\right],\\ (\chi)_{\text{landau}} &= \frac{(M)_{\text{landau}}}{B_0} = -\frac{\mu_B^2}{3k_BT},\end{aligned} \tag{11.331}$$

where, as usual, μ_B stands for a Bohr magneton.

For a quantum gas of free electrons at very high temperature, in addition to Landau diamagnetism there is also the Pauli paramagnetism—i.e., $(\chi_\infty)_{\text{pauli}}$, given in (11.317). Adding these two contributions together yields

$$\begin{aligned}\chi_{\text{electron gas at high temeperature}} &= \big(\chi(T)\big)_{\text{pauli}} + (\chi)_{\text{landau}}\\ &\approx (\chi_\infty)_{\text{pauli}} + (\chi)_{\text{landau}} \approx \frac{(g\mu_B)^2}{4k_BT} - \frac{\mu_B^2}{3k_BT}.\end{aligned} \tag{11.332}$$

Observe that we have not used the obvious simplification—namely, if we set $g = 2$ the last row would equal $\frac{2\mu_B^2}{3k_BT}$. We have not done this because depending upon the source of the free electrons being considered, g may not be exactly equal to 2. Indeed, because the Pauli and Landau processes have quite different roots, the effective mass and therefore the μ_B that occurs in the two terms in the above equation may also be slightly different.

11.50 Richardson Effect: Thermionic Emission

At normal temperature, few if any electrons are observed to escape from metals. Clearly, while electrons are known to roam around within the so-called free-electron metals, most of them do not spontaneously stray outside. The metallic ions provide an attractive force for the electrons within the metal. Therefore, in order to escape the metal, they need more kinetic energy than they normally have at such temperatures. Schematically we can express this behavior by bestowing the electrons with negative potential energy, whereby they can be thought to be moving around at the bottom of a potential well of some depth, say, W.

If the temperature of the system should rise, there will result an increase in the number of electrons whose kinetic energy exceeds the depth W. Accordingly, some electrons will begin to escape the well and flow out of the surface of the metal. Unless the temperature is very large, the number of electrons flowing out will be a very small fraction of the total number of "free electrons" in the metal. Therefore, to a very good approximation we can still use equations valid for the original total number of electrons. Thus (11.318) can be used for estimating the number of escaping electrons (per unit time, per unit area of the metal surface). For instance in the z-direction, the expression for the number, $d\wp$, of such escaping electrons whose x- and y-components of the momenta lie between (p_x, p_y) and $(p_x + dp_x, p_y + dp_y)$ is readily seen to be the following:

$$d\wp_{x,y} = \frac{2}{h^3}\int_{p_z=\sqrt{(2mW)}}^{\infty}\left(\frac{p_z}{m}\right)dp_z\frac{dp_x dp_y}{\exp(\frac{\varepsilon_p-\mu}{k_B T})+1}. \tag{11.333}$$

Here $\varepsilon_p = \frac{p_x^2+p_y^2+p_z^2}{2m}$, the spin S has been set at $\frac{1}{2}$, and μ as usual is the chemical potential.

The integrations over the two variables p_x and p_y is more easily done by transforming to the cylindrical coordinates whereby $p_x^2 + p_y^2 = p_r^2$, and

$$\int_{p_x=-\infty}^{+\infty}\int_{p_y=-\infty}^{+\infty}\cdots dp_x dp_y = 2\pi\int_{p_r=0}^{\infty}\cdots p_r dp_r .$$

Thus, the total rate of emission for the Fermi–Dirac gas, i.e., $\wp_{F-D}$, is

$$\begin{aligned}\wp_{F-D} &= \frac{4\pi}{h^3}\int_{p_z=\sqrt{(2mW)}}^{\infty}\left(\frac{p_z}{m}\right)dp_z\int_{p_r=0}^{\infty}\frac{p_r dp_r}{\exp\{\frac{(\frac{p_z^2+p_r^2}{2m})-\mu}{k_B T}\}+1}\\ &= \frac{4\pi m k_B T}{h^3}\int_{p_z=\sqrt{(2mW)}}^{\infty}\left(\frac{p_z}{m}\right)dp_z\\ &\times\int_{p_r=0}^{\infty}\frac{\exp\{\frac{-(\frac{p_z^2+p_r^2}{2m})+\mu}{k_B T}\}p_r dp_r/(mk_B T)}{\exp\{\frac{-(\frac{p_z^2+p_r^2}{2m})+\mu}{k_B T}\}+1}\end{aligned}$$

$$= \frac{4\pi m k_B T}{h^3} \int_{p_z=\sqrt{(2mW)}}^{\infty} \left(\frac{p_z}{m}\right) dp_z \ln\left[\exp\left\{\frac{-(\frac{p_z^2+p_r^2}{2m}) + \mu}{k_B T}\right\} + 1\right]_{p_r=0}^{\infty}$$

$$= \frac{4\pi m k_B T}{h^3} \int_{p_z=\sqrt{(2mW)}}^{\infty} \left(\frac{p_z}{m}\right) dp_z \ln\left[\exp\left\{\frac{-\frac{p_z^2}{2m} + \mu}{k_B T}\right\} + 1\right]. \quad (11.334)$$

Let us set $\frac{p_z^2}{2m} = \epsilon$. Then $(\frac{p_z}{m})dp_z = \epsilon$ and the limits on the relevant integral are $\epsilon = W$ and $\epsilon = \infty$. In this fashion (11.334) can be written as

$$\wp_{F-D} = \frac{4\pi m k_B T}{h^3} \int_W^{\infty} \ln\left[\exp\left\{\frac{-\epsilon + \mu}{k_B T}\right\} + 1\right] d\epsilon. \quad (11.335)$$

Note that the exponent in this equation varies between $\frac{-W+\mu}{k_B T}$ and $-\infty$. At laboratory temperatures and below, μ is essentially equal to $\mu_0 = E_F$. Also the measured value of W is generally a few electron-volts higher than the Fermi energy E_F.[55] Therefore, in practice in (11.335),

$$\exp\left\{\frac{-\epsilon + \mu_0}{k_B T}\right\} \ll 1.$$

As such, we can expand the logarithm and to a good approximation write:

$$\wp_{F-D} \approx \frac{4\pi m k_B T}{h^3} \int_W^{\infty} \exp\left\{\frac{-\epsilon + \mu_0}{k_B T}\right\} d\epsilon$$

$$= \left[\frac{4\pi m (k_B T)^2}{h^3}\right] \exp\left\{\frac{-W + E_F}{k_B T}\right\}. \quad (11.336)$$

It is convenient to define the so-called work function Ø as the difference between the potential depth and Fermi energy. Therefore, the Fermi–Dirac thermionic current density, J_{F-D}, may be represented as follows:

$$\emptyset = (W - E_F),$$

$$J_{F-D} = e\wp_{F-D} = e\left[\frac{4\pi m (k_B)^2}{h^3}\right] T^2 \exp\left\{\frac{-\emptyset}{k_B T}\right\}. \quad (11.337)$$

As before, e stands for the electron charge.

The above prediction agrees with the experiment in the sense that at low temperature there is practically no spontaneous emission of electrons. An external stimulus that adds to, or subtracts from, the work function Ø—as, for example, is the case with an applied field—would affect the rate of thermionic emission. Similarly, with rise in temperature, both T^2 and $\exp\{\frac{-\emptyset}{k_B T}\}$ would increase. In this fashion, an increase in temperature would result in increasing, what then would properly be called, thermionic emission.

[55] For instance, the experimental value of the work function Ø for nickel is ≈ 5 eV and for tungsten it is ≈ 4.5 eV.

11.50.1 Richardson Effect: Quasiclassical Statistics

Equations (11.336) and (11.337) refer to a system of Fermi–Dirac free electrons. For quasiclassical free electrons the concept of Fermi energy has no relevance. As such, $\exp\left(\frac{\mu}{k_B T}\right)$ can no longer be replaced by $\exp\left(\frac{E_B}{k_B T}\right)$. Rather, it must be replaced by the relevant quasiclassical expression described earlier—as, for example, in (11.217), i.e.,

$$\exp\left(\frac{\mu}{k_B T}\right) \rightarrow \frac{N\lambda^3}{V}.$$

Note that N is the number of noninteracting, quasiclassical electrons, V is the system volume, and λ is as given in (11.212), that is,

$$\lambda = \left(\frac{2\pi m k_B T}{h^2}\right)^{-\frac{1}{2}}.$$

Accordingly, the quasiclassical version of the Richardson thermionic emission, $J_{\text{quasiclassical}}$, is obtained by replacing, in (11.336) and (11.337), $\exp\left(\frac{\mu}{k_B T}\right)$ by $\frac{N\lambda^3}{gV}$, and the subscript $F - D$ by quasiclassical. In this fashion we get

$$\begin{aligned} J_{\text{quasiclassical}} &\approx e\left[\frac{4\pi m (k_B)^2}{h^3}\right]\left(\frac{N\lambda^3}{V}\right)T^2 \exp\left\{\frac{-W}{k_B T}\right\} \\ &= e\left(\frac{2k_B N}{gV(2\pi m k_B)^{\frac{1}{2}}}\right)T^{\frac{1}{2}} \exp\left\{\frac{-W}{k_B T}\right\}. \end{aligned} \tag{11.338}$$

Remark 11.8 As noted above, the physics leading to (11.337) and (11.338) is completely different. Therefore, it would not be surprising if the corresponding results for thermionic current density should also turn out to be substantially different. On close examination one finds that the most striking difference lies in their different dependence upon the temperature. However, it is not so much the fact that the two results have different power dependence on T—namely one proportional to T^2 and the other proportional to $T^{\frac{1}{2}}$. Rather, the major part of this difference is rooted in the exponents. Indeed, as noted by Pathria, whether we plot $\ln\left(\frac{J_{F-D}}{T^{\frac{1}{2}}}\right)$ versus $\frac{1}{T}$, or plot $\ln\left(\frac{J_{\text{quasiclassical}}}{T^{\frac{1}{2}}}\right)$ against $\frac{1}{T}$, in each case we get a fairly good straight line. Therefore, this information by itself does not help much in choosing between the two results.

But it turns out that the numerical value of the slope—rather than just the requirement that for each of the results the slope should essentially be constant for different values of $\frac{1}{T}$—gives information as to whether the experiment favors $-\emptyset$ or just $-W$. Because the value of W can be determined independently—as, for instance, was done by Davisson and Germer[56]—and with that information, the essential superiority of the $F - D$ version of the result given in (11.337) can be established.

[56] Clinton Joseph Davisson and Lester Halbert Germer studied electron diffraction from a number of metals with various values of the initial kinetic energy [29]. In this manner they were able to estimate the relevant values of W.

11.51 Bose–Einstein Gas: Low Density, High Temperature

In the preceding sections it was found that in a system with large number of particles that obey the Fermi–Dirac (F–D) statistics, the quantum statistical effects are greatly diminished at very high temperatures. This is especially true when the gas density is relatively low and/or the particle mass is relatively large. It turns out that under the same conditions, a quantum Bose–Einstein (B–E) gas also behaves in somewhat similar manner. And the system approaches the status of a classical ideal gas.

Yet, despite this apparent similarity of behavior at high temperature, there are fundamental and significant differences in the thermodynamics of the F–D and B–E quantum gases at low temperatures.

11.51.1 Bose–Einstein Gas: Grand Potential

As explained earlier, (11.234)[57] with $\Delta = 1$ and $a = -1$ may be used to calculate the grand potential $Q(z, V, T)$ for the B–E system. And the knowledge of $Q(z, V, T)$ provides information about other thermodynamic functions. For instance, (11.235) may be written as

$$\frac{VP(z,V,T)}{k_B T} = Q(z,V,T) = -\sum_{\mathrm{p}} \ln\big[1 - z\exp(-\beta\varepsilon_{\mathrm{p}})\big]. \tag{11.339}$$

Similarly, we can write (11.237) and (11.236) as follows:

$$\bar{N} = z\left(\frac{\partial Q(z,V,T)}{\partial z}\right)_{\mathrm{V,T}} = -\sum_{\mathrm{p}}\big[-z^{-1}\exp(\beta\varepsilon_{\mathrm{p}}) + 1\big]^{-1}, \tag{11.340}$$

$$\begin{aligned} U(z,V,T) &= -\left(\frac{\partial Q(z,V,T)}{\partial \beta}\right)_{\mathrm{z,V}} = k_B T^2\left(\frac{\partial\big[\frac{VP(z,V,T)}{k_B T}\big]}{\partial T}\right)_{\mathrm{z,V}} \\ &= -\sum_{\mathrm{p}} \varepsilon_{\mathrm{p}}\big[-z^{-1}\exp(\beta\varepsilon_{\mathrm{p}}) + 1\big]^{-1}. \end{aligned} \tag{11.341}$$

11.51.2 Bose–Einstein Gas: Three Dimensions

Because for large N the single particle states are essentially continuous, for an ideal B–E gas, the sum $\sum_{\mathrm{p}}$ may be replaced by an integral over the relevant volume V and the momentum $\boldsymbol{p}$. That is, $\sum_{\mathrm{p}} = \frac{1}{h^3}\int \Downarrow \boldsymbol{x} \int \mathrm{d}\boldsymbol{p}\cdots = V\int(4\pi p^2)\mathrm{d}p\cdots$. Further we can write ε_{p}, which is the free particle translational energy for momentum $\bar{p}$, as $\frac{\bar{p}.\bar{p}}{2m} = \frac{p^2}{2m}$. As a result, we have $p^2\mathrm{d}p = m\sqrt{2m\varepsilon_{\mathrm{p}}}\mathrm{d}\varepsilon_{\mathrm{p}}$ and (11.340) may be

[57] Note that (11.234) leads to (11.235), ..., (11.237).

written as

$$
\begin{aligned}
N &= -\sum_{\mathrm{p}}\left[-z^{-1}\exp(\beta\varepsilon_{\mathrm{p}})+1\right]^{-1} \\
&= O\left(-\left[-z^{-1}+1\right]^{-1}\right)+\left(\frac{V}{h^3}\right)\int_{p>0}^{\infty}\left[z^{-1}\exp(\beta\varepsilon_{\mathrm{p}})-1\right]^{-1}4\pi p^2\mathrm{d}p \\
&= 2\pi V\left(\frac{2mk_{\mathrm{B}}T}{h^2}\right)^{\frac{3}{2}}\int_{x>0}^{\infty}\frac{(x)^{\frac{1}{2}}}{z^{-1}\exp(x)-1}\mathrm{d}x\,, \quad \text{if} \quad z\neq 1\,, \\
&= N_0+2\pi V\left(\frac{2mk_{\mathrm{B}}T}{h^2}\right)^{\frac{3}{2}}\int_{x>0}^{\infty}\frac{(x)^{\frac{1}{2}}}{\exp(x)-1}\mathrm{d}x\,, \quad \text{if} \quad z=1\,.
\end{aligned}
\tag{11.342}
$$

Here we have arranged the sum over p so that a possible contribution from momentum p exactly equal to zero[58] is separated from the rest of the sum over "nonzero" values of momentum p. Also two other things have been done: First, a variable $x = \beta\varepsilon_{\mathrm{p}} = \frac{\beta p^2}{2m}$ has been introduced. Second, for $z = 1$ the contribution from the zero momentum term has been denoted as N_0. When N_0 is nonzero, there is macroscopic occupancy of the zero energy state.

As shown in (11.230) and (11.231), for the B–E system being considered here, the physically acceptable values of z and μ lie within the ranges $1 \geq z \geq 0$ and $0 \geq \mu \geq -\infty$, respectively. And because the behavior of the number of particles in the system is quite different for the cases $z \to 1$ and $0 \leq z < 1$, it is best to separate (11.342) into two parts. For instance, setting $z \to 1$ leads to the result for N_{excited} which—being the number of particles in the noncondensed excited states—is equal to $N - N_0$, where

$$
N - N_0 = N_{\mathrm{excited}} = V\left(\frac{2\pi mk_{\mathrm{B}}T}{h^2}\right)^{\frac{3}{2}} g_{\frac{3}{2}}(1)\,. \tag{11.343}
$$

Similarly, for general z—meaning when $0 \leq z < 1$—(11.342) gives a relationship between z and the total number of particles N, where

$$
N = V\left(\frac{2\pi mk_{\mathrm{B}}T}{h^2}\right)^{\frac{3}{2}} g_{\frac{3}{2}}(z)\,. \tag{11.344}
$$

The relevant value of z for any given temperature T can be calculated from (11.344).

The differences between (11.343) and (11.344) suggest that some important milestone is being reached at the "point" $z = 1$. Let us call the temperature at this transition point, T_{c}. In the limit when z is infinitesimally close to 1—without actually being exactly equal to 1—the temperature is infinitesimally close to T_{c}. Equation (11.344) still applies here and we get

$$
N = V\left(\frac{2\pi mk_{\mathrm{B}}T_{\mathrm{c}}}{h^2}\right)^{\frac{3}{2}} g_{\frac{3}{2}}(1)\,. \tag{11.345}
$$

[58]Clearly, any such zero momentum contribution, denoted as $N_0 \sim O(-[-z^{-1}+1]^{-1})$, is a function of the temperature, T, and is significant only when $z = 1$.

This relationship determines the temperature T_c as

$$T_c = \left(\frac{h^2}{2\pi m k_B}\right)\left[\frac{N}{V\varsigma(\frac{3}{2})}\right]^{\frac{2}{3}}. \quad (11.346)$$

Note that, for any n, $g_n(1) = \varsigma(n)$.

Equation (11.346) specifies how the temperature T_c is a function of both the particle mass m and the number density $\frac{N}{V}$.

11.51.3 Bose–Einstein Gas: Condensation Temperature $T \leq T_c$

Subtract (11.343) from (11.345) and divide the result by (11.345). We get

$$\frac{N-(N-N_0)}{N} = \frac{V(\frac{2\pi m k_B T_c}{h^2})^{\frac{3}{2}} g_{\frac{3}{2}}(1) - (V(\frac{2\pi m k_B T}{h^2})^{\frac{3}{2}} g_{\frac{3}{2}}(1))}{V(\frac{2\pi m k_B T_c}{h^2})^{\frac{3}{2}} g_{\frac{3}{2}}(1)}$$
$$= \frac{N_0}{N} = 1 - \left(\frac{T}{T_c}\right)^{\frac{3}{2}}. \quad (11.347)$$

In other words, in order for N_0 to be nonzero and positive—which means, in order for the Bose–Einstein condensation to occur—the system temperature has to be lower than T_c.

Below T_c the system is a mixture of two apparent phases: one consisting of N_{excited} "excited-particles" with momentum—or equivalently, the kinetic energy—greater than zero and the other comprised of N_0 "condensed particles" that are all in the single, zero-momentum—equivalent to zero kinetic energy—quantum state.

11.51.4 Bose–Einstein Gas: Pressure and Internal Energy

To determine the pressure and internal energy, we need to reexpress (11.339) and (11.341) in a form suitable for calculation. We get

$$\frac{P(z,V,T)}{k_B T} = -\left(\frac{1}{h^3}\right)\int_{p=0}^{\infty} \ln\left[1 - z\exp(-\beta\varepsilon_p)\right]4\pi p^2 \mathrm{d}p$$
$$= -2\pi\left(\frac{2mk_B T}{h^2}\right)^{\frac{3}{2}}\int_{x>0}^{\infty} \ln\left[1 - z\exp(-x)\right]x^{\frac{1}{2}}\mathrm{d}x$$
$$+\left(\frac{N}{V}\right)O\left[\frac{\ln(N_0+1)}{N}\right] = g_{\frac{5}{2}}(z)\left(\frac{2\pi m k_B T}{h^2}\right)^{\frac{3}{2}}. \quad (11.348)$$

(Because $(\frac{N}{V})O[\frac{\ln(N_0+1)}{N}] \ll 1$, it has been ignored in the final result above.)

$$U(z,V,T) = -\left(\frac{\partial Q(z,V,T)}{\partial\beta}\right)_{z,V} = k_B T^2\left(\frac{\partial[\frac{VP(z,V,T)}{k_B T}]}{\partial T}\right)_{z,V}$$
$$= \frac{3}{2}PV. \quad (11.349)$$

The last expression, namely $U = \frac{3}{2}PV$, applies to all three-dimensional, nonrelativistic ideal gases whether they be classical—see, for example, (11.108)—or quantum (see (11.246) and (11.292)). As noted earlier, it contrasts with the corresponding result for extremely relativistic three-dimensional ideal gases which, according to (11.107), gives $U = 3PV$.

Comment In (11.344), and what follows, $g_n(z)$ signifies

$$g_n(z) = \frac{1}{\Gamma(n)} \int_0^\infty x^{n-1}\left[z^{-1}\exp(x) - 1\right]^{-1} dx\,. \tag{11.350}$$

As described in (11.230) and (11.231), for the B–E gas under consideration, the physically acceptable values of z lie within the range $1 \geq z \geq 0$. Therefore we can appropriately expand $g_n(z)$ in powers of z.[59] Using (11.350), we have[60]

$$\begin{aligned} g_n(z) &= \frac{1}{\Gamma(n)} \int_0^\infty x^{n-1}\left[\frac{z\exp(-x)}{1 - z\exp(-x)}\right] dx \\ &= \frac{1}{\Gamma(n)} \int_0^\infty x^{n-1} dx \sum_{l=1}^{\infty} [z\exp(-x)]^l \\ &= \sum_{l=1}^{\infty} \frac{z^l}{l^n} = z + \frac{z^2}{2^n} + \frac{z^3}{3^n} + o(z^4)\,. \end{aligned} \tag{11.351}$$

For small z, $g_n(z)$ tends to z. But as z approaches unity, $g_n(z)$ becomes equal to the well known Riemann zeta function, $\varsigma(n)$.

Throughout the range $1 \geq z \geq 0$, $g_n(z)$ increases monotonically.[61] In other words, the maximum value of $g_n(z)$, that is appropriate to the present physical system,[62] occurs for $z = 1$:

$$g_n(z) \leq g_n(1) = \varsigma(n) = \sum_{l=1}^{\infty} \frac{1}{l^n}\,. \tag{11.352}$$

[59] For $\Gamma(n)$, see (11.245). In particular, $\Gamma(\frac{3}{2}) = \frac{\sqrt{\pi}}{2}$, $\Gamma(\frac{5}{2}) = \frac{3}{2}\frac{\sqrt{\pi}}{2}$.

[60] Compare and contrast $g_n(z)$ with $f_n(z)$: the latter was used for the F–D gas and was defined in (11.241). Especially, contrast the small z expansions for $g_n(z)$—given in (11.351) above—and $f_n(z)$ that was given in (11.249). All terms in the $g_n(z)$ and $f_n(z)$ expansions look similar except that those with even powers of z in the $f_n(z)$ expansion have negative sign.

[61] Next let us look at the chemical potential μ. Recalling (11.231), we note that because $z = \exp(\beta\mu)$, the physically acceptable values of μ are negative and they range between minus-zero and minus-infinity.

[62] Here we shall need $g_{\frac{3}{2}}(1) = \varsigma(\frac{3}{2})$ and $g_{\frac{5}{2}}(1) = \varsigma(\frac{5}{2})$ which are ≈ 2.61238 and 1.34149, respectively.

Later on[63] in this chapter, we shall need to know if, and where, $\varsigma(n)$ diverges. This question is readily answered by using the so-called integral test. That is, by examining the convergence of the integral $\int^{\infty} \frac{dx}{x^n}$. For $n \neq 1$, we have

$$\int^{\infty} \frac{dx}{x^n} = \left(\frac{x^{1-n}}{1-n} \right)|^{\infty}, \tag{11.353}$$

which diverges for $n < 1$. For n exactly equal to 1, the integral

$$\int^{\infty} \frac{dx}{x} = \ln(x)|^{\infty} \tag{11.354}$$

also diverges.

In other words, $\varsigma(n)$ diverges for $n \leq 1$.

11.51.5 Bose–Einstein Gas: Degenerate Ideal Gas

At temperatures lower than T_c, the B–E gas is "degenerate." Its physical properties are starkly different from those of a classical noninteracting gas. In particular, in defiance of the traditional classical Boltzmann–Maxwell–Gibbs prescription, even as the temperature rises above zero the macroscopic occupancy of the zero energy state persists. For instance, let us begin with the equality (11.347). Because $\frac{T}{T_c}$ is positive, the largest value of N_0—which is the number of particles in the degenerate state—consists of all the N particles and it occurs at zero temperature. As the temperature is raised above zero, the number of degenerate particles decreases until it reaches zero at $T = T_c$. This behavior is the opposite of that of particles in the excited states. $N_{\text{excited}} = N - N_0$, is vanishing at zero temperature and reaches its maximum, N, at the critical temperature.

11.52 Degenerate Regime: Specific Heat

As mentioned before, throughout the degenerate regime, z is infinitesimally close to being equal to unity. Therefore, the pressure and the internal energy in the degenerate regime are readily found from (11.348) and (11.349) by setting $z = 1$. We get

$$P = \varsigma\left(\frac{5}{2}\right) k_B T \left(\frac{2\pi m k_B T}{h^2} \right)^{\frac{3}{2}}. \tag{11.355}$$

Note that the pressure in the degenerate state is independent of the system volume. Could it be the result of the macroscopic presence of the B–E condensed state in which all particles have zero momentum and thus do not contribute to the pressure?

[63] $\varsigma(n)$ diverges for $n \leq 1$.

According to (11.349), the internal energy within the degenerate regime is, as usual, directly proportional to PV, namely

$$U = V\left(\frac{3}{2}\right)\varsigma\left(\frac{5}{2}\right)k_B T\left(\frac{2\pi m k_B T}{h^2}\right)^{\frac{3}{2}} = \frac{3}{2}\left[\frac{\varsigma(\frac{5}{2})}{\varsigma(\frac{3}{2})}\right] N k_B T\left(\frac{T}{T_c}\right)^{\frac{3}{2}}. \quad (11.356)$$

(The expression for the critical temperature T_c that was used in (11.356) above is recorded in (11.346).)

The specific heat, C_v, within the degenerate regime is

$$C_v(T) = \left(\frac{\partial U}{\partial T}\right)_V = \frac{15}{4}\left[\frac{\varsigma(\frac{5}{2})}{\varsigma(\frac{3}{2})}\right] N k_B \left(\frac{T}{T_c}\right)^{\frac{3}{2}}. \quad (11.357)$$

Notice that as T approaches T_c from below, the specific heat approaches the value $C_v(T_c) = \frac{15}{4}[\frac{\varsigma(\frac{5}{2})}{\varsigma(\frac{3}{2})}]Nk_B = 1.92567Nk_B$. This is precisely the value of the specific heat when T_c is approached from above (see (11.364)). Therefore, the specific heat is continuous at all temperatures.

Next examine the derivative of the specific heat. We have

$$\left(\frac{\partial C_v}{\partial T}\right)_V = \frac{45}{8}\left[\frac{\varsigma(\frac{5}{2})}{\varsigma(\frac{3}{2})}\right]\frac{N k_B}{T}\left(\frac{T}{T_c}\right)^{\frac{3}{2}}. \quad (11.358)$$

Clearly, at low temperature, the slope of the "specific heat-versus-the temperature" curve keeps rising at the rate $T^{\frac{1}{2}}$. When the temperature T reaches T_c from below, the slope of the specific heat curve reaches the value, $\frac{45}{8}[\frac{\varsigma(\frac{5}{2})}{\varsigma(\frac{3}{2})}] \times \frac{Nk_B}{T_c} = 2.88851\frac{Nk_B}{T_c}$.

11.52.1 Degenerate Regime: State Functions

Recalling that in the degenerate regime $z = 1$, or equivalently, the chemical potential $\mu = 0$, other state functions are readily found as

$$\begin{aligned}
G &= n\mu = 0\,,\\
F &= G - PV = -\frac{2}{3}U = -\left[\frac{\varsigma(\frac{5}{2})}{\varsigma(\frac{3}{2})}\right] N k_B T\left(\frac{T}{T_c}\right)^{\frac{3}{2}},\\
S &= \frac{U-F}{T} = \frac{5}{3}\frac{U}{T} = \frac{5}{2}\left[\frac{\varsigma(\frac{5}{2})}{\varsigma(\frac{3}{2})}\right] N k_B \left(\frac{T}{T_c}\right)^{\frac{3}{2}},\\
H &= U + PV = \frac{5}{3}U = \frac{5}{2}\left[\frac{\varsigma(\frac{5}{3})}{\varsigma(\frac{3}{2})}\right] N k_B T\left(\frac{T}{T_c}\right)^{\frac{3}{2}}.
\end{aligned} \quad (11.359)$$

The notation used is as in (11.7). That is, n is the number of moles, μ is the chemical potential per mole, G is Gibbs potential, F is Helmholtz free energy, S is the entropy, and H is the enthalpy.

11.53 Bose–Einstein Gas: Nondegenerate Regime Specific Heat

At zero temperature, essentially all the particles are in the condensed state. As the temperature begins to rise, the number of condensed particles begins to decrease. Their number approaches zero as the temperature reaches T_c. And, as the temperature rises beyond T_c, the value of the parameter z starts to decrease below unity. To make use of this information, let us start with (11.344) and (11.348) both of which refer to $T \geq T_c$:

$$N = V \frac{g_{\frac{3}{2}}(z)}{\lambda^3}, \quad P = k_B T \frac{g_{\frac{5}{2}}(z)}{\lambda^3}. \tag{11.360}$$

On combining the two relationships given above, we get

$$P = \frac{N k_B T}{V} \left[\frac{g_{\frac{5}{2}}(z)}{g_{\frac{3}{2}}(z)} \right]. \tag{11.361}$$

And, of course, we also have the usual relationship between the internal energy and the product of the pressure and the volume. In three dimensions we have,

$$U = \frac{3}{2} P V = \frac{3}{2} N k_B T \left[\frac{g_{\frac{5}{2}}(z)}{g_{\frac{3}{2}}(z)} \right]. \tag{11.362}$$

(Compare and contrast this with the corresponding result for the F–D ideal gas—that is, $U = \frac{3}{2} N k_B T [\frac{f_{\frac{5}{2}}(z)}{f_{\frac{3}{2}}(z)}]$.)

Given the internal energy, the calculation of the specific heat $(\frac{\partial U}{\partial T})_V = C_v(T)$ for temperatures $T \geq T_c$ is straightforward, but requires a little bit of algebra. The relevant algebra and the final result are given in an appendix—see the argument that leads to (N.6). For convenience that result is restated below:

$$C_v(T) = N k_B \left[\frac{15}{4} \left(\frac{g_{\frac{5}{2}}(z)}{g_{\frac{3}{2}}(z)} \right) - \frac{9}{4} \left(\frac{g_{\frac{3}{2}}(z)}{g_{\frac{1}{2}}(z)} \right) \right]. \tag{11.363}$$

In the limit that the temperature approaches T_c (from above), $z \rightarrow 1$, and $g_{\frac{1}{2}}(1) \rightarrow \infty$, the specific heat reaches the value given below:

$$C_v(T) =_{T \rightarrow T_c + 0} N k_B \left[\frac{15}{4} \left(\frac{g_{\frac{5}{2}}(1)}{g_{\frac{3}{2}}(1)} \right) \right] = N k_B \left[\frac{15}{4} \left(\frac{\varsigma(\frac{5}{2})}{\varsigma(\frac{3}{2})} \right) \right]$$
$$\approx 1.92567 N k_B . \tag{11.364}$$

This result and that given in (11.357) indicate that as the temperature is lowered from ∞—where the specific heat is at its classical limit $\frac{3}{2}Nk_{\rm B}$—down to $T_{\rm c}$, the specific heat increases monotonically with decrease in temperature. It reaches the value $1.92567Nk_{\rm B}$ at $T = T_{\rm c}$. On the other hand, as also noted before, at zero temperature the specific heat is vanishing. But it increases monotonically with increase in temperature. And as the temperature approaches $T_{\rm c}$ from below, $C_{\rm v}(T \rightarrow T_{\rm c})$ reaches precisely the same value as reached from above: that is $1.92567Nk_{\rm B}$—see (11.357). Thus the specific heat is continuous and its highest value is reached at $T = T_{\rm c}$.

However, while the specific heat itself is continuous throughout, it turns out that the same is not true for its slope. As shown in (11.358), as the temperature rises above zero, the slope begins to increase slowly with increase in temperature. And when T reaches $T_{\rm c}$, the slope has achieved the value $\frac{45}{8}[\frac{\zeta(\frac{5}{2})}{\zeta(\frac{3}{2})}]\frac{Nk_{\rm B}}{T_{\rm c}} = 2.88851\frac{Nk_{\rm B}}{T_{\rm c}}$.

This is to be contrasted with what happens as the temperature increases from $T_{\rm c} - 0$ across to $T_{\rm c} + 0$. Upon crossing $T_{\rm c}$ the slope immediately changes to $\{\frac{45}{8}[\frac{\zeta(\frac{5}{2})}{\zeta(\frac{3}{2})}] - \frac{27}{16\pi}[\zeta(\frac{3}{2})]^2\}\frac{Nk_{\rm B}}{T_{\rm c}} = -0.77807$ and stays negative beyond $T_{\rm c}$ until it reaches the limiting value of -0 at $T = \infty$. Thus the slope has a discontinuity at $T = T_{\rm c}$ [30]. The difference between the rising and the falling slopes is $\{\frac{27}{16\pi}[\zeta(\frac{3}{2})]^2\}\frac{Nk_{\rm B}}{T_{\rm c}} = 3.66658\frac{Nk_{\rm B}}{T_{\rm c}}$.

In the degenerate regime z is unity and N_0, the number of particles in the condensed state, is nonzero. In contrast, in a state of nondegeneracy, the number of particles in the condensed state is vanishing and z ranges between zero and a value that approaches unity from below. Yet, much like the F–D gas studied earlier, rather than z, we are interested in results in terms of the temperature. To this end we need to invert the series expansion implicit in (11.348) and (11.349). The needed inversion will change powers of z into powers of ρ. The inversion procedure is almost identical to that followed for the F–D gas.

The relevant difference in the grand potential for the F–D and the B–E systems lies in the different values of the parameters Δ and a. (This difference is demonstrated in (11.234).) In particular, $\Delta = 2$ and $a = 1$ for the $S = \frac{1}{2}$ F–D gas while $\Delta = 1$ and $a = -1$ for the $S = 0$ B–E system. Therefore, the high temperature, small z, results can immediately be transformed from one system to the other. For example, both $\frac{PV}{k_{\rm B}T}$ and the internal energy U, given in (11.256) and (11.257), are readily reexpressed as the corresponding results for the B–E gas by changing the signs of the terms with odd powers of ρ.

We get

$$PV = k_{\rm B}TN\left\{1 - \left(\frac{1}{2^{\frac{5}{2}}}\right)\rho + \left(\frac{1}{8} - \frac{2}{3^{\frac{5}{2}}}\right)\rho^2 - o(\rho^3)\right\}$$

where, as per (11.250),

$$\rho = \frac{N\lambda^3}{V\Delta} = \left(\frac{Nh^3}{V\Delta}\right)(2\pi mk_{\rm B}T)^{\frac{-3}{2}}. \tag{11.365}$$

Note that in the above equation, and in the following, $\Delta = 1$. Comparing with a classical gas, the above equation appears to represent the behavior of a gas with "attractive" interparticle force. Notice that this apparent attractive force is being produced by a completely noninteracting quantum gas! We recall that if the quantum particles obey, rather than the B–E, the F–D statistics, the apparent interparticle interaction is "repulsive." See (11.256).

As usual, the internal energy is found from (11.365) by using the identity $U = \frac{3}{2}PV$. We get

$$U = \frac{3k_B TN}{2}\left\{1 - \left(\frac{1}{2^{\frac{5}{2}}}\right)\rho + \left(\frac{1}{8} - \frac{2}{3^{\frac{5}{2}}}\right)\rho^2 - o\left(\rho^3\right)\right\}. \quad (11.366)$$

Because $(\frac{\partial\rho}{\partial T})_{N,V} = -\frac{3\rho}{2T}$, the system specific heat is

$$C_V = \left(\frac{\partial U}{\partial T}\right)_{N,V} = \frac{3k_B N}{2}\left\{1 + \left(\frac{1}{2^{\frac{7}{2}}}\right)\rho + \left(\frac{4}{3^{\frac{5}{2}}} - \frac{1}{4}\right)\rho^2 + o\left(\rho^3\right)\right\}. \quad (11.367)$$

For a B–E gas at very high temperature, the specific heat, C_V, approaches its classical limit, $1.5k_B N$. But as the temperature is lowered below ∞ to a finite value, the specific heat starts to rise[64] above its classical value. However, such increase in the specific heat with decreasing temperature cannot continue all the way down to zero temperature. This is the case because for all thermodynamically stable systems the specific heat is vanishing at absolute zero. Therefore at some finite temperature the increase in the specific heat C_V must stop and a process of decrease must begin. (Because of such dramatic change in thermodynamic behavior we expect this temperature to be T_c.) Indeed, coming from the opposite direction—as was shown earlier in (11.357)—the specific heat approaches the value $C_V(T_c) = \frac{15}{4}[\frac{\zeta(\frac{5}{2})}{\zeta(\frac{3}{2})}]Nk_B \approx 1.92567Nk_B$ as the temperature approaches T_c from below. Clearly, therefore, it is important to determine what happens as the temperature approaches T_c from above.

11.54 Bose–Einstein Condensation: In δ-Dimensions

As stated before, for large N the single particle states for an ideal B–E gas are essentially continuous. Therefore, the sum $\sum_p$ may be replaced by an integral. Following the procedure employed for treating a three-dimensional system—see (11.342)—first we determine the integration factor. In δ dimensions the integration factor is

$$\sum_p = \frac{1}{h^\delta}\int d\boldsymbol{x}\int d\boldsymbol{p}\cdots = \left(\frac{v^\delta}{h^\delta}\right)\delta\left[\frac{\pi^{\frac{\delta}{2}}}{(\frac{\delta}{2})!}\right]p^{\delta-1}dp\cdots, \quad (11.368)$$

where v^δ is the δ-dimensional volume. For its calculation, see [30]. The energy ε_p for momentum p is the free particle translational energy equal to $\frac{p^2}{2m}$.

[64] This means the slope of the specific heat versus the temperature curve is negative here.

The number of particles are given by the equality

$$N = -\sum_{\mathrm{p}}\left[-z^{-1}\exp(\beta\varepsilon_{\mathrm{p}})+1\right]^{-1} = O\left(-\left[-z^{-1}+1\right]^{-1}\right)$$

$$+\left(\frac{v^{\delta}}{h^{\delta}}\right)\cdot\delta\cdot\left[\frac{\pi^{\frac{\delta}{2}}}{(\frac{\delta}{2})!}\right]\int_{p>0}^{\infty}\left[z^{-1}\exp(\beta\varepsilon_{\mathrm{p}})-1\right]^{-1}p^{\delta-1}\mathrm{d}p\,.$$

$$= N_0 + \left(\frac{v^{\delta}\cdot\delta}{2h^{\delta}}\right)\cdot\left[\frac{\pi^{\frac{\delta}{2}}}{(\frac{\delta}{2})!}\right](2mk_{\mathrm{B}}T)^{\frac{\delta}{2}}\left[\int_{x>0}^{\infty}\frac{x^{(\frac{\delta}{2}-1)}}{z^{-1}\exp(x)-1}\mathrm{d}x\right]$$

$$= N_0 + \left(\frac{v^{\delta}\cdot\delta}{2h^{\delta}}\right)\cdot\left[\frac{\pi^{\frac{\delta}{2}}}{(\frac{\delta}{2})!}\right](2mk_{\mathrm{B}}T)^{\frac{\delta}{2}}\left[g_{(\frac{\delta}{2})}(z)\Gamma\left(\frac{\delta}{2}\right)\right], \tag{11.369}$$

where N_0 is the number of particles in the zero momentum condensed state. In the above, the contribution from momentum p exactly equal to zero—i.e., $O(-[-z^{-1}+1]^{-1}) = N_0$—is separated from the rest of the sum over “nonzero” values of the momentum p. Also a variable $x = \beta\varepsilon_{\mathrm{p}} = \frac{\beta p^2}{2m}$ has been introduced. Above $T = T_{\mathrm{c}}$, $N_0 = 0$ but it becomes nonzero below T_{c}.

As the critical temperature is approached from above T_{c}, and while N_0 is still zero, the parameter z tends to unity. That is, at $T = (T_{\mathrm{c}} + \text{zero})$, $N_0 = 0$ and $z = (1 - \text{zero})$. Therefore, (11.369) predicts that the total number of particles is given by the relationship

$$N = \left(\frac{v^{\delta}\cdot\delta}{2h^{\delta}}\right)\cdot\left[\frac{\pi^{\frac{\delta}{2}}}{(\frac{\delta}{2})!}\right](2mk_{\mathrm{B}}T_{\mathrm{c}})^{\frac{\delta}{2}}\varsigma\left(\frac{\delta}{2}\right)\Gamma\left(\frac{\delta}{2}\right), \tag{11.370}$$

where $\varsigma(\frac{\delta}{2})$ and $\Gamma(\frac{\delta}{2})$, respectively, are the Riemann zeta, and the Gamma, functions of argument $\frac{\delta}{2}$. The requirement that the number of particles, N, is finite places constraints on (11.370). In this regard, the only component of this equation that needs double-checking is $\varsigma(\frac{\delta}{2})\Gamma(\frac{\delta}{2})$.

To this purpose, we recall the results of (11.353) and (11.354) where it was shown that $\varsigma(n)$ diverges whenever $n \le 1$. Therefore the Riemann zeta function $\varsigma(\frac{\delta}{2})$, and the predicted result for N, tend to ∞ when $\delta \le 2$. Clearly, such a result for N would be completely unphysical.

Hence, there can be no Bose–Einstein condensation in noninteracting Bose gases with dimensionality less than or equal to two.

11.55 Critical Temperature: Bose–Einstein Gas

The critical temperature T_{c} can be found by rearranging and inverting (11.370),

$$T_{\mathrm{c}} = \left[N\left\{\left(\frac{v^{\delta}\cdot\delta}{2h^{\delta}}\right)\cdot\left[\frac{\pi^{\frac{\delta}{2}}}{(\frac{\delta}{2})!}\right](2mk_{\mathrm{B}})^{\frac{\delta}{2}}\varsigma\left(\frac{\delta}{2}\right)\Gamma\left(\frac{\delta}{2}\right)\right\}^{-1}\right]^{\frac{2}{\delta}}. \tag{11.371}$$

Decreasing the temperature below T_{c} brings about Bose–Einstein condensation.

11.55.1 Temperature Dependence of Condensate: Bose–Einstein Gas

Now that T_c has been calculated, we introduce $T_c^{-\frac{\delta}{2}}$ into (11.369)—where z must now be set equal to 1—and after a little algebra arrive at the following relationship for the number N_0 of particles in the B–E condensate:

$$N_0 = N\left[1 - \left(\frac{T}{T_c}\right)^{\frac{\delta}{2}}\right]. \tag{11.372}$$

Outwardly the above relationship is in disagreement with the previous result that predicts that B–E condensation cannot occur if δ is less than or equal to 2. In other words, it appears to disavow the need for δ to be greater than two. That appearance, however, is in the nature of an optical mirage. When $\delta \leq 2$, then according to (11.371) the so-called critical temperature, T_c, goes to zero and (11.372) truly and, indeed visibly, fails to have any meaning.

11.55.2 Pressure and Internal Energy of Condensate: Bose–Einstein Gas

In order that (11.339) may be transformed to apply to δ dimensions, we set $z = 1$ and follow the procedure used in the preceding section for deriving the corresponding results for three dimensions. In this fashion we get the following expression for the system pressure at temperature T:

$$\begin{aligned}\frac{P}{k_B T} &= -\left(\frac{\delta}{2h^\delta}\right)\left[\frac{\pi^{\frac{\delta}{2}}}{(\frac{\delta}{2})!}\right](2mk_B T)^{\frac{\delta}{2}}\left[\int_{x>0}^{\infty} x^{(\frac{\delta}{2}-1)} \ln\{1 - \exp(-x)\}\mathrm{d}x\right] \\ &= \left(\frac{1}{h^\delta}\right)\left[\frac{\pi^{\frac{\delta}{2}}}{(\frac{\delta}{2})!}\right](2mk_B T)^{\frac{\delta}{2}}\left[\varsigma\left(\frac{\delta}{2}+1\right)\Gamma\left(\frac{\delta}{2}+1\right)\right].\end{aligned} \tag{11.373}$$

The internal energy, U, and the specific heat, $C_V(T)$, are also readily found as

$$\begin{aligned}U &= k_B T^2 v^\delta \left(\frac{\partial(\frac{P}{k_B T})}{\partial T}\right)_{z, v^\delta} \\ &= \frac{\delta}{2}\left(\frac{v^\delta}{h^\delta}\right)\left[\frac{\pi^{\frac{\delta}{2}}}{(\frac{\delta}{2})!}\right](2m)^{\frac{\delta}{2}}(k_B T)^{(\frac{\delta}{2}+1)}\left[\varsigma\left(\frac{\delta}{2}+1\right)\Gamma\left(\frac{\delta}{2}+1\right)\right]\end{aligned} \tag{11.374}$$

and

$$\begin{aligned}C_V(T) &= \left(\frac{\partial U}{\partial T}\right)_{v^\delta} = k_B\left(\frac{\delta}{2}\right)\left(\frac{\delta}{2}+1\right)\left(\frac{v^\delta}{h^\delta}\right)\left[\frac{\pi^{\frac{\delta}{2}}}{(\frac{\delta}{2})!}\right](2mk_B T)^{(\frac{\delta}{2})} \\ &\times \left[\varsigma\left(\frac{\delta}{2}+1\right)\Gamma\left(\frac{\delta}{2}+1\right)\right].\end{aligned} \tag{11.375}$$

Because in order for the B–E condensation to occur δ has to be greater than 2, the specific heat always goes to zero faster than the first power of the temperature.

11.56 Black Body Radiation: Thermodynamic Consideration

The preceding treatment referred to boson systems that have a given number, N, of particles. Any possibility that some particles may be "destroyed" or additional particles "created" was not considered. In other words, in the preceding analyses, the particle number was conserved.

In contrast, there exist boson quasiparticles—such as "photons" in electromagnetic radiation, "phonons" produced by atomic vibrational motion in solids, spin-waves in magnetic systems, etc.—whose number is variable, dependent on the temperature and other relevant parameters.

Consider a macroscopic collection of photons. In order for these photons to reach equilibrium, assume them to be enclosed in an opaque container of volume V that is constructed of diathermal walls which are in thermal contact with a heat energy reservoir at a fixed temperature T. The container also has a microscopic opening which allows a small amount of electromagnetic radiation to escape every second.

The containing walls absorb and reemit radiation. Generally, in the process, photons experience multiple reflections from the walls, and a state of dynamic equilibrium—to which a temperature may be assigned—is achieved thereby. Clearly, the number of such photons must be linearly proportional to the volume V and must depend on the system temperature. Note the system temperature is the same as the temperature T of the heat energy reservoir.

Let Υ stand for the internal energy, per unit volume, at temperature T, for frequencies that lie in the narrow range between ν and $\nu + \mathrm{d}\nu$. Clearly, Υ is a function only of the temperature T and the frequency ν. That is, $\Upsilon = \Upsilon(T, \nu)$. Summing it over all the possible frequencies yields the internal energy density, u, which is dependent only on the temperature,

$$u(T) = \int_0^\infty \Upsilon(T, \nu)\mathrm{d}\nu \,. \tag{11.376}$$

The total internal energy, U, is equal to the volume V multiplied by the internal energy density $u(T)$,

$$U = U(T, V) = V u(T) \,. \tag{11.377}$$

Maxwell's theory predicts that the pressure, P, exerted by electromagnetic radiation, is directly related to the internal energy density u,

$$P = \frac{1}{3} u(T) \,. \tag{11.378}$$

In an earlier chapter we proved—see (5.16)—the identity

$$T\left(\frac{\partial P}{\partial T}\right)_{\mathrm{V}} = P + \left(\frac{\partial U}{\partial V}\right)_{\mathrm{T}} .$$

Here it translates into

$$T\left(\frac{1}{3}\right)\frac{\mathrm{d}u(T)}{\mathrm{d}T}=\frac{1}{3}u(T)+u(T)=\frac{4}{3}u(T)\,, \tag{11.379}$$

which gives

$$T\left(\frac{1}{3}\right)\frac{\mathrm{d}u(T)}{\mathrm{d}T}=\frac{1}{3}u(T)+u(T)=\frac{4}{3}u(T)\,,$$

$$\text{that is}\qquad \frac{\mathrm{d}u}{u}=4\frac{\mathrm{d}T}{T}\,. \tag{11.380}$$

Integrating the above yields the so-called Stefan–Boltzmann T^4-law:

$$\ln(u)=4\ln(T)+\text{const.}\,,\qquad u=\sigma T^4\,. \tag{11.381}$$

Here σ is the integration constant.

In the years since the experimental observation of this result by Josef Stefan, it has received ample reconfirmation. The Stefan constant σ has been found to be [31]

$$\sigma\approx 7.561*10^{-16}\,\mathrm{J\,m^{-3}\,K^{-4}}\,. \tag{11.382}$$

The pressure and the total internal energy are also readily found from (11.378) and (11.377) as

$$P=\frac{1}{3}\sigma T^4\,,\qquad U=V\sigma T^4\,. \tag{11.383}$$

(It is important to note that unlike all nonrelativistic classical or quantum perfect gases, here $U\neq\frac{3}{2}PV$. Rather, $U=3PV$. This result is consistent with that of a completely relativistic perfect gas. Compare, for example, with (11.107).)

The calculation of entropy is also straightforward. Recall that the first–second law—see (5.6)—states $T\mathrm{d}S=\mathrm{d}U+P\mathrm{d}V$. At constant volume it becomes $T(\mathrm{d}S)_{\mathrm{V}}=(\mathrm{d}U)_{\mathrm{V}}$. Division by T and integration leads to

$$\begin{aligned}\int_{\mathrm{at\,T=0}}^{\mathrm{at\,T=T}}(\mathrm{d}S)_{\mathrm{V}}&=S|_{\mathrm{T=0}}^{\mathrm{T=T}}=S\\ &=\int_{\mathrm{T=0}}^{\mathrm{T=T}}\frac{(\mathrm{d}U)_{\mathrm{V}}}{T}=V\sigma\int_{\mathrm{T=0}}^{\mathrm{T=T}}4T^3\mathrm{d}T\\ &=V\sigma\left(\frac{4}{3}\right)T^3\,.\end{aligned} \tag{11.384}$$

Now, using (11.7), the Helmholtz free energy, F, the Gibbs potential, G and the enthalpy, H, can also be found as

$$F=U-TS=-V\sigma\left(\frac{1}{3}\right)T^4\,,$$

$$G = F + PV = -V\sigma\left(\frac{1}{3}\right)T^4 + V\sigma\left(\frac{1}{3}\right)T^4 = 0\,, \tag{11.385}$$
$$H = U + PV = \; V\sigma\left(\frac{4}{3}\right)T^4\,.$$

11.56.1 Black Body Radiation: Chemical Potential Zero

Radiation photons experience multiple reflections from the containing walls, resulting in absorption and reemission. This means that a photon ideal-gas is fundamentally different from an ideal gas composed of ordinary particles because the total number of ordinary particles is a constant. Here the number of moles n of the given quasiparticles—namely the photons—is not fixed. Rather, n is a variable dependent upon the temperature and the volume. The determining factor, of course, must be the requirement that the Helmholtz free energy, F, is a minimum for given fixed values of the volume, V, and the temperature, T. That means the differential $\mathrm{d}F$,

$$\mathrm{d}F = \left(\frac{\partial F}{\partial V}\right)_{\mathrm{T,n}} \mathrm{d}V + \left(\frac{\partial F}{\partial T}\right)_{\mathrm{V,n}} \mathrm{d}T + \left(\frac{\partial F}{\partial n}\right)_{\mathrm{T,V}} \mathrm{d}n\,, \tag{11.386}$$

must be vanishing when V and T are constants equal to the given values of the volume and the temperature. In other words, when in (11.386) both $\mathrm{d}V$ and $\mathrm{d}T$ are equal to zero, the following equality must hold:

$$\mathrm{d}F = \left(\frac{\partial F}{\partial n}\right)_{\mathrm{T,V}} \mathrm{d}n = 0\,. \tag{11.387}$$

This means that for a photon gas in thermodynamic equilibrium we have

$$\left(\frac{\partial F}{\partial n}\right)_{\mathrm{T,V}} = 0\,. \tag{11.388}$$

As we recall from (11.7), the chemical potential, μ—for one mole of a given system—is defined as $(\frac{\partial F}{\partial n})_{\mathrm{T,V}}$. Therefore it follows that the chemical potential for a photon gas in thermal equilibrium is equal to zero.

11.56.2 Black Body Radiation: Energy

Following Bose [32] and Einstein, the electromagnetic radiation is treated as a gas of noninteracting, quasiparticles called photons. In quantum-mechanical description, the energy of a photon of momentum p is $\varepsilon_{\mathrm{p}} = cp = \hbar\omega$, where c is the velocity of light in vacuum and ω is the rotational frequency of the photon.

Because an extensive analysis of the B–E gas already exists in the preceding part of this chapter, we can simply make use of it here. In particular, for calculating the system pressure, the relevant equation to use is (11.235). Similarly, the internal energy $U(z, V, T)$ is available from (11.236).

11.56.3 Black Body Radiation: Pressure

For convenience we reproduce (11.235) below:

$$\frac{VP(z,V,T)}{k_B T} = Q(z,V,T) = \frac{\Delta}{a}\sum_{p} \ln\left[1 + az\exp(-\beta\varepsilon_p)\right]$$
$$= \frac{\Delta}{a}\left(\frac{V}{h^3}\right)\int_0^\infty \ln\left[1 + az\exp\left(-\frac{\beta p^2}{2m}\right)\right]4\pi p^2 dp\,. \qquad (11.389)$$

For a system of noninteracting photons being considered here, we need to introduce appropriate changes, namely set $z = 1$, $\varepsilon_p = cp$, and because photons follow Bose–Einstein statistics, set $a = -1$. Orthogonal to the electromagnetic radiation—meaning normal to the direction of the velocity c—photons have two mutually perpendicular polarizations. Therefore, Δ has to be set equal to 2. Accordingly, (11.389) can be written as

$$PV = -V\left(\frac{2k_B T}{h^3}\right)\int_0^\infty \ln[1 - \exp(-\beta cp)]4\pi p^2 \mathrm{d}p$$
$$= -V\left[\frac{8\pi(k_B T)^4}{(hc)^3}\right]\int_0^\infty \ln[1-\exp(-x)]x^2 \mathrm{d}x$$
$$= -V\left[\frac{8\pi(k_B T)^4}{(hc)^3}\right]\left(\frac{-1}{3}\right)\int_0^\infty \frac{x^3\mathrm{d}x}{\exp(x)-1}$$
$$= V\left[\frac{8\pi^5}{45(hc)^3}\right](k_B T)^4 = V\sigma\left(\frac{1}{3}\right)T^4\,, \qquad (11.390)$$

where the Stefan constant σ, referred to in (11.381)–(11.383), is

$$\sigma = \left[\frac{8\pi^5}{15(hc)^3}\right](k_B)^4\,. \qquad (11.391)$$

11.56.4 Black Body Radiation: Internal Energy

Similarly,[65] we can use (11.236) for the internal energy—which is also, for convenience, reproduced below:

$$U(z,V,T) = -\left(\frac{\partial Q(z,V,T)}{\partial\beta}\right)_{z,V}$$
$$= \frac{\Delta}{a}\sum_{p}\varepsilon_p\left[a^{-1}z^{-1}\exp(\beta\varepsilon_p)+1\right]^{-1} = \sum_{p}\varepsilon_p N_p$$

[65] The integral above is well known, that is, $\int_0^\infty \frac{x^3\mathrm{d}x}{\exp(x)-1} = \frac{\pi^4}{15}$.

$$= \frac{\Delta}{a}\left(\frac{V}{h^3}\right)\int_0^\infty \left(\frac{p^2}{2m}\right)\left[a^{-1}z^{-1}\exp\left(\frac{\beta p^2}{2m}\right)+1\right]^{-1} 4\pi p^2 \mathrm{d}p \,. \tag{11.392}$$

Again, in order to convert this equation to apply to a gas of noninteracting bosons, we need to set $z = 1$, $\varepsilon_\mathrm{p} = cp$, $a = -1$, and $\Delta = 2$. This gives

$$\begin{aligned} U &= -\left(\frac{2V}{h^3}\right)\int_0^\infty (cp)[-\exp(\beta cp)+1]^{-1} 4\pi p^2 \mathrm{d}p \\ &= -V\left[\frac{8\pi (k_\mathrm{B}T)^4}{(hc)^3}\right]\int_0^\infty [-\exp(x)+1]^{-1} x^3 \mathrm{d}x \\ &= V\left[\frac{8\pi (k_\mathrm{B}T)^4}{(hc)^3}\right]\int_0^\infty \frac{x^3 \mathrm{d}x}{\exp(x)-1} \\ &= V\left[\frac{8\pi (k_\mathrm{B}T)^4}{(hc)^3}\right]\left(\frac{\pi^4}{15}\right) = 3PV\, V\sigma T^4 \,. \end{aligned} \tag{11.393}$$

Again we note that the equality $U = 3PV$ is consistent with that yielded by an extremely relativistic monatomic ideal gas. Compare, for example, with (11.107).

11.56.5 Black Body Radiation: Other Thermodynamic Potentials

All other thermodynamic potentials can now be related to PV. For example, we have already learned that the chemical potential μ is equal to 0. Accordingly, the Gibbs free energy must also be vanishing. This tells us that the Helmholtz free energy F is

$$\begin{aligned} F &= G - PV = n\mu - PV = 0 - PV \\ &= -V\left[\frac{8\pi^5}{45(hc)^3}\right](k_\mathrm{B}T)^4 = -V\sigma\left(\frac{1}{3}\right)T^4 \,. \end{aligned} \tag{11.394}$$

But $F = U - TS$ and $U = 3PV$. Therefore the entropy, S, is

$$S = 4\left(\frac{PV}{T}\right) = V\left[\frac{32\pi^5 (k_\mathrm{B}T)^3}{45(hc)^3}\right]k_\mathrm{B} = V\sigma\left(\frac{4}{3}\right)T^3 \,. \tag{11.395}$$

Finally, the enthalpy H is

$$H = G + TS = 0 + TS = V\sigma\left(\frac{4}{3}\right)T^4 \,. \tag{11.396}$$

11.57 Phonons: In a Continuum

At finite temperature, atoms vibrate in the vicinity of stable mechanical equilibrium. These vibrations emit sound waves. At sufficiently low temperature, sound waves have low frequencies and long wave-lengths. If the wave-length is long compared to the lattice spacing, the lattice may be approximated as a continuum. Such continuum would have (continuous) vibrational modes. These modes—much like those of the electromagnetic field that gives rise to photons—would give rise to elementary excitations that are called phonons. While one might imagine the frequencies of such elementary excitations to extend all the way from zero to infinity, in fact, because the lattice spacing is not zero, the phonon wave-length and the phonon frequency must have finite limits. Such limits would depend on the interatomic spacing of the solid—or equivalently, on the number of atoms per unit volume.[66] But, as shown below, at low enough temperatures, where the continuum approximation is valid, the actual magnitude, ν_{max}, of the maximum value of the phonon frequency has little effect on the physics of the phonon system.

A simple model of phonons assumes they are noninteracting bosons whose number—much like that of the photons, which were studied in the preceding section—depends on the system volume and the temperature. Therefore—as was also the case for photons—the number of phonons is not conserved. Another important assumption about the phonon gas is that—again, like the photon gas—it is not subject to quantum condensation. Indeed, it is clear that all the preceding calculations for the photon gas can readily be transplanted to apply to the phonon gas. Therefore, in the following we shall look for the key that helps transplant one set of results into the other.

For low frequencies and long wavelengths, phonons have—also see the succeeding subsection—the following dispersion relation:

$$\varepsilon_p = \nu h = \hbar\omega = \left(\frac{c_0 k}{2\pi}\right) h = c_0 \left(\frac{h}{\lambda}\right) = c_0 p\,. \tag{11.397}$$

Here ν is the frequency, ω the angular frequency, and k is the magnitude of the wave-vector. The wave-length is λ, and c_0 is the velocity of sound—assumed here to be independent of the temperature as well as the direction. Also, as in quantum mechanics, it is convenient to define a momentum p, that is, $p = \hbar k = \frac{h}{\lambda}$.

In (11.393) that refers to the photon gas, ε_p was equivalent to cp where c is the velocity of light in vacuum. Therefore, for transforming the photon results into those for phonons, we need to replace the velocity of light, wherever it occurs, by the velocity of sound, that is, $c \to c_0$.

Additionally, the photons have two mutually perpendicular polarizations orthogonal to their direction of motion. In equations for the internal energy and the pressure, that fact requires the choice $\Delta = 2$. In contrast, phonons have three modes, one longitudinal and two transverse. Thus Δ has to be equal to 3 here.

Finally, while the integration limit for the photon frequency is correctly set at ∞, for phonons the limit should necessarily be finite. Still it is expected that the limit would be high enough that the additional contribution to the relevant integrals would be very small. (This is ensured by the integrand decreasing exponentially with increase in the momentum.)

---

[66] An estimate of the maximum value of the phonon frequency can be obtained from (11.417) to (11.422).

With these changes, the thermodynamics of phonons that are produced by a continuum of atoms, can be obtained directly from that of photons. The two trivial numerical changes needed are $c \rightarrow c_0$ and $\Delta = 2 \rightarrow \Delta = 3$. As a result, for instance, we multiply the right-hand side of (11.393) by $\frac{3}{2}$ and change c to c_0. This gives

$$\begin{aligned} U &= -\left(\frac{3}{2}\right)\left(\frac{2V}{h^3}\right)\int_0^{p_{\max}}(c_0 p)\left[-\exp(\beta c_0 p)+1\right]^{-1}4\pi p^2 \mathrm{d}p \\ &= -V\left(\frac{3}{2}\right)\left[\frac{8\pi(k_B T)^4}{(hc_0)^3}\right]\int_0^{x_{\max}}\left[-\exp(x)+1\right]^{-1}x^3\mathrm{d}x \\ &\approx V\left(\frac{3}{2}\right)\left[\frac{8\pi(k_B T)^4}{(hc_0)^3}\right]\int_0^{\infty}\frac{x^3\mathrm{d}x}{\exp(x)-1} \\ &= V\left[\frac{12\pi(k_B T)^4}{(hc_0)^3}\right]\left(\frac{\pi^4}{15}\right), \end{aligned} \tag{11.398}$$

where

$$x_{\max} = \beta h\nu_{\max} = \beta\hbar\omega_{\max} = \beta c_0 p_{\max}\,. \tag{11.399}$$

At low to moderate temperatures, $x_{\max}$ should be $\gg 1$.

The specific heat of phonons in the continuum approximation is

$$\begin{aligned} (c_V)_{\text{continuum}} &= \left(\frac{\partial U}{\partial T}\right)_V \approx V\left[\frac{12\pi k_B{}^4}{(hc_0)^3}\right]\left(\frac{\pi^4}{15}\right)4T^3 \\ &= \frac{12\pi^4 N k_B}{5}\left(\frac{T}{\Theta_{\text{continuum}}}\right)^3, \end{aligned} \tag{11.400}$$

$$\text{where}\quad \Theta_{\text{continuum}} = \left(\frac{hc_0}{k_B}\right)\left(\frac{3N}{4\pi V}\right)^{\frac{1}{3}}. \tag{11.401}$$

Exercise 11.8 By following the procedure used for the calculation of the internal energy U, or otherwise, calculate all the remaining thermodynamic potentials for phonons in the continuum approximation.

11.57.1 Phonons in Lattices

Consider N classical atoms. Denote their three-dimensional Cartesian position vectors as $\sum_{\mathrm{i}=1}^{\mathrm{N}}(x_{\mathrm{i},1}, x_{\mathrm{i},2}, x_{\mathrm{i},3})$. Assume that at low temperature these atoms form a solid. As an effect of the temperature, their positions undergo small vibrations in the vicinity of stable mechanical equilibrium. Let the equilibrium value of the position vectors be $\sum_{\mathrm{i}=1}^{\mathrm{N}}(\overline{x}_{\mathrm{i},1}, \overline{x}_{\mathrm{i},2}, \overline{x}_{\mathrm{i},3})$. Denote the displacement from the equilibrium of the ith atom in the α direction—note, $\alpha = 1, 2$, or, 3—as $q_{\mathrm{i},\alpha}$. In other words, set

$$q_{\mathrm{i},\alpha} = x_{\mathrm{i},\alpha} - \overline{x}_{\mathrm{i},\alpha}\,. \tag{11.402}$$

The kinetic energy of the ith atom, of mass $m_{\rm i}$, is $\frac{1}{2}\sum_{\alpha=1}^{3} m_{\rm i}(\dot{x_{\rm i,\alpha}})^2$. Because $(\dot{x_{\rm i,\alpha}}) = (\dot{q_{\rm i,\alpha}})$, the total kinetic energy, E, of the N atoms,

$$E = \frac{1}{2}\sum_{\rm i=1}^{\rm N}\sum_{\alpha=1}^{3} m_{\rm i}(\dot{x_{\rm i,\alpha}})^2\,, \tag{11.403}$$

can be written as

$$E = \frac{1}{2}\sum_{\rm i=1}^{\rm N}\sum_{\alpha=1}^{3} m_{\rm i}(\dot{q_{\rm i,\alpha}})^2\,. \tag{11.404}$$

Let us denote the potential energy of the system of (N atoms) as

$$V = V\sum_{\rm i=1}^{\rm N}\sum_{\alpha=1}^{3} x_{\rm i,\alpha}\,. \tag{11.405}$$

Because displacements from the equilibrium are small, V can be expanded in Taylor series:

$$\begin{aligned} V = V_0 &+ \sum_{\rm i=1}^{\rm N}\sum_{\alpha=1}^{3}\left(\frac{\partial V}{\partial x_{\rm i,\alpha}}\right)_{\rm x_{i,\alpha}=\overline{x}_{i,\alpha}} \cdot q_{\rm i,\alpha} \\ &+ \frac{1}{2}\sum_{\rm i,j=1}^{\rm N}\sum_{\alpha,\eta=1}^{3}\left(\frac{\partial^2 V}{\partial x_{\rm i,\alpha}\partial x_{\rm j,\eta}}\right)_{(\rm x_{i,\alpha}=\overline{x}_{i,\alpha};x_{j,\eta}=\overline{x}_{j,\eta})} \cdot q_{\rm i,\alpha}q_{\rm j,\eta} + \cdots\,. \end{aligned} \tag{11.406}$$

Here V_0 is the potential energy when the system is in equilibrium at rest. And because the potential energy is at its minimum when $x_{\rm i,\alpha} = \overline{x}_{\rm i,\alpha}$, therefore, its first derivative $(\frac{\partial V}{\partial x_{\rm i,\alpha}})_{\rm x_{i,\alpha}=\overline{x}_{i,\alpha}}$ is vanishing. Accordingly, to the leading—meaning, to the second—order in the small variables $q_{\rm i,\alpha}$, $q_{\rm j,\eta}$, etc., the system Hamiltonian $\mathcal{H}$ can be written as

$$\begin{aligned} \mathcal{H} &= E + V \\ &\approx V_0 + \sum_{\rm i,j=1}^{\rm N}\sum_{\alpha,\eta=1}^{3}\gamma_{\rm i,\alpha;j,\eta}\cdot(q_{\rm i,\alpha}q_{\rm j,\eta}) + \frac{1}{2}\sum_{\rm i=1}^{\rm N}\sum_{\alpha=1}^{3} m_{\rm i}(\dot{q_{\rm i,\alpha}})^2\,, \end{aligned} \tag{11.407}$$

where we have used the notation

$$\gamma_{\rm i,\alpha;j,\eta} = \frac{1}{2}\left(\frac{\partial^2 V}{\partial x_{\rm i,\alpha}\partial x_{\rm j,\eta}}\right)_{(\rm x_{i,\alpha}=\overline{x}_{i,\alpha};x_{j,\eta}=\overline{x}_{j,\eta})} = \gamma_{\rm j,\eta;i,\alpha}\,. \tag{11.408}$$

Clearly, the given Hamiltonian, $\mathcal{H}$, is an homogeneous, symmetric quadratic form in $3N$ variables like $q_{\rm i,\alpha}$, etc. As shown in (H.1)–(H.12), a change of variable can always be devised to reduce an homogeneous, symmetric quadratic form into a sum of squares—in this case, $3N$ different squares. Indeed, the reduction to a sum of squares may be done in many different ways. See, for example, [33]. An appropriate change of variables here would be the so-called normal coordinates ξ_i where $i = 1, 2, 3, \ldots, 3N$, whereby the Hamiltonian given in (11.407) would get transformed into

$$\mathcal{H} = V_0 + \sum_{i=1}^{3N} \frac{1}{2} m_i \left[(\dot{\xi}_i)^2 + \omega_i^2 \xi_i^2 \right] = V_0 + \sum_{i=1}^{3N} \mathcal{H}_i \,. \tag{11.409}$$

Here the ω_i—all $3N$ of them—are the[67] "characteristic angular frequencies" of the so-called normal modes of the system.

The above Hamiltonian looks similar to that in a problem studied earlier: namely, one that related to a collection of N, quasiclassical, distinguishable, mutually noninteracting, one-dimensional simple-harmonic oscillators—compare (11.72)–(11.77) where one sets $a \equiv b = 0$. There is, however, a fundamental difference between the distinguishable oscillators studied earlier and the system of indistinguishable phonons that obey Bose statistics. Regarding the earlier case, we note a statement by Pathria: these oscillators are assumed to be "distinguishable" because they are merely a (classical) representation of the (quantum) distinguishable energy levels available in the system. Simple harmonic oscillators themselves are not particles, or even so-called quasiparticles such as photons or phonons that have to be treated as being indistinguishable.

The Hamiltonian $\mathcal{H}_i$—where $i = 1, 2, 3, \ldots, 3N$—has (quantum) eigenvalues of the form:[68]

$$\begin{aligned} \langle n_l | \mathcal{H}_i | n_j \rangle &= \hbar\omega_i \left(n_l + \frac{1}{2} \right) \delta_{n_l, n_j} \,, \\ \langle n_l | \exp(-\beta \mathcal{H}_i) | n_j \rangle &= \exp\left\{ -\beta\hbar\omega_i \left(n_l + \frac{1}{2} \right) \right\} \delta_{n_l, n_j} \,, \\ \text{where} \qquad n_l &= 0, 1, 2, \ldots, \infty \,. \end{aligned} \tag{11.410}$$

The subscript i ranges over the normal modes, of which there are $3N$. Also, each of these normal modes has an integral number—ranging between 0 and ∞—of phonons. Thus, while the number of phonon energy levels is fixed at $3N$, the number of phonons is not conserved.[69] And similarly to (11.126)–(11.129), for all statistical analyses, the convenient partition function to use is the canonical partition function. Following the procedure for calculating canonical partition function, Ξ, for quasi-classical quantum systems—see (11.126)—we can write the partition function as

$$\begin{aligned} \Xi &= \mathrm{Tr}\left[\exp\left(-\beta \mathcal{H} \right) \right] \\ &= \exp(-\beta V_0) \prod_{i=1}^{3N} \left[\sum_{n_l=0}^{\infty} \langle n_l | \exp\left(-\beta \mathcal{H}_i \right) | n_l \rangle \right] \end{aligned}$$

[67]Clearly, these frequencies depend on the system potential energy, which itself—in view of what has been assumed above—would depend on the $\frac{N(N-1)}{2}$ two-body interatomic potentials.

[68]Compare (11.160)–(11.161).

[69]Rather, the total number of phonons depends upon the system temperature as well as its volume, etc.

$$= \exp(-\beta E_0) \prod_{i=1}^{3N} \left[\sum_{n_l=0}^{\infty} \exp(-\beta \hbar \omega_i n_l) \right]$$

$$= \exp(-\beta E_0) \prod_{i=1}^{3N} \left[1 - \exp(-\beta \hbar \omega_i) \right]^{-1} , \qquad (11.411)$$

where

$$E_0 = V_0 + \frac{\hbar \sum_{j=1}^{3N} \omega_j}{2} \qquad (11.412)$$

is the energy of the system at zero temperature. Notice that it includes both the sum of the zero-point energies of the $3N$ normal modes and the total interparticle potential V_0. Because E_0 represents the binding energy of the lattice (at zero temperature), it is necessarily negative.

Once the partition function is known, various thermodynamic potentials can be calculated.[70] To this purpose, we need first the logarithm of the partition function,

$$\ln\{\Xi\} = -\beta E_0 - \sum_{i=1}^{3N} \ln\left[1 - \exp(-\beta \hbar \omega_i)\right] . \qquad (11.413)$$

The system internal energy, U, and the specific heat, $c_V(T)$, are readily found as

$$U = -\left(\frac{\partial \ln\{\Xi\}}{\partial \beta} \right)_V = E_0 + \sum_{i=1}^{3N} \frac{\hbar \omega_i}{\exp(\beta \hbar \omega_i) - 1} , \qquad (11.414)$$

$$c_V(T) = \left(\frac{\partial U}{\partial T} \right)_V = k_B \sum_{i=1}^{3N} \frac{(\frac{\hbar \omega_i}{k_B T})^2 \exp(\beta \hbar \omega_i)}{[\exp(\beta \hbar \omega_i) - 1]^2} . \qquad (11.415)$$

11.57.2 Phonons: Einstein Approximation

Einstein, who was the first to apply quantum-mechanical description to the specific heat of solids, suggested an approximation whereby the angular frequencies of the $3N$ normal modes are treated as though they are all equal, say $= \omega_E$. Then (11.415) gives

$$\left[c_V(T)\right]_{\text{Einstein}} = 3Nk_B \frac{(\frac{\hbar \omega_E}{k_B T})^2 \exp(\beta \hbar \omega_E)}{[\exp(\beta \hbar \omega_E) - 1]^2} . \qquad (11.416)$$

At high temperature, where $\frac{\hbar \omega_E}{k_B T} \ll 1$, (11.416) yields the well-known classical result of Dulong and Petit, namely $c_V(T) \approx 3Nk_B$.

The result is unsatisfactory at very low temperatures. Unlike the experiment, which suggests a T^3 behavior, the Einstein approximation leads to a very rapid, exponential fall off with decreasing temperature.

[70] Compare (11.129) and the description that followed.

11.57.3 Phonons: Debye Approximation

While the high temperature limit is trivial, in order to actually calculate the thermodynamics of the system at intermediate temperatures, knowledge of the normal modes spectrum is needed. This requires either theoretically solving an often time laborious, quantum-mechanical problem with interparticle coupling or physically doing the relevant experiments.

The angular frequencies, ω_i, of the $3N$ normal modes are generally closely spaced. Therefore, it is often convenient to define a frequency spectrum, $g(\omega)$, where $g(\omega)\mathrm{d}\omega$ is equal to the number of normal modes whose angular frequencies lie between ω and $\omega + \mathrm{d}\omega$. In other words, rather than working directly with different normal modes, we work with their frequency spectrum

$$\int_0^{\omega_D} g(\omega)\mathrm{d}\omega = 3N\,. \tag{11.417}$$

The upper limit of the angular frequency, ω_D, is so chosen that the total number of normal modes is equal to the actual number $3N$.

As is implicit in the definition of the partition function, the area of quantum statistical phase space per normal mode is equal to h. Phonon modes occur in triplicate, one longitudinal and two transverse. Therefore, the number of normal modes in a given volume, $V\mathrm{d}p_x\mathrm{d}p_y\mathrm{d}p_z$, of the phase space is the following:

$$\begin{aligned} g(\omega)\mathrm{d}\omega &= 3\left(\frac{V\mathrm{d}p_x\mathrm{d}p_y\mathrm{d}p_z}{h^3}\right) = 3\left(\frac{V}{h^3}\right)4\pi p^2\mathrm{d}p \\ &\rightarrow \frac{V}{2\pi^2}\left(\frac{3}{c_0^3}\right)\omega^2\mathrm{d}\omega\,. \end{aligned} \tag{11.418}$$

In the last term[71] much like (11.397), the magnitude of the momentum has been expressed as being equal to the ratio of $\hbar\omega$ and the velocity of sound,

$$p = \frac{\hbar\omega}{c_0}\,. \tag{11.419}$$

Furthermore, in (11.418) and (11.419) the velocity of sound, c_0, has been assumed to be isotropic so that it is the same for the longitudinal as well as the transverse modes. In practise, this is not a good approximation. The two velocities—c_L, for the longitudinal, and c_T, for the transverse modes—are generally different. This means that in (11.418) the factor, $\frac{3}{c_0^3}$, that represents all three modes having the same velocity c_0, should be replaced by the corresponding expression $(\frac{1}{c_L^3}) + (\frac{2}{c_T^3})$.

[71]Using the fact that in a three-dimensional lattice with N atoms the total number of normal modes is equal to $3N$, from (11.418) we can determine the value of ν_{max} that is relevant to the case where the lattice is treated as a continuum—see the reference to ν_{max} in (11.399). In other words, the requirement that the integral $\int_0^{\omega_{max}} g(\omega)\mathrm{d}\omega$ be equal to $3N$ gives $\omega_{max} = c_0(\frac{6\pi^2 N}{V})^{\frac{1}{3}}$. Note that $\omega_{max} = 2\pi\nu_{max}$.

The latter expression properly sets the velocity c_L for the longitudinal mode and c_T for the two transverse modes.[72] In order to do this, (11.418) must be reexpressed in the proper Debye form:

$$g(\omega)\mathrm{d}\omega = \frac{V}{2\pi^2}\left[\left(\frac{1}{c_L^3}\right)+\left(\frac{2}{c_T^3}\right)\right]\omega^2\mathrm{d}\omega\,. \tag{11.420}$$

Integrating both sides of (11.420)—over the physically allowed limits $\omega = 0$ and $\omega = \omega_D$—gives

$$3N = \int_0^{\omega_D} g(\omega)\mathrm{d}\omega = \frac{V}{6\pi^2}\left[\left(\frac{1}{c_L^3}\right)+\left(\frac{2}{c_T^3}\right)\right]\omega_D^3\,, \tag{11.421}$$

which determines ω_D as

$$\omega_D^3 = 18\pi^2\left(\frac{N}{V}\right)\left[\left(\frac{1}{c_L^3}\right)+\left(\frac{2}{c_T^3}\right)\right]^{-1}. \tag{11.422}$$

Eliminating the factor $(\frac{1}{c_L^3}) + (\frac{2}{c_T^3})$ from (11.422) and (11.420) leads to the second, alternate expression for the Debye frequency spectrum. For $0 \le \omega \le \omega_D$, the two equivalent expressions for $g(\omega)\mathrm{d}\omega$ are

$$g(\omega)D\omega = \frac{V}{2\pi^2}\left[\left(\frac{1}{c_L^3}\right)+\left(\frac{2}{c_T^3}\right)\right]\omega^2\mathrm{d}\omega = \left(\frac{9N}{{\omega_D}^3}\right)\omega^2\mathrm{d}\omega\,. \tag{11.423}$$

Either form of (11.423) may be used for calculating the various thermodynamic potentials for the Debye system of phonons.

Let us begin with (11.414) for the internal energy. Because $3N$, the number of normal modes, is large and the modes are generally closely spaced, the discrete sum $\sum_{i=1}^{3N}$ in (11.413)–(11.415), may, with reasonable approximation, be replaced by an integral. Furthermore, we shall make the assumption that the upper limit for the integral is large enough that it can be approximated as being of order ∞. Then

$$\int_0^{x_D}\frac{x^3}{\exp(x)-1}\mathrm{d}x \approx \int_0^{\infty}\frac{x^3}{\exp(x)-1}\mathrm{d}x$$

and

$$\begin{aligned} U &= E_0 + \int_0^{\omega_D}\left[\frac{\hbar\omega}{\exp(\beta\hbar\omega)-1}\right]g(\omega)\mathrm{d}\omega \\ &= E_0 + \left(\frac{V}{2\pi^2}\right)\left[\left(\frac{1}{c_L^3}\right)+2\left(\frac{1}{c_T^3}\right)\right]\int_0^{\omega_D}\left[\frac{\hbar\omega^3}{\exp(\beta\hbar\omega)-1}\right]\mathrm{d}\omega \\ &= E_0 + \left(\frac{V}{2\pi^2}\right)\left[\left(\frac{1}{c_L^3}\right)+2\left(\frac{1}{c_T^3}\right)\right]\left(\frac{(k_BT)^4}{\hbar^3}\right)\int_0^{x_D}\left[\frac{x^3}{\exp(x)-1}\right]\mathrm{d}x \end{aligned}$$

[72] Only if $c_l = c_T = c_0$ is the last expressions the same as $\frac{3}{c_0^3}$.

$$\approx E_0 + \left(\frac{4V\pi^5 k_B^4}{15h^3}\right)\left[\left(\frac{1}{c_L^3}\right) + 2\left(\frac{1}{c_T^3}\right)\right] T^4$$

$$= E_0 + \left(\frac{24Nk_B^4\pi^7}{5h^3}\right)\frac{T^4}{(\omega_D)^3} = E_0 + \frac{3Nk_B\pi^4}{5}\frac{T^4}{(\Theta_D)^3}, \quad (11.424)$$

where

$$x_D = \beta\hbar\omega_D = \frac{\Theta_D}{T},$$

$$\int_0^{x_D} \frac{x^3}{\exp(x)-1}\mathrm{d}x \approx \int_0^{\infty} \frac{x^3}{\exp(x)-1}\mathrm{d}x = \frac{\pi^4}{15}. \quad (11.425)$$

Therefore we can write

$$\frac{1}{T}\left(\frac{\partial U}{\partial T}\right)_V = \frac{c_V(T)}{T} = \left(\frac{16V\pi^5 k_B^4}{15h^3}\right)\left[\left(\frac{1}{c_L^3}\right) + 2\left(\frac{1}{c_T^3}\right)\right] T^2$$

$$= \left(\frac{96Nk_B^4\pi^7}{5h^3}\right)\left(\frac{1}{\omega_D}\right)^3 T^2$$

$$= \frac{12Nk_B\pi^4}{5}\left(\frac{1}{\Theta_D}\right)^3 T^2. \quad (11.426)$$

If the speeds c_L and c_T were the same, say equal to c_0, then Θ_D, in (11.424)–(11.426), would be equal to $\Theta_{continuum}$ defined in (11.401). Consequently, for isotropic velocity, the above result for $c_V(T)$ is identical to the $(c_V)_{continuum}$ that was obtained for a lattice approximated as a continuum—see (11.400).

While in many cases the Debye theory does fairly well in representing the temperature dependence of the specific heat over a wide range of temperatures, at very low temperature—which in practise means T being of order $0.02\Theta_D$ or lower—the phonon specific heat is indeed observed—for example, see Fig. 11.19—to closely follow the T^3 law.[73] In fact, the low temperature measurements are often used to estimate the magnitude of Θ_D.

An alternate route to estimating Θ_D is through first calculating ω_D—as in (11.422)—and next relating it to Θ_D—as per (11.425). If the parameters c_L, c_T, and the ratio $\frac{N}{V}$ are available—or, are measured—then (11.422) determines ω_D. Next, (11.425)—that is, the relationship $\Theta_D = \frac{\hbar\omega_D}{k_B}$—readily leads to Θ_D.

The fact that the two different routes to estimating Θ_D lead to essentially the same result—see Table 11.2—provides convincing physical support for the Debye ideas.

[73]This is certainly the case in nonmetallic solids. As shown in (11.295), in metals free electrons contribute specific heat that is proportional to the first power of the temperature. Clearly, at very low temperatures, T^1 wins over T^3.

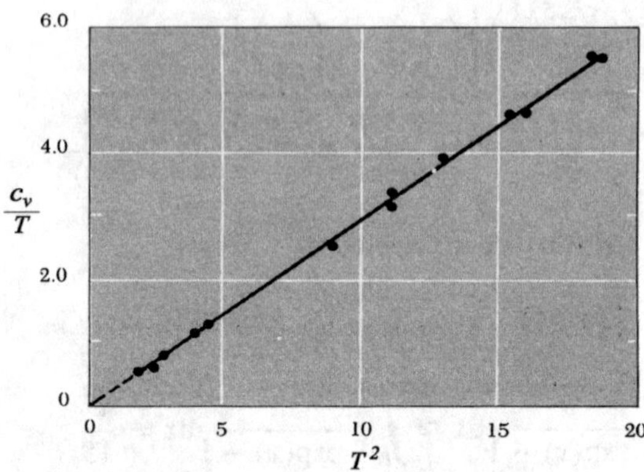

Fig. 11.19 As reported in [34], the plot presents the ratio $\frac{C_V}{T}$ for KCl as a function of T^2. Measurements are from Keesom, P.H., and Pearlman, N. Notice that there is no intercept. This indicates that at very low temperature, the specific heat varies as T^3 rather than as T. (The drawing is copied with permission from [35, Fig. 10.2.3])

Table 11.2 The Debye temperature for a number of crystals as quoted in [23, p. 198]

Crystal	Θ_D from specific heat	Θ_D from ω_D
Pb	88	73
Ag	215	214
Zn	308	305
Cu	345	332
Al	398	402
C	$\sim$ 1850	–
NaCl	308	320
KCl	233	240
MgO	$\sim$ 850	$\sim$ 950

11.57.4 Phonons: Other Thermodynamic Potentials

In order to calculate other thermodynamic functions of interest, begin with (11.413) which represents the logarithm of the relevant partition function Ξ. Much like what was done to get to (11.424), approximate the sum by an integral, meaning employ the approximation

$$\sum_{i=1}^{3N} f(\omega_i) \approx \int_0^{\omega_D} f(\omega) g(\omega) d\omega \,. \tag{11.427}$$

Next, use the expression for $g(\omega)d\omega$ given in (11.420). In this manner, write

$$\ln\{\Xi\} = -\beta E_0 - \int_0^{\omega_D} \ln[1 - \exp(-\beta\hbar\omega)] \frac{V}{2\pi^2}\left[\left(\frac{1}{c_L^3}\right) + \left(\frac{2}{c_T^3}\right)\right]\omega^2 d\omega$$

$$= -\beta E_0 - \int_0^{\omega_D} \ln[1 - \exp(-x)] \frac{V}{2\pi^2} \left[\left(\frac{1}{c_L^3} \right) + \left(\frac{2}{c_T^3} \right) \right] \left(\frac{1}{\beta \hbar} \right)^3 x^2 \mathrm{d}x$$

$$= -\beta E_0 + \frac{V}{2\pi^2} \left[\left(\frac{1}{c_L^3} \right) + \left(\frac{2}{c_T^3} \right) \right] \left(\frac{k_B T}{\hbar} \right)^3 \left(\frac{\pi^4}{45} \right). \tag{11.428}$$

As in (11.424) the approximation implicit in the derivation of the final result in (11.428) is the assumption that the upper limit, ω_D, in the integral may be replaced by ∞, meaning

$$\int_0^{\omega_D} \ln[1 - \exp(-x)] x^2 \mathrm{d}x \approx \int_0^{\infty} \ln[1 - \exp(-x)] x^2 \mathrm{d}x$$

$$= - \int_0^{\infty} \frac{x^3}{3} \left(\frac{1}{\exp(x) - 1} \right) \mathrm{d}x = -\frac{1}{3} \left(\frac{\pi^4}{15} \right). \tag{11.429}$$

Knowing $\ln\{\Xi\}$—see (11.428)—thermodynamic potentials are readily determined as

$$\begin{aligned}
F &= -k_B T \ln\{\Xi\} = E_0 - V \left[\left(\frac{1}{c_L^3} \right) + \left(\frac{2}{c_T^3} \right) \right] \left(\frac{4\pi^5 k_B^4}{45 h^3} \right) T^4, \\
P &= -\left(\frac{\partial F}{\partial V} \right)_T = \left[\left(\frac{1}{c_L^3} \right) + \left(\frac{2}{c_T^3} \right) \right] \left(\frac{4\pi^5 k_B^4}{45 h^3} \right) T^4, \\
S &= -\left(\frac{\partial F}{\partial T} \right)_V = V \left[\left(\frac{1}{c_L^3} \right) + \left(\frac{2}{c_T^3} \right) \right] \left(\frac{16 k_B^4 \pi^5}{45 h^3} \right) T^3, \\
G &= F + PV = E_0, \\
H &= U + PV = V \left[\left(\frac{1}{c_L^3} \right) + \left(\frac{2}{c_T^3} \right) \right] \left(\frac{16\pi^5 k_B^4}{45 h^3} \right) T^4,
\end{aligned} \tag{11.430}$$

where F is the Helmholtz potential, P is the pressure, S is the entropy, G is the Gibbs Potential, and H is the enthalpy.

References

1. M. Abramowitz, I. Stegun, *Handbook of Mathematical Functions* (2002)
2. R.B. Paris, D. Kaminsky, *Asymptotics and the Mellin–Barnes Integrals* (Cambridge University Press, New York, 2001)
3. E.T. Whittaker, G.N. Watson, *A Course in Modern Analysis*, 4th edn. (Cambridge University Press, New York, 1996)
4. J. Stirling. Asymptotic expansion. Am. Math. Mon. **107**(6)
5. M. Abramowitz, I. Stegun (2002)
6. J.B. Marion, S.T. Thornton, *Classical Dynamics of Particles and Systems*, 4th edn. (Saunders College Publishing/Harcourt Brace, Philadelphia, 1995), p. 34, equation (1.100)
7. F.W. Sears, G.L. Salinger, Thermodynamics, Kinetic Theory, and Statistical Thermodynamics, p. 378
8. R.K. Pathria, *Statistical Mechanics* (Pergamon Press, Oxford, 1977), pp. 255–278

9. J.E. Mayer, M.G. Mayer, *Statistical Mechanics* (John Wiley, New York, 1940)
10. F.W. Sears, G.L. Salinger, op. cit., p. 379
11. F. London, Z. Phys. **60**, 245 (1930)
12. F. London, Z. Phys. Chem. B **11**, 222 (1930)
13. C. Cohen-Tannoudji, B. Diu, F. Laloë, *Quantum Mechanics* (Wiley-Interscience, New York, 1977)
14. E. Merzbacher, *Quantum Mechanics*, 2nd edn. (Wiley, New York, 1970), pp. 178–190
15. D.M. Dennison, Proc. R. Soc. Lond. A **115**, 483 (1927)
16. R.K. Pathria, op. cit., p. 167
17. W. Nernst, *The New Heat Theorem* (Dover, New York, 1926). 1969
18. K. Denbigh, *The Principles of Chemical Equilibrium* (Cambridge University Press, Cambridge, 1971)
19. F.W. Sears, G.L. Salinger, p. 197
20. K. Denbigh, *Principles of Chemical, Equilibrium* (Cambridge University Press, Cambridge, 1969)
21. M. Goldstein, I.F. Goldstein, *Refrigerator and Universe* (Harvard University Press, Cambridge, 1993)
22. E.M. Purcell, R.V. Pound, Phys. Rev. **81**, 279 (1951)
23. R.K. Pathria, *Statistical Mechanics* (Pergamon Press, Oxford, 1980)
24. M.L. Boas (John Wiley, 1966)
25. E. Cornell, K. Wieman, W. Ketterle, Nobel Lectures (2001)
26. J.D. Clinton, H.G. Lester, Nature **119**, 558 (1927)
27. R.K. Pathria, op. cit., p. 183, equation (7.1.38)
28. L.M. Roberts, Proc. Phys. Soc. **70B**, 744 (1957)
29. F.W. Sears, G.L. Salinger, op. cit., p. 227
30. S.N. Bose, Z. Phys. **26**, 178 (1924)
31. A. Einstein, Berl. Ber. **22**, 261 (1924)
32. A. Einstein, Berl. Ber. **1**, 3 (1925)
33. H. Jeffrey, B.S. Jeffrey, op. cit.
34. Phys. Rev. **91**, 1354 (1953)
35. F. Reif, *Fundamentals of Statistical and Thermal Physics* (Waveland Press, Long Grove, 2009), p. 417

Second-Order Phase Transitions 12

Conduction electrons in metals are nearly free. That is largely due to the fact that the positive ion-core in a metal can be approximated by smearing out so that it can be represented by a uniform background charge. Then in a mean-field type of approximation the ion-core interaction with the electrons almost cancels out the electron–electron interaction. This cancelation leaves the metal effectively to have nearly free, weakly-interacting, electrons. In Sect. 12.1 we describe the Landau theory. It originated with his desire to formulate a theory that is both analytic and obeys the symmetry of the Hamiltonian. Section 12.1.1 deals with Ginzburg's work who followed Landau and recommended additions should be made to the Landau free energy. He suggested, the free energy Φ_T, rather than being given by (12.3), should also include additional terms.

12.1 Landau Theory

Lanau asserted that in the vicinity of a second-order phase transition the free energy, Φ, of the system should be expressed as a truncated Taylor expansion of a complex order-parameter, φ, which is expected to be small. That is,

$$\Phi = g_0 + \alpha|\varphi|^2 + \beta|\varphi|^4 + \cdots . \tag{12.1}$$

The parameters g_0, α, and β are to be temperature-dependent. Note also that the parameter g_0 represents the ground state free energy. Because the ground state energy does not affect the physics of the second-order phase transition, therefore in analyzing the phase transition it is traditional to ignore g_0. To that purpose, it is often set equal to zero. That is,

$$g_0 = 0 . \tag{12.2}$$

Because the order parameter is expected to be small, except for the second- and the fourth-power of the order parameter, additional terms in the expansion may be ignored. In other words, at temperature T close to the critical temperature T_c, Landau

R. Tahir-Kheli, *General and Statistical Thermodynamics*,
https://doi.org/10.1007/978-3-030-20700-7_12

worked with the relation

$$\Phi_{\mathrm{T}} = \alpha_{\mathrm{T}}|\varphi|_{\mathrm{T}}^{2} + \beta_{\mathrm{T}}|\varphi|_{\mathrm{T}}^{4}\,. \tag{12.3}$$

The Landau second-order phase transition occurs abruptly at a critical temperature T_{c}. Therefore, for thermodynamic stability, the free energy Φ at $T = T_{\mathrm{c}}$ is a minimum. Accordingly, its first derivative with respect to the order parameter $|\varphi|_{\mathrm{T=T_c}}$ is zero. That is,

$$\begin{aligned}\left(\frac{\partial \Phi}{\partial|\varphi|}\right)_{\mathrm{T=T_c}} &= 2\alpha_{\mathrm{T=T_c}}|\varphi|_{\mathrm{T=T_c}} + 4\beta_{\mathrm{T=T_c}}|\varphi|_{\mathrm{T=T_c}}^{3} \\ &= 2|\varphi|_{\mathrm{T=T_c}}\left[\alpha_{\mathrm{T=T_c}} + 2\beta_{\mathrm{T=T_c}}|\varphi|_{\mathrm{T=T_c}}^{2}\right] = 0\,.\end{aligned} \tag{12.4}$$

Equation (12.4) is solved by setting

$$2|\varphi|_{\mathrm{T=T_c}} = 0\,. \tag{12.5}$$

Here we have a conundrum because (12.4) and (12.5) do not teach us anything about the parameter α_{T}. Therefore, to proceed further, we must require that for stability the minimum of the free energy must also occur at general temperatures. In order to investigate that issue, it is convenient to rewrite (12.4) for general temperatures as

$$\begin{aligned}\left(\frac{\partial \Phi}{\partial|\varphi|}\right)_{\mathrm{T}} &= 2\alpha_{\mathrm{T}}|\varphi|_{\mathrm{T}} + 4\beta_{\mathrm{T}}|\varphi|_{\mathrm{T}}^{3} \\ &= 2|\varphi|_{\mathrm{T}}\left[\alpha_{\mathrm{T}} + 2\beta_{\mathrm{T}}|\varphi|_{\mathrm{T}}^{2}\right] = 0\,.\end{aligned} \tag{12.6}$$

Landau resolved the above stated conundrum by arbitrarily making the following choice that is to apply to temperatures T equal to or immediately below the critical temperature T_{c}:

$$\alpha_{\mathrm{T}} = -c_0(T_{\mathrm{c}} - T)\,. \tag{12.7}$$

Here c_0 is a temperature-independent constant that presumably can be measured by an appropriate experiment.

Note that, at general temperatures, the order parameter, $|\varphi|_{\mathrm{T}}$, is not zero. As such we proceed by inserting (12.7) into (12.6) and, as a result, for temperatures T immediately below the critical temperature T_{c}, we must have

$$\alpha_{\mathrm{T}} + 2\beta_{\mathrm{T}}|\varphi|_{\mathrm{T}}^{2} = 0\,,$$

or equivalently,

$$|\varphi|_{\mathrm{T}}^{2} = \frac{c_0(T_{\mathrm{c}} - T)}{2\beta_{\mathrm{T}}}\,, \tag{12.8}$$

with the result

$$|\varphi|_{\mathrm{T}} = \sqrt{\frac{c_0(T_{\mathrm{c}} - T)}{2\beta_{T}}}\,. \tag{12.9}$$

For temperatures above T_c, the system is in normal state, meaning its order parameter is zero. Below T_c, $|\varphi|_T^2$ represents the fraction of the total number of electrons that form the superconducting state. Clearly, above $T = T_c$ this fraction is vanishing. Equation (12.8) indicates that at temperatures below T_c, the fraction of the total number of electrons that form the superconducting state increases linearly with decrease in temperature.

12.1.1 Ginzburg Contribution

As mentioned above, according to Ginzburg, additions should be made to Landau free energy. The free energy, Φ_T, rather than being given by (12.3), should also include additional terms. That is,

$$\Phi_T = \alpha_T|\varphi|_T^2 + \beta_T|\varphi|_T^4 + \left(\frac{1}{2m_0}\right)|(-I\hbar\nabla - 2e_0 M_T)\varphi|_T^2 + \frac{|F|_T^2}{2\mu_0}. \tag{12.10}$$

Here m_0 is the effective mass, e_0 is the effective charge of an electron, M is the magnetic vector potential, and $F = \nabla \times M$ is the magnetic field. As before, it is necessary to minimize Φ_T with respect to the order parameter φ_T. As such its first derivative with respect to φ_T must be set equal to zero. This process leads to what are termed Landau–Ginzburg equations:

$$0 = \left(\frac{\partial \Phi}{\partial |\varphi|}\right)_T = 2\alpha_T\varphi_T + 4\beta_T\varphi_T|\varphi_T|^2 + \left(\frac{1}{m_0}\right)(-I\hbar\nabla - 2e_0 M_T)\varphi_T\,, \tag{12.11}$$

$$\nabla \times F = \mu_0 j\,, \quad j = \left(\frac{2e}{m}\right) Re\{\varphi^*(-I\hbar\nabla - 2eM)\varphi\}. \tag{12.12}$$

In (12.11), j denotes the dissipation-less electric current density, Re represents the real part of whatever follows it, and I stands for $\sqrt{-1}$. The order parameter $|\varphi_T|$ is determined by (12.11). With that information in hand, (12.12) can be used to determine j as well as the penetration depth that was first calculated by the London brothers.[1]

References

1. Ginzburg–Landau Theory, Wikipedia

[1]For details, see [1]. Compare with (15.24).

Cooper Pair

13

Section 13.1 covers the Fermi sea. Sections 13.1.1 and 13.1.2 describe the system Hamiltonians, both classical and quantum. Results in center of mass coordinates are expressed in Sect. 13.1.3. Schrödinger equation and its solution are presented in Sects. 13.2 and 13.3. Cooper pair with no center-of-mass motion and its binding energy are studied in Sects. 13.3.1 and 13.4.

13.1 Fermi Sea

Despite the fact that conduction electrons in a metal are free to move, they do interact with each other via Coulomb forces. However, if the positive ion-core in the metal can be approximated by smearing out so that the system is represented by a uniform background charge, then in a mean-field type of approximation the ion core interaction with the electrons cancels out the electron–electron interaction. This cancelation leaves the metal effectively to have a system of noninteracting electrons.

Consider a large number of such electrons, each of mass m, at absolute zero. Being fermions, electrons follow the Pauli exclusion principle. Accordingly, the collection of electrons fill-up the lowest available energy state in an energy ordered fashion. When this process is complete, these electrons have built-up a so-called Fermi sea, that is, a band of energies starting at zero and rising to the highest energy level equal to the Fermi energy E_{F}. If we know the volume υ, energy ϵ, and the number of wave-vector q-states per unit volume of real space, or equivalently, the number of energy states $N(\epsilon)$, we can determine the relationship between the Fermi energy, E_{F}, and the total number, $N(E_{\mathrm{F}})$, of electrons in the Fermi sea. That is,[1]

[1]For help with the derivation of (13.1) see [1]. In particular, read the description of their equations (3.164) and (9.123). Note that the indicated Patterson–Bailey results have been multiplied by 2 to take account of the two possible spin-states per electron.

R. Tahir-Kheli, *General and Statistical Thermodynamics*,
https://doi.org/10.1007/978-3-030-20700-7_13

$$\frac{\mathrm{d}^3 q}{(2\pi)^3} \equiv N(\epsilon)\mathrm{d}\epsilon\,, \qquad \frac{\mathrm{d}N(\epsilon)}{\mathrm{d}\epsilon}\mathrm{d}\epsilon = 2\frac{m^{\frac{3}{2}}\sqrt{\epsilon}}{\hbar^3\pi^2\sqrt{2}}\mathrm{d}\epsilon\,,$$

$$\int_0^{E_\mathrm{F}} \frac{\mathrm{d}N(\epsilon)}{\mathrm{d}\epsilon}\mathrm{d}\epsilon = N(E_\mathrm{F}) = 2\int_0^{E_\mathrm{F}} \frac{m^{\frac{3}{2}}\sqrt{\epsilon}}{\hbar^3\pi^2\sqrt{2}}\mathrm{d}\epsilon = \frac{(2mE_\mathrm{f})^{\frac{3}{2}}}{3\hbar^3\pi^2}\,, \tag{13.1}$$

$$E_\mathrm{F} = \frac{\hbar^2}{2m}\left[3\pi^2 N(E_\mathrm{F})\right]^{\frac{2}{3}}.$$

The Fermi energy, E_F, is the kinetic energy of the fastest moving electron whose velocity v_F is equal to

$$v_\mathrm{F} = \sqrt{\frac{2E_\mathrm{F}}{m}} = \frac{\hbar}{m}\left[3\pi^2 N(E_\mathrm{F})\right]^{\frac{1}{3}}. \tag{13.2}$$

13.1.1 Classical Hamiltonian

Consider a large collection of electrons, each of mass m. Position a pair of these well above the Fermi sea. The Hamiltonian representing the pair consists of the sum of the kinetic energies of the two electrons and an interaction term that depends on the physical separation between the two electrons in the pair. In classical representation, the Hamiltonian H has the form

$$H = \frac{m}{2}\left[(\dot{x}_1)^2 + (\dot{y}_1)^2 + (\dot{z}_1)^2\right] + \frac{m}{2}\left[(\dot{x}_2)^2 + (\dot{y}_2)^2 + (\dot{z}_2)^2\right] + V\left(|\boldsymbol{r}_1 - \boldsymbol{r}_2|\right) \tag{13.3}$$

$$= \frac{m}{2}\dot{\boldsymbol{r}}_1 \cdot \dot{\boldsymbol{r}}_1 + \frac{m}{2}\dot{\boldsymbol{r}}_2 \cdot \dot{\boldsymbol{r}}_2 + V\left(\left|\boldsymbol{i}(x_1 - x_2) + \boldsymbol{j}(y_1 - y_2) + \boldsymbol{k}(z_1 - z_2)\right|\right). \tag{13.4}$$

The usual notation is used here whereby $\boldsymbol{i}$, $\boldsymbol{j}$, and $\boldsymbol{k}$ are unit vectors along positive direction of the x, y, and z axes. Additionally, for electrons labeled 1 and 2, the variables $\boldsymbol{r}_1 = \boldsymbol{i}x_1 + \boldsymbol{j}y_1 + \boldsymbol{k}z_1$ and $\boldsymbol{r}_2 = \boldsymbol{i}x_2 + \boldsymbol{j}y_2 + \boldsymbol{k}z_2$ are their position vectors; $\dot{\boldsymbol{r}}_1$ and $\dot{\boldsymbol{r}}_2$ their velocities; $m\dot{\boldsymbol{r}}_1$ and $m\dot{\boldsymbol{r}}_2$ their momenta, and $V(|\boldsymbol{r}_1 - \boldsymbol{r}_2|)$ is the interaction between the pair. Rather than depending upon the interparticle interaction the center of mass motion would depend only on outside forces that may be exerted on the pair. There being no such force present, the center of mass motion is free. In contrast, both members of the pair are affected by the interparticle interaction which is nondirectional, meaning the interaction between the two electrons is not dependent on the direction of their separation $\boldsymbol{r}_1 - \boldsymbol{r}_2$, but rather on the distance, $|\boldsymbol{r}_1 - \boldsymbol{r}_2|$, of the separation.

13.1.2 Quantum Hamiltonian

Before the Schrödinger equation can properly be constructed, the classical Hamiltonian H must be transformed into quantum Hamiltonian $\mathcal{H}$. For instance, regarding electron 1, the essence of this transformation is the conversion of the classical momentum variable $m\dot{x_1}$ into quantum momentum variable $-I\hbar\frac{d}{dx_1}$. Here h stands for the Planck constant and

$$\hbar = \frac{h}{2\pi}, \qquad m\dot{y_1} \to -I\hbar\frac{\mathrm{d}}{\mathrm{d}y_1}, \qquad m\dot{z_1} \to -I\hbar\frac{\mathrm{d}}{\mathrm{d}z_1}, \qquad \text{etc.} \tag{13.5}$$

Also

$$I = \sqrt{-1}. \tag{13.6}$$

More compactly (13.5) can be written as

$$m\dot{\boldsymbol{r}_1} \to I\hbar\left[\boldsymbol{i}\frac{\mathrm{d}}{\mathrm{d}x_1} + \boldsymbol{j}\frac{\mathrm{d}}{\mathrm{d}y_1} + \boldsymbol{k}\frac{\mathrm{d}}{\mathrm{d}z_1}\right]. \tag{13.7}$$

In view of (13.7) and the fact that $I \cdot I = -1$, $\boldsymbol{i} \cdot \boldsymbol{i} = 1$, and $\boldsymbol{i} \cdot \boldsymbol{j} = 0$, etc., we have

$$\frac{1}{2m}\left[m\dot{\boldsymbol{r}_1} \cdot m\dot{\boldsymbol{r}_1}\right] = -\frac{(\hbar)^2}{2m}\left[\frac{\mathrm{d}^2}{\mathrm{d}x_1^2} + \frac{\mathrm{d}^2}{\mathrm{d}y_1^2} + \frac{\mathrm{d}^2}{\mathrm{d}z_1^2}\right], \tag{13.8}$$

$$\frac{1}{2m}\left[m\dot{\boldsymbol{r}_2} \cdot m\dot{\boldsymbol{r}_2}\right] = -\frac{(\hbar)^2}{2m}\left[\frac{\mathrm{d}^2}{\mathrm{d}x_2^2} + \frac{\mathrm{d}^2}{\mathrm{d}y_2^2} + \frac{\mathrm{d}^2}{\mathrm{d}z_2^2}\right]. \tag{13.9}$$

These changes help transform the classical Hamiltonian H, given in (13.3) and (13.4), into the quantum Hamiltonian $\mathcal{H}$ given below in (13.10), namely

$$\begin{aligned}\mathcal{H} = &-\frac{(\hbar)^2}{2m}\left[\frac{\mathrm{d}^2}{\mathrm{d}x_1^2} + \frac{\mathrm{d}^2}{\mathrm{d}y_1^2} + \frac{\mathrm{d}^2}{\mathrm{d}z_1^2}\right] - \frac{(\hbar)^2}{2m}\left[\frac{\mathrm{d}^2}{\mathrm{d}x_2^2} + \frac{\mathrm{d}^2}{\mathrm{d}y_2^2} + \frac{\mathrm{d}^2}{\mathrm{d}z_2^2}\right] \\ &+ V\left(\left|\boldsymbol{i}(x_1 - x_2) + \boldsymbol{j}(y_1 - y_2) + \boldsymbol{k}(z_1 - z_2)\right|\right).\end{aligned} \tag{13.10}$$

13.1.3 Center-of-Mass Coordinates

To study the dynamics of the quantum Hamiltonian $\mathcal{H}$ given in (13.10), and its wave function $\Psi(\boldsymbol{r}_1, \boldsymbol{r}_2)$, it is important to recognize that we have access to what are essentially two sets of coordinates and their associated variables. For instance, there are the Cartesian coordinates for the two electrons in the pair, namely x_1, y_1, z_1 and x_2, y_2, z_2. In addition, there are the center-of-mass coordinates, $\boldsymbol{R}, X, Y, Z$, for the pair.

Therefore it is helpful to have, in addition to the usual and previously used notation, also some other notation that could be needed later:

$$\begin{aligned}\boldsymbol{R} &= \left(\frac{\boldsymbol{r}_1 + \boldsymbol{r}_2}{2}\right) \\ &= \boldsymbol{i}\left(\frac{x_1 + x_2}{2}\right) + \boldsymbol{j}\left(\frac{y_1 + y_2}{2}\right) + \boldsymbol{k}\left(\frac{z_1 + z_2}{2}\right), \\ \boldsymbol{R} &= \boldsymbol{i}X + \boldsymbol{j}Y + \boldsymbol{k}Z, \\ |\boldsymbol{R}| &= R,\end{aligned} \tag{13.11}$$

$$\begin{aligned}
\boldsymbol{r_0} &= \boldsymbol{r_1} - \boldsymbol{r_2} \\
&= \boldsymbol{i}(x_1 - x_2) + \boldsymbol{j}(y_1 - y_2) + \boldsymbol{k}(z_1 - z_2) \\
&= \boldsymbol{i}(x_0) + \boldsymbol{j}(y_0) + \boldsymbol{k}(z_0)\,, \\
|\boldsymbol{r_0}| &= r_0\,, \\
\boldsymbol{r} &= \boldsymbol{i}x + \boldsymbol{j}y + \boldsymbol{k}z\,, \\
|\boldsymbol{r}| &= r\,, \\
\boldsymbol{D} &= \boldsymbol{i}\frac{\mathrm{d}}{\mathrm{d}x} + \boldsymbol{j}\frac{\mathrm{d}}{\mathrm{d}y} + \boldsymbol{k}\frac{\mathrm{d}}{\mathrm{d}z}\,, \\
\boldsymbol{D}\cdot\boldsymbol{D} = D^2 &= \frac{\mathrm{d}^2}{\mathrm{d}^2x} + \frac{\mathrm{d}^2}{\mathrm{d}^2y} + \frac{\mathrm{d}^2}{\mathrm{d}^2z}\,, \\
D_{\mathrm{R}}^2 &= \frac{\mathrm{d}^2}{\mathrm{d}^2X} + \frac{\mathrm{d}^2}{\mathrm{d}^2Y} + \frac{\mathrm{d}^2}{\mathrm{d}^2Z}\,, \\
\boldsymbol{k} &= \boldsymbol{i}k_{\mathrm{x}} + \boldsymbol{j}k_{\mathrm{y}} + \boldsymbol{k}k_{\mathrm{z}}\,, \\
k^2 &= \boldsymbol{k}\cdot\boldsymbol{k} = k_{\mathrm{x}}^2 + k_{\mathrm{y}}^2 + k_{\mathrm{z}}^2\,, \\
\boldsymbol{k}.\boldsymbol{r} &= xk_{\mathrm{x}} + yk_{\mathrm{y}} + zk_{\mathrm{z}}\,,
\end{aligned} \tag{13.12}$$

$$\begin{aligned}
\int\{\ldots\}d^3\boldsymbol{r} &\equiv \left(\frac{1}{2L}\right)^3 \int_{-\mathrm{L}}^{\mathrm{L}}\mathrm{d}x\int_{-\mathrm{L}}^{\mathrm{L}}\mathrm{d}y\int_{-\mathrm{L}}^{\mathrm{L}}\mathrm{d}z\{\ldots\}\,, \\
\varphi(\boldsymbol{k}) &= \int\{\varphi(\boldsymbol{r})\exp(I\boldsymbol{k}.\boldsymbol{r})\}\mathrm{d}^3\boldsymbol{r}\,, \\
V(\boldsymbol{r}) &= \left(\frac{1}{2\pi}\right)^3 \int_{-\pi}^{\pi}\int_{-\pi}^{\pi}\int_{-\pi}^{\pi} V(\boldsymbol{w})\exp(I\boldsymbol{w}\cdot\boldsymbol{r})\mathrm{d}^3w\,.
\end{aligned} \tag{13.13}$$

13.2 Schrödinger Equation

Because the Hamiltonian is already available, in order to set up Schrödinger equation for the Cooper pair, one needs an eigenfunction. To that end, any appropriate eigenfunction, say $\Psi(\boldsymbol{r}_1, \boldsymbol{r}_2)$, should be represented also as a possible function of R—the position coordinate relating to the center of mass—as well as the interparticle separation r_0. That is,

$$\Psi(\boldsymbol{r}_1, \boldsymbol{r}_2) \rightarrow \Phi(\boldsymbol{r_0}, \boldsymbol{R})\,. \tag{13.14}$$

Additionally, the following information is needed. According to (13.12) and (13.11), we have

$$\begin{aligned}
x_0 &= x_1 - x_2, & X &= \frac{x_1 + x_2}{2}\,, \\
y_0 &= y_1 - y_2\,, & Y &= \frac{y_1 + y_2}{2}\,, \\
z_0 &= z_1 - z_2\,, & Z &= \frac{z_1 + z_2}{2}\,,
\end{aligned} \tag{13.15}$$

or equivalently,

$$\begin{aligned} x_1 &= \frac{x_0}{2} + X\,, \qquad & x_2 &= -\frac{x_0}{2} + X\,, \\ y_1 &= \frac{y_0}{2} + Y\,, \qquad & y_2 &= -\frac{y_0}{2} + Y\,, \\ z_1 &= \frac{z_0}{2} + Z\,, \qquad & z_2 &= -\frac{z_0}{2} + Z\,. \end{aligned} \tag{13.16}$$

Our next task is to begin with one of the Cartesian axes—say, the x-axis—and start using the chain rule for derivatives. According to (13.16), the x-coordinates of both electrons in the Cooper pair depend on the variables x_0 and X. That is,

$$\begin{aligned} x_1 &\equiv x_1(x_0, X)\,, \\ x_2 &\equiv x_2(x_0, X)\,. \end{aligned} \tag{13.17}$$

Therefore by using (13.15), (13.16), and (13.17), one can write

$$\frac{\mathrm{d}}{\mathrm{d}x_1} = \left(\frac{\partial x_0}{\partial x_1}\right)\frac{\mathrm{d}}{\mathrm{d}x_0} + \left(\frac{\partial X}{\partial x_1}\right)\frac{\mathrm{d}}{\mathrm{d}X} = \frac{\mathrm{d}}{\mathrm{d}x_0} + \left(\frac{1}{2}\right)\frac{\mathrm{d}}{\mathrm{d}X}\,. \tag{13.18}$$

Similarly, for the derivative with respect to x_2, we have

$$\frac{\mathrm{d}}{\mathrm{d}x_2} = \left(\frac{\partial x_0}{\partial x_2}\right)\frac{\mathrm{d}}{\mathrm{d}x_0} + \left(\frac{\partial X}{\partial x_2}\right)\frac{\mathrm{d}}{\mathrm{d}X} = -\frac{\mathrm{d}}{\mathrm{d}x_0} + \left(\frac{1}{2}\right)\frac{\mathrm{d}}{\mathrm{d}X}\,. \tag{13.19}$$

Next we calculate the second derivative $\frac{\mathrm{d}^2}{\mathrm{d}x_1{}^2}$ as

$$\begin{aligned} \frac{\mathrm{d}^2}{\mathrm{d}x_1{}^2} &= \frac{\mathrm{d}}{\mathrm{d}x_1}\cdot\frac{\mathrm{d}}{\mathrm{d}x_1} = \left[\frac{\mathrm{d}}{\mathrm{d}x_0} + \left(\frac{1}{2}\right)\frac{\mathrm{d}}{\mathrm{d}X}\right]\cdot\left[\frac{\mathrm{d}}{\mathrm{d}x_0} + \left(\frac{1}{2}\right)\frac{\mathrm{d}}{\mathrm{d}X}\right] \\ &= \frac{\mathrm{d}^2}{\mathrm{d}^2x_0} + \frac{\mathrm{d}}{\mathrm{d}x_0}\cdot\frac{\mathrm{d}}{\mathrm{d}X} + \left(\frac{1}{4}\right)\frac{\mathrm{d}^2}{\mathrm{d}X^2}\,. \end{aligned} \tag{13.20}$$

Similarly,

$$\begin{aligned} \frac{\mathrm{d}^2}{\mathrm{d}x_2{}^2} &= \frac{\mathrm{d}}{\mathrm{d}x_2}\cdot\frac{\mathrm{d}}{\mathrm{d}x_2} = \left[-\frac{\mathrm{d}}{\mathrm{d}x_0} + \left(\frac{1}{2}\right)\frac{\mathrm{d}}{\mathrm{d}X}\right]\cdot\left[-\frac{\mathrm{d}}{\mathrm{d}x_0} + \left(\frac{1}{2}\right)\frac{\mathrm{d}}{\mathrm{d}X}\right] \\ &= \frac{\mathrm{d}^2}{\mathrm{d}^2x_0} - \frac{\mathrm{d}}{\mathrm{d}x_0}\cdot\frac{\mathrm{d}}{\mathrm{d}X} + \left(\frac{1}{4}\right)\frac{\mathrm{d}^2}{\mathrm{d}X^2}\,. \end{aligned} \tag{13.21}$$

Summing (13.20) and (13.21) yields

$$\frac{\mathrm{d}^2}{\mathrm{d}x_1{}^2} + \frac{\mathrm{d}^2}{\mathrm{d}x_2{}^2} = 2\frac{\mathrm{d}^2}{\mathrm{d}^2x_0} + \left(\frac{1}{2}\right)\frac{\mathrm{d}^2}{\mathrm{d}X^2}\,. \tag{13.22}$$

Similarly, for the y- and the z-axes, we would get

$$\frac{\mathrm{d}^2}{\mathrm{d}y_1{}^2} + \frac{\mathrm{d}^2}{\mathrm{d}y_2{}^2} = 2\frac{\mathrm{d}^2}{\mathrm{d}^2y_0} + \left(\frac{1}{2}\right)\frac{\mathrm{d}^2}{\mathrm{d}Y^2}\,, \tag{13.23}$$

$$\frac{d^2}{dz_1{}^2} + \frac{d^2}{dz_2{}^2} = 2\frac{d^2}{d^2z_0} + \left(\frac{1}{2}\right)\frac{d^2}{dZ^2}. \tag{13.24}$$

Adding (13.22), (13.23), and (13.24) gives

$$\left\{\left(\frac{d^2}{dx_1{}^2} + \frac{d^2}{dx_1{}^2} + \frac{d^2}{dx_1{}^2}\right) + \left(\frac{d^2}{dx_2{}^2} + \frac{d^2}{dy_2{}^2} + \frac{d^2}{dz_2{}^2}\right)\right\}$$
$$= \left\{2\left(\frac{d^2}{d^2x_0} + \frac{d^2}{d^2y_0} + \frac{d^2}{d^2z_0}\right) + \left(\frac{1}{2}\right)\left(\frac{d^2}{dX^2} + \frac{d^2}{dY^2} + \frac{d^2}{dZ^2}\right)\right\}. \tag{13.25}$$

Rearrange slightly the quantum Hamiltonian given in (13.10) to get

$$\mathcal{H} = -\frac{(\hbar)^2}{2m}\left\{\left(\frac{d^2}{d^2x_1}+\frac{d^2}{d^2y_1}+\frac{d^2}{d^2z_1}\right)+\left(\frac{d^2}{d^2x_2}+\frac{d^2}{d^2y_2}+\frac{d^2}{d^2z_2}\right)\right\}+V(r_0). \tag{13.26}$$

Interchange relationships in (13.25) with the relevant ones in (13.26). This changes the quantum Hamiltonian $\mathcal{H}$ in (13.26) to that in (13.27), namely

$$\mathcal{H} = -\frac{(\hbar)^2}{2m}\left\{2\left(\frac{d^2}{d^2x_0} + \frac{d^2}{d^2y_0} + \frac{d^2}{d^2z_0}\right) + \left(\frac{1}{2}\right)\left(\frac{d^2}{\partial X^2} + \frac{d^2}{dY^2} + \frac{d^2}{dZ^2}\right)\right\}$$
$$+ V(r_0). \tag{13.27}$$

Most noteworthy feature of $\mathcal{H}$ in (13.27) is it has two distinct parts, one that depends only on the Cartesian coordinates x_0, y_0, and z_0, which are the components of the vector variable $\boldsymbol{r_0}$, and the other that depends on X, Y, and Z, which are components of the vector variable $\boldsymbol{R}$. Therefore we can expect the eigenfunction to display the same features, meaning the eigenfunction should be a function of the form $\Phi(\boldsymbol{r_0}, \boldsymbol{R})$.

Now multiply both the left- and right-hand sides of (13.27) from the right by the wave-function $\Phi(\boldsymbol{r_0}, \boldsymbol{R})$, work with the notation D^2 and D_{R}^2 given in (13.11) and (13.12), use the symbol E_{cpair} for the eigenvalue, and write the resultant (13.27) in the form of a Schrödinger equation:

$$\mathcal{H}\Phi(\boldsymbol{r_0}, \boldsymbol{R}) = -\frac{(\hbar)^2}{2m}\left\{2\left(\frac{d^2}{d^2x_0} + \frac{d^2}{d^2y_0} + \frac{d^2}{d^2z_0}\right) + \left(\frac{1}{2}\right)D_R^2\right\}\Phi(\boldsymbol{r_0}, \boldsymbol{R})$$
$$+ V(r_0)\Phi(\boldsymbol{r_0}, \boldsymbol{R}) = E_{\mathrm{cpair}}\Phi(\boldsymbol{r_0}, \boldsymbol{R}). \tag{13.28}$$

The variable E_{cpair}, the eigenvalue of the Hamiltonian $\mathcal{H}$, represents the energy of the Cooper pair. In addition, because $\boldsymbol{r_0}$ and $\boldsymbol{R}$ are linearly independent, we can write

$$\Phi(\boldsymbol{r_0}, \boldsymbol{R}) \equiv \varphi(\boldsymbol{r_0}) \cdot \phi(\boldsymbol{R}). \tag{13.29}$$

Now use the expression $\Phi(\boldsymbol{r_0}, \boldsymbol{R})$ given in (13.29). For sake of convenience, change the notation from r_0 to r and also do some slight reorganization. Thereby, as shown below, (13.28) may be transformed to the following form:

$$
\begin{aligned}
\mathcal{H}\Phi(\boldsymbol{r_0}, \boldsymbol{R}) &= -\frac{(\hbar)^2}{2m}\left\{2\left(\frac{d^2}{d^2x_0}+\frac{d^2}{d^2y_0}+\frac{d^2}{d^2z_0}\right)+\left(\frac{1}{2}\right)D_R^2\right\}\Phi(\boldsymbol{r_0}, \boldsymbol{R}) \\
&\quad + V(r_0)\Phi(\boldsymbol{r_0}, \boldsymbol{R}) = E_{\text{cpair}}\Phi(\boldsymbol{r_0}, \boldsymbol{R}) \rightarrow \\
\mathcal{H}\phi(\boldsymbol{R})\cdot\varphi(\boldsymbol{r}) &= \left\{-\left(\frac{\hbar^2}{m}\right)D^2\right\}\phi(\boldsymbol{R})\cdot\varphi(\boldsymbol{r}) + \left\{-\left(\frac{\hbar^2}{4m}\right)D_{\text{R}}^2\right\}\phi(\boldsymbol{R})\cdot\varphi(\boldsymbol{r}) \\
&\quad + V(r)\phi(\boldsymbol{R})\cdot\varphi(\boldsymbol{r}) = E_{\text{cpair}}\phi(\boldsymbol{R})\cdot\varphi(\boldsymbol{r})\,.
\end{aligned}
\tag{13.30}
$$

As indicated in (13.30), the operator D_{R}^2 does not operate on $\varphi(r)$. Therefore, $\varphi(r)$ can be transferred across this operator to the left-hand side. Similarly, D^2 cannot operate on $\phi(\boldsymbol{R})$. Therefore $\phi(\boldsymbol{R})$ can be transferred across it to the left-hand side. After doing these transfers, (13.30) becomes

$$
\begin{aligned}
\mathcal{H}\phi(\boldsymbol{R})\cdot\varphi(\boldsymbol{r}) &= \phi(\boldsymbol{R})\left\{-\left(\frac{\hbar^2}{m}\right)D^2\right\}\varphi(\boldsymbol{r}) + \varphi(\boldsymbol{r})\left\{-\left(\frac{\hbar^2}{4m}\right)D_{\text{R}}^2\right\}\phi(\boldsymbol{R}) \\
&\quad + V(r)\phi(\boldsymbol{R})\cdot\varphi(\boldsymbol{r}) = E_{\text{cpair}}\phi(\boldsymbol{R})\cdot\varphi(\boldsymbol{r})\,.
\end{aligned}
\tag{13.31}
$$

Let us deal first with the component in (13.31) that depends only on the center-of-mass vector $\boldsymbol{R}$. That is,

$$
\varphi(\boldsymbol{r})\left\{-\left(\frac{\hbar^2}{4m}\right)D_{\text{R}}^2\right\}\phi(\boldsymbol{R})\,. \tag{13.32}
$$

According to (13.11), $\boldsymbol{R}$ in (13.32) is expressed as

$$
\boldsymbol{R} = |\boldsymbol{R}| = |\boldsymbol{i}X + \boldsymbol{j}Y + \boldsymbol{k}Z|\,.
$$

In (13.32), the presence of derivatives with respect to X, Y, and Z, in the given simple format suggests we can try representing $\phi(\boldsymbol{R})$ in (13.31) as

$$
\phi(\boldsymbol{R}) = \phi(0)\exp\big[I(\boldsymbol{k_x}\cdot\boldsymbol{X} + \boldsymbol{k_y}\cdot\boldsymbol{Y} + \boldsymbol{k_z}\cdot\boldsymbol{Z})\big]\,. \tag{13.33}
$$

Inserting the $\phi(\boldsymbol{R})$ given in (13.33) into (13.32) leads to

$$
\begin{aligned}
&\varphi(\boldsymbol{r})\left\{-\left(\frac{\hbar^2}{4m}\right)D_{\text{R}}^2\right\}\phi(\boldsymbol{R}) \\
&= \varphi(\boldsymbol{r})\left\{-\left(\frac{\hbar^2}{4m}\right)D_{\text{R}}^2\right\}\phi(0)\exp\big[I(\boldsymbol{k_x}\cdot\boldsymbol{X} + \boldsymbol{k_y}\cdot\boldsymbol{Y} + \boldsymbol{k_z}\cdot\boldsymbol{Z})\big] \\
&= \varphi(\boldsymbol{r})\left(\frac{\hbar^2}{4m}\right)(k_{\text{x}}^2 + k_{\text{y}}^2 + k_{\text{z}}^2)\phi(\boldsymbol{R}) = \varphi(\boldsymbol{r})\left(\frac{\hbar^2 k^2}{4m}\right)\phi(\boldsymbol{R})\,.
\end{aligned}
\tag{13.34}
$$

In (13.34) use was made of the notation introduced in (13.11)–(13.13). Note that on the right-hand side of (13.34) the quantity $\frac{\hbar^2 k^2}{4m}$ represents the kinetic energy of the center-of-mass. The center-of-mass kinetic energy is zero in two limits. First, the trivial one that is obtained if both electrons of the pair are stationary, and the second when the two electrons are moving in opposite directions with the same speed. The wave-function $\varphi(\boldsymbol{r})$ is spatially symmetric if $\varphi(\boldsymbol{r}) = \varphi(-\boldsymbol{r})$ and spatially antisymmetric if $\varphi(\boldsymbol{r}) = -\varphi(-\boldsymbol{r})$. However, in order to satisfy the demands of the Pauli principle, the total wave-function has to be antisymmetric, requiring the spins of the pair to form either a singlet or a triplet state.

Our next task is to make use of the result derived in (13.34) whereby we can replace $\varphi(\boldsymbol{r})\{-(\frac{\hbar^2}{4m})D_{\mathrm{R}}^2\}\phi(\boldsymbol{R})$ by $\varphi(\boldsymbol{r})(\frac{\hbar^2 k^2}{4m})\phi(\boldsymbol{R})$. We make that replacement in (13.31) and rewrite it as

$$\begin{aligned}\mathcal{H}\phi(\boldsymbol{R}) \cdot \varphi(\boldsymbol{r}) &= \phi(\boldsymbol{R})\left\{\left(-\frac{\hbar^2}{m}\right)D^2 + V(\boldsymbol{r})\right\}\varphi(\boldsymbol{r}) \\ &= \phi(\boldsymbol{R})\left(E_{\mathrm{cpair}} - \frac{\hbar^2 k^2}{4m}\right)\varphi(\boldsymbol{r}), \end{aligned} \tag{13.35}$$

where E_{cpair} is the energy of the Cooper pair. The Hamiltonian $\mathcal{H}$ in (13.35) no longer depends on $\boldsymbol{R}$. Therefore $\phi(\boldsymbol{R})$ can safely be extracted out of both sides of[2] (13.35) leading to (13.36), namely

$$\mathcal{H} \cdot \varphi(\boldsymbol{r}) = \left\{\left(-\frac{\hbar^2}{m}\right)D^2 + V(\boldsymbol{r})\right\}\varphi(\boldsymbol{r}) = \left(E_{\mathrm{cpair}} - \frac{\hbar^2 k^2}{4m}\right)\varphi(\boldsymbol{r}). \tag{13.36}$$

While attempting to proceed further, one notices that there is no need to retain the term on the left-hand side of (13.36) as long as its equivalent on the right-hand side is fully retained. Indeed, the two terms on the right-hand side of (13.36), expressed as (13.37), contain the entirety of the Schrödinger differential equation,

$$\left\{\left(-\frac{\hbar^2}{m}\right)D^2 + V(\boldsymbol{r})\right\}\varphi(\boldsymbol{r}) = \left(E_{\mathrm{cpair}} - \frac{\hbar^2 k^2}{4m}\right)\varphi(\boldsymbol{r}). \tag{13.37}$$

13.3 Solution of Schrödinger Equation

In order conveniently to solve the Schrödinger equation (13.37), it is helpful to work with the Fourier transform of $\varphi(\boldsymbol{r})$ rather than directly with $\varphi(\boldsymbol{r})$ itself and rewrite (13.37) as

$$\begin{aligned}&\left\{\left(-\frac{\hbar^2}{m}\right)D^2\right\}\int\{\varphi(\boldsymbol{r})\exp(I\boldsymbol{k}\cdot\boldsymbol{r})\}D^3\boldsymbol{r} + \int\{V(\boldsymbol{r})\varphi(\boldsymbol{r})\exp(I\boldsymbol{k}\cdot\boldsymbol{r})\}D^3\boldsymbol{r} \\ &= \left(E_{\mathrm{cpair}} - \frac{\hbar^2 k^2}{4m}\right)\int\{\varphi(\boldsymbol{r})\exp(I\boldsymbol{k}\cdot\boldsymbol{r})\}D^3\boldsymbol{r}. \end{aligned} \tag{13.38}$$

[2]This process is equivalent to dividing both sides of (13.35) by $\phi(\boldsymbol{R})$.

Now carry out the double-differentiation implicit in D^2, as indicated on the extreme left-hand side of (13.38), and use (13.13) for treating two of the integrals that appear in this differential equation. These integrals are $\{(-\frac{\hbar^2}{m})\cdot D^2\}\int\{\varphi(\boldsymbol{r})\exp(I\boldsymbol{k}\cdot\boldsymbol{r})\}D^3\boldsymbol{r}$ and $(E_{\text{cpair}} - \frac{\hbar^2k^2}{4m})\int\{\varphi(\boldsymbol{r})\exp(I\boldsymbol{k}\cdot\boldsymbol{r})\}D^3\boldsymbol{r}$. When that is done, (13.38) becomes

$$\left(\frac{\hbar^2k^2}{m}\right)\varphi(\boldsymbol{k}) + \int\left\{V(\boldsymbol{r})\varphi(\boldsymbol{r})\exp(I\boldsymbol{k}\cdot\boldsymbol{r})\right\}D^3\boldsymbol{r} = \left(E_{\text{cpair}} - \frac{\hbar^2k^2}{4m}\right)\varphi(\boldsymbol{k})\,. \quad (13.39)$$

At the extreme left-hand side of (13.39), the parameter $\frac{\hbar^2k^2}{m}$ refers to the sum of the kinetic energies of two independent electrons. Henceforth we shall display this parameter as 2ϵ where ϵ is the kinetic energy of a single free electron at zero temperature above the Fermi surface. That is,

$$\frac{\hbar^2k^2}{m} = 2\left(\frac{\hbar^2k^2}{2m}\right) \equiv 2\epsilon\,,$$

and (13.39) becomes

$$2\epsilon\varphi(\boldsymbol{k}) + \int\left\{V(\boldsymbol{r})\varphi(\boldsymbol{r})\exp(I\boldsymbol{k}\cdot\boldsymbol{r})\right\}D^3\boldsymbol{r} = \left(E_{\text{cpair}} - \frac{\hbar^2k^2}{4m}\right)\varphi(\boldsymbol{k})\,. \quad (13.40)$$

13.3.1 No Center-of-Mass Motion

Assume the two electrons in the Cooper pair are moving with equal, but opposite momentum. As a result, the center-of-mass kinetic energy $\frac{\hbar^2k^2}{4m}$ is vanishing. Accordingly for such "no center-of-mass motion" Cooper pair the differential (13.39) reduces to

$$2\epsilon\,\varphi(\boldsymbol{k}) + \int V(\boldsymbol{r}) \times \left\{\varphi(\boldsymbol{r})\exp(I\boldsymbol{k}\cdot\boldsymbol{r})\right\}D^3\boldsymbol{r} = E_{\text{cpair}}\varphi(\boldsymbol{k})\,, \quad (13.41)$$

where E_{cpair}, as before, represents the energy of the Cooper pair.

Transfer $2\epsilon\,\varphi(\boldsymbol{k})$ to the right-hand side, lift the expression for $V(\boldsymbol{r})$ given at the end of (13.13), and insert it into (13.41). We get

$$\begin{aligned}&\int\left(\frac{1}{2\pi}\right)^3\int_{-\pi}^{\pi}\int_{-\pi}^{\pi}\int_{-\pi}^{\pi} V(\boldsymbol{w})\exp(I\boldsymbol{w}\cdot\boldsymbol{r})d^3w \times \left\{\varphi(\boldsymbol{r})\exp(I\boldsymbol{k}.\boldsymbol{r})\right\}D^3\boldsymbol{r}\\ &= \left(\frac{1}{2\pi}\right)^3\int_{-\pi}^{\pi}\int_{-\pi}^{\pi}\int_{-\pi}^{\pi} V(\boldsymbol{w})d^3w\int\left\{\varphi(\boldsymbol{r})\exp\left[I(\boldsymbol{k}+\boldsymbol{w})\cdot\boldsymbol{r}\right]\right\}D^3\boldsymbol{r}\\ &= (E_{\text{cpair}} - 2\epsilon)\varphi(\boldsymbol{k})\,. \end{aligned} \quad (13.42)$$

As in (13.13), the integral $\int\{\varphi(\boldsymbol{r})\exp[I(\boldsymbol{k}+\boldsymbol{w})\cdot\boldsymbol{r}]\}D^3\boldsymbol{r}$ can be expressed as $\varphi(\boldsymbol{k}+\boldsymbol{w})$. As such (13.41) and (13.42) take the form

$$\left(\frac{1}{2\pi}\right)^3\int_{-\pi}^{\pi}\int_{-\pi}^{\pi}\int_{-\pi}^{\pi} V(\boldsymbol{w})\varphi(\boldsymbol{k}+\boldsymbol{w})d^3w = (E_{\text{cpair}} - 2\epsilon)\varphi(\boldsymbol{k})\,. \quad (13.43)$$

In (13.43), it is helpful to change some symbols, and reset some notation. First, let us change the variable in the integral and use the notation $k + w = q$. That means setting $w = q - k$ and $d^3w = d^3q$. As a result, (13.43) becomes

$$\int_{-\pi}^{\pi}\int_{-\pi}^{\pi}\int_{-\pi}^{\pi} V(\boldsymbol{q}-\boldsymbol{k})\varphi(\boldsymbol{q})\frac{d^3q}{(2\pi)^3} = (E_{\text{cpair}} - 2\epsilon)\varphi(\boldsymbol{k})\,. \tag{13.44}$$

Note that the two electrons in the Cooper pair would be in a bound-state if the Cooper pair energy[3] E_{cpair} is less than the energy of two free electrons each moving with kinetic energy ϵ. Thus for a bound Cooper pair, the following inequality holds: $E_{\text{cpair}} < 2\epsilon$.

Next, let us start using the notation

$$(E_{\text{cpair}} - 2\epsilon)\varphi(\boldsymbol{k}) = \lozenge(\boldsymbol{k})\,. \tag{13.45}$$

By interchanging k and q, (13.45) would lead to

$$\varphi(\boldsymbol{q}) = \frac{\lozenge(\boldsymbol{q})}{E_{\text{cpair}} - 2\epsilon}\,. \tag{13.46}$$

Exchange the $\varphi(\boldsymbol{q})$ given in (13.46) with that in (13.44) to obtain

$$\int_{-\pi}^{\pi}\int_{-\pi}^{\pi}\int_{-\pi}^{\pi} V(\boldsymbol{q}-\boldsymbol{k})\frac{\lozenge(\boldsymbol{q})}{(E_{\text{cpair}} - 2\epsilon)}\frac{d^3q}{(2\pi)^3} = \lozenge(\boldsymbol{k})\,. \tag{13.47}$$

Finally, by using the equivalency $\frac{d^3q}{(2\pi)^3} \equiv N(\epsilon)d\epsilon$ that is recorded in (13.1), and summing the resultant energy integral from the Fermi energy, E_F, all the way up to Fermi + Debye energy, i.e., E_F to $E_F + \hbar\Omega_D$, (13.47) can be reexpressed as

$$\int_{E_F}^{F_f+\hbar\Omega_D} V(\boldsymbol{q}-\boldsymbol{k})\frac{\lozenge(\boldsymbol{q})}{(E_{\text{cpair}} - 2\epsilon)}N(\epsilon)d\epsilon = \lozenge(\boldsymbol{k})\,. \tag{13.48}$$

13.4 Binding Energy

Finally, we are able to calculate the binding energy of a Cooper pair. Assume the interparticle potential $V(r)$ is attractive, meaning it has a negative value. Further consider that the related parameter $V(\boldsymbol{q}-\boldsymbol{k})$, that appears in (13.48), can be approximated by a constant $-V_0$ as long as the Cooper pair is positioned above the Fermi sea and, what is highly likely, the kinetic energy of either of the electrons is much less than the Debye energy. Furthermore, let us make another assumption whereby $\lozenge(\boldsymbol{k})$ and $\lozenge(\boldsymbol{q})$ are independent of k and q, and can be approximated simply as $\lozenge$. This approximation clearly implies that the spatial wave-function is even and as a consequence the spins of the two electrons in the pair are antiparallel. Furthermore,

[3]Refer to the description of Cooper pair energy described just below (13.28).

as a result of this approximation, (13.48) leads to

$$\int_{E_F}^{E_F+\hbar\Omega_D} [-V_0] \frac{\lozenge}{E_{cpair} - 2\epsilon} N(\epsilon) d\epsilon = \lozenge\,, \tag{13.49}$$

or equivalently,

$$-V_0 \int_{E_F}^{E_F+\hbar\Omega_D} \frac{N(\epsilon) d\epsilon}{E_{cpair} - 2\epsilon} = 1\,. \tag{13.50}$$

The parameter, $N(\epsilon)$, in (13.50) is a slowly varying function of the electron kinetic energy, ϵ. As such, within the energy range between the Fermi energy to the Debye energy, that is, E_F to $E_F + \hbar\Omega_D$, it can be assumed to have a nearly constant value approximately equal to $N(E_F)$. And as a constant, $N(E_F)$ can then be moved out of the integral. As a result, (13.50) would become

$$1 = V_0 N(E_F) \int_{E_F}^{E_F+\hbar\Omega_D} \frac{d\epsilon}{(2\epsilon - E_{cpair})}\,. \tag{13.51}$$

After computing the integral in (13.51), we can represent the result in the form

$$\frac{2}{V_0 N(E_F)} = \ln\left[\frac{2E_F - E_{cpair} + 2\hbar\Omega_D}{2E_F - E_{cpair}}\right]. \tag{13.52}$$

The Cooper pair sits somewhat above the Fermi sea, but much lower than the Debye point, that is, $2\hbar\Omega_D \gg 2E_F - E_{cpair}$. Therefore $2\hbar\Omega_D + (2E_F - E_{cpair}) \approx 2\hbar\Omega_D$. As such (13.52) can be approximated as

$$\frac{2}{V_0 N(E_F)} \approx \ln\left(\frac{2\hbar\Omega_D}{2E_F - E_{cpair}}\right). \tag{13.53}$$

Exponentiate both sides,

$$\exp\left(\frac{2}{V_0 N(E_F)}\right) \approx \frac{2\hbar\Omega_D}{2E_F - E_{cpair}}\,, \tag{13.54}$$

and invert (13.54) to reach the following result:

$$2E_F - E_{cpair} \equiv E_B \approx 2\hbar\Omega_D \exp\left[-\left(\frac{2}{V_0 N(E_F)}\right)\right], \tag{13.55}$$

where E_B in (13.55) represents the binding energy of the Cooper pair. It is directly proportional to the Debye frequency Ω_D. Notice that even if the interparticle attraction represented by V_0 is extremely weak—meaning $V_0 N(E_F) \ll 1$—the binding energy E_B remains positive. However, its strength wanes exponentially fast as the interparticle attraction decreases.

References

1. J. Patterson, B. Bailey, *Solid State Physics*, 2nd edn. (Springer, Berlin, 2010)

Bogolyubov Representation

14

The system Hamiltonian is presented in Sect. 14.1. Quasiparticles are discussed in Sect. 14.2, followed by a remark about Bogulyubov theory.

14.1 Hamiltonian and Solution

The analysis presented in the chapter titled "Cooper Pair" demonstrates that two free electrons, with attractive interpair interaction, positioned somewhere above the Fermi sea, end up forming a bound Cooper pair. As such it is reasonable to expect that the presence of a large number of free electrons above the Fermi sea, and the attractiveness—no matter how weak—of the net interparticle interaction, would cause the formation of a large number of bound Cooper pairs, leading to the formation of quasiparticles called bogolyubons. Equally, when equilibrium is reached, it is reasonable to expect under favorable temperature conditions that the large number of quasiparticles would condense into an ordered manybody bound state of bogolyubons [1, 2].

The system energy consists of kinetic and potential energies. The latter is directly dependent on the interparticle interaction. Solving the manybody problem with the totality of interparticle interaction is extremely complicated if not well-nigh impossible. But because the attractive part of the interparticle interaction is central to inducing Cooper pairing, it must be kept intact as best as possible. Therefore, while seeking simplicity, yet maintaining accuracy, Bogolyubov analyzed only an appropriately truncated version of the attractive interaction [3]. The interaction that is used, according to Bogolyubov theories [4–6], is fully adequate for describing the physics of bogolyubons.

Bogolyubov used the Hamiltonian given below

$$\mathcal{H}_{\mathrm{bogo}} = \sum_{\mathrm{k}} \xi_{\mathrm{k}} \left(c^{\dagger}_{\mathrm{k},\uparrow} c_{\mathrm{k},\uparrow} + c^{\dagger}_{-\mathrm{k},\downarrow} c_{-\mathrm{k},\downarrow} \right) + \left(\frac{1}{N} \right) \sum_{\mathrm{k},\mathrm{q}} V_{\mathrm{k},\mathrm{q}} . c^{\dagger}_{\mathrm{k},\uparrow} c^{\dagger}_{-\mathrm{k},\downarrow} c_{-\mathrm{q},\downarrow} c_{\mathrm{q},\uparrow} \, . \tag{14.1}$$

R. Tahir-Kheli, *General and Statistical Thermodynamics*,
https://doi.org/10.1007/978-3-030-20700-7_14

The variable ξ_{k} is defined as

$$\xi_{\mathrm{k}} \equiv \epsilon_{\mathrm{k}} - E_{\mathrm{F}} , \tag{14.2}$$

where ϵ_{k} is the kinetic energy of an electron and E_{F} is the Fermi energy. Above the superconducting state, ϵ_{k} is $> E_{\mathrm{F}}$; below the superconducting state, ϵ_{k} is $< E_{\mathrm{F}}$, and exactly at the superconducting state, $\epsilon_{\mathrm{k}} = E_{\mathrm{F}}$.

Equation (14.1) uses both up- and down-spin creation and destruction operators for the electrons, namely $c^{\dagger}_{\mathrm{k},\uparrow}$, $c^{\dagger}_{\mathrm{k},\downarrow}$, $c_{\mathrm{k},\uparrow}$, and $c_{\mathrm{k},\downarrow}$.

The assumption here is that the electrons that are important have energies ξ_{k} above the Fermi sea where ξ_{k} is much less than the Debye energy $\hbar\Omega_{\mathrm{D}}$. Note that $\xi_{\mathrm{k}} \equiv \xi_{-\mathrm{k}} \equiv \xi^{\dagger}{}_{\mathrm{k}}$.

The analysis of interaction that leads to bogolyubons is helped by using the following expressions:

$$\begin{aligned} c^{\dagger}_{\mathrm{k},\uparrow} c^{\dagger}_{-\mathrm{k},\downarrow} &= \langle c^{\dagger}_{\mathrm{k},\uparrow} c^{\dagger}_{-\mathrm{k},\downarrow} \rangle + \left[c^{\dagger}_{\mathrm{k},\uparrow} c^{\dagger}_{-\mathrm{k},\downarrow} - \langle c^{\dagger}_{\mathrm{k},\uparrow} c^{\dagger}_{-\mathrm{k},\downarrow} \rangle \right] \\ &\equiv X_{\mathrm{k}} + \left[c^{\dagger}_{\mathrm{k},\uparrow} c^{\dagger}_{-\mathrm{k},\downarrow} - X_{\mathrm{k}} \right] \end{aligned} \tag{14.3}$$

and

$$\begin{aligned} c_{-\mathrm{q},\downarrow} c_{\mathrm{q},\uparrow} &= \langle c_{-\mathrm{q},\downarrow} c_{\mathrm{q},\uparrow} \rangle + \left[c_{-\mathrm{q},\downarrow} c_{\mathrm{q},\uparrow} - \langle c_{-\mathrm{q},\downarrow} c_{\mathrm{q},\uparrow} \rangle \right] \\ &\equiv X^{\dagger}_{\mathrm{q}} + \left[c_{-\mathrm{q},\downarrow} c_{\mathrm{q},\uparrow} - X^{\dagger}_{\mathrm{q}} \right] , \end{aligned} \tag{14.4}$$

where

$$X_{\mathrm{k}} = \langle c^{\dagger}_{\mathrm{k},\uparrow} c^{\dagger}_{-\mathrm{k},\downarrow} \rangle, \qquad X^{\dagger}_{\mathrm{q}} = \langle c_{-\mathrm{q},\downarrow} c_{\mathrm{q},\uparrow} \rangle . \tag{14.5}$$

The operators $c^{\dagger}_{\mathrm{k},\uparrow} c^{\dagger}_{-\mathrm{k},\downarrow}$, and $c_{-\mathrm{q},\downarrow} c_{\mathrm{q},\uparrow}$ fluctuate around their thermodynamic averages by amounts $[c^{\dagger}_{\mathrm{k},\uparrow} c^{\dagger}_{-\mathrm{k},\downarrow} - X_{\mathrm{k}}]$ and $[c_{-\mathrm{q},\downarrow} c_{\mathrm{q},\uparrow} - X^{\dagger}_{\mathrm{q}}]$. These are tiny fluctuations such that their squares can be assumed to be too small to be relevant.

With the use of expressions (14.3)–(14.5), the Hamiltonian $\mathcal{H}_{\mathrm{bogo}}$—see (14.1)—can be written as

$$\begin{aligned} \mathcal{H}_{\mathrm{bogo}} - \sum_{\mathrm{k}} \xi_{\mathrm{k}} \left(c^{\dagger}_{\mathrm{k},\uparrow} c_{\mathrm{k},\uparrow} + c^{\dagger}_{-\mathrm{k},\downarrow} c_{-\mathrm{k},\downarrow} \right) \\ = \frac{1}{N} \sum_{\mathrm{k},\mathrm{q}} V_{\mathrm{k},\mathrm{q}} \left\{ X_{\mathrm{k}} + \left[c^{\dagger}_{\mathrm{k},\uparrow} c^{\dagger}_{-\mathrm{k},\downarrow} - X_{\mathrm{k}} \right] \right\} \times \left\{ X^{\dagger}_{\mathrm{q}} + \left[c_{-\mathrm{q},\downarrow} c_{\mathrm{q},\uparrow} - X^{\dagger}_{\mathrm{q}} \right] \right\} \\ = \frac{1}{N} \sum_{\mathrm{k},\mathrm{q}} V_{\mathrm{k},\mathrm{q}} \left\{ X_{\mathrm{k}} c_{-\mathrm{q},\downarrow} c_{\mathrm{q},\uparrow} + X^{\dagger}_{\mathrm{q}} c^{\dagger}_{\mathrm{k},\uparrow} c^{\dagger}_{-\mathrm{k},\downarrow} - X_{\mathrm{k}} X^{\dagger}_{\mathrm{q}} \right\} \\ + \frac{1}{N} \sum_{\mathrm{k},\mathrm{q}} V_{\mathrm{k},\mathrm{q}} \left\{ \left[c^{\dagger}_{\mathrm{k},\uparrow} c^{\dagger}_{-\mathrm{k},\downarrow} - X_{\mathrm{k}} \right] \cdot \left[c_{-\mathrm{q},\downarrow} c_{\mathrm{q},\uparrow} - X^{\dagger}_{\mathrm{q}} \right] \right\} . \end{aligned} \tag{14.6}$$

The last line on the right-hand side of (14.6) represents a product of two fluctuations. Because both these fluctuations are expected to be small, their product will be much smaller. Consequently, Bogolyubov theory ignores this product. In other words, it sets

$$\frac{1}{N}\sum_{\mathrm{k,q}} V_{\mathrm{k,q}}\{[c^{\dagger}_{\mathrm{k},\uparrow}c^{\dagger}_{-\mathrm{k},\downarrow} - X_{\mathrm{k}}]\cdot[c_{-\mathrm{q},\downarrow}c_{\mathrm{q},\uparrow} - X^{\dagger}_{\mathrm{q}}]\} \approx 0\,. \tag{14.7}$$

Using the approximation (14.7), the Hamiltonian $\mathcal{H}_{\mathrm{bogo}}$ becomes

$$\begin{aligned}\mathcal{H}_{\mathrm{bogo}} &- \sum_{\mathrm{k}} \xi_{\mathrm{k}}\big(c^{\dagger}_{\mathrm{k},\uparrow}c_{\mathrm{k},\uparrow} + c^{\dagger}_{-\mathrm{k},\downarrow}c_{-\mathrm{k},\downarrow}\big)\\ &= \frac{1}{N}\sum_{\mathrm{k,q}} V_{\mathrm{k,q}}\{X_{\mathrm{k}}c_{-\mathrm{q},\downarrow}c_{\mathrm{q},\uparrow} + X^{\dagger}_{\mathrm{q}}c^{\dagger}_{\mathrm{k},\uparrow}c^{\dagger}_{-\mathrm{k},\downarrow} - X_{\mathrm{k}}X^{\dagger}_{\mathrm{q}}\}\,.\end{aligned} \tag{14.8}$$

Given the fact $V_{\mathrm{k,q}} = V_{\mathrm{q,k}}$, it is helpful to make a variable change from q to k and vice-versa in the first-term on the right-hand side of (14.8). Accordingly, we can write (14.8) as

$$\begin{aligned}\mathcal{H}_{\mathrm{bogo}} &- \sum_{\mathrm{k}} \xi_{\mathrm{k}}\big(c^{\dagger}_{\mathrm{k},\uparrow}c_{\mathrm{k},\uparrow} + c^{\dagger}_{-\mathrm{k},\downarrow}c_{-\mathrm{k},\downarrow}\big)\\ &= \frac{1}{N}\sum_{\mathrm{k,q}} V_{\mathrm{k,q}}\{X_{\mathrm{q}}c_{-\mathrm{k},\downarrow},c_{\mathrm{k},\uparrow} + X^{\dagger}_{\mathrm{q}}c^{\dagger}_{\mathrm{k},\uparrow}c^{\dagger}_{-\mathrm{k},\downarrow} - X_{\mathrm{k}}X^{\dagger}_{\mathrm{q}}\}\,.\end{aligned} \tag{14.9}$$

For simplicity of display, let us introduce a notational change in (14.9):

$$\nabla_{\mathrm{k}} = \frac{1}{N}\sum_{\mathrm{q}} V_{\mathrm{k,q}}X_{\mathrm{q}}\,, \qquad \nabla_{\mathrm{k}}{}^{\dagger} = \frac{1}{N}\sum_{\mathrm{q}} V_{\mathrm{k,q}}X^{\dagger}_{\mathrm{q}}\,. \tag{14.10}$$

As a result, $\mathcal{H}_{\mathrm{bogo}}$ given in (14.9) can be rewritten in a more compact form, namely

$$\begin{aligned}\mathcal{H}_{\mathrm{bogo}} &- \sum_{\mathrm{k}} \xi_{\mathrm{k}}\big(c^{\dagger}_{\mathrm{k},\uparrow}c_{\mathrm{k},\uparrow} + c^{\dagger}_{-\mathrm{k},\downarrow}c_{-\mathrm{k},\downarrow}\big)\\ &= \sum_{\mathrm{k}}\{\nabla_{\mathrm{k}}c_{-\mathrm{k},\downarrow}c_{\mathrm{k},\uparrow} + \nabla_{k}{}^{\dagger}c^{\dagger}_{\mathrm{k},\uparrow}c^{\dagger}_{-\mathrm{k},\downarrow}\} + \Delta\,.\end{aligned} \tag{14.11}$$

The parameter Δ stands for the variable-free function

$$\Delta = -\frac{1}{N}\sum_{\mathrm{k,q}} V_{\mathrm{k,q}}X_{\mathrm{k}}X^{\dagger}_{\mathrm{q}} \equiv \sum_{\mathrm{k}}\langle c^{\dagger}_{\mathrm{k},\uparrow}c^{\dagger}_{-\mathrm{k},\downarrow}\rangle\Lambda_{\mathrm{k}}\,. \tag{14.12}$$

Here Λ_{k}, the so-called gap-function, which is defined as follows:

$$\Lambda_{\mathrm{k}} = -\frac{1}{N}\sum_{\mathrm{q}} V_{\mathrm{k,q}}\langle c_{-\mathrm{q},\downarrow}c_{\mathrm{q},\uparrow}\rangle\,. \tag{14.13}$$

Although the Hamiltonian $\mathcal{H}_{\mathrm{bogo}}$ does not conserve particle number, it can be diagonalized by using Bogolyubov fermionic quasiparticle operators. These operators create—e.g., $\gamma^{\dagger}_{\mathrm{k},\sigma}$—and destroy—e.g., $\gamma_{\mathrm{k},\sigma}$—Bogolyubov fermionic quasiparticles.

In the following we study a representation of such quasiparticle operators as function of the electron-creation and electron-destruction operators $c^{\dagger}_{\mathrm{k},\sigma}$ and $c_{\mathrm{k},\sigma}$,

$$\begin{aligned}\gamma_{\mathrm{k},\uparrow} &\equiv u_{\mathrm{k}},\, c_{\mathrm{k},\uparrow} - v_{\mathrm{k}} c^{\dagger}_{-\mathrm{k},\downarrow}\,,\\ \gamma_{-\mathrm{k},\downarrow} &\equiv u_{\mathrm{k}} c_{-\mathrm{k},\downarrow} + v_{\mathrm{k}} c^{\dagger}_{\mathrm{k},\uparrow}\,.\end{aligned} \tag{14.14}$$

Hermitian conjugate of (14.14) is

$$\begin{aligned}\gamma^{\dagger}_{\mathrm{k},\uparrow} &\equiv u^{*}_{\mathrm{k}} c^{\dagger}_{\mathrm{k},\uparrow} - v^{*}_{\mathrm{k}} c_{-\mathrm{k},\downarrow}\,,\\ \gamma^{\dagger}_{-\mathrm{k},\downarrow} &\equiv u^{*}_{\mathrm{k}} c^{\uparrow}_{-\mathrm{k},\downarrow} + v^{*}_{\mathrm{k}} c_{\mathrm{k},\uparrow}\,.\end{aligned} \tag{14.15}$$

We shall also need the inverse of (14.14) and (14.15) so that electron creation and destruction operators are represented as functions of the fermionic quasiparticle creation and destruction operators. That is,

$$\begin{aligned}c_{\mathrm{k},\uparrow} &\equiv u^{*}_{\mathrm{k}} \gamma_{\mathrm{k},\uparrow} + v_{\mathrm{k}} \gamma^{\dagger}_{-\mathrm{k},\downarrow}\,,\\ c_{-\mathrm{k},\downarrow} &\equiv -v_{\mathrm{k}} \gamma^{\dagger}_{\mathrm{k},\uparrow} + u^{*}_{\mathrm{k}} \gamma_{-\mathrm{k},\downarrow}\,,\end{aligned} \tag{14.16}$$

and

$$\begin{aligned}c^{\dagger}_{\mathrm{k},\uparrow} &\equiv u_{\mathrm{k}} \gamma^{\dagger}_{\mathrm{k},\uparrow} + v^{*}_{\mathrm{k}} \gamma_{-\mathrm{k},\downarrow}\,,\\ c^{\dagger}_{-\mathrm{k},\downarrow} &\equiv -v^{*}_{\mathrm{k}} \gamma_{\mathrm{k},\uparrow} + u_{\mathrm{k}} \gamma^{\dagger}_{-\mathrm{k},\downarrow}\,.\end{aligned} \tag{14.17}$$

For help with this inversion, see below.

Generally, for a 2×2 matrix A,

$$\begin{pmatrix} a & b \\ c & d \end{pmatrix} = A\,, \tag{14.18}$$

the inverse is

$$\begin{pmatrix} d & -b \\ -c & a \end{pmatrix} /(ad - bc) = (A)^{-1}\,. \tag{14.19}$$

Compare, for instance, [7]. Details of the inversion procedure that was used to derive (14.16) and (14.17) are the following. Consider the equations

$$\begin{aligned}c_{\mathrm{k},\uparrow} &= u^{*}_{\mathrm{k}} \gamma_{\mathrm{k},\uparrow} + v_{\mathrm{k}} \gamma^{\dagger}_{-\mathrm{k},\downarrow}\,,\\ c^{\dagger}_{-\mathrm{k},\downarrow} &= -v^{*}_{\mathrm{k}} \gamma_{\mathrm{k},\uparrow} + u_{\mathrm{k}} \gamma^{\dagger}_{-\mathrm{k},\downarrow}\,.\end{aligned}$$

The relevant matrix is

$$\begin{pmatrix} u^{*}_{\mathrm{k}} & v_{\mathrm{k}} \\ -v^{*}_{\mathrm{k}} & u_{\mathrm{k}} \end{pmatrix} = M\,. \tag{14.20}$$

And the inverse matrix, M^{-1}, is the following:

$$\begin{pmatrix} u_{\mathrm{k}} & -v_{\mathrm{k}} \\ v_{\mathrm{k}}^* & u_{\mathrm{k}}^* \end{pmatrix} / \left(|u_{\mathrm{k}}|^2 + |v_{\mathrm{k}}|^2\right) = (M)^{-1}, \tag{14.21}$$

which leads to the result

$$\begin{aligned} \gamma_{\mathrm{k},\uparrow} &\equiv \left[u_{\mathrm{k}} c_{\mathrm{k},\uparrow} - v_{\mathrm{k}} c^{\dagger}_{-\mathrm{k},\downarrow}\right] / \left(|u_{\mathrm{k}}|^2 + |v_{\mathrm{k}}|^2\right), \\ \gamma^{\dagger}_{-\mathrm{k},\downarrow} &\equiv \left[v_{\mathrm{k}}^* c_{\mathrm{k},\uparrow} + u_{\mathrm{k}}^* c^{\dagger}_{-\mathrm{k},\downarrow}\right] / \left(|u_{\mathrm{k}}|^2 + |v_{\mathrm{k}}|^2\right). \end{aligned} \tag{14.22}$$

14.2 Quasiparticles

Because the electron operators obey the anticommutation rules,

$$\begin{aligned} &\left[c_{\mathrm{k},\uparrow}, c^{\dagger}_{\mathrm{k},\uparrow}\right]_+ = 1, \quad &&\left[c_{\mathrm{k},\uparrow}, c^{\dagger}_{-\mathrm{k},\downarrow}\right]_+ = 0, \quad &&\left[c_{-\mathrm{k},\downarrow}, c^{\dagger}_{-\mathrm{k},\downarrow}\right]_+ = 1, \\ &\left[c_{-\mathrm{k},\downarrow}, c_{\mathrm{k},\uparrow}\right]_+ = 0, \quad &&\left[c^{\dagger}_{\mathrm{k},\uparrow}, c^{\dagger}_{-\mathrm{k},\downarrow}\right]_+ = 0, \quad &&\left[c_{-\mathrm{k},\uparrow}, c^{\dagger}_{-\mathrm{k},\downarrow}\right]_+ = 0, \end{aligned} \tag{14.23}$$

it follows that the fermionic—i.e., the Bogulyubov quasiparticle—operators, $\gamma_{\mathrm{k},\sigma}$ and $\gamma^{\dagger}_{\mathrm{k},\sigma}$, also do the same. For instance, for $\sigma \equiv \uparrow$, we have

$$\begin{aligned} \left[\gamma_{\mathrm{k},\uparrow}, \gamma^{\dagger}_{\mathrm{k},\uparrow}\right]_+ &= \left[u_{\mathrm{k}} c_{\mathrm{k},\uparrow} - v_{\mathrm{k}} c^{\dagger}_{-\mathrm{k},\downarrow}, u_{\mathrm{k}}^* c^{\dagger}_{\mathrm{k},\uparrow} - v_{\mathrm{k}}^* c_{-\mathrm{k},\downarrow}\right]_+ \\ &= |u_k|^2\left[c_{\mathrm{k},\uparrow}, c^{\dagger}_{\mathrm{k},\uparrow}\right]_+ + |v_k|^2\left[c^{\dagger}_{-\mathrm{k},\downarrow}, c_{-\mathrm{k},\downarrow}\right]_+ + 0 + 0 \\ &= |u_k|^2 + |v_k|^2. \end{aligned} \tag{14.24}$$

And the same is the case for $\sigma \equiv \downarrow$. Importantly, the satisfaction of both these cases requires u_{k} and v_{k} obey the normalization requirement

$$|u_{\mathrm{k}}|^2 + |v_{\mathrm{k}}|^2 = 1. \tag{14.25}$$

Thus the normalization requirement (14.25) is needed both for the inversions (14.16)–(14.17) to hold and for the fermionic quasiparticle operators, $\gamma_{\mathrm{k},\sigma}$ and $\gamma^{\dagger}_{\mathrm{k},\sigma}$, to be canonical. It is easy to check that in addition to satisfying the normalization requirement the fermionic quasiparticle operators also obey all the needed anticommutation rules. For instance, from (14.14) we have

$$\begin{aligned} [\gamma_{\mathrm{k},\uparrow}, \gamma_{-\mathrm{k},\downarrow}]_+ &\equiv \gamma_{\mathrm{k},\uparrow}\gamma_{-\mathrm{k},\downarrow} + \gamma_{-\mathrm{k},\downarrow}\gamma_{\mathrm{k},\uparrow} \\ &= u_{\mathrm{k}}^2 \times 0 + u_{\mathrm{k}} v_{\mathrm{k}} - v_{\mathrm{k}} u_{\mathrm{k}} - v_{\mathrm{k}}^2 \times 0 = 0, \quad \text{etc.} \end{aligned} \tag{14.26}$$

In order to represent $\mathcal{H}_{\mathrm{bogo}}$ in terms of the fermionic operators, we need to insert the relevant contents of (14.16) and (14.17) into (14.11). Thereby $\mathcal{H}_{\mathrm{bogo}}$ becomes

$$
\begin{aligned}
&\mathcal{H}_{\text{bogo}} \\
&= \sum_{\mathrm{k}} \xi_{\mathrm{k}} \big(u_{\mathrm{k}} \gamma^{\dagger}_{\mathrm{k},\uparrow} + v^{*}_{\mathrm{k}} \gamma_{-\mathrm{k},\downarrow}\big)\big(u^{*}_{\mathrm{k}} \gamma_{\mathrm{k},\uparrow} + v_{\mathrm{k}} \gamma^{\dagger}_{-\mathrm{k},\downarrow}\big) \\
&\quad + \sum_{\mathrm{k}} \xi_{\mathrm{k}} \big(-v^{*}_{\mathrm{k}} \gamma_{\mathrm{k},\uparrow} + u_{\mathrm{k}} \gamma^{\dagger}_{-\mathrm{k},\downarrow}\big)\big(-v_{\mathrm{k}} \gamma^{\dagger}_{\mathrm{k},\uparrow} + u^{*}_{\mathrm{k}} \gamma_{-\mathrm{k},\downarrow}\big) \\
&\quad + \sum_{\mathrm{k}} \nabla^{\dagger}_{\mathrm{k}} \big\{(u_{\mathrm{k}})^{2} \gamma^{\dagger}_{\mathrm{k},\uparrow} \gamma^{\dagger}_{-\mathrm{k},\downarrow} - u_{\mathrm{k}} v^{*}_{\mathrm{k}} \gamma^{\dagger}_{\mathrm{k},\uparrow} \gamma_{\mathrm{k},\uparrow} + v^{*}_{\mathrm{k}} u_{\mathrm{k}} \gamma_{-\mathrm{k},\downarrow} \gamma^{\dagger}_{-\mathrm{k},\downarrow} - \big(v^{*}_{\mathrm{k}}\big)^{2} \gamma_{-\mathrm{k},\downarrow} \gamma_{\mathrm{k},\uparrow}\big\} \\
&\quad + \sum_{\mathrm{k}} \nabla_{\mathrm{k}} \big\{\big(u^{*}_{\mathrm{k}}\big)^{2} \gamma_{-\mathrm{k},\downarrow} \gamma_{\mathrm{k},\uparrow} + u^{*}_{\mathrm{k}} v_{\mathrm{k}} \gamma_{-\mathrm{k},\downarrow} \gamma^{\dagger}_{-\mathrm{k},\downarrow} - v_{\mathrm{k}} u^{*}_{\mathrm{k}} \gamma^{\dagger}_{\mathrm{k},\uparrow} \gamma_{\mathrm{k},\uparrow} - (v_{\mathrm{k}})^{2} \gamma^{\dagger}_{\mathrm{k},\uparrow} \gamma^{\dagger}_{-\mathrm{k},\downarrow}\big\} \\
&\quad + \Delta\,.
\end{aligned}
\tag{14.27}
$$

After some reorganization of the ξ_{k} terms, (14.27) leads to (14.28), that is,

$$
\begin{aligned}
&\mathcal{H}_{\text{bogo}} \\
&= \sum_{\mathrm{k}} \xi_{\mathrm{k}} \big\{\big(|u_{\mathrm{k}}|^{2} - |v_{\mathrm{k}}|^{2}\big)\big(\gamma^{\dagger}_{\mathrm{k},\uparrow} \gamma_{\mathrm{k},\uparrow} + \gamma^{\dagger}_{-\mathrm{k},\downarrow} \gamma_{-\mathrm{k},\downarrow}\big)\big\} \\
&\quad + \sum_{\mathrm{k}} \xi_{\mathrm{k}} \big\{2u_{\mathrm{k}} v_{\mathrm{k}} \big(\gamma^{\dagger}_{\mathrm{k},\uparrow} \gamma^{\dagger}_{-\mathrm{k}\downarrow}\big) + 2u^{*}_{\mathrm{k}} v^{*}_{\mathrm{k}} (\gamma_{-\mathrm{k},\downarrow} \gamma_{\mathrm{k},\uparrow})\big\} \\
&\quad + \sum_{\mathrm{k}} \xi_{\mathrm{k}} \big\{2|v_{\mathrm{k}}|^{2}\big\} \\
&\quad + \sum_{\mathrm{k}} \nabla^{\dagger}_{\mathrm{k}} \big\{(u_{\mathrm{k}})^{2} \gamma^{\dagger}_{\mathrm{k},\uparrow} \gamma^{\dagger}_{-\mathrm{k},\downarrow} - u_{\mathrm{k}} v^{*}_{\mathrm{k}} \gamma^{\dagger}_{\mathrm{k},\uparrow} \gamma_{\mathrm{k},\uparrow} + v^{*}_{\mathrm{k}} u_{\mathrm{k}} \gamma_{-\mathrm{k},\downarrow} \gamma^{\dagger}_{-\mathrm{k},\downarrow} - \big(v^{*}_{\mathrm{k}}\big)^{2} \gamma_{-\mathrm{k},\downarrow} \gamma_{\mathrm{k},\uparrow}\big\} \\
&\quad + \sum_{\mathrm{k}} \nabla_{\mathrm{k}} \big\{\big(u^{*}_{\mathrm{k}}\big)^{2} \gamma_{-\mathrm{k},\downarrow} \gamma_{\mathrm{k},\uparrow} + u^{*}_{\mathrm{k}} v_{\mathrm{k}} \gamma_{-\mathrm{k},\downarrow} \gamma^{\dagger}_{-\mathrm{k},\downarrow} - v_{\mathrm{k}} u^{*}_{\mathrm{k}} \gamma^{\dagger}_{\mathrm{k},\uparrow} \gamma_{\mathrm{k},\uparrow} - (v_{k})^{2} \gamma^{\dagger}_{\mathrm{k},\uparrow} \gamma^{\dagger}_{-\mathrm{k},\downarrow}\big\} \\
&\quad + \Delta\,.
\end{aligned}
\tag{14.28}
$$

It is helpful to represent (14.28) in a more symmetrical form such as in (14.29) given below:

$$
\begin{aligned}
\mathcal{H}_{\text{bogo}} &\equiv \mathcal{H}_{\text{diagonal}} + \mathcal{H}_{\text{nondiagonal}} \\
&= \sum_{\mathrm{k}} \big[\xi_{\mathrm{k}} \big(|u_{\mathrm{k}}|^{2} - |v_{\mathrm{k}}|^{2}\big) - \nabla_{\mathrm{k}} u^{*}_{\mathrm{k}} v_{\mathrm{k}} - \nabla^{\dagger}_{\mathrm{k}} v^{*}_{\mathrm{k}} u_{\mathrm{k}}\big]\big(\gamma^{\dagger}_{\mathrm{k},\uparrow} \gamma_{\mathrm{k},\uparrow}\big) \\
&\quad + \sum_{\mathrm{k}} \big[\xi_{\mathrm{k}} \big(|u_{\mathrm{k}}|^{2} - |v_{\mathrm{k}}|^{2}\big) - \nabla_{\mathrm{k}} u^{*}_{\mathrm{k}} v_{\mathrm{k}} - \nabla^{\dagger}_{\mathrm{k}} v^{*}_{\mathrm{k}} u_{\mathrm{k}}\big]\big(\gamma^{\dagger}_{-\mathrm{k},\downarrow} \gamma_{-\mathrm{k},\downarrow}\big) \\
&\quad + \sum_{\mathrm{k}} \big(\gamma^{\dagger}_{\mathrm{k},\uparrow} \gamma^{\dagger}_{-\mathrm{k},\downarrow}\big)\big\{\nabla^{\dagger}_{\mathrm{k}} (u_{\mathrm{k}})^{2} - \nabla_{\mathrm{k}} (v_{\mathrm{k}})^{2} + 2\xi_{\mathrm{k}} u_{\mathrm{k}} v_{\mathrm{k}}\big\} \\
&\quad + \sum_{\mathrm{k}} (\gamma_{-\mathrm{k},\downarrow} \gamma_{\mathrm{k},\uparrow})\big\{\nabla_{\mathrm{k}} \big(u^{*}_{\mathrm{k}}\big)^{2} - \nabla^{\dagger}_{\mathrm{k}} \big(v^{*}_{\mathrm{k}}\big)^{2} + 2\xi_{\mathrm{k}} u^{*}_{\mathrm{k}} v^{*}_{\mathrm{k}}\big\} \\
&\quad + \sum_{\mathrm{k}} \big\{2\xi_{\mathrm{k}} |v_{\mathrm{k}}|^{2} + \nabla^{\dagger}_{\mathrm{k}} u_{\mathrm{k}} v^{*}_{\mathrm{k}} + \nabla_{\mathrm{k}} u^{*}_{\mathrm{k}} v_{\mathrm{k}}\big\} + \Delta\,.
\end{aligned}
\tag{14.29}
$$

On the right-hand side of (14.29), the functions on the first and second lines are diagonal and constitute the Hamiltonian $\mathcal{H}_{\text{diagonal}}$. Those on the third and fourth lines are equal in the sense that they are complex-conjugates of each other. And they are nondiagonal. The functions on the fifth line represent the ground-state energy and do not contain any operators. These three functions—meaning those on the third, fourth, and fifth lines—constitute the Hamiltonian $\mathcal{H}_{\text{nondiagonal}}$.

Clearly, in order to eliminate the two nondiagonal terms that are mutually conjugate and are recorded on the third and fourth lines of (14.29), we need to set only one of them equal to zero. And that arrangement is fully ensured by (14.30):

$$\nabla_{\mathrm{k}}^{\dagger}(u_{\mathrm{k}})^2 - \nabla_{\mathrm{k}}(v_{\mathrm{k}})^2 + 2\xi_{\mathrm{k}} u_{\mathrm{k}} v_{\mathrm{k}} = 0 . \tag{14.30}$$

Dividing (14.30) by $-\nabla_{\mathrm{k}}(u_{\mathrm{k}})^2$ leads to the quadratic (14.31), that is,

$$\left(\frac{v_{\mathrm{k}}}{u_{\mathrm{k}}}\right)^2 - 2\left(\frac{\xi_{\mathrm{k}}}{\nabla_{\mathrm{k}}}\right)\left(\frac{v_{\mathrm{k}}}{u_{\mathrm{k}}}\right) - \left(\frac{\nabla_{\mathrm{k}}^{\dagger}}{\nabla_{\mathrm{k}}}\right) = 0 . \tag{14.31}$$

Of the two roots, namely

$$\frac{v_{\mathrm{k}}}{u_{\mathrm{k}}} = \frac{1}{\nabla_{\mathrm{k}}}\left[\xi_{\mathrm{k}} \pm \sqrt{(\xi_{\mathrm{k}})^2 + |\nabla_{\mathrm{k}}|^2}\right], \tag{14.32}$$

of this quadratic we choose the one that ensures thermodynamic stability, meaning the ground state energy is a minimum. That is,

$$\frac{v_{\mathrm{k}}}{u_{\mathrm{k}}} = \frac{1}{\nabla_{\mathrm{k}}}\left[\xi_{\mathrm{k}} - \sqrt{(\xi_{\mathrm{k}})^2 + |\nabla_{\mathrm{k}}|^2}\right]. \tag{14.33}$$

Given ${\xi^{\dagger}}_{\mathrm{k}} = \xi_{\mathrm{k}} = \xi_{-\mathrm{k}}$, the Hermitian-conjugate of (14.33) is

$$\frac{v_{\mathrm{k}}^*}{u_{\mathrm{k}}^*} = \frac{1}{\nabla_{\mathrm{k}}^{\dagger}}\left[\xi_{\mathrm{k}} - \sqrt{(\xi_{\mathrm{k}})^2 + |\nabla_{\mathrm{k}}|^2}\right]. \tag{14.34}$$

Now multiply each against the other the left-hand sides of (14.33) and (14.34), and equate them to the corresponding multiple of the right-hand sides. One gets

$$\frac{|v_{\mathrm{k}}|^2}{|u_{\mathrm{k}}|^2} = \frac{1}{|\nabla_{\mathrm{k}}|^2}\left[\xi_{\mathrm{k}} - \sqrt{(\xi_{\mathrm{k}})^2 + |\nabla_{\mathrm{k}}|^2}\right]^2 . \tag{14.35}$$

Add unity to both sides of (14.35), invert both sides, and make use of (14.25) whereby $|u_{\mathrm{k}}^2| + |v_{\mathrm{k}}^2| = 1$. The result is

$$\frac{1}{1 + \frac{|v_{\mathrm{k}}|^2}{|u_{\mathrm{k}}|^2}} = \frac{|u_{\mathrm{k}}|^2}{|u_{\mathrm{k}}|^2 + |v_{\mathrm{k}}|^2} = |u_{\mathrm{k}}|^2 = \frac{1}{1 + \frac{1}{|\nabla_{\mathrm{k}}|^2}[\xi_{\mathrm{k}} - \sqrt{{\xi_{\mathrm{k}}}^2 + |\nabla_{\mathrm{k}}|^2}]^2} . \tag{14.36}$$

Completing the squaring process in the denominator of (14.36) leads to

$$
\begin{aligned}
|u_{\mathrm{k}}|^2 &= \frac{1}{2}\left(\frac{|\nabla_{\mathrm{k}}|^2}{\xi_{\mathrm{k}}^2 + |\nabla_{\mathrm{k}}|^2 - \xi_{\mathrm{k}}\sqrt{\xi_{\mathrm{k}}^2 + |\nabla_{\mathrm{k}}|^2}}\right) \\
&= \frac{1}{2}\frac{1}{\sqrt{\xi_{\mathrm{k}}{}^2 + |\nabla_{\mathrm{k}}|^2}}\left[\frac{|\nabla_{\mathrm{k}}|^2}{\sqrt{\xi_{\mathrm{k}}{}^2 + |\nabla_{\mathrm{k}}|^2} - \xi_{\mathrm{k}}}\right]\cdot\left(\frac{\sqrt{\xi_{\mathrm{k}}{}^2 + |\nabla_{\mathrm{k}}|^2} + \xi_{\mathrm{k}}}{\sqrt{\xi_{\mathrm{k}}{}^2 + |\nabla_{\mathrm{k}}|^2} + \xi_{\mathrm{k}}}\right) \\
&= \frac{1}{2}\left(\frac{\sqrt{\xi_{\mathrm{k}}{}^2 + |\nabla_{\mathrm{k}}|^2} + \xi_{\mathrm{k}}}{\sqrt{\xi_{\mathrm{k}}{}^2 + |\nabla_{\mathrm{k}}|^2}}\right)\cdot\left[\frac{|\nabla_{\mathrm{k}}|^2}{\xi_{\mathrm{k}}{}^2 + |\nabla_{\mathrm{k}}|^2 - \xi_{\mathrm{k}}{}^2}\right] \\
&= \frac{1}{2}\left(1 + \frac{\xi_{\mathrm{k}}}{\sqrt{\xi_{\mathrm{k}}{}^2 + |\nabla_{\mathrm{k}}|^2}}\right)\cdot\left[\frac{|\nabla_{\mathrm{k}}|^2}{|\nabla_{\mathrm{k}}|^2}\right] \\
&= \frac{1}{2}\left(1 + \frac{\xi_{\mathrm{k}}}{\sqrt{\xi_{\mathrm{k}}{}^2 + |\nabla_{\mathrm{k}}|^2}}\right).
\end{aligned}
\tag{14.37}
$$

Now that $|u_{\mathrm{k}}|^2$ has been determined, we can use it to calculate $|v_{\mathrm{k}}|^2$. To that end, write

$$
|v_{\mathrm{k}}|^2 = 1 - |u_{\mathrm{k}}|^2\,, \tag{14.38}
$$

and insert $|u_{\mathrm{k}}|^2$ from (14.37) into (14.38). We get

$$
|v_{\mathrm{k}}|^2 = \frac{1}{2}\left(1 - \frac{\xi_{\mathrm{k}}}{\sqrt{\xi_{\mathrm{k}}{}^2 + |\nabla_{\mathrm{k}}|^2}}\right). \tag{14.39}
$$

The Hamiltonian $\mathcal{H}_{\mathrm{bogo}}$ has the following diagonal terms:

$$
H_{\mathrm{diagonal}} = \sum_{\mathrm{k},\sigma}\left[\xi_{\mathrm{k}}\left(|u_{\mathrm{k}}|^2 - |v_{\mathrm{k}}|^2\right) - \nabla_{\mathrm{k}}u_{\mathrm{k}}^*v_{\mathrm{k}} - \nabla_{\mathrm{k}}^\dagger v_{\mathrm{k}}^*u_{\mathrm{k}}\right]\cdot\left(\gamma_{\mathrm{k},\sigma}^\dagger\gamma_{\mathrm{k},\sigma}\right). \tag{14.40}
$$

Using (14.37) for $|u_{\mathrm{k}}|^2$ and (14.39) for $|v_{\mathrm{k}}|^2$, we can write the first term of (14.40) as

$$
\sum_{\mathrm{k},\sigma}\xi_{\mathrm{k}}\left(|u_{\mathrm{k}}|^2 - |v_{\mathrm{k}}|^2\right)\cdot\left(\gamma_{\mathrm{k},\sigma}^\dagger\gamma_{\mathrm{k},\sigma}\right) = \sum_{\mathrm{k},\sigma}\left(\frac{\xi_{\mathrm{k}}^2}{\sqrt{\xi_{\mathrm{k}}^2 + |\nabla_{\mathrm{k}}|^2}}\right)\cdot\left(\gamma_{\mathrm{k},\sigma}^\dagger\gamma_{\mathrm{k},\sigma}\right). \tag{14.41}
$$

The second and third terms of (14.40) require knowledge of $\nabla_{\mathrm{k}}u_{\mathrm{k}}^*v_{\mathrm{k}}$ and $\nabla_{\mathrm{k}}^\dagger v_{\mathrm{k}}^*u_{\mathrm{k}}$. These two terms are Hermitian conjugates of each other. Therefore only one of them needs to be evaluated. To that end, consider (14.33) and multiply both sides by $u_{\mathrm{k}}^*u_{\mathrm{k}}$. The result is

$$
\begin{aligned}
u_{\mathrm{k}}^*u_{\mathrm{k}}\frac{v_{\mathrm{k}}}{u_{\mathrm{k}}} &= u_{\mathrm{k}}^*v_{\mathrm{k}} \\
&= u_{\mathrm{k}}^*u_{\mathrm{k}}\left(\frac{1}{\nabla_{\mathrm{k}}}\right)\left[\xi_{\mathrm{k}} - \sqrt{(\xi_{\mathrm{k}})^2 + |\nabla_{\mathrm{k}}|^2}\right].
\end{aligned}
\tag{14.42}
$$

The second term on the right-hand side of (14.42) is

$$u_{\mathrm{k}}^* v_{\mathrm{k}} = u_{\mathrm{k}}^* u_{\mathrm{k}} \left(\frac{1}{\nabla_{\mathrm{k}}^{\dagger}} \right) \left[\xi_{\mathrm{k}} - \sqrt{(\xi_{\mathrm{k}})^2 + |\nabla_{\mathrm{k}}|^2} \right]. \tag{14.43}$$

Using (14.42) and (14.43), we write sequentially the second and third terms of (14.40).

The second term is

$$\begin{aligned} \sum_{\mathrm{k},\sigma} \{-\nabla_{\mathrm{k}} u_{\mathrm{k}}^* v_{\mathrm{k}}\} (\gamma_{\mathrm{k},\sigma}^{\dagger} \gamma_{\mathrm{k},\sigma}) &= \left\{ -\frac{\nabla_{\mathrm{k}} u_{\mathrm{k}}^* u_{\mathrm{k}}}{\nabla_{\mathrm{k}}} \right\} \left[\xi_{\mathrm{k}} - \sqrt{\xi_{\mathrm{k}}{}^2 + |\nabla_{\mathrm{k}}|^2} \right] (\gamma_{\mathrm{k},\sigma}^{\dagger} \gamma_{\mathrm{k},\sigma}) \\ &= \sum_{k,\sigma} \left\{ -|u_{\mathrm{k}}|^2 \left[\xi_{\mathrm{k}} - \sqrt{\xi_{\mathrm{k}}{}^2 + |\nabla_{\mathrm{k}}|^2} \right] \right\} (\gamma_{\mathrm{k},\sigma}^{\dagger} \gamma_{\mathrm{k},\sigma}). \end{aligned} \tag{14.44}$$

At this stage we replace $|u_{\mathrm{k}}|^2$ by $\frac{1}{2}(1 + \frac{\xi_{\mathrm{k}}}{\sqrt{\xi_{\mathrm{k}}{}^2 + |\nabla_{\mathrm{k}}|^2}})$ as suggested by (14.37). We get for the second term in H_{diagonal}:

$$\begin{aligned} &-\sum_{\mathrm{k},\sigma} \left(\frac{1}{2} \right) \left\{ \frac{(\xi_{\mathrm{k}} + \sqrt{\xi_{\mathrm{k}}{}^2 + |\nabla_{\mathrm{k}}|^2})}{\sqrt{\xi_{\mathrm{k}}{}^2 + |\nabla_{\mathrm{k}}|^2}} \left[\xi_{\mathrm{k}} - \sqrt{\xi_{\mathrm{k}}{}^2 + |\nabla_{\mathrm{k}}|^2} \right] \right\} (\gamma_{\mathrm{k},\sigma}^{\dagger} \gamma_{\mathrm{k},\sigma}) \\ &= \sum_{\mathrm{k},\sigma} \left(\frac{1}{2} \right) \left\{ \frac{|\nabla_{\mathrm{k}}|^2}{\sqrt{\xi_{\mathrm{k}}{}^2 + |\nabla_{\mathrm{k}}|^2}} \right\} (\gamma_{\mathrm{k},\sigma}^{\dagger} \gamma_{\mathrm{k},\sigma}). \end{aligned} \tag{14.45}$$

And finally, the third term in H_{diagonal}, that is,

$$\begin{aligned} \sum_{\mathrm{k},\sigma} \{-\nabla_{\mathrm{k}}^{\dagger} v_{\mathrm{k}}^* u_{\mathrm{k}}\} (\gamma_{\mathrm{k},\sigma}^{\dagger} \gamma_{\mathrm{k},\sigma}) &= -\sum_{\mathrm{k},\sigma} \frac{\nabla_{\mathrm{k}}^{\dagger} \Delta_{\mathrm{k}} v_{\mathrm{k}}^* v_{\mathrm{k}}}{[\xi_{\mathrm{k}} - \sqrt{\xi_{\mathrm{k}}{}^2 + |\nabla_{\mathrm{k}}|^2}]} (\gamma_{\mathrm{k},\sigma}^{\dagger} \gamma_{\mathrm{k},\sigma}) \\ &= -\sum_{\mathrm{k},\sigma} \frac{\nabla_{\mathrm{k}}^{\dagger} \Delta_{\mathrm{k}} (\frac{1}{2})(1 - \frac{\xi_{\mathrm{k}}}{\sqrt{\xi_{\mathrm{k}}{}^2 + |\nabla_{\mathrm{k}}|^2}})}{[\xi_{\mathrm{k}} - \sqrt{\xi_{\mathrm{k}}{}^2 + |\nabla_{\mathrm{k}}|^2}]} (\gamma_{\mathrm{k},\sigma}^{\dagger} \gamma_{\mathrm{k},\sigma}) \\ &= \sum_{\mathrm{k},\sigma} \left(\frac{1}{2} \right) \left\{ \frac{|\nabla_{\mathrm{k}}|^2}{\sqrt{\xi_{\mathrm{k}}{}^2 + |\nabla_{\mathrm{k}}|^2}} \right\} (\gamma_{\mathrm{k},\sigma}^{\dagger} \gamma_{\mathrm{k},\sigma}). \end{aligned} \tag{14.46}$$

Notice that the third term is exactly the same as the second. Summing (14.41), (14.45), and (14.46) leads to Bogolyubov representation of the diagonal terms. That is,

$$\begin{aligned} H_{\text{diagonal}} &= \sum_{\mathrm{k},\sigma} \{ \sqrt{\xi_{\mathrm{k}}{}^2 + |\nabla_{\mathrm{k}}|^2} \} (\gamma_{\mathrm{k},\sigma}^{\dagger} \gamma_{\mathrm{k},\sigma}) \\ &\equiv \sum_{\mathrm{k},\sigma} \Upsilon_{\mathrm{k}} (\gamma_{\mathrm{k},\sigma}^{\dagger} \gamma_{\mathrm{k},\sigma}). \end{aligned} \tag{14.47}$$

In (14.47), $(\gamma^{\dagger}_{k,\sigma}\gamma_{k,\sigma})$ is the number operator for bogolyubons with wave-vector k and spin σ, and Υ_k is their dispersion relation given below as

$$\Upsilon_k \equiv \sqrt{\xi_k{}^2 + |\nabla_k|^2} . \tag{14.48}$$

As recorded in (14.2), there is the relationship

$$\xi_k = \epsilon_k - E_F = \xi_{-}k , \tag{14.49}$$

where ϵ_k is the kinetic energy of an electron and E_F is the Fermi energy. Also, according to (14.37), (14.39), and (14.48), we have

$$|u_k|^2 = \frac{1}{2}\left(1 + \frac{\xi_k}{|\Upsilon_k|}\right), \qquad |v_k|^2 = \frac{1}{2}\left(1 - \frac{\xi_k}{|\Upsilon_k|}\right), \tag{14.50}$$

where $\Upsilon_k \equiv \sqrt{\xi_k{}^2 + |\nabla_k|^2}$. As such, for $\xi_k \gg |\nabla_k|$, which basically means when $|\nabla_k| \to 0$, we get $\xi_k \to |\Upsilon_k|$, $|u_k|^2 \to 1$, and $|v_k|^2 \to 0$. Accordingly, in this limit the Bogulyubov quasiparticle is essentially an ordinary particle and the system is close to its normal state. Therefore, close to the normal state, creating a Bogolyubov excitation amounts to creating an electron at energies above the Fermi sea and destroying an electron of opposite spin and momentum within the Fermi sea. On the other hand, in the opposite limit where $\xi_k \ll -|\nabla_k|$, $\xi_k \to -|\Upsilon_k|$, $|u_k|^2 \to 0$, and $|v_k|^2 \to 1$. As such, in this limit, the Bogulyubov quasiparticle is essentially a hole. The energy separation between the higher and the lower energy-states is $\geq 2|\nabla_k|$. In other words, the system needs a minimum of $2|\nabla_k|$ in energy to excite the Bogulyubov quasiparticles.

To sum up, for energies much lower than the Fermi energy, the Bogolyubov quasiparticles act as "holes." On the other hand, for energies much higher than the Fermi energy, the quasiparticles act as normal electrons. In-between these energy levels, where $1 > |u|^2 > 0$ and $1 > |v|^2 > 0$, the Bogolyubov quasiparticles are a mixture of particles and holes. And as the kinetic energy of an electron approaches the Fermi energy,[1] $\xi_k \to 0$ and (14.50) lead to

$$|u_k|^2 = \frac{1}{2}, \qquad |v_k|^2 = \frac{1}{2}, \tag{14.51}$$

or equivalently, $u_k = v_k = \frac{1}{\sqrt{2}}$. Therefore, in the limit $\xi_k \to 0$, (14.14) and (14.15) give

$$\begin{aligned} \gamma_{k,\uparrow} &=_{\xi_k \to 0} \frac{1}{\sqrt{2}} c_{k,\uparrow} - \frac{1}{\sqrt{2}} c^{\dagger}_{-k,\downarrow} , \\ \gamma^{\dagger}_{k,\uparrow} &=_{\xi_k \to 0} \frac{1}{\sqrt{2}} c^{\dagger}_{k,\uparrow} - \frac{1}{\sqrt{2}} c_{-k,\downarrow} , \end{aligned} \tag{14.52}$$

[1] This means as ϵ_k approaches E_F.

and

$$\gamma_{-\mathrm{k},\downarrow} =_{\xi_\mathrm{k}\to 0} \frac{1}{\sqrt{2}} c_{-\mathrm{k},\downarrow} + \frac{1}{\sqrt{2}} c^{\dagger}_{\mathrm{k},\uparrow}\,,$$
$$\gamma^{\dagger}_{-\mathrm{k},\downarrow} =_{\xi_\mathrm{k}\to 0} \frac{1}{\sqrt{2}} c^{\uparrow}_{-\mathrm{k},\downarrow} + \frac{1}{\sqrt{2}} c_{\mathrm{k},\uparrow}\,. \tag{14.53}$$

Equations (14.52) and (14.53) indicate that exactly at the Fermi surface, i.e., when $\xi_\mathrm{k} \to 0$, the quasiparticles act as electrons and holes with equal strength, leading to electrical neutrality.

Consequent to the truncation implicit in (14.29) and (14.30), $H_{\mathrm{nondiagonal}}$ is equivalent to $H_0 + \Delta$, where

$$H_{\mathrm{nondiagonal}} \equiv H_0 + \Delta = \sum_{\mathrm{k}} \{2\xi_\mathrm{k}|v_\mathrm{k}|^2 + \nabla_\mathrm{k} u^*_\mathrm{k} v_\mathrm{k} + \nabla^{\dagger}_\mathrm{k} u_\mathrm{k} v^*_\mathrm{k}\} + \Delta\,. \tag{14.54}$$

The Bogolyubov representation of the set of three terms that have variables—and are given on the right-hand side of (14.54)—is the following. The first term is

$$\sum_{\mathrm{k}} \{2\xi_\mathrm{k}|v_\mathrm{k}|^2\} = \sum_{\mathrm{k}} 2\xi_\mathrm{k}\left(\frac{1}{2}\right)\left(1 - \frac{\xi_\mathrm{k}}{\sqrt{\xi_\mathrm{k}{}^2 + |\nabla_\mathrm{k}|^2}}\right)$$
$$= \sum_{\mathrm{k}} \left(\xi_\mathrm{k} - \frac{\xi^2_\mathrm{k}}{\sqrt{\xi_\mathrm{k}{}^2 + |\nabla_\mathrm{k}|^2}}\right). \tag{14.55}$$

The second term, $\sum_\mathrm{k}\{\nabla_\mathrm{k} u^*_\mathrm{k} v_\mathrm{k}\}$, has already been evaluated–see (14.45):

$$\sum_{\mathrm{k}} \{\nabla_\mathrm{k} u^*_\mathrm{k} v_\mathrm{k}\} = -\frac{1}{2}\sum_{\mathrm{k}} \left\{\frac{|\nabla_\mathrm{k}|^2}{\sqrt{\xi_\mathrm{k}{}^2 + |\nabla_\mathrm{k}|^2}}\right\}. \tag{14.56}$$

Similarly, the third term, $\sum_\mathrm{k}\{\nabla^{\dagger}_\mathrm{k} u_\mathrm{k} v^*_\mathrm{k}\}$, was evaluated in (14.46) and again we have the same result as in (14.45), namely

$$\sum_{\mathrm{k}} \{\nabla^{\dagger}_\mathrm{k} u_\mathrm{k} v^*_\mathrm{k}\} = -\frac{1}{2}\sum_{\mathrm{k},\sigma} \left\{\frac{|\nabla_\mathrm{k}|^2}{\sqrt{\xi_\mathrm{k}{}^2 + |\nabla_\mathrm{k}|^2}}\right\}. \tag{14.57}$$

Combining (14.54)–(14.56) leads to the result

$$H_{\mathrm{non-diagonal}} = \sum_{\mathrm{k}} \left\{\xi_\mathrm{k} - \frac{(\xi_\mathrm{k})^2}{\sqrt{\xi_\mathrm{k}{}^2 + |\nabla_\mathrm{k}|^2}} - \frac{|\nabla_\mathrm{k}|^2}{\sqrt{\xi_\mathrm{k}{}^2 + |\nabla_\mathrm{k}|^2}}\right\} - \Delta$$
$$= \sum_{\mathrm{k}} \left(\xi_\mathrm{k} - \sqrt{\xi_\mathrm{k}{}^2 + |\nabla_\mathrm{k}|^2}\right) + \Delta\,. \tag{14.58}$$

The parameter Δ stands for the variable-free function given below

$$\begin{aligned}\Delta &= \frac{1}{N}\sum_{\mathrm{k,q}} V_{\mathrm{k,q}} X_{\mathrm{k}} X_{\mathrm{q}}^{\dagger} = \frac{1}{N}\sum_{\mathrm{k}}\sum_{\mathrm{q}} V_{\mathrm{k,q}} \langle c_{-\mathrm{q},\downarrow} c_{\mathrm{q},\uparrow}\rangle \langle c_{\mathrm{k},\uparrow}^{\dagger} c_{-\mathrm{k},\downarrow}^{\dagger}\rangle \\ &\equiv \sum_{\mathrm{k}} \Lambda_{\mathrm{k}} \langle c_{\mathrm{k},\uparrow}^{\dagger} c_{-\mathrm{k},\downarrow}^{\dagger}\rangle ,\end{aligned} \tag{14.59}$$

where

$$\Lambda_{\mathrm{k}} = \frac{1}{N}\sum_{\mathrm{q}} V_{\mathrm{k,q}} \langle c_{-\mathrm{q},\downarrow} c_{\mathrm{q},\uparrow}\rangle . \tag{14.60}$$

According to (14.29), the total Hamiltonian is the sum of nondiagonal and diagonal Hamiltonians. In the following we use a convenient representation of these Hamiltonians. That is,

$$\mathcal{H}_{\mathrm{bogo}} = H_{\mathrm{nondiagonal}} + H_{\mathrm{diagonal}} \tag{14.61}$$

with

$$H_{\mathrm{nondiagonal}} = \sum_{\mathrm{k}} \left[(\xi_{\mathrm{k}} - \Upsilon_{\mathrm{k}}) + \Lambda_{\mathrm{k}} \langle c_{\mathrm{k},\uparrow}^{\dagger} c_{-\mathrm{k},\downarrow}^{\dagger}\rangle \right] , \tag{14.62}$$

$$H_{\mathrm{diagonal}} = \sum_{\mathrm{k},\sigma} \Upsilon_{\mathrm{k}} \left(\gamma_{\mathrm{k},\sigma}^{\dagger} \gamma_{\mathrm{k},\sigma} \right) , \tag{14.63}$$

where $(\gamma_{\mathrm{k},\sigma}^{\dagger}\gamma_{\mathrm{k},\sigma})$ is the number of bogolyubons with wave-vector k and spin σ. An important point to note is that the nondiagonal Hamiltonian can more informatively be referred to as the ground-state energy of the system, heretofore to be denoted E_{ground}. Also that, according to (14.2), the variable ξ_{k} stands for the parameter $(\epsilon_{\mathrm{k}} - E_{\mathrm{F}})$, where ϵ_{k} is the kinetic energy of an electron and E_{F} is the Fermi energy. Similarly, according to (14.47), the dispersion relation for a bogolyubon is Υ_{k},

$$\Upsilon_{\mathrm{k}} \equiv \sqrt{\xi_{\mathrm{k}}{}^{2} + |\nabla_{\mathrm{k}}|^{2}} . \tag{14.64}$$

Remark 14.1 Despite the elegance of the Bogolyubov theory, it failed to predict the onset of superconductivity.

References

1. A.L. Fetter, J.D. Walecka, *Quantum Theory of Many Particle Systems* (McGraw-Hill, New York, 1971)
2. G. Rickayzen, *Theory of Superconductivity* (Wiley, New York, 1965)
3. P.-G. de Gennes, *Superconductivity of Metals and Alloys* (Perseus Books, Reading, 1999). Advanced Book Program

4. M. Tinkham, *Introduction to Superconductivity*, 2nd edn. (Dover, New York, 2004)
5. N.N. Bogoliubov, Sov. Phys. JETP **7**(1), 41–46 (1958)
6. N.N. Bogoliubov, Sov. Phys. JETP **34**(7), 51–55 (1958)
7. G.B. Thomas, *Calculus and Analytic Geometry*, 4th edn., vol. 15 (Addison-Wesley, Reading, 1969), p. 441, Exercise 15

London and London Theory 15

Electricity and magnetism are interconnected. In the following, i.e., Sect. 15.1, we study first the flow of electricity and its effect on magnetic properties of the system. Electric resistivity is an important parameter in the choice of a current carrying wire. Conductivity is the physical opposite of resistivity. In Sect. 15.2 causes of perfect conductivity are analyzed. Section 15.3 refers to one-dimensional theory and Sect. 15.4 deals with arbitrary change of theory, and the Meissner effect.

15.1 Electricity and Magnetism

Kamerlingh Onnes [1], while examining transport properties of solid mercury, noticed a strange phenomenon. As the temperature was lowered, there occurred a great decrease in electrical resistivity. The reduction in resistivity was both sudden and total. Indeed, at 4.2 K the resistance appeared to have become zero. The conductivity had become infinite. In other words, solid mercury had become a superconductor.

Electrical resistivity in ordinary metals—other than having a constant value due to impurities—decreases monotonically as the square-root of the temperature. This occurrence is credited to interparticle interaction. There is also a much faster decrease, due to phonon scattering, that is proportional to the fifth-power of temperature. These expectations contrasted with the observed behavior of solid mercury. Clearly, any reliable explanation of Kamerlingh Onnes observations needed new physics.

In addition to conductivity becoming infinite at a critical low temperature, another important characteristic of superconductors is that in its super-conducting state an externally imposed magnetic field does not fully penetrate the system. This phenomenon is called Meissner effect [2]. In other words, since the superconducting state allows only a limited penetration of the externally applied magnetic field, the metal has become diamagnetic. The penetration depth of the field is usually short compared to the depth of the sample and does not affect the inner workings of the system. There is, however, a special strength of the magnetic field to completely

R. Tahir-Kheli, *General and Statistical Thermodynamics*,
https://doi.org/10.1007/978-3-030-20700-7_15

destroy superconductivity. The critical strength of the magnetic field at which complete destruction of superconductivity occurs is dependent on the temperature and the chemistry of the material under consideration.

Many superconductors behave similarly to solid mercury and show dramatic drop in resistance at a particular transition temperature. These are labeled type-I superconductors. When moving up the temperature scale, these systems cross back into the normal state at the same critical temperature. There are, however, other superconductors—called type-II—where the drop in resistance is gradual rather than sudden. An intermediate state, labeled mixed state, appears before transition to the superconducting state. On the way back, the system again passes through the mixed state before normalcy recurs. And more than one temperature is involved in this process.

15.2 Perfect Conductivity

Consider an electric field $\boldsymbol{F}_{\mathbf{r}}(t)$ that is a function of time and space and assume that, at any given spatial location $\boldsymbol{r}$, it does not vary with time t. As such

$$\frac{\partial \boldsymbol{F}_{\mathbf{r}}(t)}{\partial t} = 0 . \tag{15.1}$$

Because of perfect conductivity, which implies vanishing resistivity, the motion of an electric charge, $-e$, of mass, m, occurs without resistance. The applied electric field, $\boldsymbol{F}_{\mathbf{r}}(t)$, accelerates the motion of the charge in its direction. That is,

$$-e\boldsymbol{F}_{\mathbf{r}}(t) = m\boldsymbol{r}^{\cdot\cdot} = m\mathbf{drt} . \tag{15.2}$$

Note that $\boldsymbol{r}^{\cdot\cdot}$ stands for the acceleration vector. Assuming the electric field acts on ν superconducting electrons, the current density, $\boldsymbol{C}_{\mathbf{r}}(t)$, is

$$\boldsymbol{C}_{\mathbf{r}}(t) = -e\nu\boldsymbol{r}^{\cdot} = -e\nu\frac{\partial \boldsymbol{r}(t)}{\partial t} . \tag{15.3}$$

Using (15.2), we can write the time differential of (15.3) as

$$\frac{\partial \boldsymbol{C}_{\mathbf{r}}(t)}{\partial t} = -e\nu\boldsymbol{r}^{\cdot\cdot} = \frac{e^2\nu\boldsymbol{F}_{\mathbf{r}}(t)}{m} . \tag{15.4}$$

Maxwell's third equation is the Faraday's law that relates the vector product of $\boldsymbol{D}$ and $\boldsymbol{F}_{\mathbf{r}}(t)$ to the rate of change of the magnetic-flux density $\boldsymbol{M}$ and shows how a magnetic field changing with time creates an electric field circulating around it. The magnetic-flux density is a function both of time, t, and position, $\boldsymbol{r}$. Its partial derivative with respect to time is used in the Faraday's law written below, namely

$$\boldsymbol{D} \times \boldsymbol{F}_{\mathbf{r}}(t) = -\frac{1}{c}\frac{\partial \boldsymbol{M}_{\mathbf{r}}(t)}{\partial t} . \tag{15.5}$$

In (15.5), c is the speed of light, $\boldsymbol{M}_{\mathbf{r}}(t)$ is the magnetic flux density, $\boldsymbol{D}$ is as defined in (15.5), i.e.,

$$\boldsymbol{D} = \boldsymbol{i}\frac{d}{\mathrm{d}x} + \boldsymbol{j}\frac{d}{\mathrm{d}y} + \boldsymbol{k}\frac{d}{\mathrm{d}z}, \qquad \boldsymbol{D}\cdot\boldsymbol{D} = D^2 = \frac{d^2}{\partial^2 x} + \frac{d^2}{\partial^2 y} + \frac{d^2}{\partial^2 z}, \tag{15.6}$$

the symbol $\times$ denotes vector-product, and $\boldsymbol{i}, \boldsymbol{j}, \boldsymbol{k}$ are unit vectors in the x-, y-, and z-directions of a Cartesian coordinate system. Inserting the result for $\boldsymbol{F}_{\mathbf{r}}(t)$, given in (15.4), into (15.5) gives

$$\boldsymbol{D} \times \frac{\partial \boldsymbol{C}_{\mathbf{r}}(t)}{\partial t} = -\left(\frac{\nu e^2}{mc}\right)\frac{\partial \boldsymbol{M}_{\mathbf{r}}(t)}{\partial t}. \tag{15.7}$$

Relevant to this subject matter is the work of Ampere and Maxwell [3]. Indeed, the Ampere–Maxwell law states that $\boldsymbol{D} \times \boldsymbol{M}_{\mathbf{r}}(\boldsymbol{\partial t}) = \frac{1}{c}(4\pi \boldsymbol{C}_{\mathbf{r}}(t) + \frac{\partial \boldsymbol{F}}{\mathrm{d}t})$. Because of (15.1), the relevant form of the Ampere–Maxwell law is

$$\boldsymbol{D} \times \boldsymbol{M}_{\mathbf{r}}(t) = \frac{1}{c}\left(4\pi \boldsymbol{C}_{\mathbf{r}}(t)\right). \tag{15.8}$$

Its time-derivative gives

$$\boldsymbol{D} \times \frac{\partial \boldsymbol{M}_{\mathbf{r}}(t)}{\partial t} = \frac{1}{c}\left(4\pi \frac{\partial \boldsymbol{C}_{\mathbf{r}}(t)}{\partial t}\right). \tag{15.9}$$

By vector-multiplying (15.9) from the left by $\boldsymbol{D}$, we get

$$\boldsymbol{D} \times \boldsymbol{D} \times \frac{\partial \boldsymbol{M}_{\mathbf{r}}(\boldsymbol{\partial t})}{\partial t} = \left(\frac{4\pi}{c}\right)\boldsymbol{D} \times \frac{\partial \boldsymbol{C}_{\mathbf{r}}(t)}{\partial t}. \tag{15.10}$$

Use (15.7) and rewrite (15.10) as

$$\boldsymbol{D} \times \boldsymbol{D} \times \frac{\partial \boldsymbol{M}_{\mathbf{r}}(t)}{\partial t} = -\left(\frac{4\pi e^2 \mathrm{v}}{mc^2}\right)\frac{\partial \boldsymbol{M}_{\mathbf{r}}(t)}{\partial t}. \tag{15.11}$$

Under the rules of vector multiplication [4], we have the identity

$$\boldsymbol{D} \times \boldsymbol{D} \times \frac{\partial \boldsymbol{M}_{\mathbf{r}}(t)}{\partial t} = \boldsymbol{D}\left(\boldsymbol{D}\cdot\frac{\partial \boldsymbol{M}_{\mathbf{r}}(t)}{\partial t}\right) - (\boldsymbol{D}\cdot\boldsymbol{D})\frac{\partial \boldsymbol{M}_{\mathbf{r}}(t)}{\partial t}. \tag{15.12}$$

Gauss' law for magnetism states that the magnetic monopoles do not exist. That is,

$$\boldsymbol{D}\cdot\boldsymbol{M}_{\mathbf{r}}(t) = 0. \tag{15.13}$$

By differentiating (15.13) with respect to time, we can write

$$\boldsymbol{D}\cdot\frac{\partial \boldsymbol{M}_{\mathbf{r}}(t)}{\partial t} = 0. \tag{15.14}$$

As a result the term $\boldsymbol{D}(\boldsymbol{D} \cdot \frac{\partial \boldsymbol{M}_{\mathbf{r}}(t)}{\partial t})$ in (15.12) can be set equal to zero, and with the help of (15.11), (15.12) can be rewritten as

$$-\boldsymbol{D} \times \boldsymbol{D} \times \frac{\partial \boldsymbol{M}_{\mathbf{r}}(t)}{\partial t} = \boldsymbol{D} \cdot \boldsymbol{D} \frac{\partial \boldsymbol{M}_{\mathbf{r}}(t)}{\partial t} = \left(\frac{4\pi e^2 \nu}{mc^2}\right) \frac{\partial \boldsymbol{M}_{\mathbf{r}}(t)}{\partial t} . \tag{15.15}$$

Noting the equalities

$$\begin{aligned} \boldsymbol{D} \cdot \boldsymbol{D} &= D^2 = \operatorname{div} \operatorname{grad} = \frac{\partial^2}{\partial^2 x} + \frac{\partial^2}{\partial^2 y} + \frac{\partial^2}{\partial^2 z} , \\ \boldsymbol{M}_{\mathbf{r}}(t) &= \boldsymbol{i} M_{\mathrm{x}}(t) + \boldsymbol{j} M_{\mathrm{y}}(t) + \boldsymbol{k} M_{\mathrm{z}}(t) , \end{aligned} \tag{15.16}$$

equation (15.15) gives

$$D^2 \frac{\partial \boldsymbol{M}_{\mathbf{r}}(t)}{\partial t} = \left(\frac{4\pi e^2 \nu}{mc^2}\right) \frac{\partial \boldsymbol{M}_{\mathbf{r}}(t)}{\partial t} , \tag{15.17}$$

or equivalently,

$$\begin{aligned} &\boldsymbol{i} \frac{\partial^2 \{\frac{\partial \boldsymbol{M}_{\mathbf{x}}(t)}{\partial t}\}}{\partial^2 x} + \boldsymbol{j} \frac{\partial^2 \{\frac{\partial \boldsymbol{M}_{\mathbf{y}}(t)}{\partial t}\}}{\partial^2 y} + \boldsymbol{k} \frac{\partial^2 \{\frac{\partial \boldsymbol{M}_{\mathbf{z}}(t)}{\partial t}\}}{\partial^2 z} \\ &= \left(\frac{4\pi e^2 \nu}{mc^2}\right) \left[\boldsymbol{i} \frac{\partial \boldsymbol{M}_{\mathbf{x}}(t)}{\partial t} + \boldsymbol{j} \frac{\partial \boldsymbol{M}_{\mathbf{y}}(t)}{\partial t} + \boldsymbol{k} \frac{\partial \boldsymbol{M}_{\mathbf{z}}(t)}{\partial t} \right] . \end{aligned} \tag{15.18}$$

15.3 One-Dimensional Theory

Rather than treating a three-dimensional system, a good physical feel can be had by studying a one-dimensional system. That means keeping only the x-dependent terms in differential (15.18):

$$\boldsymbol{i} \frac{\partial^2 \{\frac{\partial \boldsymbol{M}_{\mathbf{x}}(t)}{\partial t}\}}{\partial^2 x} = \boldsymbol{i} \left(\frac{4\pi e^2 \nu}{mc^2}\right) \left\{ \frac{\partial \boldsymbol{M}_{\mathbf{x}}(t)}{\partial t} \right\} . \tag{15.19}$$

Assume the system is a perfect conductor for all locations $x \geq 0$.

A possible solution of the of the one-dimensional differential (15.19) is (15.20), namely

$$\begin{aligned} \frac{\partial \boldsymbol{M}_{\mathbf{x}}(t)}{\partial t} &= \text{const.}_1 \cdot \exp\left[-x \cdot \sqrt{\left(\frac{4\pi e^2 \nu}{mc^2}\right)} \right] \\ &\quad + \text{const.}_2 \cdot \exp\left[x \cdot \sqrt{\left(\frac{4\pi e^2 \nu}{mc^2}\right)} \right] . \end{aligned} \tag{15.20}$$

Clearly, the second term in (15.20), that increases exponentially with position, is unacceptable. Therefore a more appropriate form of (15.20) is (15.21), that is,

$$\frac{\partial \boldsymbol{M}_{\mathbf{x}}(\boldsymbol{t})}{\partial t} = \text{const.}_1 \cdot \exp\left[-x \cdot \sqrt{\left(\frac{4\pi e^2 \nu}{mc^2}\right)}\right]. \tag{15.21}$$

In order to get a feel for the behavior of $\boldsymbol{M}_{\mathbf{x}}(\boldsymbol{t})$, we need to integrate (15.21) with respect to time to get

$$\boldsymbol{M}_{\mathbf{x}}(\boldsymbol{t}) = \text{const.}_1 \cdot \exp\left[-x \cdot \sqrt{\left(\frac{4\pi e^2 \nu}{mc^2}\right)}\right] \cdot t + \text{const.}_3 \,. \tag{15.22}$$

One might wonder why there is more than one unknown constant. It is so because we have inherited this result from a higher-order differential equation (15.19). More importantly, one might ask why is the final result in physical disagreement with Meissner's experimental observations [2], namely that while a material is in its superconducting state magnetic-flux density, $\boldsymbol{M}_{\mathbf{x}}(\boldsymbol{t})$, penetrates it for only a tiny distance of order 10^{-8} meters, vanishes within the semiconductor, and this behavior does not change with time. The present theory predicts something entirely different, namely that while the magnetic-flux density inside a material decreases exponentially with position—which would be not too-bad just by itself—at any given location, the magnetic-flux density forever keeps increasing with time. Clearly, this unhappy result is owed entirely to the unphysical nature of differential equation (15.19). Therefore, for achieving a physically correct solution, the differential equation (15.19) needs to be modified.

15.4 Arbitrary Change of Theory: Meissner Effect

Presented with the unphysical results described in (15.22), London and London proposed a change to the theory and suggested eliminating the time-derivative from the one dimensional master-differential equation (15.19). That means using $\boldsymbol{M}_{\mathbf{x}}(\boldsymbol{t})$ in place of $\frac{\partial \boldsymbol{M}_{\mathbf{x}}(\boldsymbol{t})}{\partial t}$, thereby arbitrarily changing (15.19) into (15.23), which is

$$\boldsymbol{i}\frac{\partial^2 \boldsymbol{M}_{\mathbf{x}}(\boldsymbol{t})}{\partial^2 x} = \boldsymbol{i}\frac{\boldsymbol{M}_{\mathbf{x}}(\boldsymbol{t})}{\lambda^2} \,. \tag{15.23}$$

Note that the new notation was introduced in (15.23) whereby

$$\sqrt{\frac{4\pi e^2 \nu}{mc^2}} \equiv \frac{1}{\lambda} \,. \tag{15.24}$$

Being second-order, (15.23) has two solutions,

$$\boldsymbol{M}_{\mathbf{x}}(\boldsymbol{t}) = A \exp\left[\pm\frac{x}{\lambda}\right]. \tag{15.25}$$

As noted before, the exponentially increasing solution is unphysical. Therefore, the London brothers used the exponentially decreasing solution

$$\boldsymbol{M}_{\mathbf{x}}(t) = A \exp\left[-\frac{x}{\lambda}\right]. \tag{15.26}$$

Upon penetrating a distance λ, the magnetic-flux density decreases by more than 62% of its original value. For this reason, the symbol λ is referred to as the London penetration depth.

Equation (15.26) suggests that the system is moderately diamagnetic. Moreover, and quite importantly, because such behavior is not affected by the passage of time, one can conclude that the theory gives a fair description of the Meissner effect in one dimension.

Similar to one-dimensional superconductor, in order to describe a three-dimensional superconductor London and London arbitrarily changed the master differential equation (15.17) to the differential equation (15.27) given below as

$$D^2 \boldsymbol{M}_{\mathbf{r}}(t) = \left(\frac{4\pi e^2 \nu}{mc^2}\right) \boldsymbol{M}_{\mathbf{r}}(t) = \frac{1}{\lambda^2} \boldsymbol{M}_{\mathbf{r}}(t). \tag{15.27}$$

Again, much like the one-dimensional differential equation (15.23), the three-dimensional London–London differential equation (15.27) gives a fair description of the Meissner effect.

References

1. H. Kamerlingh Onnes, Leiden, Comm., 120 b, 122 b, 124 c (1911)
2. W. Meissner, R. Ochsenfeld, Naturwissenchaften **21**(787) (1933)
3. J.C. Maxwell, Op. cit.
4. M.-L. Boaz, Mathematical Methods in Physical Sciences, 3rd edn., p. 297, Example 1

Superconductivity 16

Transition from a state of normal conductivity to a superconducting state is a second-order phase transition. Landau's theory of second-order phase transitions is presented in Chap. 12 ("Second-Order Phase Transitions").

In Chap. 13 two free electrons, with attractive interpair interaction, positioned somewhere above the Fermi sea are shown to form a bound Cooper pair. As such it is reasonable to expect the presence of many free electrons above the Fermi sea and the attractiveness—no matter how weak—of the net interparticle interaction would cause formation of a manybody bound-state of Cooper pairs, leading to greatly reduced resistivity. That happenstance provides a rationale [1–6] for superconductivity. An important part of Cooper's theory is described in Chap. 13 ("Cooper Pair").

Chapter 14, titled "Bogolyubov Representation," chronicles the rather substantial effort Bogolyubov made toward explaining a manybody ordered state beginning with a specified Hamiltonian.

London and London's work regarding perfect semiconductors is discussed in Chap. 15 titled "London and London Theory." A notable subject studied there is the Meissner [7] effect. Compare (15.23)–(15.27). Despite the arbitrariness of the London and London theory, it provides a plausible description of the Meissner effect which is a notable feature of the superconducting state. Beginning with Sect. 16.1 some important aspects of Bardeen, Cooper, and Schrieffer (BCS) theory are presented in Chap. 16. In Sects. 16.2–16.17 the following subjects are studied:

Section 16.2 treats $[\mathcal{H}_{\mathrm{BCS}}]_{\mathrm{MFA}}$ Hamiltonian, while Sect. 16.3 deals with fermionic quasiparticles. Section 16.4 describes the inversion of (16.14) and (16.15). Section 16.5 studies elimination of terms in the nondiagonal Hamiltonian. Section 16.6 refers to the transformation of $\mathcal{H}_{\mathrm{nondiagonal}}$. Section 16.7 deals with the evaluation of diagonal terms. Section 16.8 provides information about Fermi surface. Section 16.9 studies the regions both above and within the Fermi sea. Section 16.10 provides information about the gap function and its solutions. Section 16.11 studies the critical temperature T_{c}. Section 16.12 describes the Debye potential, $\hbar\Omega_{\mathrm{D}}$. Section 16.13 discusses the isotope effect. Section 16.14 looks at the specific heat near T_{c}.

R. Tahir-Kheli, *General and Statistical Thermodynamics*,
https://doi.org/10.1007/978-3-030-20700-7_16

Section 16.15 describes the calculation of jump in specific heat, Sect. 16.16 presents a universal ratio, Sect. 16.17 another universal ratio, while Sect. 16.17 introduces the result at zero temperature.

As noted in (16.74) and (16.75), exactly at the Fermi surface, the fermionic quasiparticles behave as electrons and holes with equal strength, leading to electrical neutrality. Such behavior portends superconductivity.

16.1 Theory

The microscopic theory of superconductivity begins with Bardeen, Cooper and Schrieffer (BCS) Hamiltonian

$$\mathcal{H}_{\mathrm{BCS}} = \sum_{\mathrm{k}} \xi_{\mathrm{k}} \left(c^{\dagger}_{\mathrm{k},\uparrow} c_{\mathrm{k},\uparrow} + c^{\dagger}_{-\mathrm{k},\downarrow} c_{-\mathrm{k},\downarrow} \right) + \frac{1}{N} \sum_{\mathrm{k,q}} V c^{\dagger}_{\mathrm{k},\uparrow} c^{\dagger}_{-\mathrm{k},\downarrow} c_{-\mathrm{q},\downarrow} c_{\mathrm{q},\uparrow} \,, \tag{16.1}$$

where symbol V stands for the interparticle attractive force.

Equation (16.1) uses both up- and down-spin creation and destruction operators for electrons. These operators are $c^{\dagger}_{\mathrm{k},\uparrow}$, $c^{\dagger}_{\mathrm{k},\downarrow}$, $c_{\mathrm{k},\uparrow}$, and $c_{\mathrm{k},\downarrow}$. The energy variable ξ_{k} is defined as

$$\xi_{\mathrm{k}} \equiv \epsilon_{\mathrm{k}} - E_{\mathrm{F}} \,, \tag{16.2}$$

where ϵ_{k} is the kinetic energy of an electron, E_{F} is the Fermi energy, and $\xi_{\mathrm{k}} \equiv \xi_{-\mathrm{k}} \equiv \xi^{\dagger}{}_{\mathrm{k}}$. At the Fermi surface, $\epsilon_{\mathrm{k}} = E_{\mathrm{F}}$. Therefore exactly at the Fermi surface, $\xi_{\mathrm{k}} = 0$. Another point to note is that the maximum possible range of the energy variable ξ_{k} is from $-\hbar\Omega_{\mathrm{D}}$ to $+\hbar\Omega_{\mathrm{D}}$ where the symbol $\hbar\Omega_{\mathrm{D}}$ stands for the Debye energy. In practice, a majority of the electrons are likely to sit closer to the Fermi surface than their maximum possible separation permits.

Expressions that are helpful for studying $\mathcal{H}_{\mathrm{BCS}}$ are the following:

$$\begin{aligned} c^{\dagger}_{\mathrm{k},\uparrow} c^{\dagger}_{-\mathrm{k},\downarrow} &= \langle c^{\dagger}_{\mathrm{k},\uparrow} c^{\dagger}_{-\mathrm{k},\downarrow} \rangle + \left[c^{\dagger}_{\mathrm{k},\uparrow} c^{\dagger}_{-\mathrm{k},\downarrow} - \langle c^{\dagger}_{\mathrm{k},\uparrow} c^{\dagger}_{-\mathrm{k},\downarrow} \rangle \right], \\ &\equiv X_{\mathrm{k}} + \left[c^{\dagger}_{\mathrm{k},\uparrow} c^{\dagger}_{-\mathrm{k},\downarrow} - X_{\mathrm{k}} \right] \end{aligned} \tag{16.3}$$

and

$$\begin{aligned} c_{-\mathrm{q},\downarrow} c_{\mathrm{q},\uparrow} &= \langle c_{-\mathrm{q},\downarrow} c_{\mathrm{q},\uparrow} \rangle + \left[c_{-\mathrm{q},\downarrow} c_{\mathrm{q},\uparrow} - \langle c_{-\mathrm{q},\downarrow} c_{\mathrm{q},\uparrow} \rangle \right], \\ &\equiv X^{\dagger}_{\mathrm{q}} + \left[c_{-\mathrm{q},\downarrow} c_{\mathrm{q},\uparrow} - X^{\dagger}_{\mathrm{q}} \right], \end{aligned} \tag{16.4}$$

where

$$X_{\mathrm{k}} = \langle c^{\dagger}_{\mathrm{k},\uparrow} c^{\dagger}_{-\mathrm{k},\downarrow} \rangle, \qquad X^{\dagger}_{\mathrm{q}} = \langle c_{-\mathrm{q},\downarrow} c_{\mathrm{q},\uparrow} \rangle \,. \tag{16.5}$$

The symbol $\langle \ldots \rangle$ means the thermodynamic average of "...", while the operators $c^{\dagger}_{\mathrm{k},\uparrow} c^{\dagger}_{-\mathrm{k},\downarrow}$ and $c_{-\mathrm{q},\downarrow} c_{\mathrm{q},\uparrow}$ fluctuate around their thermodynamic averages by amounts $[c^{\dagger}_{\mathrm{k},\uparrow} c^{\dagger}_{-\mathrm{k},\downarrow} - X_{\mathrm{k}}]$ and $[c_{-\mathrm{q},\downarrow} c_{\mathrm{q},\uparrow} - X^{\dagger}_{\mathrm{q}}]$. These fluctuations are tiny and their squares can be assumed to be too small to be relevant.

With the use of expressions (16.3)–(16.5), the BCS Hamiltonian $[\mathcal{H}_{\mathrm{BCS}}]$ may be written as

$$
\begin{aligned}
&[\mathcal{H}_{\mathrm{BCS}}] - \sum_{\mathrm{k}} \xi_{\mathrm{k}}\left(c^{\dagger}_{\mathrm{k},\uparrow} c_{\mathrm{k},\uparrow} + c^{\dagger}_{-\mathrm{k},\downarrow} c_{-\mathrm{k},\downarrow}\right) \\
&= \frac{1}{N} \sum_{\mathrm{k,q}} V\left\{X_{\mathrm{k}} + \left[c^{\dagger}_{\mathrm{k},\uparrow} c^{\dagger}_{-\mathrm{k},\downarrow} - X_{\mathrm{k}}\right]\right\} \times \left\{X^{\dagger}_{\mathrm{q}} + \left[c_{-\mathrm{q},\downarrow} c_{\mathrm{q},\uparrow} - X^{\dagger}_{\mathrm{q}}\right]\right\} \\
&= \frac{1}{N} \sum_{\mathrm{k,q}} V\left\{X_{\mathrm{k}} c_{-\mathrm{q},\downarrow} c_{\mathrm{q},\uparrow} + X^{\dagger}_{\mathrm{q}} c^{\dagger}_{\mathrm{k},\uparrow} c^{\dagger}_{-\mathrm{k},\downarrow} - X_{\mathrm{k}} X^{\dagger}_{\mathrm{q}}\right\} \\
&+ \frac{1}{N} \sum_{\mathrm{k,q}} V\left\{\left[c^{\dagger}_{\mathrm{k},\uparrow} c^{\dagger}_{-\mathrm{k},\downarrow} - X_{\mathrm{k}}\right] \cdot \left[c_{-\mathrm{q},\downarrow} c_{\mathrm{q},\uparrow} - X^{\dagger}_{\mathrm{q}}\right]\right\}. \qquad (16.6)
\end{aligned}
$$

The last line on the right-hand side of (16.6) represents a product of two fluctuations. Because both these fluctuations are expected to be small, their product should be much smaller. Consequently, in a mean-field type approximation (MFA) this product may be ignored. That is,

$$
\frac{1}{N} \sum_{\mathrm{k,q}} V\left\{\left[c^{\dagger}_{\mathrm{k},\uparrow} c^{\dagger}_{-\mathrm{k},\downarrow} - X_{\mathrm{k}}\right] \cdot \left[c_{-\mathrm{q},\downarrow} c_{\mathrm{q},\uparrow} - X^{\dagger}_{\mathrm{q}}\right]\right\} \approx 0. \qquad (16.7)
$$

16.2 $[\mathcal{H}_{\mathrm{BCS}}]_{\mathrm{MFA}}$ Hamiltonian

Using the approximation (16.7), the Hamiltonian $[\mathcal{H}_{\mathrm{BCS}}]$ becomes $[\mathcal{H}_{\mathrm{BCS}}]_{\mathrm{MFA}}$, that is,

$$
\begin{aligned}
[\mathcal{H}_{\mathrm{BCS}}]_{\mathrm{MFA}} &= \sum_{\mathrm{k}} \xi_{\mathrm{k}}\left(c^{\dagger}_{\mathrm{k},\uparrow} c_{\mathrm{k},\uparrow} + c^{\dagger}_{-\mathrm{k},\downarrow} c_{-\mathrm{k},\downarrow}\right) \\
&+ \frac{1}{N} \sum_{\mathrm{k,q}} V\left\{X_{\mathrm{k}} c_{-\mathrm{q},\downarrow} c_{\mathrm{q},\uparrow} + X^{\dagger}_{\mathrm{q}} c^{\dagger}_{\mathrm{k},\uparrow} c^{\dagger}_{-\mathrm{k},\downarrow} - X_{\mathrm{k}} X^{\dagger}_{\mathrm{q}}\right\}. \qquad (16.8)
\end{aligned}
$$

Let us make a notational change from q to k and vice-versa in the first term on the right-hand side of (16.8). Whereby we can write (16.8) as

$$
\begin{aligned}
[\mathcal{H}_{\mathrm{BCS}}]_{\mathrm{MFA}} &= \sum_{\mathrm{k}} \xi_{\mathrm{k}}\left(c^{\dagger}_{\mathrm{k},\uparrow} c_{\mathrm{k},\uparrow} + c^{\dagger}_{-\mathrm{k},\downarrow} c_{-\mathrm{k},\downarrow}\right) \\
&+ \frac{1}{N} \sum_{\mathrm{k,q}} V\left\{X_{\mathrm{q}} c_{-\mathrm{k},\downarrow} c_{\mathrm{k},\uparrow} + X^{\dagger}_{\mathrm{q}} c^{\dagger}_{\mathrm{k},\uparrow} c^{\dagger}_{-\mathrm{k},\downarrow} - X_{\mathrm{k}} X^{\dagger}_{\mathrm{q}}\right\}. \qquad (16.9)
\end{aligned}
$$

For simplicity of current and future display, we need to introduce another notational change, this time in (16.9), namely

$$\nabla = \frac{1}{N}\sum_{q} V X_q = \left(\frac{1}{N}\right)\sum_{q} V\langle c^{\dagger}_{q\uparrow} c^{\dagger}_{-q\downarrow}\rangle$$
$$\nabla^{\dagger} = \frac{1}{N}\sum_{q} V X^{\dagger}_q = \frac{1}{N}\sum_{q} V\langle c_{-q\downarrow} c_{q\uparrow}\rangle\,. \tag{16.10}$$

As a result, $[\mathcal{H}_{BCS}]_{MFA}$ given in (16.9) can be rewritten in a more compact form as

$$[\mathcal{H}_{BCS}]_{MFA} - \sum_{k} \xi_k\left(c^{\dagger}_{k,\uparrow}c_{k,\uparrow} + c^{\dagger}_{-k,\downarrow}c_{-k,\downarrow}\right)$$
$$= \sum_{k}\left\{\nabla c_{-k,\downarrow}c_{k,\uparrow} + \nabla^{\dagger} c^{\dagger}_{k,\uparrow}c^{\dagger}_{-k,\downarrow}\right\} - N\frac{\nabla\nabla^{\dagger}}{V}$$
$$= \sum_{k}\left\{\nabla c_{-k,\downarrow}c_{k,\uparrow} + \nabla^{\dagger} c^{\dagger}_{k,\uparrow}c^{\dagger}_{-k,\downarrow}\right\} + \Delta\,. \tag{16.11}$$

The parameter Δ stands for the variable-free function,

$$\Delta = -N\frac{\nabla\nabla^{\dagger}}{V} = -\frac{V}{N}\sum_{k,q} X_k X^{\dagger}_q$$
$$\equiv \sum_{k}\langle c^{\dagger}_{k,\uparrow}c^{\dagger}_{-k,\downarrow}\rangle\Lambda\,. \tag{16.12}$$

Here Λ is the so called gap-function that is defined as follows:

$$\Lambda = -\frac{V}{N}\sum_{q}\langle c_{-q,\downarrow}c_{q,\uparrow}\rangle = -\nabla^{\dagger}\,. \tag{16.13}$$

We have used (16.10) to insert $-\nabla^{\dagger}$ at the end of (16.13). Note that Λ, Λ†, ∇, and $\nabla^{\dagger}$ are not operators.

16.3 Fermionic Quasiparticles

Electron creation and destruction operators obey anticommutation rules. Our requirement for the fermionic quasiparticle operators, $\gamma_{k,\sigma}$ and $\gamma^{\dagger}_{k,\sigma}$, is to do the same. This requirement is confirmed by the subject matter given below.

Although the Hamiltonian $[\mathcal{H}_{BCS}]_{MFA}$ does not conserve particle number, it can be diagonalized by using an appropriate set, i.e.,

$$c^{\dagger}_{k,\uparrow} \equiv u\gamma^{\dagger}_{k,\uparrow} + v\gamma_{-k,\downarrow}\,,$$
$$c^{\dagger}_{-k,\downarrow} \equiv -v\gamma_{k,\uparrow} + u\gamma^{\dagger}_{-k,\downarrow}\,, \tag{16.14}$$

$$c_{k,\uparrow} \equiv u\gamma_{k,\uparrow} + v\gamma^{\dagger}_{-k,\downarrow}\,,$$
$$c_{-k,\downarrow} \equiv -v\gamma^{\dagger}_{k,\uparrow} + u\gamma_{-k,\downarrow}\,, \tag{16.15}$$

of fermionic quasiparticle operators. Such operators create—e.g., $\gamma^{\dagger}_{k,\sigma}$—and destroy—e.g., $\gamma_{k,\sigma}$—fermionic quasiparticles. Furthermore, these operators follow the same anticommutation rules that are obeyed by electron creation—i.e., $c^{\dagger}_{k,\uparrow}$—and destruction—i.e., $c_{k,\uparrow}$—operators. To double-check on this statement, we need to construct an inverse relationship that precisely inverts[1] (16.14) and (16.15) into the desired fermionic quasiparticle operators. Doing that would give $\gamma_{k,\uparrow}$ and $\gamma^{\dagger}_{-k,\downarrow}$ in terms of $c_{k,\uparrow}$ and $c^{\dagger}_{-k,\downarrow}$. For the reader's convenience, relevant parts of the results are recorded below:

$$\gamma_{k,\uparrow} \equiv uc_{k,\uparrow} - vc^{\dagger}_{-k,\downarrow}\,,$$
$$\gamma^{\dagger}_{-k,\downarrow} \equiv vc_{k,\uparrow} + uc^{\uparrow}_{-k,\downarrow}\,. \tag{16.16}$$

The complex conjugate of (16.16) is

$$\gamma^{\dagger}_{k,\uparrow} \equiv \left[uc^{\dagger}_{k,\uparrow} - vc_{-k,\downarrow}\right],$$
$$\gamma_{-k,\downarrow} \equiv \left[vc^{\dagger}_{k,\uparrow} + uc_{-k,\downarrow}\right]. \tag{16.17}$$

Electron creation and destruction operators obey anticommutation rules. Our requirement for the fermionic quasi-particle operators, $\gamma_{k,\sigma}$ and $\gamma^{\dagger}_{k,\sigma}$, is to do the same. That is,

$$\begin{aligned}
&\left[\gamma_{-k,\uparrow}, \gamma^{\dagger}_{-k,\downarrow}\right]_{+} = 0\,, \quad &&[\gamma_{-k,\downarrow}, \gamma_{k,\uparrow}]_{+} = 0\,, \quad &&\left[\gamma^{\dagger}_{k,\uparrow}, \gamma^{\dagger}_{-k,\downarrow}\right]_{+} = 0\,,\\
&\left[\gamma_{k,\uparrow}, \gamma^{\dagger}_{k,\uparrow}\right]_{+} = 1\,, &&\left[\gamma_{k,\uparrow}, \gamma^{\dagger}_{-k,\downarrow}\right]_{+} = 0\,, &&\left[\gamma_{-k,\downarrow}, \gamma^{\dagger}_{-k,\downarrow}\right]_{+} = 1\,,\\
&\left[\gamma_{k,\downarrow}, \gamma^{\dagger}_{k,\downarrow}\right]_{+} = 1\,, &&\left[\gamma_{k,\downarrow}, \gamma^{\dagger}_{-k,\uparrow}\right]_{+} = 0, &&\left[\gamma_{-k,\uparrow}, \gamma^{\dagger}_{-k,\uparrow}\right]_{+} = 1\,.
\end{aligned} \tag{16.18}$$

In this regard, let us check how well that works out for $[\gamma_{k,\uparrow}, \gamma^{\dagger}_{k,\uparrow}]_{+}$. We have

$$\begin{aligned}
\left[\gamma_{k,\uparrow}, \gamma^{\dagger}_{k,\uparrow}\right]_{+} &= \left[uc_{k,\uparrow} - vc^{\dagger}_{-k,\downarrow}, uc^{\dagger}_{k,\uparrow} - v^{*}c_{-k,\downarrow}\right]_{+}\\
&= u^{2}\left[c_{k,\uparrow}, c^{\dagger}_{k,\uparrow}\right]_{+} + v^{2}\left[c^{\dagger}_{-k,\downarrow}, c_{-k,\downarrow}\right]_{+} + 0 + 0\\
&= u^{2} + v^{2}\,.
\end{aligned} \tag{16.19}$$

And the same is obtained for $\sigma \equiv\downarrow$. Importantly, the satisfaction of both these commutation relationships requires u and v obey the so-called normalization requirement,

$$u^{2} + v^{2} = 1\,. \tag{16.20}$$

[1]This inversion is worked out in detail in (16.26), (16.27), (16.28) and (16.29).

Clearly, the normalization requirement (16.20) is needed for the fermionic quasiparticle operators to be canonical. It is easy to check that in addition to satisfying the normalization requirement the fermionic quasiparticle operators also obey all the commutation rules.

In order to represent $\mathcal{H}_{\text{BCS}}$ in terms of the fermionic quasiparticle operators, we need to insert the relevant contents of (16.14) and (16.15) into (16.11). That is,

$$
\begin{aligned}
&[\mathcal{H}_{\text{BCS}}]_{\text{MFA}} \\
&\quad = \sum_{\mathrm{k}} \xi_{\mathrm{k}}\big(c^{\dagger}_{\mathrm{k},\uparrow} c_{\mathrm{k},\uparrow} + c^{\dagger}_{-\mathrm{k},\downarrow} c_{-\mathrm{k},\downarrow}\big) \\
&\qquad + \sum_{\mathrm{k}} \big\{\nabla c_{-\mathrm{k},\downarrow} c_{\mathrm{k},\uparrow} + \nabla^{\dagger} c^{\dagger}_{\mathrm{k},\uparrow} c^{\dagger}_{-\mathrm{k},\downarrow}\big\} + \Delta\,. \\
&\quad = \sum_{\mathrm{k}} \xi_{\mathrm{k}}\big(u\gamma^{\dagger}_{\mathrm{k},\uparrow} + v\gamma_{-\mathrm{k},\downarrow}\big)\big(u\gamma_{\mathrm{k},\uparrow} + v\gamma^{\dagger}_{-\mathrm{k},\downarrow}\big) \\
&\qquad + \sum_{\mathrm{k}} \xi_{\mathrm{k}}\big(-v\gamma_{\mathrm{k},\uparrow} + u\gamma^{\dagger}_{-\mathrm{k},\downarrow}\big)\big(-v\gamma^{\dagger}_{\mathrm{k},\uparrow} + u\gamma_{-\mathrm{k},\downarrow}\big) \\
&\qquad + \sum_{\mathrm{k}} \nabla\big\{-v^2\gamma^{\dagger}_{\mathrm{k},\uparrow}\gamma^{\dagger}_{-\mathrm{k},\downarrow} - uv\gamma^{\dagger}_{\mathrm{k},\uparrow}\gamma_{\mathrm{k},\uparrow} + vu\gamma_{-\mathrm{k},\downarrow}\gamma^{\dagger}_{-\mathrm{k},\downarrow} + u^2\gamma_{-\mathrm{k},\downarrow}\gamma_{\mathrm{k},\uparrow}\big\} \\
&\qquad + \sum_{\mathrm{k}} \nabla^{\dagger}\big\{-v^2\gamma_{-\mathrm{k},\downarrow}\gamma_{\mathrm{k},\uparrow} + uv\gamma_{-\mathrm{k},\downarrow}\gamma^{\dagger}_{-\mathrm{k},\downarrow} - vu\gamma^{\dagger}_{\mathrm{k},\uparrow}\gamma_{\mathrm{k},\uparrow} + u^2\gamma^{\dagger}_{\mathrm{k},\uparrow}\gamma^{\dagger}_{-\mathrm{k},\downarrow}\big\} \\
&\qquad + \Delta\,.
\end{aligned}
\tag{16.21}
$$

Note that Δ is as defined in (16.12) and (16.13). After some reorganization of the ξ_{k} terms, (16.21) leads to (16.22), namely

$$
\begin{aligned}
&[\mathcal{H}_{\text{BCS}}]_{\text{MFA}} \\
&\quad = \sum_{\mathrm{k}} \xi_{\mathrm{k}}\big\{(u^2 - v^2)\big(\gamma^{\dagger}_{\mathrm{k},\uparrow}\gamma_{\mathrm{k},\uparrow} + \gamma^{\dagger}_{-\mathrm{k},\downarrow}\gamma_{-\mathrm{k},\downarrow}\big)\big\} \\
&\qquad + \sum_{\mathrm{k}} \xi_{\mathrm{k}}\big\{2uv\big(\gamma^{\dagger}_{\mathrm{k},\uparrow}\gamma^{\dagger}_{-\mathrm{k}\downarrow}\big) + 2uv(\gamma_{-\mathrm{k},\downarrow}\gamma_{\mathrm{k},\uparrow})\big\} \\
&\qquad + \sum_{\mathrm{k}} \xi_{\mathrm{k}}\big\{2v^2\big\} \\
&\qquad + \sum_{\mathrm{k}} \nabla\big\{-v^2\gamma^{\dagger}_{\mathrm{k},\uparrow}\gamma^{\dagger}_{-\mathrm{k},\downarrow} - uv\gamma^{\dagger}_{\mathrm{k},\uparrow}\gamma_{\mathrm{k},\uparrow} + vu\gamma_{-\mathrm{k},\downarrow}\gamma^{\dagger}_{-\mathrm{k},\downarrow} + u^2\gamma_{-\mathrm{k},\downarrow}\gamma_{\mathrm{k},\uparrow}\big\} \\
&\qquad + \sum_{\mathrm{k}} \nabla^{\dagger}\big\{-v^2\gamma_{-\mathrm{k},\downarrow}\gamma_{\mathrm{k},\uparrow} + uv\gamma_{-\mathrm{k},\downarrow}\gamma^{\dagger}_{-\mathrm{k},\downarrow} - vu\gamma^{\dagger}_{\mathrm{k},\uparrow}\gamma_{\mathrm{k},\uparrow} + u^2\gamma^{\dagger}_{\mathrm{k},\uparrow}\gamma^{\dagger}_{-\mathrm{k},\downarrow}\big\} \\
&\qquad + \Delta\,.
\end{aligned}
\tag{16.22}
$$

It is helpful to represent (16.22) in a more symmetrical form as in (16.23), (16.24) and (16.25):

$$
[\mathcal{H}_{\text{BCS}}]_{\text{MFA}} \equiv \mathcal{H}_{\text{diagonal}} + \mathcal{H}_{\text{nondiagonal}}\,, \tag{16.23}
$$

where

$$\mathcal{H}_{\text{diagonal}} = \sum_{\mathrm{k}}\left[\xi_{\mathrm{k}}\left(u^2 - v^2\right) - \nabla uv - \nabla^{\dagger} vu\right]\left(\gamma^{\dagger}_{\mathrm{k},\uparrow}\gamma_{\mathrm{k},\uparrow}\right)$$
$$+ \sum_{\mathrm{k}}\left[\xi_{\mathrm{k}}\left(u^2 - v^2\right) - \nabla uv - \nabla^{\dagger} vu\right]\left(\gamma^{\dagger}_{-\mathrm{k},\downarrow}\gamma_{-\mathrm{k},\downarrow}\right) \qquad (16.24)$$

and

$$\mathcal{H}_{\text{nondiagonal}} = \sum_{\mathrm{k}}\left(\gamma^{\dagger}_{\mathrm{k},\uparrow}\gamma^{\dagger}_{-\mathrm{k},\downarrow}\right)\left\{\nabla^{\dagger} u^2 - \nabla v^2 + 2\xi_{\mathrm{k}} uv\right\}$$
$$+ \sum_{\mathrm{k}}\left(\gamma_{-\mathrm{k},\downarrow}\gamma_{\mathrm{k},\uparrow}\right)\left\{\nabla u^2 - \nabla^{\dagger} v^2 + 2\xi_{\mathrm{k}} uv\right\}$$
$$+ \sum_{\mathrm{k}}\left\{2\xi_{\mathrm{k}} v^2 + \nabla^{\dagger} uv + \nabla uv\right\} + \Delta\,. \qquad (16.25)$$

In $\mathcal{H}_{\text{diagonal}}$, the two functions on the right-hand side of (16.24) contain operators and are both diagonal. In contrast, while the top two functions on the right-hand side of (16.25) also contain operators they are both nondiagonal. Interestingly, however, these two functions are equivalent in the sense that they are complex-conjugates of each other. The functions on the third line of (16.25) represent the ground-state energy and do not contain any operators. These three functions—meaning those on the first, second, and the third lines of (16.25)—constitute the Hamiltonian $\mathcal{H}_{\text{nondiagonal}}$.

16.4 Procedure to Invert

The Hamiltonian, $[\mathcal{H}_{\text{BCS}}]_{\text{MFA}}$, described in (16.23), (16.24), and (16.25) contains variables. Unfortunately, these variables are functions of the fermionic quasiparticle creation and destruction operators. In contrast, what one actually measures is a function of the electron creation and destruction operators. In order to be able to do that, we need to invert (16.14) and (16.15).

As mentioned in (14.18) and (14.19), generally, for a 2×2 matrix A,

$$\begin{pmatrix} a & b \\ c & d \end{pmatrix} = A\,, \qquad (16.26)$$

the inverse is

$$\begin{pmatrix} d & -b \\ -c & a \end{pmatrix} / (ad - bc) = (A)^{-1}\,. \qquad (16.27)$$

Compare, for instance, [8]. Details of the needed inversion are the following. Consider the equations

$$c_{\mathrm{k},\uparrow} = u\gamma_{\mathrm{k},\uparrow} + v\gamma^{\dagger}_{-\mathrm{k},\downarrow}\,,$$
$$c^{\dagger}_{-\mathrm{k},\downarrow} = -v\gamma_{\mathrm{k},\uparrow} + u\gamma^{\dagger}_{-\mathrm{k},\downarrow}\,.$$

The relevant matrix is

$$\begin{pmatrix} u & v \\ -v & u \end{pmatrix} = M\,. \tag{16.28}$$

And the inverse matrix, M^{-1}, is the following

$$\begin{pmatrix} u & -v \\ v & u \end{pmatrix} / \left(u^2 + v^2\right) = (M)^{-1} \tag{16.29}$$

which leads to the result

$$\begin{aligned} \gamma_{k,\uparrow} &\equiv \left[u c_{k,\uparrow} - v c^{\dagger}_{-k,\downarrow}\right] / \left(u^2 + v^2\right), \\ \gamma^{\dagger}_{-k,\downarrow} &\equiv \left[v c_{k,\uparrow} + u c^{\uparrow}_{-k,\downarrow}\right] / \left(u^2 + v^2\right). \end{aligned} \tag{16.30}$$

Because $u^2 + v^2 = 1$, (16.30) gives

$$\begin{aligned} \gamma_{k,\uparrow} &\equiv \left[u c_{k,\uparrow} - v c^{\dagger}_{-k,\downarrow}\right], \\ \gamma^{\dagger}_{-k,\downarrow} &\equiv \left[v c_{k,\uparrow} + u c^{\uparrow}_{-k,\downarrow}\right]. \end{aligned} \tag{16.31}$$

Complex conjugate of (16.31) is also helpful:

$$\begin{aligned} \gamma^{\dagger}_{k,\uparrow} &\equiv \left[u c^{\dagger}_{k,\uparrow} - v c_{-k,\downarrow}\right], \\ \gamma_{-k,\downarrow} &\equiv \left[v c^{\dagger}_{k,\uparrow} + u c_{-k,\downarrow}\right]. \end{aligned} \tag{16.32}$$

16.5 Nondiagonal Hamiltonian: Elimination of Terms

Note that to help the reader, references to the wave-vector are explicitly indicated in the current and the following two sections.

Consequent to the truncation described in (16.34), the first and second terms on the right-hand side of (16.25) are vanishing. Therefore $\mathcal{H}_{\text{nondiagonal}}$ becomes

$$H_{\text{nondiagonal}} \equiv H_0 + \Delta = \sum_{k} \left\{2\xi_k v^2 + \nabla u v + \nabla^{\dagger} u v\right\}_k + \Delta\,. \tag{16.33}$$

Here Δ is as defined in (16.12) and (16.13), and H_0 is available in (16.33).

Clearly, in order to eliminate the two nondiagonal terms that are mutually conjugate and are recorded on the first and the second lines of (16.25), we need to set only one of them equal to zero. And that arrangement is fully ensured by (16.34) given below:

$$\left[\nabla^{\dagger} u^2 - \nabla v^2 + 2\xi_k u v\right]_k = 0\,. \tag{16.34}$$

Dividing (16.34) by $-\nabla u^2$ leads to the quadratic equation (16.35), that is,

$$\left[\left(\frac{v}{u}\right)^2 - 2\left(\frac{\xi_k}{\nabla}\right)\left(\frac{v}{u}\right) - \left(\frac{\nabla^{\dagger}}{\nabla}\right)\right]_k = 0\,. \tag{16.35}$$

Of the two roots,

$$\left(\frac{v}{u}\right)_{\mathrm{k}} = \frac{1}{\nabla}\left[\xi_{\mathrm{k}} \pm \sqrt{(\xi_{\mathrm{k}})^2 + |\nabla|^2}\right], \tag{16.36}$$

of the quadratic we choose the one that ensures thermodynamic stability, meaning the ground state energy is a minimum. That is,

$$\left(\frac{v}{u}\right)_{\mathrm{k}} = \frac{1}{\nabla}\left[\xi_{\mathrm{k}} - \sqrt{(\xi_{\mathrm{k}})^2 + |\nabla|^2}\right]. \tag{16.37}$$

Given $\xi^{\dagger}{}_{\mathrm{k}} = \xi_{\mathrm{k}} = \xi_{-\mathrm{k}}$, the Hermitian-conjugate of (16.37) is

$$\left(\frac{v}{u}\right)_{\mathrm{k}} = \frac{1}{\nabla^{\dagger}}\left[\xi_{\mathrm{k}} - \sqrt{(\xi_{\mathrm{k}})^2 + |\nabla|^2}\right]. \tag{16.38}$$

Now multiply each against the other the left-hand sides of (16.37) and (16.38), and equate them to the corresponding multiple of the right-hand sides. One gets

$$\left(\frac{v^2}{u^2}\right)_{\mathrm{k}} = \frac{1}{|\nabla|^2}\left[\xi_{\mathrm{k}} - \sqrt{(\xi_{\mathrm{k}})^2 + |\nabla|^2}\right]^2. \tag{16.39}$$

Add unity to both sides of (16.39), invert both sides, and make use of (16.20) whereby $(u^2)_{\mathrm{k}} + (v^2)_{\mathrm{k}} = 1$. The result is

$$\frac{1}{1 + (\frac{v^2}{u^2})_{\mathrm{k}}} = \left(\frac{u^2}{u^2 + v^2}\right)_{\mathrm{k}} = \left(u^2\right)_{\mathrm{k}} = \frac{1}{1 + \frac{1}{|\nabla|^2}[\xi_{\mathrm{k}} - \sqrt{\xi_{\mathrm{k}}{}^2 + |\nabla|^2}]^2}. \tag{16.40}$$

Completing the squaring process in the denominator of (16.40) leads to

$$\begin{aligned}
\left(u^2\right)_{\mathrm{k}} &= \frac{1}{2}\left(\frac{|\nabla|^2|}{\xi_{\mathrm{k}}^2 + |\nabla|^2 - \xi_{\mathrm{k}}\sqrt{\xi_{\mathrm{k}}^2 + |\nabla|^2|}}\right) \\
&= \frac{1}{2}\frac{1}{\sqrt{\xi_{\mathrm{k}}{}^2 + |\nabla|^2|}}\left[\frac{|\nabla|^2}{\sqrt{\xi_{\mathrm{k}}{}^2 + \nabla^2} - \xi_{\mathrm{k}}}\right]\cdot\left(\frac{\sqrt{\xi_{\mathrm{k}}{}^2 + |\nabla|^2|} + \xi_{\mathrm{k}}}{\sqrt{\xi_{\mathrm{k}}{}^2 + |\nabla|^2} + \xi_{\mathrm{k}}}\right) \\
&= \frac{1}{2}\left(\frac{\sqrt{\xi_{\mathrm{k}}{}^2 + |\nabla|^2|} + \xi_{\mathrm{k}}}{\sqrt{\xi_{\mathrm{k}}{}^2 + |\nabla|^2}}\right)\cdot\left[\frac{|\nabla|^2}{\xi_{\mathrm{k}}{}^2 + |\nabla|^2 - \xi_{\mathrm{k}}{}^2}\right] \\
&= \frac{1}{2}\left(1 + \frac{\xi_{\mathrm{k}}}{\sqrt{\xi_{\mathrm{k}}{}^2 + |\nabla|^2}}\right)\cdot\left[\frac{|\nabla|^2}{|\nabla|^2}\right] \\
&= \frac{1}{2}\left(1 + \frac{\xi_{\mathrm{k}}}{\sqrt{\xi_{\mathrm{k}}{}^2 + |\nabla|^2}}\right).
\end{aligned} \tag{16.41}$$

Now that $(u^2)_{\mathrm{k}}$ has been determined, we can use it to calculate $(v^2)_{\mathrm{k}}$. To that end, write

$$\left(v^2\right)_{\mathrm{k}} = 1 - \left(u^2\right)_{\mathrm{k}}, \tag{16.42}$$

and insert $(u^2)_\text{k}$ from (16.41) into (16.42). We get

$$\left(v^2\right)_\text{k} = \frac{1}{2}\left(1 - \frac{\xi_\text{k}}{\sqrt{{\xi_\text{k}}^2 + |\nabla|^2}}\right). \tag{16.43}$$

It is interesting to calculate $(u\,v)_\text{k}$ here. We shall also have another occasion to do the same, but from a different perspective.

Multiply $(u^2)_\text{k}$ and $(v^2)_\text{k}$ given in (16.41) and (16.43) and take the square-root of the result. The possible solutions are

$$2(uv)_\text{k} = \pm\sqrt{1 - \left(\frac{\xi_\text{k}}{\sqrt{{\xi_\text{k}}^2 + |\nabla|^2}}\right)^2} = \pm\frac{\sqrt{|\nabla|^2}}{\sqrt{{\xi_\text{k}}^2 + |\nabla|^2}}. \tag{16.44}$$

16.6 Transformation of $\mathcal{H}_{\text{nondiagonal}}$

The Hamiltonian $\mathcal{H}_{\text{nondiagonal}}$ is defined in (16.25). Let us transform it into a more useful form. To that purpose, multiply (16.37) by $(u^2)_\text{k}\nabla$ and (16.38) by $(u^2)_\text{k}\nabla^\dagger$ and thereby get (16.45) and (16.46), that is,

$$(uv)_\text{k}(\nabla) = \left(u^2\right)_\text{k}\left[\xi_\text{k} - \sqrt{(\xi_\text{k})^2 + |\nabla|^2}\right], \tag{16.45}$$

$$(uv)_\text{k}\left(\nabla^\dagger\right) = \left(u^2\right)_\text{k}\left[\xi_\text{k} - \sqrt{(\xi_\text{k})^2 + |\nabla|^2}\right]. \tag{16.46}$$

Use the equivalence,

$$\nabla^\dagger \equiv \nabla, \tag{16.47}$$

that follows from (16.45) and (16.46). And multiply (16.43) by $2\xi_\text{k}$ to obtain

$$2\xi_\text{k}\left(v^2\right)_\text{k} = \xi_\text{k} - \frac{{\xi_\text{k}}^2}{\sqrt{{\xi_\text{k}}^2 + |\nabla|^2}}. \tag{16.48}$$

By employing (16.45), (16.46), and (16.48), $H_{\text{nondiagonal}}$ given in (16.33) becomes

$$\begin{aligned}
H_{\text{nondiagonal}} &\equiv H_0 + \Delta \\
&= \sum_\text{k}\left\{2\xi_\text{k}\left(v^2\right)_\text{k} + \nabla(uv)_\text{k} + \nabla^\dagger(uv)_\text{k}\right\} + \Delta. \\
&= \sum_\text{k}\left\{\left(\xi_\text{k} - \frac{\xi_\text{k}^2}{\sqrt{{\xi_\text{k}}^2 + |\nabla|^2}}\right)\right\} \\
&\quad + 2\sum_\text{k}\left\{\left(u^2\right)_\text{k}\left[\xi_\text{k} - \sqrt{(\xi_\text{k})^2 + |\nabla|^2}\right]\right\} + \Delta.
\end{aligned} \tag{16.49}$$

Next, as suggested by (16.41), replace $2(u^2)_k$—that occurs in the last line of (16.49)—by $1+\frac{\xi_k}{\sqrt{\xi_k{}^2+|\nabla|^2}}$ which is equal to $\frac{1}{\sqrt{\xi_k{}^2+|\nabla|^2}}(\sqrt{\xi_k{}^2+|\nabla|^2}+\xi_k)$. Therefore (16.49) becomes

$$
\begin{aligned}
H_{\text{nondiagonal}} &\equiv H_0 + \Delta \\
&= \sum_k \left\{2\xi_k \left(v^2\right)_k + \nabla (uv)_k + \nabla^\dagger (uv)_k\right\} + \Delta . \\
&= \sum_k \left(\xi_k - \frac{\xi_k^2}{\sqrt{\xi_k{}^2 + |\nabla|^2}}\right) \\
&\quad + \sum_k \left(\frac{1}{\sqrt{\xi_k{}^2 + |\nabla|^2}}\right)\left(\sqrt{\xi_k{}^2 + |\nabla|^2} + \xi_k\right)\left(\xi_k - \sqrt{(\xi_k)^2 + |\nabla|^2}\right) \\
&\quad + \Delta .
\end{aligned} \tag{16.50}
$$

The last term in (16.50) can be organized further. For instance.

$$
\begin{aligned}
&\frac{1}{\sqrt{\xi_k{}^2 + |\nabla|^2}}\left(\sqrt{\xi_k{}^2 + |\nabla|^2} + \xi_k\right)\left(\xi_k - \sqrt{(\xi_k)^2 + |\nabla|^2}\right) + \Delta \\
&= \frac{1}{\sqrt{\xi_k{}^2 + |\nabla|^2}}\left[\xi_k^2 - \left(\xi_k^2 + |\nabla|^2\right)\right] + \Delta = \Delta - \frac{|\nabla|^2}{\sqrt{\xi_k{}^2 + |\nabla|^2}} .
\end{aligned} \tag{16.51}
$$

Inserting into (16.50), the result derived in (16.51) leads to

$$
\begin{aligned}
H_{\text{nondiagonal}} &\equiv H_0 + \Delta \\
&= \sum_k \left\{2\xi_k \left(v^2\right)_k + \nabla (uv)_k + \nabla^\dagger (uv)_k\right\} + \Delta . \\
&= \sum_k \left\{\left(\xi_k - \frac{\xi_k^2}{\sqrt{\xi_k{}^2 + |\nabla|^2}}\right) - \left(\frac{|\nabla|^2}{\sqrt{\xi_k{}^2 + |\nabla|^2}}\right)\right\} + \Delta .
\end{aligned} \tag{16.52}
$$

A more compact form of (16.52) is the following:

$$
H_{\text{nondiagonal}} = \sum_k \left(\xi_k - \sqrt{\xi_k{}^2 + |\nabla|^2}\right) + \Delta . \tag{16.53}
$$

The thermodynamic average of $H_{\text{nondiagonal}}$ is

$$
\langle H_{\text{nondiagonal}} \rangle = \left\langle \sum_k \left(\xi_k - \sqrt{\xi_k{}^2 + |\nabla|^2}\right) \right\rangle + \Delta . \tag{16.54}
$$

Note that all measurements are of thermodynamic averages.

16.7 Evaluation of Diagonal Terms

According to (16.23), (16.24), (16.25), and (16.70), Hamiltonian, $[\mathcal{H}_{\mathrm{BCS}}]_{\mathrm{MFA}}$, is the sum of the diagonal Hamiltonian,

$$\begin{aligned} H_{\text{diagonal}} &= \sum_{\mathrm{k},\sigma} \mho_{\mathrm{k}} \big(\gamma_{\mathrm{k},\sigma}^{\dagger} \gamma_{\mathrm{k},\sigma}\big) = \sum_{\mathrm{k}} \mho_{\mathrm{k}} \sum_{\sigma} \big(\gamma_{\mathrm{k},\sigma}^{\dagger} \gamma_{\mathrm{k},\sigma}\big) \\ &= \sum_{\mathrm{k}} \mho_{\mathrm{k}} \big[\big(\gamma_{\mathrm{k},\uparrow}^{\dagger} \gamma_{\mathrm{k},\uparrow}\big) + \big(\gamma_{\mathrm{k},\downarrow}^{\dagger} \gamma_{\mathrm{k},\downarrow}\big) \big], \end{aligned} \tag{16.55}$$

and the nondiagonal Hamiltonian, $H_{\text{nondiagonal}}$. In order to correspond with thermodynamic representation of the nondiagonal Hamiltonian given in (16.54), we determine the thermodynamic average also of H_{diagonal} as given in (16.56). That is,

$$\langle H_{\text{diagonal}} \rangle = \sum_{\mathrm{k}} \mho_{\mathrm{k}} \sum_{\sigma} \big\langle \big(\gamma_{\mathrm{k},\sigma}^{\dagger} \gamma_{\mathrm{k},\sigma}\big) \big\rangle . \tag{16.56}$$

In (16.55), $\mho_{\mathrm{k}}$, which is given by the expression

$$\mho_{\mathrm{k}} \equiv \sqrt{\xi_{\mathrm{k}}{}^2 + |\nabla|^2}, \tag{16.57}$$

is the quasiparticle dispersion relation. Because here the quasiparticles are fermions, the thermodynamic average of the number N_{k} of quasiparticles with wave-vector k and spins up or down, namely $\langle(\gamma_{\mathrm{k},\sigma}^{\dagger} \gamma_{\mathrm{k},\sigma})\rangle$, is given by the relationship

$$\begin{aligned} \big\langle \big(\gamma_{\mathrm{k},\uparrow}^{\dagger} \gamma_{\mathrm{k},\uparrow}\big) \big\rangle = \big\langle \big(\gamma_{\mathrm{k},\downarrow}^{\dagger} \gamma_{\mathrm{k},\downarrow}\big) \big\rangle &= \big[\exp(\beta \mho_{\mathrm{k}}) + 1\big]^{-1} \\ &= \Big[\exp\Big(\beta \sqrt{\xi_{\mathrm{k}}{}^2 + |\nabla|^2}\Big) + 1\Big]^{-1} \equiv N_{\mathrm{k}}, \end{aligned} \tag{16.58}$$

where N_{k} is the canonical representation of the number of weakly interacting fermions with interparticle interaction $\mho_{\mathrm{k}}$ at temperature $T = \frac{1}{k_B \beta}$ and appropriately carries the up–down symmetry. As such (16.55) becomes

$$\begin{aligned} \langle H_{\text{diagonal}} \rangle &= \sum_{\mathrm{k}} \mho_{\mathrm{k}} \left[\frac{2}{\exp\big(\beta \sqrt{\xi_{\mathrm{k}}{}^2 + |\nabla|^2}\big) + 1} \right] \\ &= \sum_{\mathrm{k}} \mho_{\mathrm{k}} \left[\frac{2 \exp\big(-(\frac{\beta}{2}) \sqrt{\xi_{\mathrm{k}}{}^2 + |\nabla|^2}\big)}{\exp\big((\frac{\beta}{2}) \sqrt{\xi_{\mathrm{k}}{}^2 + |\nabla|^2}\big) + \exp\big(-(\frac{\beta}{2}) \sqrt{\xi_{\mathrm{k}}{}^2 + |\nabla|^2}\big)} \right] \\ &= \sum_{\mathrm{k}} \mho_{\mathrm{k}} \left[\frac{\exp\big((\frac{\beta}{2}) \sqrt{\xi_{\mathrm{k}}{}^2 + |\nabla|^2}\big) + \exp\big(-(\frac{\beta}{2}) \sqrt{\xi_{\mathrm{k}}{}^2 + |\nabla|^2}\big)}{\exp\big((\frac{\beta}{2}) \sqrt{\xi_{\mathrm{k}}{}^2 + |\nabla|^2}\big) + \exp\big(-(\frac{\beta}{2}) \sqrt{\xi_{\mathrm{k}}{}^2 + |\nabla|^2}\big)} \right] \\ &\quad - \sum_{\mathrm{k}} \mho_{k} \left[\frac{\exp\big((\frac{\beta}{2}) \sqrt{\xi_{\mathrm{k}}{}^2 + |\nabla|^2}\big) - \exp\big(-(\frac{\beta}{2}) \sqrt{\xi_{\mathrm{k}}{}^2 + |\nabla|^2}\big)}{\exp\big((\frac{\beta}{2}) \sqrt{\xi_{\mathrm{k}}{}^2 + |\nabla|^2}\big) + \exp\big(-(\frac{\beta}{2}) \sqrt{\xi_{\mathrm{k}}{}^2 + |\nabla|^2}\big)} \right]. \end{aligned} \tag{16.59}$$

A more compact form of (16.59) is given in the following equation, namely

$$\langle H_{\text{diagonal}} \rangle = \sum_{\text{k}} \mho_{\text{k}} \left[1 - \tanh\left(\frac{\beta}{2} \sqrt{\xi_{\text{k}}{}^2 + |\nabla|^2} \right) \right]. \tag{16.60}$$

Also note that (16.55)–(16.60) identify the relationship

$$\begin{aligned} &\langle (\gamma^{\dagger}_{\text{k},\uparrow} \gamma_{\text{k},\uparrow}) \rangle + \langle (\gamma^{\dagger}_{\text{k},\downarrow} \gamma_{\text{k},\downarrow}) \rangle = 2 \langle (\gamma^{\dagger}_{\text{k},\uparrow} \gamma_{\text{k},\uparrow}) \rangle \\ &= \left[\frac{2}{\exp(\beta \sqrt{\xi_{\text{k}}{}^2 + |\nabla|^2}) + 1} \right] = \left[1 - \tanh\left(\frac{\beta}{2} \sqrt{\xi_{\text{k}}{}^2 + |\nabla|^2} \right) \right]. \end{aligned} \tag{16.61}$$

The Hamiltonian $[\mathcal{H}_{\text{BCS}}]_{\text{MFA}}$ that was given in (16.23) and (16.24) has the following diagonal terms:

$$H_{\text{diagonal}} = \sum_{\text{k},\sigma} [(\xi)_{\text{k}} \{ (u^2)_{\text{k}} - (v^2)_{\text{k}} \} - \{ \nabla (uv)_{\text{k}} - \nabla^{\dagger} (vu)_{\text{k}} \}] \cdot (\gamma^{\dagger}{}_{\text{k},\sigma} \gamma_{\text{k},\sigma}) . \tag{16.62}$$

Using (16.41) for $(u^2)_{\text{k}}$ and (16.43) for $(v^2)_{\text{k}}$, we can write the first term of (16.62) as

$$\sum_{\text{k},\sigma} (\xi)_{\text{k}} \{ (u^2)_{\text{k}} - (v^2)_{\text{k}} \} \cdot (\gamma^{\dagger}_{\text{k},\sigma} \gamma_{\text{k},\sigma}) = \sum_{\text{k},\sigma} \left(\frac{\xi_{\text{k}}^2}{\sqrt{\xi_{\text{k}}^2 + |\nabla|^2}} \right) \cdot (\gamma^{\dagger}_{\text{k},\sigma} \gamma_{\text{k},\sigma}) . \tag{16.63}$$

The second and third terms of (16.62) require knowledge of $\nabla (uv)_{\text{k}}$ and $\nabla^{\dagger} (vu)_{\text{k}}$. These two terms are Hermitian conjugates of each other. Therefore only one of them needs to be evaluated. To that end, consider (16.37) and multiply both sides by $(uu)_{\text{k}}$. The result is

$$(uu)_{\text{k}} \left(\frac{v}{u} \right)_{\text{k}} = (uv)_{\text{k}} = (uu)_{\text{k}} \left(\frac{1}{\nabla} \right) [\xi_{\text{k}} - \sqrt{(\xi_{\text{k}})^2 + |\nabla|^2}] . \tag{16.64}$$

We also need the Hermitian conjugate of (16.64), which is

$$(vu)_{\text{k}} = (uu)_{\text{k}} \left(\frac{1}{\nabla^{\dagger}_{\text{k}}} \right) [\xi_{\text{k}} - \sqrt{(\xi_{\text{k}})^2 + |\nabla|^2}] . \tag{16.65}$$

By combining (16.64), (16.65), and (16.62), we can commence the evaluation of the second term of (16.62). For convenience this second term is to be denoted $[H_{\text{diagonal}}]_{\text{2nd term}}$, which is

$$\begin{aligned} &[H_{\text{diagonal}}]_{\text{2nd term}} \\ &= \sum_{\text{k},\sigma} \{ -\nabla (uv)_{\text{k}} \} (\gamma^{\dagger}_{\text{k},\sigma} \gamma_{\text{k},\sigma}) = \left\{ -\frac{\nabla (uu)_{\text{k}}}{\nabla} \right\} [\xi_{\text{k}} - \sqrt{\xi_{\text{k}}{}^2 + |\nabla|^2}] (\gamma^{\dagger}_{\text{k},\sigma} \gamma_{\text{k},\sigma}) \\ &= \sum_{\text{k},\sigma} \{ -(u^2)_{\text{k}} [\xi_{\text{k}} - \sqrt{\xi_{\text{k}}{}^2 + |\nabla|^2|}] \} (\gamma^{\dagger}_{\text{k},\sigma} \gamma_{\text{k},\sigma}) . \end{aligned} \tag{16.66}$$

At this stage, we replace $(u^2)_k$ as suggested by (16.41). That is,

$$(u^2)_k = \frac{1}{2}\left(1 + \frac{\xi_k}{\sqrt{{\xi_k}^2 + \nabla^2}}\right) = \frac{1}{2}\frac{\xi_k + \sqrt{{\xi_k}^2 + |\nabla|^2|}}{\sqrt{{\xi_k}^2 + \nabla^2}}. \tag{16.67}$$

Accordingly, (16.66) leads to the following result for the second term:

$$\begin{aligned}
&[H_{\text{diagonal}}]_{\text{2nd term}} \\
&= \sum_{k,\sigma}\{-\nabla(uv)_k\}(\gamma^\dagger_{k,\sigma}\gamma_{k,\sigma}) \\
&= \sum_{k,\sigma}\{-(u^2)_k[\xi_k - \sqrt{{\xi_k}^2 + |\nabla|^2|}]\}(\gamma^\dagger_{k,\sigma}\gamma_{k,\sigma}) \\
&= -\sum_{k,\sigma}\left(\frac{1}{2}\right)\left\{\frac{\xi_k + \sqrt{{\xi_k}^2 + |\nabla|^2}}{\sqrt{{\xi_k}^2 + \nabla^2}}[\xi_k - \sqrt{{\xi_k}^2 + |\nabla|^2|}]\right\}(\gamma^\dagger_{k,\sigma}\gamma_{k,\sigma}) \\
&= \sum_{k,\sigma}\left(\frac{1}{2}\right)\left\{\frac{|\nabla|^2}{\sqrt{{\xi_k}^2 + |\nabla|^2|}}\right\}(\gamma^\dagger_{k,\sigma}\gamma_{k,\sigma}).
\end{aligned} \tag{16.68}$$

And finally, let us determine the third term in H_{diagonal}. As mentioned above, the third term is simply the complex conjugate of the second term. But because the second term is real, both the second and third terms are exactly equal. Therefore summing (16.63) and twice (16.68) determines the diagonal terms:

$$\begin{aligned}
&H_{\text{diagonal}} \\
&= \sum_{k,\sigma}\left(\frac{\xi_k^2}{\sqrt{\xi_k^2 + |\nabla|^2}}\right)\cdot(\gamma^\dagger_{k,\sigma}\gamma_{k,\sigma}) + \sum_{k,\sigma}\left\{\frac{|\nabla|^2|}{\sqrt{{\xi_k}^2 + |\nabla|^2|}}\right\}(\gamma^\dagger_{k,\sigma}\gamma_{k,\sigma}) \\
&= \sum_{k,\sigma}\sqrt{{\xi_k}^2 + |\nabla|^2|}\cdot(\gamma^\dagger_{k,\sigma}\gamma_{k,\sigma}).
\end{aligned} \tag{16.69}$$

With the preceding information in hand, more organized version of $\langle[\mathcal{H}_{\text{BCS}}]_{\text{MFA}}\rangle$ can be written. Using (16.23), (16.24), (16.25), (16.54), and (16.69), we can right the thermodynamic average $\langle[\mathcal{H}_{\text{BCS}}]_{\text{MFA}}\rangle$ as

$$\begin{aligned}
\langle[\mathcal{H}_{\text{BCS}}]_{\text{MFA}}\rangle &\equiv \langle\mathcal{H}_{\text{diagonal}}\rangle + \langle\mathcal{H}_{\text{nondiagonal}}\rangle \\
&= \sum_{k,\sigma}\sqrt{{\xi_k}^2 + |\nabla|^2|}\cdot\langle(\gamma^\dagger_{k,\sigma}\gamma_{k,\sigma})\rangle + \sum_k(\xi_k - \sqrt{{\xi_k}^2 + |\nabla|^2}) + \Delta \\
&\equiv \sum_{k,\sigma}\mho_k\langle(\gamma^\dagger_{k,\sigma}\gamma_{k,\sigma})\rangle + \sum_k(\xi_k - \sqrt{{\xi_k}^2 + |\nabla|^2}) + \Delta.
\end{aligned} \tag{16.70}$$

16.8 Fermi Surface

As recorded in (16.2), there is the relationship

$$\xi_k = \epsilon_k - E_F = \xi_{-}k\,, \tag{16.71}$$

where ϵ_k is the kinetic energy of an electron and E_F is the Fermi energy. As the kinetic energy of an electron approaches the Fermi energy, or equivalently, when $[\epsilon_k \to E_F]$, we get $\xi_k \to 0$ and (16.76) gives

$$u^2 = \frac{1}{2}, \qquad v^2 = \frac{1}{2}, \tag{16.72}$$

or equivalently,

$$u = v = \frac{1}{\sqrt{2}}. \tag{16.73}$$

Inserting (16.73) into (16.31) and (16.32) leads to

$$\begin{aligned} \gamma_{k,\uparrow} &=_{\xi_k \to 0} \frac{1}{\sqrt{2}} c_{k,\uparrow} - \frac{1}{\sqrt{2}} c^{\dagger}_{-k,\downarrow}\,, \\ \gamma^{\dagger}_{k,\uparrow} &=_{\xi_k \to 0} \frac{1}{\sqrt{2}} c^{\dagger}_{k,\uparrow} - \frac{1}{\sqrt{2}} c_{-k,\downarrow}\,, \end{aligned} \tag{16.74}$$

and

$$\begin{aligned} \gamma_{-k,\downarrow} &=_{\xi_k \to 0} \frac{1}{\sqrt{2}} c_{-k,\downarrow} + \frac{1}{\sqrt{2}} c^{\dagger}_{k,\uparrow}\,, \\ \gamma^{\dagger}_{-k,\downarrow} &=_{\xi_k \to 0} \frac{1}{\sqrt{2}} c^{\uparrow}_{-k,\downarrow} + \frac{1}{\sqrt{2}} c_{k,\uparrow}\,. \end{aligned} \tag{16.75}$$

Equations (16.74) and (16.75) indicate that exactly at the Fermi surface, i.e., when $\xi_k \to 0$, the quasiparticles act as electrons and holes with equal strength, leading to electrical neutrality. Consequently, this position is the ground state of any theory of superconductivity.

16.9 Fermi Sea: Above and Within

Much above the Fermi sea, according to (16.41) and (16.43), we have

$$u^2 = \frac{1}{2}\left(1 + \frac{\xi_k}{\sqrt{{\xi_k}^2 + |\nabla|^2}}\right), \qquad v^2 = \frac{1}{2}\left(1 - \frac{\xi_k}{\sqrt{{\xi_k}^2 + |\nabla|^2}}\right). \tag{16.76}$$

As such, for $\xi_k \gg \nabla$, i.e., when $\nabla \approx 0$, one gets $u^2 \approx 1$ and $v^2 \approx 0$. Accordingly, in this limit, the fermionic quasiparticle is essentially an ordinary particle and the system is in its normal state. Therefore, in the normal state, creating a fermionic quasiparticle excitation amounts to creating an electron at energies above the Fermi sea and destroying an electron of opposite spin and momentum within the Fermi sea.

On the other hand, much below the Fermi sea we have $\xi_{\mathrm{k}} \ll -\nabla$, $u^2 \to 0$, and $v^2 \to 1$. As such, in this limit, the fermionic quasiparticle is a hole. The minimum energy separation between the higher and the lower energy-states is 2∇. In other words, in this limit, the system needs a minimum of 2∇ in energy to excite fermionic quasiparticles.

To sum up, for energies much lower than the Fermi energy, the fermionic quasiparticles act as "holes." On the other hand, for energies much higher than the Fermi energy, the fermionic quasiparticles act as normal electrons. In-between these energy levels, where $1 > u^2 > 0$ and $1 > v^2 > 0$, these quasiparticles are a mixture of particles and holes. But exactly at the Fermi surface, where $\xi_{\mathrm{k}} = 0$, we have

$$u^2 = v^2 = \frac{1}{2} . \tag{16.77}$$

Accordingly, exactly at the Fermi surface, a fermionic quasiparticle is both a hole and particle with equal strength. As such, exactly at the Fermi surface, there are no fermionic quasiparticles.

16.10 Gap Function and Solutions

The gap function Λ was defined in (16.13) as

$$\Lambda = -\frac{V}{N} \sum_{\mathrm{q}} \langle c_{-\mathrm{q},\downarrow} c_{\mathrm{q},\uparrow} \rangle = -\nabla^\dagger . \tag{16.78}$$

To express it in terms of the fermionic quasiparticle operators, we need (16.15). That is,

$$\begin{aligned} c_{-\mathrm{q},\downarrow} &\equiv -v\gamma^\dagger_{\mathrm{q},\uparrow} + u\gamma_{-\mathrm{q},\downarrow} , \\ c_{\mathrm{q},\uparrow} &\equiv u\gamma_{\mathrm{q},\uparrow} + v\gamma^\dagger_{-\mathrm{q},\downarrow} . \end{aligned} \tag{16.79}$$

Picking the expressions for c_{q} and $c_{-\mathrm{q}}$ from (16.79) and introducing them into (16.78) yields the result

$$\begin{aligned} \Lambda &= -\frac{1}{N} \sum_{\mathrm{q}} V \langle \left(-v\gamma^\dagger_{\mathrm{q},\uparrow} + u\gamma_{-\mathrm{q},\downarrow}\right)\left(u\gamma_{\mathrm{q},\uparrow} + v\gamma^\dagger_{-\mathrm{q},\downarrow}\right) \rangle \\ &= \frac{1}{N} \sum_{\mathrm{q}} V \langle vu\left(\gamma^\dagger_{\mathrm{q},\uparrow}\gamma_{\mathrm{q},\uparrow} - \gamma_{-\mathrm{q},\downarrow}\gamma^\dagger_{-\mathrm{q},\downarrow}\right) \rangle \\ &\quad - \frac{1}{N} \sum_{\mathrm{q}} V \langle -v^2 \gamma^\dagger_{\mathrm{q},\uparrow}\gamma^\dagger_{-\mathrm{q},\uparrow} + u^2 \gamma_{-\mathrm{q},\downarrow}\gamma_{\mathrm{q},\uparrow} \rangle . \end{aligned} \tag{16.80}$$

Because u^2 and v^2 are constants, they can be withdrawn out of the averaging process. As a result, the nondiagonal terms that appear in the last term on the right-hand side of (16.80) are vanishing. To see this, proceed as follows:

$$\begin{aligned} & -\frac{V}{N}\sum_{\mathbf{q}}\langle -v^2\gamma^{\dagger}_{\mathbf{q},\uparrow}\gamma^{\dagger}_{-\mathbf{q},\uparrow}+u^2\gamma_{-\mathbf{q},\downarrow}\gamma_{\mathbf{q},\uparrow}\rangle \\ &= \frac{V}{N}v^2\sum_{\mathbf{q}}\langle\gamma^{\dagger}_{\mathbf{q},\uparrow}\gamma^{\dagger}_{-\mathbf{q},\uparrow}\rangle-\frac{V}{N}u^2\sum_{\mathbf{q}}\langle\gamma_{-\mathbf{q},\downarrow}\gamma_{\mathbf{q},\uparrow}\cdot\rangle \\ &= \frac{V}{N}v^2\times\sum_{\mathbf{q}}0-\frac{V}{N}u^2\times\sum_{\mathbf{q}}0=0\,. \end{aligned} \tag{16.81}$$

As such (16.80) becomes

$$\begin{aligned} \Lambda &= \frac{V}{N}\sum_{\mathbf{q}}\left[vu\langle\gamma^{\dagger}_{\mathbf{q},\uparrow}\gamma_{\mathbf{q},\uparrow}\rangle-vu\langle\gamma_{-\mathbf{q},\downarrow}\gamma^{\dagger}_{-\mathbf{q},\downarrow}\rangle\right] \\ &= \frac{V}{N}\sum_{\mathbf{q}}vu\left[\langle\gamma^{\dagger}_{\mathbf{q},\uparrow}\gamma_{\mathbf{q},\uparrow}\rangle+\langle\gamma^{\dagger}_{-\mathbf{q},\downarrow}\gamma_{-\mathbf{q},\downarrow}-1\rangle\right]. \end{aligned} \tag{16.82}$$

The up–down symmetry of the Cooper pair leads to the fact that in the superconducting state, which has a large number of electrons, the number of up-spin electrons is essentially equal to the number of down-spin electrons. Therefore the two thermodynamic averages given in (16.83) below are essentially equal,

$$\langle\gamma^{\dagger}_{\mathbf{q},\uparrow}\gamma_{\mathbf{q},\uparrow}\rangle=\langle\gamma^{\dagger}_{\mathbf{q},\downarrow}\gamma_{\mathbf{q},\downarrow}\rangle\,. \tag{16.83}$$

Also because of dynamical symmetry,

$$\langle\gamma^{\dagger}_{\mathbf{q},\downarrow}\gamma_{\mathbf{q},\downarrow}\rangle=\langle\gamma^{\dagger}_{-\mathbf{q},\downarrow}\gamma_{-\mathbf{q},\downarrow}\rangle\,. \tag{16.84}$$

Combining (16.83) and (16.84) gives

$$\langle\gamma^{\dagger}_{-\mathbf{q},\downarrow}\gamma_{-\mathbf{q},\downarrow}\rangle=\langle\gamma^{\dagger}_{\mathbf{q},\uparrow}\gamma_{\mathbf{q},\uparrow}\rangle\,. \tag{16.85}$$

As a result, (16.82) becomes

$$\Lambda=\frac{V}{N}\sum_{\mathbf{q}}vu\left[2\langle\gamma^{\dagger}_{\mathbf{q},\uparrow}\gamma_{\mathbf{q},\uparrow}\rangle-1\right]=-\nabla^{\dagger}\,. \tag{16.86}$$

From (16.60), (16.61), and (16.62), we glean following information:

$$\begin{aligned} 2\langle(\gamma^{\dagger}_{\mathbf{q},\uparrow}\gamma_{\mathbf{q},\uparrow})\rangle-1 &= \frac{2}{\exp(\beta\sqrt{\xi_{\mathbf{q}}{}^2+|\nabla|^2})}-1 \\ &= -\left[1-\tanh\left(\frac{\beta}{2}\sqrt{\xi_{\mathbf{q}}{}^2+|\nabla|^2}\right)\right]-1 \\ &= -\tanh\left(\frac{\beta}{2}\sqrt{\xi_{\mathbf{q}}{}^2+|\nabla|^2}\right) \end{aligned} \tag{16.87}$$

and

$$vu = \frac{1}{2}\left\{\frac{\nabla^\dagger}{\sqrt{\xi_q{}^2 + |\nabla|^2|}}\right\}. \tag{16.88}$$

Combining (16.86), (16.87), and (16.88) gives

$$\nabla^\dagger = \frac{V}{N}\sum_q \left(\frac{1}{2}\right)\left\{\frac{\nabla^\dagger}{\sqrt{\xi_q{}^2 + |\nabla|^2|}}\right\}\tanh\left(\frac{\beta}{2}\sqrt{\xi_q{}^2 + |\nabla|^2}\right). \tag{16.89}$$

Now divide both sides of (16.89) by the nonzero parameter $\nabla^\dagger$ and arrive at the equality

$$1 = \left(\frac{V}{N}\right)\left(\frac{1}{2}\right)\sum_q \left\{\frac{1}{\sqrt{\xi_q{}^2 + |\nabla|^2|}}\right\}\tanh\left(\frac{\beta}{2}\sqrt{\xi_q{}^2 + |\nabla|^2}\right). \tag{16.90}$$

According to (16.2), the energy variable ξ_q is defined as

$$\xi_q \equiv \epsilon_q - E_F\,, \tag{16.91}$$

where ϵ_q is the kinetic energy of an electron, E_F is the Fermi energy, and $\xi_q \equiv \xi_{-q} \equiv \xi^\dagger{}_q$. At the Fermi surface, $\epsilon_q = E_F$. Therefore exactly at the Fermi surface, $\xi_q = 0$. Another point to note is that the maximum possible range of the energy variable ξ_q is motivated by phonons which essentially cut off at the Debye energy $\hbar\Omega_D$. In practice, however, a majority of the electrons are likely to sit closer to the Fermi surface than their maximum possible separation permits. For this reason, it is reasonable to assume the density of states per unit volume does not vary much for the relevant group of electrons, meaning

$$\text{density of states per unit volume} \approx \Theta\,, \tag{16.92}$$

where Θ is a constant.

Equation (16.90) involves a sum over the entirety of the phonon wave-vector space. Calculating this sum in the form given would require a tiresome numerical computation. However, because the phonon wave vectors are both continuous and very large in number, the sum is well approximated by an integral. Integration requires knowledge of the density of states for all the relevant energies. For simplicity, in (16.92), we assumed such density of states can be approximated by a constant Θ.

With these approximations in place, the relationship (16.90) can be represented as

$$\begin{aligned} 1 &= \frac{V}{2}\int_{-\hbar\Omega_D}^{\hbar\Omega_D} d\xi_q \left\{\frac{\Theta}{\sqrt{\xi_q{}^2 + |\nabla|^2|}}\right\}\tanh\left(\frac{\beta}{2}\sqrt{\xi_q{}^2 + |\nabla|^2}\right) \\ &= V\int_0^{\hbar\Omega_D} d\xi_q \left\{\frac{\Theta}{\sqrt{\xi_q{}^2 + |\nabla|^2|}}\right\}\tanh\left(\frac{\beta}{2}\sqrt{\xi_q{}^2 + |\nabla|^2}\right). \end{aligned} \tag{16.93}$$

It should be noted that (16.93) is the master relationship that can be used for calculating some basic results for the BCS Hamiltonian. Also remember that V stands for the strength of the interparticle coupling.

16.11 Critical Temperature T_c

Use the fact that at $T = T_c$, $\nabla = 0$. Additionally, for $T = T_c$, introduce the symbol $\beta_c \equiv \frac{1}{k_b T_c}$. Then (16.93) gives

$$1 = V \int_0^{\hbar\Omega_D} d\xi_q \left\{ \frac{\Theta}{\xi_q} \right\} \tanh\left(\frac{\beta_c}{2} \xi_q \right) . \tag{16.94}$$

Now change the variable and set

$$\xi_q \equiv \frac{2x}{\beta_c} . \tag{16.95}$$

Therefore

$$d\xi_q = \frac{2dx}{\beta_c} , \qquad \frac{\beta_c}{2}\xi_q = x , \tag{16.96}$$

and

$$\frac{d\xi_q}{\xi_q} = \frac{dx}{x} . \tag{16.97}$$

Regarding the limits:

$$\begin{aligned} &\text{For} \quad \xi_q = 0 , \quad \text{we have} \quad x = 0 , \\ &\text{and for} \quad \xi_q = \hbar\Omega_D , \quad \text{we have} \quad x = \frac{\beta_c \hbar\Omega_D}{2} . \end{aligned} \tag{16.98}$$

Given (16.95)–(16.98), (16.94) becomes

$$1 = V\Theta \int_0^{\frac{\beta_c\hbar\Omega_D}{2}} dx \left\{ \frac{\tanh(x)}{x} \right\} = V\Theta \log\left(2.2677318 \frac{\beta_c \hbar\Omega_D}{2} \right) , \tag{16.99}$$

or equivalently,

$$k_B T_c = 1.1338659(\hbar\Omega_D) \exp\left[\frac{-1}{V\Theta} \right] . \tag{16.100}$$

The critical temperature, given in (16.100), has a somewhat weak dependence on the density of states and the interparticle attractive force V. Yet, the relevant relationship contains important information. For instance, the critical temperature remains finite for arbitrarily small interparticle attractive forces as long it is > 0. On the other hand, (16.99) confirms that superconductivity cannot occur for repulsive—i.e., for negative V—interparticle forces. This fact can readily be checked because the left-hand side of (16.99), being equal to $+1$, is positive while a negative V would make the right-hand side negative.

In contrast, the critical temperature depends directly on the Debye potential energy $\hbar\Omega_{\mathrm{D}}$.

16.12 Debye Potential $\hbar\Omega_{\mathrm{D}}$

Assume phonons motivate much of the interparticle attraction that is the basis of superconductivity. In this regard, they affect the Debye integral. Importantly, the Debye integral excludes consideration of those phonons that have reached wavelengths equal to the width of the unit cell, meaning the Debye integral cuts off at that wave-length. Such truncated computation is generally assumed to lead to a Debye potential $\hbar\Omega_{\mathrm{D}}$ which is directly proportional to the number density, N_{d}, as well as the speed of sound, v_{s}, in the appropriate solid at the given temperature T. Note that N_{d} is the number of atoms in unit volume and

$$\hbar\Omega_{\mathrm{D}} \propto N_{\mathrm{d}} v_{\mathrm{s}} , \tag{16.101}$$

where

$$v_{\mathrm{s}} = \sqrt{\frac{\text{bulk modulus}}{\text{mass density}}} . \tag{16.102}$$

Clearly, the mass density is directly proportional to the ionic mass M_{ion}. Therefore (16.101) and (16.102) lead to the following result:

$$\hbar\Omega_{\mathrm{D}} \propto \sqrt{\frac{1}{\text{mass density}}} \propto \sqrt{\frac{1}{M_{\mathrm{ion}}}} . \tag{16.103}$$

16.13 Isotope Effect

And finally, because the critical temperature T_{c} is directly proportional to $\hbar\Omega_{\mathrm{D}}$, (16.103) provides an important prediction that the critical temperature depends directly on the inverse square root of the ionic mass,

$$T_{\mathrm{c}} \propto \sqrt{\frac{1}{M_{\mathrm{ion}}}} . \tag{16.104}$$

There is ample experimental verification of this prediction.

16.14 Specific Heat Near T_c

The specific heat is defined as rate of change of the internal energy with respect to temperature. Equivalently, it can be related to the rate of change of the entropy. This is helpful because unlike the internal energy, depending as it does on the interparticle forces, the entropy can readily be calculated. At low temperatures, entropy for free electrons varies as the temperature T. And in the superconducting state the entropy, S, can be approximated by the relationship

$$S \approx -2k_B \sum_k \left[N_k \log(N_k) + (1 - N_k)\log(1 - N_k)\right], \tag{16.105}$$

where N_k is as defined in (16.58). As such, N_k is the canonical representation of the number of weakly interacting fermions with interparticle interaction $\mho_k$ at temperature $T = \frac{1}{k_B\beta}$. That is,

$$N_k \equiv \left[\exp(\beta\mho_k) + 1\right]^{-1}, \tag{16.106}$$

where

$$\mho_k \equiv \sqrt{\xi_k{}^2 + |\nabla|^2}\,. \tag{16.107}$$

In thermodynamics there is a well-known relationship between the specific heat, C_v, at a given temperature, T, under constant volume, v, to the temperature derivative of the entropy, namely

$$C_v = T\left(\frac{dS}{dT}\right)_v = T\left(\frac{dS}{d\beta}\cdot\frac{d\beta}{dT}\right)_v = -\beta\left(\frac{dS}{d\beta}\right)_v. \tag{16.108}$$

In (16.108) use was made of the identity

$$T\frac{d\beta}{dT} = -\frac{1}{k_B T} = \beta\,. \tag{16.109}$$

Next we need to calculate the derivative $(\frac{dS}{d\beta})_V$. Let us do that in parts. First consider the derivative of $N_k \log(N_k)$, which is

$$\begin{aligned}\frac{d(N_k\log(N_k))}{d\beta} &= \log(N_k)\frac{d(N_k)}{d\beta} + N_k\frac{d\log(N_k)}{d\beta}\\ &= \log(N_k)\frac{d(N_k)}{d\beta} + N_k.\frac{1}{N_k}\frac{d(N_k)}{d\beta} = \frac{d(N_k)}{d\beta}\left[\log(N_k) + 1\right],\end{aligned} \tag{16.110}$$

Next the derivative

$$\begin{aligned}\frac{d[(1-N_k)\log(1-N_k)]}{d\beta} &= -\log(1-N_k)\frac{d(N_k)}{d\beta} - \frac{d(N_k)}{d\beta}\\ &= -\frac{d(N_k)}{d\beta}\left[\log(1-N_k) + 1\right].\end{aligned} \tag{16.111}$$

Now add the results in (16.110) and (16.111) and insert the sum into the derivative of (16.105). We get

$$\begin{aligned} C_v &= -\beta\left(\frac{\mathrm{d}S}{\mathrm{d}\beta}\right)_V \\ &= 2\beta k_B\left(\frac{\sum_k \mathrm{d}[N_k \log(N_k) + (1-N_k)\log(1-N_k)]}{\mathrm{d}\beta}\right)_V \\ &= 2\beta k_B \sum_k \frac{\mathrm{d}(N_k)}{\mathrm{d}\beta}\big[\log(N_k) + 1 - \log(1-N_k) - 1\big] \\ &= 2\beta k_B \sum_k \frac{\mathrm{d}(N_k)}{\mathrm{d}\beta}\big[\log(N_k) - \log(1-N_k)\big]. \end{aligned} \tag{16.112}$$

Now

$$\begin{aligned} &\log(N_k) - \log(1-N_k) \\ &\quad = \log\big[\left(\exp(\beta\mho_k)+1\right)^{-1}\big] - \log\big[1 - \left(\exp(\beta\mho_k)+1\right)^{-1}\big] \\ &\quad = \log\big[\left(\exp(\beta\mho_k)+1\right)^{-1}\big] - \log\big[\exp(\beta\mho_k)\left(\exp(\beta\mho_k)+1\right)^{-1}\big] \\ &\quad = -\beta\mho_k\,, \end{aligned} \tag{16.113}$$

where

$$\mho_k \equiv \sqrt{\xi_k{}^2 + |\nabla|^2}\,. \tag{16.114}$$

Therefore the heat capacity C_V given in (16.112) is

$$C_v = -2\beta^2 k_B \sum_k \mho_k \left(\frac{\mathrm{d}N_k}{\mathrm{d}\beta}\right)_V. \tag{16.115}$$

Because N_k depends both on β and $\mho_k$, the total derivative being used in (16.115) can be broken up as

$$\left(\frac{\mathrm{d}N_k}{\mathrm{d}\beta}\right)_V = \left(\frac{\partial N_k}{\partial\beta}\right)_V + \left(\frac{\partial N_k}{\partial\mho_k}\right)_V \cdot \left(\frac{\partial\mho_k}{\partial\beta}\right)_V. \tag{16.116}$$

Given $N_k \equiv [\exp(\beta\mho_k)+1]^{-1}$, we can write

$$\left(\frac{\partial N_k}{\partial\beta}\right)_V = \left(\frac{\partial N_k}{\partial\mho_k}\right)_V \cdot \frac{\mho_k}{\beta}. \tag{16.117}$$

Similarly, given $\mho_k \equiv \sqrt{\xi_k{}^2 + |\nabla|^2}$, we have

$$\left(\frac{\partial\mho_k}{\partial\beta}\right)_V = \left(\frac{\partial\mho_k}{\partial|\nabla|^2}\right)_V \cdot \left(\frac{\partial|\nabla|^2}{\partial\beta}\right)_V = \left(\frac{1}{2\mho_k}\right) \cdot \left(\frac{\partial|\nabla|^2}{\partial\beta}\right)_V. \tag{16.118}$$

By combining (16.116), (16.117), and (16.118), we get

$$\left(\frac{\mathrm{d}N_{\mathrm{k}}}{\mathrm{d}\beta}\right)_{\mathrm{v}} = \left(\frac{\partial N_{\mathrm{k}}}{\partial \mho_{\mathrm{k}}}\right)_{\mathrm{v}} \cdot \left[\frac{\mho_{\mathrm{k}}}{\beta} + \left(\frac{1}{2\mho_{\mathrm{k}}}\right) \cdot \left(\frac{\partial |\nabla|^2}{\partial \beta}\right)_{\mathrm{v}}\right]. \tag{16.119}$$

Inserting (16.119) into (16.115) gives

$$C_{\mathrm{v}} = -2\beta k_{\mathrm{B}} \sum_{\mathrm{k}} \left(\frac{\partial N_{\mathrm{k}}}{\partial \mho_{\mathrm{k}}}\right)_{\mathrm{v}} \cdot \left[\mho_{\mathrm{k}}{}^2 + \left(\frac{\beta}{2}\right) \cdot \left(\frac{\partial |\nabla|^2}{\partial \beta}\right)_{\mathrm{v}}\right]. \tag{16.120}$$

At $T = T_{\mathrm{c}} + 0$, the parameter $|\nabla|^2 = 0$. Therefore (16.120) gives

$$C_{\mathrm{v+}} = -2\beta k_{\mathrm{B}} \sum_{\mathrm{k}} \left(\frac{\partial N_{\mathrm{k}}}{\partial \mho_{\mathrm{k}}}\right)_{\mathrm{v}} \cdot \left[\mho_{\mathrm{k}}{}^2\right]. \tag{16.121}$$

To begin simplifying (16.120), consider the first expression on the right-hand side, that is, $(\frac{\partial N_{\mathrm{k}}}{\partial \mho_{\mathrm{k}}})_{\mathrm{v}}$. From (16.106) we know $N_{\mathrm{k}} = [\exp(\beta\mho_{\mathrm{k}}) + 1]^{-1}$. Differentiating N_{k} with respect to $\mho_{\mathrm{k}}$ gives

$$\left(\frac{\partial N_{\mathrm{k}}}{\partial \mho_{\mathrm{k}}}\right)_{\mathrm{v}} = -\beta \exp(\beta\mho_{\mathrm{k}})\left[\exp(\beta\mho_{\mathrm{k}}) + 1\right]^{-2} = -\beta N_{\mathrm{k}}(1 - N_{\mathrm{k}}). \tag{16.122}$$

Inserting (16.122) into (16.120) leads to

$$C_{\mathrm{v}} = 2\beta^2 k_{\mathrm{B}} \sum_{\mathrm{k}} N_{\mathrm{k}}(1 - N_{\mathrm{k}}) \cdot \left[\mho_{\mathrm{k}}{}^2 + \left(\frac{\beta}{2}\right) \cdot \left(\frac{\partial |\nabla|^2}{\partial \beta}\right)_{\mathrm{v}}\right]. \tag{16.123}$$

Subtracting (16.121) from (16.123) provides an expression that represents the specific heat jump at the critical temperature, which is

$$C_{\mathrm{v-}} - C_{\mathrm{v+}} = \beta^3 k_{\mathrm{B}} \sum_{\mathrm{k}} \left[N_{\mathrm{k}}(1 - N_{\mathrm{k}}) \cdot \left(\frac{\partial |\nabla|^2}{\partial \beta}\right)_{\mathrm{v}}\right]_{\mathrm{T=T_c}}. \tag{16.124}$$

In order to calculate the specific heat jump, we need the parameter N_{k} as well as the expression $(\frac{\partial |\nabla|^2}{\partial \beta})_{\mathrm{v}}$ which requires knowledge of $|\nabla|^2$ at temperatures close to, but just below T_{c}. For convenience we introduce the notation $|\nabla|^2 \equiv |\nabla_0|^2$ that applies at these temperatures.

Calculation of $|\nabla_0|^2$

To that purpose, return to the gap function master equation (16.93). Use the fact that at $T = T_{\mathrm{c}}$, $\nabla = 0$. Additionally, for $T = T_{\mathrm{c}}$, introduce the symbol $\beta_{\mathrm{c}} \equiv \frac{1}{k_{\mathrm{b}} T_{\mathrm{c}}}$. Then (16.93) reduces to the following equality:

$$1 - V\Theta \int_0^{\hbar\Omega_{\mathrm{D}}} \left\{\frac{\mathrm{d}\xi_{\mathrm{q}}}{\xi_{\mathrm{q}}}\right\} \tanh\left(\frac{\beta_{\mathrm{c}}}{2}\xi_{\mathrm{q}}\right) = 0. \tag{16.125}$$

Whereas at and above the transition temperature $\nabla = 0$, just below the transition temperature it is equal to ∇_0 which is nonzero, but small. Similarly, at such temperatures it is convenient to work also with another small parameter δ_T defined below by

$$T = T_c - (T_c - T) \equiv T_c - \delta_T\,, \quad \frac{\delta_T}{T_c} \ll 1\,, \quad \nabla_0 \ll k_B T_c\,. \tag{16.126}$$

By changing ∇ to ∇_0 the master equation (16.93) becomes (16.127).

$$\begin{aligned} 1 &= \frac{V}{2}\int_{-\hbar\Omega_D}^{\hbar\Omega_D} d\xi_q \left\{\frac{\Theta}{\sqrt{\xi_q{}^2 + |\nabla_0|^2|}}\right\} \tanh\left(\frac{\beta_c}{2}\sqrt{\xi_q{}^2 + |\nabla_0|^2}\right) \\ &= V\int_{0}^{\hbar\Omega_D} d\xi_q \left\{\frac{\Theta}{\sqrt{\xi_q{}^2 + |\nabla_0|^2|}}\right\} \tanh\left(\frac{\beta_c}{2}\sqrt{\xi_q{}^2 + |\nabla_0|^2}\right). \end{aligned} \tag{16.127}$$

Here $\beta_c = \frac{1}{k_B T_c}$.

Expanding (16.127) as a power series in the small parameters δ_T and ∇_0 leads to

$$\begin{aligned} 0 = 1 &- V\Theta\int_0^{\hbar\Omega_D} d\xi_q \left\{\frac{1}{\xi_q}\right\} \tanh\left(\frac{\beta_c}{2}\xi_q\right) + \left(\frac{V\Theta\delta_T}{2k_B T_c{}^2}\right)\int_0^{\hbar\Omega_D} \frac{d\xi_q}{\cosh^2\left(\frac{\xi_q}{2k_B T_c}\right)} \\ &+ \left(\frac{V\Theta(\nabla_0)^2}{4k_B T_c}\right)\int_0^{\hbar\Omega_D} \frac{d\xi_q}{\xi_q^2}\left\{\frac{1}{\cosh^2\left(\frac{\xi_q}{2k_B T_c}\right)} - \frac{2k_B T_c \tanh\left(\frac{\xi_q}{2k_B T_c}\right)}{\xi_q}\right\} \\ &+ \text{higher order terms}\,. \end{aligned} \tag{16.128}$$

Inserting the information contained in (16.125) into (16.128) makes the two terms on the first line of (16.128) add up to zero. As a reasonable approximation, we make two assumptions: first, beyond the leading order, the power series of the small parameters can be ignored; and second, because of weak interparticle coupling, $\hbar\Omega_D \gg k_b T_c$. Therefore the upper limit in the integrals may safely be extended from $\hbar\Omega_D$ to ∞.

With above two assumptions in place, (16.128) reduces to

$$\begin{aligned} 0 \approx &\left(\frac{V\Theta\delta_T}{2k_B T_c{}^2}\right)\int_0^{\infty} \frac{d\xi_q}{\cosh^2\left(\frac{\xi_q}{2k_B T_c}\right)} \\ &+ \left(\frac{V\Theta(\nabla_0)^2}{4k_B T_c}\right)\int_0^{\infty} \frac{d\xi_q}{\xi_q^2}\left\{\frac{1}{\cosh^2\left(\frac{\xi_q}{2k_B T_c}\right)} - \frac{2k_B T_c \tanh\left(\frac{\xi_q}{2k_B T_c}\right)}{\xi_q}\right\}. \end{aligned} \tag{16.129}$$

The integrals in (16.129) are

$$\int_0^{\infty} \frac{d\xi_q}{\cosh^2\left(\frac{\xi_q}{2k_B T_c}\right)} = 2k_B T_c \int_0^{\infty} \frac{dx}{\cosh^2 x} = 2k_B T_c\big[\tanh(x)\big]_0^{\infty} = 2k_B T_c \tag{16.130}$$

and

$$\int_0^\infty \frac{d\xi_q}{\xi_q^2}\left\{\frac{1}{\cosh^2(\frac{\xi_q}{2k_BT_c})} - \frac{2k_BT_c\tanh(\frac{\xi_q}{2k_BT_c})}{\xi_q}\right\}$$
$$= \left(\frac{1}{2k_BT_c}\right)\left[\int_0^\infty \frac{dx}{x^2\cosh^2(x)} - \int_0^\infty \frac{\tanh(x)}{x^3}dx\right] = -\left(\frac{1}{2k_BT_c}\right)\left(\frac{7\zeta(3)}{\pi^2}\right). \tag{16.131}$$

Insert the results in (16.130) and (16.131) into the appropriate locations in (16.129). This process leads to

$$0 \approx \left(\frac{V\Theta\delta_T}{2k_BT_c{}^2}\right)2k_BT_c - \left(\frac{V\Theta(|\nabla_0)^2|}{(k_BT_c)^2}\right)\frac{7\zeta(3)}{8\pi^2}$$
$$= (V\Theta)\left[\left(\frac{\delta_T}{T_c}\right) - \frac{|(\nabla_0)^2|}{(k_BT_c)^2}\frac{7\zeta(3)}{8\pi^2}\right]. \tag{16.132}$$

Equation (16.132) records the equality

$$\left|(\nabla_0)^2\right| = \left(\frac{8\pi^2(k_B)^2T_c}{7\zeta(3)}\right)\delta_T = \left(\frac{8\pi^2(k_B)^2T_c}{7\zeta(3)}\right)(T_c - T). \tag{16.133}$$

Differentiation of the equality with respect to T gives

$$\frac{d(|\nabla_0)^2|}{dT} = -\frac{8\pi^2(k_B)^2T_c}{7\zeta(3)}, \tag{16.134}$$

or equivalently,

$$\frac{d(|\nabla_0)^2|}{d\beta} = -k_BT_c^2\frac{d|(\nabla_0)^2|}{dT} = \frac{8\pi^2(k_BT_c)^3}{7\zeta(3)}. \tag{16.135}$$

16.15 Specific Heat: Calculation of Jump in

A relationship expressing the jump in specific heat at the critical temperature was noted in (16.124). Inserting (16.135) into (16.124) leads to

$$C_{v^-} - C_{v^+} = (\beta^3k_B)\cdot\left(\frac{\partial|\nabla_0|^2}{\partial\beta}\right)_v\cdot\sum_k[N_k(1-N_k)]_{T=T_c}$$
$$= (\beta^3k_B)\cdot\left(\frac{8\pi^2(k_BT_c)^3}{7\zeta(3)}\right)_v\cdot\sum_k[N_k(1-N_k)]_{T=T_c}$$
$$= \left(\frac{8\pi^2k_B}{7\zeta(3)}\right)\cdot\sum_k[N_k(1-N_k)]_{T=T_c}\cdot \tag{16.136}$$

Finally, we consider the infinite sum

$$\sum_{\mathrm{k}} [N_{\mathrm{k}}(1 - N_{\mathrm{k}})]_{\mathrm{T=T_c}} , \tag{16.137}$$

where $[N_{\mathrm{k}}]_{\mathrm{T=T_C}}$ in (16.137) is the canonical representation of the number of weakly interacting fermions with interparticle interaction $[\mho_{\mathrm{k}} \equiv \sqrt{\xi_{\mathrm{k}}{}^2}]_{\mathrm{T=T_c}}$ at temperature $T_{\mathrm{c}} = \frac{1}{k_{\mathrm{B}}\beta_{\mathrm{c}}}$ given as

$$[N_{\mathrm{k}}]_{\mathrm{T=T_c}} \equiv \left[\exp(\beta\sqrt{\xi_{\mathrm{k}}{}^2}) + 1\right]^{-1}_{\mathrm{T=T_c}} . \tag{16.138}$$

Therefore the infinite sum over k-space gives

$$\sum_{\mathrm{k}} [N_{\mathrm{k}}(1 - N_{\mathrm{k}})]_{\mathrm{T=T_c}} = \Theta \int_{-\infty}^{\infty} V \mathrm{d}\xi_{\mathrm{k}} N_{\mathrm{k}}(1 - N_{\mathrm{k}}) = \frac{\Theta V}{\beta_{\mathrm{c}}} , \tag{16.139}$$

where, as defined in (16.92), Θ is the density of states per unit volume.

Inserting (16.138) into (16.139) leads to

$$C_{\mathrm{v}^-} - C_{\mathrm{v}^+} = \left(\frac{8\pi^2 k_{\mathrm{B}}}{7\zeta(3)}\right) \cdot \sum_{\mathrm{k}} [N_{\mathrm{k}}(1 - N_{\mathrm{k}})]_{\mathrm{T=T_c}} = \left(\frac{8\pi^2 T_{\mathrm{c}} (k_{\mathrm{B}})^2}{7\zeta(3)}\right) \cdot \Theta V . \tag{16.140}$$

16.16 A Universal Ratio

At $T = T_{\mathrm{c}} + 0$, the parameter $|\nabla_0|^2 = 0$. Therefore (16.120) gives

$$C_{\mathrm{v}^+} = -2\beta k_{\mathrm{B}} \sum_{\mathrm{k}} \left(\frac{\partial N_{\mathrm{k}}}{\partial \mho_{\mathrm{k}}}\right)_{\mathrm{v}} \cdot [\mho_{\mathrm{k}}{}^2] \Theta V . \tag{16.141}$$

By using the so-called Sommerfeld expansion, the sum on the right-hand side of (16.141) has been calculated [9, 10]. The result is

$$C_{\mathrm{v}^+} = \left(\frac{2\pi^2 T_{\mathrm{c}} (k_{\mathrm{B}})^2}{3}\right) \Theta V . \tag{16.142}$$

Dividing (16.140) by (16.142) yields the ratio of the jump in specific heat at the critical temperature and the specific heat of the normal state above the critical temperature, which is

$$\frac{C_{\mathrm{v}^-} - C_{\mathrm{v}^+}}{C_{\mathrm{v}^+}} = \frac{12}{7\zeta(3)} . \tag{16.143}$$

This ratio is just a number unaffected by the substance of the superconducting material. As such it is universal. Indeed, it agrees well with experimental results of many BCS superconductors.

16.17 Another Universal Ratio and Result at Zero Temperature

Beginning with results at zero temperature, consider yet other universal ratio.

To that end let us return to the master equation (16.93) that governs the gap function at arbitrary temperature and set $T \rightarrow 0$, i.e., set $\beta \rightarrow \infty$. At $\beta = \infty$, $\tanh\left(\frac{\beta}{2}\sqrt{\xi_q{}^2 + |\nabla_0|^2}\right) \rightarrow 1$. As a result, at $T = 0$ (16.93) gives

$$\frac{1}{V\Theta} = \int_0^{\hbar\Omega_D} \frac{d\xi_q}{\sqrt{\xi_q{}^2 + |\nabla_0|^2|}} = \log\left[\sqrt{\hbar\Omega_D{}^2 + |\nabla_0|^2} + \hbar\Omega_D\right] - \log \nabla_0 \,, \tag{16.144}$$

where ∇_0 signifies ∇ at $T = 0$. Because in BCS superconductors $|\nabla_0| \ll \hbar\Omega_D$, (16.100) can be well approximated by the relationship

$$\frac{1}{V\Theta} \approx \log\left[2\hbar\Omega_D\right] - \log \nabla_0 = -\log\left[\frac{\nabla_0}{2\hbar\Omega_D}\right]. \tag{16.145}$$

After transferring the minus sign and exponentiating both sides, (16.145) leads to

$$\nabla_0 = 2\hbar\Omega_D \exp\left[-\frac{1}{V\Theta}\right]. \tag{16.146}$$

Divide ∇_0 given in (16.146) by $k_B T_c$ given in (16.100) to obtain

$$\frac{\nabla_0}{k_B T_c} = \frac{2\hbar\Omega_D \exp[-\frac{1}{V\Theta}]}{1.1338659(\hbar\Omega_D)\exp[\frac{-1}{V\Theta}]} = \frac{2}{1.1338659} = 1.7638770\,. \tag{16.147}$$

This ratio is just a number. As such it is universal. Indeed, in conventional superconductors, experiment appears to be in moderate agreement with this result. That fact provides support to the theory.

References

1. J. Bardeen, L.N. Cooper, J.R. Schrieffer, Phys. Rev. **106**(1), 162–164 (1957)
2. J. Bardeen, L.N. Cooper, J.R. Schrieffer, Phys. Rev. **108**(5), 1175–1204 (1957) = BCS
3. A.L. Fetter, J.D. Walecka, *Quantum Theory of Many Particle Systems* (McGraw-Hill, New York, 1971)
4. G. Rickayzen, *Theory of Superconductivity* (Wiley, New York, 1965)
5. C. Timm, Theory of Superconductivity. www.physik.tu.dresden.de
6. R.M. Fernandes, Lecture Notes: BCS theory of superconductivity
7. W. Meissner, R. Ochsenfeld, Naturwissenchaften **21**(787) (1933)
8. G.B. Thomas, *Calculus and Analytic Geometry*, 4th edn. (Addison-Wesley, Reading, 1969), p. 441, Exercise 15
9. R.M. Fernandes, Lecture Notes: BCS theory of superconductivity
10. Carston.Timm@tu-dresden.de, Theory of superconductivity in FreeBookCentre.net

Large Numbers. The Most Probable State

A

Thermodynamics deals with systems with very large number of atoms. For instance, four grams of helium have approximately 6×10^{23} molecules.[1] Considering that the age of the Universe is only about 5×10^{17} s, this is a very large number.

Interparticle interactions make exact analysis of most thermodynamic systems well-nigh impossible. Indeed, when theoretical formulation is used—as, for example, is done within the framework of statistical mechanics—approximations are often needed for its evaluation. Therefore, rather than getting involved with a priori calculations, thermodynamics generally deals with interrelationships of physical properties of macroscopic systems. Because such knowledge can often help relate easily measurable properties to those that are hard to measure, thermodynamic plays an important role in scientific disciplines.

In this appendix, we show how large numbers, that are central to the validity of thermodynamic relationships, possess some simplifying properties. This fact is best demonstrated by analyzing an idealized model. Frequency moments of the exact distribution function as well as those of the relevant Gaussian approximation are worked out and the impressive validity of the Gaussian approximation is pointed out. Also the helpful use of binomial expansion is noted. It is demonstrated that in a macroscopically large system, the most probable configuration is overwhelmingly so. Therefore, the result of any "macroscopic" measurement is well described by an accurate calculation of the most probable state.

A.1 Random Number Generator

Random number generators are available in most mathematical software packages. When a "perfect random number generator" (PRNG) is set to return values within a "specified range" that extends, let us say, from 0 to 1, it does so with equal probability for all values that lie within the range. And, if the PRNG were called an infinite number of times, the number density of returned values per unit microscopic length would be identically the same throughout the specified range $0 \rightarrow 1$.

[1]Note that one He atom is a molecule.

R. Tahir-Kheli, *General and Statistical Thermodynamics*,
https://doi.org/10.1007/978-3-030-20700-7

Therefore, let us consider the following scenario. Some person wishes to use a PRNG to fill a group of N locations—to be called N "sites"—with a total of exactly $N\mu$ "occupied" sites[2] and $N(1-\mu)$ "unoccupied" sites.[3] A person might imagine that all that he/she needs to do is the following:

Catalogue every call returned by the PRNG into one of two possible statements, "occupied" or "unoccupied."

Record that the ith site is "occupied" if the ith call returns a value somewhere between 0 and μ. Denote this fact by setting an occupancy variable $\sigma_{\mathrm{i}} = 1$.

Otherwise, if the ith call returns a value that lies between μ and 1, the ith site is to be considered "unoccupied." And this fact is to be denoted by setting the occupancy variable $\sigma_{\mathrm{i}} = 0$.

For extreme simplicity, let us deal first with a trivial case. Assume that the total number of sites $N = 2$ and $\mu = 0.5$. So the number of occupied sites is equal to the number of unoccupied sites, both being equal to 1. Assume that we want to save effort and decide to make only 2 PRNG calls in the belief that the number of calls need not be much larger than the number of sites available. We ask the question: What is wrong with a person's belief that the proposed arrangement, based on only 2 PRNG calls, will successfully result in fitting the two sites with exactly $N\mu = 1$ occupied, and $N(1-\mu) = 1$ unoccupied, sites.

An incorrect answer would claim that, inasmuch as the statements about the PRNG returning calls between 0, $\frac{1}{2}$, and 1 do not specify what happens exactly at 0, $\frac{1}{2}$, and 1, the proposal given to the computer—that runs the PRNG—is vague. Hence, the difficulty!

The correct answer is that, unlike in a thermodynamic system, the number of calls is not "large." Therefore, the fluctuations in the result are significant. And, hence there is a sizeable probability that the PRNG result for the number of occupied sites will turn out to equal 2 or 0, both very different from the actual number that is equal to 1.

A.2 Binomial Expansion

Determining the results of a large number of calls of the PRNG requires some effort. Fortunately, a little help from the binomial expansion does the trick. The binomial equality given below holds for finite values of the two given variables μ and μ_0:

$$(\mu + \mu_0)^N = \sum_{p=0}^{N} \left[\frac{N!}{(p)!(N-p)!} \right] \mu^p \mu_0^{N-p} ; \quad p = 0, 1, 2, \ldots, N . \tag{A.1}$$

[2]Important notice: The symbol μ, which is mostly used for denoting the chemical potential, is temporarily being appropriated for use as the relative concentration of occupied sites.

[3]The present formulation also applies to a group of N noninteracting spins. Out of such a group, $N\mu$ spins are supposed to be pointing up so that each can be said to have a spin $s = +1$, and $N(1-\mu)$ spins are pointing down so that each of their spins has $s = -1$.

The above equality can conveniently be used to study the case where the total number of sites—and also the total number of the PRNG calls—is equal to N. To do that:

Set the variables μ and μ_0 to represent the desired concentrations of the occupied and the unoccupied sites.[4] Further, denote 100% concentration as equalling 1. Then we have,

$$0 \leq \mu \leq 1\,, \qquad 0 \leq \mu_0 \leq 1\,, \qquad \mu + \mu_0 = 1\,. \tag{A.2}$$

As a result, the left-hand side of (A.1) is equal to unity. Therefore, $[\frac{N!}{p!(N-p)!}]\,\mu^p(1-\mu)^{N-p}$ represents the probability that out of N sites, a number (p) will be found to be occupied and $(N-p)$ will be found unoccupied. Note that according to (A.1) and (A.2), the sum of all such probabilities, i.e., where the number p of occupied sites ranges from 0 to N, is equal to unity, that is, it is 100%.

Two Sites: A Trivial Example As mentioned before, it is helpful to treat first the rather trivial case where the total number of sites, i.e., N, is very small, that is, $N = 2$. The two sites can be occupied in four different ways, meaning four different combinations of the occupancy variables σ_1 and σ_2 are possible. For example, we can have:

$$\begin{aligned} &\sigma_1 = 0\,, \quad \sigma_2 = 0\,; \qquad \sigma_1 = 1\,, \quad \sigma_2 = 1\,; \\ &\sigma_1 = 1\,, \quad \sigma_2 = 0\,; \qquad \sigma_1 = 0\,, \quad \sigma_2 = 1\,. \end{aligned} \tag{A.3}$$

For simplicity, let us choose the desired value of the concentration, μ, of the occupied sites to be equal to 0.5.[5]

According to (A.1), the probability that two calls to the PRNG will actually lead to zero number of occupied sites—i.e., $p = 0$—is,

$$\left[\frac{N!}{p!(N-p)!}\right]\mu^p(1-\mu)^{N-p} = \left[\frac{2!}{0!\,2!}\right](0.5)^0(0.5)^2 = \frac{1}{4}\,.$$

Similarly, if we set $p = 1$, we get the probability that the two calls to the PRNG will lead to only a single occupied site:

$$\left[\frac{N!}{p!(N-p)!}\right]\mu^p(1-\mu)^{N-p} = \left[\frac{2!}{1!\,1!}\right](0.5)^1(0.5)^1 = \frac{1}{2}\,.$$

And finally, the probability that the two calls to the PRNG will lead to exactly two occupied sites is found by setting $p = 2$:

$$\left[\frac{N!}{p!(N-p)!}\right]\mu^p(1-\mu)^{N-p} = \left[\frac{2!}{2!\,0!}\right](0.5)^2(0.5)^0 = \frac{1}{4}\,.$$

[4]Remember: We can immediately apply the present formulation and its results to the field-free, noninteracting, up–down spin problem. All that needs to be done is to replace the "occupied sites" by "up-spins" and the "unoccupied" sites by "down-spins."

[5]When this is the case, these four combinations are all equally likely to occur. As such, they can be referred to as the four possible "microstates" of the system. Note that the system consists of only two sites.

By the way, in order to get the relevant number of states, the above results for the probabilities need to be multiplied by $N^2 = 4$.[6]

We notice that in the small system being studied above, that has only $N = 2$ sites and $2^N = 4$ states, the results given by the PRNG do not correspond well with our expressed desire. We had wanted the PRNG to return exactly 50% occupancy. This would have happened only if all the four states had contained exactly one occupied, and one unoccupied, site. Instead, what really happened is that out of four possible states, only two are of the desired variety. Of the other two states, the first contains no occupied sites, while the second has both sites occupied. Accordingly, there is only a half-chance that, in practice, our expressed desire will actually be realized.

A physically more instructive description of the above happenstance is the following: The desired result is the most likely result.[7] For a system that is not "large,"[8] the desired result is still not overwhelmingly probable.[9] One expects that things will improve if the number of calls, N, is "large." Hopefully then the desired result will be overwhelmingly probable.[10] This matter is investigated below.

A.3 Large Number of Calls

Let us, for the moment, continue to treat the simple case where the desired concentration, μ, of the occupied sites equals that of the unoccupied sites, $1 - \mu$. That is, $\mu = \frac{1}{2}$.

Consider a system with a very large number of sites, i.e., $N \gg 1$. As per the described procedure, this entails making a large number of calls—equal to N. To deal with this situation—see (A.1)—we need factorials of large numbers.

When N is very large compared to unity, according to Stirling's zeroth-order approximation, the factorial of N can be approximated as follows:

$$N! \sim (N/e)^N . \tag{A.4}$$

Insert this value of $N!$ into (A.1) and (A.2) and study the case where the actual number of occupied, p, and the unoccupied, $N - p$, sites is exactly equal to that suggested by the desired concentration of the occupied and the unoccupied sites, that is, $p = \mu N$ and $N - p = (1 - \mu)N$. Then, because μ has been chosen to be $\frac{1}{2}$

[6]Compare this prediction with the demonstration of the four possible states shown in (A.3).

[7]Note that while two states corresponded to the desired result, only one refers to each undesired state.

[8]A system with only two sites is a "small" system.

[9]An event that has only half a chance of occurring is not considered to be a highly probable event.

[10]In other words, then the actual value returned by the PRNG will correspond much more closely to that which was desired.

here, the probability of such occupancy is

$$\left[\frac{N!}{p!(N-p)!}\right]\mu^p(1-\mu)^{N-p} = \left[\frac{N!}{(N/2)!(N/2)!}\right](1/2)^{N/2}(1/2)^{N/2}$$
$$\approx \left[\frac{(N/e)^N}{(N/2e)^{N/2}(N/2e)^{N/2}}\right](1/2)^{N/2}(1/2)^{N/2}$$
$$= \left[\frac{(N/e)^N}{(N/2e)^N}\right](1/2)^N = 1\,. \quad \text{(A.5)}$$

What a fantastic outcome! The result is exactly as was desired. Its probability of occurrence is 100% and it occurs exactly at the desired place, namely, where the number of occupied and unoccupied sites is equal, i.e., they are both $N/2$. Accordingly, an infinitely narrow region must contain all the 2^N states of the model!

Is this really true? Or has the crudeness of the approximation for $N!$—given in (A.4)—deceived us?

To investigate this matter further, it is necessary first to use a more accurate version of the Stirling asymptotic series. That is,

$$N! = (2\pi N)^{1/2}(N/e)^N\left[1 + \frac{1}{12N} + \frac{1}{288N^2} + O\left(\frac{1}{N^3}\right)\right]. \quad \text{(A.6)}$$

Inserting the above approximation for $N!$ into the right-hand side of (A.5), readily yields the following result:

$$\left[\frac{N!}{(N/2)!(N/2)!}\right](1/2)^{N/2}(1/2)^{N/2} = \sqrt{(2/\pi N)}\big[1 + O(1/N)\big]. \quad \text{(A.7)}$$

Exercise A.1 Derive (A.7).

Unlike the fantastic statement made by (A.5), (A.7) is quite sensible. All the states do not reside, exactly as desired, at $p = N/2$. Rather, in order to collect most of them, one would need to sum over a range of p values—i.e., the occupied sites—that are narrowly spread around the desired occupancy number $N\mu = N/2$. As a rough guess, as the above equation indicates, the width of such a region should be approximately equal to $\sqrt{(\pi N/2)}$. A more precise estimate of the size of the width is discussed below.

A.3.1 Remark: A Gaussian Distribution

An adequate estimate of the width of the distribution can be had by making a simple assumption that around the most probable location, i.e., at the desired value of the occupancy, the distribution of states has roughly a Gaussian shape (see Fig. A.1).

As a physical test, for general value of the desired concentration μ, we carry out an exact calculation of several frequency moments of the distribution function. We find that these moments are well represented by a Gaussian approximation for the exact distribution function.

Moreover, as indicated below, the functional form of the density of states—for $\mu = \frac{1}{2}$ and very large N—also yields a result that is close to a Gaussian.

Gaussian Approximation for a Region Around Half-Concentration

In order to calculate the probability that the occupancy closely ranges around the desired value, one needs to employ (A.1), or equivalently, the left-hand side of (A.5). Next, one sets $N \gg 1$, and chooses the desired concentration, μ, of the occupied sites. For instance, here we have chosen $\mu = 1 - \mu = 0.5$. In order to examine the region that lies immediately around the relevant occupancy number—that is, $\frac{N}{2}$ here—one needs to set $p = \frac{N}{2} + n$ and $N - p = \frac{N}{2} - n$ and remember to choose $n \ll N$.

Then, according to (A.1), the result for the probability distribution function is

$$\frac{N!}{(\frac{N}{2}+n)!(\frac{N}{2}-n)!}\left(\frac{1}{2}\right)^{\frac{N}{2}+n}\left(\frac{1}{2}\right)^{\frac{N}{2}-n}. \tag{A.8}$$

Notice that the distribution is symmetric for the interchange $\pm n \rightarrow \mp n$. Moreover, when $\frac{n^2}{N} \ll 1$, the above can be expanded in powers of $\frac{n^2}{N}$. That is,

$$\begin{aligned}
&\frac{N!}{(\frac{N}{2}+n)!(\frac{N}{2}-n)!}\left(\frac{1}{2}\right)^{\frac{N}{2}+n}\left(\frac{1}{2}\right)^{\frac{N}{2}-n}\\
&=\sqrt{\frac{2}{\pi N}}\cdot\left[1-\left(\frac{2}{N}n^2\right)+O\left\{\left(\frac{n^2}{N}\right)^2\right\}\right]\\
&\approx\sqrt{\frac{2}{\pi N}}\cdot\exp\left(-\frac{2}{N}n^2\right).
\end{aligned} \tag{A.9}$$

Exercise A.2 Derive the relationships implicit in (A.9).

Gaussian Approximation—Continued

There were three good reasons why the expansion on the right-hand side of (A.9) was approximated by the exponential $\exp(-\frac{2}{N}n^2)$. First, the two leading terms agree. Second, the quadratic dependence on n is consistent with the symmetry, $n\pm \rightarrow n\mp$, of the left-hand side of (A.9). Third, the requirement

$$\left(\frac{1}{2}+\frac{1}{2}\right)^N = \sum_{p=0}^{N}\left[\frac{N!}{p!(N-p)!}\right]\left(\frac{1}{2}\right)^p\left(\frac{1}{2}\right)^{N-p} = 1, \tag{A.10}$$

when translated by setting $p = \frac{N}{2} + n$—as in (A.11) below—

$$\begin{aligned}
&\sum_{n=-\frac{N}{2}}^{n=+\frac{N}{2}}\frac{N!}{(\frac{N}{2}+n)!(\frac{N}{2}-n)!}\left(\frac{1}{2}\right)^{\frac{N}{2}+n}\left(\frac{1}{2}\right)^{\frac{N}{2}-n} = 1\\
&\approx\sqrt{\frac{2}{\pi N}}\cdot\int_{-\frac{N}{2}}^{+\frac{N}{2}}\exp\left(-\frac{2}{N}n^2\right)\cdot dn
\end{aligned} \tag{A.11}$$

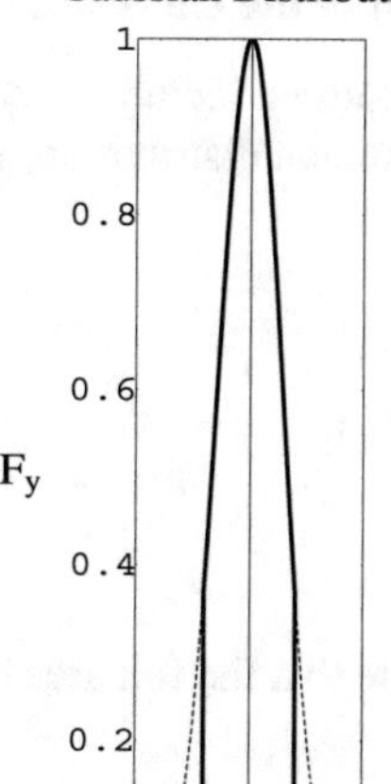

Fig. A.1 Gaussian distribution function

is valid when $N \gg 1$. Compare (A.1) and (A.9). Note that, when the above width is described by the relation

$$n = W_{\text{width}} = \pm\sqrt{\frac{N}{2}}, \tag{A.12}$$

the exponential falls off to $\exp(-1) \approx 0.37$ of its maximum value. (The maximum occurs when $n = 0$.) Henceforth, we shall call W_{width} the half-width of the distribution function.

To an untutored eye, the estimated size of the half-width, i.e., $\sqrt{(N/2)}$, would seem to be inordinately large. To alleviate such concerns, all we need to remember is that the full distribution, in principle, can extend from $p = 0$ all the way to $p = N$. (Recall that N is very large.) Therefore, a more meaningful measure of the width is the "relative width," which we shall denote by the symbol ϖ_{rel}. Here

$$\varpi_{\text{rel}} = \frac{\text{twice the half-width}}{\text{full range over which the distribution could extend}}$$
$$= \frac{2W_{\text{width}}}{N} = \left(\frac{2}{N}\right)^{1/2}. \tag{A.13}$$

For example, when $N \sim 10^{24}$, the width $\sim 10^{12}$ appears to be very large. Yet, the relative width, $\varpi_{\text{rel}} \sim 10^{-12}$, is extremely small.

Despite the finding that the distribution is not infinitely peaked at the desired concentration[11] but has a finite width, the above discussion shows that for large N the desired state is overwhelmingly probable.

[11]Note that here the desired concentration would have led to the midpoint $p = \frac{N}{2}$.

Plot of the Gaussian Distribution

Because a picture is worth many words, appended above is a plot of the normalized Gaussian distribution, F_y, as a function of the variable y,

$$y = n\sqrt{\frac{2}{N}}\,. \tag{A.14}$$

That is,

$$F_y = \frac{\exp(-y^2)}{\sqrt{\pi}}\,. \tag{A.15}$$

Note that the full area of the curve is normalized to unity, i.e.,

$$\int_{-\infty}^{+\infty} F_y \mathrm{d}y = \int_{-\infty}^{+\infty} \frac{\exp(-y^2)}{\sqrt{\pi}} \mathrm{d}y = 1\,. \tag{A.16}$$

The two dark vertical lines, and the thick dark area of the curve, enclose that part of the distribution which lies within what we have called the width of the distribution. In terms of the abscissa y, these lines fall at positions $y = \pm 1$. It is interesting to note that this narrow region—consisting only of the width of the distribution—covers $\approx 84.3\%$ of the total weight of the distribution, i.e.,

$$\int_{-1}^{+1} F_y \mathrm{d}y = 0.84271\,. \tag{A.17}$$

Indeed, if we extend this region to twice the width, that is, $y = \pm 2$, we recover almost all, that is 99.53%, of the full weight of the distribution.

Moments of the Distribution Function: Remark B

Moments of appropriate distribution functions can often yield valuable information about thermodynamic states. To make use of this fact, in the following the so-called normalized moments of the Gaussian distribution function are calculated. Next, the exact second normalized moment is calculated and is set to agree with the corresponding result of the Gaussian distribution function for general concentration.[1] It is interesting to note that for half-concentration, the Gaussian approximation—without any fiddling—yields exact results for the second normalized moment.

B.1 Moments of the Gaussian Distribution Function

B.1.1 Unnormalized Moments: Gaussian Distribution

The first several unnormalized moments of the Gaussian distribution function are calculated below. That is,

$$I(a,n) = \int_{-N/2}^{N/2} \mathrm{d}x \exp\left(-ax^2\right)x^n\,, \quad n = 0,\ldots,6 \quad \text{for } N \gg 1\,. \tag{B.1}$$

Integration

It is clear that because of symmetry, $I(a,n)$ is zero for all odd values of n. To evaluate the integrals for even n, consider doing the following:

$$\begin{aligned} I(a,0)\cdot I(a,0) &= \int_{-N/2}^{N/2} \mathrm{d}x \exp\left(-ax^2\right) \int_{-N/2}^{N/2} \mathrm{d}y \exp\left(-ay^2\right) \\ &= \int_{-N/2}^{N/2}\int_{-N/2}^{N/2} \mathrm{d}x\mathrm{d}y \exp\left[-a\left(x^2+y^2\right)\right], \quad N \gg 1\,. \end{aligned} \tag{B.2}$$

[1]Recall that general concentration refers to the occupancy departing from the midpoint. This means that the probability that the ith call on the PRNG returns an occupied site, i.e., $\sigma_i = 1$, is not necessarily equal to that which returns $\sigma_i = 0$.

R. Tahir-Kheli, *General and Statistical Thermodynamics*,
https://doi.org/10.1007/978-3-030-20700-7

Changing over to polar coordinates, r and θ,

$$x = r\cos\theta\,, \qquad y = r\sin\theta\,,$$

and noting that

$$\mathrm{d}x\mathrm{d}y = \begin{vmatrix} (\frac{\partial x}{\partial r})_\theta & (\frac{\partial x}{\partial \theta})_\mathrm{r} \\ (\frac{\partial y}{\partial r})_\theta & (\frac{\partial y}{\partial \theta})_\mathrm{r} \end{vmatrix} \mathrm{d}r\mathrm{d}\theta = r\mathrm{d}r\mathrm{d}\theta\,,$$

the double integral over a square of size $N \times N$ in the Cartesian plane is transformed into one over a circular, planar disk of radius $N/2$:

$$\begin{aligned} \left[I(a,0)\right]^2 &= \int_0^{2\pi} \mathrm{d}\theta \int_0^{N/2} \mathrm{d}r \exp(-ar^2)r \\ &= 2\pi\left(\frac{1}{-2a}\right)\left[\exp\left\{-a\left(\frac{N}{2}\right)^2\right\} - 1\right] \\ &\approx \pi/a\,, \quad N \gg 1. \end{aligned} \tag{B.3}$$

Regarding the limits, we note that even though the square and the circle do not exactly fit over each other, for $N/2 \gg 1$ the discrepant region is so very far away from the origin that the exponential in the integral makes the discrepancy vanishingly small. Therefore, as long as a is positive, without any loss in accuracy, integrals $I(a,n)$ an be evaluated by replacing N by ∞. Consequently, the following result is close to being exact:

$$I(a,0) = \sqrt{\frac{\pi}{a}}. \tag{B.4}$$

For $n > 0$, the nth moment $I(a,n)$ can be evaluated by repeated differentiation with respect to a. For instance, the second, fourth and sixth moments of the Gaussian distribution, $\exp(-ax^2)$, are,

$$I(a,2) = -\frac{\mathrm{d}I(a,0)}{\mathrm{d}a} = \int_{-\infty}^{\infty} \mathrm{d}x \exp(-ax^2)x^2 = \frac{1}{2}\sqrt{\frac{\pi}{a^3}}\,, \tag{B.5}$$

$$\begin{aligned} I(a,4) &= -\frac{\mathrm{d}I(a,2)}{\mathrm{d}a} = \int_{-\infty}^{\infty} \mathrm{d}x \exp(-ax^2)x^4 = \frac{1}{2}\cdot\frac{3}{2}\cdot\sqrt{\frac{\pi}{a^5}}\,, \\ I(a,6) &= -\frac{\mathrm{d}I(a,4)}{\mathrm{d}a} = \int_{-\infty}^{\infty} \mathrm{d}x \exp(-ax^2)x^6 = \frac{1}{2}\cdot\frac{3}{2}\cdot\frac{5}{2}\cdot\sqrt{\frac{\pi}{a^7}}. \end{aligned} \tag{B.6}$$

B.1.2 Normalized Moments of the Gaussian Distribution Function

Let us define the jth normalized Gaussian moment for a general choice of the parameter a as follows:

$$\Delta_\mathrm{j}^\mathrm{G}(a) = \frac{I(a,j)}{I(a,0)}. \tag{B.7}$$

Clearly, by definition, the 0th normalized moment, $\Delta_0^G(a)$, is unity. And the other normalized Gaussian moments are:

$$\Delta_2^G(a) = \frac{1}{2a}, \qquad \Delta_4^G(a) = \frac{3}{4a^2}, \qquad \Delta_6^G(a) = \frac{15}{8a^3}. \tag{B.8}$$

C Normalized Moments of the Exact Distribution Function for General Occupancy

The overall average value of the occupancy variable $\sigma_{\rm i}$, written as $\langle\sigma_{\rm i}\rangle$, is equal to the concentration μ of the occupied sites for which $\sigma_{\rm i} = 1$. Note that the unoccupied sites, for which $\sigma_{\rm i} = 0$ and whose concentration is $1 - \mu$, contribute nothing to $\langle\sigma_{\rm i}\rangle$. Thus,

$$\langle\sigma_{\rm i}\rangle = \mu \times 1 + (1 - \mu) \times 0 = \mu \,, \tag{C.1}$$

where

$$1 \geq \mu \geq 0 \,. \tag{C.2}$$

The present notation for μ is the same as originally used in (A.2).

For the exact distribution, a convenient formal representation for the nth normalized moment, $\Delta_{\rm n}^{\rm exact}(\mu)$, is the following:

$$\Delta_{\rm n}^{\rm exact}(\mu) = \left\langle \left[\sum_{\rm i=1}^{N} (\sigma_{\rm i} - \mu) \right]^{n} \right\rangle . \tag{C.3}$$

As shown below, the zeroth order normalized moment, $\Delta_0^{\rm exact}(\mu)$, is unity. Similarly, because of the requirement that the thermodynamic average of $\sigma_{\rm i}$—to be denoted as $\langle\sigma_{\rm i}\rangle$—be equal to the density μ of the occupied sites, the first moment $\Delta_1^{\rm exact}(\mu)$ is equal to zero:

$$\begin{aligned} \Delta_0^{\rm exact}(\mu) &= \left\langle \left[\sum_{\rm i}^{N} (\sigma_{\rm i} - \mu) \right]^{n=0} \right\rangle = \langle 1 \rangle = 1 \,, \\ \Delta_1^{\rm exact}(\mu) &= \left\langle \left[\sum_{\rm i}^{N} (\sigma_{\rm i} - \mu) \right]^{n=1} \right\rangle \\ &= \left\langle \sum_{\rm i}^{N} \sigma_{\rm i} \right\rangle - \langle N\mu \rangle \\ &= \sum_{\rm i}^{N} \langle\sigma_{\rm i}\rangle - N\mu = \sum_{\rm i}^{N} \mu - N\mu = 0 \,. \end{aligned} \tag{C.4}$$

R. Tahir-Kheli, *General and Statistical Thermodynamics*,
https://doi.org/10.1007/978-3-030-20700-7

C.1 Exact Second Normalized Moment

In order to calculate

$$\Delta_2^{\text{exact}}(\mu) = \left\langle \left[\sum_{\mathrm{i}}^{N} (\sigma_{\mathrm{i}} - \mu) \right]^2 \right\rangle , \tag{C.5}$$

expand the square, and note that i is a dummy index. Therefore

$$\begin{aligned} \Delta_2^{\text{exact}}(\mu) &= \left\langle \sum_{\mathrm{i}}^{N} (\sigma_{\mathrm{i}} - \mu) \cdot \sum_{\mathrm{j}}^{N} (\sigma_{\mathrm{j}} - \mu) \right\rangle \\ &= \left\langle \sum_{\mathrm{i}}^{N} \sum_{\mathrm{j}}^{N} (\sigma_{\mathrm{i}} \sigma_{\mathrm{j}}) \right\rangle - N\mu \left[\sum_{\mathrm{i}}^{N} \langle \sigma_{\mathrm{i}} \rangle + \sum_{\mathrm{j}}^{N} \langle \sigma_{\mathrm{j}} \rangle \right] + N^2 \mu^2 . \end{aligned} \tag{C.6}$$

Because

$$\sum_{\mathrm{i}}^{N} \langle \sigma_{\mathrm{i}} \rangle = \sum_{\mathrm{j}}^{N} \langle \sigma_{\mathrm{j}} \rangle = \sum_{\mathrm{i}}^{N} \mu = N\mu , \tag{C.7}$$

the only term that remains to be evaluated is

$$\begin{aligned} \gamma_2 &= \left\langle \sum_{\mathrm{i}}^{N} \sum_{\mathrm{j}}^{N} \sigma_{\mathrm{i}} \sigma_{\mathrm{j}} \right\rangle \\ &= \sum_{\mathrm{i}}^{N} \sum_{\mathrm{j} \neq \mathrm{i}}^{N} \langle \sigma_{\mathrm{i}} \sigma_{\mathrm{j}} \rangle + \left[\sum_{\mathrm{i}}^{N} \langle \sigma_{\mathrm{i}} \sigma_{\mathrm{j}} \rangle (j = i) \right] \\ &= \sum_{\mathrm{i}}^{N} \sum_{\mathrm{j} \neq \mathrm{i}}^{N} \langle \sigma_{\mathrm{i}} \sigma_{\mathrm{j}} \rangle + \left[\sum_{\mathrm{i}}^{N} \langle (\sigma_{\mathrm{i}})^2 \rangle \right] . \end{aligned} \tag{C.8}$$

Because σ_{i} can take on only the two possible values 1 and 0,

$$(\sigma_{\mathrm{i}})^2 = \sigma_{\mathrm{i}} .$$

This property will be invoked and also use will be made of the fact that any two independent calls to the perfect random number are completely uncorrelated. In other words,

$$\langle \sigma_{\mathrm{i}} \sigma_{\mathrm{j}} \rangle = \langle \sigma_{\mathrm{i}} \rangle \langle \sigma_{\mathrm{j}} \rangle = \mu^2 , \quad \text{for } j \neq i .$$

Therefore, (C.8) becomes

$$\begin{aligned}\gamma_2 &= \sum_{\mathrm{i}}^{N}\sum_{\mathrm{j}\neq\mathrm{i}}^{N}\langle\sigma_{\mathrm{i}}\rangle\langle\sigma_{\mathrm{j}}\rangle + \sum_{\mathrm{i}}^{N}\langle\sigma_{\mathrm{i}}\rangle \\ &= \sum_{\mathrm{i}}^{N}\sum_{\mathrm{j}\neq\mathrm{i}}^{N}\mu^2 + \sum_{\mathrm{i}}^{N}\mu \\ &= N(N-1)\mu^2 + N\mu\,. \end{aligned} \tag{C.9}$$

Equations (C.9), (C.8), (C.7), and (C.3) lead to the following exact result for the second order normalized frequency moment:

$$\begin{aligned}\Delta_2^{\mathrm{exact}}(\mu) &= \gamma_2 - N\mu(N\mu + N\mu) + N^2\mu^2 \\ &= N\mu - N\mu^2 = N\mu(1-\mu)\,. \end{aligned} \tag{C.10}$$

Note that the half-concentration $\Delta_2^{\mathrm{exact}}(\mu = \frac{1}{2})$, as confirmed by the right-hand side of (C.10), is equal to $N/4$. This result is identical to that given by the Gaussian approximation, $\Delta_2^{\mathrm{G}}(a_{\frac{1}{2}})$. See (C.10). Clearly, therefore, for general concentration, μ, (C.10) suggests that, in view of (C.11), the dependence of the variable a_μ on μ should be as follows:

$$\frac{1}{2a_\mu} = \Delta_2^{\mathrm{exact}}(\mu) = N\mu(1-\mu)\,. \tag{C.11}$$

Consequently, the Gaussian approximation yields an exact result for the second order normalized moment. This holds true not only for half-concentration, $\mu = \frac{1}{2}$, but also for general (concentration) μ. That is,

$$\Delta_2^{\mathrm{G}}(a_\mu) = \Delta_2^{\mathrm{exact}}(\mu)\,. \tag{C.12}$$

Indeed, as will be shown later, with $a_\mu = \frac{1}{2N\mu(1-\mu)}$, the Gaussian approximation yields results for the normalized moments that are exact to the leading order in $\frac{1}{N}$.

C.2 Exact Third–Sixth Normalized Moments

For brevity, only the essential steps will be given below.

C.2.1 The Third Moment

The third-order normalized moment is

$$\Delta_3^{\mathrm{exact}}(\mu) = \left\langle\left[\sum_{\mathrm{i}}^{N}(\sigma_{\mathrm{i}} - \mu)\right]^3\right\rangle = \gamma_3 - 3\mu N\gamma_2 + 2(\mu N)^3\,, \tag{C.13}$$

where γ_2 is as given in (C.8) and

$$\gamma_3 = \sum_{\mathrm{i}}^{N} \sum_{\mathrm{j}}^{N} \sum_{\mathrm{k}}^{N} \langle \sigma_{\mathrm{i}} \sigma_{\mathrm{j}} \sigma_{\mathrm{k}} \rangle$$
$$= N(N-1)(N-2)\mu^3 + 3N(N-1)\mu^2 + N\mu \,. \tag{C.14}$$

Thus the exact result for the third-order normalized moment for the general concentration is the following:

$$\Delta_3^{\mathrm{exact}}(\mu) = N\mu\big(1 - 3\mu + 2\mu^2\big) \,. \tag{C.15}$$

Note that, when the concentration μ of occupied sites is $\frac{1}{2}$—i.e., the half-concentration case—the third moment, like all the odd-order moments, is equal to zero.

Exercise C.1 Using the procedure described for calculating γ_2, derive (C.14). (See (C.8) and (C.9)) Indeed, a sufficiently motivated student may even want to derive (C.16), (C.18), and (C.19) that are given below.

C.2.2 The Fourth Moment

In addition to γ_2 and γ_3, the evaluation of the fourth moment also requires the knowledge of γ_4, which is

$$\gamma_4 = \sum_{\mathrm{i}}^{N} \sum_{\mathrm{j}}^{N} \sum_{\mathrm{k}}^{N} \sum_{\mathrm{l}}^{N} \langle \sigma_{\mathrm{i}} \sigma_{\mathrm{j}} \sigma_{\mathrm{k}} \sigma_{\mathrm{l}} \rangle$$
$$= N(N-1)(N-2)(N-3)\mu^4 + 6N(N-1)(N-2)\mu^3$$
$$+ 7N(N-1)\mu^2 + N\mu \,. \tag{C.16}$$

Combination of this with the results of the averages of the two- and three-site sums, γ_2 and γ_3, leads to the following exact result for the fourth moment:

$$\Delta_4^{\mathrm{exact}}(\mu) = \gamma_4 - 4\mu N\gamma_3 + 6\mu^2 N^2 \gamma_2 - 3\mu^4 N^4$$
$$= 3N^2\mu^2(1-\mu)^2 + N\mu\big(1 - 7\mu + 12\mu^2 - 6\mu^3\big) \,. \tag{C.17}$$

While the procedure for calculating the higher order moments is similar, the effort involved rapidly increases with the rise of the order.

C.2.3 Calculation of the Fifth and Sixth Normalized Moments

With somewhat more than the usual amount of effort, one finds:

$$\begin{aligned}\gamma_5 &= \sum_{\mathrm{i}}^{N}\sum_{\mathrm{j}}^{N}\sum_{\mathrm{k}}^{N}\sum_{\mathrm{l}}^{N}\sum_{\mathrm{m}}^{N}\langle\sigma_{\mathrm{i}}\sigma_{\mathrm{j}}\sigma_{\mathrm{k}}\sigma_{\mathrm{l}}\sigma_{\mathrm{m}}\rangle \\ &= N(N-1)(N-2)(N-3)(N-4)\mu^5 \\ &\quad + 10N(N-1)(N-2)(N-3)\mu^4 \\ &\quad + 25N(N-1)(N-2)\mu^3 + 15N(N-1)\mu^2 + N\mu \end{aligned} \tag{C.18}$$

and

$$\begin{aligned}\gamma_6 &= \sum_{\mathrm{i}}^{N}\sum_{\mathrm{j}}^{N}\sum_{\mathrm{k}}^{N}\sum_{\mathrm{l}}^{N}\sum_{\mathrm{m}}^{N}\sum_{\mathrm{n}}^{N}\langle\sigma_{\mathrm{i}}\sigma_{\mathrm{j}}\sigma_{\mathrm{k}}\sigma_{\mathrm{l}}\sigma_{\mathrm{m}}\sigma_{\mathrm{n}}\rangle \\ &= N(N-1)(N-2)(N-3)(N-4)(N-5)\mu^6 \\ &\quad + 15N(N-1)(N-2)(N-3)(N-4)\mu^5 \\ &\quad + 65N(N-1)(N-2)(N-3)\mu^4 + 90N(N-1)(N-2)\mu^3 \\ &\quad + 31N(N-1)\mu^2 + N\mu\,. \end{aligned} \tag{C.19}$$

The rest of the task is easy. We have

$$\begin{aligned}\Delta_5^{\mathrm{exact}}(\mu) &= \left(\gamma_5 - 5N\mu\gamma_4 + 10N^2\mu^2\gamma_3\right) - \left(10N^3\mu^3\gamma_2 + 4N^5\mu^5\right) \\ &= 10N^2\mu^2\left(1 - 4\mu + 5\mu^2 - 2\mu^3\right) \\ &\quad + N\mu\left(1 - 15\mu + 50\mu^2 - 60\mu^3 + 24\mu^4\right) \end{aligned} \tag{C.20}$$

and

$$\begin{aligned}\Delta_6^{\mathrm{exact}}(\mu) &= \left(\gamma_6 - 6N\mu\gamma_5 + 15N^2\mu^2\gamma_4\right) - \left(20N^3\mu^3\gamma_3 + 15N^4\mu^4\gamma_2 - 5N^6\mu^6\right) \\ &= 15\mu^3N^3\left(1 - 3\mu + 3\mu^2 - \mu^3\right) \\ &\quad + 5N^2\mu^2\left(5 - 36\mu + 83\mu^2 - 78\mu^3 + 26\mu^4\right) \\ &\quad + N\mu\left(1 - 31\mu + 180\mu^2 - 390\mu^3 + 360\mu^4 - 120\mu^5\right). \end{aligned} \tag{C.21}$$

For the normalized moments, it is instructive to compare the exact results with those given by the corresponding Gaussian approximation. Remember that the appropriate choice for the parameter a_μ—that is to be used instead of a in the Gaussian expansion valid for general concentration of occupied sites—is as recorded in (C.11).

Consequently, one has

$$
\begin{aligned}
\frac{\Delta_2^{\mathrm{G}}(a_\mu)}{N^{2/2}} &= \frac{\Delta_2^{\mathrm{exact}}(\mu)}{N^{2/2}} = \frac{\left(\frac{1}{2a_\mu}\right)}{N^{2/2}} = \mu(1-\mu)\,, \\
\frac{\Delta_4^{\mathrm{G}}(a_\mu)}{N^{4/2}} &= \frac{\left(\frac{3}{4a_\mu^2}\right)}{N^{4/2}} = 3\mu^2(1-\mu)^2\,, \\
\frac{\Delta_4^{\mathrm{exact}}(\mu)}{N^{4/2}} &= \frac{\Delta_4^{\mathrm{G}}(a_\mu)}{N^{4/2}} + \frac{\mu(1-7\mu+12\mu^2-6\mu^3)}{N}\,, \\
\frac{\Delta_6^{\mathrm{G}}(a_\mu)}{N^{6/2}} &= \frac{\left(\frac{15}{8a_\mu^3}\right)}{N^{6/2}} = 15\mu^3(1-\mu)^3\,, \\
\frac{\Delta_6^{\mathrm{exact}}(\mu)}{N^{6/2}} &= \frac{\Delta_6^{\mathrm{G}}(a_\mu)}{N^{6/2}} + \frac{5\mu^2(5-36\mu+83\mu^2-78\mu^3+26\mu^4)}{N} \\
&\quad + \frac{\mu}{N^2}\left[1-31\mu+180\mu^2-390\mu^3+360\mu^4-120\mu^5\right].
\end{aligned}
\tag{C.22}
$$

$$
\begin{aligned}
\frac{\Delta_3^{\mathrm{G}}(a_\mu)}{N^{3/2}} &= 0\,, \\
\frac{\Delta_3^{\mathrm{exact}}(\mu)}{N^{3/2}} &= \frac{\Delta_3^{\mathrm{G}}(a_\mu)}{N^{3/2}} + \mu\left[\frac{(1-3\mu+2\mu^2)}{N^{1/2}}\right], \\
\frac{\Delta_5^{\mathrm{G}}(a_\mu)}{N^{3/2}} &= 0\,, \\
\frac{\Delta_5^{\mathrm{exact}}(\mu)}{N^{5/2}} &= \frac{\Delta_5^{\mathrm{G}}(a_\mu)}{N^{3/2}} + 10\mu^2\left[\frac{(1-4\mu+5\mu^2-2\mu^3)}{N^{1/2}}\right] \\
&\quad + \mu\left[\frac{(1-15\mu+50\mu^2-60\mu^3+24\mu^4)}{N^{3/2}}\right].
\end{aligned}
\tag{C.23}
$$

Note that, for $\mu = \frac{1}{2}$, the odd-order moments are vanishing.

As mentioned before, while the second moment is exactly given by the Gaussian approximation, the third, fourth, fifth, and the sixth moments are "exact" only to the leading order in the large N limit. Because in thermodynamic systems N is extraordinarily large, the Gaussian approximation is impressively accurate.

C.3 Concluding Remark

For $\mu = 1$ or 0, all exact, normalized, moments are vanishing. Also, despite the fact that the Gaussian approximation—which is symmetric—cannot be expected to be accurate over the whole concentration range, in the immediate vicinity of the most probable state the exact distribution function appears to be very nearly symmetric and is close to being a Gaussian. This is testified to by the following facts:

(1) The Gaussian approximation exactly reproduces the zeroth and second moment of the distribution.

(2) The approximation predicts that the ratio of the fourth normalized moment to the square of the second normalized moment is equal to 3 for general values of μ. That is,

$$\frac{\Delta_4^{\mathrm{G}}(a_\mu)}{[\Delta_2^{\mathrm{G}}(a_\mu)]^2} = 3\,. \tag{C.24}$$

The same is also true, to the leading order in N, for the exact moments: that is,

$$\frac{\Delta_4^{\mathrm{exact}}(\mu)}{[\Delta_2^{\mathrm{exact}}(\mu)]^2} = 3 + \left(N^{-1}\right)\left[\frac{1}{\mu(1-\mu)}\right]^2 \left[1 - 7\mu + 12\mu^2 - 6\mu^3\right]. \tag{C.25}$$

(3) The Gaussian approximation predicts that the ratio of the sixth normalized moment to the cube of the second normalized moment is equal to 15 for general values of μ. That is,

$$\frac{\Delta_6^{\mathrm{G}}(a_\mu)}{[\Delta_2^{\mathrm{G}}(a_\mu)]^3} = 15\,. \tag{C.26}$$

The same is also true, to the leading order in N, for the exact moments, i.e.,

$$\frac{\Delta_6^{\mathrm{exact}}(\mu)}{[\Delta_2^{\mathrm{exact}}(\mu)]^3} = 15 + \left(N^{-1}\right)\left[\frac{5\mu^2(5 - 36\mu + 83\mu^2 - 78\mu^3 + 26\mu^4)}{\mu^3(1-\mu)^3}\right] + O\left(N^{-2}\right). \tag{C.27}$$

(4) While the Gaussian distribution is symmetric and thus leads to vanishing odd-order normalized moments, the exact distribution is asymmetric and its odd-order normalized moments are nonvanishing. Such asymmetry is very small in the neighborhood of the most probable state. This is demonstrated by the size of the normalized third- and fifth-order moments for very large N. Here, instead of $\Delta_3^{\mathrm{exact}}(\mu)$ and $\Delta_5^{\mathrm{exact}}(\mu)$ having the usual, canonical size, namely

$$\left[\Delta_{\mathrm{n}}^{\mathrm{exact}}(\mu)\right]^{1/n} \to O\left(N^{1/2}\right), \quad n = 2, 3, \ldots, \tag{C.28}$$

their size is actually much smaller, i.e.,

$$\left[\Delta_3^{\mathrm{exact}}(\mu)\right]^{1/3} \to O\left(N^{1/3}\right) \tag{C.29}$$

and

$$\left[\Delta_5^{\mathrm{exact}}(\mu)\right]^{1/5} \to O\left(N^{2/5}\right). \tag{C.30}$$

Thus, as long as a is chosen to be equal to $a_\mu = \frac{1}{2N\mu(1-\mu)}$, the line-shape, $\exp\left(-a_\mu z^2\right)$, retains its physical validity for large N. Another feature to note is the size of the relative fluctuation[1] as a function of the concentration μ:

$$\varpi_{\mathrm{rel}} = \sqrt{2\Delta_2^{\mathrm{exact}}(\mu)/(N\mu)} = \sqrt{2(1-\mu)/(N\mu)}\,. \tag{C.31}$$

[1] Compare with (A.13).

Not unexpectedly when μ approaches unity, almost all the calls to the random number generator return the value 1. And, additionally the total number of *occupied* sites becomes large, approaching N. Thus the relative width of the distribution narrows still further. Opposite is the case when μ approaches zero because now there may be some calls to the generator that actually return the value unity while the average of the total number of occupied sites, $N\mu$, has become very small. This fact merely restates the obvious: the relative fluctuation in a small sample is large.

C.4 Summary

In a macroscopically large system, the most probable configuration is overwhelmingly so. Therefore, the result of any "macroscopic" measurement is well described by an accurate calculation of the most probable state.

Perfect Gas Revisited

D

As mentioned before, a perfect gas consists of N identical molecules, each of the same mass. The number of molecules is very large, that is, $N \gg 1$. The gas is enclosed in a vessel of arbitrary shape. The volume of the vessel is V. There are no intermolecular interactions, the size of the molecules is vanishingly small, the containing walls of the vessel are smooth and featureless. All collisions between the molecules and the walls are perfectly elastic; effects of gravity are absent; no other external forces are present. Further, the molecules are in a state of random motion.

Here, in this appendix, we first revisit the standard thermodynamics treatment for a qualitative derivation of the equation of state. This time, somewhat greater detail is provided than was done previously. Next we use an elementary statistical mechanical procedure, that employs Boltzmann–Maxwell–Gibbs distribution, to precisely and quantitatively derive the equation of state.

D.1 Monatomic Perfect Gas

All the molecules are monatomic and each has mass m. The molecules—meaning the atoms—are all of zero-size and in three dimensions each has only three possible degrees of freedom related to its translational motion. The atomic size being zero forbids any meaningful possibility of self-rotation. Furthermore, zero interatomic interaction disallows any interparticle coupling.

D.1.1 Monatomic Perfect Gas: Pressure

Consider a vessel of arbitrary shape. The walls of the vessel are smooth and their shape can be represented in terms of nonsingular equations. Consider a small, but finite volume of gas inside the vessel. For simplicity assume that the small volume is shaped as a parallelepiped whose three, mutually perpendicular, imaginary walls[1] lie

[1]Although we have called the walls imaginary, we do not treat them as representing an open boundary. Indeed, it is not unreasonable to impose boundary conditions on these "imaginary" walls which are more restrictive than the conditions for an open boundary. For instance, the passage of

R. Tahir-Kheli, *General and Statistical Thermodynamics*,
https://doi.org/10.1007/978-3-030-20700-7

along the x-, y-, and z-axes of a Cartesian coordinate system and are of length Δx, Δy, and Δz. Assume the number of molecules, N_{pp}, within such a parallelepiped is large compared with unity, i.e., $N \gg N_{\text{pp}} \gg 1$.

Set the origin of the Cartesian coordinates at the bottom left-hand corner of the parallelepiped and the positive direction of the axes along the three edges. As such, the top corner diagonally opposite to the origin is at the point $(\Delta x, \Delta y, \Delta z)$.

Examine the course of events involved in molecular collisions against the two walls of the parallelepiped that are perpendicular to the x-axis. Denote the x-component of the velocity of the ith molecule—$i = 1, 2, \ldots, N_{\text{pp}}$—as $v_{\text{i,x}}$.

Perfect elasticity of collisions requires that upon striking the right-hand side wall—at $x = +\Delta x$—with x component of momentum $m \cdot v_{\text{i,x}}$ the molecule gets reflected and the x component of its momentum becomes $-m \cdot v_{\text{i,x}}$. Accordingly, the change in x component of momentum of the molecule after one collision is,

$$\begin{aligned} &\text{final momentum of colliding molecule} - \text{its initial momentum} \\ &= [-m \cdot v_{\text{i,x}}] - [m \cdot v_{\text{i,x}}] = -2m \cdot v_{\text{i,x}} \,. \end{aligned} \tag{D.1}$$

Because there are no external forces, the total momentum in any direction is conserved. Invoking this fact for the x-direction leads to the requirement:

$$\begin{aligned} &\text{change in total momentum} \\ &= \text{change in particle momentum} + \text{change in momentum of wall} \\ &= -2m \cdot v_{\text{i,x}} + \delta(\text{mom wall})_{\text{i,x}} = 0 \,. \end{aligned} \tag{D.2}$$

That is, a single collision of the wall perpendicular to the x-axis causes an increase in the x-component of the momentum of the wall equal to

$$\Delta(\text{mom wall})_{\text{i,x}} = 2m \cdot v_{\text{i,x}} \,. \tag{D.3}$$

The absence of slowing-down mechanisms insures that after traversing across the parallelepiped to the left-hand side wall placed at $x = 0$ this molecule returns for another collision against the original wall at $x = +\Delta x$. Such a round-trip—from the right-hand side wall to the wall on the left and then back to the wall on the right—is of length $2\Delta x$. Further, it is traversed at constant speed $|v_{\text{i,x}}|$. Therefore, the time, Δt, taken by the molecule for the round trip travel is

$$\Delta t = \frac{\text{distance traveled}}{\text{speed of travel}} = \frac{2\Delta x}{|v_{\text{i,x}}|} . \tag{D.4}$$

As a result, the rate of transfer of momentum by one molecule to this wall,

$$[\text{momentum transferred in one collision/time taken between the collisions}] \,,$$

a sufficiently long interval of time ensures that equal number of particles with roughly the same energy have been incident from opposite directions on a given "imaginary wall." This behavior is not unlike that which results from specular boundary conditions and results in collisions that can be treated as being perfectly elastic.

can be written as

$$\frac{\Delta(\text{mom wall})_{\text{i,x}}}{\Delta t} = \frac{2m \cdot |v_{\text{i,x}}|}{\{2\Delta x/|v_{\text{i,x}}|\}} = \left(\frac{m}{\Delta x}\right) v_{\text{i,x}}^2 . \tag{D.5}$$

Summing this over all the molecules—that is, from $i = 1$ to $i = N_{\text{pp}}$—within the parallelepiped gives us the total transfer rate of momentum to the right-hand side wall. According to Newton's second law of motion, this is equal to the force, F_{pp}, exerted by the gas on the relevant wall of the parallelepiped,

$$F_{\text{pp}} = \sum_{\text{i=1}}^{N_{\text{pp}}} \frac{\Delta(\text{mom wall})_{\text{i,x}}}{\Delta t} = \left(\frac{m}{\Delta x}\right) \sum_{\text{i=1}}^{N_{\text{pp}}} v_{\text{i,x}}^2 . \tag{D.6}$$

The force F_{pp} acts normal to the wall under examination. Accordingly, it exerts pressure P_{pp}, which is defined as the perpendicular force on the wall per unit area,

$$P_{\text{pp}} = F_{\text{pp}}/(\text{area of the wall}) = \frac{F_{\text{pp}}}{\Delta y \Delta z} . \tag{D.7}$$

Combining this with (D.6) yields

$$P_{\text{pp}} = \left(\frac{m}{\Delta x \Delta y \Delta z}\right) \sum_{\text{i}}^{N_{\text{pp}}} v_{\text{i,x}}^2 = \left(\frac{m}{V_{\text{pp}}}\right) \sum_{\text{i}}^{N_{\text{pp}}} v_{\text{i,x}}^2 = \left(\frac{m}{V_{\text{pp}}}\right) N_{\text{pp}} \langle v_{\text{x}}^2 \rangle_{\text{pp}} . \tag{D.8}$$

Here $V_{\text{pp}} = (\Delta x \Delta y \Delta z)$ is the volume of the elementary[2] parallelepiped under consideration and the pointed brackets with suffix pp, i.e., $\langle v_{\text{x}}^2 \rangle_{\text{pp}}$ signifies the average of $v_{\text{i,x}}^2$ over all—i.e., for $i = 1, \ldots, N_{\text{pp}}$—molecules in the parallelepiped,

$$\langle v_{\text{x}}^2 \rangle_{\text{pp}} = \frac{\sum_{\text{i}}^{N_{\text{pp}}} v_{\text{i,x}}^2}{N_{\text{pp}}} . \tag{D.9}$$

The gas is isotropic. Therefore

$$\langle v_{\text{x}}^2 \rangle_{\text{pp}} = \langle v_{\text{y}}^2 \rangle_{\text{pp}} = \langle v_{\text{z}}^2 \rangle_{\text{pp}} = \frac{1}{3}\left[\langle v_{\text{x}}^2 \rangle_{\text{pp}} + \langle v_{\text{y}}^2 \rangle_{\text{pp}} + \langle v_{\text{z}}^2 \rangle_{\text{pp}}\right] = \frac{1}{3}\langle \boldsymbol{v}^2 \rangle_{\text{pp}} . \tag{D.10}$$

Thus, (D.8) can be recast as

$$P_{\text{pp}} V_{\text{pp}} = m N_{\text{pp}} \langle v_{\text{x}}^2 \rangle_{\text{pp}} = \frac{m}{3} N_{\text{pp}} \langle \boldsymbol{v}^2 \rangle_{\text{pp}} . \tag{D.11}$$

In the above equation all the three quantities V_{pp}, N_{pp}, and $\langle \boldsymbol{v}^2 \rangle_{\text{pp}}$ are independent of the direction x, y, or z. Therefore, the pressure P_{pp} is also independent of the direction.

[2] Yet macroscopic, because $N_{\text{pp}} \gg 1$.

Now sum the above equation over all the parallelepipeds designated by the index pp—or equivalently, over all the molecules—in the vessel,

$$\sum_{\text{all pp}} P_{\text{pp}} V_{\text{pp}} = \left(\frac{m}{3}\right) \sum_{\text{all pp}} N_{\text{pp}} \langle \boldsymbol{v}^2 \rangle_{\text{pp}} = \left(\frac{m}{3}\right) N \langle \boldsymbol{v}^2 \rangle . \tag{D.12}$$

In the above, the pointed brackets without any suffix, i.e., $\langle \boldsymbol{v}^2 \rangle$, represent the average over all the N molecules. That is,

$$\langle \boldsymbol{v}^2 \rangle = \sum_{\text{all pp}} \frac{N_{\text{pp}} \langle \boldsymbol{v}^2 \rangle_{\text{pp}}}{N} . \tag{D.13}$$

Define the pressure P inside the vessel from the relationship

$$\sum_{\text{all pp}} P_{\text{pp}} V_{\text{pp}} = P V . \tag{D.14}$$

Given that the surface enclosing the container does not possess pathological singularities, the sum $\sum_{\text{all pp}}$ over a very large number of appropriately small parallelepipeds of volumes V_{pp} reproduces the actual volume V that encloses the container. That is,

$$\sum_{\text{all pp}} V_{\text{pp}} = V .$$

Multiplying both sides by the constant P gives

$$P \sum_{\text{all pp}} V_{\text{pp}} = \sum_{\text{all pp}} P V_{\text{pp}} = P V . \tag{D.15}$$

Subtraction of (D.14) from (D.15) leads to the equality

$$\sum_{\text{all pp}} (P - P_{\text{pp}}) V_{\text{pp}} = 0 . \tag{D.16}$$

The above sum is over an arbitrary number of different parallelepipeds whose volumes sum to the total volume V. Because these volumes, V_{pp}, are all arbitrary—other than each being very small but still containing a moderate number of molecules—therefore, the above equation can be satisfied only if

$$P_{\text{pp}} = P \tag{D.17}$$

for all parallelepipeds. This result is in agreement with that implicit in Pascal's law, which asserts that the pressure is constant throughout the vessel. Note also that (D.12) and (D.14) lead to an important relationship

$$P V = \left(\frac{m}{3}\right) N \langle \boldsymbol{v}^2 \rangle . \tag{D.18}$$

Consider a monatomic perfect gas in three-dimensions that obeys classical Boltzmann–Maxwell–Gibbs statistics.

D.2 Classical Statistics: Boltzmann–Maxwell–Gibbs Distribution

The gas has N atoms whose total volume is V. The location and momentum of an infinitesimal sized atom i is specified by three position coordinates, i.e., $q_{\mathrm{i,x}}, q_{\mathrm{i,y}}, q_{\mathrm{i,z}}$, and three vector-components, i.e., $p_{\mathrm{i,x}}, p_{\mathrm{i,y}}, p_{\mathrm{i,z}}$, of the momentum $\boldsymbol{p}_{\mathbf{i}}$. Denote the 3-$N$ position coordinates of the N atoms as Q, the 3-N components of the momentum vectors divided by h^{3N} as P, and use the notation

$$\begin{aligned} \mathrm{d}Q &= \mathrm{d}q_{1,\mathrm{x}}\mathrm{d}q_{1,\mathrm{y}}\mathrm{d}q_{1,\mathrm{z}}\cdots \mathrm{d}q_{\mathrm{N,x}}\mathrm{d}q_{\mathrm{N,y}}\mathrm{d}q_{\mathrm{N,z}}\,, \\ \mathrm{d}P &= \mathrm{d}p_{1,\mathrm{x}}\mathrm{d}p_{1,\mathrm{y}}\mathrm{d}p_{1,\mathrm{z}}\cdots \mathrm{d}p_{\mathrm{N,x}}\mathrm{d}p_{\mathrm{N,y}}\mathrm{d}p_{\mathrm{N,z}}/\left(h^{3N}\right). \end{aligned} \tag{D.19}$$

Next define the BMG distribution factor $f(Q, P)$,

$$f(Q,P) = \frac{\exp\left(-\beta\mathcal{H}\right)}{\int_{\mathrm{Q}}\cdots\int_{\mathrm{P}}\exp\left(-\beta\mathcal{H}\right)\cdot\mathrm{d}Q\cdot\mathrm{d}P}\,. \tag{D.20}$$

Here

$$\beta = \frac{1}{k_{\mathrm{B}}T} = \frac{N_{\mathrm{A}}}{RT} = \frac{N}{nRT}\,, \tag{D.21}$$

$\mathcal{H}$ is the Hamiltonian—i.e., the functional form of the system energy in terms of the $6N$ variables Q and P—and T represents the statistical-mechanical temperature—usually called the Kelvin temperature and labeled as K. Constants n and N_{A} have already been defined in (2.12). Additionally, k_{B}, and therefore R, are also constants. That is,

$$\begin{aligned} R &= 8.3144\,72(15)\ \mathrm{J\,mol^{-1}\,K^{-1}}\,, \\ k_{\mathrm{B}} &= 1.38065\,04(24)\times10^{-23}\ \mathrm{J\,K^{-1}}\,. \end{aligned} \tag{D.22}$$

The above R is called the "molar gas constant" and k_{B} is known as the Boltzmann constant.

In accordance with the Boltzmann–Maxwell–Gibbs (BMG) postulates in thermodynamic equilibrium, the normalized average (i.e., the observed value $\langle\Omega\rangle$) of any thermodynamic function, $\Omega(Q, P)$, is given by the following integral:[3]

$$\langle\Omega\rangle = \int_{\mathrm{Q}}\cdots\int_{\mathrm{P}}\left[\Omega(Q,P)\right]\cdot f(Q,P)\cdot\mathrm{d}Q\cdot\mathrm{d}P\,. \tag{D.23}$$

Note that $f(Q, P)$ here is the same as defined in (D.20). Further, that the integrations over the $3N$ position variables, Q, occur over the maximum (three-dimensional) volume V available to each and all of the N atoms. The integration over the 3-N momentum variables is over the infinite range from $-\infty$ to $+\infty$.

[3]Note that the normalized average of any constant, say α, is equal to itself, that is, $\langle\alpha\rangle = \alpha$.

The denominator that appears in (D.20), i.e.,

$$\int_{\mathrm{Q}} \cdots \int_{\mathrm{P}} \exp\left(-\beta \mathcal{H}\right) \cdot \mathrm{d}Q \cdot \mathrm{d}P\,, \tag{D.24}$$

is of great importance. (Remember $\beta = \frac{1}{k_{\mathrm{B}}t}$.) Except for a multiplying constant, this denominator is proportional to the so-called "partition function," $\Xi(N, V, T)$, which will be described in detail later. The partition function is fundamental to the use of statistical mechanics. We shall have occasion to expand on this statement in Chap. 11.

Because for a perfect gas interatomic interaction is assumed to be completely absent, the Hamiltonian $\mathcal{H}$ in (D.20), (D.23), and (D.24) contains only the kinetic energy and depends on just the momenta of the N monatoms,[4] i.e.,

$$\mathcal{H} = \frac{1}{2m} \sum_{\mathrm{j}=1}^{N} \left(p_{\mathrm{j,x}}^2 + p_{\mathrm{j,x}}^2 + p_{\mathrm{j,x}}^2\right). \tag{D.25}$$

D.2.1 Energy in a Monatomic Perfect Gas

According to (D.25), the average value of the energy—here to be called the internal energy and denoted as U—in a monatomic perfect gas of N, non-interacting infinitesimal sized atoms each of mass m, is given by the relation

$$\langle \mathcal{H} \rangle = \frac{1}{2m} \sum_{\mathrm{j}=1}^{N} \left[\langle (p_{\mathrm{j,x}}^2)\rangle + \langle (p_{\mathrm{j,y}}^2)\rangle + \langle (p_{\mathrm{j,z}}^2)\rangle\right]. \tag{D.26}$$

Because there are no direction dependent forces[5] present, the gas is isotropic. Therefore, the above can be written as

$$\langle \mathcal{H} \rangle = \frac{1}{2m} \sum_{\mathrm{j}=1}^{N} \left[3\langle (p_{\mathrm{j,x}}^2)\rangle\right] = \sum_{\mathrm{j}=1}^{N} \frac{\langle \boldsymbol{p}_{\mathrm{j}}^2 \rangle}{2m}. \tag{D.27}$$

Let us now use the BMG procedure, given in (D.20), (D.23), and (D.24), and calculate the thermodynamic average $\langle p_{\mathrm{i,x}}^2 \rangle$ for any arbitrary atom i,

$$\langle p_{\mathrm{i,x}}^2 \rangle = \frac{\int_{\mathrm{Q}} \cdots \int_{\mathrm{P}} \cdots [p_{\mathrm{i,x}}^2] \cdot \exp\left(-\beta \mathcal{H}\right) \cdot \mathrm{d}Q \cdot \mathrm{d}P}{\int_{\mathrm{Q}} \cdots \int_{\mathrm{P}} \cdots \exp\left(-\beta \mathcal{H}\right) \cdot \mathrm{d}Q \cdot \mathrm{d}P}. \tag{D.28}$$

In the above equation, the integral over Q is trivial because $\mathcal{H}$ and $p_{\mathrm{i,x}}^2$ do not depend on any of the $3N$ position coordinates, $Q = \cdots, q_{\mathrm{i,x}}, q_{\mathrm{i,y}}, q_{\mathrm{i,z}}, \ldots$, etc. Therefore, each of the N atoms simply contributes a factor V equal to the maximum volume

[4] Diatomic perfect gas is treated in Chap. 11.

[5] For example, such as the gravity.

available to it. That is, for any atom j, we have

$$\int_{\mathrm{Q}} \ldots \mathrm{d}Q = \left[\int_{\mathrm{q_{j,x}}} \int_{\mathrm{q_{j,y}}} \int_{\mathrm{q_{j,z}}} \mathrm{d}q_{\mathrm{j,x}} \mathrm{d}q_{\mathrm{j,y}} \mathrm{d}q_{\mathrm{j,z}}\right]^N = V^N ,$$

and as a result we get

$$\langle (p_{\mathrm{i,x}}^2) \rangle = \frac{V^N \int_{\mathrm{P}} \cdots [p_{\mathrm{i,x}}^2] \cdot \exp(-\beta \mathcal{H}) \cdot \mathrm{d}P}{V^N \int_{\mathrm{P}} \cdots \exp(-\beta \mathcal{H}) \cdot \mathrm{d}P} . \tag{D.29}$$

The remaining integrals in (D.29) are of a standard form and are worked out in detail in (B.1)–(B.5), which, in particular, say,

$$\begin{aligned} \int_{-\infty}^{+\infty} \exp(-\alpha p^2) \mathrm{d}p &= \sqrt{\frac{\pi}{\alpha}} , \\ \int_{-\infty}^{+\infty} p^2 \exp(-\alpha p^2) \mathrm{d}p &= \frac{1}{2} \sqrt{\frac{\pi}{\alpha^3}} . \end{aligned} \tag{D.30}$$

Therefore, in the following only a brief description is provided. First, let us look at the denominator in (D.28) and (D.29). Although we need to calculate only a part of this integral for the present purposes—see below—with a view to using it later it is worked out in toto here.

The denominator of the right-hand side of (D.29) is the following:

$$\begin{aligned} &\left(\frac{V}{h^3}\right)^N \int_{-\infty}^{\infty} \cdots \int_{-\infty}^{\infty} \mathrm{d}p_{1,\mathrm{x}} \mathrm{d}p_{1,\mathrm{y}} \mathrm{d}p_{1,\mathrm{z}} \cdots \mathrm{d}p_{\mathrm{N,x}} \mathrm{d}p_{\mathrm{N,y}} \mathrm{d}p_{\mathrm{N,z}} \\ &\times \exp\left(\frac{-\beta\{p_{1,\mathrm{x}}^2 + p_{1,\mathrm{y}}^2 + p_{1,\mathrm{z}}^2 + \cdots + p_{\mathrm{N,x}}^2 + p_{\mathrm{N,y}}^2 + p_{\mathrm{N,z}}^2\}}{2m}\right) . \end{aligned} \tag{D.31}$$

The $3N$ seemingly different integrals in (D.31) that are being multiplied together are all equal. Therefore, their product can be written very simply as follows:

$$\left(\frac{V}{h^3}\right)^N \left[\int_{-\infty}^{\infty} \exp\left(\frac{-\beta p^2}{2m}\right) \mathrm{d}p\right]^{3N} = \left(\frac{V}{h^3}\right)^N \left(\frac{2m\pi}{\beta}\right)^{\frac{3N}{2}} . \tag{D.32}$$

To deal with the numerator of (D.29), let us separate the integral that is taken over the variable $p_{\mathrm{i,x}}$, i.e.,

$$\int_{-\infty}^{\infty} [p_{\mathrm{i,x}}^2] \exp\left(\frac{-\beta\{p_{\mathrm{i,x}}^2\}}{2m}\right) \mathrm{d}p_{\mathrm{i,x}} ,$$

from the rest of the $3N - 1$ integrals. We get

$$\begin{aligned}
&\left(\frac{V}{h^3}\right)^N \int_Q \cdots \int_P [p_{i,x}^2] \cdot \exp(-\beta\mathcal{H}) \cdot dP \\
&= \left(\frac{V}{h^3}\right)^N \left(\int_{-\infty}^{\infty} [p_{i,x}^2] \exp\left(\frac{-\beta\{p_{i,x}^2\}}{2m}\right) dp_{i,x}\right) \cdot \left[\int_{-\infty}^{\infty} \exp\left(\frac{-\beta p^2}{2m}\right) dp\right]^{3N-1} \\
&= \left(\frac{V}{h^3}\right)^N \left(\frac{1}{2}\sqrt{\frac{\pi}{(\frac{\beta}{2m})^3}}\right) \cdot \left[\frac{\pi}{(\frac{\beta}{2m})}\right]^{\frac{3N-1}{2}} \\
&= \left(\frac{V}{h^3}\right)^N \left(\frac{1}{2}\sqrt{\pi}\left(\frac{2m}{\beta}\right)^{\frac{3}{2}}\right) \cdot \left[\frac{2m\pi}{\beta}\right]^{\frac{3N-1}{2}} .
\end{aligned} \tag{D.33}$$

Equation (D.33) gives the numerator of the right-hand side of (D.29). As is clear from (D.29), to determine the thermodynamic average $\langle p_{i,x}^2 \rangle$, we need to divide the result obtained in (D.33) by that found in (D.31) and (D.32). Further, because the system is isotropic, we get

$$\begin{aligned}
\langle p_{i,x}^2 \rangle &= \frac{\langle p_{i,x}^2 \rangle + \langle p_{i,y}^2 \rangle + \langle p_{i,z}^2 \rangle}{3} = \frac{\langle \boldsymbol{p}_i^2 \rangle}{3} \\
&= \frac{(\frac{V}{h^3})^N \cdot \frac{1}{2}\sqrt{\pi}(\frac{2m}{\beta})^{\frac{3}{2}} \cdot (\frac{2m\pi}{\beta})^{\frac{3N-1}{2}}}{(\frac{V}{h^3})^N \cdot (\frac{2m\pi}{\beta})^{\frac{3N}{2}}} = \frac{m}{\beta} = k_B T m .
\end{aligned} \tag{D.34}$$

We notice that $\langle \boldsymbol{p}_i^2 \rangle$ is independent of the position i of the ith particle. Thus we can write

$$3\langle p_{i,x}^2 \rangle = \langle \boldsymbol{p}_i^2 \rangle = \langle \boldsymbol{p}^2 \rangle . \tag{D.35}$$

Therefore, according to (D.27) and (D.34), the internal energy of a perfect gas consisting of N atoms is equal to

$$U = \langle \mathcal{H} \rangle = N\frac{\langle \boldsymbol{p}^2 \rangle}{2m} = \frac{3N}{2} k_B T . \tag{D.36}$$

Second Law: Carnot Version Leads to Clausius Version

E

Carnot's ideas revolutionized the physics of heat engines. The formulation of the necessary ingredients for achieving maximum efficiency, inspired Lord Kelvin to attempt to understand the true meaning of "temperature." Indeed, some would say that "absolute temperature" and the relevant "Kelvin scale" were owed directly to this understanding.

Another milestone in the history of thermodynamics is the second law. It turns out, however, that the Carnot law, the absolute temperature scale, and the second law are all closely related. And the Carnot statement, in fact, leads to the second law.

To analyze these issues, we consider two engines in tandem, one a perfect Carnot engine and the other an engine of ordinary variety. In this fashion we prove that violation of the Carnot version of the second law leads to a physically unacceptable conclusion, namely that without external assistance a positive amount of heat energy can be extracted from a cold dump and all of it transferred to a hot reservoir. Clearly, therefore, a violation of the Carnot version of the second law necessarily results in a violation of the Clausius version of the second law.

E.1 A Carnot and an Ordinary Engine in Tandem

Let T_{H} and T_{C} be the temperatures of the hot reservoir and the cold dump, respectively.

Consider an "ordinary" cyclic engine. As usual, each cycle comprises four legs. But, unlike a perfect Carnot engine, here at least one—but possibly all—of the four legs, are traversed either wholly or partially irreversibly.

Arrange the ordinary engine so that it withdraws heat energy $Q'(T_{\mathrm{H}})$ from the hot reservoir at temperature T_{H}, and $Q'(T_{\mathrm{C}})$ from the cold dump at the lower temperature T_{C}. As usual, set this engine up so that it works in the forward direction, and does positive amount of work, W'. The work done per cycle is equal to the total heat energy input into the working substance during the two isothermal legs. Therefore,

$$W' = Q'(T_{\mathrm{H}}) + Q'(T_{\mathrm{C}})\,. \tag{E.1}$$

R. Tahir-Kheli, *General and Statistical Thermodynamics*,
https://doi.org/10.1007/978-3-030-20700-7

Also get hold of a perfect Carnot engine but arrange it to work "backwards"! Realize that a perfect Carnot engine, much like any other engine, can be run forwards or backwards.

A backward running perfect Carnot engine withdraws negative amount of heat energy equal to $-Q^{rev}(T_H)$, from the hot reservoir during the isothermal leg at temperatures T_H. Similarly, it withdraws a negative amount of heat energy $-Q^{rev}(T_C)$, from the cold dump during the isothermal leg at temperature T_C. Arrange things so that as a result of withdrawing negative amount of heat energy during both the isothermal legs, the work W^{rev} done, per cycle, by the perfect Carnot engine, i.e.,

$$W^{rev} = -Q^{rev}(T_H) - Q^{rev}(T_C) \,,$$

is equal to the negative of the work, W', done per cycle by the ordinary engine described in (E.1) above.[1] That is,

$$W^{rev} = -W' \,. \tag{E.2}$$

Using (E.1), we can represent the efficiency, ϵ', of the ordinary engine as follows:

$$\epsilon' = \frac{W'}{Q'(T_H)} \,. \tag{E.3}$$

Similarly, (E.2) leads to the following expression for the efficiency, ϵ_{carnot}, of the perfect Carnot engine:

$$\epsilon_{carnot} = \left[\frac{W^{rev}}{-Q^{rev}(T_H)}\right] = \frac{-W^{rev}}{Q^{rev}(T_H)} = \frac{W'}{Q^{rev}(T_H)} \,. \tag{E.4}$$

Now, if the following inequality were ever true,

$$\epsilon' > \epsilon_{carnot} \,, \tag{E.5}$$

it would violate the Carnot version of the second law. Using (E.3) and (E.4), the disallowed inequality (E.5) can also be represented as

$$\frac{W'}{Q'(T_H)} > \frac{W'}{Q^{rev}(T_H)} \,. \tag{E.6}$$

Multiplying both sides by $Q^{rev}(T_H) \cdot Q'(T_H)$, the inequality (E.6) becomes

$$Q^{rev}(T_H) - Q'(T_H) > 0 \,. \tag{E.7}$$

(Although different looking, in fact the inequality (E.7) is a restatement of the inequality (E.5).) Thus, if the inequality (E.7) were ever true, the Carnot version of the second law would be violated.

[1] In general, it is always possible to arrange for this to happen. For instance, according to (4.7), (4.8), and (4.14), when the working substance is ideal gas, for prescribed values of T_H and T_C, V_1 and V_2 determine the values of the heat energy exchanges and therefore the amount of work done.

We recall that the two sums,

$$Q'(T_H) + Q'(T_C)$$

and

$$Q^{rev}(T_H) + Q^{rev}(T_C)\,,$$

have been set—see (E.1) and (E.2)—to be equal, i.e.,

$$Q'(T_H) + Q'(T_C) = Q^{rev}(T_H) + Q^{rev}(T_C)\,.$$

Let us represent this fact in the following form:

$$Q'(T_C) - Q^{rev}(T_C) = Q^{rev}(T_H) - Q'(T_H)\,. \tag{E.8}$$

Using (E.8), the inequality given in (E.7) also leads to the following inequality:

$$Q'(T_C) - Q^{rev}(T_C) > 0\,. \tag{E.9}$$

The inequalities (E.7) and (E.9) are of fundamental interest and are needed in the discussion that follow later.

Let us now enclose the two engines inside an isolating, adiabatic chamber and run the two in tandem. Remember, the two engines consist of one ordinary engine—described by (E.1)—that is working in the forward direction, and one perfect Carnot engine working backwards as described by (E.2).

The tandem mode of operation is described by the sum of (E.1) and (E.2). It is important to note that the two engines working in tandem in the manner described, manage to do no work at all! That is,

$$\left[Q'(T_H) + Q'(T_C)\right] - \left[Q^{rev}(T_H) + Q^{rev}(T_C)\right] = W' - W' = 0\,. \tag{E.10}$$

Rearranging (E.10)—or equivalently, (E.8)—gives

$$Q'(T_C) - Q^{rev}(T_C) = -\left[Q'(T_H) - Q^{rev}(T_H)\right]. \tag{E.11}$$

Let us pause to consider the message contained in (E.11) with some care.

The left-hand side represents the heat energy isothermally "added by the dump"—which is at temperature T_C—"into the working substance of the two engines operating in tandem."[2] And, if the inequality (E.9) holds—meaning, if the Carnot statement of the second law is violated—then this heat energy is positive. So where did this positive amount of heat energy go? Because no work has been done, and the fact that the tandem engine is operating within an adiabatic enclosure, all of this positive amount of heat energy, supplied by the low temperature dump, must have been "transferred to the hot reservoir maintained at the higher

[2]Do not forget that the tandem operation consists of a forwards working ordinary engine and a backwards operating Carnot engine.

temperature T_H." And this is exactly what is demonstrated by the right-hand side of (E.11). (To double-check on this last statement, see the inequality (E.7).) The above findings may be summarized as follows:

Violation of the Carnot version of the second law leads to a physically unacceptable conclusion, namely that without external assistance a positive amount of heat energy can be extracted from a cold dump and all of it transferred to a hot reservoir. Thus, a violation of the Carnot version of the second law necessarily results in a violation of the Clausius version of the second law.

Positivity of the Entropy Increase: Equation (4.71)

F

According to (4.71), in order to demonstrate the positivity of the total increase of the entropy for general values of $M_{\rm H}$ and $M_{\rm S}$, one needs to show that the following is true:

$$\begin{aligned}\Delta S_{\rm total} &= \Delta S_{\rm c} + \Delta S_{\rm h} = \ln\left(\frac{T_{\rm f}}{T_{\rm c}}\right)^{M_{\rm c}} + \ln\left(\frac{T_{\rm f}}{T_{\rm h}}\right)^{M_{\rm h}} \\ &= \ln\left[\left(\frac{T_{\rm f}}{T_{\rm c}}\right)^{M_{\rm c}} \times \left(\frac{T_{\rm f}}{T_{\rm h}}\right)^{M_{\rm h}}\right] \geq 0 \,. \end{aligned} \tag{F.1}$$

Equivalently, the validity of the following inequality needs to be proven:

$$\left(\frac{T_{\rm f}}{T_{\rm c}}\right)^{M_{\rm c}} \times \left(\frac{T_{\rm f}}{T_{\rm h}}\right)^{M_{\rm h}} \geq 1 \,. \tag{F.2}$$

In what follows in the present appendix, we prove the validity of the inequality (F.2).

Proof of Validity of Given Inequality

Transferring $T_{\rm c}$ and $T_{\rm h}$ to the right-hand side, we get a convenient form for this inequality:

$$T_{\rm f} \geq (T_{\rm c})^{\frac{M_{\rm c}}{M_{\rm c}+M_{\rm h}}} \times (T_{\rm h})^{\frac{M_{\rm h}}{M_{\rm c}+M_{\rm h}}} \,. \tag{F.3}$$

In order to demonstrate the validity of the above inequality, it is helpful to introduce some notational changes. For convenience we shall use the following notation:

$$\beta = \frac{M_{\rm h}}{M_{\rm c}} > 0 \,, \qquad \alpha = \frac{T_{\rm h}}{T_{\rm c}} \geq 1 \,, \qquad z = \frac{M_{\rm h}}{M_{\rm c} + M_{\rm h}} = \frac{\beta}{1+\beta} \,. \tag{F.4}$$

Note that z given above lies within the range $1 > z > 0$. Also, that because both $M_{\rm h}$ and $M_{\rm c}$ are positive, so is their ratio β.

R. Tahir-Kheli, *General and Statistical Thermodynamics*,
https://doi.org/10.1007/978-3-030-20700-7

Using the notation introduced above in (F.4), (4.68) can be written as

$$T_{\mathrm{f}} = \frac{M_{\mathrm{c}}T_{\mathrm{c}} + M_{\mathrm{h}}T_{\mathrm{h}}}{M_{\mathrm{c}} + M_{\mathrm{h}}} = M_{\mathrm{c}}T_{\mathrm{c}}\left(\frac{1 + \frac{M_{\mathrm{h}}T_{\mathrm{h}}}{M_{\mathrm{c}}T_{\mathrm{c}}}}{M_{\mathrm{c}} + M_{\mathrm{h}}}\right) = T_{\mathrm{c}}\left(\frac{1+\beta\alpha}{1+\beta}\right)$$
$$= T_{\mathrm{c}}\left[\frac{1}{1+\beta} + z\alpha\right] = T_{\mathrm{c}}(1 - z + z\alpha)\,. \tag{F.5}$$

Similarly,

$$(T_{\mathrm{c}})^{\frac{M_{\mathrm{c}}}{M_{\mathrm{c}}+M_{\mathrm{h}}}} \times (T_{\mathrm{h}})^{\frac{M_{\mathrm{h}}}{M_{\mathrm{c}}+M_{\mathrm{h}}}}$$

can be recast as

$$T_{\mathrm{c}}^{\frac{1}{1+\beta}} \times T_{\mathrm{h}}^{\frac{\beta}{1+\beta}} = T_{\mathrm{c}}\left(\frac{T_{\mathrm{h}}}{T_{\mathrm{c}}}\right)^{\frac{\beta}{1+\beta}} = T_{\mathrm{c}}\alpha^{z}\,. \tag{F.6}$$

We can now restate the positivity requirement for $\Delta S_{\mathrm{total}}$, last represented in (F.3), in a very compact form as follows:

$$1 - z + z\alpha \geq \alpha^{z},$$

or, equivalently, as[1]

$$f(\alpha, z) = 1 - z + z\alpha - \alpha^{z} \geq 0\,. \tag{F.7}$$

We notice that when $\alpha = 1$, the equality holds and there is no change in the total entropy. This, of course, is the trivial case where T_{h} is equal to T_{c}. Similarly trivial are the cases for $z = 0$ and $z \to 1$ which arise when either the mass or the specific heat is zero in the expressions M_{h} or M_{c}. Another fact to remember is that when $z = 0.5$ the general inequality reduces to the one already treated above in (4.74).

To begin the demonstration of the validity of the general inequality given in (F.7), let us consider first the case where $\alpha = 1 + \epsilon$ with $\epsilon \ll 1$. Then for all values of z lying within its allowed domain, the inequality must hold because

$$f(1+\epsilon, z) = z(1-z)\frac{\epsilon^2}{2} > 0\,. \tag{F.8}$$

Next we look at the rate of change of $f(\alpha, z)$ with respect to α. That is,

$$\frac{\mathrm{d}f(\alpha, z)}{\mathrm{d}\alpha} = z - \frac{z}{\alpha^{1-z}}\,. \tag{F.9}$$

Now, because $\alpha > 1$ and $1 - z$ is positive

$$\alpha^{1-z} > 1\,, \tag{F.10}$$

[1] To see this, first replace the left-hand side of (F.3) by the right-hand side of (F.5). Next, replace the right-hand side of (F.3) by the right-hand side of (F.6). Finally, cancel the multiplying factor T_c from both sides.

within the specified domains for α and z. This fact ensures the positivity of the slope $\frac{df(\alpha,z)}{d\alpha}$. And beginning with $\alpha = 1 + \epsilon$, where $f(\alpha, z)$ is positive, the positivity of the slope insure the positivity of $f(\alpha, z)$ itself. That is, of course, within the specified domain for α and z, namely, $\alpha \geq 1$ and $1 > z > 0$.

Mixture of Van der Waals Gases

G

Because of the great historical importance of the Van der Waals theory of imperfect gases, and the impetus it provided to the development of thermodynamics, it is in order to ask how would the equation of state change for a mixture of different Van der Waals gases. In particular, would the mixture preserve the Dalton's law of partial pressures?

In this appendix we show that the equation of state for a mixture of Van der Waals gases remains unchanged in form. However, despite its similarity to the equation of state of an unmixed Van der Waals fluid, Dalton's law of partial pressures is not necessarily valid for a mixture of dissimilar gases.

Analysis: Mixture of Van der Waals Gases

If the gases being mixed have no intermolecular interaction, the equation of state for the mixture is simply found as

$$PV = (N_1 + N_2)k_B T = (n_1 + n_2)RT\,, \tag{G.1}$$

where N_1, N_2, or n_1, n_2 represent the number of molecules, or the number of moles, in the two gases. As for a single gas, when interactions are taken into account for Van der Waals gas, we expect both the pressure P and the volume V to get modified.

Let us treat first the attractive part of the potential. As for a single type of molecular pair, we assume the range of interaction to be practically infinite for all molecular pairs. Thus the mutual potential energy of any pair of molecules, separated by more than the hard core radius, is independent of their separation. Accordingly, the total mutual potential energy, E_{11}, of the N_1 molecules of the first Van der Waals gas is proportional to the number of distinct pairs of the first type of molecules. That is,

$$E_{11} \propto -\frac{N_1(N_1 - 1)}{2}\,. \tag{G.2}$$

As noted earlier, this negative potential energy results in an attractive force between molecules that leads to a reduction in the pressure which in the limit $N_1 \gg 1$ can be expressed as

$$Z_{11}(V)N_1^2/2\,, \tag{G.3}$$

R. Tahir-Kheli, *General and Statistical Thermodynamics*,
https://doi.org/10.1007/978-3-030-20700-7

where $N_1^2/2$ is the number of distinct molecular pairs of the first type of molecules and $Z_{11}(V)$ is a function of V and positive.

Of course, the mutual potential energy of the second type of molecules would also cause a correspondingly similar reduction in pressure, namely

$$Z_{22}(V)N_2^2/2\,, \tag{G.4}$$

where $Z_{22}(V)$ is a function of V and positive. Also we have made the approximation $N_2 \gg 1$.

Similarly, the attractive interaction between the two different type of molecules would be proportional to the number of distinct pairs that can be formed. Note that this number is $N_1 \times N_2$. The corresponding reduction in pressure would therefore be

$$Z_{12}(V)N_1N_2\,. \tag{G.5}$$

Because P is an intensive state variable, in the limit $(N_1+N_2) \gg 1$ it is independent of N_1+N_2. Similarly, V is an extensive state variable, therefore in this limit it scales linearly with N_1+N_2. Clearly, therefore all three $Z(V)$'s must scale as

$$Z(V) \propto \frac{1}{V^2}\,, \tag{G.6}$$

leading to the following expression for the change in pressure:

$$\begin{aligned}\delta P &= -Z_{11}(V)N_1^2/2 - Z_{22}(V)N_2^2/2 - Z_{12}(V)N_1N_2\\ &= -z_{11}\frac{N_1^2}{V^2} - z_{22}\frac{N_2^2}{V^2} - 2z_{12}\frac{N_1N_2}{V^2}\,,\end{aligned} \tag{G.7}$$

where we introduced the notation

$$\frac{Z_{11}(V)}{2} = \frac{z_{11}}{V^2}\,, \qquad \frac{Z_{22}(V)}{2} = \frac{z_{22}}{V^2}\,, \qquad \frac{Z_{12}(V)}{2} = \frac{z_{12}}{V^2}\,. \tag{G.8}$$

Similar to the case where all molecules were identical, the phenomenological constants z_{11}, z_{22}, z_{12} are all positive.

As before, it is convenient to work in molal units, and use the notation that includes the Avogadro's number N_A:

$$\begin{aligned} N_1 &= n_1N_A\,, & N_2 &= n_2N_A,\\ z_{11}N_A^2 &= a_{11}\,, & z_{22}N_A^2 &= a_{22},\\ z_{12}N_A^2 &= a_{12}\,, & V &= (n_1+n_2)v\,.\end{aligned} \tag{G.9}$$

Therefore, in complete analogy with the Van der Waals gas for only one type of molecules with interaction parameter a, for the mixed gas we can also define an—effective—interaction parameter a',

$$\Delta P = -\frac{a'}{v^2}\,, \tag{G.10}$$

where

$$a' = \frac{a_{11}n_1^2 + a_{22}n_2^2 + 2a_{12}n_1n_2}{(n_1+n_2)^2}. \tag{G.11}$$

The hard-core part of the potential can also be treated in an analogous manner. As before, the two types of molecules are assumed incompressible, spherical hard-balls of radii r_1 and r_2.

Consider a pair of type 1 molecules. The nearest distance that their centers can get to is $2r_1$. Accordingly, for each molecule, the equivalent of half of the spherical volume—that is, $\frac{1}{2} \cdot \frac{4\pi}{3}(2r_1)^3$—is excluded. Therefore, the contribution to the total excluded volume due to short range repulsion *only between type* 1 *molecules* is given by

$$N_1\left[\frac{1}{2} \times \frac{4\pi}{3}(2r_1)^3\right] \times \left(\frac{N_1}{N_1+N_2}\right). \tag{G.12}$$

Note that when there are only a single type of molecules present, i.e., $N_2 = 0$, we retrieve the earlier result. When the second type of molecules are also there, we do need the weighting factor described by the second term, namely $\frac{N_1}{N_1+N_2}$, which determines the probability for a chosen molecule to actually be of type 1.

Following the same argument, the contribution to the total excluded volume due to the short range repulsion only between type 2 molecules is given by

$$N_2\left[\frac{1}{2} \times \frac{4\pi}{3}(2r_2)^3\right] \times \left(\frac{N_2}{N_1+N_2}\right). \tag{G.13}$$

The excluded volume due to the avoidance of hard cores overlap for a type 1–type 2 pair is found as follows. The diameter of the excluded sphere is now $r_1 + r_2$. Also, as mentioned above, rather than being $N_1^2/2$ or $N_2^2/2$, for this case the number of distinct pairs is a function of both N_1 and N_2. Thus we may represent this part of the excluded volume as

$$\alpha(N_1, N_2) \times \left[\frac{1}{2} \times \frac{4\pi}{3}(r_1+r_2)^3\right]. \tag{G.14}$$

The constant $\alpha(N_1, N_2)$ can be determined either by careful argument or a dimensional approach. Below we pursue the latter course because it is both shorter and easier to understand.

Adding the three contributions to the excluded volume, we get

$$\begin{aligned} &N_1\left[\frac{1}{2} \times \frac{4\pi}{3}(2r_1)^3\right] \times \left(\frac{N_1}{N_1+N_2}\right) \\ &+ N_2\left[\frac{1}{2} \times \frac{4\pi}{3}(2r_2)^3\right] \times \left(\frac{N_2}{N_1+N_2}\right) \\ &+ \alpha(N_1, N_2) \times \left[\frac{1}{2} \times \frac{4\pi}{3}(r_1+r_2)^3\right]. \end{aligned} \tag{G.15}$$

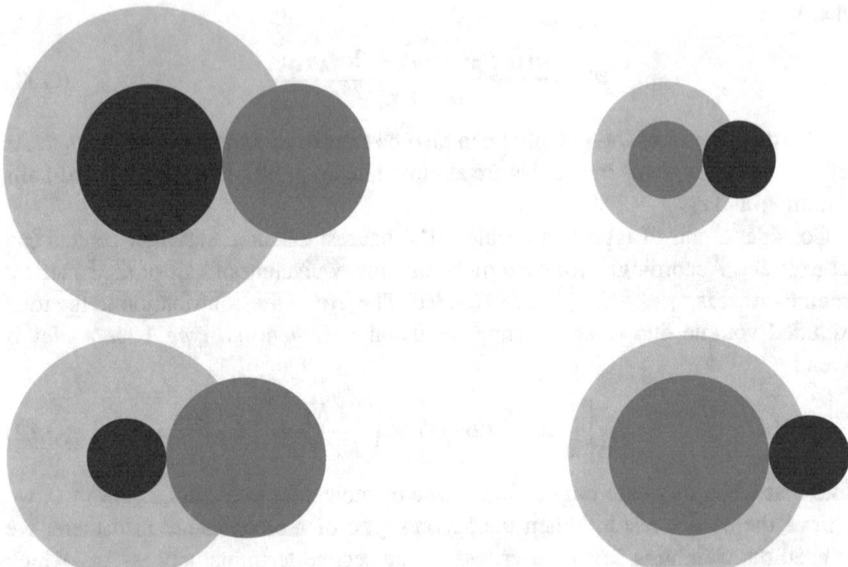

Fig. G.1 **(a)**–**(b)** Hard-core sphere of exclusion for homogeneous pairs. In a mixed gas composed of molecules with radii r_1 and r_2, the size of the sphere of exclusion depends on the nature of the neighboring pairs. For pairs of type 1 molecules, the sphere of exclusion has radius $2r_1$. Similarly, for pairs of type 2 molecules, the radius of the sphere of exclusion is $2r_2$. The shading for the molecules is either dark or grey. The excluded volume is shaded very light-grey. **(c)**–**(d)** Hard-core sphere of exclusion for heterogeneous pairs. In a mixed gas composed of molecules with radii r_1 and r_2, the size of the sphere of exclusion depends on the nature of the neighboring pair. When the pair consists of different types of molecules, the excluded volume is a sphere of radius $r_1 + r_2$. Such a sphere—shaded very light-grey—can equivalently be considered to reside on either of the two molecules

Clearly, when $r_1 = r_2 = r_0$, the distinction between the two types of molecules disappears. Accordingly, this entire expression should reduce to that obtained earlier for molecules of a single type which number $N = N_1 + N_2$. This requirement is satisfied if

$$\frac{N_1^2 + N_2^2}{N_1 + N_2} + \alpha(N_1, N_2) = N_1 + N_2 \,. \tag{G.16}$$

Thus we are immediately led to the result

$$\alpha(N_1, N_2) = \frac{2N_1 N_2}{N_1 + N_2} \,. \tag{G.17}$$

Therefore, to correct for hard core repulsion, the volume V in the ideal gas equation of state has to be changed to $(V - \Delta V)$ where

$$\begin{aligned} \Delta V &= (n_1 + n_2)b' \,, \\ b' &= \frac{b_{11}n_1^2 + b_{22}n_2^2 + 2b_{12}n_1 n_2}{(n_1 + n_2)^2} \,, \end{aligned} \tag{G.18}$$

and

$$b_{11} = N_{\mathrm{A}}\frac{2\pi}{3}(2r_1)^3\,, \qquad b_{22} = N_{\mathrm{A}}\frac{2\pi}{3}(2r_2)^3\,, \qquad b_{12} = N_{\mathrm{A}}\frac{2\pi}{3}(r_1 + r_2)^3\,. \tag{G.19}$$

See Figs. G.1.a–d above.

To sum up: The equation of state for a mixture of Van der Waals fluids remains unchanged in form. That is, it can still be represented as

$$\left(P + \frac{a'}{v^2}\right)(v - b') = RT\,, \tag{G.20}$$

where a' and b' are the effective interaction and hard core exclusion parameters of the mixture. It is easy to check that despite this similarity, unless $a_{11} = a_{12} = a_{22}$ and $b_{11} = b_{12} = b_{22}$, Dalton's law of partial pressures is not necessarily valid for the mixture.

Positive-Definite Homogeneous Quadratic Form

A change of variable can always be devised to reduce an homogeneous, symmetric quadratic form Q_n of the type

$$Q_n = \tilde{\boldsymbol{x}}\hat{\boldsymbol{a}}\boldsymbol{x} = \sum_{i,j}^{n} x_i a_{i,j} x_j \tag{H.1}$$

to a sum of squares.[1] In (H.1), x_i's are assumed to be real and $a_{ij} = a_{ji}$.

Analysis: Positive-Definite Homogeneous Quadratic Form

To see how this may be done, let us introduce a variable ξ_1

$$\xi_1 = x_1 + \frac{a_{12}}{a_{11}} x_2 + \cdots + \frac{a_{1n}}{a_{11}} x_n . \tag{H.2}$$

Now, subtract $a_{11}\xi_1^2$ from Q_n. This eliminates all terms that involve x_1. In a similar fashion, a second variable ξ_2 can then be introduced to eliminate all terms containing x_2. Following this process, eventually one obtains

$$Q_n = a_{11}\xi_1^2 + b_2\xi_2^2 + \cdots + b_{n-1}\xi_{n-1}^2 + b_n x_n^2 , \tag{H.3}$$

where ξ_j contains terms that involve x_i's for $i \geq j$.

[1]Note that, by using a linear transformation, the vector $\tilde{\boldsymbol{x}}$ can be transformed to another set of variables. Thus, the reduction to sum of squares can be done in innumerable ways. This does not affect the substance of the requirement for positive-definiteness as long as the transformation is nonsingular. Indeed, irrespective of the choice of variables, the number of positive, zero, or negative coefficients does not change. Note that the objective of the current exercise is to affect a transformation that reduces the quadratic form to a sum of squares. For positivity of the form, every one of the coefficients multiplying the square terms must therefore be positive. See, for instance, Harold and B.S. Jeffreys, Methods of Mathematical Physics, p. 137, Cambridge University Press (1957) and references cited there.

R. Tahir-Kheli, *General and Statistical Thermodynamics*,
https://doi.org/10.1007/978-3-030-20700-7

H.1 Positive Definiteness of 3 × 3 Quadratic Form

It is helpful to demonstrate in detail the procedure described above. To this end, we append below the case for $n = 3$.

Let[2] us look at Q_3, which is

$$
\begin{aligned}
Q_3 &= \sum_{\mathrm{i,j}}^{3} x_{\mathrm{i}} a_{\mathrm{i,j}} x_{\mathrm{j}} \\
&= \left[a_{11}x_1^2 + 2a_{12}x_1x_2 + 2a_{13}x_1x_3\right] + 2a_{23}x_2x_3 + a_{22}x_2^2 + a_{33}x_3^2 .
\end{aligned}
$$

To eliminate x_1 from the above equation, define

$$
\xi_1 = x_1 + \left(\frac{2a_{12}}{2a_{11}}\right)x_2 + \left(\frac{2a_{13}}{2a_{11}}\right)x_3 . \tag{H.4}
$$

Next calculate $a_{11}\xi_1^2$ and subtract it from Q_3. This leads to the following, where x_1 has been eliminated:

$$
\begin{aligned}
Q_3 - a_{11}\xi_1^2 = &\left[\frac{(a_{11}a_{22} - a_{12}^2)}{a_{11}}x_2^2 + \frac{(a_{11}a_{23} - a_{12}a_{13})}{a_{11}}2x_2x_3\right] \\
&+ \frac{(a_{11}a_{33} - a_{13}^2)}{a_{11}}x_3^2 .
\end{aligned} \tag{H.5}
$$

Next we need to eliminate x_2. To this end, much like what we did earlier, we choose:

$$
\xi_2 = x_2 + \frac{2\left(\frac{a_{11}a_{23} - a_{12}a_{13}}{a_{11}}\right)}{2\left(\frac{a_{11}a_{22} - a_{12}^2}{a_{11}}\right)}x_3 . \tag{H.6}
$$

Then, much as was done before, we calculate $\frac{(a_{11}a_{22} - a_{12}^2)}{a_{11}}\xi_2^2$ and write

$$
\begin{aligned}
Q_3 = {} & a_{11}\xi_1^2 + \frac{(a_{11}a_{22} - a_{12}^2)}{a_{11}}\xi_2^2 \\
&+ \frac{(a_{11}a_{22}a_{33} - a_{12}^2a_{33} - a_{13}^2a_{22} - a_{11}a_{23}^2 + 2a_{12}a_{13}a_{23})}{(a_{11}a_{22} - a_{12}^2)}x_3^2 .
\end{aligned}
$$

Note that, in addition to x_1, we have now also excluded x_2. There is no need to bother about excluding x_3 because it occurs only as a square which is exactly the form we are looking for. Note that the last variable—which is x_3 here—will always occur in the squared format.

---

[2]Note that square brackets will be inserted here to assist the reader in recognizing the pattern used for introducing the variables ξ_{i}.

For Q_3 to be positive definite, the coefficients of the square terms have to be positive definite. That is,

$$a_{11} > 0\,, \tag{H.7}$$

$$\frac{a_{11}a_{22} - a_{12}^2}{a_{11}} > 0\,, \tag{H.8}$$

and

$$\frac{a_{11}a_{22}a_{33} - a_{12}^2 a_{33} - a_{13}^2 a_{22} - a_{11}a_{23}^2 + 2a_{12}a_{13}a_{23}}{a_{11}a_{22} - a_{12}^2} > 0\,. \tag{H.9}$$

Note that the denominators of (H.8) and (H.9) are positive because of the preceding inequalities given in (H.7) and (H.8), respectively.

All the three inequalities (H.7), (H.8), and (H.9) can more conveniently be displayed as determinants:

$$|\boldsymbol{A_1}| = a_{11} > 0\,, \tag{H.10}$$

$$|\boldsymbol{A_2}| = \begin{vmatrix} a_{11} & a_{12} \\ a_{21} & a_{22} \end{vmatrix} > 0\,, \tag{H.11}$$

and

$$|\boldsymbol{A_3}| = \begin{vmatrix} a_{11} & a_{12} & a_{13} \\ a_{21} & a_{22} & a_{23} \\ a_{31} & a_{32} & a_{33} \end{vmatrix} > 0\,. \tag{H.12}$$

Note that $|\boldsymbol{A_1}|$, $|\boldsymbol{A_2}|$, and $|\boldsymbol{A_3}|$ are the principal minors of $\hat{\boldsymbol{a}}$. Indeed, it turns out[3] that positive definiteness of such a quadratic form for general n is assured if all the principal minors of the determinant—of the $n \times n$ matrix of the quadratic form—are positive.

H.2 Helpful Surprise

In the following we study a thermodynamic system that generates a 3×3 homogeneous, symmetric quadratic form. For this system, the substance of the inequalities (H.10) and (H.11)—that involve differentiation with respect to only the two variables x_1 and x_2—is readily unraveled and we are led to two eminently simple physical dictates for thermodynamic stability, namely that the specific heat C_V and the isothermal compressibility χ_T must both be positive. On the other hand, the analysis of the rather fierce looking inequality (H.12)—which involves nine terms—is much more involved. It is, however, noted that:

[3] See footnote 1.

(i) The given quadratic form can just as easily be represented as

$$Q_3 = \{x_3, x_2, x_1\} \cdot \hat{\boldsymbol{b}} \cdot \begin{Bmatrix} x_3 \\ x_2 \\ x_1 \end{Bmatrix} > 0\,, \tag{H.13}$$

where

$$\hat{\boldsymbol{b}} = \begin{pmatrix} a_{33} & a_{32} & a_{31} \\ a_{23} & a_{22} & a_{21} \\ a_{13} & a_{12} & a_{11} \end{pmatrix}. \tag{H.14}$$

According to this representation, positive definiteness of Q_3 is assured if

$$|\boldsymbol{B_1}| = a_{33} > 0\,, \tag{H.15}$$

$$|\boldsymbol{B_2}| = \begin{vmatrix} a_{33} & a_{32} \\ a_{23} & a_{22} \end{vmatrix} > 0\,, \tag{H.16}$$

and

$$|\boldsymbol{B_3}| = \begin{vmatrix} a_{33} & a_{32} & a_{31} \\ a_{23} & a_{22} & a_{21} \\ a_{13} & a_{12} & a_{11} \end{vmatrix} = |\boldsymbol{A_3}| > 0\,. \tag{H.17}$$

(ii) Note that the corresponding terms equivalent to the current inequality (H.15) appear to irreducibly involve only the third linearly independent variable x_3. Thus, a complete description of the positive definiteness of Q_3 is provided by the already disentangled two inequalities (H.10) and (H.11), and the simple third inequality given in (H.15). (See (I.18) for details.)

Thermodynamic Stability: Three Extensive Variables

I

In an earlier analysis—see (9.39)—the requirement that under appropriate conditions the system energy is a minimum was used to predict the stability criteria of a simple system. We found that for intrinsic thermodynamic stability, both the specific heat at constant volume and the isothermal compressibility must be positive. That is, the following must hold:

$$C_{\mathrm{v}} > 0\,, \qquad \chi_{\mathrm{t}} > 0\,. \tag{I.1}$$

The first requirement, namely, $C_{\mathrm{v}} > 0$, has an obvious physical basis: When heat energy is not allowed to be used for expanding the volume, all that any addition of heat energy can do is increase the system temperature.

The positivity of the isothermal compressibility χ_{t} is testified to by observation. When the temperature is held constant, increase in compression shrinks the volume of the object being compressed.

In this appendix we consider an isolated system composed of a single chemical constituent with variable mole number. The variability of the mole number introduces n as an additional extensive variable. Accordingly, as shown below, the conditions for intrinsic stability also include a third requirement, namely

$$\left(\frac{\partial \mu}{\partial n}\right)_{\mathrm{S,V}} > 0\,. \tag{I.2}$$

Much like the first two requirements, the third requirement also has an obvious physical basis. In order to maintain thermodynamic equilibrium, addition of molecules, to an otherwise isolated system in equilibrium, must increase their chemical potential. A hint of this phenomenon was already noted in Chap. 9—see (9.17)—where we observed that the chemical potential is higher in a region of higher particle density.

I.1 Energy Minimum Procedure: Intrinsic Stability

To satisfy the energy minimum extremum, two requirements must be satisfied. The first requirement is the satisfaction

$$(\mathrm{d}U)_{\mathrm{S,V,n}} = 0\,. \tag{I.3}$$

R. Tahir-Kheli, *General and Statistical Thermodynamics*,
https://doi.org/10.1007/978-3-030-20700-7

I.2 The First Requirement for Intrinsic Stability

The first requirement, given in (I.3), has already been fully explored and exploited in Chap. 9 titled "Le Châtelier Principle" and Sects. 9.4 and 9.4.1.

Note that the internal energy U is being considered to be a function of three independent variables S, V, and n. That is,

$$U = U(S, V, n) . \tag{I.4}$$

According to the first–second law,

$$\begin{aligned} dU(S, V, n) &= \left(\frac{\partial U}{\partial S}\right)_{V,n} dS + \left(\frac{\partial U}{\partial V}\right)_{S,n} dV + \left(\frac{\partial U}{\partial n}\right)_{V,S} dn \\ &= T dS - P dV + \mu dn . \end{aligned} \tag{I.5}$$

I.3 The Second Requirement for Intrinsic Stability

The second requirement is the satisfaction of

$$(d^2U)_{S,V,n} > 0 . \tag{I.6}$$

To study the second requirement, proceed as follows. In order to analyze the second requirement for the energy to be a minimum—see (I.6),—it is convenient to utilize a matrix procedure and introduce compact notation. To that end, define the matrix

$$\hat{J} = \begin{pmatrix} U_{SS} & U_{SV} & U_{Sn} \\ U_{VS} & U_{VV} & U_{Vn} \\ U_{nS} & U_{nV} & U_{nn} \end{pmatrix} \tag{I.7}$$

where

$$\begin{aligned} U_{SS} &= dUSV, n , & U_{SV} &= \left(\frac{\partial^2 U}{\partial S \partial V}\right)_n , & U_{Sn} &= \left(\frac{\partial^2 U}{\partial S \partial n}\right)_V , \\ U_{VS} &= \left(\frac{\partial^2 U}{\partial V \partial S}\right)_n , & U_{VV} &= dUVS, n , & U_{V,n} &= \left(\frac{\partial^2 U}{\partial V \partial n}\right)_S , \\ U_{nS} &= \left(\frac{\partial^2 U}{\partial n \partial S}\right)_V , & U_{nV} &= \left(\frac{\partial^2 U}{\partial n \partial V}\right)_S , & U_{nn} &= dUnV, S . \end{aligned} \tag{I.8}$$

Then write

$$d^2U(S, V, n) = \{dS, dV, dn\} \cdot \hat{J} \cdot \begin{Bmatrix} dS \\ dV \\ dn \end{Bmatrix} > 0 . \tag{I.9}$$

For this inequality to hold, the principal minors of the determinant $|\hat{J}|$ have to be positive. Accordingly,

$$|J_1| = U_{SS} > 0\,, \tag{I.10}$$

$$|J_2| = \begin{vmatrix} U_{SS} & U_{SV} \\ U_{VS} & U_{VV} \end{vmatrix} > 0\,, \tag{I.11}$$

and

$$|J_3| = |\hat{J}| = \begin{vmatrix} U_{SS} & U_{SV} & U_{Sn} \\ U_{VS} & U_{VV} & U_{Vn} \\ U_{nS} & U_{nV} & U_{nn} \end{vmatrix} > 0\,. \tag{I.12}$$

Now,

$$\begin{aligned} |J_1| &= \left(\frac{\partial((\frac{\partial U}{\partial S})_{V,n})}{\partial S}\right)_{V,n} = \left(\frac{\partial T}{\partial S}\right)_{V,n} \\ &= \left(\frac{\partial T}{\partial U}\right)_{V,n} \cdot \left(\frac{\partial U}{\partial S}\right)_{V,n} = \left(\frac{1}{C_V}\right) \cdot T > 0\,. \end{aligned} \tag{I.13}$$

Again, much like the entropy maximum principle, the energy minimum principle requires the positivity of the specific heat C_V.

Next, let us for convenience use the Jacobian form for the inequality (I.11),

$$|J_2| = \left[\frac{\partial((\frac{\partial U}{\partial S})_V, (\frac{\partial U}{\partial V})_S)}{\partial(S, V)}\right]_n > 0\,. \tag{I.14}$$

Because

$$\left(\frac{\partial U}{\partial S}\right)_{V,n} = T$$

and

$$\left(\frac{\partial U}{\partial V}\right)_{S,n} = -P\,,$$

we can write

$$\begin{aligned} |J_2| &= \left[\frac{\partial(T, -P)}{\partial(S, V)}\right]_n = \left[\frac{\partial(-P, T)}{\partial(V, S)}\right]_n \\ &= \left[\frac{\partial(-P, T)}{\partial(V, T)} \cdot \frac{\partial(V, T)}{\partial(V, S)}\right]_n = \left[\frac{\partial(-P, T)}{\partial(V, T)} \cdot \frac{\partial(T, V)}{\partial(S, V)}\right]_n \\ &= -\left[\left(\frac{\partial P}{\partial V}\right)_T \cdot \left(\frac{\partial T}{\partial S}\right)_V\right]_n = \left[\left(\frac{1}{V\chi_T}\right) \cdot \left(\frac{T}{C_V}\right)\right]_n > 0\,. \end{aligned} \tag{I.15}$$

Noting the already established requirement that C_V be > 0, this inequality is satisfied only if χ_T is also > 0.

Thus the physical requirements for intrinsic stability in a simple isolated system are the same irrespective of whether we use the principle of entropy maximum or energy minimum.

I.4 The Third Requirement For Intrinsic Stability

The first two requirements were handled with relative ease. Indeed, the foregoing analysis of the positivity of $|\boldsymbol{J}_1|$ and $|\boldsymbol{J}_2|$ has not been hard work. It appears that the same may not be true for the third term. Because analyzing the positivity of the determinant $|\boldsymbol{J}_3|$—which has nine terms, each of which is a multiple of three different double derivatives—is time consuming. Therefore, in order to sail this route, we take a different track.

As noted in Appendix G, a quadratic form of the type found in $\mathrm{d}^2U(S, V, n)$ that is given in (I.9) can just as well be represented as follows:

$$\mathrm{d}^2U(S, V, n) = \{\mathrm{d}n, \mathrm{d}V, \mathrm{d}S\} \cdot \hat{\boldsymbol{K}} \cdot \begin{Bmatrix} \mathrm{d}n \\ \mathrm{d}V \\ \mathrm{d}S \end{Bmatrix} > 0\,, \tag{I.16}$$

where

$$\hat{\boldsymbol{K}} = \begin{pmatrix} U_{\mathrm{nn}} & U_{\mathrm{nV}} & U_{\mathrm{nS}} \\ U_{\mathrm{Vn}} & U_{\mathrm{VV}} & U_{\mathrm{VS}} \\ U_{\mathrm{Sn}} & U_{\mathrm{SV}} & U_{\mathrm{SS}} \end{pmatrix} > 0\,. \tag{I.17}$$

As a result, its positivity requires that the principal minors $|\boldsymbol{K}_1|$, $|\boldsymbol{K}_2|$, and $|\boldsymbol{K}_3|$ be positive. Here

$$|\boldsymbol{K}_1| = U_{\mathrm{nn}} = \mathrm{d}UnS, V = \left(\frac{\partial(\frac{\partial U}{\partial n})_{\mathrm{S,V}}}{\partial n}\right)_{\mathrm{S,V}} = \left(\frac{\partial \mu}{\partial n}\right)_{\mathrm{S,V}} > 0\,. \tag{I.18}$$

Because for given S and V, the above requirement involves only the rate of change of the chemical potential with respect to the occupancy n (of the relevant molecule), and the earlier two requirements explicitly depended only on the rates of change of the other two extensive parameters, S and V, therefore this must be the third linearly independent requirement for intrinsic stability.

A clear display of the important results is given below.

For intrinsic thermodynamic stability, the following three physical requirements must be satisfied:

$$\begin{aligned} C_{\mathrm{V}} &> 0\,, \\ \chi_{\mathrm{T}} &> 0\,, \quad \text{and} \\ \left(\frac{\partial \mu}{\partial n}\right)_{\mathrm{S,V}} &> 0\,. \end{aligned} \tag{I.19}$$

Much like the first two requirements noted above, the third requirement also has an obvious physical basis: to maintain thermodynamic equilibrium, addition of molecules, to an otherwise isolated system in equilibrium, must increase their chemical potential. A hint of this phenomenon has already been noted in Chap. 9—see (9.17)—where we observed the equivalent of the fact that the chemical potential is higher in a region of higher particle density.[1]

I.5 Solved Problems

Problem I.1 Represent a row-vector of intensive variables, $\{dT, -dP, d\mu\}$, in terms of its conjugate column-vector. In other words, find the matrix $\hat{\boldsymbol{M}}$ that is defined by the following relationship:

$$\{dT, -dP, d\mu\} = \hat{\boldsymbol{M}} \cdot \begin{Bmatrix} dS \\ dV \\ dn \end{Bmatrix} . \tag{I.20}$$

Solution I.1 Begin with the first intensive variable on the left-hand side and represent it as a function of all the extensive variables. That is,

$$T = T(S, V, n) . \tag{I.21}$$

Exploit the fact that dT is an exact differential,

$$dT(S, V, n) = \left(\frac{\partial T}{\partial S}\right)_{V,n} dS + \left(\frac{\partial T}{\partial V}\right)_{S,n} dV + \left(\frac{\partial T}{\partial n}\right)_{S,V} dn . \tag{I.22}$$

Now invoke the first–second law,

$$TdS = dU + PdV - \mu dn , \tag{I.23}$$

and note that for constant V and n,

$$T(dS)_{V,n} = (dU)_{V,n} . \tag{I.24}$$

Therefore

$$T = \left(\frac{\partial U}{\partial S}\right)_{V,n} . \tag{I.25}$$

Introduce this expression for T in the three terms on the right-hand side of (I.22) to get

$$\begin{aligned} dT(S, V, n) &= \left(\frac{\partial(\frac{\partial U}{\partial S})_{V,n}}{\partial S}\right)_{V,n} dS + \left(\frac{\partial(\frac{\partial U}{\partial S})_{V,n}}{\partial V}\right)_{S,n} dV + \left(\frac{\partial(\frac{\partial U}{\partial S})_{V,n}}{\partial n}\right)_{S,V} dn \\ &= U_{SS}dS + U_{SV}dV + U_{Sn}dn . \end{aligned} \tag{I.26}$$

[1]For instance, compare with that statement: Simply put, the requirement that in an isolated system the entropy must increase in an isothermal spontaneous process mandates the molecular flow, at constant temperature, to occur away from a region of higher chemical potential towards a region of lower chemical potential.

We follow the same procedure for the remaining two intensive variables—P and μ—on the left-hand side of (I.20). Without comment, salient steps of this exercise are recorded below:

$$\begin{aligned}
P &= P(S, V, n)\,, \\
\mathrm{d}P &= \left(\frac{\partial P}{\partial S}\right)_{\mathrm{V,n}} \mathrm{d}S + \left(\frac{\partial P}{\partial V}\right)_{\mathrm{S,n}} \mathrm{d}V + \left(\frac{\partial P}{\partial n}\right)_{\mathrm{S,V}} \mathrm{d}n\,, \\
P &= -\left(\frac{\partial U}{\partial V}\right)_{\mathrm{S,n}}, \\
-\mathrm{d}P &= U_{\mathrm{S,V}}\mathrm{d}S + U_{\mathrm{VV}}\mathrm{d}V + U_{\mathrm{nV}}\mathrm{d}n\,;
\end{aligned} \tag{I.27}$$

$$\begin{aligned}
\mu &= \mu(S, V, n)\,, \\
\mathrm{d}\mu &= \left(\frac{\partial \mu}{\partial S}\right)_{\mathrm{V,n}} \mathrm{d}S + \left(\frac{\partial \mu}{\partial V}\right)_{\mathrm{S,n}} \mathrm{d}V + \left(\frac{\partial \mu}{\partial n}\right)_{\mathrm{S,V}} \mathrm{d}n\,, \\
\mu &= \left(\frac{\partial U}{\partial n}\right)_{\mathrm{S,V}} \\
\mathrm{d}\mu &= U_{\mathrm{nS}}\mathrm{d}S + U_{\mathrm{nV}}\mathrm{d}V + U_{\mathrm{nn}}\mathrm{d}n\,.
\end{aligned} \tag{I.28}$$

Equations (I.26)–(I.28) are readily seen to be represented by (I.20) where

$$\hat{\boldsymbol{M}} = \begin{pmatrix} U_{\mathrm{SS}} & U_{\mathrm{SV}} & U_{\mathrm{Sn}} \\ U_{\mathrm{SV}} & U_{\mathrm{VV}} & U_{\mathrm{nV}} \\ U_{\mathrm{nS}} & U_{\mathrm{nV}} & U_{\mathrm{nn}} \end{pmatrix}. \tag{I.29}$$

Problem I.2 Find matrix $\hat{\boldsymbol{H}}$ where

$$\{\mathrm{d}T, -\mathrm{d}V, \mathrm{d}\mu\} = \hat{\boldsymbol{H}} \cdot \begin{Bmatrix} \mathrm{d}S \\ -\mathrm{d}P \\ \mathrm{d}n \end{Bmatrix}. \tag{I.30}$$

Solution I.2

$$\begin{aligned}
T &= T(S, P, n)\,, \\
\mathrm{d}T &= \left(\frac{\partial T}{\partial S}\right)_{\mathrm{P,n}} \mathrm{d}S + \left(\frac{\partial T}{\partial P}\right)_{\mathrm{S,n}} \mathrm{d}P + \left(\frac{\partial T}{\partial n}\right)_{\mathrm{S,P}} \mathrm{d}n\,, \\
T\mathrm{d}S &= \mathrm{d}H - V\mathrm{d}P - \mu\mathrm{d}n\,, \\
T &= \left(\frac{\partial H}{\partial S}\right)_{\mathrm{P,n}}, \\
\mathrm{d}T &= H_{\mathrm{SS}}\mathrm{d}S + U_{\mathrm{PS}}\mathrm{d}P + H_{\mathrm{nS}}\mathrm{d}n\,;
\end{aligned} \tag{I.31}$$

$$
\begin{aligned}
V &= V(S, P, n)\,, \\
\mathrm{d}V &= \left(\frac{\partial V}{\partial S}\right)_{\mathrm{P,n}} \mathrm{d}S + \left(\frac{\partial V}{\partial P}\right)_{\mathrm{S,n}} \mathrm{d}P + \left(\frac{\partial V}{\partial n}\right)_{\mathrm{S,P}} \mathrm{d}n\,, \\
V &= \left(\frac{\partial H}{\partial P}\right)_{\mathrm{S,n}}\,, \\
\mathrm{d}V &= H_{\mathrm{SP}}\mathrm{d}S + H_{\mathrm{PP}}\mathrm{d}P + H_{\mathrm{nP}}\mathrm{d}n\,;
\end{aligned}
\tag{I.32}
$$

$$
\begin{aligned}
\mu &= \mu(S, P, n)\,, \\
\mathrm{d}\mu &= \left(\frac{\partial \mu}{\partial S}\right)_{\mathrm{P,n}} \mathrm{d}S + \left(\frac{\partial \mu}{\partial P}\right)_{\mathrm{S,n}} \mathrm{d}P + \left(\frac{\partial \mu}{\partial n}\right)_{\mathrm{S,P}} \mathrm{d}n\,, \\
\mu &= \left(\frac{\partial H}{\partial n}\right)_{\mathrm{S,V}}\,, \\
\mathrm{d}\mu &= H_{\mathrm{Sn}}\mathrm{d}S + H_{\mathrm{Pn}}\mathrm{d}P + H_{\mathrm{nn}}\mathrm{d}n\,.
\end{aligned}
\tag{I.33}
$$

Thus

$$
\hat{\boldsymbol{H}} = \begin{pmatrix} H_{\mathrm{SS}} & H_{\mathrm{PS}} & H_{\mathrm{nS}} \\ H_{\mathrm{SP}} & H_{\mathrm{PP}} & H_{\mathrm{nP}} \\ H_{\mathrm{Sn}} & H_{\mathrm{Pn}} & H_{\mathrm{nn}} \end{pmatrix}\,.
\tag{I.34}
$$

Massieu Transforms: The Entropy Representation

J

To determine the Massieu transforms, one needs to use the entropy representation.[1]

In this representation, the entropy s acts as the central thermodynamic potential.

J.1 Massieu Potential, $M\{v, u\}$

Let us consider, in the entropy representation, the fundamental equation for a simple thermodynamic system that is recorded in (8.18). For convenience, it is reproduced in an equivalent form below:

$$s = \left(\frac{p}{t}\right)v + \left(\frac{1}{t}\right)u - \left(\frac{\mu}{t}\right)n + \left(\frac{\mathcal{Y}}{t}\right)\mathcal{X}. \tag{J.1}$$

This equation describes how the extensive variable, the entropy s, depends on other extensive variables: the volume v, internal energy u, mole numbers n, and extensive parameters such as $\mathcal{X}$. The simple system being treated here has constant number of atoms and does not have any dependence on the term $\mathcal{Y}\mathcal{X}$. For such a system, a convenient appropriate relationship is provided by the statement of the first–second law. That statement was originally given in (5.6). Again, for convenience, we reproduce it below (also compare with (10.13)):

$$\mathrm{d}s = \left(\frac{p}{t}\right)\mathrm{d}v + \left(\frac{1}{t}\right)\mathrm{d}u. \tag{J.2}$$

Clearly, for the entropy s, the "canonical"—i.e., the characteristic—independent variables are the volume v and the internal energy u, that is,

$$s = s(v, u). \tag{J.3}$$

The entropy $s(v, u)$, therefore, is itself the Massieu potential $M\{v, u\}$. That is, $s(v, u) = M\{v, u\}$.

[1] Corresponding work in the energy representation—called the Legendre transformations—is discussed in Sect. 10.11.

R. Tahir-Kheli, *General and Statistical Thermodynamics*,
https://doi.org/10.1007/978-3-030-20700-7

J.1.1 Massieu Potential, $M\{v, \frac{1}{t}\}$

Rather than the internal energy u—as is the case in (J.2) above—the inverse temperature, $\frac{1}{t}$, is greatly preferred as an independent variable. Therefore, in (J.2), the variable of interest—that we should transform out of—is the internal energy u which occurs in the form $+\frac{1}{t}\mathrm{d}u$. In order to eliminate $\mathrm{d}u$, we transcribe $+\frac{1}{t}\mathrm{d}u$ as follows.

Add $-\frac{1}{t}u$ to the primary potential s. This introduces an alternate Massieu potential $M\{v, \frac{1}{t}\}$,

$$M\left\{v, \frac{1}{t}\right\} = s - \left(\frac{1}{t}\right)u = -\frac{f}{t}. \tag{J.4}$$

Note that $-tM\{v, \frac{1}{t}\}$ is equal to the Helmholtz free energy f. Thus $M\{v, \frac{1}{t}\}$ is really a close relative of an old, but important, thermodynamic potential. Differentiate $M\{v, \frac{1}{t}\}$ to get

$$\mathrm{d}M\left\{v, \frac{1}{t}\right\} = \mathrm{d}s - \left(\frac{1}{t}\right)\mathrm{d}u - u\mathrm{d}\left(\frac{1}{t}\right). \tag{J.5}$$

And, cancel $-\frac{1}{t}\mathrm{d}u$ by inserting the original relationship for $\mathrm{d}s$ given in (J.2). That is,

$$\mathrm{d}M\left\{v, \frac{1}{t}\right\} = \left(\frac{p}{t}\right)\mathrm{d}v + \left(\frac{1}{t}\right)\mathrm{d}u - \left(\frac{1}{t}\right)\mathrm{d}u - u\mathrm{d}\left(\frac{1}{t}\right) = \left(\frac{p}{t}\right)\mathrm{d}v - u\mathrm{d}\left(\frac{1}{t}\right). \tag{J.6}$$

The inverse temperature $\frac{1}{t}$ is the new independent variable that is conjugate to the previous independent variable the internal energy u. Note that both the characteristic independent variables here, v and $\frac{1}{t}$, are easy to measure.

Remark J.1 The simple thermodynamic system being considered here provides only four possible choices for the pairs of independent characteristic variables. These are: (v, u), $(v, \frac{1}{t})$, $(\frac{p}{t}, \frac{1}{t})$, and $(\frac{p}{t}, u)$. So far in this appendix, only two of these four pairs have been utilized, namely (v, u) in $M\{v, u\}$ and $(v, \frac{1}{t})$ in $M\{v, \frac{1}{t}\}$. To make use of the last two pairs, namely $(\frac{p}{t}, \frac{1}{t})$ and $(\frac{p}{t}, u)$, we need to proceed, in the usual fashion, as follows:

J.1.2 Massieu Potential, $M\{\frac{p}{t}, u\}$

The appropriate new thermodynamic potential is

$$M\left\{\frac{p}{t}, u\right\} = s - \left(\frac{p}{t}\right)v. \tag{J.7}$$

Take its derivative to get

$$\mathrm{d}\left(M\left\{\frac{p}{t} u\right\}\right) = \mathrm{d}s - \left(\frac{p}{t}\right)\mathrm{d}v - v\mathrm{d}\left(\frac{p}{t}\right). \tag{J.8}$$

Now, following (J.2), replace ds by $(\frac{p}{t})\mathrm{d}v + (\frac{1}{t})\mathrm{d}u$. This yields the following relationship:

$$\begin{aligned} \mathrm{d}\left(M\left\{\frac{p}{t}u\right\}\right) &= \mathrm{d}s - \left(\frac{p}{t}\right)\mathrm{d}v - v\mathrm{d}\left(\frac{p}{t}\right) \\ &= \left(\frac{p}{t}\right)\mathrm{d}v + \left(\frac{1}{t}\right)\mathrm{d}u - \left(\frac{p}{t}\right)\mathrm{d}v - v\mathrm{d}\left(\frac{p}{t}\right) \\ &= \left(\frac{1}{t}\right)\mathrm{d}u - v\mathrm{d}\left(\frac{p}{t}\right). \end{aligned} \tag{J.9}$$

Note that the Massieu potential, $M\{\frac{p}{t}u\}$, whose independent characteristic variables are $\frac{p}{t}$ and u, is not a "close relative" of previous thermodynamic potentials. Rather, it is a new thermodynamic potential.

J.1.3 Massieu Potential, $M\{\frac{p}{t}, \frac{1}{t}\}$

Choose the following Massieu potential:

$$M\left\{\frac{p}{t}, \frac{1}{t}\right\} = M\left\{v, \frac{1}{t}\right\} - \left(\frac{p}{t}\right)v = s - \left(\frac{1}{t}\right)u - \left(\frac{p}{t}\right)v = -\frac{g}{t}. \tag{J.10}$$

Note that $-t\, M\{\frac{p}{t}, \frac{1}{t}\}$ is equal to the Gibbs free energy g. Thus $M\{\frac{p}{t}, \frac{1}{t}\}$ is not a new function. Rather, it is a close relative of an old, but important, thermodynamic potential, the Gibbs free energy.

Then differentiate both sides to obtain

$$\mathrm{d}\left(M\left\{\frac{p}{t}, \frac{1}{t}\right\}\right) = \mathrm{d}\left(M\left\{v, \frac{1}{t}\right\}\right) - \left(\frac{p}{t}\right)\cdot \mathrm{d}v - v \cdot \mathrm{d}\left(\frac{p}{t}\right). \tag{J.11}$$

Now, using (J.6), replace $\mathrm{d}M\{v, \frac{1}{t}\}$ by $(\frac{p}{t})\mathrm{d}v - u\mathrm{d}(\frac{1}{t})$. This yields the desired relationship

$$\mathrm{d}\left(M\left\{\frac{p}{t}, \frac{1}{t}\right\}\right) = -v \cdot \mathrm{d}\left(\frac{p}{t}\right) - u \cdot \mathrm{d}\left(\frac{1}{t}\right). \tag{J.12}$$

As mentioned above, the Massieu potential $M\{\frac{p}{t}, \frac{1}{t}\}$ is not new. Rather, it is closely related to the Gibbs free energy g.

Integral (11.83) K

We need to calculate the following:

$$
\begin{aligned}
u_{\mathrm{i}} &= u \\
&= \frac{\int_0^{\pi} \sin(\theta_1)\mathrm{d}\theta_1 \int_0^{2\pi} \mathrm{d}\phi_1 \int_0^{\pi} \sin(\theta_2)\mathrm{d}\theta_2 \int_0^{2\pi} \mathrm{d}\phi_2 F(1,2) \exp[-\beta F(1,2)]}{\int_0^{\pi} \sin(\theta_1)\mathrm{d}\theta_1 \int_0^{2\pi} \mathrm{d}\phi_1 \int_0^{\pi} \sin(\theta_2)\mathrm{d}\theta_2 \int_0^{2\pi} d\phi_2 \exp[-\beta F(1,2)]},
\end{aligned}
\tag{K.1}
$$

where

$$
F(1,2) = -\left(\frac{C\mu_1\mu_2}{R^3}\right)\left[2\cos\theta_1\cos\theta_2 - \sin\theta_1\sin\theta_2\cos(\phi_1-\phi_2)\right]. \tag{K.2}
$$

Normally, the needed integrals would be evaluated by numerical methods. However, if the system is at high enough temperature such that $(\frac{C\mu_1\mu_2}{k_B T R^3}) \ll 1$, the exponential can be expanded in powers of the exponent, i.e.,

$$
\exp\left[-\beta F(1,2)\right] = 1 - \beta F(1,2) + \frac{1}{2}\left[\beta F(1,2)\right]^2 + \cdots, \tag{K.3}
$$

and the resultant integrals are evaluated by standard analytical methods.

Let us look first at the denominator of (K.1). We have

$$
\begin{aligned}
&\int_0^{\pi} \sin(\theta_1)\mathrm{d}\theta_1 \int_0^{2\pi} \mathrm{d}\phi_1 \int_0^{\pi} \sin(\theta_2)\mathrm{d}\theta_2 \int_0^{2\pi} \mathrm{d}\phi_2 \exp\left[-\beta F(1,2)\right] \\
&= \int_0^{\pi} \sin(\theta_1)\mathrm{d}\theta_1 \int_0^{2\pi} \mathrm{d}\phi_1 \int_0^{\pi} \sin(\theta_2)\mathrm{d}\theta_2 \int_0^{2\pi} \mathrm{d}\phi_2 \\
&\quad - \beta \int_0^{\pi} \sin(\theta_1)\mathrm{d}\theta_1 \int_0^{2\pi} \mathrm{d}\phi_1 \int_0^{\pi} \sin(\theta_2)\mathrm{d}\theta_2 \int_0^{2\pi} \mathrm{d}\phi_2 F(1,2) \\
&\quad + \frac{\beta^2}{2} \int_0^{\pi} \sin(\theta_1)\mathrm{d}\theta_1 \int_0^{2\pi} \mathrm{d}\phi_1 \int_0^{\pi} \sin(\theta_2)\mathrm{d}\theta_2 \int_0^{2\pi} \mathrm{d}\phi_2 \left[F(1,2)\right]^2 + \cdots \\
&= A_1 - \beta A_2 + \frac{(\beta)^2}{2} A_3 + \cdots .
\end{aligned}
\tag{K.4}
$$

R. Tahir-Kheli, *General and Statistical Thermodynamics*,
https://doi.org/10.1007/978-3-030-20700-7

Then

$$\begin{aligned} A_1 &= \int_0^{\pi} \sin(\theta_1)\mathrm{d}\theta_1 \int_0^{2\pi} \mathrm{d}\phi_1 \int_0^{\pi} \sin(\theta_2)\mathrm{d}\theta_2 \int_0^{2\pi} \mathrm{d}\phi_2 \\ &= \left[-\cos(\theta_1)\right]_0^{\pi}(2\pi)\left[-\cos(\theta_2)\right]_0^{\pi}(2\pi) \\ &= 2(2\pi)2(2\pi) = (4\pi)^2 \end{aligned} \tag{K.5}$$

and

$$\begin{aligned} A_2 &= \int_0^{\pi} \sin(\theta_1)\mathrm{d}\theta_1 \int_0^{2\pi} \mathrm{d}\phi_1 \int_0^{\pi} \sin(\theta_2)\mathrm{d}\theta_2 \int_0^{2\pi} \mathrm{d}\phi_2 F(1,2) \\ &= -\left(\frac{C\mu_1\mu_2}{R^3}\right) \int_0^{\pi} \sin(\theta_1)\mathrm{d}\theta_1 \int_0^{2\pi} \mathrm{d}\phi_1 \int_0^{\pi} \sin(\theta_2)\mathrm{d}\theta_2 \int_0^{2\pi} \mathrm{d}\phi_2 \\ &\quad \times \left[2\cos\theta_1\cos\theta_2 - \sin\theta_1\sin\theta_2\cos(\phi_1-\phi_2)\right] = 0\,. \end{aligned} \tag{K.6}$$

To convince ourselves that $A_2 = 0$, all we need to notice is that the following two integrals are vanishing. That is,

$$\begin{aligned} &2\int_0^{\pi} \sin(\theta_1)\cos(\theta_1)\mathrm{d}\theta_1 \\ &= \int_0^{\pi} \sin(2\theta_1)\mathrm{d}\theta_1 = -\left[\frac{\cos(2\theta_1)}{2}\right]_0^{\pi} = -\left[\frac{1-1}{2}\right] = 0 \end{aligned} \tag{K.7}$$

and

$$\begin{aligned} &\int_0^{2\pi} \cos(\phi_1-\phi_2)\mathrm{d}\phi_1 \\ &= \cos(\phi_2)\int_0^{2\pi} \cos(\phi_1)\mathrm{d}\phi_1 + \sin(\phi_2)\int_0^{2\pi} \sin(\phi_1)\mathrm{d}\phi_1 \\ &= \cos(\phi_2)\left[\sin(\phi_1)\right]_0^{2\pi} - \sin(\phi_2)\left[\cos(\phi_1)\right]_0^{2\pi} \\ &= \cos(\phi_2)[0-0] - \sin(\phi_2)[1-1] = 0\,. \end{aligned} \tag{K.8}$$

The calculation of A_3 requires a little more effort. We have

$$\begin{aligned} \frac{A_3}{[-(\frac{C\mu_1\mu_2}{R^3})]^2} &= \int_0^{\pi} \sin(\theta_1)\mathrm{d}\theta_1 \int_0^{2\pi} \mathrm{d}\phi_1 \int_0^{\pi} \sin(\theta_2)\mathrm{d}\theta_2 \int_0^{2\pi} \mathrm{d}\phi_2 \left[\frac{F(1,2)}{-(\frac{C\mu_1\mu_2}{R^3})}\right]^2 \\ &= \int_0^{\pi} \sin(\theta_1)\mathrm{d}\theta_1 \int_0^{2\pi} \mathrm{d}\phi_1 \int_0^{\pi} \sin(\theta_2)\mathrm{d}\theta_2 \int_0^{2\pi} \mathrm{d}\phi_2 \\ &\quad \times \left[2\cos\theta_1\cos\theta_2 - \sin\theta_1\sin\theta_2\cos(\phi_1-\phi_2)\right]^2 \\ &\equiv \text{Part}_1 + \text{Part}_2\,, \end{aligned} \tag{K.9}$$

where

$$
\begin{aligned}
\mathrm{Part}_1 &= \int_0^{\pi} \sin(\theta_1)\mathrm{d}\theta_1 \int_0^{2\pi} \mathrm{d}\phi_1 \int_0^{\pi} \sin(\theta_2)\mathrm{d}\theta_2 \int_0^{2\pi} \mathrm{d}\phi_2 \\
&\quad \times \left[4\cos\theta_1^2 \cos\theta_2^2 + \sin\theta_1^2 \sin\theta_2^2 \cos(\phi_1 - \phi_2)^2\right], \\
\mathrm{Part}_2 &= \int_0^{\pi} \sin(\theta_1)\mathrm{d}\theta_1 \int_0^{2\pi} \mathrm{d}\phi_1 \int_0^{\pi} \sin(\theta_2)\mathrm{d}\theta_2 \int_0^{2\pi} \mathrm{d}\phi_2 \\
&\quad \times \left[-4\cos\theta_1 \cos\theta_2 \cdot \sin\theta_1 \sin\theta_2 \cos(\phi_1 - \phi_2)\right] \\
&= 0 .
\end{aligned}
\tag{K.10}
$$

We have asserted in the above that Part_2 is vanishing. This is assured by the fact that one of its multiplying factors is the integral $\int_0^{2\pi} \cos(\phi_1 - \phi_2)\mathrm{d}\phi_1$. That integral was shown, in (K.8), to be equal to zero.

Regarding Part_1, it is convenient to separate it into two subsidiary parts. That is,

$$
\mathrm{Part}_1 \equiv \mathrm{Part}_1\,(\mathrm{I}) + \mathrm{Part}_1\,(\mathrm{II}) , \tag{K.11}
$$

where

$$
\begin{aligned}
\mathrm{Part}_1\,(\mathrm{I}) &= \int_0^{\pi} \sin(\theta_1)\mathrm{d}\theta_1 \int_0^{2\pi} \mathrm{d}\phi_1 \int_0^{\pi} \sin(\theta_2)\mathrm{d}\theta_2 \int_0^{2\pi} \mathrm{d}\phi_2 \left[4\cos\theta_1^2 \cos\theta_2^2\right] \\
&= 4 \int_{\cos(\theta_1)=1}^{\cos(\theta_1)=-1} \cos\theta_1^2 \{-\mathrm{d}\cos(\theta_1)\} \int_0^{2\pi} \mathrm{d}\phi_1 \\
&\quad \times \int_{\cos(\theta_2)=1}^{\cos(\theta_2)=-1} \cos\theta_2^2 \{-\mathrm{d}\cos(\theta_2)\} \int_0^{2\pi} \mathrm{d}\phi_2 \\
&= 4 \int_{-1}^{1} \eta^2 \mathrm{d}\eta\,(2\pi) \times \int_{-1}^{1} \gamma^2 \mathrm{d}\gamma\,(2\pi) = 4\left(\frac{4\pi}{3}\right)^2
\end{aligned}
\tag{K.12}
$$

and

$$
\begin{aligned}
\mathrm{Part}_1\,(\mathrm{II}) &= \int_0^{\pi} \sin(\theta_1)\mathrm{d}\theta_1 \int_0^{2\pi} \mathrm{d}\phi_1 \int_0^{\pi} \sin(\theta_2)\mathrm{d}\theta_2 \int_0^{2\pi} \mathrm{d}\phi_2 \\
&\quad \times \left[\sin\theta_1^2 \sin\theta_2^2 \cos(\phi_1 - \phi_2)^2\right] \\
&= \int_0^{\pi} \sin(\theta_1)^3 \mathrm{d}\theta_1 \int_0^{\pi} \sin(\theta_2)^3 \mathrm{d}\theta_2 \\
&\quad \times \int_0^{2\pi} \mathrm{d}\phi_1 \int_0^{2\pi} \mathrm{d}\phi_2 \cos(\phi_1 - \phi_2)^2 .
\end{aligned}
\tag{K.13}
$$

Relevant integrals in (K.13) are found as follows:

$$
\begin{aligned}
\int_0^{\pi} \sin(\theta_1)^3 \mathrm{d}\theta_1 &= \int_0^{\pi} \sin(\theta_2)^3 \mathrm{d}\theta_2 = \int_0^{\pi} \left[\frac{3\sin(\theta_1) - \sin(3\theta_1)}{4}\right] \mathrm{d}\theta_1 \\
&= \frac{3}{4}[1+1] - \frac{1}{4}\frac{[1+1]}{3} = \left(\frac{4}{3}\right),
\end{aligned}
$$

$$\int_0^{2\pi} d\phi_1 \int_0^{2\pi} d\phi_2 \cos(\phi_1 - \phi_2)^2 \tag{K.14}$$
$$= \int_0^{2\pi} d\phi_1 \int_0^{2\pi} d\phi_2 \left[\cos(\phi_1)^2 \cos(\phi_2)^2 + \sin(\phi_1)^2 \sin(\phi_2)^2\right]$$
$$+ 2 \int_0^{2\pi} \sin(\phi_1)\cos(\phi_1) d\phi_1 \int_0^{2\pi} \sin(\phi_2)\cos(\phi_2) d\phi_2 \,.$$

The integrals in the last row in (K.14) are clearly both equal to zero. And the relevant integrals in the row before are of the form:

$$\int_0^{2\pi} d\phi_1 \cos(\phi_1)^2 = \int_0^{2\pi} \left[\frac{1+\cos(2\phi_1)}{2}\right] d\phi_1 = \frac{2\pi}{2},$$
$$\int_0^{2\pi} d\phi_1 \sin(\phi_1)^2 = \int_0^{2\pi} \left[\frac{1-\cos(2\phi_1)}{2}\right] d\phi_1 = \frac{2\pi}{2}.$$

As a result

$$\int_0^{2\pi} d\phi_1 \int_0^{2\pi} d\phi_2 \cos(\phi_1 - \phi_2)^2$$
$$= \int_0^{2\pi} d\phi_1 \int_0^{2\pi} d\phi_2 \left[\cos(\phi_1)^2 \cos(\phi_2)^2 + \sin(\phi_1)^2 \sin(\phi_2)^2\right]$$
$$+ 2 \int_0^{2\pi} \sin(\phi_1)\cos(\phi_1) d\phi_1 \int_0^{2\pi} \sin(\phi_2)\cos(\phi_2) d\phi_2$$
$$= \left(\frac{2\pi}{2}\right)^2 + \left(\frac{2\pi}{2}\right)^2 + 2 \times 0 \times 0 = 2\pi^2 \,. \tag{K.15}$$

Combining (K.13)–(K.15) gives

$$\text{Part}_1\,(\text{II}) = \left(\frac{4}{3}\right) \times \left(\frac{4}{3}\right) \times \left(2\pi^2\right).$$

Having calculated Part_1 (I) and Part_1 (II) and recalling that Part_2 is vanishing—see (K.10)—we are now able to write the complete result for (K.4), namely the denominator of (K.1). Note this denominator is needed for the calculation of, u, the average energy per mol:

$$\int_0^{\pi} \sin(\theta_1) d\theta_1 \int_0^{2\pi} d\phi_1 \int_0^{\pi} \sin(\theta_2) d\theta_2 \int_0^{2\pi} d\phi_2 \exp\left[-\beta F(1,2)\right]$$
$$= A_1 - \beta A_2 + \frac{(\beta)^2}{2} A_3 + \cdots$$
$$= (4\pi)^2 + 0 + \frac{\beta^2}{2}\left[\left(\frac{C\mu_1\mu_2}{R^3}\right)\right]^2 \left[\text{Part}_1\,(\text{I}) + \text{Part}_1\,(\text{II}) + \text{Part}_2\right] + \cdots$$
$$= (4\pi)^2 + \frac{\beta^2}{2}\left[\left(\frac{C\mu_1\mu_2}{R^3}\right)\right]^2 \left[4\left(\frac{4\pi}{3}\right)^2 + \left(\frac{4}{3}\right)\left(\frac{4}{3}\right)2\pi^2 + 0\right] + \cdots$$

$$= (4\pi)^2\left[1+\frac{\beta^2}{3}\left(\frac{C\mu_1\mu_2}{R^3}\right)^2\right]+\cdots \tag{K.16}$$

Having thus calculated the denominator on the right-hand side of (K.1), we deal next with the numerator. Expanding the exponential as given in (K.3), the numerator of (K.1) is the following:

$$\begin{aligned}
&\int_0^{\pi}\sin(\theta_1)\mathrm{d}\theta_1\int_0^{2\pi}\mathrm{d}\phi_1\int_0^{\pi}\sin(\theta_2)\mathrm{d}\theta_2\int_0^{2\pi}\mathrm{d}\phi_2 F(1,2)\exp\left[-\beta F(1,2)\right]\\
&=\int_0^{\pi}\sin(\theta_1)\mathrm{d}\theta_1\int_0^{2\pi}\mathrm{d}\phi_1\int_0^{\pi}\sin(\theta_2)\mathrm{d}\theta_2\int_0^{2\pi}\mathrm{d}\phi_2 F(1,2)\\
&\quad-\beta\int_0^{\pi}\sin(\theta_1)\mathrm{d}\theta_1\int_0^{2\pi}d\phi_1\int_0^{\pi}\sin(\theta_2)\mathrm{d}\theta_2\int_0^{2\pi}\mathrm{d}\phi_2\left[F(1,2)\right]^2+\cdots\\
&=0-\beta A_3\\
&=-\beta\frac{2}{3}(4\pi)^2\left[\frac{C\mu_1\mu_2}{R^3}\right]^2 .
\end{aligned} \tag{K.17}$$

Here we have made use of the following information: As was demonstrated earlier—see (K.6) and the discussion that followed—we have

$$\int_0^{\pi}\sin(\theta_1)\mathrm{d}\theta_1\int_0^{2\pi}\mathrm{d}\phi_1\int_0^{\pi}\sin(\theta_2)\mathrm{d}\theta_2\int_0^{2\pi}d\phi_2 F(1,2)=0 .$$

Also the last integral in (K.17) has already been evaluated as A_3—see (K.4)–(K.16).

The average value, U, of the energy of N dipole pairs is N times the average energy, u, of a single dipole pair. Note that u is the ratio of the results given in (K.17) and (K.16). That is,

$$\begin{aligned}
U=Nu&=\frac{-N\beta A_3}{(4\pi)^2[1+\frac{\beta^2}{3}(\frac{C\mu_1\mu_2}{R^3})^2]}+\cdots\\
&=-N\beta\frac{2}{3}\left(\frac{C\mu_1\mu_2}{R^3}\right)^2\left[1-\frac{\beta^2}{3}\left(\frac{C\mu_1\mu_2}{R^3}\right)^2\right]+\cdots\\
&\approx-N\beta\frac{2}{3}\left(\frac{C\mu_1\mu_2}{R^3}\right)^2 .
\end{aligned} \tag{K.18}$$

K.1 Average Force Between a Pair

The force Λ between any pair of dipoles can be determined by differentiating their intrapair potential energy $F(1,2)$ that was specified in (K.2). That is,

$$\begin{aligned}
F(1,2)&=-\left(\frac{C\mu_1\mu_2}{R^3}\right)\left[2\cos\theta_1\cos\theta_2-\sin\theta_1\sin\theta_2\cos(\phi_1-\phi_2)\right],\\
\Lambda&=-\left(\frac{\partial F(1,2)}{\partial R}\right)_{\mu_1,\mu_2,\theta_1,\theta_2,\phi_1,\phi_2}=\left(\frac{3}{R}\right)F(1,2) .
\end{aligned} \tag{K.19}$$

Similar to the average energy of a pair, $\langle F(1,2)\rangle$—see (11.83) and (11.84)—we can also represent the average value of the intradipolar force, Λ, as $\langle\Lambda\rangle$. Very conveniently, in this case they are related! Indeed,

$$\langle\Lambda\rangle = \left(\frac{3}{R}\right)\langle F(1,2)\rangle = \left(\frac{3}{RN}\right)U$$
$$= -\beta\frac{2}{R}\left(\frac{C\mu_1\mu_2}{R^3}\right)^2. \tag{K.20}$$

Note that U is as given in (K.18).

Clearly, the thermodynamic average of the force between any single pair of dipoles is attractive.

Indistinguishable, Noninteracting Quantum Particles

L

For a gas of n noninteracting free particles with mass m and momentum p, the Hamiltonian $\mathcal{H}_{\rm n}$ is

$$\mathcal{H}_{\rm n} = \sum_{\rm p} n_{\rm p} \varepsilon_{\rm p} , \tag{L.1}$$

where $n_{\rm p}$ is the number of free quantum particles with momentum p,

$$n = \sum_{\rm p} n_{\rm p} , \qquad \varepsilon_{\rm p} = \frac{p^2}{2m} . \tag{L.2}$$

Therefore the partition function, given in (11.207), can be written as

$$\begin{aligned}
\Psi(z, V, T) &= \sum_{\rm n=0}^{\infty} {\rm Tr}\big[\exp\{-\beta(\mathcal{H}_{\rm n} - \mu n)\}\big] \\
&= \sum_{\rm n=0}^{\infty} {\rm Tr}\left[\exp\left\{-\beta\left(\sum_{\rm p}(\varepsilon_{\rm p} - \mu) n_{\rm p}\right)\right\}\right] \\
&= \sum_{\rm n=0}^{\infty} {\sum_{\rm n_p}}' \prod_{\rm p} \big[\exp\{-\beta(\varepsilon_{\rm p} - \mu) n_{\rm p}\}\big] ,
\end{aligned} \tag{L.3}$$

where the primed sum over $n_{\rm p}$, i.e., $\sum'_{\rm n_p}$, includes only those values of $n_{\rm p}$ for which $\sum_{\rm p} n_{\rm p} = n$.

Note that in deference to the physical requirements of the grand canonical ensemble, the double summation in the last row of (L.3) must occur in the following manner:

First, we must sum over $n_{\rm p}$ in such a way that the total number of atoms are kept equal to n, that is, $\sum_{\rm p} n_{\rm p} = n$.

Second, the next summation is over n, which must include all physically allowed values.

R. Tahir-Kheli, *General and Statistical Thermodynamics*,
https://doi.org/10.1007/978-3-030-20700-7

An important mathematical result is that such double summation—in which the summation over n occurs after the n_p summation—is in fact equivalent to a product of independent, single sums over all values of the numbers n_p—and not just those that satisfy the sum-rule,[1] $\sum_p n_p = n$. As a result we can reexpress the last row of (L.3) as follows:

$$\begin{aligned}\Psi(z, V, T) &= \sum_{n_0=0}^{n_{max}} \left[\exp\{-\beta(\varepsilon_0 - \mu)n_0\}\right] \\ &\times \sum_{n_1=0}^{n_{max}} \left[\exp\{-\beta(\varepsilon_1 - \mu)n_1\}\right] \times \sum_{n_2=0}^{n_{max}} \left[\exp\{-\beta(\varepsilon_2 - \mu)n_2\}\right] \times \cdots .\end{aligned} \tag{L.4}$$

For the Bose–Einstein (B–E) gas, n_{max} can be as large as ∞. On the other hand, for a Fermi–Dirac (F–D) gas, the occupancy of each state is limited to two. Therefore, for an F–D gas, $n_0, n_1, n_2, \ldots$, etc., may take on only two values, 0 and 1.

L.1 Quantum Statistics: Grand Canonical Partition Function

The so-called Bose–Einstein particles obey the Bose statistics in which there is no restriction on the occupation number of any state, so it can range between zero and infinity. Therefore, $n_{max} = \infty$ and any of the sums in (L.4)—for example, the ith—can be written as follows:

$$\sum_{n_i=0}^{n_{max}=\infty} \left[\exp\{-\beta(\varepsilon_i - \mu)n_i\}\right] = \frac{1}{1 - \exp\{-\beta(\varepsilon_i - \mu)\}} . \tag{L.5}$$

As a result, (L.4) becomes

$$\begin{aligned}\left[\Psi(z, V, T)\right]_{B-E} &= \Psi(z, V, T) = \prod_i \left(\frac{1}{1 - \exp\{-\beta(\varepsilon_i - \mu)\}}\right), \\ \left[Q(z, V, T)\right]_{B-E} &= \Delta_{B-E} \ln\left[\Psi(z, V, T)\right]_{B-E} \\ &= -\Delta_{B-E} \sum_i \ln\left[1 - \exp\{-\beta(\varepsilon_i - \mu)\}\right] \\ &= -\Delta_{B-E} \sum_i \ln\left[1 - z\exp(-\beta\varepsilon_i)\right] .\end{aligned} \tag{L.6}$$

Recall that $z = \exp(\beta\mu)$. Also that Δ_{B-E} is the spin degeneracy weight factor which is equal to unity for the B–E gas. Both ε_i and the sum $\sum_i$ are explained in (L.9).

[1] A motivated student would want to test this assertion by working through the sum over n from $n = 0$ to $n = 2$.

In contrast with Bose statistics, for Fermi particles with spin $\frac{1}{2}$, the sum over n_i is restricted to only two cases, $n_i = 0$ and 1. This limits n_{max} to the number 1. Therefore, for the so called Fermi–Dirac gas, any of the sums in (L.4)—say, the ith—can be written as

$$\sum_{n_i=0}^{n_{max}=1} \left[\exp\left\{-\beta(\varepsilon_i - \mu)n_i\right\}\right] = 1 + \exp\left\{-\beta(\varepsilon_i - \mu)\right\}. \tag{L.7}$$

As a result (L.4) becomes

$$\begin{aligned}
\left[\Psi(z, V, T)\right]_{F-D} &= \Psi(z, V, T) = \prod_i \left(1 + \exp\left\{-\beta(\varepsilon_i - \mu)\right\}\right), \\
\left[Q(z, V, T)\right]_{F-D} &= \Delta_{F-D} \ln\left[\Psi(z, V, T)\right]_{F-D} \\
&= \Delta_{F-D} \sum_i \ln\left[1 + \exp\left\{-\beta(\varepsilon_i - \mu)\right\}\right] \\
&= \Delta_{F-D} \sum_i \ln\left[1 + z\exp(-\beta\varepsilon_i)\right],
\end{aligned} \tag{L.8}$$

where the spin-degeneracy factor $\Delta_{F-D} = 2$.

Following a suggestion by Pathria, we can combine (L.6) and (L.8) and thus describe in a single equation the quantum noninteracting particles of both the B–E and F–D varieties. While we do not necessarily need it, such a description also simultaneously contains the corresponding representation for a classical noninteracting gas,

$$\begin{aligned}
Q(z, V, T) &= \frac{\Delta}{a} \sum_i \ln\left[1 + az\exp(-\beta\varepsilon_i)\right], \quad \text{where} \\
a &= -1, \quad \text{for B–E gas}, \\
&= +1, \quad \text{for F–D gas}, \\
&\to 0, \quad \text{for classical gas, and} \\
\Delta &= (2S + 1) = 2, \quad \text{for spin } \tfrac{1}{2} \text{ F–D gas}, \\
&= 1, \quad \text{otherwise}.
\end{aligned} \tag{L.9}$$

Landau Diamagnetism

M

In addition to usual paramagnetism, conduction electrons are also subject to a sort of negative paramagnetism whereby the effect of applied field in a given direction produces a magnetic moment that faces in the opposite direction. Such behavior, resulting in negative susceptibility, is called diamagnetic.

So far in our study of the conduction electrons, we have not considered any effects of the orbital motion. In the presence of applied magnetic field B_0—say, along the z-axis—an electron follows a helical path around the z-axis. The rotational motion is caused by the so-called Lorentz force. Its projection on the x–y plane is completely circular. While a single rotating electron may be diamagnetic, when a large number N of electrons are present, and they are subject to reflection from the enclosing walls, classical statistics predicts that there is "complete"—i.e., to the leading order in N—cancelation of diamagnetic effect. Therefore, as first noted by Bohr, van Leeuwen, and others, classical statistical description of this motion cannot lead to diamagnetism.

Unlike classical statistics, L.D. Landau has shown that quantum statistics do lead to nonzero diamagnetism. While the details of Landau's work are somewhat involved, a relatively simple general comment can still be made. Due to reflection from the bounding walls, the quantum electrons near the boundary—unlike the classical, Maxwell–Boltzmann electrons—have on average different quantized velocities from the electrons that have not been reflected. Therefore, complete compensation of the diamagnetic effect that occurs for perfectly reflecting classical electrons does not take place for quantum electrons.

Analysis: Landau Diamagnetism

For the electrons being studied, we assume there is no interparticle interaction. Therefore, we can work with a collection of separate single particles.

A particle's circular motion—say, with angular velocity ω—is quantized so that the relevant rotational energy levels are not continuous but are equal, say, to $[\ell + \frac{1}{2}](\hbar\omega)$, where $\ell = 0, 1, 2, 3, \ldots$ (Here $\omega = \frac{eB_0}{mc}$, e is the electronic charge, m its mass, and c is the velocity of light in vacuum.) While the energy for the longitudinal motion, i.e., $\frac{p_z^2}{2m}$, is also quantized, due to the large size of the system the levels lie

R. Tahir-Kheli, *General and Statistical Thermodynamics*,
https://doi.org/10.1007/978-3-030-20700-7

very close together. As a result, quantization along the z-direction can be ignored and p_z can be treated as a continuous variable. Therefore, the energy spectrum to treat is

$$\epsilon = \left[\frac{p_x^2}{2m} + \frac{p_y^2}{2m} + \frac{p_z^2}{2m}\right] = \left[\ell + \frac{1}{2}\right](\hbar\omega) + \left[\frac{p_z^2}{2m}\right]$$
$$= \left[\ell + \frac{1}{2}\right]\left(\frac{\hbar e B_0}{mc}\right) + \left[\frac{p_z^2}{2m}\right] = -B_0 \hat{M}_l + \left[\frac{p_z^2}{2m}\right]. \tag{M.1}$$

Much as was done in (11.234), we work out the grand potential

$$Q(z, V, T, B_0) = \sum_{\epsilon} \ln\{1 + z\exp(-\beta\epsilon)\}. \tag{M.2}$$

Then the total magnetic moment, M, is

$$M = \sum_{l=0}^{N} \langle \hat{M}_l \rangle = \frac{1}{\beta}\left(\frac{\partial Q(z, V, T, B_0)}{\partial B_0}\right)_{z,V,T}. \tag{M.3}$$

Also, as noted in (11.237), the observed value of the number of particles $\bar{N}$ is

$$\bar{N} = z\left(\frac{\partial Q(z, V, T, B_0)}{\partial z}\right)_{B_0,V,T}. \tag{M.4}$$

M.1 Multiplicity Factor: Landau Diamagnetism

We consider separately the contribution due to the rotational and the translational states. The first part of the integral then leads to the "multiplicity factor."

The quantized, rotational energy levels of a particle, that refer to the x–y space, are necessarily degenerate due to the "coalescing together" of the almost continuous set of zero-field levels. As a result, essentially all those levels, that lie between any nearest two eigenvalues of the x–y component of the Hamiltonian, coalesce together into a single level that may be characterized by the quantum index ℓ. The difference in energy between any nearest pair of levels is therefore independent of their quantum index ℓ. The number of these levels is the "multiplicity factor." According to Pathria,[1] this factor is the "quantum-mechanical measure of the freedom available to the particle for the center of its orbit to be 'located' in the total area X, Y of the physical space." (In the present notation, the x- and the y-components of the total physical space are denoted as X and Y, respectively.) Pathria tells us that the multiplicity factor may be found to be the following:

$$\left(\frac{XY}{h^2}\right) \times \left\{\text{relevant area:} \int \mathrm{d}p_x \int \mathrm{d}p_y\right\}$$
$$= \left(\frac{XY}{h^2}\right) 2\pi(m\hbar\omega) = XY\left(\frac{eB_0}{hc}\right). \tag{M.5}$$

[1] See Pathria's equation (8.2.29), Op. cit.

As a result the grand potential—see (M.1) and (M.2)—is

$$
\begin{aligned}
Q(z, V, T, B_0) &= \sum_{\epsilon} \ln\{1 + z\exp(-\beta\epsilon)\} \\
&= XY\left(\frac{eB_0}{hc}\right)\sum_{\ell=0}^{\infty}\int_{-\infty}^{+\infty}\left(\frac{Z}{h}\right)\mathrm{d}p_z \ln\{1 + z\exp(-\beta\epsilon)\} \\
&= \left(\frac{eVB_0}{h^2c}\right)\sum_{\ell=0}^{\infty}\int_{-\infty}^{+\infty}\mathrm{d}p_z \\
&\times \ln\left\{1 + z\exp\left[-\beta\left(\left[\ell + \frac{1}{2}\right]\left(\frac{\hbar eB_0}{mc}\right) + \left[\frac{p_z^2}{2m}\right]\right)\right]\right\}, \qquad \text{(M.6)}
\end{aligned}
$$

where we have replaced XYZ by the system volume V. At general temperatures, the integral above is best evaluated numerically. However, in the limit of high and low temperature, it readily yields to analytical evaluation.

M.2 High Temperature: Landau Diamagnetism

Here z and therefore $z\exp(-\beta\epsilon)$ are small compared to unity. Therefore

$$
\ln\{1 + z\exp(-\beta\epsilon)\} = z\exp(-\beta\epsilon) + O[z\exp(-\beta\epsilon)]^2 . \qquad \text{(M.7)}
$$

Accordingly, at high temperatures, (M.6), (M.4), and (M.3) lead to the following results for the grand potential, average number of particles, and the total diamagnetic moment:

$$
\begin{aligned}
Q(z, V, T, B_0) &\approx \left(\frac{eVB_0}{h^2c}\right) z \int_{-\infty}^{+\infty}\exp\left(-\frac{\beta p_z^2}{2m}\right)\mathrm{d}p_z \\
&\times \sum_{\ell=0}^{\infty}\exp\left\{-\beta\left[\ell + \frac{1}{2}\right](\hbar\omega)\right\} \\
&= \left(\frac{zeVB_0}{2h^2c}\right)(2\pi m k_B T)^{\frac{1}{2}}\left\{\sinh\left(\frac{\beta\hbar eB_0}{2mc}\right)\right\}^{-1} . \qquad \text{(M.8)}
\end{aligned}
$$

In (M.8) above, use was made of the following identity:

$$
\begin{aligned}
\sum_{\ell=0}^{\infty}\exp\left\{-\beta\left[\ell + \frac{1}{2}\right](\hbar\omega)\right\} &= \frac{\exp\{-\beta(\frac{\hbar\omega}{2})\}}{1 - \exp\{-\beta(\hbar\omega)\}} \\
&= \left\{2\sinh\left(\frac{\beta\hbar\omega}{2}\right)\right\}^{-1} . \qquad \text{(M.9)}
\end{aligned}
$$

Also, ω was replaced by its value $\frac{eB_0}{mc}$.

In order to determine the diamagnetic susceptibility per particle, we need to find the total magnetization as a function of the applied field and the measured value $\bar{N}$—namely, the thermodynamic average—of the number of particles present. The latter, according to (M.4), involves differentiation with respect to z and then multiplication by z. Similarly, the high temperature result for the Landau diamagnetic moment, $(M)_{\text{landau}}$, can be found from (M.3) and (M.8). In this manner, we readily find

$$
\begin{aligned}
\bar{N} &= z\left(\frac{\partial Q(z, V, T, B_0)}{\partial z}\right)_{V,T,B_0} = Q(z, V, T, B_0)\,, \\
(M)_{\text{landau}} &= \frac{1}{\bar{N}\beta}\left(\frac{\partial Q(z, V, T, B_0)}{\partial B_0}\right)_{z,V,T} \\
&= \left(\frac{1}{\beta B_0}\right)\left\{1 - \left(\frac{\beta\hbar e B_0}{2mc}\right)\coth\left(\frac{\beta\hbar e B_0}{2mc}\right)\right\} \\
&= -\left(\frac{e\hbar}{2mc}\right)\left[\coth\left(\frac{\beta\hbar e B_0}{2mc}\right) - \left(\frac{2mc}{\beta\hbar e B_0}\right)\right] \\
&= -\mu_B L(\Upsilon)\,,
\end{aligned}
\tag{M.10}
$$

where $\mu_B = (\frac{e\hbar}{2mc})$ is the Bohr magneton and

$$
\begin{aligned}
\Upsilon &= \left(\frac{\beta\hbar e B_0}{2mc}\right) = \beta B_0 \mu_B, \\
L(\Upsilon) &= \coth(\Upsilon) - (1/\Upsilon)\,.
\end{aligned}
\tag{M.11}
$$

At high temperatures, unless the applied field, B_0, is exceptionally large, $\Upsilon \ll 1$. As a result $L(\Upsilon) = (\Upsilon/3) - O[(\Upsilon)^3]$, and the thermodynamic average of the diamagnetic moment per particle and the resultant Landau susceptibility, $(\chi)_{\text{landau}}$, are as follows:

$$
\begin{aligned}
(M)_{\text{landau}} &= -\frac{B_0\mu_B^2}{3k_B T} + O\left[\mu_B(\beta B_0 \mu_B)^3\right], \\
(\chi)_{\text{landau}} &= \frac{(M)_{\text{landau}}}{B_0} = -\frac{\mu_B^2}{3k_B T}\,.
\end{aligned}
\tag{M.12}
$$

Except for the negative sign, other features of this result are reminiscent of those of Langevin's classical theory result for paramagnetism! (For example, compare (11.98) where we found $(\chi)_{\text{langevin classical}} = \mu_c^2/(3k_B T)$. Recall that μ_c was the magnetic moment of a single, classical Langevin-particle.) We hasten to add, however, that this is merely a happenstance and has no fundamental significance (that, at least the current author is aware of).

At such high temperature, the Pauli result for the paramagnetic susceptibility of a gas of free, quantum electrons appears to be equal to the corresponding quasiclassical estimate given in (11.296). The equality holds only when the Landé g factor is equal to 2. Then we can legitimately replace the "intrinsic magnetic moment per electron," $\frac{g\mu_B}{2}$, by the Bohr magneton μ_B.

Adding that result to Landau diamagnetism yields

$$\chi_{\text{electronathightemeperature}} = (\chi)_{\text{quasiclassical}} + (\chi)_{\text{landau}}$$
$$= \frac{\mu_B{}^2}{k_B T} - \frac{\mu_B^2}{3k_B T}. \tag{M.13}$$

Observe that we have not used the obvious simplification—namely, $\frac{2\mu_B^2}{3k_B T}$—for the sum of the above two terms. The reason is that depending upon the source of the free electrons being considered, μ_B occurring in the two terms in the above equation may be slightly different. This is because of its dependence on the effective electronic mass and/or the Landé g-factor, which are not necessarily the same for the two processes being represented.

Specific Heat for the B–E Gas

N

Knowing the internal energy in the form of (11.362), that is, $U = \frac{3}{2}Nk_BT[\frac{g_{\frac{5}{2}}(z)}{g_{\frac{3}{2}}(z)}]$, the objective of the present appendix is to derive the general expression for the specific heat that is given in (11.363).

Analysis

Briefly we proceed as follows. First, we determine the derivative of $g_n(z)$ as

$$
\begin{aligned}
z\left(\frac{\partial g_n(z)}{\partial z}\right)_V &= \frac{z}{\Gamma(n)}\int_0^\infty \exp(x)x^{n-1}\big[\exp(x)-z\big]^{-2}dx \\
&= \frac{z}{\Gamma(n)}\left[-\frac{x^{n-1}}{\exp(x)-z}|_0^\infty\right] \\
&\quad + \frac{z(n-1)}{\Gamma(n)}\left[\int_0^\infty x^{n-2}\big[\exp(x)-z\big]^{-1}dx\right] \\
&= 0 + \frac{1}{\Gamma(n-1)}\int_0^\infty x^{n-2}\big[z^{-1}\exp(x)-1\big]^{-1}dx \\
&= g_{n-1}(z)\,. \qquad\qquad (N.1)
\end{aligned}
$$

(Note that here V is just a dummy index.) Then we write

$$
\begin{aligned}
\frac{C_V(T)}{Nk_B} &= \frac{1}{Nk_B}\left(\frac{\partial U}{\partial T}\right)_V = \frac{3}{2}\left[\frac{g_{\frac{5}{2}}(z)}{g_{\frac{3}{2}}(z)}\right] \\
&\quad + \left(\frac{3T}{2g_{\frac{3}{2}}^2(z)}\right)\left[g_{\frac{3}{2}}(z)\left(\frac{\partial g_{\frac{5}{2}}(z)}{\partial z}\right)_V - g_{\frac{5}{2}}(z)\left(\frac{\partial g_{\frac{3}{2}}(z)}{\partial z}\right)_V\right]\left(\frac{\partial z}{\partial T}\right)_V .
\end{aligned}
\qquad (N.2)
$$

R. Tahir-Kheli, *General and Statistical Thermodynamics*,
https://doi.org/10.1007/978-3-030-20700-7

Next we need to determine $(\frac{\partial z}{\partial T})_V$. To this purpose, we begin with (11.360), namely

$$g_{\frac{3}{2}}(z) = \left(\frac{N}{V}\right)\lambda^3, \quad \text{where, as always,} \quad \lambda = \left(\frac{2\pi m k_B T}{h^2}\right)^{-\frac{1}{2}}. \tag{N.3}$$

Then

$$\begin{aligned}\left(\frac{\partial g_{\frac{3}{2}}(z)}{\partial T}\right)_V &= \left(\frac{\partial g_{\frac{3}{2}}(z)}{\partial z}\right)_V \left(\frac{\partial z}{\partial T}\right)_V = \left[\frac{g_{\frac{1}{2}}(z)}{z}\right]\left(\frac{\partial z}{\partial T}\right)_V \\ &= \left(\frac{-3N}{2TV}\right)\lambda^3 = \left(-\frac{3}{2T}\right) g_{\frac{3}{2}}(z),\end{aligned} \tag{N.4}$$

which leads to

$$\left(\frac{\partial z}{\partial T}\right)_V = -\left(\frac{3z}{2T}\right)\frac{g_{\frac{3}{2}}(z)}{g_{\frac{1}{2}}(z)}. \tag{N.5}$$

Now we insert $(\frac{\partial z}{\partial T})_V$ given above into (N.2). This finally leads to the specific heat $C_V(T)$ that is valid for temperatures $T \geq T_c$,

$$C_V(T) = Nk_B\left[\frac{15}{4}\left(\frac{g_{\frac{5}{2}}(z)}{g_{\frac{3}{2}}(z)}\right) - \frac{9}{4}\left(\frac{g_{\frac{3}{2}}(z)}{g_{\frac{1}{2}}(z)}\right)\right]. \tag{N.6}$$

Bogolyubov Transformation: Inversion Procedure

O

Equations (16.26) and (16.27) were inverted to arrive at (16.28) and (16.29). Generally, for a 2×2 matrix A,

$$\begin{pmatrix} a & b \\ c & d \end{pmatrix} = A \,, \tag{O.1}$$

the inverse is[1]

$$\begin{pmatrix} d & -b \\ -c & a \end{pmatrix} /(ad - bc) = (A)^{-1} \,. \tag{O.2}$$

Therefore for the matrix M,

$$\begin{pmatrix} u_{\mathrm{k}}^* & v_{\mathrm{k}} \\ -v_{\mathrm{k}}^* & u_{\mathrm{k}} \end{pmatrix} = M \,, \tag{O.3}$$

the inverse matrix is M^{-1} given below:

$$\begin{pmatrix} u_{\mathrm{k}} & -v_{\mathrm{k}} \\ v_{\mathrm{k}}^* & u_{\mathrm{k}}^* \end{pmatrix} /|u_{\mathrm{k}}|^2 + |v_{\mathrm{k}}|^2 = (M)^{-1} \,. \tag{O.4}$$

[1]Compare, for instance, George B. Thomas, "Calculus and Analytic Geometry," 4th Ed., p. 441, Exercise 15: Addison-Wesley (1969).

R. Tahir-Kheli, *General and Statistical Thermodynamics*,
https://doi.org/10.1007/978-3-030-20700-7

List of Problems and Proofs

R. Tahir-Kheli, *General and Statistical Thermodynamics*,
https://doi.org/10.1007/978-3-030-20700-7

Scientists

Ampère, André-Marie (1775 Jan 20–1836 Jun 10), see p. 535

Avogadro, Lorenzo Romano Amadeo Carlo (1776 Aug 9–1856 Jun 9), see p. 4

Berthelot, Pierre Eugène Marcellin (1827 Oct 25–1907 Mar 18), see p. 428

Bohr, Niels Henrik David (1885 Oct 7–1962 Nov 18), see p. 419

Boltzmann, Ludwig Eduard (1844 Feb 20–1906 Sep 5), see p. 22

Boyle, Robert (1627 Jan 25–1691 Dec 31), see p. 26

Brillouin, Léon Nicolas (1889 Aug 27–1969 Oct 4), see p. 421

de Broglie, Louis V.P.R (1892 Aug 15–1987 Mar 19), see p. 438

Callen, Herbert B (1919–1993 May 22), see p. xiii

Carnot, N.L. Sadi (1796 Jun 1–1832 Aug 24), see p. 26

Celsius, Anders (1701 Nov 27–1744 Apr 25), see p. 27

Charles, Jacques Alexander César (1746 Nov 12–1823 Apr 7), see p. 27

Clausius, Rudolf Julius Emanuel (1822 Jan 2–1888 Aug 24), see p. 124

Dalton, John (1766 Sep 6–1844 Jul 27), see p. 31, 228

Davisson, Clinton Joseph (1881 Oct 22–1958 Feb 1), see p. 476

Debye, Peter Joseph William (1984 Mar 24–1966 Nov 2), see p. 497

Dirichlet, Johann Peter Gustav (1805 Feb 13–1859 May 5), see p. 405

Drude, Paul Karl Ludwig (1863 Jul 12–1906 Jul 5), see p. 461

Euler, Leonhard (1707 Apr 15–1783 Sep 18), see p. 410

Faraday, Michael (1791 Sep 22–1867 Aug 25), see p. 534

Gay-Lussac, Joseph Louis (1778 Dec 6–1850 May 9), see p. 266

Germer, Lester Halbert (1896 Oct 10–1971 Oct 3), see p. 476

Gibbs, Josiah Willard (1839 Feb 11–1903 Apr 28), see p. 22, 373

Hess, Germain Henri (1802 Aug 7–1850 Nov 30), see p. 77

Jacobian, Carl Gutav Jacob J (1804 Dec 10–1851 Feb 18), see p. 221

R. Tahir-Kheli, *General and Statistical Thermodynamics*,
https://doi.org/10.1007/978-3-030-20700-7

Jeffreys, Harold (1891 Apr 22–1989 Mar 18), see p. 609

Joule, James Prescott (1818 Dec 24–1889 Oct 11), see p. 50

Kac, Mark (1914 Aug 3–1984 Oct 26), see p. 226

Kamerlingh-Onnes, Heike (1853 Sep 21–1926 Feb 21), see p. 228

Kelvin, Lord: born William Thomson (1824 Jun 26–1907 Dec 17), see p. 114

Lagrange, Joseph-Louis (1736 Jan 25–1813 Apr 10), see p. 381

Landau, Lev Davidovich (1908 Jan 22–1968 Apr 1), see p. 472

Lande, Alfred (1888 Dec 13–1976 Oct 30), see p. 419

Laplace, Pierre-Simon (1749 Mar 23–1827 Mar 5), see p. 223

Lennard-Jones, John Edward (1894 Oct 27–1954 Nov 1), see p. 400

London, Fritz Wolfgang (1900 Mar 7–1954 Mar 30), see p. 401, 537

London, Heinz (1907 Nov 7–1970 Aug 3), see p. 537

Lorentz, Hendrik Antoon (1863 Jul 18–1928 Feb 4), see p. 461

Maclaurin, Colin (1698 Feb–1746 Jun 14), see p. 410

Massieu, François Jacques Dominique (1832 Aug 4–1896 Feb 5), see p. 621

Maxwell, James Clerk (1831 Jun 13–1879 Nov 5), see p. 22, 225, 534

Mayer, Joseph Edward (1904 Feb 5–1983 Oct 15), see p. 229

Nernst, Walther Hermann (1864 Jun 25–1941 Nov 18), see p. 428

Newton, Isaac (1643 Jan 4–1727 Mar 31), see p. 105

Otto, Nicolaus (1832 Jun 14–1891 Jan 26), see p. 176

Pascal, Blaise (1623 Jun 19–1662 Aug 19), see p. 313

Pauli, Wolfgang (1900 Apr 25–1958 Dec 15), see p. 462

Pippard, Alfred Brian (1920 Sep 7–2008 Sep 21), see p. 4

Planck, Max (1858 Apr 23–1947 Oct 4), see p. 126

Ramsey, Norman Foster (1915 Aug 27–1993 Dec 7), see p. 182

Richardson, Owen Willans (1879 Apr 26–1959 Feb 15), see p. 473

Riemann, Bernhard Georg Friedrich (1826 Sep 17–1866 Jul 20), see p. 454

Rumford, Benjamin Thompson, Count of Bavaria (1753 Mar 26–1814 Aug 21), see p. 50

Schrödinger, Erwin Rudolf Josef (1887 Aug 12, 1961 Jan 9), see p. 404

Stefan, Josef (1835 Mar 24–1893 Jan 7), see p. 488

Stirling, James (1692 May–1770 Dec 5), see p. 571

Stirling, Robert (1790 Oct 25–1878 Jun 6), see p. 171

ter Haar, Dirk (1919 Apr 22–2002 Sep 3), see p. xiii

Thomsen, H. Julius (1826 Feb 16–1909 Feb 13), see p. 428

Uhlenbeck, George Eugene (1900 Dec 6–1988 Oct 31), see p. 226

Van der Waals, Johannes Diderik (1837 Nov 23–1923 Mar 18), see p. 223

Index

R. Tahir-Kheli, *General and Statistical Thermodynamics*,
https://doi.org/10.1007/978-3-030-20700-7

Q

R

S

Printed in the United States
by Baker & Taylor Publisher Services